S0-AKW-687

Practical Heating Technology

Practical Heating Technology

Second Edition

William M. Johnson

Kevin Standiford

Australia • Brazil • Japan • Korea • Mexico • Singapore • Spain • United Kingdom • United States

Practical Heating Technology, 2nd Edition
William M. Johnson, Kevin Standiford

Vice President, Career and Professional Editorial: Dave Garza

Director of Learning Solutions: Sandy Clark

Senior Acquisitions Editor: James Devoe

Managing Editor: Larry Main

Senior Product Manager: John Fisher

Editorial Assistant: Thomas Best

Vice President, Career and Professional Marketing: Jennifer McAvey

Marketing Director: Deborah S. Yarnell

Marketing Manager: Jimmy Stephens

Marketing Coordinator: Mark Pierro

Production Director: Wendy Troeger

Production Manager: Mark Bernard

Content Project Manager: David Plagenza

Senior Art Director: David Arsenault

Technology Project Manager: Christopher Catalina

Production Technology Analyst: Thomas Stover

Library of Congress Control Number: 2008926660

ISBN-13: 978-1-418-08039-6

ISBN-10: 1-418-08039-X

Delmar
5 Maxwell Drive
Clifton Park, NY 12065-2919
USA

Cengage Learning is a leading provider of customized learning solutions with office locations around the globe, including Singapore, the United Kingdom, Australia, Mexico, Brazil, and Japan. Locate your local office at: **international.cengage.com/region**

Cengage Learning products are represented in Canada by Nelson Education, Ltd.

For your lifelong learning solutions, visit **delmar.cengage.com**

Visit our corporate website at **www.cengage.com.**

Printed in the United States of America
2 3 4 5 XX 10 09

Brief Contents

Contents

Preface

Practical Heating Technology was written to provide students and technicians extended coverage of the practical aspects of heating equipment and systems as well as more professional installation and service techniques. Installation and service techniques that are more professional result in fewer problems and help to create a more satisfying profession. The more professional technicians also have the best career opportunities.

The student or technician should have studied the fundamentals of heating before starting this text. Each type of equipment is discussed in detail with the assumption that the basics of heating—including Btu, temperature, and pressure—have been studied previously. This text discusses how a furnace or boiler is constructed and what the manufacturers' intentions are when the various features are designed into the equipment.

Pipe sizing for gas piping systems for low-pressure and 2-pound-per-square-inch systems is discussed to broaden the technician's knowledge of subjects that are very seldom discussed in other textbooks. Venting practices and sizing are also discussed for gas and oil.

An effort has been made to address the more common systems, whether old or new. Many older types of heating systems that are still in use as well as many of the newer types of systems are included. Because literature for older systems is out of print, the technician may benefit from this information when troubleshooting these systems.

Features of This Text

Objectives are listed at the beginning of each unit. The objective statements are kept clear and concise to give the students direction. Summary and review questions appear at the end of each unit. Students should answer the questions while reviewing what they have read; instructors can use the summaries and review questions to stimulate class discussion and for classroom unit review. Practical troubleshooting procedures are an important feature of this text. There are practical component and system troubleshooting suggestions and techniques. In many units, practical examples of service technician calls are presented in a down-to-earth situational format.

Safety precautions and techniques are discussed throughout the text. Each unit cautions the technician about the particular safety aspect that should be known and followed at this point of study.

Illustrations and photos are used generously throughout the text. An effort has been made to illustrate each component and its relationship to the system function in a manner that the student can get a word picture of the event.

Supplements to This Text

To further enhance the training of the technician, the following supplements are also available to accompany the text.

Practical Heating Technology Lab Manual A series of practical exercises help guide the student through many types of equipment used in the field. When the student has successfully completed these exercises, a better understanding of how to disassemble the common components for faster service will be accomplished. The student will have handled all of the components and performed many tasks on these components for the purpose of diagnostics and repair.

Practical Heating Technology Instructor Guide Guide of suggested activities and methods for presenting the text material in each unit, including material from other sources that have been used. It includes the answers to all questions in the text and the lab manual.

e.resource This is an educational resource that creates a truly electronic classroom. It is a CD-ROM containing tools and instructional resources that enrich your classroom and make your preparation time shorter. The elements of e.resource link directly to the text and tie together to provide a unified instructional system. Spend your time teaching, not preparing to teach.

Features contained in e.resource include the following:

- PowerPoint® Presentations: These slides provide the basis for a lecture outline that helps you to present concepts and material. Key points and concepts can be graphically highlighted for student retention.
- Test Questions: More than 800 questions of varying levels of difficulty are provided in true/false and multiple-choice formats. These questions can be used to assess student comprehension or can be made available to the student for self-evaluation.

Acknowledgments

A special acknowledgment is due the instructors who reviewed certain chapters in detail.

Dr. Clyde Perry, Gateway Community College, Phoenix, Arizona

Terry Rogers, Midlands Technical College, Columbia, South Carolina

About the Authors

Bill Johnson graduated from Southern Polytechnic in Marietta, Georgia with an associates degree in gas fuel technology and refrigeration. He owned and operated an air-conditioning, heating, and refrigeration business for 10 years. He has unlimited licenses for North Carolina in heating, air conditioning, and refrigeration. Bill taught heating, air conditioning, and refrigeration installation, service, and design for 15 years at Central Piedmont Community College in Charlotte, North Carolina, and was instrumental in standardizing the heating, air conditioning, and refrigeration curriculum for the state community college system. He is a member of Refrigeration Service Engineers Society (RSES). He has a series of articles on the website for the *Air Conditioning, Heating & Refrigeration News*. See achrnews.com, Special Edition BTU Buddy articles. These articles are service situation calls for technicians.

Kevin Standiford, author/contributor/consultant has been in the technology fields of manufacturing processes, HVACR, process piping and robotics for more than 20 years. While attending college to obtain his bachelor of science in mechanical engineering technology, he worked for McClelland Consulting Engineers as a mechanical designer, designing HVAC, complex processing piping, and cogeneration systems for commercial and industrial applications. During his college years, he became a student member of the American Society of Heating, Refrigerating and Air-Conditioning Engineers where he developed and later wrote a paper on a computer application that enabled the user to simulate, design, and draw heating and cooling systems by using AutoCAD. The paper was entered into a student design competition and became the first-place winner for the region and state. After graduation, Kevin worked for one of the leading HVAC engineering firms in the state of Arkansas, Pettit and Pettit Consulting Engineers, as a mechanical design engineer. While working for Pettit and Pettit, Kevin designed and selected equipment for large commercial and government projects by using manual design techniques and computer simulations. In addition to working at Pettit and Pettit, Kevin started teaching part-time evening engineering and design courses for Garland County Community College in Hot Springs, Arkansas. Subsequently he stopped working full time in the engineering field and started teaching technology classes, which included heat transfer, duct design, and properties of air. It was also at this time that Kevin started writing textbooks for Cengage Delmar Publishers. The first textbook was a descriptive geometry book, which included a section on sheet metal design. Today Kevin is a full-time consultant working for both the publishing and engineering industries, and a part-time instructor. In the publishing industry, Kevin has worked on numerous e-resource products, mapping, and custom publications for Delmar's HVAC, CAD, and plumbing titles.

1 Basic Electricity and Magnetism

Objectives

After studying this unit, you should be able to

- **describe the structure of an atom.**
- **identify atoms with a positive charge and atoms with a negative charge.**
- **explain the characteristics that make certain materials good conductors.**
- **describe how magnetism is used to produce electricity.**
- **state the differences between alternating current and direct current.**
- **list the units of measurement for electricity.**
- **explain the differences between series and parallel circuits.**
- **state Ohm's Law.**
- **state the formula for determining electrical power.**
- **describe a solenoid.**
- **explain inductance.**
- **describe the construction of a transformer and the way a current is induced in a secondary circuit.**
- **describe how a capacitor works.**
- **describe a sine wave.**
- **state the reasons for using proper wire sizes.**
- **describe the physical characteristics and the function of several semiconductors.**
- **describe procedures for making electrical measurements.**

Safety Checklist

- Do not make any electrical measurements without specific instructions from a qualified person.
- Use only electrical conductors of the proper size to avoid overheating and possibly fire.
- Electrical circuits must be protected from current overloads. These circuits are normally protected with fuses or circuit breakers.
- Extension cords used by technicians to provide electrical power for portable power tools and other devices should be protected with ground fault circuit interrupters.
- When servicing equipment, the electrical service should be shut off at a disconnect panel whenever possible, the disconnect panel locked, and the only key kept by the technician.

1.1 Structure of Matter

To understand the theory of how an electric current flows, you must understand something about the structure of matter. Matter is made up of atoms. Atoms are made up of protons, neutrons, and electrons. Protons and neutrons are located at the center (or nucleus) of the atom. Protons have a positive charge. Neutrons have no charge and have little or no effect as far as electrical characteristics are concerned. Electrons have a negative charge and travel around the nucleus in orbits. The number of electrons in an atom is the same as the number of protons. Electrons in the same orbit are the same distance from the nucleus but do not follow the same orbital paths, Figure 1-1.

The hydrogen atom is a simple atom to illustrate because it has only one proton and one electron, Figure 1-2. Not

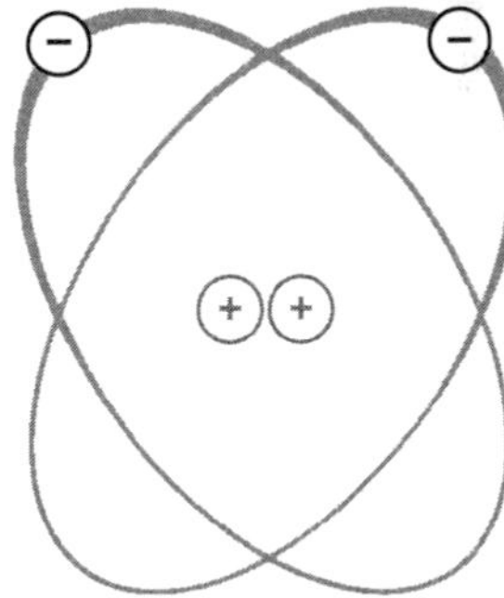

Figure 1-1. The orbital paths of electrons.

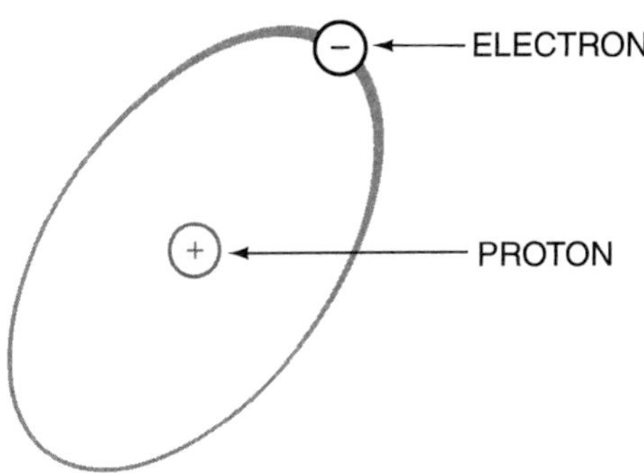

Figure 1-2. A hydrogen atom with one electron and one proton.

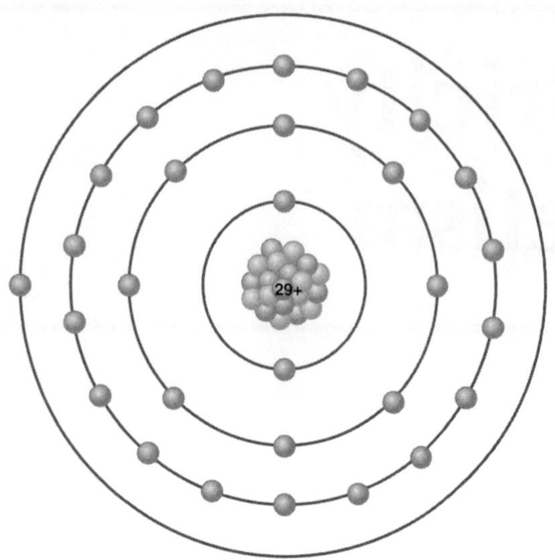

Figure 1-3. A copper atom with twenty-nine protons and twenty-nine electrons.

all atoms are as simple as the hydrogen atom. Most wiring used to conduct an electrical current is made of copper. Figure 1-3 illustrates a copper atom, which has twenty-nine protons and twenty-nine electrons. Some electron orbits are farther away from the nucleus than others. As can be seen, two travel in an inner orbit, eight in the next, eighteen in the next, and 1 in the outer orbit. It is this single electron in the outer orbit that makes copper a good conductor.

1.2 Movement of Electrons

When sufficient energy or force is applied to an atom, the outer electron (or electrons) becomes free and moves. If it leaves the atom, the atom will contain more protons than electrons. Protons have a positive charge. This means that this atom will have a positive charge, Figure 1-4(A).

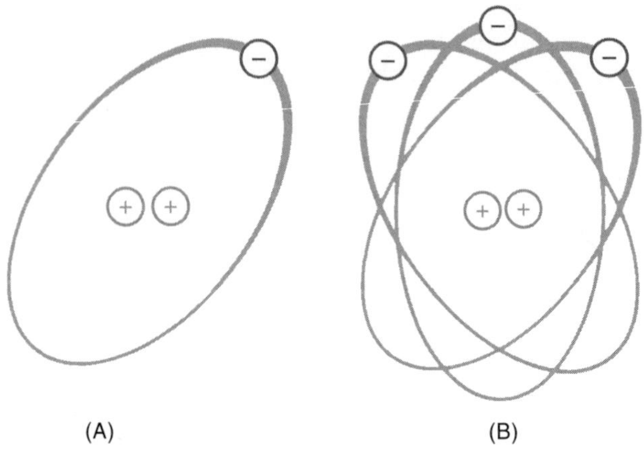

Figure 1-4. (A) This atom has two protons and one electron. It has a shortage of electrons and thus a positive charge. (B) This atom has two protons and three electrons. It has an excess of electrons and thus a negative charge.

The atom the electron joins will contain more electrons than protons, so it will have a negative charge, Figure 1-4(B).

Like charges repel each other, and unlike charges attract each other. An electron in an atom with a surplus of electrons (negative charge) will be attracted to an atom with a shortage of electrons (positive charge). An electron entering an orbit with a surplus of electrons will tend to repel an electron already there and cause it to become a free electron.

1.3 Conductors

Good conductors are those with few electrons in the outer orbit. Three common metals—copper, silver, and gold—are good conductors, and each has one electron in the outer orbit. These are considered to be free electrons because they move easily from one atom to another.

1.4 Insulators

Atoms with several electrons in the outer orbit are poor conductors. These electrons are difficult to free, and materials made with these atoms are considered to be insulators. Glass, rubber, and plastic are examples of good insulators.

1.5 Electricity Produced from Magnetism

Electricity can be produced in many ways, for example, from chemicals, pressure, light, heat, and magnetism. The electricity that air-conditioning and heating technicians are more involved with is produced by a generator using magnetism.

Magnets are common objects with many uses. Magnets have poles usually designated as the north (N) pole and the south (S) pole. They also have fields of force. Figure 1-5 shows the lines of the field of force around a permanent bar magnet. This field causes the like poles of two magnets to repel each other and the unlike poles to attract each other.

If a conductor, such as a copper wire, is passed through this field and crosses these lines of force, the outer electrons in the atoms in the wire are freed and begin to move

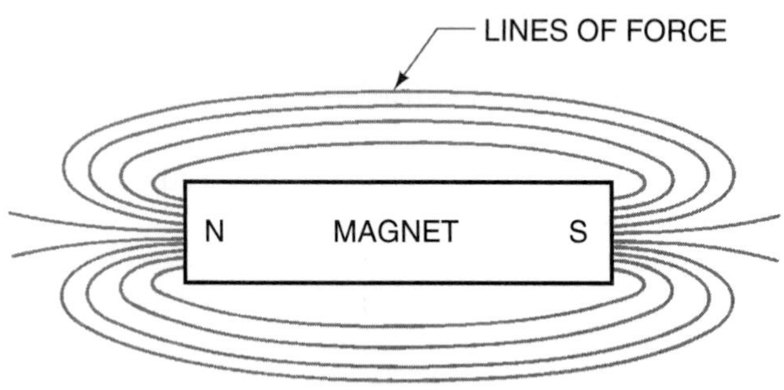

Figure 1-5. A permanent magnet with lines of force.

from atom to atom. This movement of electrons is considered to be an electrical current. They will move in one direction. It does not matter if the wire moves or if the magnetic field moves. It is only necessary that the conductor cross through the lines of force, Figure 1-6.

This movement of electrons in one direction produces the electric current. The current is an impulse transferred from one electron to the next. If you pushed a golf ball into a tube already filled with golf balls, one would be ejected instantly from the other end, Figure 1-7. Electric current travels in a similar manner at a speed of 186,000 miles/sec. The electrons do not travel through the wire at this speed, but the repelling and attracting effect causes the current to do so.

An electrical generator has a large magnetic field and many turns of wire crossing the lines of force. A large magnetic field will produce more current than a smaller magnetic field, and many turns of wire passing through a magnetic field will produce more current than a few turns of wire. The magnetic force field for generators is usually produced by electromagnets. Electromagnets have similar characteristics to permanent magnets and are discussed later in this unit. Figure 1-8 shows a simple generator.

For practical use, the generator must be rotated with enough force to pass the wires through the lines of force. This takes considerable energy. It actually takes more energy than the generator puts out. The energy used to rotate the generator is steam made by natural gas, coal, or nuclear power. Water generated power is also available. These sources of power are not practical to use where the energy is needed, so they are used to generate electrical power that can be wired to home and industry.

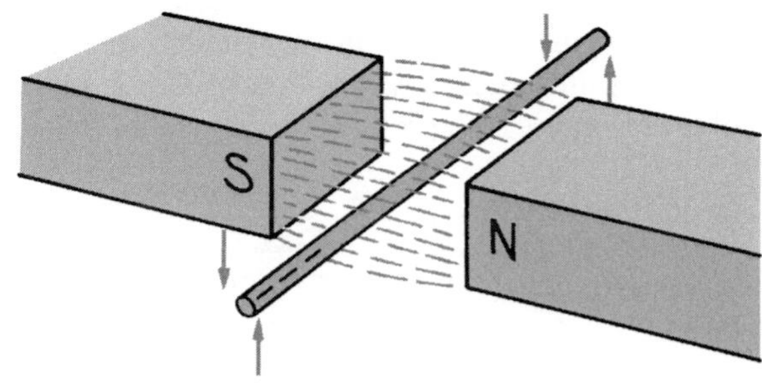

Figure 1-6. The movement of wire up and down cuts the lines of force, causing an electric current to flow in the wire.

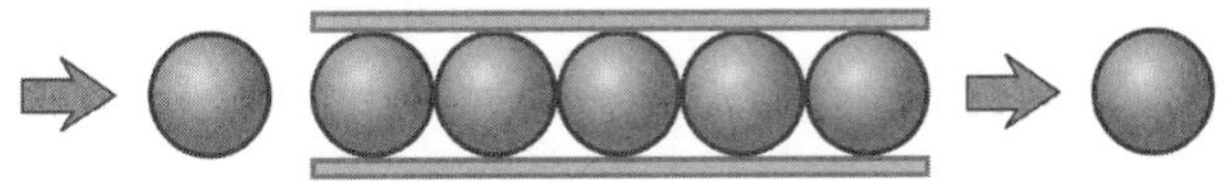

Figure 1-7. A tube filled with golf balls; when one ball is pushed in, one ball is pushed out.

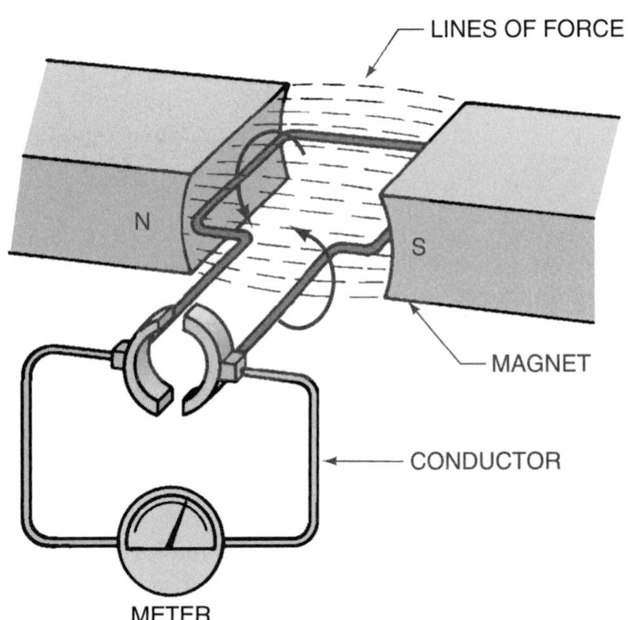

Figure 1-8. A simple generator.

1.6 Direct Current

Direct current (DC) travels in one direction. Because electrons have a negative charge and travel to atoms with a positive charge, DC is considered to flow from negative to positive.

1.7 Alternating Current

Alternating current (AC) is continually and rapidly reversing. The charge at the power source (generator) is continually changing direction; thus the current continually reverses itself. For several reasons, most electrical energy generated for public use is AC. It is much more economical to transmit electrical energy long distances in the form of AC. The voltage of this type can be readily changed so that it has many more uses. DC still has many applications, but it is usually obtained by changing AC to DC or by producing the DC locally where it is to be used.

1.8 Electrical Units of Measurement

Electromotive force (emf) or voltage (V) is used to indicate the difference of potential in two charges. When an electron surplus builds up on one side of a circuit and a shortage of electrons exists on the other side, a difference of potential or emf is created. The unit used to measure this force is the *volt.*

The *ampere* is the unit used to measure the quantity of electrons moving past a given point in a specific period of time (electron flow rate).

All materials oppose or resist the flow of an electrical current to some extent. In good conductors this opposition or resistance is very low. In poor conductors the resistance is high. The unit used to measure resistance is the ohm. A conductor has a resistance of 1 ohm when a force of 1 volt causes a current of 1 ampere to flow.

Volt = Electrical force or pressure (V)

Ampere = Quantity of electron flow rate (A)

Ohm = Resistance to electron flow (Ω)

1.9 The Electrical Circuit

An electrical circuit must have a power source, a conductor to carry the current, and a load or device to use the current. There is also generally a means for turning the electrical current flow on and off. Figure 1-9 shows an electrical generator for the source, a wire for the conductor, a light bulb for the load, and a switch for opening and closing the circuit.

The generator produces the current by passing many turns of wire through a magnetic field. If it is a DC generator, the current will flow in one direction. If it is an AC generator, the current will continually reverse itself. However, the effect on this circuit will generally be the same whether it is AC or DC.

The wire or conductor provides the path for the electricity to flow to the bulb and complete the circuit. The electrical energy is converted to heat and light energy at the bulb element.

The switch is used to open and close the circuit. When the switch is open, no current will flow. When it is closed, the bulb element will produce heat and light because current is flowing through it.

1.10 Making Electrical Measurements

In the circuit illustrated in Figure 1-9, electrical measurements can be made to determine the voltage (emf) and current (amperes). In making the measurements, Figure 1-10, the voltmeter is connected across the terminals of the bulb

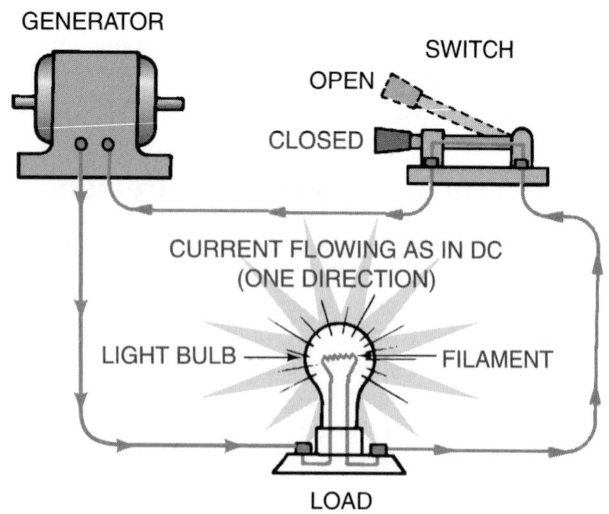

Figure 1-9. An electric circuit.

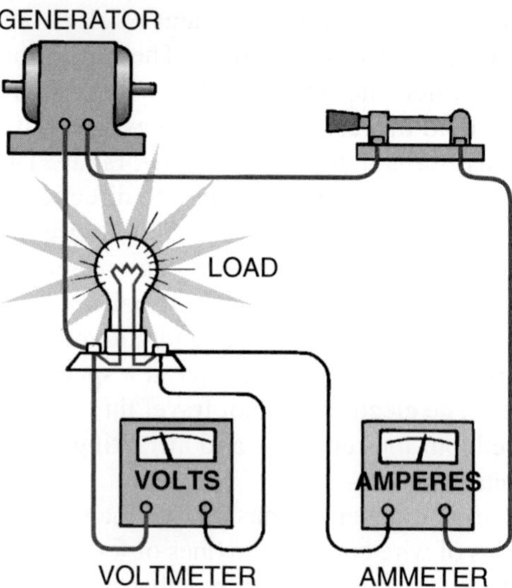

Figure 1-10. Voltage is measured across the load. Amperage is measured in series.

without interrupting the circuit. The ammeter is connected directly into the circuit so that all the current flows through it. Figure 1-11 illustrates the same circuit using symbols.

Often a circuit will contain more than one resistance or load. These resistances may be wired in series or in parallel, depending on the application or use of the circuit. Figure 1-12 shows three loads in series. This is shown pictorially and by symbols. Figure 1-13 illustrates three loads wired in parallel.

In circuits where devices are wired in series, all of the current passes through each load. When two or more loads are wired in parallel, the current is divided among the loads. This is explained in more detail later. Power-passing devices such as switches are wired in series. Most resistances or loads (power-consuming devices) that air-conditioning and heating technicians work with are wired in parallel.

Figure 1-14 illustrates how a voltmeter is connected for each of the resistances. The voltmeter is in parallel with each

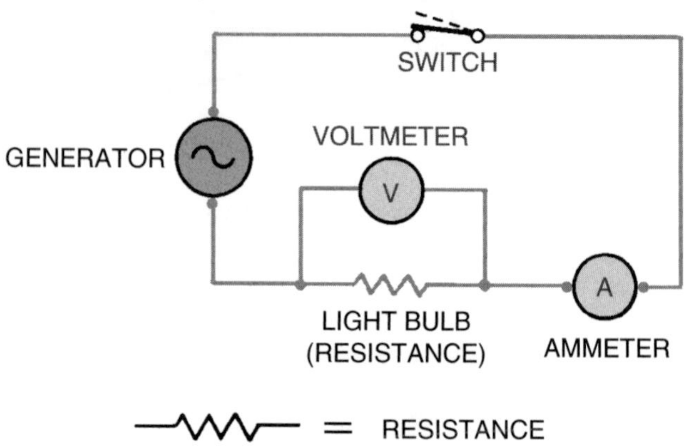

Figure 1-11. The same circuit as **Figure 1-10,** illustrated with symbols.

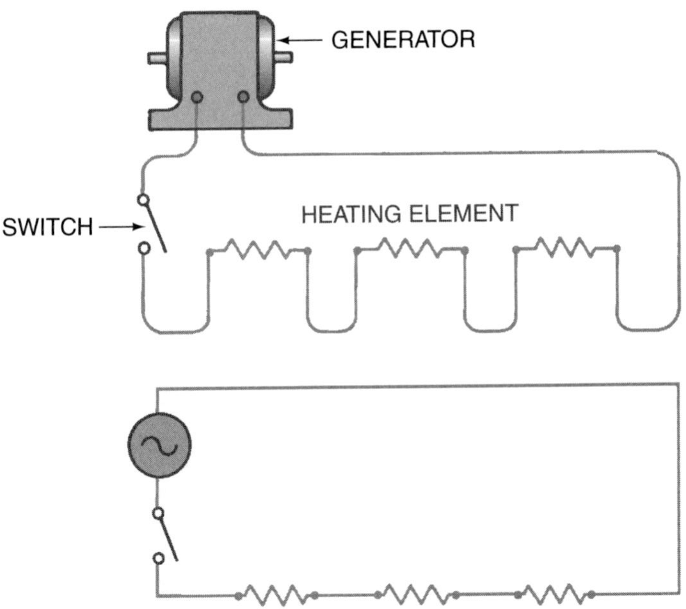

Figure 1-12. Multiple resistances (small heating elements) in series.

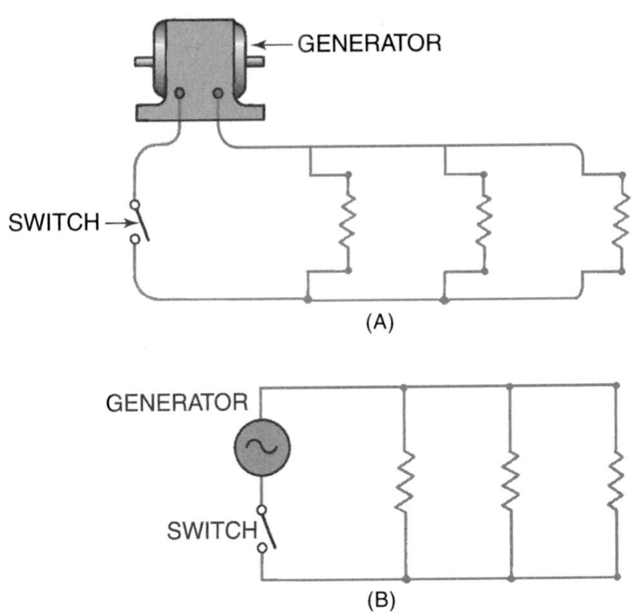

Figure 1-13. (A) Multiple resistances in parallel. (B) Three resistances in parallel using symbols.

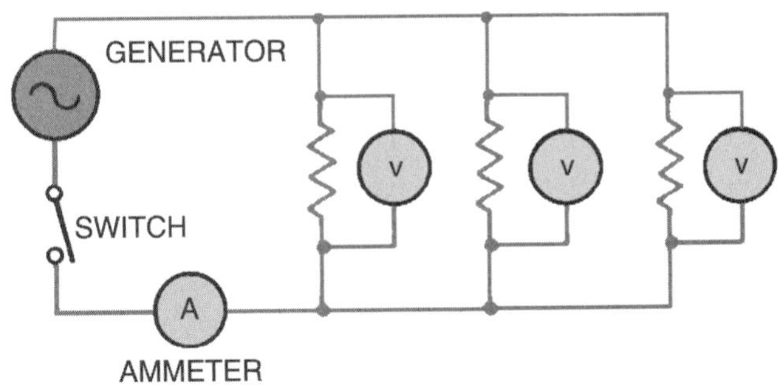

Figure 1-14. Voltage readings are taken across the resistances in the circuit.

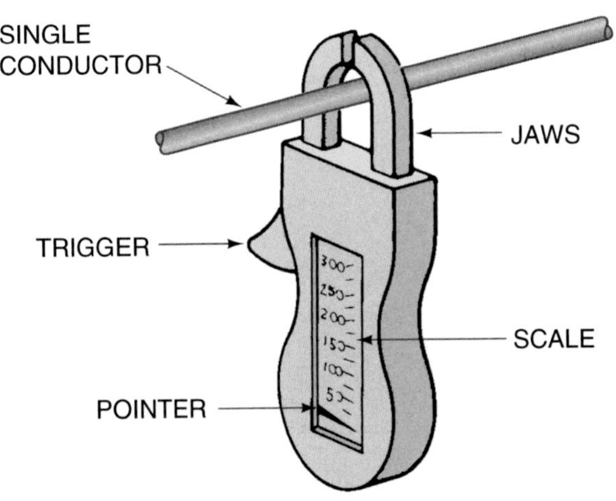

Figure 1-15. A clamp-on ammeter.

resistance. The ammeter is also shown and is wired in series in the circuit. An ammeter has been developed that can be clamped around a single conductor to measure amperes, Figure 1-15. This is convenient because it is often difficult to disconnect the circuit to connect the ammeter in series. This type of ammeter, usually called a clamp-on type, is discussed in Section 1.19, "Electrical Measuring Instruments."

1.11 Ohm's Law

During the early 1800s the German scientist Georg Simon Ohm experimented with electrical circuits and particularly with resistances in these circuits. He determined that there is a relationship between each of the factors in an electrical circuit. This relationship is called *Ohm's Law*, which is described next. Abbreviations are used to represent the different electrical factors:

E = Voltage (electromotive force)

I = Amperage (current)

R = Resistance (the electrical load where the power is consumed; the symbol for ohms is Ω)

When electrical circuits are defined, they are often compared to water circuits. In a water circuit, pressure pushes the water through the circuit, Figure 1-16. The water flows at a rate of gallons per minute, and it flows against the resistance of the friction loss in the pipe and the components in the circuit, Figure 1-17. In electrical circuits, the water pressure is compared to the voltage, the water volume in gallons per minute (GPM) is compared to the amperage, and friction loss in the pipe and components is compared to the resistance to electrical flow in a wire or electrical load, Figure 1-18.

Ohm determined that the voltage equals the amperage times the resistance:

$$E = I \times R$$

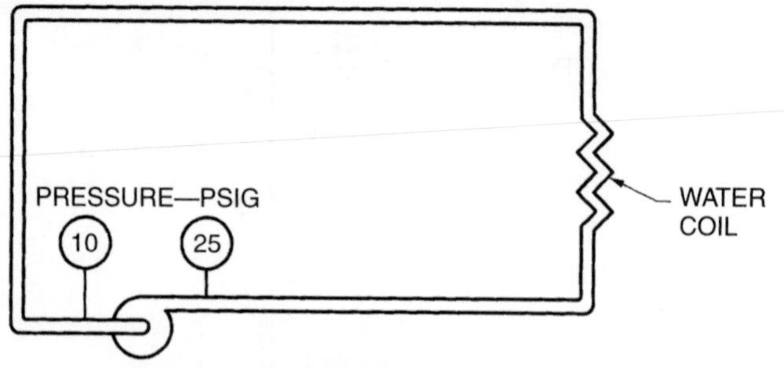

Figure 1-16. Electrical circuits may be compared to water circuits.

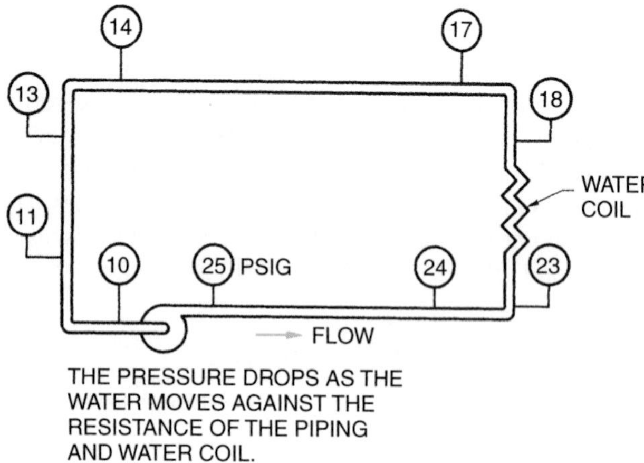

Figure 1-17. The rate of water flow meets the resistance of the friction in the pipe and components.

The amperage equals the voltage divided by the resistance:

$$I = \frac{E}{R}$$

The resistance equals the voltage divided by the amperage:

$$R = \frac{E}{I}$$

Figure 1-19 shows a convenient way to remember these formulas. These formulas apply only to circuits that have resistances as the load and do not apply to motor circuits, which involve magnetism. Motor circuits are discussed later in Unit 2.

The formulas can be used to calculate the circuit performance. For example, suppose that you can measure the resistance (60 Ω) and the current flow (2 A) of the heating coil in a circuit, and you want to know what the supply voltage should be. See Figure 1-20 for an illustration of the problem. You can calculate the correct applied voltage as follows:

$$E = I \times R = 2 \times 60 = 120 \text{ V}$$

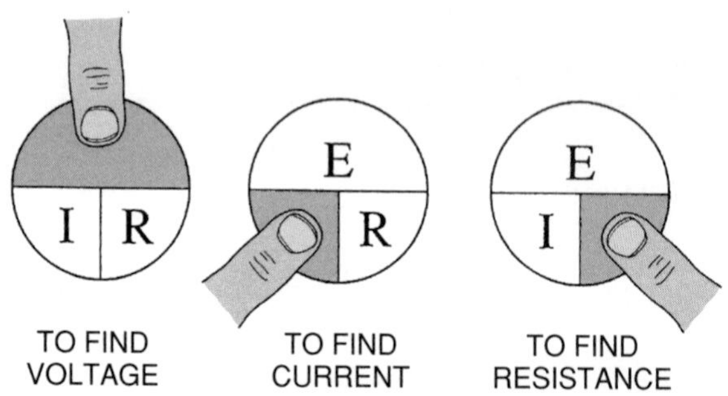

Figure 1-19. To determine the formula for the unknown quantity, cover the letter representing the unknown.

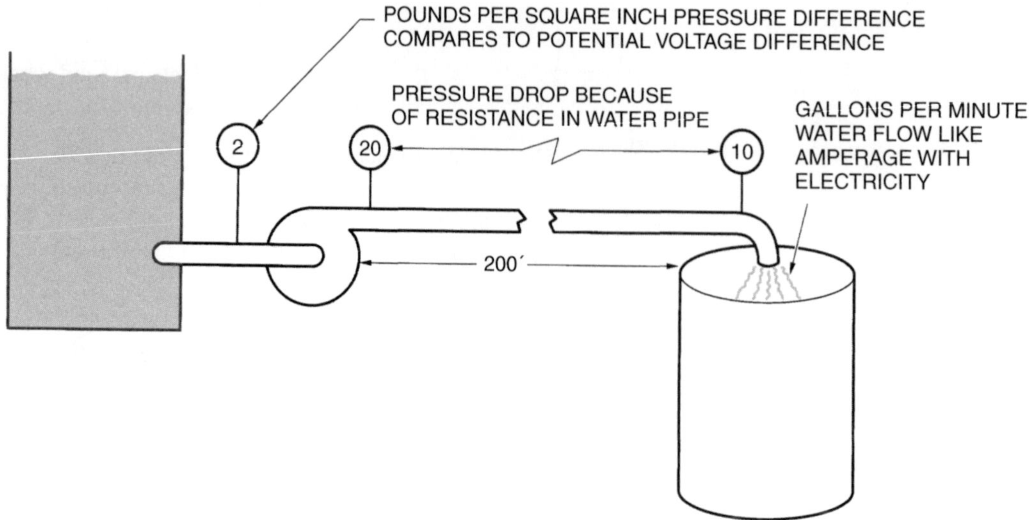

Figure 1-18. Water flow in gallons per minute compares to current flow in amperes. Water pressure in pounds per square inch compares to voltage. Friction loss in pipes and components compares to resistance in the conductors and power consuming components.

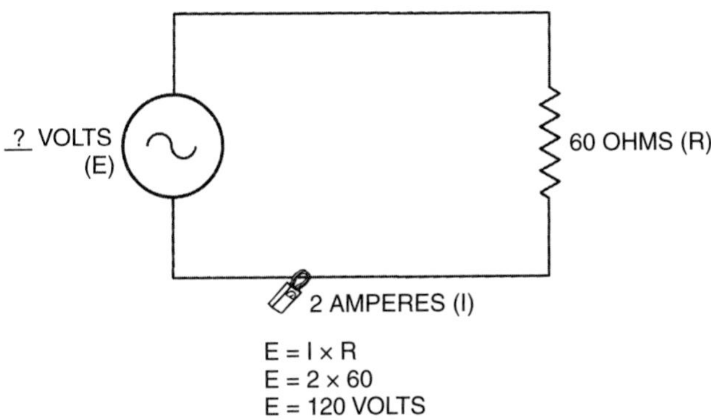

Figure 1-20. Calculating the correct applied voltage for a heating coil when the amperage and the resistance of the coil are known.

Suppose that the voltage to a heater is 220 volts and that the resistance in the heater is 10 ohms. What should the amperage be when voltage is applied to the unit? (See Figure 1-21.)

$$I = E/R = 220/10 = 22 \text{ A}$$

It is often necessary to measure the resistance in a circuit with a test instrument. The common test instrument used is the volt-ohm-milliammeter (VOM). The ohm portion of the meter is used for determining the resistance of a component or circuit, Figure 1-22. This is accomplished by placing the meter leads across the resistance to be measured with no power on the circuit, Figure 1-23. The component must be isolated from the circuit for a correct ohm reading, or there will be feedback through other components in the circuit, Figure 1-24.

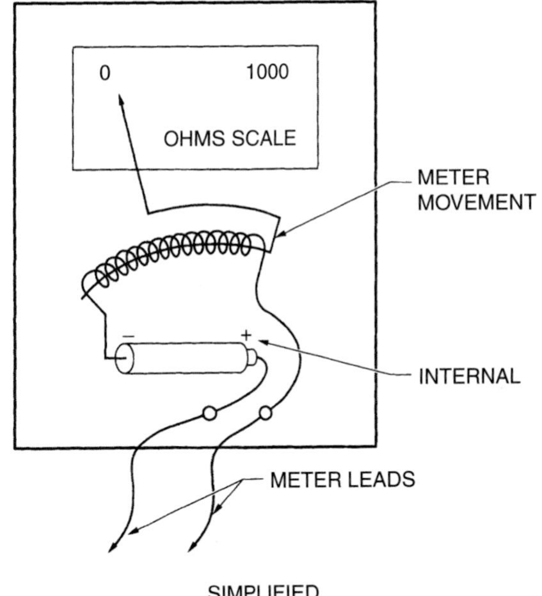

Figure 1-22. Ohmmeter for measuring resistance has a battery power supply.

Caution: Do not use electrical-measuring instruments without specific instructions from a qualified person.

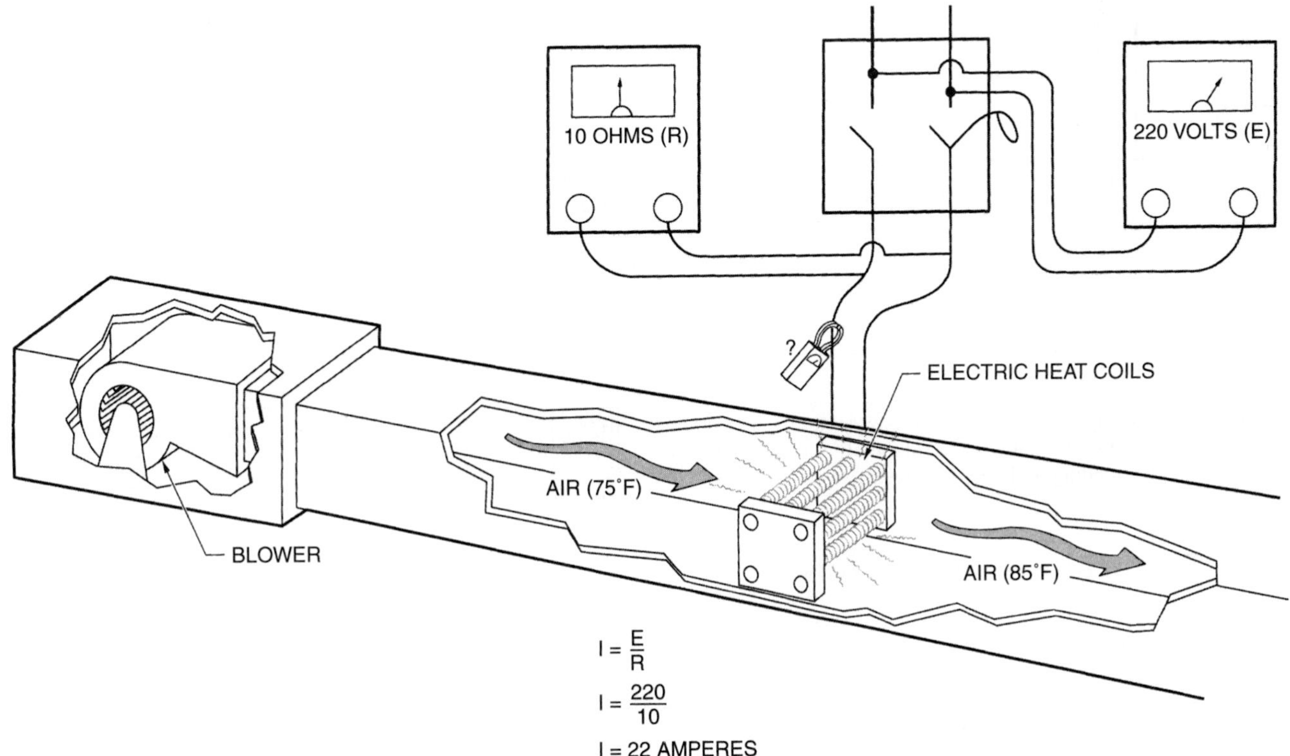

Figure 1-21. Calculating the correct amperage when the voltage and resistance of a heater are known.

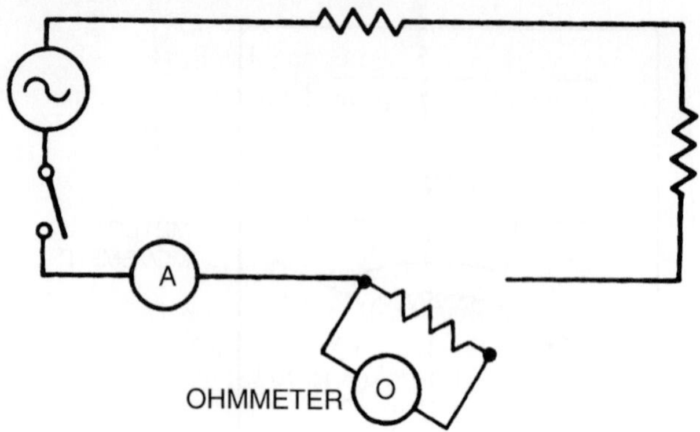

Figure 1-23. The component to be checked must be out of the circuit.

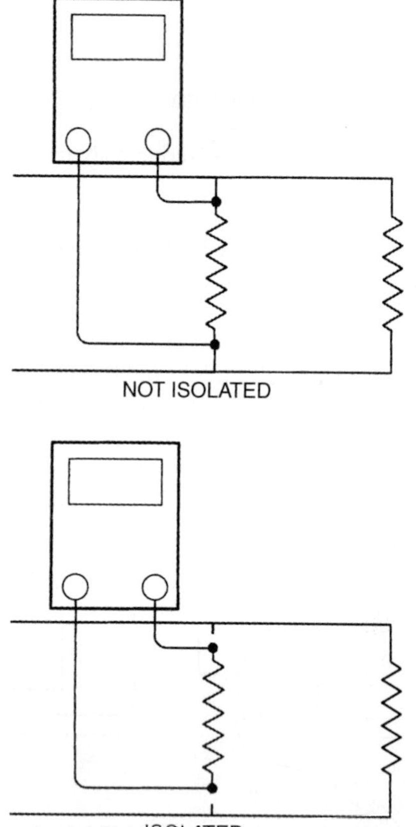

Figure 1-24. If the component to be checked is not isolated from the rest of the circuit, the meter will read the other components in the circuit and the reading will not be correct.

1.12 Characteristics of Circuits

As explained in section 1.10, there are two kinds of circuits: series and parallel. Three characteristics make a series circuit different from a parallel circuit:

1. The voltage is divided across the electrical loads in the series circuit. For example, if there are three resistances of equal value in a circuit, and the voltage applied to the circuit is 120 volts, the voltage is equally divided across each resistance, Figure 1-25. If the resistances are not equal, the voltage is divided across each according to its resistance, Figure 1-26.
2. The total current for the circuit flows through each electrical load in the circuit. With one power supply and three resistances, the current must flow through each to reach the others.
3. The resistances of the circuit can be added for the sum of the resistance of the total circuit. For example, when three resistances are in a circuit, the total resistance of the circuit is the sum of the three resistances, Figure 1-27.

Three characteristics of a parallel circuit make it different from a series circuit:

1. The total voltage for the circuit is applied across each circuit resistance. The power supply feeds each power-consuming device (load) directly, Figure 1-28.

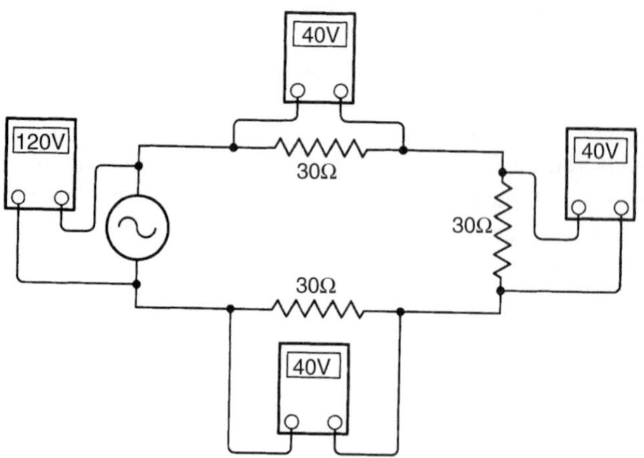

Figure 1-25. Three resistors of equal value divide the voltage equally in a series circuit.

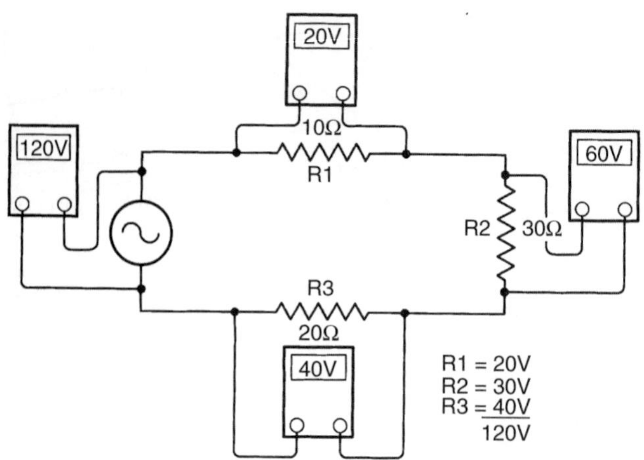

Figure 1-26. Three resistors of unequal value divide the voltage according to their values.

2. The current is divided among the different loads, or the total current is equal to the sum of the currents in each branch, Figure 1-29.
3. The total resistance is less than the value of the smallest resistance in the circuit.

Calculating the resistances in a parallel circuit requires a different procedure than simply adding them. A parallel circuit allows current along two or more paths at the same time. This type of circuit applies equal voltage to all loads. The general formula used to determine total resistance in a parallel circuit is as follows:

$$R_{\text{TOTAL}} = \frac{1}{\frac{1}{R_1} + \frac{1}{R_2} + \frac{1}{R_3} + \cdots}$$

For example, the total resistance of the circuit in Figure 1-30 is determined as follows:

$$R_{\text{TOTAL}} = \frac{1}{\frac{1}{R_1} + \frac{1}{R_2} + \frac{1}{R_3}}$$

$$= \frac{1}{\frac{1}{40} + \frac{1}{30} + \frac{1}{20}}$$

$$= \frac{1}{0.025 + 0.0333 + 0.05}$$

$$= \frac{1}{0.1083}$$

$$= 9.234\ \Omega$$

To determine the total current draw, Ohm's Law can be used. For example, in the previous problem, the voltage is known, the current flow for each component can be calculated from the total resistance, as shown in Figure 1-31. Again, notice that the total resistance is less than that of the smallest resistor.

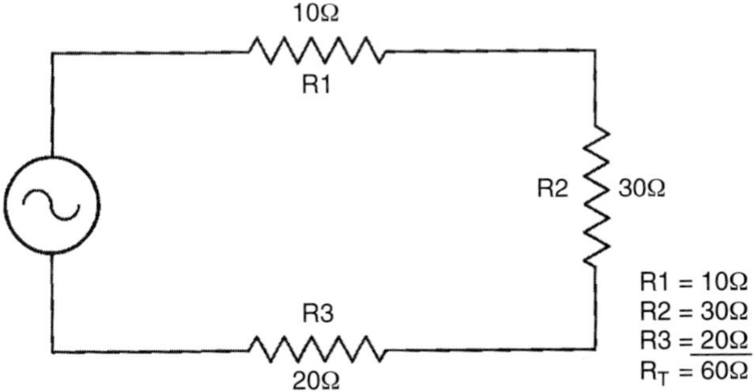

THE VOLTAGE AT EACH COMPONENT IN FIGURE 11.36 WAS CALCULATED USING THIS KNOWLEDGE IN THE FOLLOWING MANNER.

$I = \frac{E}{R_T} = \frac{120V}{60\Omega} = 2$ AMPERES

$E = I \times R_1 = 2A \times 10\Omega = 20$ VOLTS
$E = I \times R_2 = 2A \times 30\Omega = 60$ VOLTS
$E = I \times R_3 = 2A \times 20\Omega = 40$ VOLTS
120 VOLTS TOTAL

Figure 1-27. The sum of the three resistors is the total resistance for the circuit.

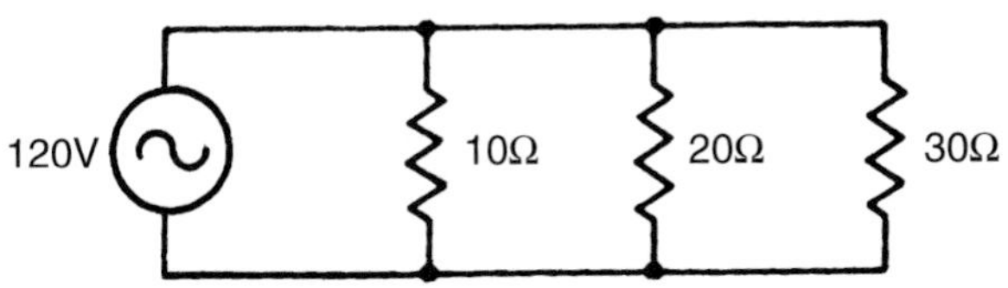

Figure 1-28. The total voltage for a parallel circuit is applied across each component.

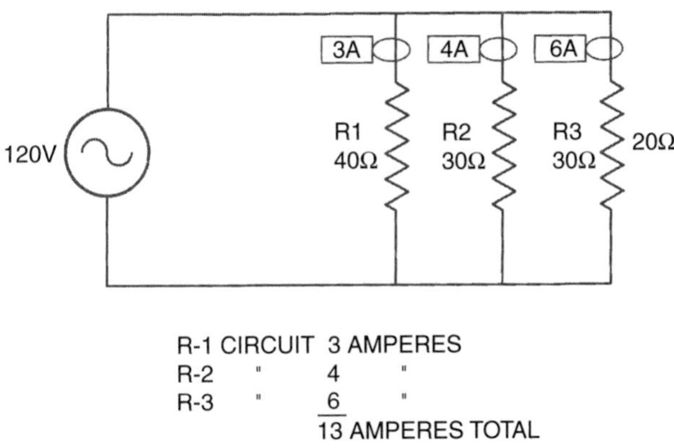

R-1 CIRCUIT 3 AMPERES
R-2 " 4 "
R-3 " 6 "
13 AMPERES TOTAL

Figure 1-29. The current is divided between the different loads in a parallel circuit.

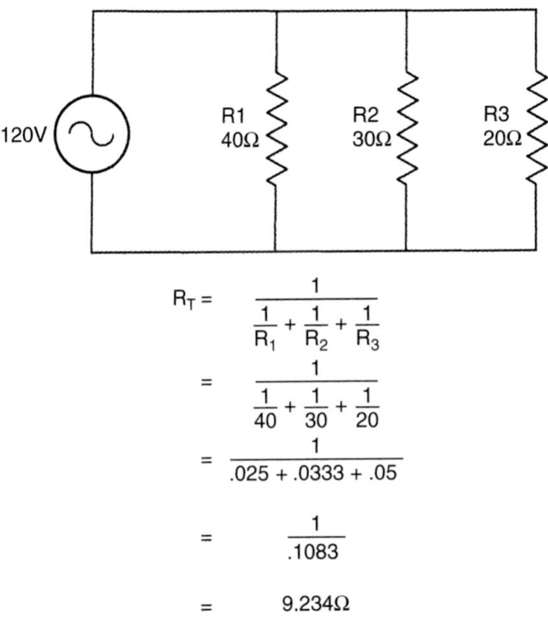

$$R_T = \frac{1}{\frac{1}{R_1} + \frac{1}{R_2} + \frac{1}{R_3}}$$

$$= \frac{1}{\frac{1}{40} + \frac{1}{30} + \frac{1}{20}}$$

$$= \frac{1}{.025 + .0333 + .05}$$

$$= \frac{1}{.1083}$$

$$= 9.234\Omega$$

Figure 1-30. The total resistance for a parallel circuit must be calculated.

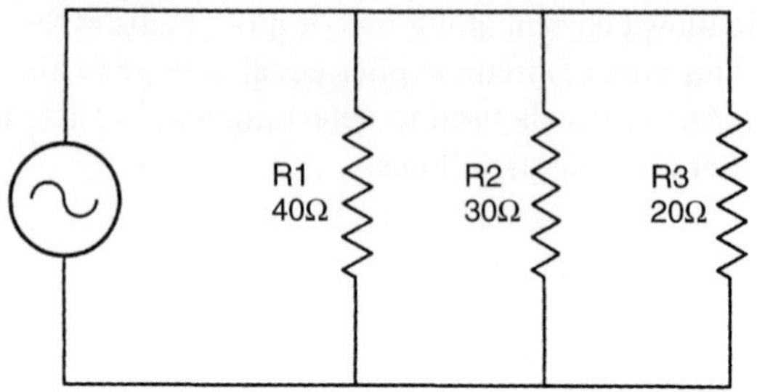

CALCULATED TOTAL RESISTANCE IS 9.234Ω
CALCULATED AMPERAGE:

$$I = \frac{E}{R}$$
$$= \frac{120}{9.234}$$
$$= 12.995 \text{ OR } 13 \text{ AMPERES}$$

Figure 1-31. The current for the circuit in **Figure 1-30** may be calculated from the total resistance when the applied voltage is known.

1.13 Electrical Power

Electrical power (P) is measured in watts. A *watt (W)* is the power used when 1 ampere flows with a potential difference of 1 volt. Therefore, power can be determined by multiplying the voltage times the amperes flowing in a circuit.

$$\text{Watts} = \text{Volts} \times \text{Amperes}$$
or
$$P = E \times I$$

The consumer of electrical power pays the electrical utility company according to the number of kilowatts (kW) used for a certain time span usually billed as kilowatt hours (kWh). A kilowatt is equal to 1000 W. To determine the power being consumed, divide the number of watts by 1000:

$$P \text{ (in kW)} = \frac{E \times I}{1000}$$

1.14 Magnetism

Magnetism was briefly discussed previously in the unit to point out how electrical generators are able to produce electricity. Magnets are classified as either permanent or temporary. Permanent magnets are used in only a few applications that air-conditioning and refrigeration technicians would work with. Electromagnets, a type of temporary magnet, are used in many electrical components of air-conditioning and refrigeration equipment.

A magnetic field exists around a wire carrying an electrical current, Figure 1-32. If the wire or conductor is formed in a loop, the magnetic field will be increased, Figure 1-33. If the wire is wound into a coil, a stronger magnetic field will be created, Figure 1-34. This coil of wire carrying an electrical current is called a *solenoid.* This solenoid or electromagnet will attract or pull an iron bar into the coil, Figure 1-35.

If an iron bar is inserted permanently in the coil, the strength of the magnetic field will be increased even more.

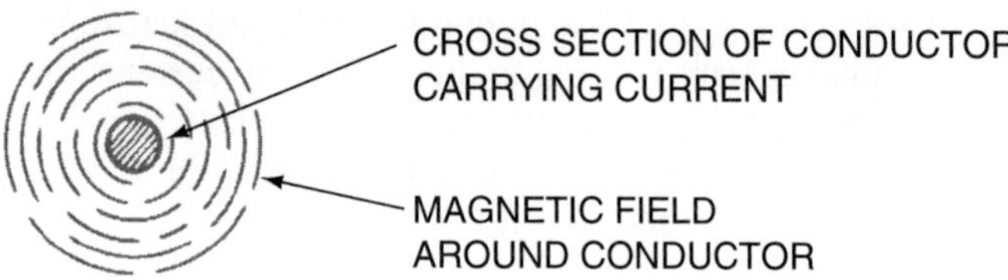

Figure 1-32. This cross section of a wire shows a magnetic field around the conductor.

Figure 1-33. The magnetic field around a loop of wire. This is a stronger field than that around a straight wire.

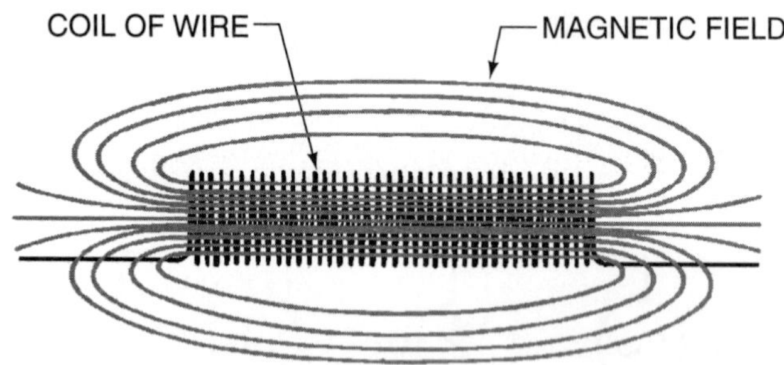

Figure 1-34. There is a stronger magnetic field surrounding wire formed into a coil.

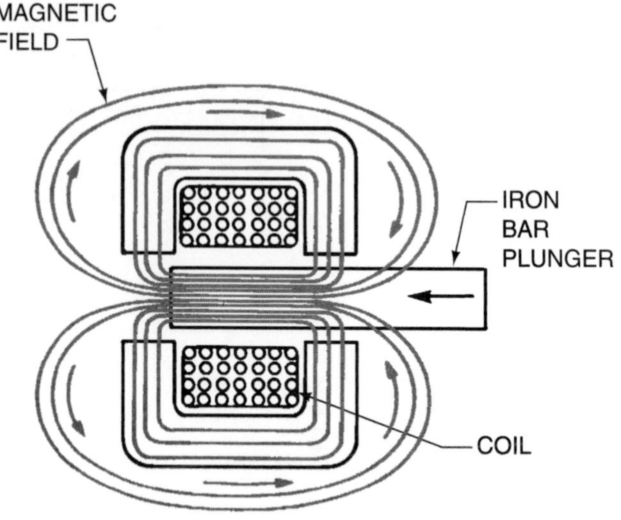

Figure 1-35. When current flows through a coil, the iron bar will be attracted into it.

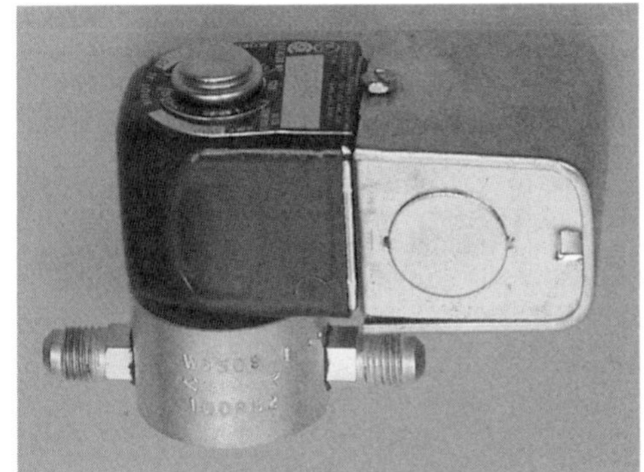

(A)

(B)

(C)

Figure 1-36 (A). (A) A solenoid. (B) A relay. (C) Contactors. *Photos (A) and (B) by Bill Johnson, (C) Courtesy Honeywell, Inc.*

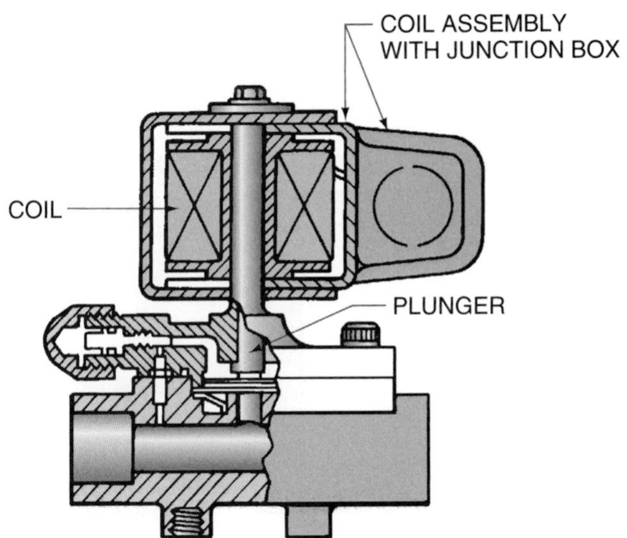

Figure 1-36 (B). A cutaway view of a solenoid. *Courtesy Parker Hannifin Corporation*

This magnetic field can be used to generate electricity and to cause electric motors to operate. The magnetic attraction can also cause motion, which is used in many controls and switching devices, such as solenoids, relays, and contactors, Figure 1-36(A). Figure 1-36(B) is a cutaway view of a solenoid.

1.15 Inductance

As mentioned previously, when voltage is applied to a conductor and current flows, a magnetic field is produced around the conductor. In an AC circuit the current is continually changing direction. This causes the magnetic field to continually build up and immediately collapse. When these lines of force build up and collapse, they cut through the wire or conductor and produce an emf or voltage. This voltage opposes the existing voltage in the conductor.

In a straight conductor this induced voltage is very small and is usually not considered, Figure 1-37. However, if a conductor is wound into a coil, these lines of force overlap and reinforce each other, Figure 1-38. This does develop an emf or voltage that is strong enough to provide opposition to the existing voltage. This opposition is called *inductive reactance* and is a type of resistance in an AC circuit. Coils, chokes, and transformers are examples of components that produce inductive reactance. Figure 1-39 shows symbols for the electrical coil.

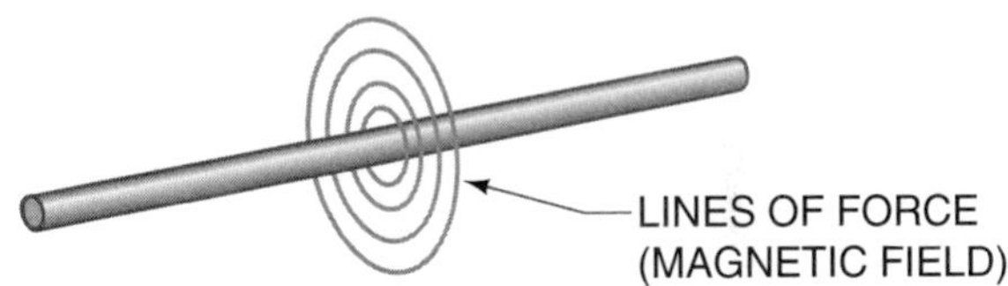

Figure 1-37. A straight conductor with a magnetic field.

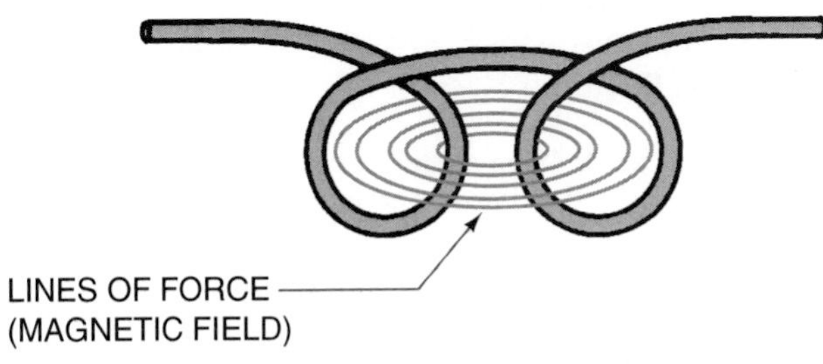

Figure 1-38. A conductor formed into a coil with lines of force.

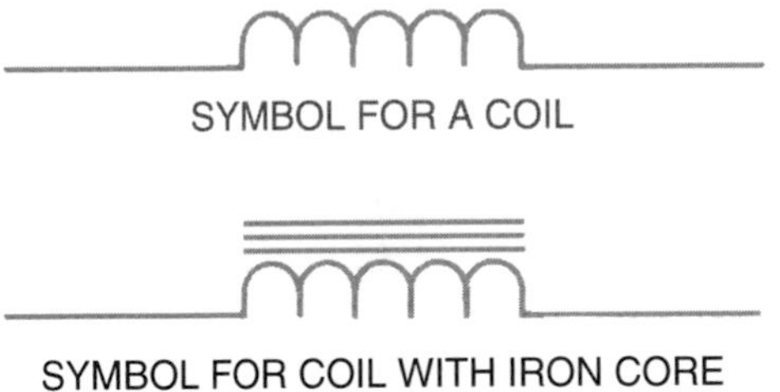

Figure 1-39. Symbols for a coil.

1.16 Transformers

Transformers are electrical devices that produce an electrical current in a second circuit through electromagnetic induction.

In Figure 1-40(A) a voltage applied across terminals A-A will produce a magnetic field around the steel or iron core. This is AC causing the magnetic field to continually build up and collapse as the current reverses. This will cause the magnetic field around the core in the second winding to cut across the conductor wound around it. An electric current is induced in the second winding.

Transformers, Figure 1-40(B), have a primary winding, a core usually made of thin plates of steel laminated together, and a secondary winding. There are step-up and step-down transformers. A step-down transformer contains more turns of wire in the primary winding than in the secondary winding. The voltage at the secondary is directly proportional to the number of turns of wire in the secondary as compared to the number of turns in the primary windings. For example, Figure 1-41 is a transformer with 1000 turns in the primary and 500 turns in the secondary. A voltage of 120 V is applied, and the voltage induced in the secondary is 60 V. Actually the voltage is slightly less due to some loss into the air of the magnetic field and because of resistance in the wire.

A step-up transformer has more windings in the secondary than in the primary. This causes a larger voltage to be induced into the secondary. In Figure 1-42, with 1000 turns in the primary, 2000 in the secondary, and an applied voltage of 120 V, the voltage induced in the secondary is double, or approximately 240 V.

The same power (watts) is available at the secondary as at the primary (except for a slight loss). If the voltage is reduced to one-half that at the primary, the current capacity nearly doubles.

Step-up transformers are used at generating stations to increase the voltage to produce more efficiency in delivering the electrical energy over long distances to substations or other distribution centers. At the substation the voltage is reduced for further distribution. To reduce the voltage, a step-down transformer is used. At a residence, the voltage may be reduced to 240 V or 120 V. Further step-down transformers may be used with air-conditioning and

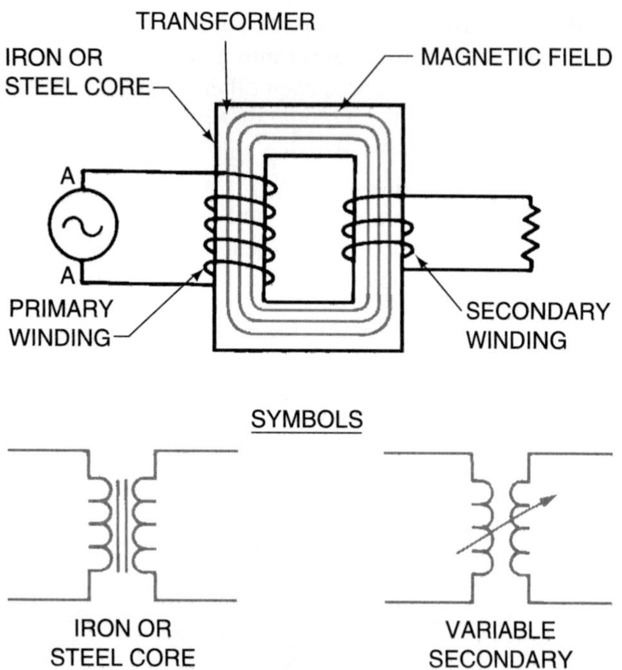

Figure 1-40(A). Voltage applied across terminals produces a magnetic field around an iron or steel core.

Figure 1-40(B). A transformer. *Photo by Bill Johnson*

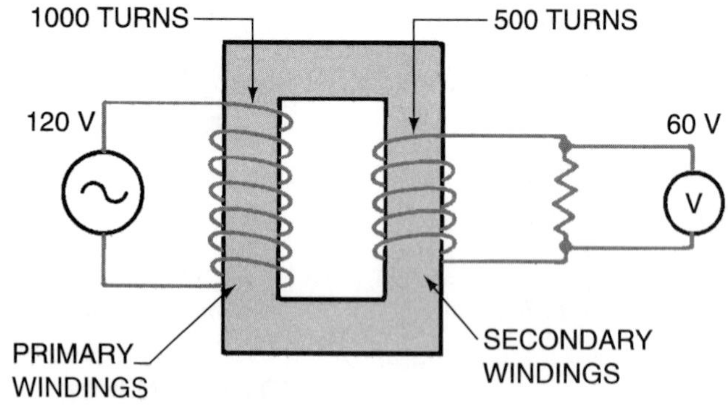

Figure 1-41. A step-down transformer.

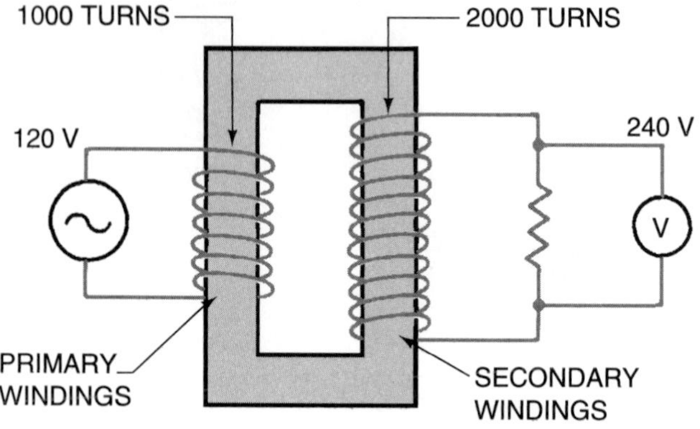

Figure 1-42. A step-up transformer.

heating equipment to produce the 24 V commonly used in thermostats and other control devices.

1.17 Capacitance

A device in an electrical circuit that allows electrical energy to be stored for later use is called a *capacitor.* A simple capacitor is composed of two plates with insulating material between them, Figure 1-43. The capacitor can store a charge of electrons on one plate. When the plate is fully charged in a DC circuit, no current will flow until there is a path back to the positive plate, Figure 1-44. When this path is available, the electrons will flow to the positive plate until the negative plate no longer has a charge, Figures 1-45 and 1-46. At this point both plates are neutral.

In an AC circuit, the voltage and current are continually changing direction. As the electrons flow in one direction, the capacitor plate on one side becomes charged. As this current and voltage reverses, the charge on the capacitor becomes greater than the source voltage, and the capacitor begins to discharge. It is discharged through the circuit, and the opposite plate becomes charged. This continues through each AC cycle.

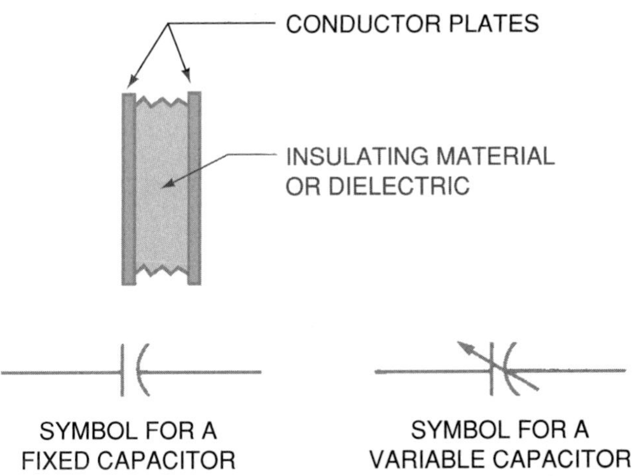

Figure 1-43. A capacitor.

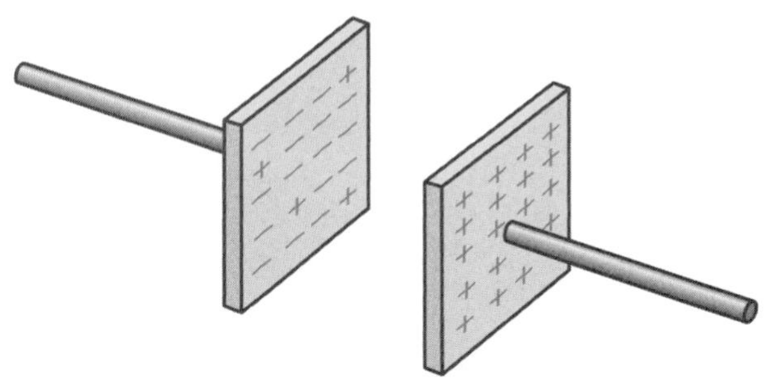

Figure 1-44. A charged capacitor.

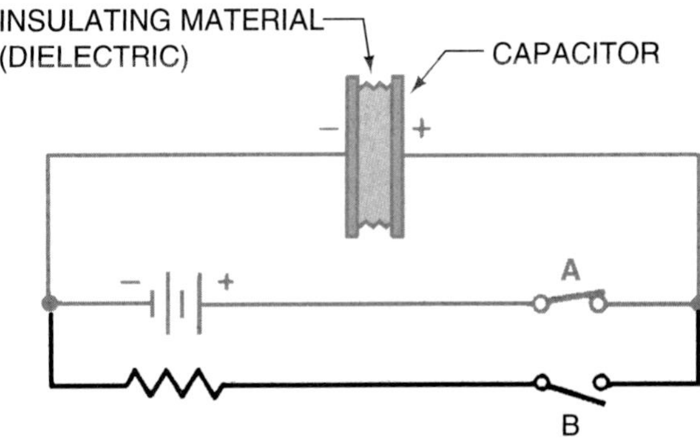

Figure 1-45. Electrons will flow to the negative plate from the battery. The negative plate will charge until the capacitor has the same potential difference as the battery.

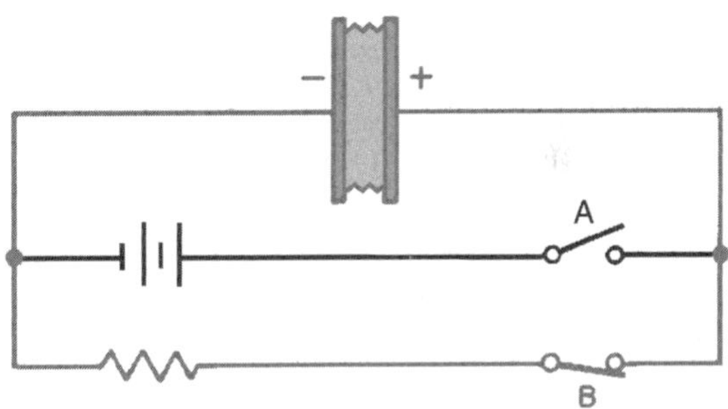

Figure 1-46. When the capacitor is charged, switch A is opened, switch B is closed, and the capacitor discharges through the resistor to the positive plate. The capacitor then has no charge.

A capacitor has *capacitance*, which is the amount of charge that can be stored. The capacitance is determined by the following physical characteristics of the capacitor:

1. Distance between the plates
2. Surface area of the plates
3. Dielectric material between the plates

Capacitors are rated in farads. However, farads represent such a large amount of capacitance that the term microfarad is normally used. A microfarad is one millionth (0.000001) of a farad. Capacitors can be purchased in ranges up to several hundred microfarads. The symbol for micro is the Greek letter μ (mu) and the symbol for farad is the capital letter F (μ F).

A capacitor opposes current flow in an AC circuit similar to a resistor or to inductive reactance. This opposition or type of resistance is called *capacitive reactance.* The capacitive reactance depends on the frequency of the voltage and the capacitance of the capacitor.

Two types of capacitors used frequently in the air-conditioning and refrigeration industry are the starting and running capacitors used on electric motors, Figure 1-47.

Figure 1-47. A starting capacitor and a running capacitor. *Photo by Bill Johnson*

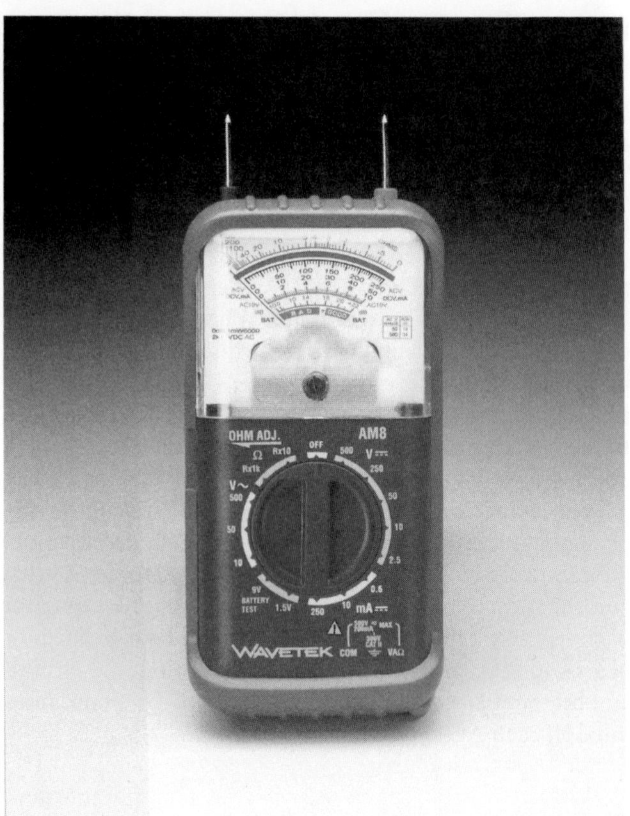

Figure 1-48. A volt-ohm-milliammeter (VOM). *Courtesy Wavetek*

1.18 Impedance

We have learned that there are three types of opposition to current flow in an AC circuit. There is pure resistance, inductive reactance, and capacitive reactance. The total effect of these three is called impedance. The voltage and current in a circuit that has only resistance are in phase with each other. The voltage leads the current across an inductor and lags behind the current across a capacitor. Inductive reactance and capacitive reactance can cancel each other. Impedance is a combination of the opposition to current flow produced by these characteristics in a circuit.

1.19 Electrical Measuring Instruments

A multimeter is an instrument that measures voltage, current, resistance, and, on some models, temperature. It is a combination of several meters and can be used for making AC or DC measurements in several ranges. It is the instrument used most often by heating, refrigeration, and air-conditioning technicians.

A multimeter often used is the *volt-ohm-milliammeter (VOM)* such as the one in Figure 1-48. This meter is used to measure AC and DC voltages, and resistance. AC amperage may be measured with some meters when used with an AC ammeter adapter, Figure 1-49. Meters have different features depending on the manufacturer. Most meters have a function switch and a range switch. We will use a typical meter with examples to explain how the technician uses the meter.

The function switch, located at the left side of the lower front panel, Figure 1-50, has –DC, +DC, and AC positions.

The range switch is in the center of the lower part of the front panel, Figure 1-50. It may be turned in either direction to obtain the desired range. It also selects the proper position for making AC measurements when using the AC clamp-on adapter.

Figure 1-49. An ammeter adapter to VOM. *Courtesy Amprobe*

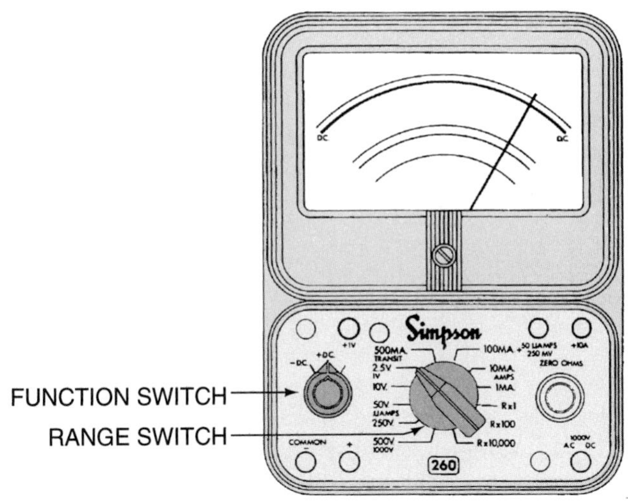

Figure 1-50. The function switch and range switch on a VOM. *Courtesy Simpson Electric Co.*

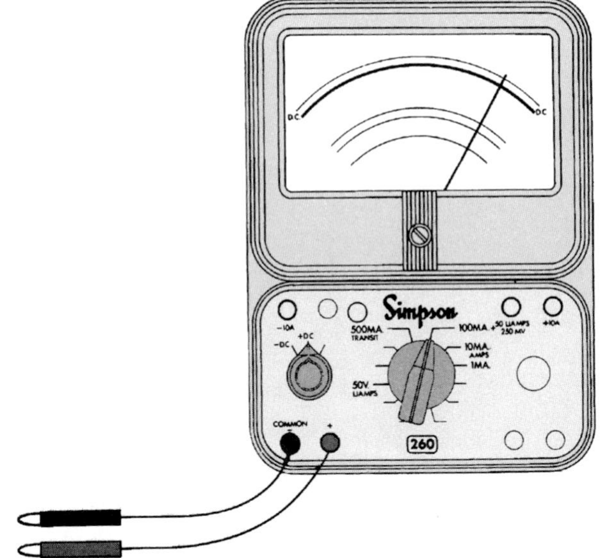

Figure 1-52. A VOM showing the – and + jacks. *Courtesy Simpson Electric Co.*

The zero ohms control on the right of the lower panel is used to adjust the meter to compensate for the aging of the meter's batteries, Figure 1-51.

When the meter is ready to operate, the pointer must read zero. If the pointer is off zero, use a screwdriver to turn the screw clockwise or counterclockwise until the pointer is set exactly at zero, Figure 1-51.

The test leads may be plugged into any of eight jacks. In this unit the common (–) and the positive (+) jacks will be the only ones used, Figure 1-52. Only a few of the basic measurements are discussed here. Other measurements are described in detail in other units.

The following instructions are for familiarization with the meter and procedure only. SAFETY PRECAUTION: Do not make any measurements without instructions and approval from an instructor or supervisor. Insert the black test lead into the common (–) jack. Insert the red test lead into the positive (+) jack.

Figure 1-53 is a DC circuit with 15 V from the battery power source. To check this voltage with the VOM, set the function switch to +DC. Set the range switch to 50 V, Figure 1-53. If you are not sure about the magnitude of the voltage, always set the range switch to the highest setting. After measuring, you can set the switch to a lower range if necessary to obtain a more accurate reading. Be

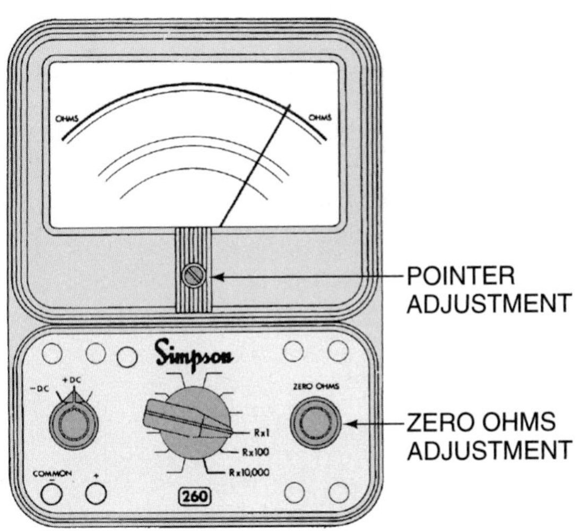

Figure 1-51. Zero ohm adjustment and pointer adjustment. *Courtesy Simpson Electric Co.*

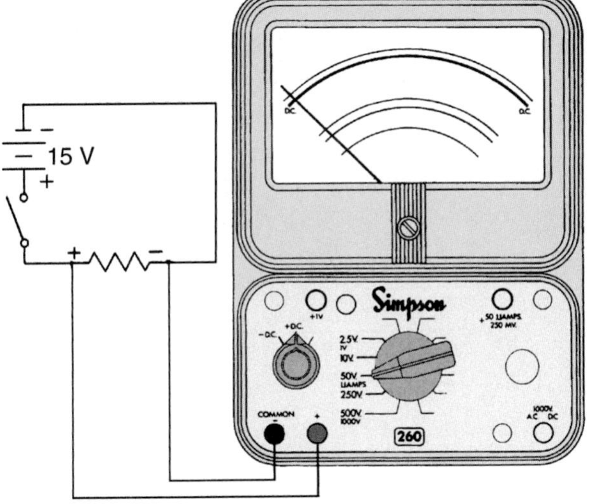

Figure 1-53. A VOM with the function switch set at +DC and range switch set at 50 V. *Courtesy Simpson Electric Co.*

sure the switch in the circuit is open. Now connect the black test lead to the negative side of the circuit, and connect the red test lead to the positive side, as indicated in Figure 1-53. Note that the meter is connected across the load (in parallel). Close the switch and read the voltage from the DC scale.

To check the voltage in the AC circuit in Figure 1-54, follow the steps listed:

1. Turn off the power.
2. Set the function switch to AC.
3. Set the range switch to 500 V.
4. Plug the black test lead into the common (–) jack and the red test lead into the positive (+) jack.
5. Connect the test leads across the load as shown in Figure 1-54.
6. Turn on the power. Read the red scale marked AC. Use the 0 to 50 figures and multiply the reading by 10.
7. Turn the range switch to 250 V. Read the red scale marked AC and use the black figures immediately above the scale.

To determine the resistance of a load, disconnect the load from the circuit. Make sure all power is off while doing this. If the load is not entirely disconnected from the circuit, the voltage from the internal battery of the ohmmeter may damage solid-state components. These solid-state components may be part of an electronic control board or microprocessor.

1. Make the zero ohms adjustment in the following manner:
   A. Turn the range switch to the desired ohms range.
      Use R × 1 for 0 to 200 Ω
      Use R × 100 for 200 to 20,000 Ω
      Use R × 10,000 for above 20,000 Ω
   B. Plug the black test lead into the common (–) jack and the red test lead into the positive (+) jack.
   C. Connect the test leads to each other.
   D. Rotate the zero ohms control until the pointer indicates zero ohms. (If the pointer cannot be adjusted to zero, replace the batteries.)

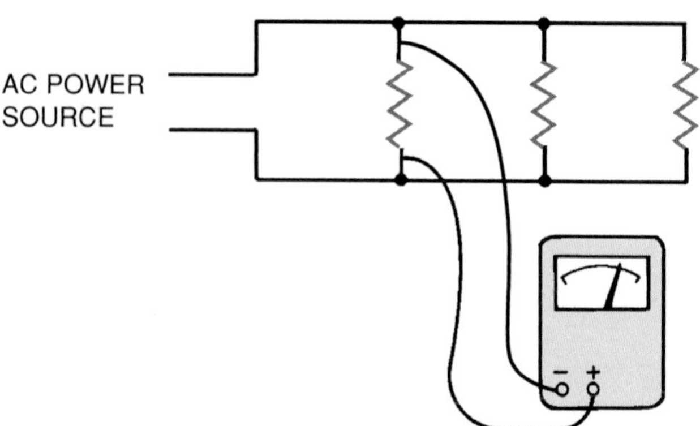

Figure 1-54. Connecting test leads across a load to check voltage.

2. Disconnect the ends of the test leads and connect them to the load being tested.
3. Set the function switch at either –DC or +DC.
4. Observe the reading on the ohms scale at the top of the dial. (Note that the ohms scale reads from right to left.)
5. To determine the actual resistance, multiply the reading by the factor at the range switch position.

The ammeter has a clamping feature that can be placed around a single wire in a circuit, and the current flowing through the wire can be read as amperage from the meter, Figure 1-55. SAFETY PRECAUTION: Do not perform any of the preceding or following tests in this unit without approval from an instructor or supervisor. These instructions are simply a general orientation to meters. Be sure to read the operator's manual for the particular meter available to you.

It is often necessary to determine voltage or amperage readings to a fraction of a volt or ampere, Figure 1-56.

Many styles and types of meters are available for making electrical measurements. Figure 1-57 shows some of these meters. Many modern meters come with digital readouts, Figure 1-58.

Electrical troubleshooting is taken one step at a time. Figure 1-59 is a partial wiring diagram of an oil burner fan circuit showing the process. The fan motor does not operate because of an open motor winding. The following

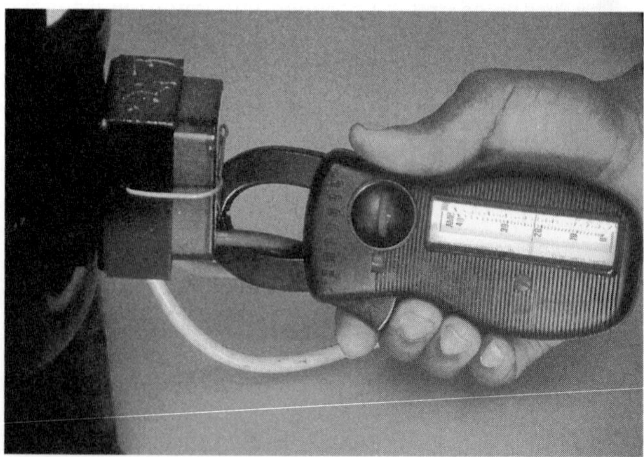

Figure 1-55. Measuring amperage by clamping the meter around the conductor. *Photo by Bill Johnson*

VOLTAGE

1 volt = 1000 millivolts (m V)
1 volt = 1000 000 microvolts (μ V)

AMPERAGE

1 ampere = 1000 milliamperes (m A)
1 ampere = 1000 000 microamperes (μ A)
Note that the symbol for micro or millionths is μ

Figure 1-56. Units of voltage and amperage.

(A)

(B)

(C)

Figure 1-57. Meters used for electrical measurements. (A) DC millivoltmeter. (B) Multimeter (VOM). (C) Digital clamp-on ammeter. *(A) Photo by Bill Johnson, (B) Courtesy Wavetek, (C) Courtesy Amprobe*

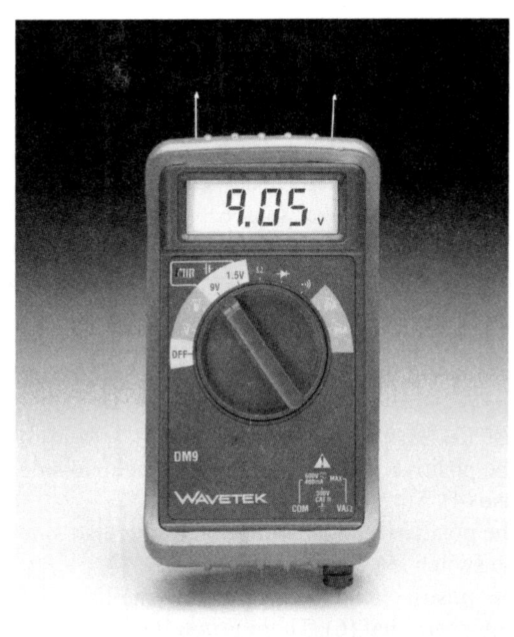

Figure 1-58. A typical VOM with digital readout. *Courtesy Wavetek*

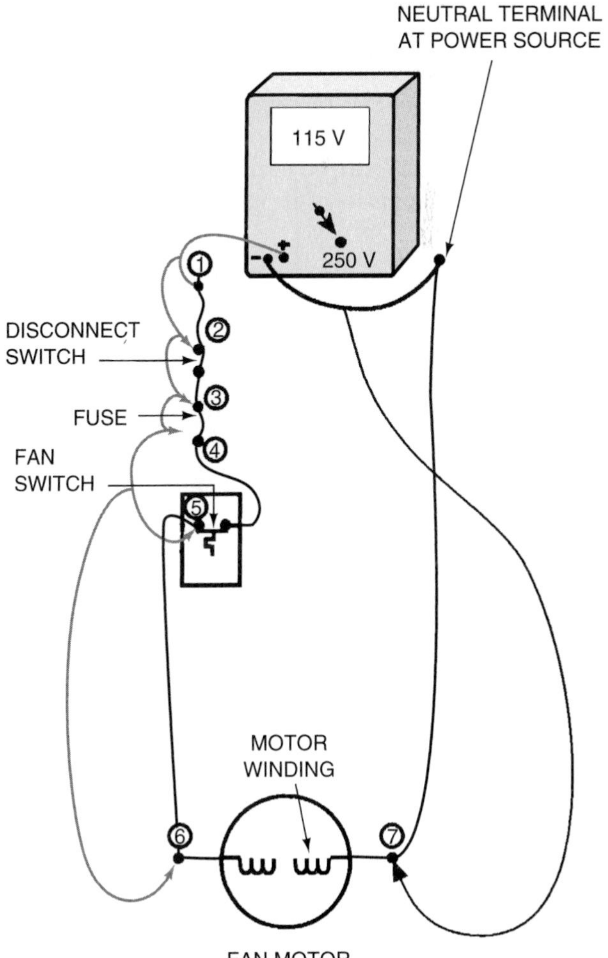

Figure 1-59. A partial diagram of an oil burner fan motor circuit.

voltage checks could be made by the technician to determine where the failure is.

1. This is a 120-V circuit. The technician would set the VOM range selector switch to 250 V AC. The neutral (– or common) meter lead is connected to a neutral terminal at the power source.
2. The positive (+) meter probe is connected to a line-side power source terminal, Figure 1-59(1). The meter should read 115 V, indicating that power is at the source.
3. Then the positive lead is connected to the line-side terminal of the disconnect switch (2). There is power.
4. The positive lead is connected to the line side of the fuse (3). There is power.
5. The positive lead is connected to the load side of the fuse (4). There is power.
6. The positive lead is connected to the load side of the fan switch (5). There is power.
7. The positive lead is connected to the line side of the motor terminal (6). There is power.
8. To ensure there is power through the conductor and terminal connections on the neutral or ground side, connect the neutral meter probe to the neutral side of the motor (7). Leave the positive probe on the line side of the motor. There should be a 115-V reading, indicating that current is flowing through the neutral conductor.

In all of the above checks, there is 115 V, but the motor does not run. It would be appropriate to conclude that the motor is defective.

The motor winding can be checked with an ohmmeter. The motor winding must have a measurable resistance for it to function properly. To check this resistance the technician may do the following, Figure 1-60:

1. Turn off the power source to the circuit.
2. Disconnect the terminals on the motor from the circuit.
3. Set the meter selector switch to ohms R × 1.
4. Touch the meter probes together and adjust the meter to 0 ohms.

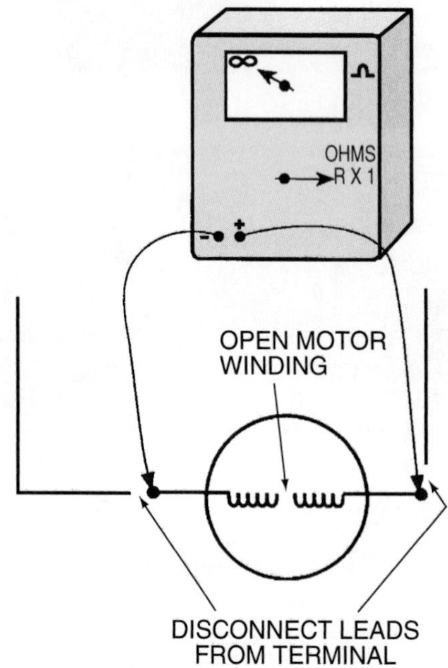

Figure 1-60. Checking an electric motor winding with an ohmmeter.

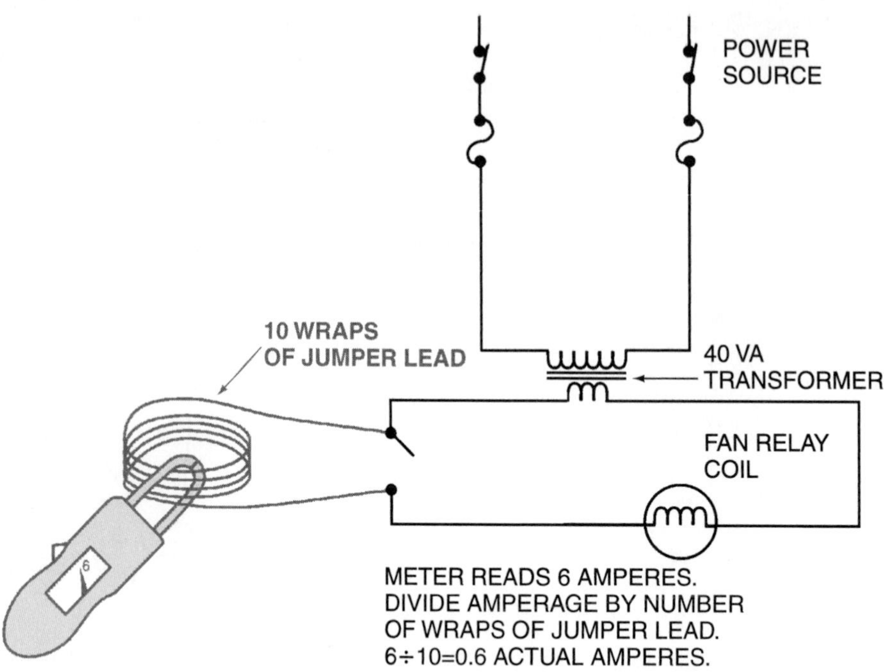

Figure 1-61. An illustration using a 10-wrap multiplier with the ammeter.

5. Touch one meter probe to one motor terminal and the other to the other terminal. The meter reads infinity (∞). This is the same reading you would get by holding the meter probes apart in the air. There is no circuit through the windings indicating that they are open.

Most clamp-on type ammeters do not read accurately in the lower amperage ranges such as in the 1 ampere or below range. However, a standard clamp-on ammeter can be modified to produce an accurate reading. For instance, transformers are rated in volt-amperes (VA). A 40-VA transformer is often used in the control circuit of combination heating and cooling systems. These transformers produce 24 V and can carry a maximum of 1.66 A. This is determined as follows:

$$\text{Output in amperes} = \frac{\text{VA rating}}{\text{Voltage}}$$

$$I = 40/24$$

$$= 1.66\text{ A}$$

Figure 1-61 illustrates how the clamp-on ammeter can be used with 10 wraps of wire to multiply the amperage reading by 10. To determine the actual amperage, divide the amperage indicated on the meter by 10.

1.20 Sine Waves

As mentioned earlier, alternating current continually reverses direction. An oscilloscope is an instrument that measures the amount of voltage over a period of time and can display this voltage on a screen. This display is called a wave form. Many types of wave forms exist, but we will discuss the one most refrigeration and air-conditioning technicians are involved with, the sine wave, Figure 1-62.

The sine wave displays the voltage of one cycle through 360°. In the United States and Canada, the standard voltage in most locations is produced with a frequency of 60 cycles per second. Frequency is measured in hertz (Hz). Therefore, the standard frequency in this country and Canada is 60 Hz.

Figure 1-62 is a sine wave as it would be displayed on an oscilloscope. At the 90° point the voltage reaches its peak (positive); at 180° it is back to 0; at 270° it reaches its negative peak; and at 360° it is back to 0. If the frequency is

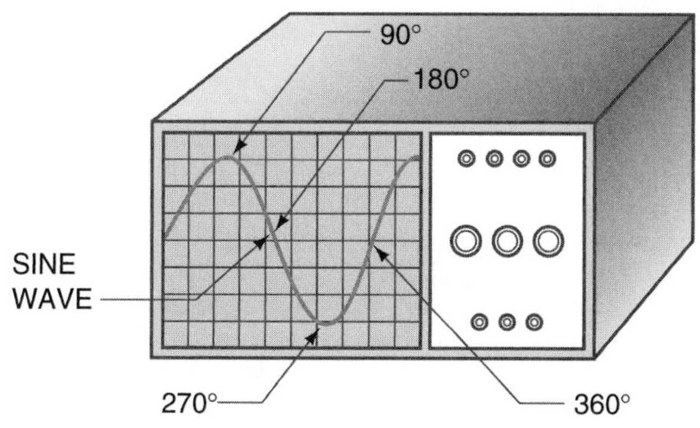

Figure 1-62. A sine wave displayed on an oscilloscope.

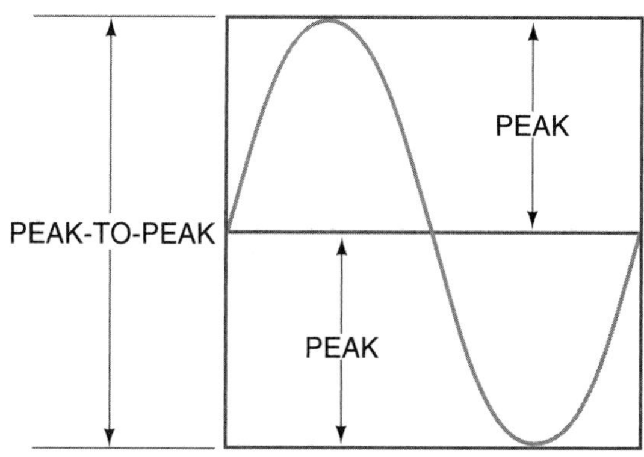

Figure 1-63. Peak and peak-to-peak AC voltage values.

60 Hz, this cycle would be repeated 60 times every second. The sine wave is a representation of a trigonometric function of an alternating current cycle.

Figure 1-63 shows the peak and peak-to-peak values. As the sine wave indicates, the voltage is at its peak value briefly during the cycle. Therefore, the peaks of the peak-to-peak values are not the effective voltage values.

The effective voltage is the RMS voltage, Figure 1-64. The RMS stands for root-mean-square value. This is the alternating current value measured by most voltmeters and ammeters. The RMS voltage is 0.707 × the peak voltage. If the peak voltage were 170 V, the effective voltage measured by a voltmeter would be 120 V (170 V × 0.707 = 120.19 V).

Sine waves can illustrate a cycle of an AC electrical circuit that contains only a pure resistance such as a circuit with electrical heaters. In a pure resistive circuit such as this, the voltage and current will be in phase. This is illustrated with sine waves such as in Figure 1-65. Notice that the voltage and current reach their negative and positive peaks at the same time.

Sine waves can also illustrate a cycle of an AC electrical circuit that contains a fan relay coil. This coil will produce an inductive reactance. The sine wave will show the current lagging the voltage in this circuit, Figure 1-66.

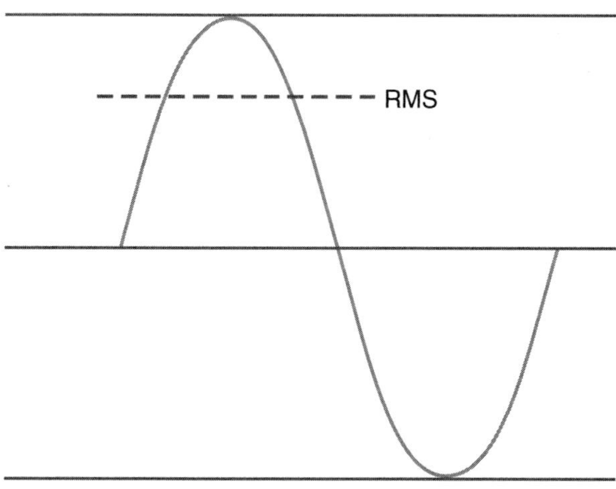

Figure 1-64. The root-mean-square (RMS) or effective voltage.

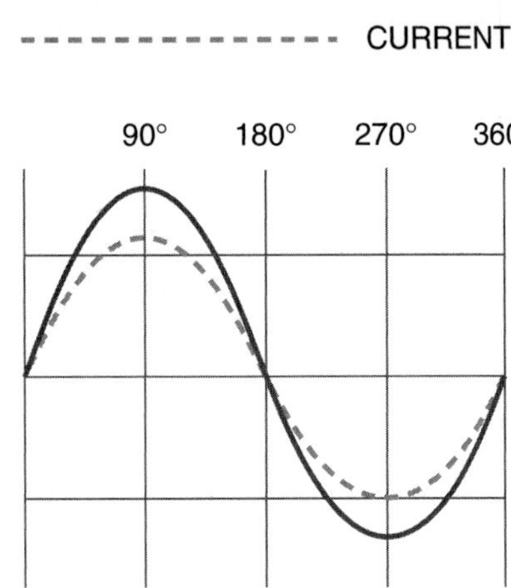

Figure 1-65. This sine wave represents both the voltage and current in phase in a pure resistive circuit.

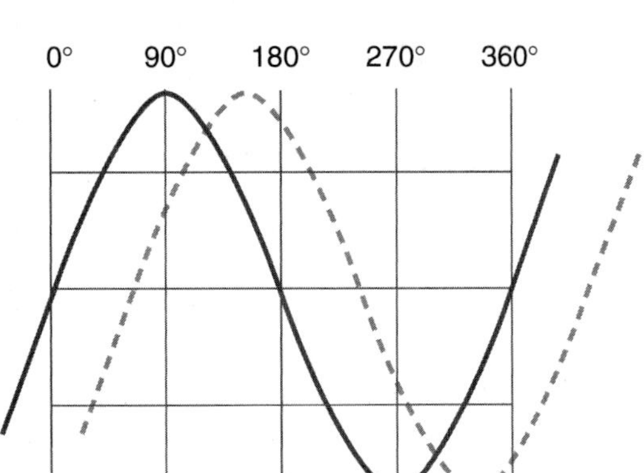

Figure 1-66. This sine wave represents the current lagging the voltage (out of phase) in an inductive circuit.

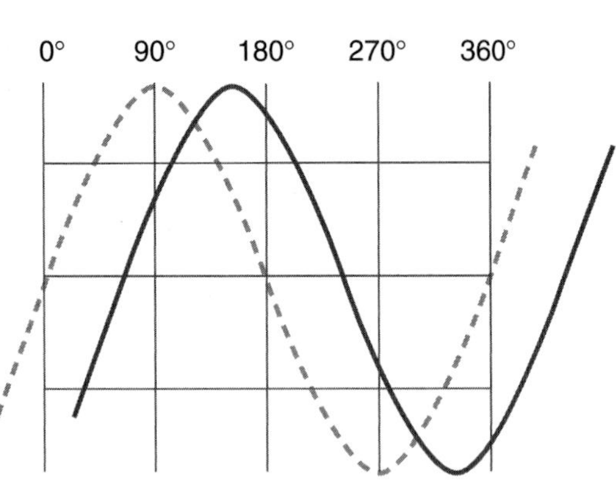

Figure 1-67. This sine wave represents a capacitive AC cycle sine with the voltage lagging the current. The voltage and current are out of phase.

Figure 1-67 shows a sine wave illustrating a cycle of an AC electrical circuit that has a capacitor producing a pure capacitive circuit. In this case the current leads the voltage.

1.21 Wire Sizes

All conductors have some resistance. The resistance depends on the conductor material, the cross-sectional area of the conductor, and the length of the conductor. A conductor with low resistance carries a current more easily than a conductor with high resistance.

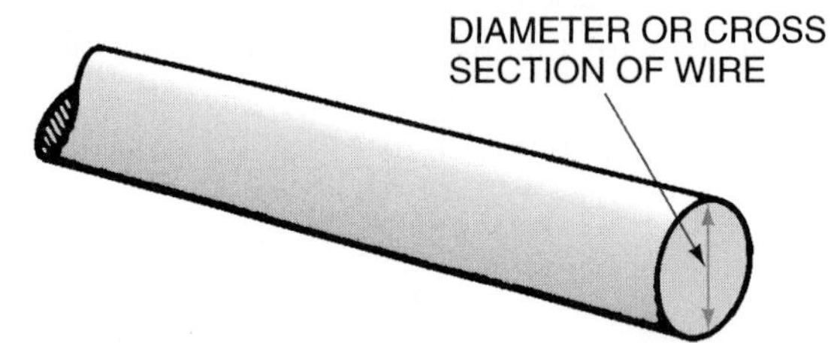

Figure 1-68. The cross section of a wire.

WIRE SIZE	
	60°C (140°F)
AWG or kcmil	Types TW, UF
COPPER	
18	—
16	—
14*	20
12*	25
10*	30
8	40
6	55
4	70
3	85
2	95
1	110

*Small Conductors. Unless specifically permitted in (e) through (g), the overcurrent protection shall not exceed 15 amperes for No. 14, 20 amperes for No. 12, and 30 amperes for No. 10 copper.

Figure 1-69. This section of the *National Electrical Code®* shows an example of how a wire is sized for only one type of conductor. It is not the complete and official position but is a representative example.

The proper wire (conductor) size must always be used. The size of a wire is determined by its diameter or cross section, Figure 1-68. A larger diameter wire has more current-carrying capacity than a smaller diameter wire. SAFETY PRECAUTION: If a wire is too small for the current passing through, it will overheat and possibly burn the insulation and could cause a fire. Standard copper wire sizes are identified by American Standard Wire Gauge numbers and measured in circular mils. A circular mil is the area of a circle 1/1000 in. in diameter. Temperature is also considered because resistance increases as temperature increases. Increasing wire size numbers indicate *smaller* wire diameters and greater resistance. For example, number 12 wire size is smaller than number 10 wire size and has less current-carrying capacity.

The technician should not determine and install a conductor of a particular size unless licensed to do so. The technician should, however, be able to recognize an undersized conductor and bring it to the attention of a qualified person. As mentioned previously, an undersized wire may cause voltage to drop, breakers or fuses to trip, and conductors to overheat.

The conductors are sized by their amperage-carrying capacity. This is called *ampacity.* Figure 1-69 contains a small part of a chart from the *National Electrical Code*® *(NEC*®*)* and a partial footnote for one type of conductor. This chart as shown and the footnote should not be used when determining a wire size. It is shown here only to familiarize you with the way in which it is presented in the *NEC*®. The footnote actually reduces the amount of amperes for number 12 and number 14 wire listed in the *NEC*® table in Figure 1-69. The reason is that number 12 and number 14 wire are used in residential houses where circuits are often overloaded unintentionally by the homeowner. The footnote exception simply adds more protection.

An example of a procedure that might be used for calculating the wire size for the outdoor unit for a heat pump follows, Figure 1-70.

General Data **4TWX4042A1000A**

OUTDOOR UNIT ①②	4TWX4042A1000A	
SOUND RATING (DECIBELS) ②⑨	77/75	
POWER CONNS. — V/PH/HZ ③	208/230/1/60	
MIN. BRCH. CIR. AMPACITY	25	25 AMPACITY
BR. CIR. PROT. RTG. MAX. (AMPS)	40	
BR. CIR. PROT. RTG. MIN. (AMPS)	40	
COMPRESSOR	CLIMATUFF® -SCROLL	
NO. USED - NO. SPEEDS	1 - 1	
VOLTS/PH/HZ	208/230/1/60	
R.L. AMPS ⑦ - L.R. AMPS	19.2-104	19.2 A
FACTORY INSTALLED		
START COMPONENTS ⑧	NO	
INSULATION/SOUND BLANKET	YES	
COMPRESSOR HEAT	YES	
OUTDOOR FAN — TYPE	PROPELLER	
DIA. (IN.) - NO. USED	27.6 - 1	
TYPE DRIVE - NO. SPEEDS	DIRECT - 2	
CFM @ 0.0 IN. W.G. ④	4200	
NO. MOTORS - HP	1 - 1/6	
MOTOR SPEED R.P.M.	825	
VOLTS/PH/HZ	200/230	
F.L. AMPS	1.4	1.4 A
OUTDOOR COIL — TYPE	SPINE FIN™	20.6 A TOTAL
ROWS - F.P.I.	1 - 24	
FACE AREA (SQ. FT.)	27.81	
TUBE SIZE (IN.)	5/16	
REFRIGERANT CONTROL	EXPANSION VALVE	
REFRIGERANT		
LBS. — R-410A (O.D. UNIT) ⑤	7/04 -LB/OZ	
FACTORY SUPPLIED	YES	
LINE SIZE - IN. O.D. GAS ⑥	3/4	
LINE SIZE - IN. O.D. LIQ. ⑥	3/8	
FCCV		
RESTRICTOR ORIFICE SIZE	0.071	
DIMENSIONS	H X W X D	
OUTDOOR UNIT - CRATED (IN.)	53.4 X 35.1 X 38.7	
UNCRATED	SEE OUTLINE DWG.	
WEIGHT		
SHIPPING (LBS.)	315	
NET (LBS.)	267	

① **CERTIFIED IN ACCORDANCE WITH THE AIR-SOURCE UNITARY HEAT PUMP EQUIPMENT CERTIFICATION PROGRAM WHICH IS BASED ON A.R.I. STANDARD 210/240.**

② **RATED IN ACCORDANCE WITH A.R.I. STANDARD 270.**

③ CALCULATED IN ACCORDANCE WITH NATIONAL ELECTRIC CODE. ONLY USE HACR CIRCUIT BREAKERS OR FUSES.

④ STANDARD AIR - DRY COIL - OUTDOOR

⑤ THIS VALUE APPROXIMATE. FOR MORE PRECISE VALUE SEE UNIT NAMEPLATE AND SERVICE INSTRUCTION.

⑥ MAX. LINEAR LENGTH: 80 FT WITH RECIPROCATING COMPRESSOR - 60 FT WITH SCROLL. MAX. LIFT - SUCTION 60 FT; MAX LIFT - LIQUID 60 FT. FOR GREATER LENGTH REFER TO REFRIGERANT PIPING SOFTWARE PUB. NO. 32-3312-01.

⑦ THE VALUE SHOWN FOR COMPRESSOR RLA ON THE UNIT NAMEPLATE AND ON THIS SPECIFICATION SHEET IS USED TO COMPUTE MINIMUM BRANCH CIRCUIT AMPACITY AND MAXIMUM FUSE SIZE. THE VALUE SHOWN IS THE BRANCH CIRCUIT SELECTION CURRENT.

⑧ NO MEANS NO START COMPONENTS
YES MEANS QUICK START KIT COMPONENTS
PTC MEANS POSITIVE TEMPERATURE COEFFICIENT STARTER.

⑨ RATED IN ACCORDANCE WITH ARI STANDARD 270/SECTION 5.3.6.

SPLIT SYSTEM

OUTDOOR UNIT WITH HEAT PUMP COILS

Figure 1-70. This specification sheet for one unit shows the electrical data that are used to size the wire to the unit. *Courtesy Trane*

This outdoor unit is a 3½-ton unit (42A means 42,000 Btu/hr or 3½ tons). The electrical data show that the unit compressor uses 19.2 FLA (full-load amperage) and the fan motor uses 1.4 FLA for a total of 20.6 FLA. The specifications round this up to an ampacity of 25. In other words, the conductor must be sized for 25 amperes. According to the *NEC*® the wire size would be 10. The specifications for the unit indicate that a maximum fuse or breaker size would be 40 A. This will give the compressor some extra amperage for the locked-rotor amperage (LRA) at start-up. Some manufacturers give the ampacity and the recommended wire size in the directions or printed on the unit nameplate.

When you find a unit that has low voltage while operating, you should first check the voltage at the entrance panel to the building. If the voltage is correct there and low at the unit, there is either a loose connection, there is undersized wire, or the wire run is too long.

1.22 Circuit Protection Devices

Safety Precaution: Electrical circuits *must* be protected from current overloads. If too much current flows through the circuit, the wires and components will overheat, resulting in damage and possible fire. Circuits are normally protected with fuses or circuit breakers.

Fuses

A *fuse* is a simple device used to protect circuits from over-loading and overheating. Most fuses contain a strip of metal that has a higher resistance than the conductors in the circuit. This strip also has a relatively low melting point. Because of its higher resistance, it will heat up faster than the conductor. When the current exceeds the rating on the fuse, the strip melts and opens the circuit.

PLUG FUSES. Plug fuses have either an Edison base or a type S base, Figure 1-71(A). Edison-base fuses are used in older installations and can be used for replacement only. Type S fuses can be used only in a type S fuse holder specifically designed for the fuse; otherwise an adapter such as the one illustrated in Figure 1-71(B) must be used. Each adapter is designed for a specific ampere rating, and these fuses cannot be interchanged. The amperage rating determines the size of the adapter. Plug fuses are rated up to 125 V and 30 A.

DUAL-ELEMENT PLUS FUSES. Many circuits have electric motors as the load or part of the load. Motors draw more current when starting and can cause a plain (single element) fuse to burn out or open the circuit. Dual-element fuses are frequently used in this situation, Figure 1-72. One element in the fuse will melt when there is a large overload, such as a short circuit. The other element will melt and open the circuit when there is a smaller current overload lasting more than a few seconds. This allows for the larger starting current of an electric motor.

CARTRIDGE FUSES. For 230-V to 600-V service up to 60 A, the ferrule cartridge fuse is used, Figure 1-73(A). From 60 A to 600 A, knife-blade cartridge fuses can be used, Figure 1-73(B). A cartridge fuse is sized according to its ampere rating to prevent a fuse with an inadequate rating from being used. Many cartridge fuses have an arc-quenching material around the element to prevent damage from arcing in severe short-circuit situations, Figure 1-74.

Circuit Breakers

A circuit breaker can function as a switch as well as a means for opening a circuit when a current overload occurs. Most modern installations in houses and many

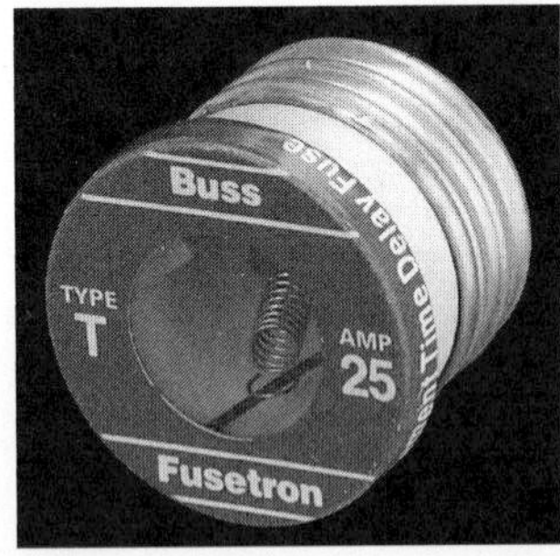

Figure 1-72. A dual-element plug fuse. *Courtesy Bussmann Division, McGraw Edison Company*

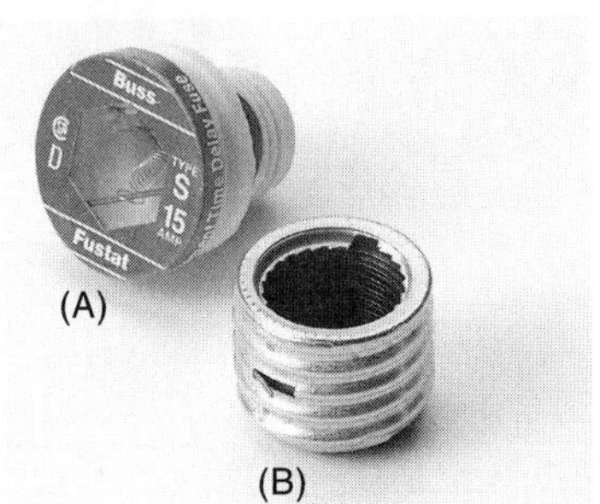

Figure 1-71. (A) A type S base plug fuse. (B) A type S fuse adapter. *Courtesy Bussmann Division, McGraw Edison Company*

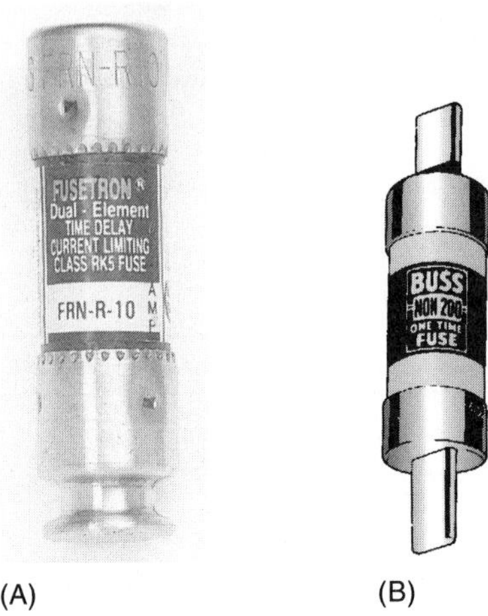

Figure 1-73. (A) A ferrule-type cartridge fuse. (B) A knife-blade cartridge fuse. *Courtesy Bussmann Division, McGraw Edison Company*

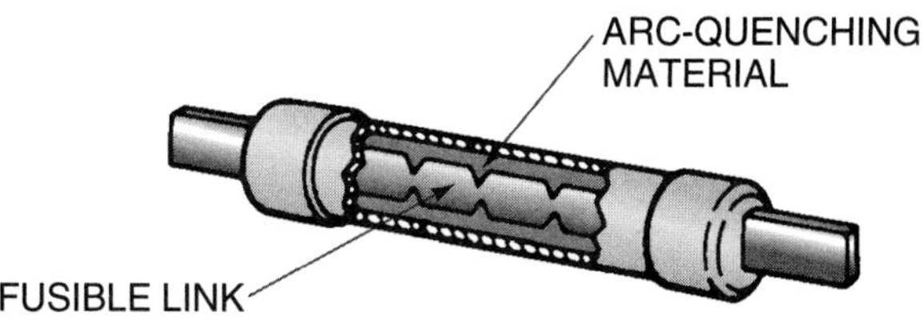

Figure 1-74. A knife-blade cartridge fuse with arc-quenching substance shown.

commercial and industrial installations use circuit breakers rather than fuses for circuit protection.

Circuit breakers use two methods to protect the circuit. One is a bimetal strip that heats up with a current overload and trips the breaker, opening the circuit. The other is a magnetic coil that causes the breaker to trip and open the circuit when there is a short circuit or other excessive current overload in a short time, Figure 1-75.

Ground Fault Circuit Interrupters

Safety Precaution: Ground fault circuit interrupters (GFCI) help protect individuals against shock, in addition to providing current overload protection. The GFCI, Figure 1-76, detects even a very small current leak to a ground. Under certain conditions this leak may cause an electrical shock. This small leak, which may not be detected by a conventional circuit breaker, will cause the GFCI to open the circuit.

Circuit protection is essential to prevent the conductors in the circuit from being overloaded. If one of the circuit power-consuming devices were to cause an overload due to a short circuit within its coil, the circuit protector would stop the current flow before the conductor became overloaded and hot. Remember, a circuit consists of a power supply, the conductor, and the power-consuming device. The conductor must be sized large enough that it does not operate beyond its rated temperature, typically 140°F (60°C) while in an ambient of 86°F (30°C). For example, a circuit may be designed to carry a load of 20 A. As long as the circuit is carrying up to its amperage, overheating is not a potential hazard. If the amperage in the circuit is gradually increased, the conductor will begin to become hot, Figure 1-77. Proper understanding of circuit protection is a lengthy process. More details can be obtained from the *NEC®* and from further study of electricity.

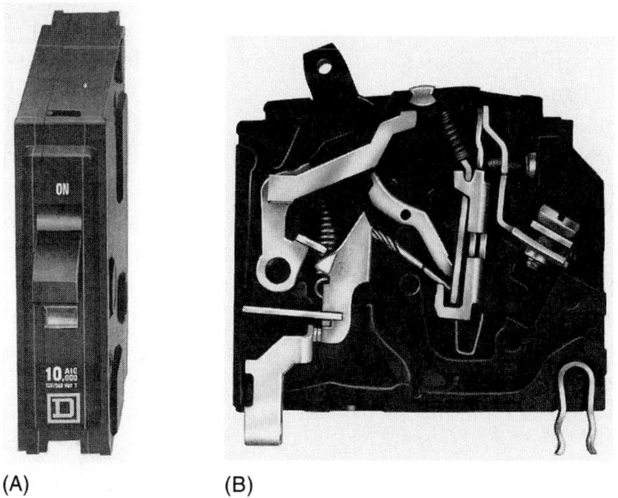

Figure 1-75. (A) A circuit breaker. (B) A cutaway. *Courtesy Square D Company*

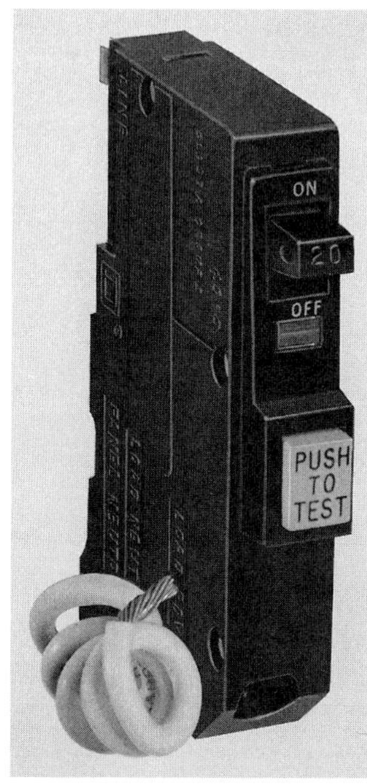

Figure 1-76. A ground fault circuit interrupter. *Courtesy Square D Company*

1.23 Semiconductors

The development of what are commonly called semiconductors or solid-state components has caused major changes in the design of electrical devices and controls.

Semiconductors are generally small and lightweight and can be mounted in circuit boards, Figure 1-78. In this unit we describe some of the individual solid-state devices and some of their uses. Refrigeration and air-conditioning technicians do not normally replace solid-state components on circuit boards. They should have some knowledge of these components, however, and should be able to determine when one or more of the circuits in which they are used are defective. In most cases, when a component is defective the entire board will need to be replaced. Often these circuit boards can be returned to the manufacturer or sent to a company that specializes in repairing or rebuilding them.

Semiconductors are usually made of silicon or germanium. Semiconductors in their pure form, as their name implies, do not conduct electricity very well. However, for semiconductors to be of value they must conduct electricity in some controlled manner. To accomplish this an additional

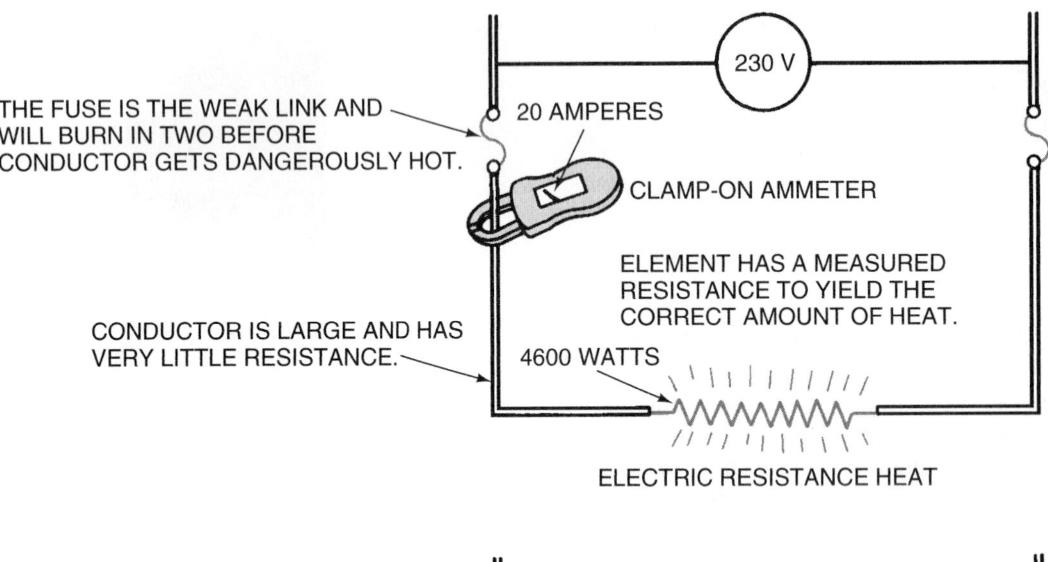

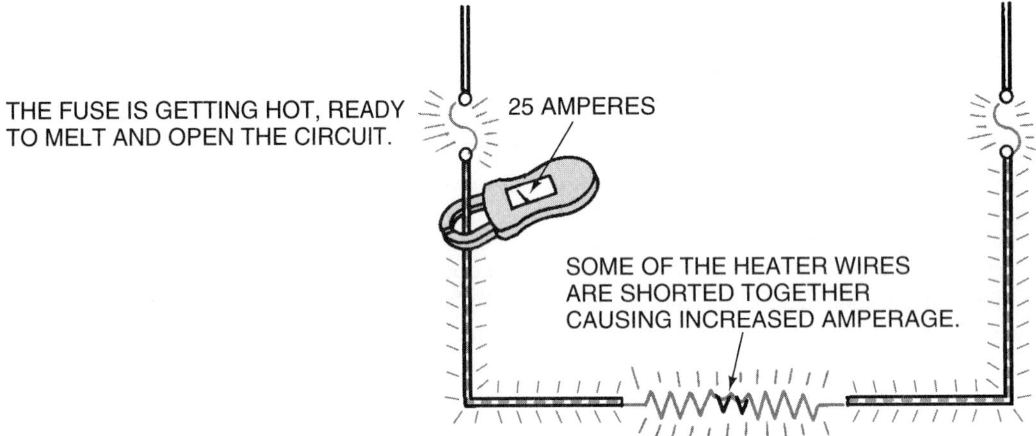

Figure 1-77. Fuses protect the circuit.

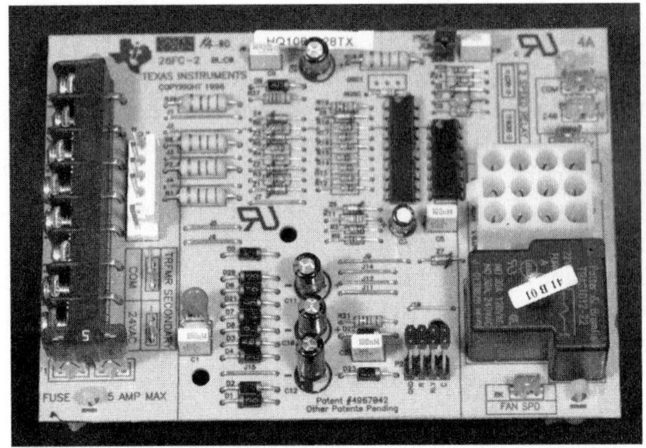

Figure 1-78. A circuit board with semiconductors. *Photo by Bill Johnson*

substance, often called an impurity, is added to the crystal-like structures of the silicon or germanium. This is called doping. One type of impurity produces a hole in the material where an electron should be. Because the hole replaces an electron (which has a negative charge), it results in the material having fewer electrons or a net result of a positive charge. This is called a P-type material. If a material of a different type is added to the semiconductor, an excess of electrons is produced; the material has a negative charge and is called an N-type material.

When a voltage is applied to a P-type material, electrons fill these holes and move from one hole to the next still moving from negative to positive. However, this makes it appear that the holes are moving in the opposite direction (from positive to negative) as the electrons move from hole to hole.

N-type material has an excess of electrons that move from negative to positive when a voltage is applied.

Solid-state components are made from a combination of N-type and P-type substances. The manner in which the materials are joined together, the thickness of the materials, and other factors determine the type of solid-state component and its electronic characteristics.

DIODES. Diodes are simple solid-state devices. They consist of P-type and N-type material connected together. When this combination of P- and N-type material is connected to a power source one way, it will allow current to flow and is said to have forward bias. When reversed it is said to have reverse bias, and no current will flow. Figure 1-79 is a drawing of a simple diode. Figure 1-80 is a photo of two types of diodes. One of the connections on the diode is called the cathode and the other the anode, Figure 1-81. If the diode is to be connected to a battery to have forward bias (current flow), the negative terminal on the battery should be connected to the cathode, Figure 1-82. Connecting the negative terminal of the battery to the anode will produce reverse bias (no current flow), Figure 1-83.

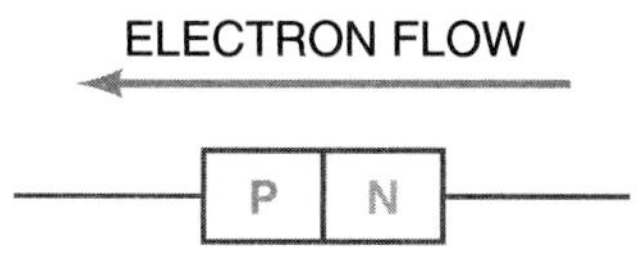

Figure 1-79. A pictorial drawing of a diode.

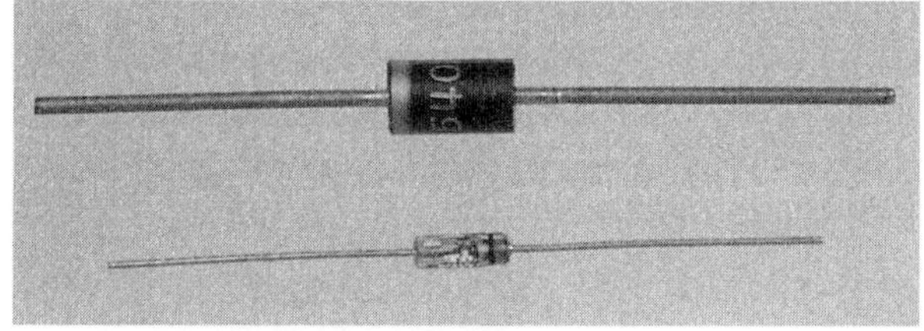

Figure 1-80. Typical diodes. *Photo by Bill Johnson*

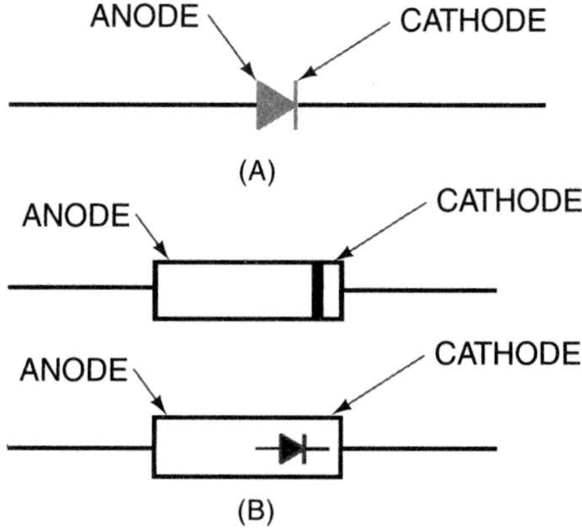

Figure 1-81. (A) The schematic symbol for a diode. (B) Identifying markings on the diode.

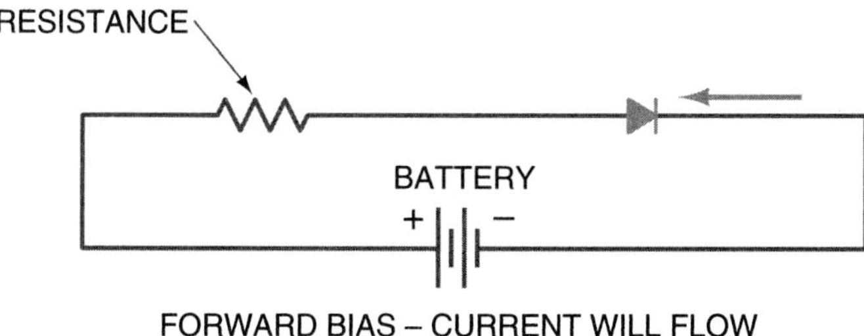

Figure 1-82. A simple diagram with a diode indicating forward bias.

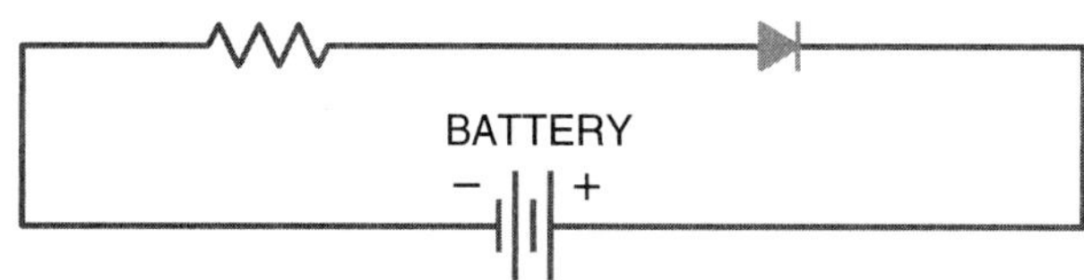

Figure 1-83. A circuit with a diode indicating reverse bias.

CHECKING A DIODE. A diode may be tested by connecting an ohmmeter across it. The diode must be removed from the circuit. The negative probe should be touched to the cathode and the positive to the anode. With the selector switch on R × 1, the meter should show a small resistance, indicating that there is continuity, Figure 1-84. Reverse the leads. The meter should show infinity, indicating there is no continuity. A diode should show continuity in one direction and not in the other. If it shows continuity in both directions, it is defective. If it does not show continuity in either direction, it is defective.

RECTIFIER. A diode can be used as a solid-state rectifier, changing AC to DC. The term diode is normally used when rated for less than 1 A. A similar component rated above 1 A is called a rectifier. A rectifier allows current to flow in one direction. Remember that AC flows in first one

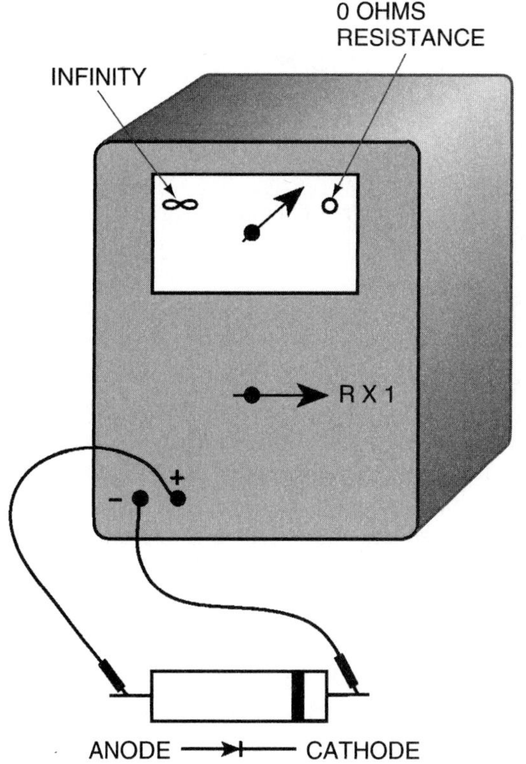

Figure 1-84. Checking a diode.

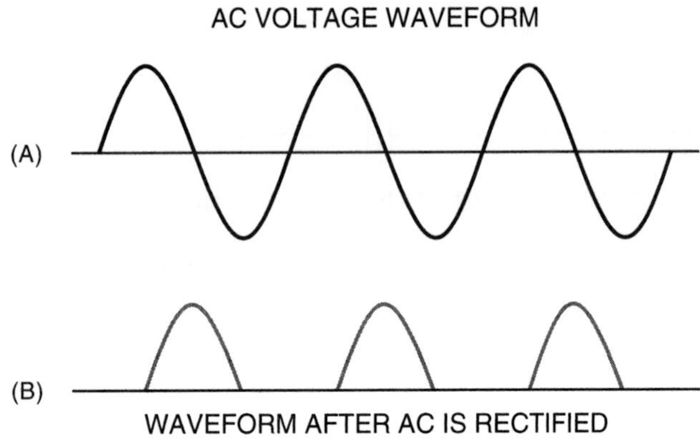

Figure 1-85. (A) A full wave AC waveform. (B) A half wave DC waveform.

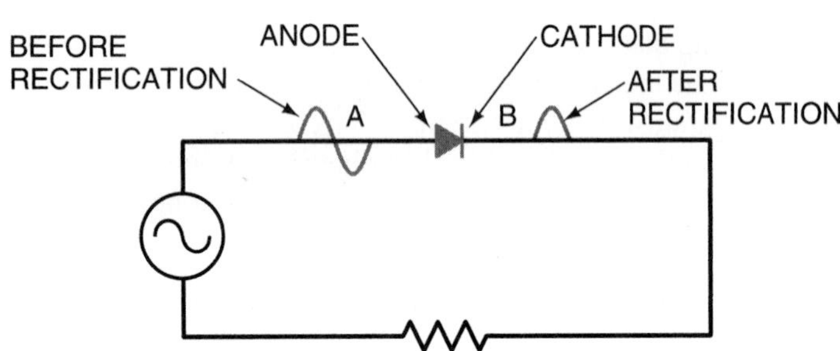

Figure 1-86. A diode rectifier circuit.

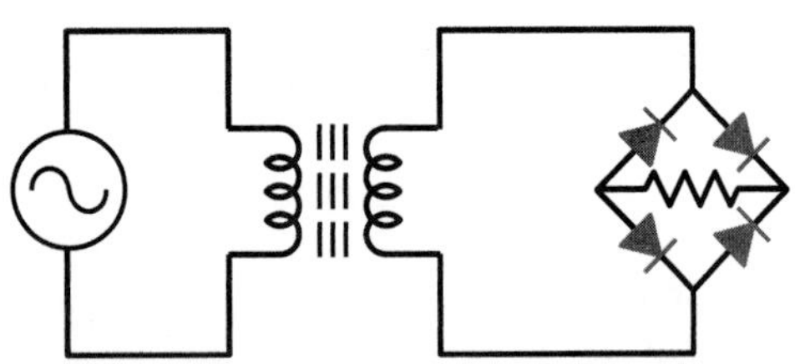

Figure 1-87. A full wave bridge rectifier.

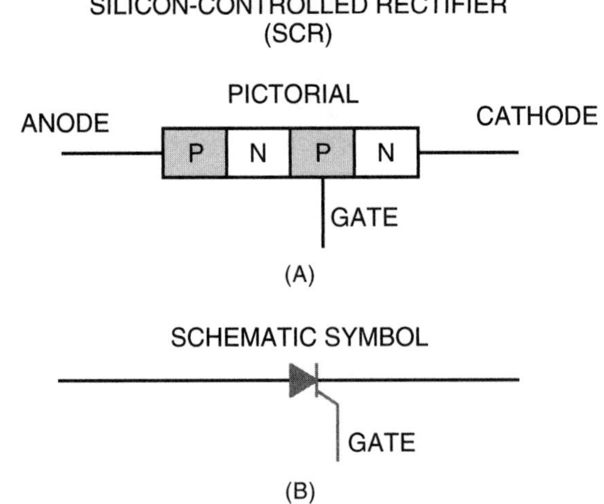

Figure 1-88. Pictorial and schematic drawings of a silicon-controlled rectifier.

Figure 1-89. A typical silicon-controlled rectifier. *Photo by Bill Johnson*

direction and then reverses, Figure 1-85(A). The rectifier allows the AC to flow in one direction but blocks it from reversing, Figure 1-85(B). Therefore, the output of a rectifier circuit is in one direction or direct current. Figure 1-86 illustrates a rectifier circuit. This is called a half wave rectifier because it allows only that part of the AC moving in one direction to pass through. Figure 1-86(A) shows the AC before it is rectified and Figure 1-86 B shows the AC after it is rectified. Full wave rectification can be achieved by using a more complicated circuit such as in Figure 1-87. This is a full wave bridge rectifier commonly used in this industry.

SILICON-CONTROLLED RECTIFIER. Silicon-controlled rectifiers (SCRs) consist of four semiconductor materials bonded together. These form a PNPN junction, Figure 1-88(A); Figure 1-88(B) illustrates the schematic symbol. Notice that the schematic is similar to the diode except for the gate. The SCR is used to control devices that may use large amounts of power. The gate is the control for the SCR. These devices may be used to control the speed of motors or to control the brightness of lights. Figure 1-89 shows a photo of a typical SCR.

Checking the Silicon-Controlled Rectifier

The SCR can also be checked with an ohmmeter. Ensure that the SCR is removed from a circuit. Set the selector switch on the meter to R × 1. Zero the meter. Fasten the negative lead from the meter to the cathode terminal of the SCR and the positive lead to the anode, Figure 1-90. If the SCR is good, the needle should not move. This is because the SCR has not fired to complete the circuit. Use a jumper to connect the gate terminal to the anode. The meter needle should now show continuity. If it does not, you may not have the cathode and anode properly identified. Reverse the leads and change the jumper to the new suspected anode. If it fires, you had the anode and cathode reversed. When the jumper is removed, the SCR will continue to conduct if the meter has enough capacity to keep the gate closed. If the meter were to show current flow without firing the gate, the SCR is defective. If the gate will not fire after the jumper and leads are attached correctly, the SCR is defective.

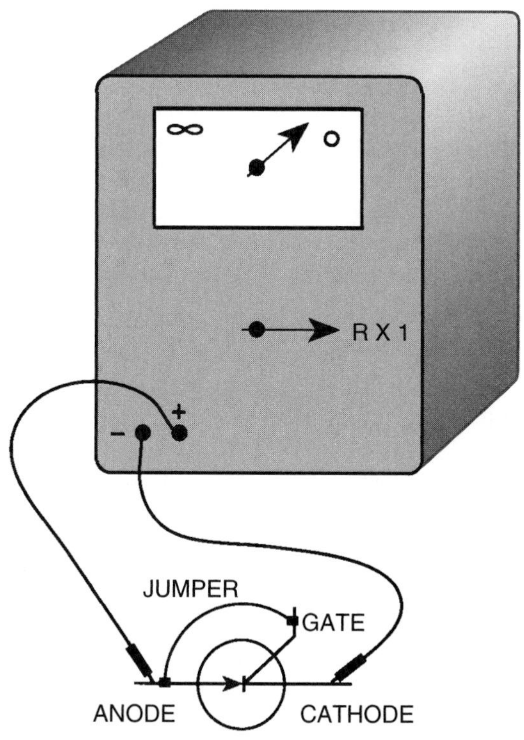

Figure 1-90. Checking a silicon-controlled rectifier.

TRANSISTORS. Transistors are also made of N-type and P-type semiconductor materials. Three pieces of these materials are sandwiched together. Transistors are either NPN or PNP types. Figure 1-91 shows diagrams, and Figure 1-92 shows the schematic symbols for the two types. As the symbols show, each transistor has a base, a collector, and an emitter. In the NPN type the collector and the base are connected to the positive; the emitter is connected to the negative. The PNP transistor has a negative base and collector connection and a positive emitter. The base must be connected to the same polarity as the collector to provide forward bias. Figure 1-93 is a photo of typical transistors.

The transistor may be used as a switch or a device to amplify or increase an electrical signal. One application of a transistor used in an air-conditioning control circuit would be to amplify a small signal to provide enough current to operate a switch or relay.

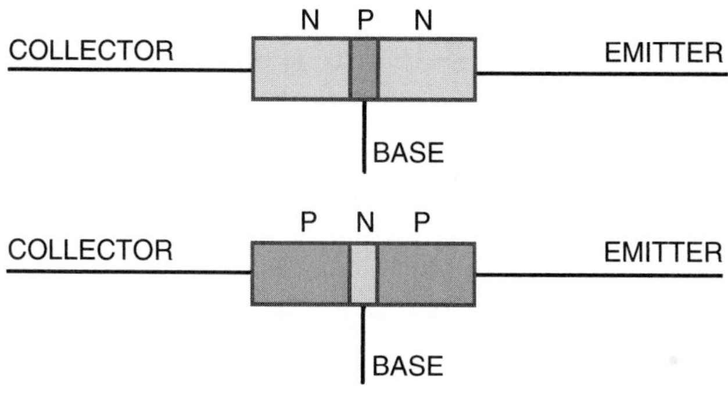

Figure 1-91. Pictorial drawings of NPN and PNP transistors.

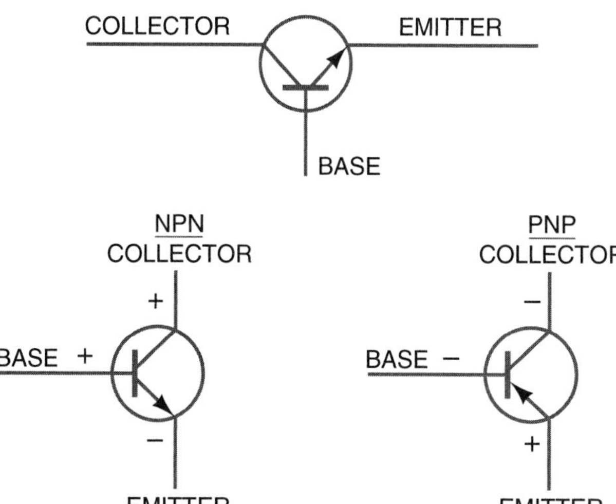

Figure 1-92. Schematic drawings of NPN and PNP transistors.

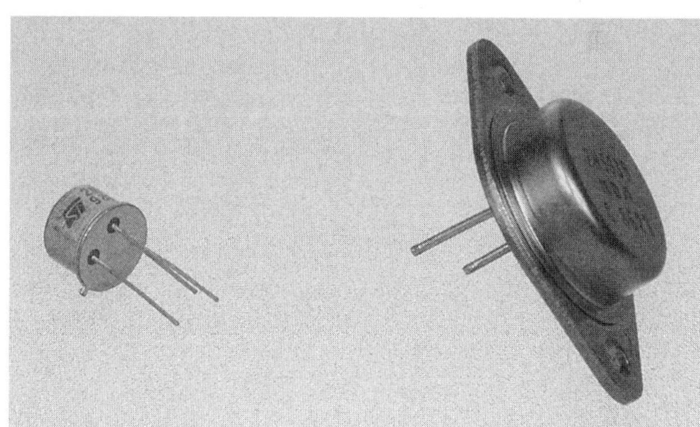

Figure 1-93. Typical transistors. *Photo by Bill Johnson*

Current will flow through the base emitter and through the collector emitter. The base-emitter current is the control, and the collector-emitter current produces the action. A very small current passing through the base emitter may allow a much larger current to pass through the collector-emitter junction. A small increase in the base-emitter junction can allow a much larger increase in the current flow through the collector emitter.

THERMISTORS. A thermistor is a type of resistor that is sensitive to temperature, Figure 1-94. The resistance of a thermistor changes with a change in temperature. Two types of thermistors exist. A positive temperature coefficient (PTC) thermistor causes the resistance of the thermistor to increase when the temperature increases. A negative temperature coefficient (NTC) causes the resistance to decrease with an increase in temperature. Figure 1-95 illustrates the schematic symbol for a thermistor.

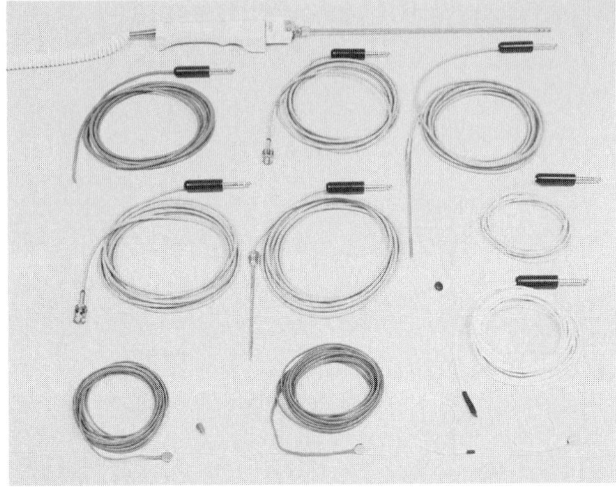

Figure 1-94. Typical thermistors. *Courtesy Omega Engineering*

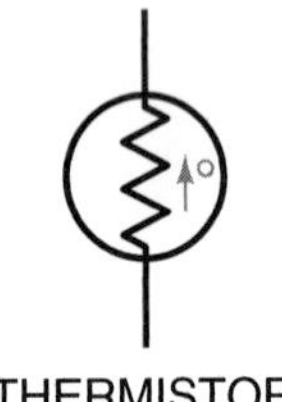

Figure 1-95. The schematic symbol for a thermistor.

An application of a thermistor is to provide motor overload protection. The thermistor is imbedded in the windings of a motor. When the winding temperature exceeds a predetermined amount, the thermistor changes in resistance. This change in resistance is detected by an electronic circuit that causes the motor circuit to open.

Another application is to provide start assistance in a PSC (permanent split-capacitor) electric motor. This thermistor is known as a positive temperature coefficient device. It allows full voltage to reach the start windings during start-up of the motor. The thermistor heats during start-up and creates resistance, turning off power to the start winding at the appropriate time. This does not give the motor the starting torque that a start capacitor does, but it is advantageous in some applications because of its simple construction and lack of moving parts.

DIACS. The diac is a two-directional electronic device. It can operate in an AC circuit and its output is AC. It is a voltage-sensitive switch that operates in both halves of the AC waveform. When a voltage is applied, it will not conduct (or will remain off) until the voltage reaches a predetermined level. Let us assume this predetermined level to be 24 V. When the voltage in the circuit reaches 24 V, the diac will begin to conduct or fire. Once it fires it will continue to conduct even at a lower voltage. Diacs are designed to have a higher cut-in voltage and lower cut-out voltage. If the cut-in voltage is 24 V, let us assume that the cut-out voltage is 12 V. In this case the diac will continue to operate until the voltage drops below 12 V at which time it will cut off.

Figure 1-96 shows two schematic symbols for a diac. Figure 1-97 illustrates a diac in a simple AC circuit. Diacs are often used as switching or control devices for triacs.

TRIACS. A triac is a switching device that will conduct on both halves of the AC waveform. The output of the triac is AC. Figure 1-98 shows the schematic symbol for a triac. Notice that it is similar to the diac but has a gate lead. A pulse supplied to the gate lead can cause the triac to fire or to conduct. Triacs were developed to provide better AC switching. As was mentioned previously, diacs often provide the pulse to the gate of the triac. A common use of a triac is as a motor speed control for AC motors.

HEAT SINKS. Some solid-state devices may appear a little different from others of the same type. This is because of the differing voltages and current they are rated to carry and the purpose for which they are designed and used. Some produce much more heat than others and will operate only at the rating specified if kept within a certain temperature range. Heat that could change the operation of or destroy the device must be dissipated. This is done by adhering the solid-state component to an object called a heat sink that has a much greater surface area, Figure 1-99. Heat will travel from the device to this object with the large

Figure 1-96. Schematic symbols for a diac.

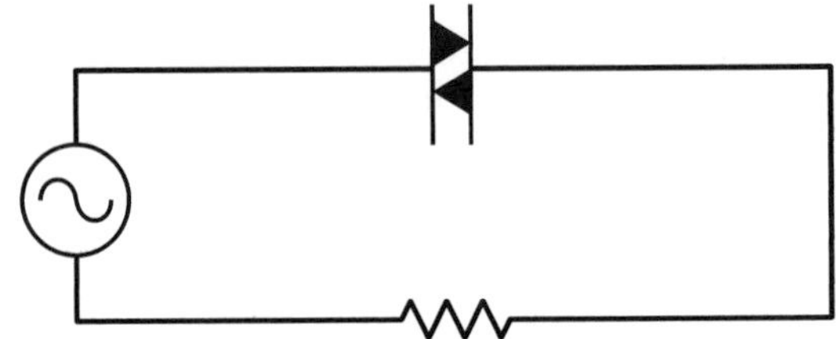

Figure 1-97. A simple diac circuit.

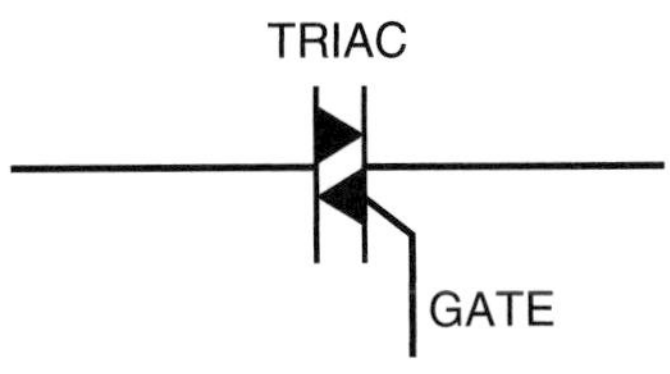

Figure 1-98. The schematic symbol for a triac.

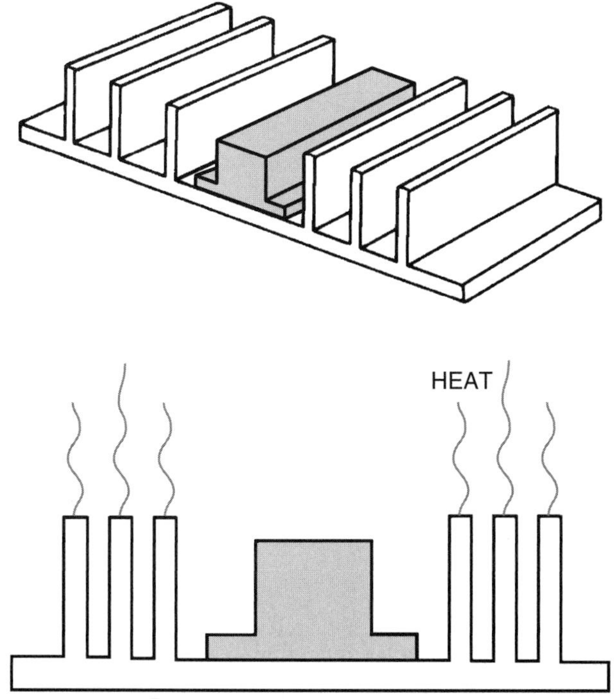

Figure 1-99. One type of heat sink.

surface area allowing the excess heat to be dissipated into the surrounding air.

Practical Semiconductor Applications for Technicians

This has been a short introduction into semiconductors. Brief descriptions in previous paragraphs outlined procedures for checking certain semiconductors. These tests were described to show how semiconductors may be checked in the field. However, it is seldom practical to remove these devices from a printed circuit board and perform the test. Other instruments may be used in the shop or laboratory to check these same components while they are still in the circuit. Refrigeration and air-conditioning technicians will be more involved with checking the input and output of circuit board circuits than with checking individual electronic components. The preceding information has been provided so that you will have some idea of the purpose of these control devices. You are encouraged to pursue the study of these components as more and more solid-state electronics will be used in the future to control the systems you will be working with. Each manufacturer will have a control sequence procedure that you must use to successfully check their controls. Make it a practice to attend seminars and factory schools in your area to increase your knowledge in all segments of this ever-changing field.

Unit 5, "Troubleshooting and Maintenance for Electric-heat Equipment" contains many exercises involving reading electrical schematic drawings.

Summary

- Atoms are made up of protons, neutrons, and electrons.
- Protons have a positive charge, and electrons have a negative charge.
- Electrons travel in orbits around the protons and neutrons.
- When a surplus of electrons is in an atom, it has a negative charge. When a deficiency of electrons exists, the atom has a positive charge.
- Electricity can be produced by using magnetism. A conductor cutting through magnetic lines of force produces electricity.
- Direct current (DC) is an electrical current moving in one direction.
- Alternating current (AC) is an electrical current that is continually reversing.
- Volt = electrical force or pressure.
- Ampere = quantity of electron flow.
- Ohm = resistance to electron flow.
- Voltage (E) = Amperage (I) × Resistance (R). This is Ohm's Law.
- In series circuits the voltage is divided across the resistances, the total current flows through each resistance, and the resistances are added together to obtain the total resistance.
- In parallel circuits the total voltage is applied across each resistance, the current is divided between the resistances, and the total resistance is less than that of the smallest resistance.
- Electrical power is measured in watts, $P = E \times I$.
- Inductive reactance is the resistance caused by the magnetic field surrounding a coil in an AC circuit.
- A step-up transformer increases the voltage and decreases the current. A step-down transformer decreases the voltage and increases the current.
- A capacitor in an AC circuit will continually charge and discharge as the current in the circuit reverses.
- A capacitor has capacitance, which is the amount of charge that can be stored.
- Impedance is the opposition to current flow in an AC circuit from the combination of resistance, inductive reactance, and capacitive reactance.
- A multimeter often used is the VOM (volt-ohm-milliammeter).
- A sine wave displays the voltage of one AC cycle through 360°.
- *Safety Precaution:* Properly sized conductors must be used. Larger wire sizes will carry more current than smaller wire sizes without overheating.
- *Safety Precaution:* Fuses and circuit breakers are used to interrupt the current flow in a circuit when the current is excessive.
- Semiconductors in their pure state do not conduct electricity well, but when they are doped with an impurity, they form an N-type or P-type material that will conduct in one direction.
- Diodes, rectifiers, transistors, thermistors, diacs, and triacs are examples of semiconductors.

Review Questions

1. The __________ is that part of an atom that moves from one atom to another.
   A. electron
   B. proton
   C. neutron
2. When this part of an atom moves to another atom, the losing atom will have a __________ charge.
   A. negative
   B. positive
   C. neutral
3. State the differences between AC and DC.
4. Describe how a meter would be connected in a circuit to measure the voltage at a light bulb.
5. Describe how an amperage reading would be taken using a clamp-on or clamp-around ammeter.
6. Describe how the total resistance in a series circuit is determined.
7. Ohm's Law for determining voltage is __________.
8. Ohm's Law for determining amperage is __________.
9. Ohm's Law for determining resistance is __________.
10. Sketch a circuit with three loads wired in parallel.
11. Describe the characteristics of the voltage, amperage, and resistances when there is more than one load in a parallel circuit.
12. If there were a current flowing of 5 amperes in a 120-volt circuit, what would the resistance be?
    A. 25 ohms
    B. 24 ohms
    C. 600 ohms
    D. 624 ohms
13. If the resistance in a 120-volt circuit was 40 ohms, what would the current be in amperes?
    A. 4800
    B. 48
    C. 4
    D. 3
14. Electrical power is measured in:
    A. amperes divided by the resistance.
    B. amperes divided by the voltage.
    C. watts.
    D. voltage divided by the amperage.
15. The formula for determining electrical power is __________.
16. Describe how a step-down transformer differs from a step-up transformer.
17. What are the three types of opposition to current flow that impedance represents?
18. Why is it important to use a properly sized wire in a particular circuit?
19. What does forward bias on a diode mean?
20. The unit of measurement for the charge a capacitor can store is the:
    A. inductive reactance.
    B. microfarad.
    C. ohm.
    D. joule.

2 Types of Electric Motors

Objectives

After studying this unit, you should be able to

- **describe the different types of open single-phase motors used to drive fans, compressors, and pumps.**
- **describe the applications of the various types of motors.**
- **state which motors have high starting torque.**
- **list the components that cause a motor to have a higher starting torque.**
- **describe a multispeed permanent split-capacitor motor and indicate how the different speeds are obtained.**
- **explain the operation of a three-phase motor.**
- **describe the use of variable-speed motors.**

2.1 Uses of Electric Motors

Electric motors are used to turn the prime movers of air, water, and refrigerant. The prime movers are the fans, pumps, and compressors, Figure 2-1. Several types of motors, each with its particular use, are available. For example, some applications need motors that will start under heavy loads and still develop their rated work horsepower under a continuous running condition. Some motors run for years in dirty operating conditions, and others operate in a refrigerant atmosphere. These are a few of the typical applications of motors in this industry. The technician must understand which motor is suitable for each job so that effective troubleshooting can be accomplished and, if necessary, the motor replaced by the proper type. But the basic operating principles of an electric motor first must be understood. Although many types of electric motors are used, most motors operate on similar principles.

2.2 Parts of an Electric Motor

Electric motors have a *stator* with windings, a *rotor, bearings, end bells, housing,* and some means to hold these parts in the proper position, Figures 2-2 and 2-3.

2.3 Electric Motors and Magnetism

Electricity and magnetism are used to create the rotation in an electric motor to drive the fans, pumps, and compressors. Magnets are known to have two different electrical poles,

(A)

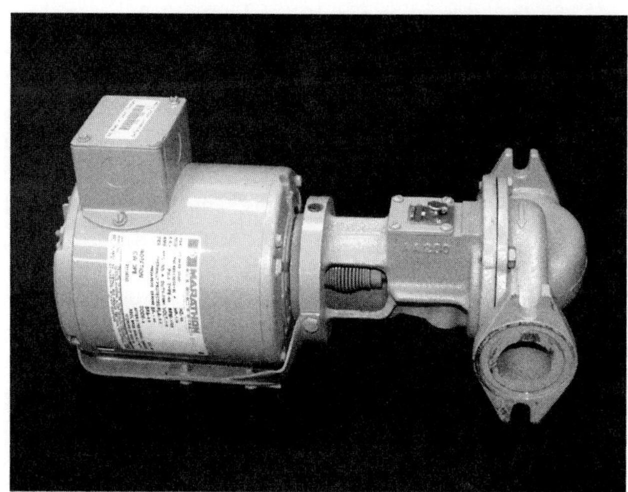

(B)

Figure 2-1. (A) Fans move air. (B) Pumps move water. *Courtesy W. W. Grainger, Inc.*

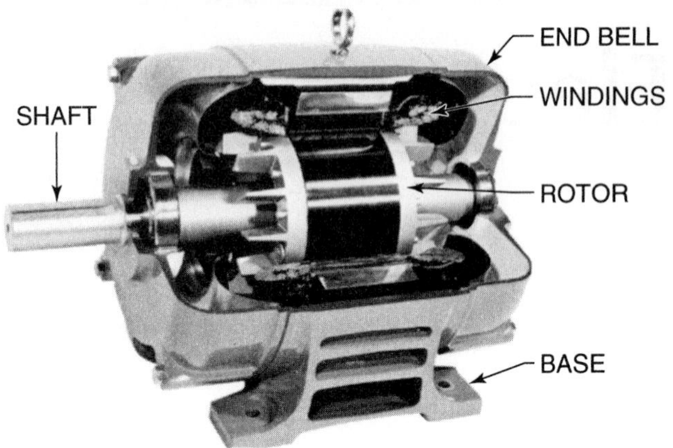

Figure 2-2. A cutaway of an electric motor. *Courtesy Century Electric, Inc.*

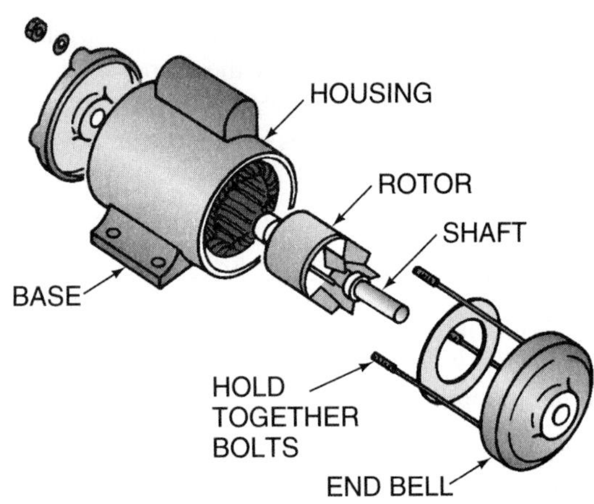

Figure 2-3. Individual electric motor parts.

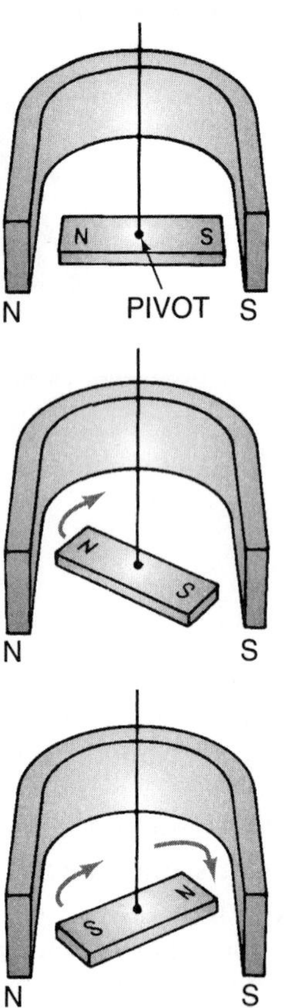

Figure 2-4. Poles (north and south) on a rotating magnet will line up with the opposite poles on a stationary magnet.

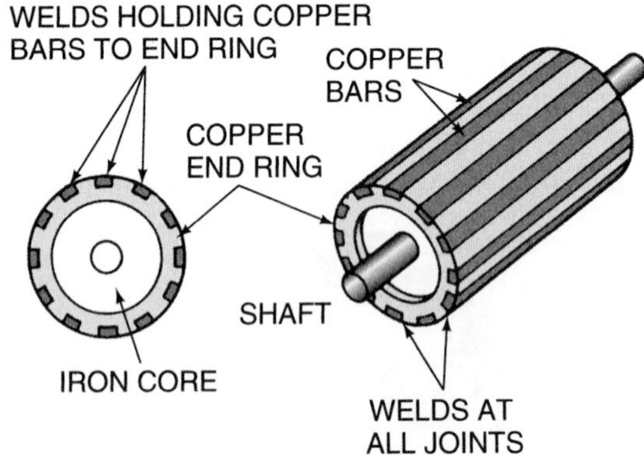

Figure 2-5. A simple sketch of a squirrel cage rotor.

north and south. Unlike poles of a magnet attract each other, and like poles repel each other. If a stationary horseshoe magnet were placed with its two poles (north and south) at either end of a free-turning magnet as in Figure 2-4, one pole of the free-rotating magnet would line up with the opposite pole of the horseshoe magnet. If the horseshoe magnet were an electromagnet and the wires on the battery were reversed, the poles of this magnet would reverse, and the poles on the free magnet would be repelled, causing it to rotate until the unlike poles again were lined up. This is the basic principle of electric motor operation. The horseshoe magnet is the stator, and the free-rotating magnet is the rotor.

In a two-pole split-phase motor the stator has two poles with insulated wire windings called the *running windings.* When an electrical current is applied, these poles become an electromagnet with the polarity changing constantly. In normal 60-cycle operation the polarity changes 60 times per second.

The rotor may be constructed of bars, Figure 2-5. This type is called a *squirrel cage rotor.* When the rotor shaft is placed in the bearings in the bell-type ends, it is positioned between the running windings. When an alternating current (AC) is applied to these windings, a magnetic field is produced in the windings and a magnetic field is also induced in the rotor. The bars in the rotor actually form a coil. This is similar to the field induced in a transformer secondary by the magnetic field in the transformer primary. The field induced in the rotor has a polarity opposite that in the run windings. The opposite poles of the run winding are wound in different directions. If one pole is wound clockwise, the opposite pole would be wound counterclockwise. This would set up opposite polarities for the attraction and repulsion forces that cause rotation.

The attracting and repelling action between the poles of the run windings and the rotor sets up a rotating magnetic field and causes the rotor to turn. Since this is AC reversing 60 times per second, the rotor turns, in effect "chasing" the changing polarity in the run windings. The motor will continue to rotate as long as power is supplied. The motor starting method determines the direction of motor rotation. The motor will run equally well in either direction.

2.4 Determining a Motor's Speed

The following formula can be used to determine the synchronous speed (without load) of motors:

$$S\ (\text{rpm}) = \frac{\text{Frequency} \times 120}{\text{Number of poles}}$$

Frequency is the number of cycles per second (also called *hertz*).

NOTE: The magnetic field builds and collapses twice each second (each time it changes direction); the 120 consists of a time conversion that changes seconds into minutes. It also consists of a conversion from the motor's pole pairs (poles/2) to the actual number of poles.

$$\text{Speed of two-pole split-phase motors} = \frac{60 \times 120}{2} = 3600$$

$$\text{Speed of four-pole split-phase motors} = \frac{60 \times 120}{4} = 1800$$

The speed under load of each motor will be approximately 3450 rpm and 1750 rpm. The difference between synchronous speed and the actual speed is called slip. Slip is caused by the load. Other combinations of the numbers of poles can yield different motor speeds. The manufacturer determines the motor speed combination.

2.5 Start Windings

The preceding explanation does not explain how to start the motor rotation, which is accomplished by using a separate motor winding called the start winding. The start winding is wound next to the run winding but is a few electrical degrees out of phase with the run winding. This design is much like the pedals on a bicycle—if both pedals were at the same angle, it would be very hard to get started. The start winding is in the circuit only for as long as it takes the rotor to get up to about 75% of its speed, and it then is electrically disconnected, Figure 2-6. The start windings have more turns than the run windings and are wound with a smaller diameter wire. Opposite poles of the start winding are wound in different directions. If one pole is wound clockwise, then the opposite pole would be wound counterclockwise. This would set up opposite polarities for the attraction and repulsion forces that cause rotation. This produces a larger magnetic field and greater resistance, which helps the rotor to start turning and determines the direction in which it will turn. This happens as a result of these windings being located between the run windings. It changes the phase angle between the voltage and the current in these windings.

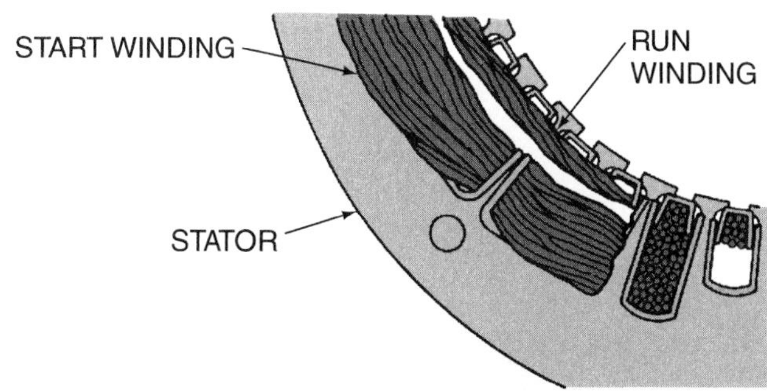

Figure 2-6. The placement of the start and run windings inside the stator.

We have just described a two-pole split-phase *induction* motor, which is rated to run at 3600 revolutions per minute (rpm). It actually turns at a slightly slower speed when running under full load. When the motor reaches approximately 75% of its normal speed, a centrifugal switch opens the circuit to the start windings and the motor continues to operate on only the run windings. Many split-phase motors have four poles and run at 1800 rpm.

2.6 Starting and Running Characteristics

Two major considerations of electric motor applications are the starting and running characteristics. A motor applied to some fans must have a high starting *torque*—it must be able to start under heavy starting loads. Torque is the twisting power of the motor shaft. The starting torque is the power to turn the shaft from the stopped position. The motor must also have enough torque to operate with the load of the application. Some motors must have a great deal of starting torque to turn the motor at the stopped position but do not need a great deal of torque to maintain speed.

Motors are rated with two different current ratings, full-load or rated-load amperes (FLA or RLA) and locked-rotor amperes (LRA). Locked-rotor amperage is often referred to as inrush current. When a motor is still, it takes a great deal of torque to get it turning, particularly if the motor has a load on it at start-up. Typically, the LRA for a motor is about five times the FLA/RLA. For example, suppose a fan operates on 25 A when at full load. This same fan will pull about 125 A on start-up for just a moment. As the motor gets up to speed, the amperage will reduce to FLA, 25 A. Since this inrush of current is of a short duration, it does not figure in the wire sizing for the motor.

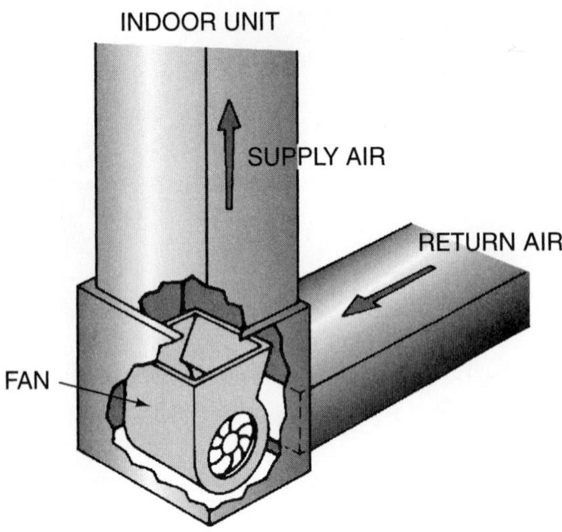

Figure 2-7. This fan has no pressure difference to overcome while starting. When it stops, the air pressure equalizes.

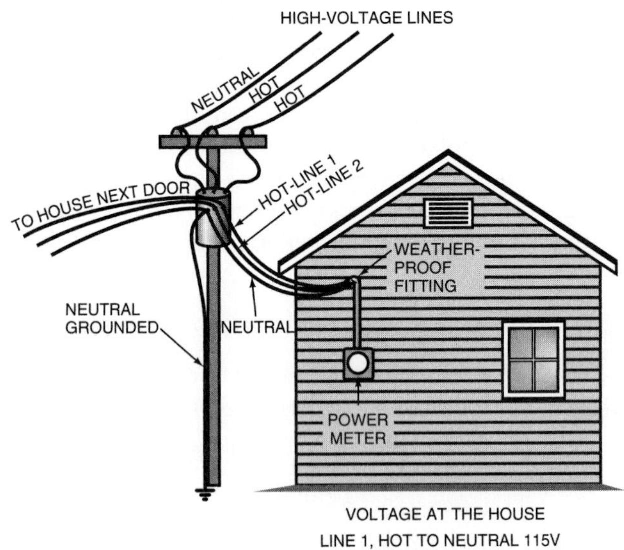

Figure 2-8. A single-phase power supply.

To start a small fan, a motor does not need as much starting torque. The motor must simply overcome the friction needed to start the fan moving. There is no pressure difference because the pressures equalize when the fan is not running, Figure 2-7. The high starting load, called starting torque, puts more load on a large fan when compared to the light load of a small fan. These are two different motor applications and require two different types of motors. These applications will be discussed one at a time as the different types of motors are discussed.

2.7 Electrical Power Supplies

The power company furnishes the power to the customer and determines what type of power supply they have. The power company moves the power across the country by means of high-voltage transmission lines to a transformer that reduces the power to the voltage needed at the consumer level. Residences are typically furnished with a single-phase (θ is the symbol often used for phase) power supply. The power pole to the house may look like the one in Figure 2-8. Several homes may be furnished from the same power-reducing transformer. The power-reducing transformer has three wires that pass through the meter base and on to the house electrical panel where the power is distributed through the circuit protectors (circuit breakers or fuses), to the various circuits in the house, and then to the power-consuming equipment, Figure 2-9.

The power panel in the house would look like the illustration in Figure 2-9. Notice that it will furnish both 115-V and 230-V service. Typically the 230-V service is used for the electric dryer, electric range, electric oven, many electric heat devices, and the air conditioning. All other appliances are typically operated from the 115-V circuits.

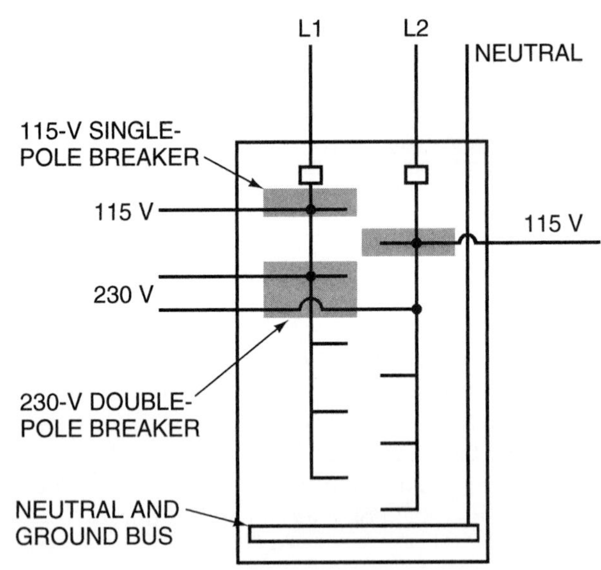

Figure 2-9. The main breaker panel for a typical residence.

Commercial buildings and factories use equipment large enough that they require three-phase power. A three-phase power supply may furnish several different voltage options:

A. 115-V single-phase for common appliances
B. 230-V single-phase for heavy-duty appliances
C. 230-V three-phase for large loads such as electric heat or motors
D. 460-V three-phase for large loads such as electric heat or motors

E. 277-V single-phase for lighting circuits (may be obtained between 460 V and neutral with certain systems)
F. 560-V three-phase for special industrial systems

A typical three-phase system wiring from the pole may look like the illustration in Figure 2-10. Again, this is furnished by the power company for this application. This building could very well have been supplied with 230-V three-phase. One of the reasons for using 460-V three-phase is that it reduces the wire size to all components in the building, which reduces the materials cost and the labor cost for installation. The same load using 230 V would use twice the current flow, so the wire would have to be sized up to twice the current-carrying capacity. Many of the components in the building will need to operate from 115 V and maybe even 230 V so a step-down transformer is used for equipment such as small fans, computers, and office equipment. The step-down transformer enables the building to take advantage of the benefits of 460 V, 277 V, 230 V, and 115 V in the same system, Figure 2-11.

Safety Precaution: A technician must be very careful with any live power circuits. 460-V circuits are particularly dangerous, and the technician must take extra care while servicing them.

Following is a description of some motors currently used in the heating, air-conditioning, and refrigeration industry. The electrical characteristics, not the working conditions, are emphasized. Some older motors are still in operation, but they are not discussed in this text.

2.8 Single-Phase Open Motors

The power supply for most *single-phase* motors is either 115 V or 208 V to 230 V. A home gas or oil furnace uses a power supply of 115 V. An electric heat furnace uses a power supply of 230 V. A commercial building may have either 208 V, 230 V, or 460 V, depending on the power company. Some single-phase motors are dual voltage. The motor has two run windings and one start winding. The two run windings have the same resistance, and the start winding has a high resistance. The motor will operate with the two run windings in parallel in the low-voltage mode. When it is required to run in the high-voltage mode, the technician changes the numbered motor leads according to the manufacturer's instructions. This wires the run windings in series with each other and delivers an effective voltage of 115 V to each winding. The motor windings are actually only 115 V because they operate only on 115 V, no matter which mode they are in. The technician can change the voltage at the motor terminal box, Figure 2-12.

Some commercial and industrial installations may use a 460-V power supply for large motors. The 460 V may be reduced to a lower voltage to operate the small motors.

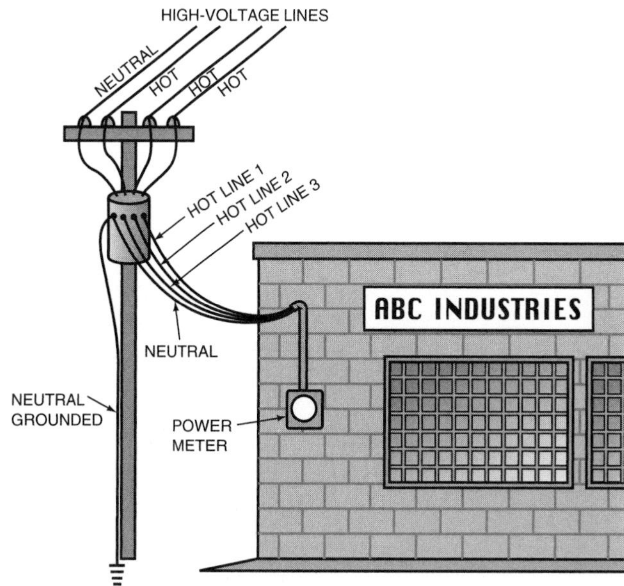

Figure 2-10. The power supply to a building, 460 V three phase.

The smaller motors may be single-phase and must operate from the same power supply, Figure 2-11.

A motor can rotate clockwise or counterclockwise. Some motors are reversible from the motor terminal box, Figure 2-13.

2.9 Split-Phase Motors

Split-phase motors have two distinctly different windings, Figure 2-14. They have a medium amount of starting torque and a good operating efficiency. The split-phase motor is normally used for operating fans in the fractional horsepower range. Its normal operating ranges are 1800 rpm and 3600 rpm. An 1800-rpm motor will normally operate at 1725 rpm to 1750 rpm under a load. The difference in the rated rpm and the actual rpm is called *slip.* If the motor is loaded to the point where the speed falls below 1725 rpm, the current draw will climb above the rated amperage. Motors rated at 3600 rpm will normally slip in speed to about 3450 rpm to 3500 rpm, Figure 2-15. Some of these motors are designed to operate at either speed, 1750 rpm or 3450 rpm. The speed of the motor is determined by the number of motor poles and by the method of wiring the motor poles. The technician can change the speed of a two-speed motor at the motor terminal box.

2.10 The Centrifugal Switch

All split-phase motors have a start and run winding. The start windings must be disconnected from the circuit

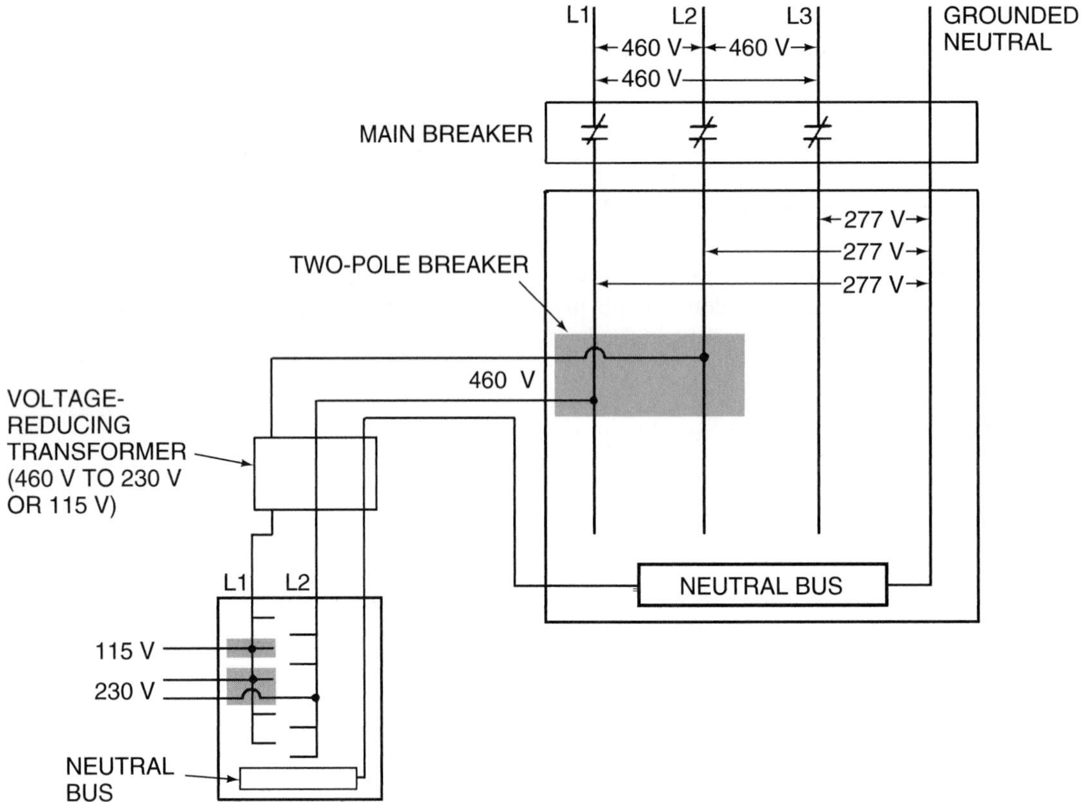

Figure 2-11. A 460 V commercial building power supply. Normally when a building has a 460 V power supply, it will have a step-down transformer to 115 V for office machines and small appliances.

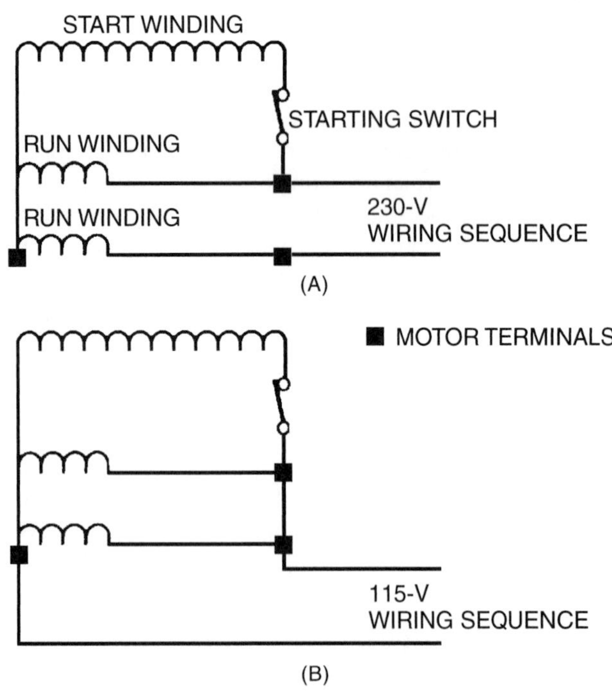

Figure 2-12. The wiring diagram of a dual-voltage motor. It is made to operate using 115 V or 230 V, depending on how the motor is wired. (A) A 230 V wiring sequence. (B) A 115 V wiring sequence.

within a very short period of time, or they will overheat. Several methods are used to disconnect the start windings. With open motors, the centrifugal switch is the most common; however, electronic start switches are sometimes used.

The *centrifugal switch* is used to disconnect the start winding from the circuit when the motor reaches approximately 75% of the rated speed. Motors described here are those that run in the atmosphere.

The centrifugal switch is a mechanical device attached to the end of the shaft with weights that will sling outward when the motor reaches approximately 75% speed. For example, if the motor has a rated speed of 1725 rpm, at 1294 rpm (1725 × 0.75) the centrifugal weights will change position and open a switch to remove the start winding from the circuit. This switch is under a fairly large current load, so a spark will occur. If the switch fails to open its contacts and remove the start winding, the motor will draw too much current, and the overload device will cause it to stop.

The more the switch is used, the more its contacts will burn from the arc. If this type of motor is started many times, the first thing that will likely fail will be the centrifugal switch, depicted in Figure 2-16. This switch makes an audible sound when the motor starts and stops.

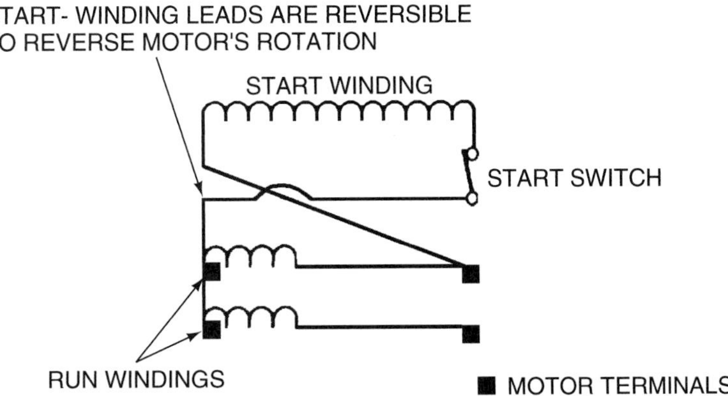

Figure 2-13. A single-phase motor can be reversed by changing the connections in the motor terminal box. The direction the motor turns is determined by the start winding. This can be shown by disconnecting the start-winding leads and applying power to the motor. It will hum and will not start. The shaft can be turned in either direction, and the motor will run in that direction.

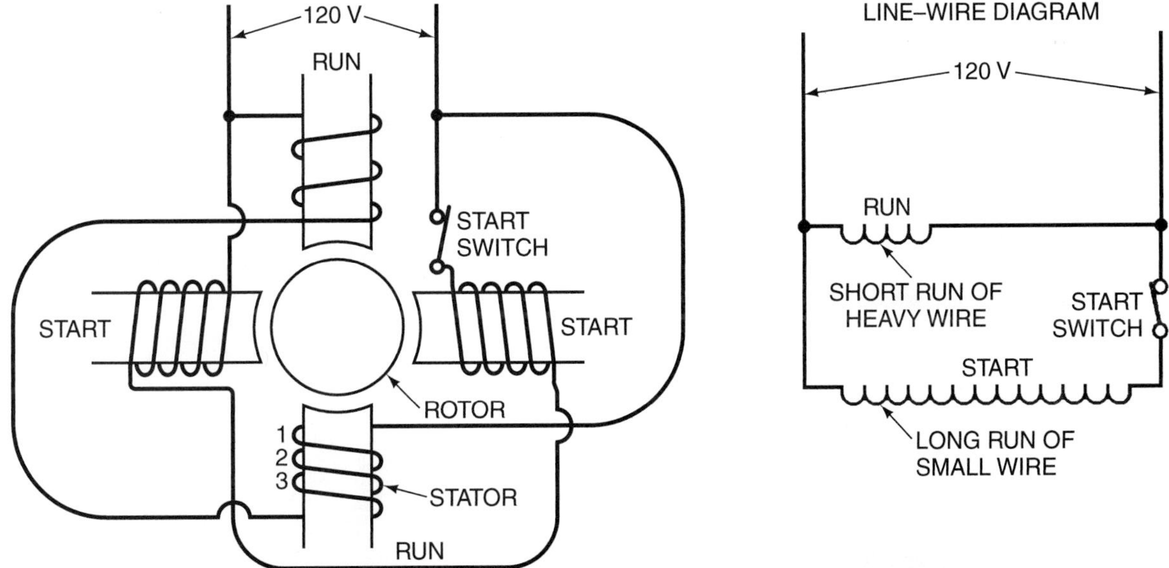

Figure 2-14. The diagram shows a difference in the resistance in start and run windings.

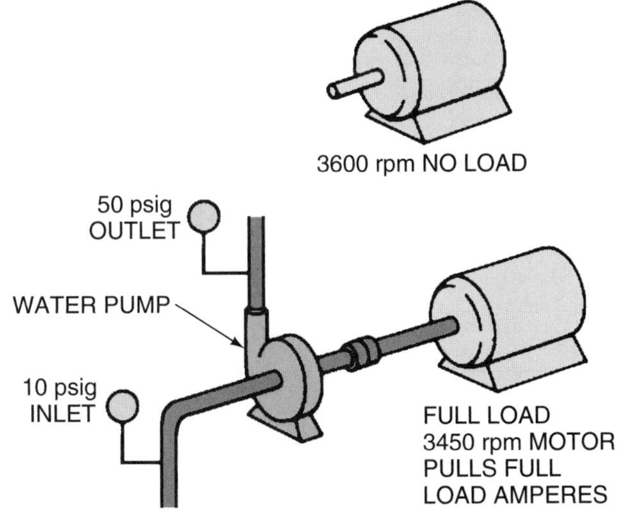

Figure 2-15. How motor revolutions per minute change under motor load.

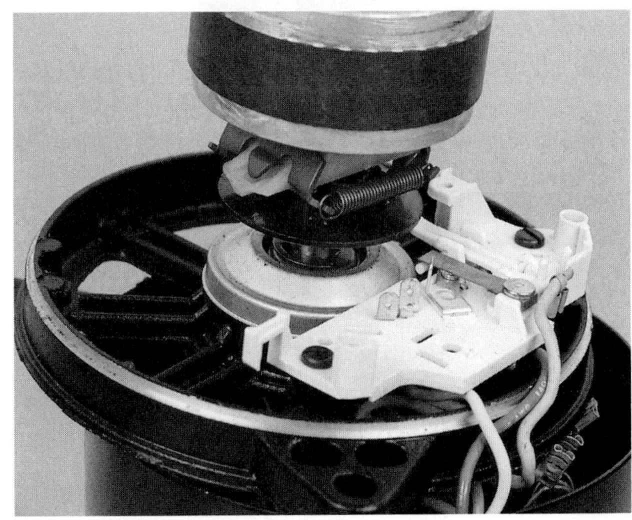

Figure 2-16. The centrifugal switch located at the end of the motor. *Photo by Bill Johnson*

2.11 The Electronic Relay

The *electronic relay* is used with some motors to open the start windings after the motor has started. This is a solid-state device designed to open the start winding circuit when the design speed has been obtained.

2.12 Capacitor-Start Motors

The capacitor-start motor is the same basic motor as the split-phase motor, Figure 2-17. It has two distinctly different windings for starting and running. The previously mentioned methods may be used to interrupt the power to the start windings while the motor is running. A start capacitor is wired in series with the start windings to give the motor more starting torque. Figure 2-18 shows voltage and current cycles in an induction motor. In an inductive circuit the current *lags* the voltage. In a capacitive circuit the current *leads* the voltage. The amount by which the current leads or lags the voltage is the *phase angle.* A capacitor is chosen to make the phase angle such that it is most efficient for starting the motor, Figure 2-19. This start capacitor is not designed to be used while the motor is running, and it must be switched out of the circuit soon after the motor starts. This is done at the same time the windings are taken out of the circuit with the same switch.

2.13 Capacitor-Start, Capacitor-Run Motors

Capacitor-start, capacitor-run motors are much the same as the split-phase motors. The run capacitor is wired into the circuit to provide the most efficient phase angle between the current and voltage when the motor is running. The run capacitor is in the circuit at any time the motor is running. Both the run and start capacitors are wired in series with the start winding but are in parallel to one another, Figure 2-20. The capacitances of two capacitors in

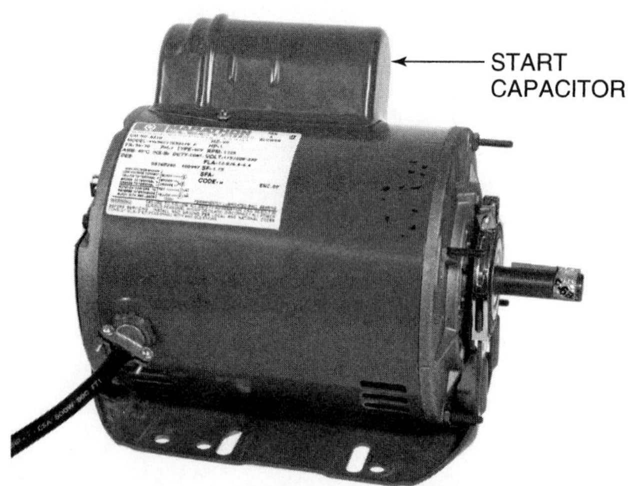

Figure 2-17. A capacitor-start motor. *Courtesy W. W. Grainger, Inc.*

Figure 2-19. A start capacitor. *Photo by Bill Johnson*

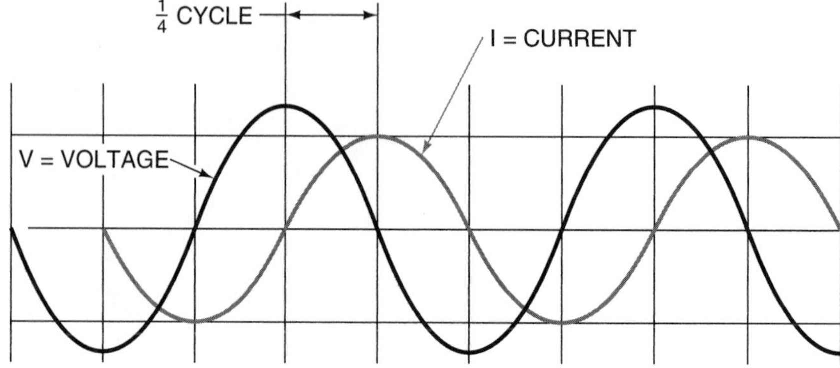

Figure 2-18. The AC cycle, voltage, and current in an induction circuit. The current lags the voltage.

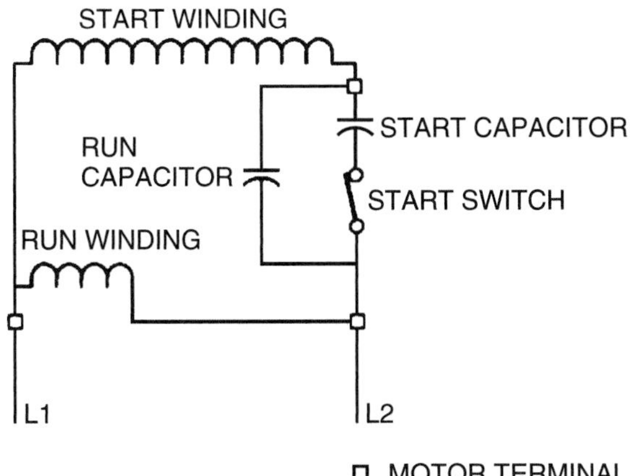

Figure 2-20. A wiring diagram of a capacitor-start, capacitor-run motor. The start capacitor is in the circuit only during starting, and the run capacitor is in the circuit during starting and running of the motor.

parallel add their values the same as resistors do in series. If the run capacitor has a capacitance of 10 microfarads and the start capacitor has a capacitance of 110 microfarads, their total capacitance would add to be 120 microfarads. During start-up, this added capacitance in series with the start winding causes a greater phase angle between the run and start winding, which gives the motor more starting torque. When the start switch opens, the start capacitor is taken out of the circuit. However, the run capacitor and start winding stay in the circuit. The run capacitor stays in series with the start winding for extra running torque. The capacitor, being in series with the start winding during the running mode, also limits the current through the start winding so the winding will not get hot. This motor is actually a permanent split-capacitor (PSC) motor when running. The start capacitor is used for nothing but added starting torque.

If a run capacitor fails because of an open circuit within the capacitor, the motor may start, but the running amperage will be about 10% too high, and the motor will get hot if operated at full load, Figure 2-20. The capacitor-start, capacitor-run motor is one of the most efficient motors used in heating and air-conditioning equipment. It is normally used with belt-drive fans and compressors.

2.14 Permanent Split-Capacitor Motors

The PSC motor has windings very similar to the split-phase motor, Figure 2-21, but it does not have a start capacitor. Instead it uses one run capacitor wired into the circuit in a similar way to the run capacitor in the capacitor-start, capacitor-run motor. This is the simplest split-phase motor. It is very efficient and has no moving parts for the starting of the motor; however, the starting torque is very low so the motor can be used only in low starting torque applications, Figure 2-22.

Figure 2-21. A permanent split-capacitor motor. *Courtesy Universal Electric Company*

Figure 2-22. This open PSC motor may be used to turn a fan. *Photo by Bill Johnson*

Figure 2-23. A multispeed PSC motor. *Photo by Bill Johnson*

A multispeed motor can be identified by the many wires at the motor electrical connections, Figures 2-23 and 2-24. As the resistance of the motor winding decreases, the speed of the motor increases. When more resistance is wired into

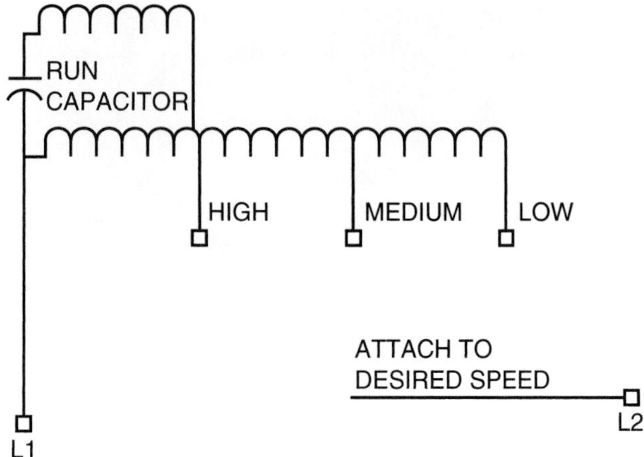

Figure 2-24. This diagram shows how a three-speed motor may be wired to run at a slow speed in the winter and at high speed in the summer.

the circuit, the motor speed decreases. Most manufacturers use this motor in the fan section in air-conditioning and heating systems. Motor speed can be changed by switching the wires. Earlier systems used a capacitor-start, capacitor-run motor, and a belt drive, and air volumes were adjusted by varying the drive pulley diameter. Many PSC motors are manufactured in 2-, 4-, 6-, and 8-pole designs. The speed in rpm of these motors depends on the frequency of the AC source and the number of poles wired in the motor. The higher the frequency, the faster the motor. The more poles there are, the slower the rpm of the motor. For 60-hertz operation, the rated or synchronous speeds would be 3600, 1800, 1200, and 900 rpm, respectively. However, due to motor slip, the actual speeds would be 3450, 1725, 1075, and 825 rpm.

The PSC motor may be used to obtain slow fan speeds during the winter heating season for higher leaving-air temperatures with gas, oil, and electric furnaces. The fan speed can be increased by switching to a different resistance in the winding using a relay. This will provide more airflow in summer to satisfy cooling requirements, Figure 2-25.

The PSC fan motor also has another advantage over the split-phase motor for fan applications. It has a very soft startup. When a split-phase motor is used for a belt-drive application, the motor, belt, and fan must get up to speed very quickly, which often creates a start-up noise. The PSC motor starts up very slowly, gradually getting up to speed. This is very desirable if the fan is close to the return air inlet to the duct system.

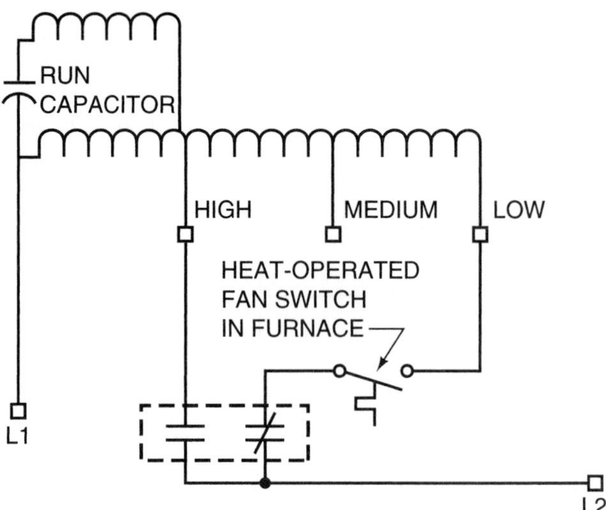

FAN RELAY: WHEN ENERGIZED, SUCH AS IN COOLING, THE FAN CANNOT RUN IN THE LOW-SPEED MODE. WHEN DE-ENERGIZED THE FAN CAN START IN THE LOW-SPEED MODE THROUGH THE CONTACTS IN THE HEAT-OPERATED FAN SWITCH.

IF THE FAN SWITCH AT THE THERMOSTAT IS ENERGIZED WHILE THE FURNACE IS HEATING, THE FAN WILL MERELY SWITCH FROM LOW TO HIGH. THIS RELAY PROTECTS THE MOTOR FROM TRYING TO OPERATE AT 2 SPEEDS AT ONCE.

Figure 2-25. This diagram of a PSC motor shows how the motor can be applied for high air volume in the summer and low air volume in the winter.

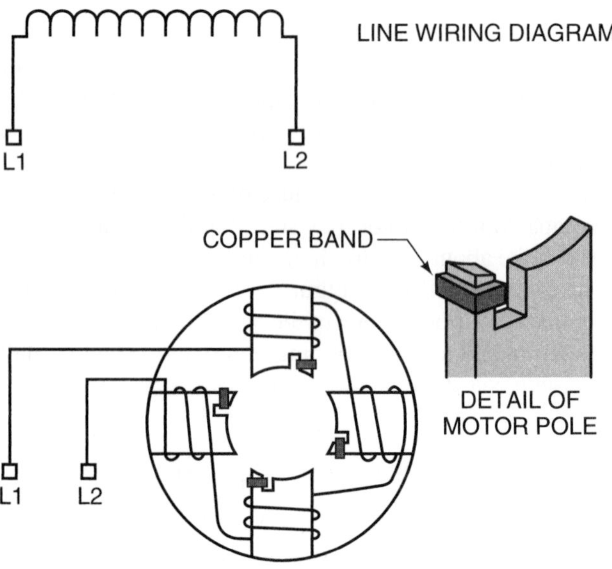

Figure 2-26. The wiring diagram of a shaded-pole motor.

2.15 Shaded-Pole Motors

The shaded-pole motor has very little starting torque and is not as efficient as the PSC motor, so it is used only for light-duty applications. These motors have small starting windings or shading coils at the corner of each pole that help the motor start by providing an induced current and a rotating field, Figure 2-26. It is an economical motor from the standpoint of initial cost. The shaded-pole motor is normally manufactured in the fractional horsepower

Figure 2-27. A shaded-pole motor. *Photo by Bill Johnson*

range. For years it has been used in some furnaces to turn the fans, Figure 2-27. The location of the shading coil on the pole face determines the direction of rotation for a shaded-pole motor. On most shaded-pole motors, rotation can be reversed by disassembling the motor and turning the stator over. This moves the shaded coil to the opposite side that the pole faces and reverses rotation.

2.16 Three-Phase Motors

Three-phase motors have some characteristics that make them popular for many applications from about 1 hp into the thousands of horsepower. These motors are very efficient and require no start assist for high-torque applications. Three-phase motors are used mainly on commercial equipment. The building power supply must have three-phase power available. (Three-phase power is seldom found in a home.) Three-phase motors have no starting windings or capacitors. They can be thought of as having three single-phase power supplies, Figure 2-28. Each of the phases can have either two or four poles. A 3600-rpm motor will have three sets, each with two poles (total of six), and an 1800-rpm motor will have three sets each with four poles (total of 12). Each phase changes direction of current flow at different times but always in the same order. A three-phase motor has high starting torque because of the three phases of current that operate the motor. At any given part of the rotation of the motor, one of the windings is in position for high torque. This makes starting large fans and compressors very easy, Figure 2-29.

The three-phase motor rpm also slips to about 1750 rpm and 3450 rpm when under full load. The motor is not normally available with dual speed; it is either an 1800-rpm or a 3600-rpm motor.

The rotation of a three-phase motor may be changed by switching any two motor leads, Figure 2-30. This rotation must be carefully observed when three-phase fans are used. If a fan rotates in the wrong direction, it will move only about half as much air. If this occurs, reverse the motor leads, and the fan will turn in the correct direction.

Motors have other characteristics that must be considered when selecting a motor for a particular application: for example, the motor mounting. Is the motor solidly mounted to a base, or is there a flexible mount to minimize noise?

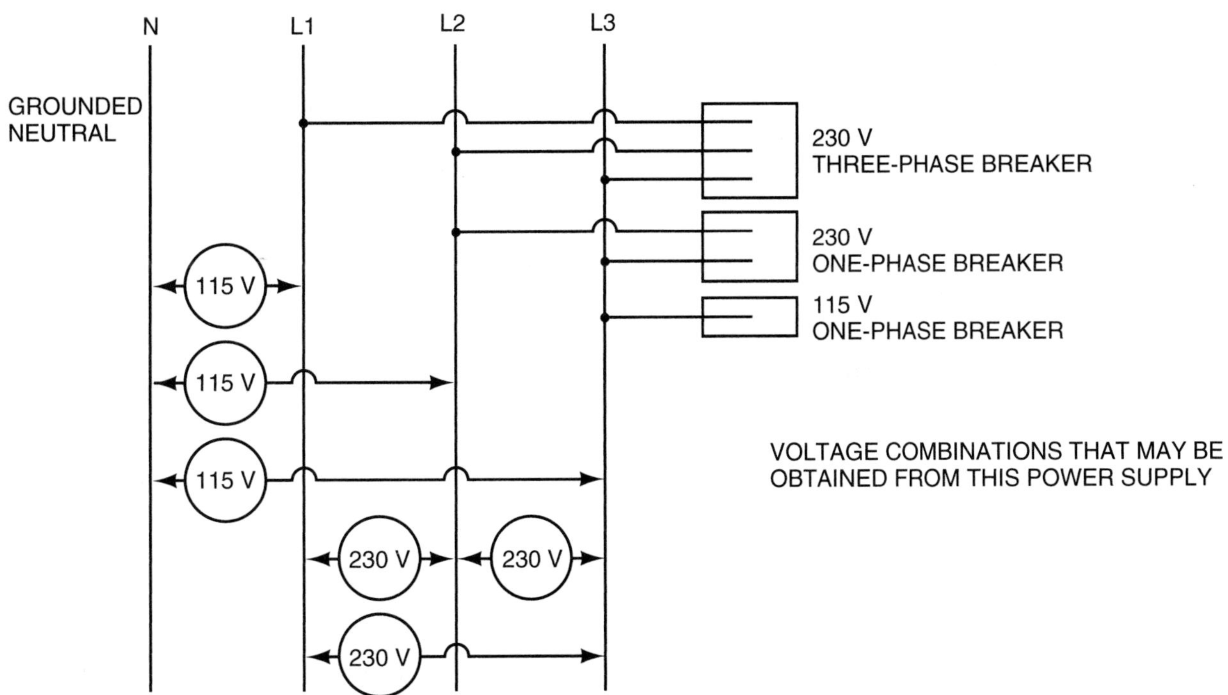

Figure 2-28. The diagram of a three-phase power supply.

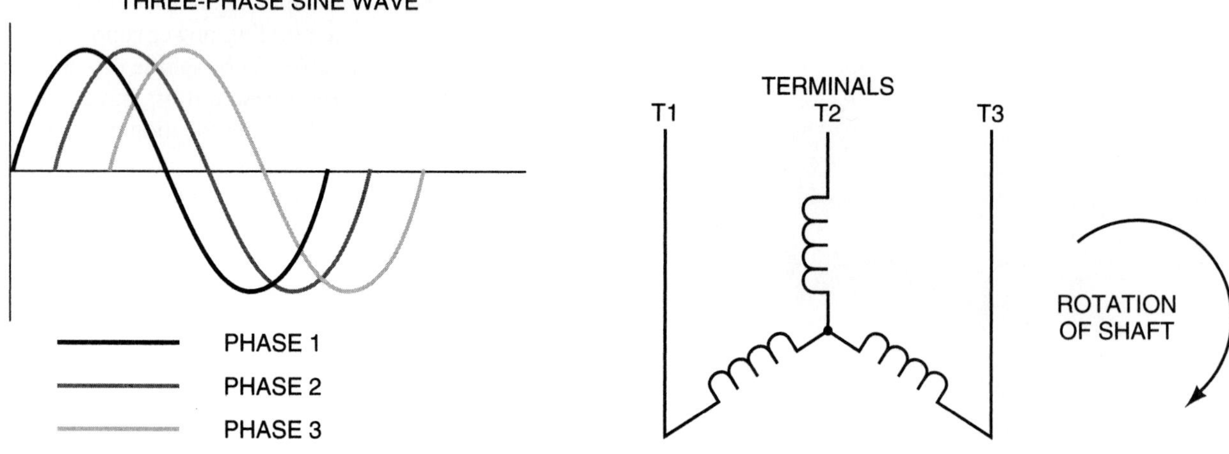

Figure 2-29. The diagram of a three-phase motor and power supply.

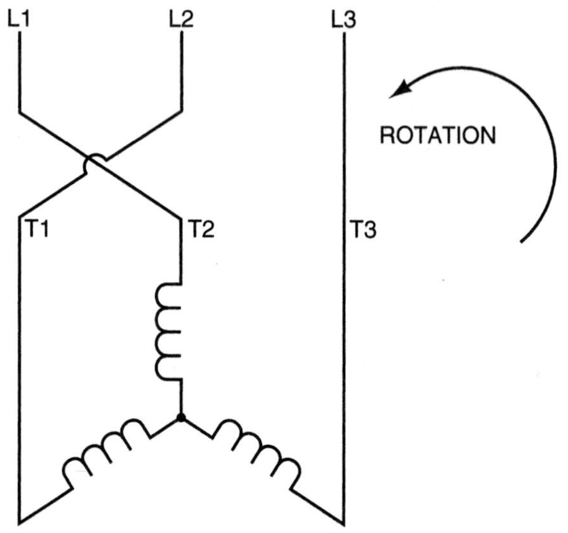

Figure 2-30. The wiring diagram of a three-phase motor. The rotation of this motor can be reversed by changing any two motor leads.

The sound level is another factor. Will the motor be used where ball bearings would make too much noise? If so, sleeve bearings should be used.

2.17 Variable-Speed Motors

The desire to control motors to provide a greater efficiency for the fans, pumps, and compressors has led the industry to explore development of and use of variable-speed motors. Most motors do not need to operate at full speed and load except during the peak temperature of the season and could easily satisfy the heating or air-conditioning load at other times by operating at a slower speed. When the motor speed is reduced, the power to operate the motor reduces proportionately. For example, if a home or building needs only 50% of the capacity of the heating unit to satisfy the space temperature, it would be advantageous to reduce the capacity of the unit rather than stop and restart the unit. When the power consumption can be reduced in this manner, the unit becomes more efficient.

The frequency (cycles per second) of the power supply and the number of poles determine the speed of a conventional motor. New motors are being used that can operate at different speeds by the use of electronic circuits. Several methods are used to vary the frequency of the power supply depending on the type of motor. Fan motors may be controlled through any number of speed combinations based on the needs.

Unit 1, "Basic Electricity and Magnetism," covered some of the fundamentals of AC electricity. All of the electricity furnished in the United States is AC and the current frequency is 60 cycles per second, or 60 hertz. AC is much more efficient to distribute than direct current (DC). AC can be converted to DC by means of rectifiers. Many applications for motors in the industry have needed variable-speed motors, and the only available variable-speed motors were DC motors. These motors are typically much more complicated to work with because they have both a field (stator) and an armature (rotor) winding and use brushes to carry power to the rotor. Brushes are carbon connectors that rub on the armature and create an arc. Variable-speed systems have been used in our industry in the past by using open-drive DC motors. There are many applications for variable-speed drive, such as pumps and fans and fractional horsepower motors.

The two types of motors that may be found in equipment today are the squirrel cage induction motor and the ECM (electronically commutated) DC motor. Instead of brushes rubbing on an armature, the motor is electronically commutated, Figure 2-31. This motor would be applied to a fan application as it is a fractional horsepower motor.

Variable-speed AC motor operation can now be accomplished with electronic components. The electronic components can switch power on and off in microseconds

Figure 2-31. An electronically commutated motor. *Courtesy General Electric*

Figure 2-32. This power transistor is basically an electronic switch that has no arc and has the ability to turn on and off very quickly.

without creating an arc; therefore, open-type contacts are not necessary for the switching purposes. The no-arc components have virtually no wear, have a long life cycle, and are reliable. Figure 2-32 shows a transistor that is an electronic switch with no contacts to arc.

The heating load on a building varies during the season and during each day. The central heating system in a house or other building would have many of the same operating characteristics. Let us use a house as an example. Starting at midnight, the outside temperature may be 0°F and the system may be required to run at full load all the time to add heat as fast as it is leaving the house. As the house warms in the afternoon, the unit may start to cycle off and then back on, based on the space temperature. Remember, every time the motor stops and restarts, there is wear at the contactor contacts and a burden is put on the motor in the form of starting up. Motor inrush current stresses the bearings and windings of the motor. Most motor bearing wear occurs in the first few seconds of start-up because the bearings are not lubricated until the motor is turning. It would be best not to ever turn the motor off and instead just keep it running at a reduced capacity.

When a system shuts off, there is normally a measurable temperature drop before it starts back up. This is very noticeable in systems that only stop and start. It starts up and runs until the thermostat is satisfied and then it shuts off. There is a measurable temperature rise before it shuts off and a measurable temperature drop before it starts back up. The actual space temperature at the thermostat location may look like the graph in Figure 2-33. The temperature graph may look more like the one in Figure 2-34 when variable-speed motor controls are used along with variable firing rate for a furnace. The same temperature

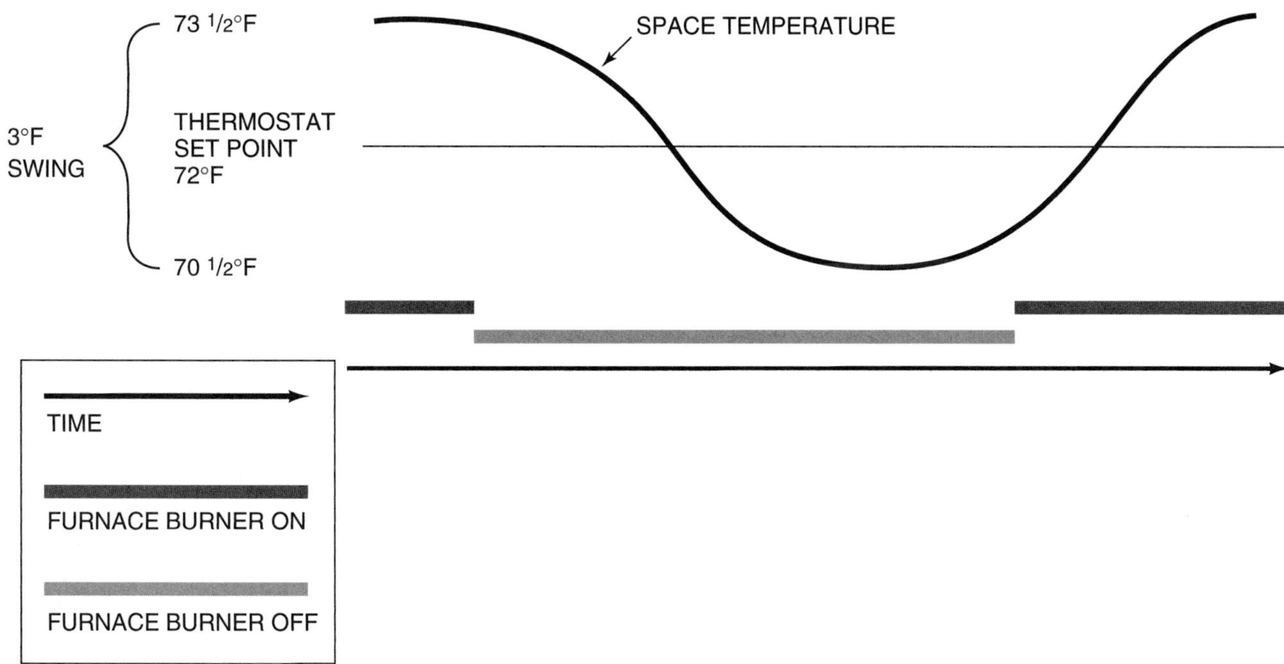

Figure 2-33. This typical on-off furnace thermostat has a 3°F temperature swing during its cycle. This is enough temperature swing to be noticeable.

Figure 2-34. This furnace is controlled by a variable-speed fan motor and firing rate. It has only a 1°F swing. This will not be noticeable.

curve profile would be true for the cooling season, a more flat profile with less temperature and humidity variations.

Variable-speed motors can level out the air-conditioner operation by running for longer periods of time. You may notice in any building when the system thermostat is satisfied and the unit shuts off. Suppose we could just keep the unit running at a reduced capacity that matched the building load. If we could gently ramp the motor speed down as the load reduces and then ramp it up as the load increases, the temperature would be more constant in the winter. This can be accomplished with modern electronics and variable-speed motor drives.

For example, a light bulb turns on and off 120 times per second when used with 60-cycle current furnished to a typical household. Figure 2-35 shows a sine wave of the current that is furnished to the home. The voltage goes from 0 V to 120 V to 0 V and back to 120 V 60 times per second; thus it is a 60-cycle current. Current to the light bulb element is interrupted 120 times in that second. The light from the bulb comes from the glowing element. It does not cool off between the cycles, so you do not notice it. Electronic devices can detect any part of the sine wave and switch off the current at any place in time.

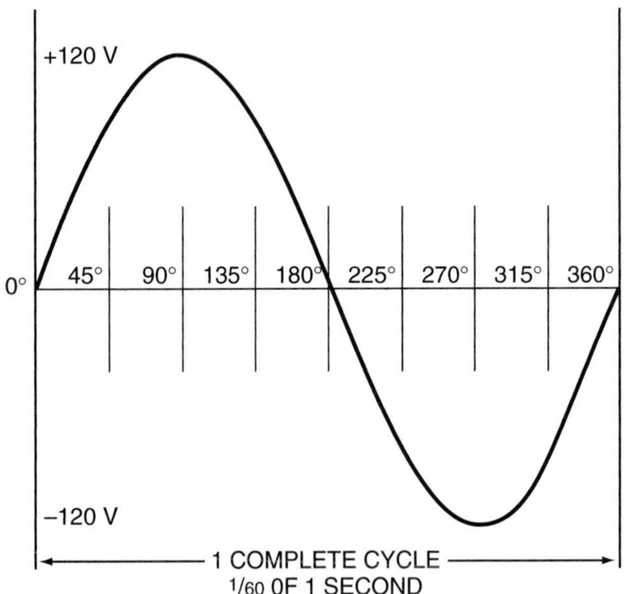

Figure 2-35. This sine wave represents 1/60th of one second.

If a switch could be installed to make even more interruptions in the current to the light bulb, it would glow at a lower light output. Figure 2-36 shows a simple rectifier in the form of a diode installed in the line that will interrupt the current flow to the light bulb on half of the sine wave cycle. The light bulb will now glow at half of its light output. This diode does not draw any current or have any moving parts but interrupts half of the current flow to the light bulb. The light bulb not only uses half power, it also gives off half as much light. The diode is like a check valve in a water circuit; it allows current flow only in one direction. The term *alternating current* tells you that current flows in two directions, one direction in one part of the cycle and the other direction in the other part of the cycle. When the current tries to reverse for the other half of the cycle, it stops it from flowing. It saves power and gives the owner some control over the light output.

AC electric motors can be controlled in a similar manner with electronic circuits. AC motor speed is directly proportional to the cycles per second, hertz. If the cycles per second are varied, the motor speed will vary. The voltage must also be varied in proportion to the cycles per second for the motor to remain efficient at all speeds. Once the voltage is converted to DC and filtered, it then goes through an inverter to change it back to AC that is controllable. Actually, this is still pulsating DC. The reason for all of this is to be able to change the frequency, cycles per second, and voltage at the same time. As the frequency is reduced, the voltage must also be reduced at the same rate. For example, if you have a 3600-rpm, 230-V, 60 cycle per second motor and you want to reduce the speed to 1800 rpm, the voltage and the cycles must be reduced in proportion. By reducing the voltage to 115 V and the frequency to 30 cycles per second, the motor will have the correct power to operate at 1800 rpm. It will not lose its torque characteristics. The motor will also operate on about half the power requirement of the full-load operation. If you reduce only the voltage, the motor will overheat.

AC that is supplied from the power company is very hard to regulate for usage at the motor, so it must be altered to make the process easier and more stable. The process involves changing the incoming AC voltage to DC. This is accomplished with a device called a converter or rectifier. This is much like a battery charger that converts

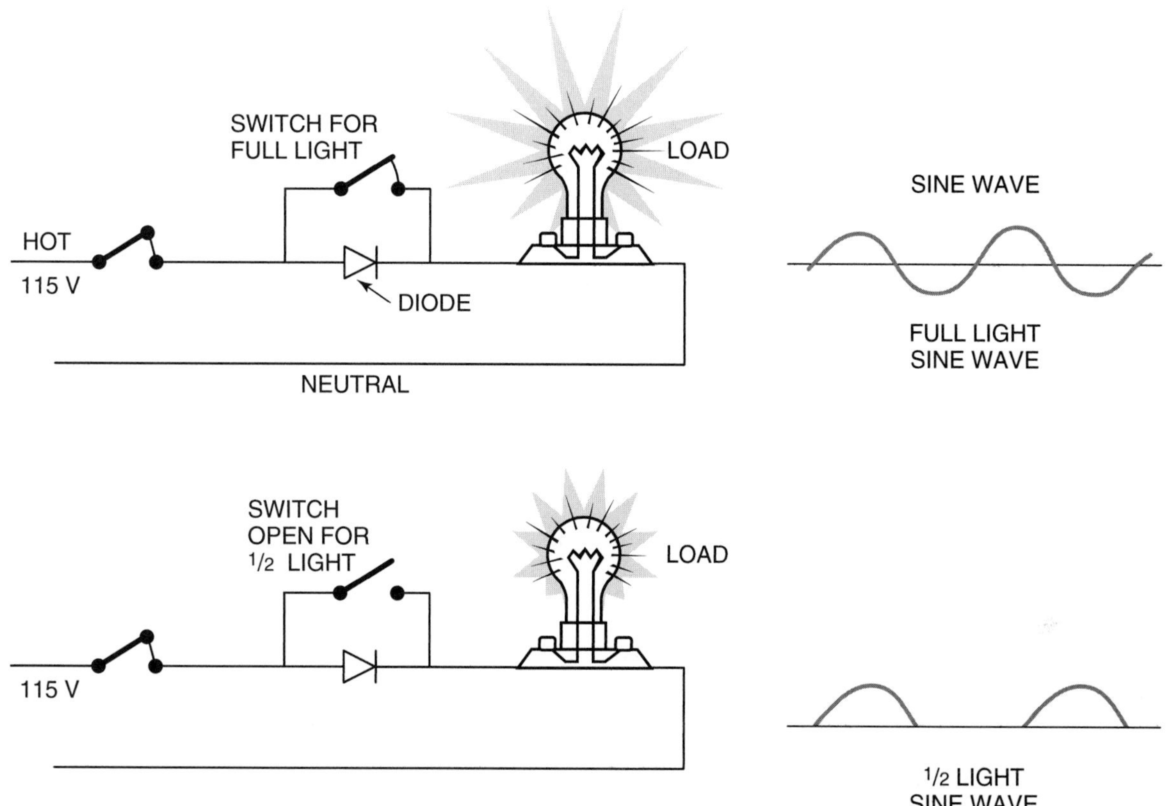

Figure 2-36. This light dimmer uses a diode to cut the voltage in half at low light.

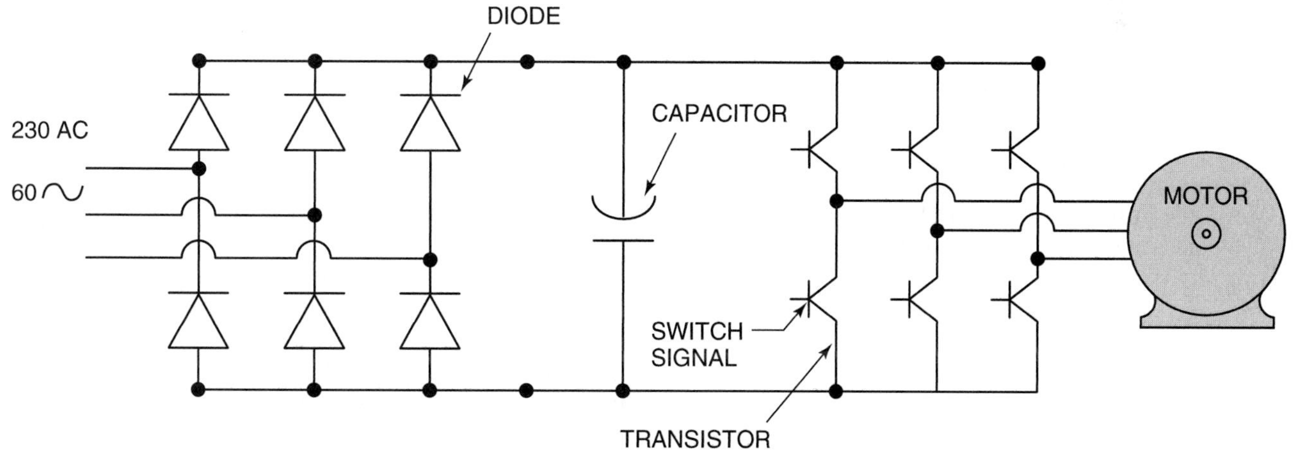

Figure 2-37. A simple diagram of a variable-speed motor drive.

AC to 14 V DC to charge an automobile battery. This DC voltage is actually known as pulsating DC voltage. The DC voltage is then filtered using capacitors to create a more pure DC voltage.

The simple diagram of a system may look like the one in Figure 2-37. There are many components in an actual system, but the technician should think about the system one component at a time for understanding and troubleshooting.

2.18 DC Converters (Rectifiers)

There are two basic types of converters, the *phase-controlled rectifier* and the *diode bridge rectifier.*

The phase-controlled rectifier receives AC from the power company and converts it to variable voltage DC. This is done by using silicon-controlled rectifiers (SCRs) and transistors that can be turned off and back on in microseconds. Figure 2-38 shows the wave form of AC that enters the

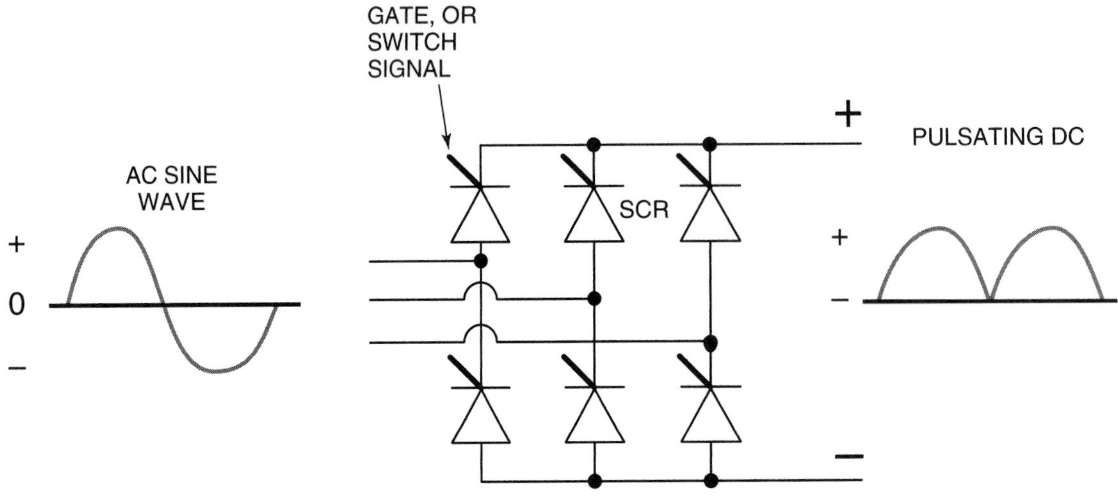

Figure 2-38. The rectifier moved all of the sine wave to one side of the graph.

phase-controlled rectifier, which is furnished by the power company, and the DC current that leaves the device. Notice the connection for turning these diodes or transistors on and off. The DC voltage leaving this rectifier is varied within the rectifier to coincide with the motor speed. The frequency of the power will be varied to the required motor speed in the inverter, which is between the converter and the motor. Remember, the voltage and the frequency must be changed for an efficient motor speed adjustment.

The other component in the system is a capacitor bank to smooth out the DC voltage. The rectifier turns AC into a pulsating DC voltage that looks like all of the AC voltage is on one side of the sine curve. The voltage looks more like pure DC voltage when it leaves the capacitor bank, Figure 2-39. This type of capacitor bank is used for any rectifier to create a better DC profile.

The diode bridge rectifier is a little different in that the DC voltage is not regulated in the rectifier. The diodes used in this rectifier are not controllable. It is a constant pure DC voltage after it has been filtered through the capacitor bank. The voltage and the frequency will be adjusted at the inverter of the system. The diode bridge rectifier has no connection for switching the diodes on and off.

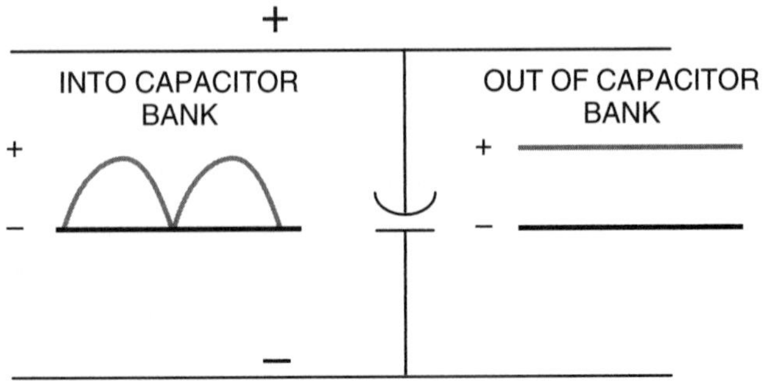

Figure 2-39. Pulsating DC enters the capacitor bank and straight line DC leaves.

2.19 Inverters

Inverters produce the correct frequency to the motor for the desired speed. Conventional motor speeds are controlled by the number of poles, and the frequency is a constant 60 Hz. Inverters can actually control motor speeds down to about 10% of their rated speed at 60 Hz and up to about 120% of their rated speed by adjusting the hertz to above the 60 Hz standard.

There are different types of inverters. A common one is a *six-step* inverter, and there are two variations. One controls voltage and the other controls current. The six-step inverter has six switching components, two for each phase of a three-phase motor. This inverter receives regulated voltage from the converter, such as the phase-controlled inverter, and the frequency is regulated in the inverter.

The voltage-controlled six-step converter has a large capacitor source at the output of the DC bus that maintains the output voltage, Figure 2-40. Notice the controllers are transistors that can be switched on and off.

The current-controlled six-step inverter also has the voltage controlled at the input. It uses a large coil, often called a choke, in the DC output bus, Figure 2-41. This helps stabilize the current flow in the system.

The *pulse width modulator* (PWM) inverter receives a fixed DC voltage from the converter, and then it pulses the voltage to the motor. At low speeds, the pulses are short; at high speeds, the pulses are longer. The PWM pulses are sine coded to where they are narrower at the part of the cycle close to the ends. This makes the pulsating signal look more like a sine wave to the motor. Figure 2-42 shows the signal the motor receives. This motor speed can be controlled very closely.

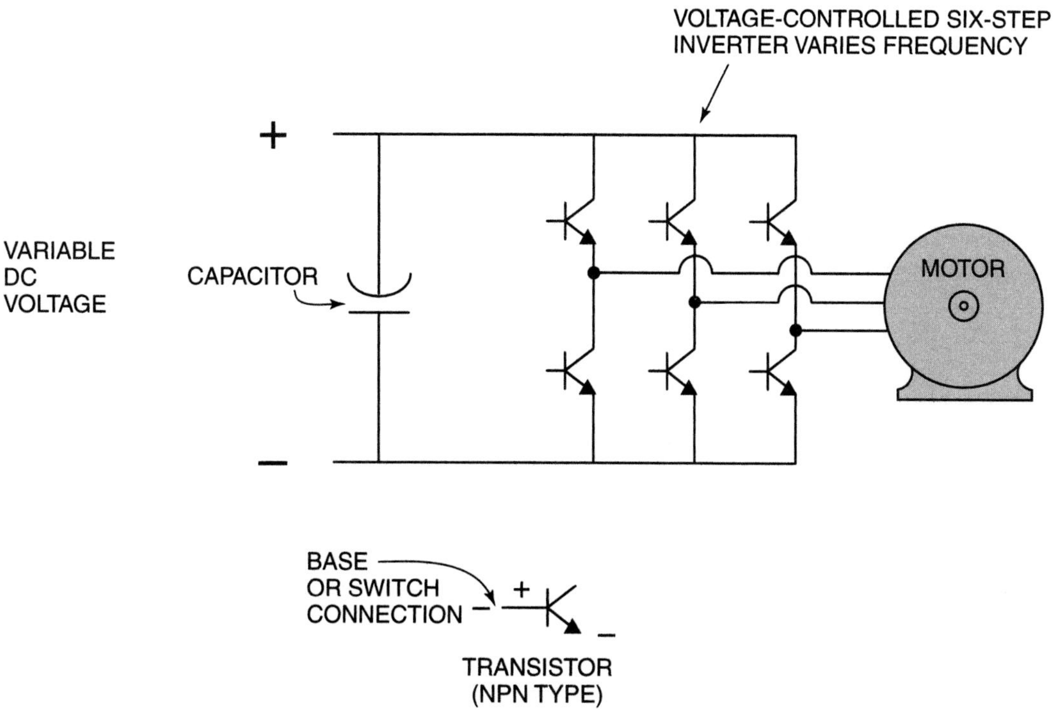

Figure 2-40. When the control system (computer) sends a signal to the base connection the transistor turns on and then it turns off when the signal is dropped.

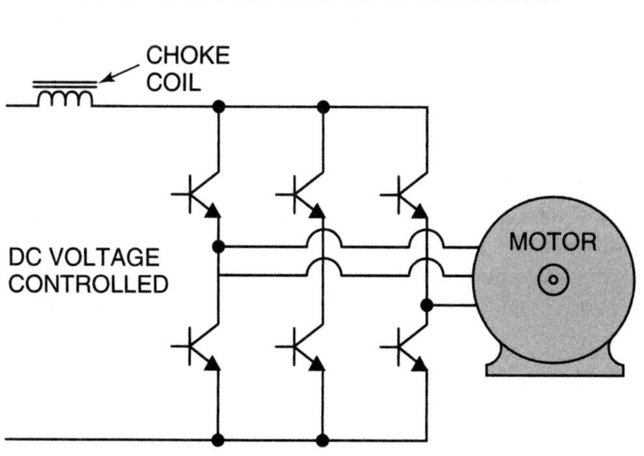

Figure 2-41. The choke coil stabilizes current flow.

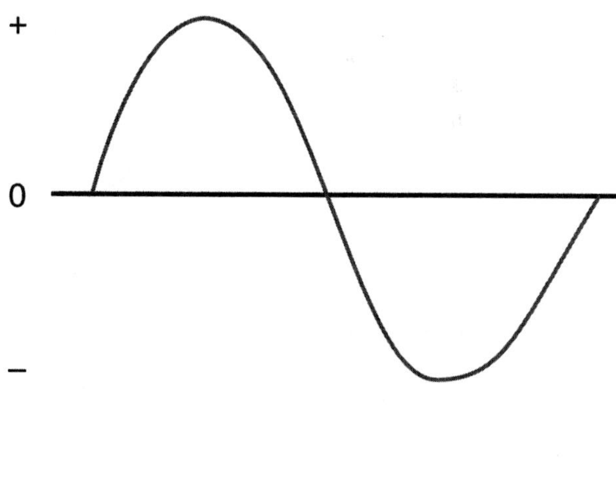

Figure 2-42. Sine coded, pulse width modulation.

2.20 Electronically Commutated Motors

The ECM is used for applications for open-drive fans in sizes 1 hp and smaller, Figure 2-31. It is reliable and has an energy savings that is worth the difference over time. The technology has been developed to electronically commutate a DC motor where there is no need for brushes, as mentioned earlier. The DC motor has always been more efficient than AC motors, but the armature had to be magnetized with DC power to the armature with the brushes.

The armature can be commutated (magnetized) now with permanent magnets attached to the armature with reliable high-tech adhesives.

This motor is factory calibrated to suit the piece of equipment it is applied to. A speed, torque, airflow, and external static pressure relationship is programmed into the motor at the factory for that particular application. This can help the air handler supply the correct airflow for different field conditions. For example, when a typical system has restricted air filters or when supply registers are closed, there is a reduced airflow. With the ECM, the motor simply speeds up to accommodate the reduced airflow and stabilizes at the correct airflow. This is accomplished with the relationship of airflow to torque, speed, and external static pressure set up by the manufacturer.

The ECM can also be applied to any system that can furnish the correct 24-V signal to the motor for load demand. This technology has provided variable-speed motors for gas furnaces, air-conditioning systems, oil burner systems, and heat pumps.

The ECM is actually a two-piece motor, the controls and the motor section. If the technician suspects that the motor is defective, the controls can be removed and the motor may be checked using an ohmmeter in the conventional manner. If the motor shows to be electrically sound, a test module (furnished by the manufacturer) can be fastened to the motor and operated manually at variable speeds, Figure 2-43. When the motor will operate correctly with the test module and will not operate without it, the ECM electronic portion of the motor can be checked for correct input and tight connections. If the motor cannot be made to operate with the correct signal to the ECM control portion, the ECM controls can be changed and the same motor used. *NOTE:* Each motor has its own ECM control package, and it is not interchangeable with just any other ECM control package from the same manufacturer. Remember, the control package is programmed for the specific piece of equipment. An exact replacement must be used.

Figure 2-43. This analyzer is connected to the motor and control board, and the technician can diagnose whether the board or the motor has a problem. *Courtesy Ferris State University, Photo by John Tomczyk*

All of these variable-speed motors must have very close control for faults. They all tend to give off a lot of electronic noise, peaks, and valleys to the input supply AC voltage. This is because they are constantly stopping and starting the input current flow to the converter. If a large current flow is stopped in the middle of the entering sine wave cycle, the power furnished by the power company will have a tendency to fluctuate. This will create a spike in voltage. The same thing will happen if a large current load is started, like the inrush current on start-up of a motor. The fluctuations must be filtered to prevent problems with computers and other electronic equipment in the vicinity. Special precautions are taken by the manufacturer of the equipment and all directions should be followed.

One of the advantages of variable-speed motors is that they can be started at a reduced speed, which will be a reduced inrush current. As discussed earlier, the inrush current is about five times the running current. When a large motor is started in an office building, the inrush current can cause the voltage to the entire building to drop to a lower value. This can cause lights to blink and computers to cycle off if it is extreme. Large motors have methods to soften the start-up. Variable-speed motors have what is called a soft start-up, at a reduced speed. Then the motor speed is ramped up to full load and full current.

Each manufacturer will have a troubleshooting procedure that should be followed. As the components become more cost effective, they will be used in smaller equipment. These systems are very efficient. Always check the incoming power supply to any device to make sure it is what it should be before you decide the electronics are the problem. It is possible to have one fuse blown on a three-phase circuit. A quick check with a voltmeter will verify whether the power supply is good.

Some benefits of variable-speed technology are

- power savings.
- load reduction based on demand.
- soft starting of the motor (no LRA).
- better space temperature and humidity control.
- solid-state motor starters without open contacts.
- oversized units for future expansion that can run at part load until the expansion.
- load and capacity matching.

2.21 Cooling Electric Motors

All motors must be cooled because part of the electrical energy input to the motor is given off as heat. Most open motors are air cooled. An air-cooled motor has fins on the surface to help give the surface more area for dissipating heat. These motors must be located in a moving airstream to cool properly.

Summary

- Motors turn fans, compressors, and pumps.
- Some of these applications need high starting torque and good running efficiencies; some need low starting torque with average or good running efficiencies.
- Small fans normally need motors that have low starting torque.
- The voltage supplied to a particular installation will determine the motor's voltage. The common voltage for furnace fans is 115 V; 230 V is the common voltage for home electric heating systems.
- Common single-phase motors are split-phase, PSC, and shaded-pole.
- When more starting torque is needed, a start capacitor is added to the motor.
- A run capacitor improves the running efficiency of the split-phase motor.
- A centrifugal switch breaks the circuit to the start winding when the motor is up to running speed. The switch changes position with the speed of the motor.
- An electronic switch may be used to interrupt power to the start winding.
- The common rated speed of a single-phase motor is determined by the number of poles or windings in the motor. The common speeds are 1800 rpm, which will slip in speed to about 1725 rpm, and 3600 rpm, which will slip to about 3450 rpm.
- The difference in 1800/3600 and the running speeds of 1750/3450 is known as the slip. Slip is due to the load imposed on the motor while operating.
- Three-phase motors are used for all large applications. They have a high starting torque and a high running efficiency. Three-phase power is not available at most residences, so these motors are limited to commercial and industrial installations.
- The PSC motor is used when high starting torque is not required. It needs no starting device other than the run capacitor.
- Variable-speed motors operate at higher efficiencies with varying loads.
- The electronically commutated motor (ECM) is a DC motor that does not have brushes.
- Variable-speed motors can even out the load on the heating and air-conditioning systems by adjusting the equipment speed to the actual load.
- Motor speed can be varied from about 10% of rated speed to about 120% using electronics.
- Electronic switches, silicon-controlled rectifiers (SCRs), and transistors can be turned on and off without making an arc. They are reliable and do not use much power.
- DC converters convert AC power to DC power.
- Capacitors even out the pulsating DC power.
- Inverters create an alternating frequency that can be varied.
- The frequency and the voltage must be changed together for the motor to perform efficiently.

Review Questions

1. The two popular operating voltages for residences are _______V and _______V.
2. When an open motor gets up to speed, the _______ switch takes the start winding out of the circuit.
3. Is the resistance in the start winding **greater** or **less** than the resistance in the run winding?
4. Which of the following devices may be wired into the starting circuit of a motor to improve the starting torque?
   A. another winding.
   B. start capacitor.
   C. thermostat.
   D. winding thermostat.
5. Which of the following devices may be wired into the running circuit of a motor to improve the running efficiency?
   A. a start capacitor.
   B. a run capacitor.
   C. both start and run capacitor.
   D. none of the above.
6. True or False: Three-phase motors have low starting torque.
7. True or False: All large motors are three-phase.
8. Why is it desirable to have a two-speed fan motor?
9. True or False: All electric motors must be cooled, or they will overheat.
10. True or False: An SCR is an electronic switch.
11. When a transistor is used as a switch:
    A. a great deal of power is consumed.
    B. it makes a lot of noise.
    C. it is subject to fail quickly.
    D. it does not create an arc.
12. Capacitors are used after the converter to:
    A. smooth out the DC current.
    B. suppress the noise.
    C. change the voltage frequency.
    D. reduce the voltage.
13. The converter:
    A. changes the frequency of the power.
    B. changes AC to DC.
    C. is a switch.
    D. is used only with DC motors.
14. The inverter:
    A. adjusts the frequency.
    B. adjusts the DC voltage.
    C. smoothes out the AC power.
    D. makes a lot of noise.

3 Electric-Heat Equipment

OBJECTIVES

Upon completion of this unit, you should be able to

- **calculate the current draw of an electric-heat system.**
- **convert the current draw to the Btu rating of an electric-heat system.**
- **estimate the current draw of an electric heating element from the resistance of the element.**

3.1 Introduction to Electric Heat

Whole-house electric heat became popular in the temperate climates in the 1960s when contractors, who began to add more insulation to houses, made a discovery. They found that when a house was insulated at every outside exposure and the infiltration from outside air was reduced, electric heat was a competitor for the fossil fuels, gas and oil. As electricity-generating costs came closer to the cost of fossil fuels, electricity began to be used more often.

Electric heat is 100% efficient because there is no flue gas loss. All the heat is used, unless the appliance is under the house or in the attic, where some heat will escape (see Figure 3-1). The cost of electric heat is still more than the cost of fossil fuels in most parts of the United States, but the high efficiency of electric heat helps offset the difference in cost.

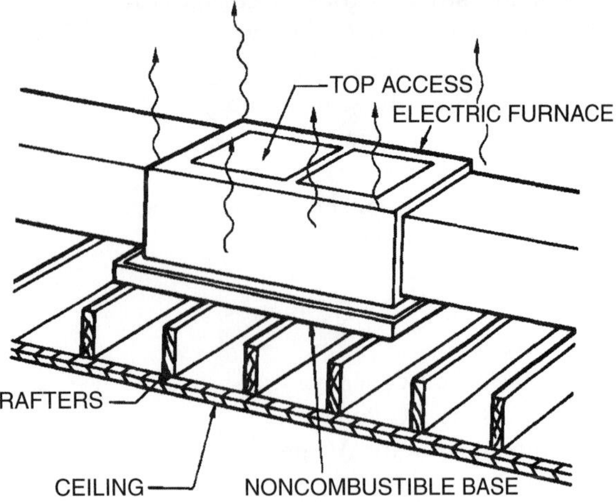

Figure 3-1. If an electric-heat appliance is in an unconditioned space, some heat will escape.

The initial cost of the electric-heat appliance is usually less than the cost of a fossil fuel appliance. This low cost makes electric heat appealing to speculative builders, the builders who construct homes for sale. The builder puts less money into the house and may make a higher profit.

The goal of electric heat is to extract the heat energy that is available from electricity. A working knowledge of Unit 2 is vital at this point in order to understand how heat is obtained from electricity.

The potential for heat in electricity is expressed in the power formula, a derivative of Ohm's law. The power formula states the following:

$$P = I \times E$$

where P = Power, expressed in watts.

I = Amperage, expressed in amperes.

E = Electromotive force, expressed in volts.

This formula can be used in the following example to calculate the amount of heat that an electric heating element will give off.

Suppose that an electric heating element has an ampere draw of 20 amperes when the applied voltage (electromotive force) is 230 volts. How many watts of heat energy will this heater give off? See Figure 3-2 for a working example.

$$\begin{aligned} P &= I \times E \\ &= 20 \times 230 \\ &= 4{,}600 \text{ Watts (W)} \end{aligned}$$

The watt method of rating heaters often yields large numbers, so another method is usually used, the kilowatt method. The amount of heat can be converted to kilowatts, or thousands of watts, by dividing watts by 1,000. In our example, 4,600 watts divided by 1,000 equals 4.6 kilowatts (kW).

The kilowatt of heat is used to rate electric-heating appliances. In the units on oil and gas heat we discuss the Btu as a rating for the fossil fuel systems. The kilo-watt heat rating can be converted to Btu by multiplying the number of kilowatts times 3,413 (there are 3,413 Btu in 1 kilowatt). The heater in our example has a rating of 4.6 kW × 3,413 Btu/kW = 15,699.8 Btu.

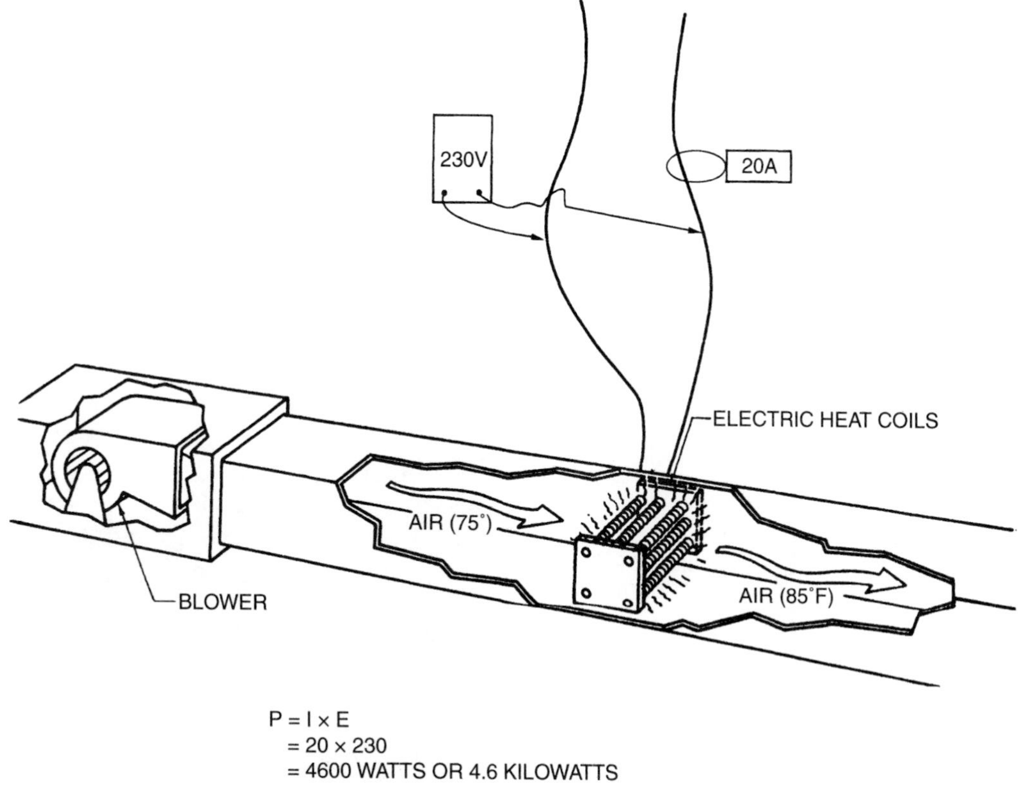

Figure 3-2. The heat output of this furnace is 100% of the electrical input.

Ohm's law can explain how heat is extracted from electricity. Remember that Ohm's law is $I = E/R$ (volts divided by the resistance of the circuit). The resistance of the circuit must be concentrated at the heating element for the heat to be given off at this point. Remember the statement that for current to flow in an electrical circuit and for heat to be produced three components must be present: a power supply, a conductor on which the power can flow, and a measurable resistance to consume the power, see Figure 3-3. The measurable resistance is the electric heating element, the load.

The heater in our example, with an ampere draw of 20 amperes at 230 volts, creates a heat rate of 4,600 watts (4.6 kW). The conductor has little resistance. For practical purposes, it has no resistance. According to Ohm's law, the heating element must have the resistance:

$$E = I \times R$$

where I = Electrical current in amperes (A).

R = Resistance in ohms (Ω).

E = Electromotive force in volts (V).

This formula can also be expressed as $R = E/I$, and the equation can then be solved for the resistance of the heater. For our example, the resistance of the heater is 11.5 ohms:

$$R = E/I$$
$$= 230/20$$
$$= 11.5\ \Omega$$

The conductor must be sized so that it gives off only a minimum of heat as the current approaches the measurable resistance. The resistance in the conductor is low. On the other hand, the heating element must have a high

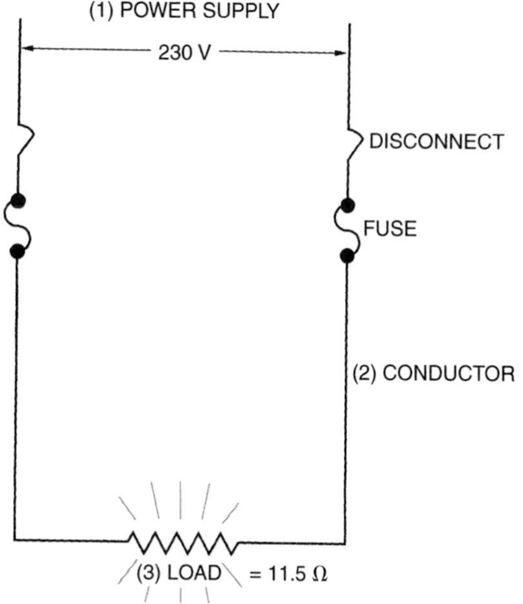

Figure 3-3. Three components must be present for electricity to flow.

resistance so that when the current reaches this point, heat will build up, Figure 3-4.

Conductors are made of copper, which has little resistance. The heaters are often made of ni–chrome (nickel–chromium) wire, which is shown in Figure 3-5. Ni–chrome wire has a fairly high resistance and does not burn away fast. It is tough and can be formed in different configurations to accomplish the type of heat desired by the manufacturer. When ni–chrome wire is formed and heated many times, it maintains its shape and does not droop. The manufacturer may want a long heater and the wire may be formed in a helix shape for this application, as shown in Figure 3-6. It may be in a ribbon form for some heaters, as shown in Figure 3-7. Some electric heaters use quartz elements for heating.

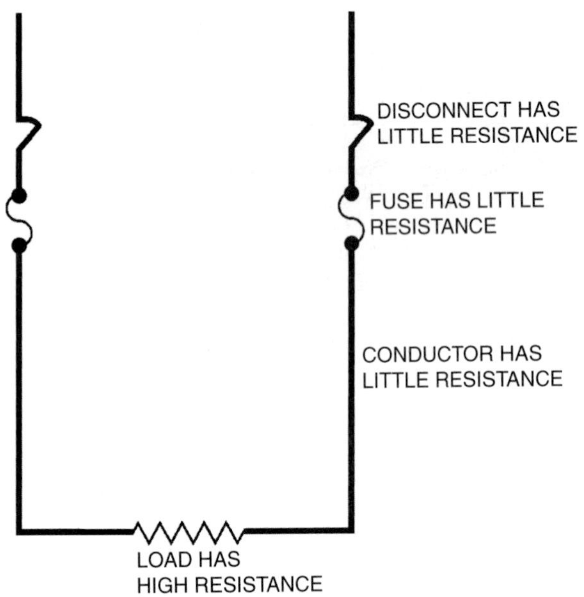

Figure 3-4. Heat is concentrated at the heating elements by the correct resistance.

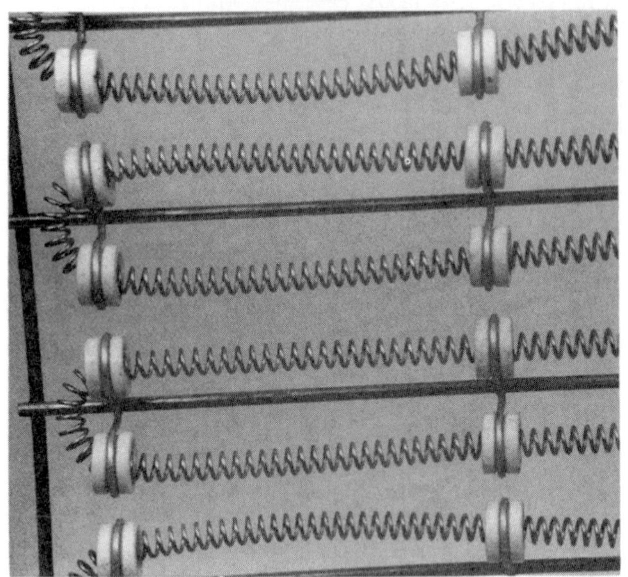

Figure 3-5. Ni–chrome heating element. *Photo by Bill Johnson*

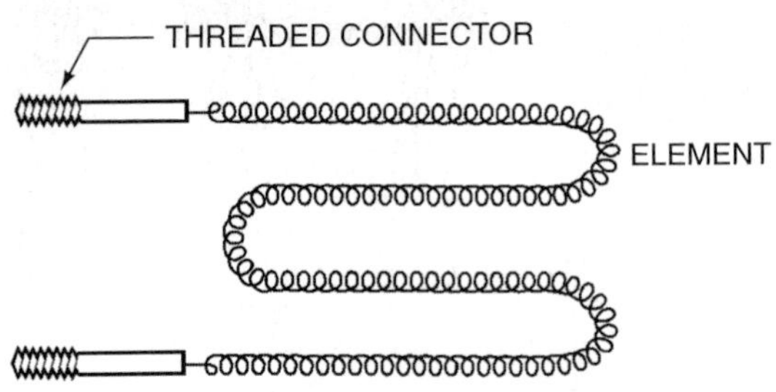

Figure 3-6. A helix shaped heating element.

Figure 3-7. Ribbon shaped heating element with ni-chrome. *Photo by Bill Johnson*

When electricity is furnished by the power company, the customer is charged by the kilowatt-hour, or kWh. When the heater in our example is operated for 1 hour at 4.6 kilowatts, the customer is charged for 4.6 kilowatt-hours. If the cost of power is $0.10 per kilowatt-hour, the customer is charged $0.46 per hour to operate the heater, Figure 3-8. At $0.46 per hour, the power bill can quickly add up.

Heat is transferred from the electric element to the surroundings in one of two ways: by conduction or by radiation. These two methods of transferring heat are used for different applications. A conduction electric heater transfers heat to the surrounding air. Figure 3-9 shows an example of a forced-air conduction heater.

The heat is transferred to the air, which is ducted to the conditioned space. Heat can be transferred into water in a boiler in the same manner Figure 3-10. The water is then circulated to the conditioned space.

Radiant heat is transferred from the heating element through air to solid objects, as shown in Figure 3-11. The radiant heating element typically does not have a fan to blow air across it, so it glows red-hot.

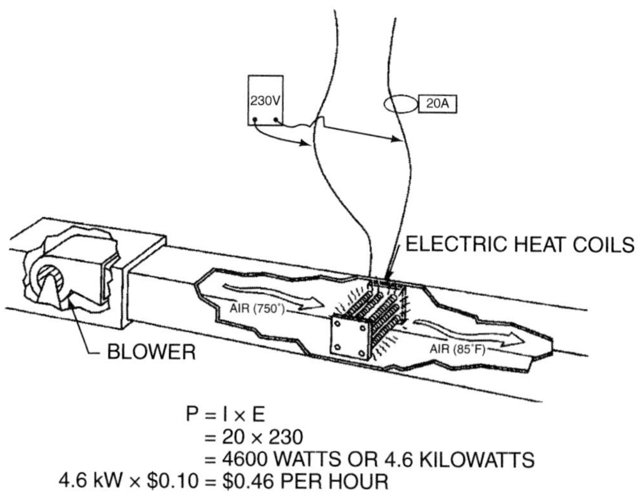

Figure 3-8. Cost of operating an electric heater.

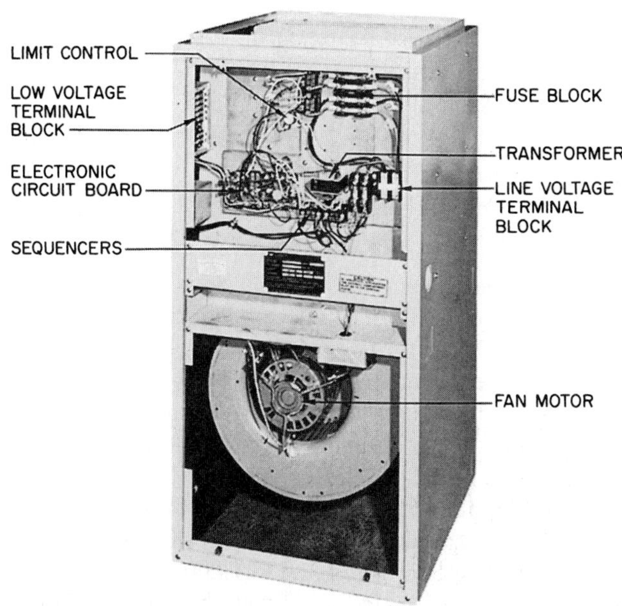

Figure 3-9. Forced–air electric conduction heater with fan. *Courtesy The Williamson Company*

Figure 3-10. Electric boiler. Heat can be transferred to water in a boiler. *Courtesy Burnham Corporation*

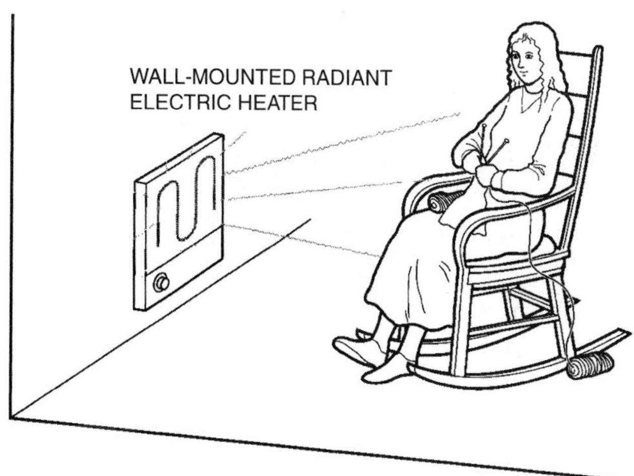

Figure 3-11. Radiant heat heats only the solid objects.

3.2 Portable Electric Heat

Portable electric heaters are purchased at department stores and hardware stores for use in the home and office as plug-in devices. These heaters are usually low in capacity, approximately 1,500 watts maximum, and can be plugged into 115-volt outlets. Portable electric heaters use radiant heat or forced-air convection for transferring heat into the conditioned space. The radiant heater pictured in Figure 3-12 has an element that glows red-hot and radiates heat to the solid objects in the space. These heaters are excellent for spot-heating in homes and offices. The element may be ni–chrome wire wound in a helix shape, like a spring, or in a ribbon form, as shown in Figure 3-13.

Warning: These electric heaters must remain upright and must be the correct distance, as recommended by the manufacturer, from combustible materials. Do not touch the front of the heater or allow children to come in contact with it, or burns may occur.

Sometimes the portable heater has a tilt switch that shuts off the heater in case it falls over, but this switch should not be relied on.

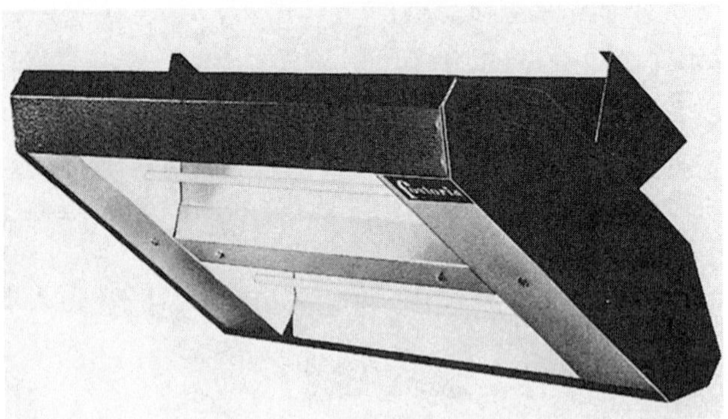

Figure 3-12. Quartz-element electric radiant heater. The element glows red-hot. *Courtesy Fostoria Industries, Inc.*

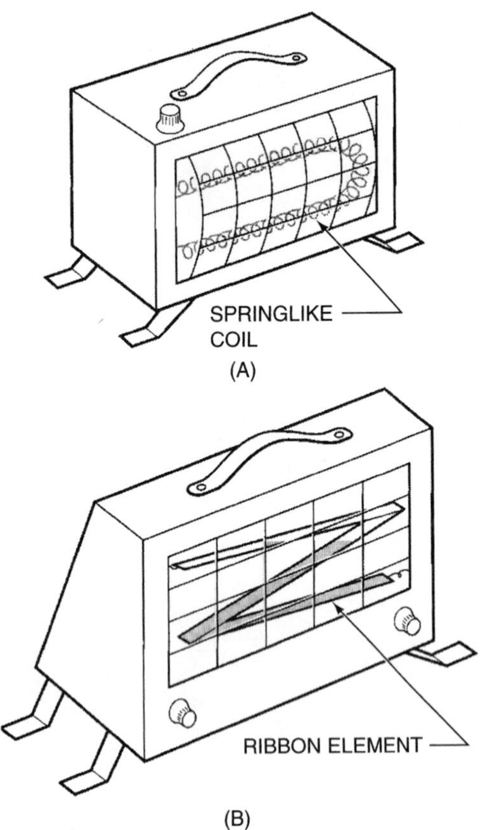

Figure 3-13. The portable electric heating element can be (A) helix wound, like a spring, or (B) ribbon like.

Forced-air convection heaters have fans to move the heat from the heaters. Heat is transferred into the air as it passes over the heating element and the heat is moved into the conditioned space, as shown in Figure 3-14. These heaters are small but emit quite a bit of heat.

This type of heater usually has a thermostat built into its casing to control the temperature somewhat, and to prevent the space from overheating. The thermostat may not maintain accurate control of the space because it is located in the heater, Figure 3-15.

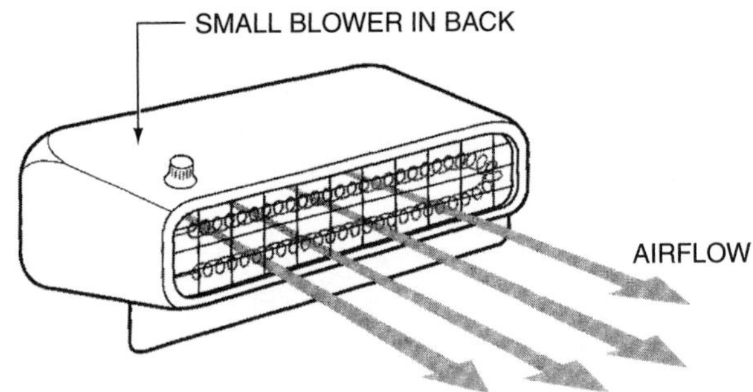

Figure 3-14. Forced-air convection heater.

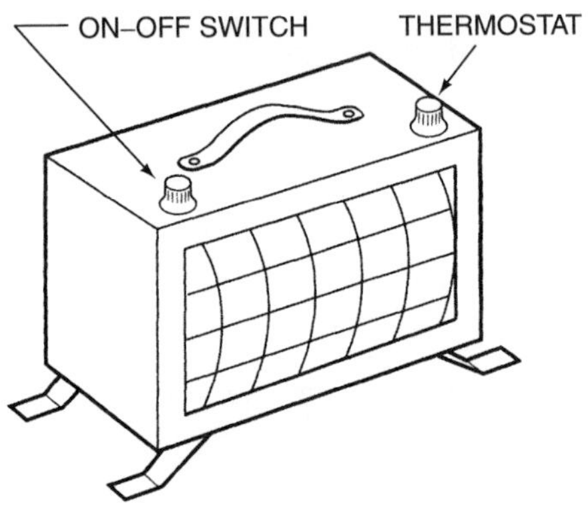

Figure 3-15. Thermostat location in an electric heater.

The portable heater often has a selector switch so that different ranges of heat can be selected. For example, a heater may have a 500-watt, 1,000-watt, and a 1,500-watt capacity output. To prevent circuit overload, always run the heater at the lowest capacity that will heat the space adequately. A 1,500-watt heater operating at 115 volts has an ampere rating of 13 amperes ($I = P/E = 1{,}500/115 = 13$ A). Most circuits in homes and businesses are 20-ampere circuits, so an electric heater operating at 13 amperes uses much of the capacity. If the heater is reduced to 1,000 watts, it requires only 8.7 amperes ($I = P/E = 1{,}000/115 = 8.7$). Reducing the load helps keep the circuits operating correctly.

Caution: Be observant of the power cord and plug. If they get hot, the connection may not be good. If the heater is allowed to operate while the plug is too hot to hold with your hands, damage or fire may occur.

Small electric plug-in heaters are not installed or serviced by heating technicians. They are mentioned only as a possible solution to spot-heating problems.

3.3 Radiant Electric Heat

Radiant heat is often used for spot-heating in places where it is not practical to add heat to the whole structure. For example, the entrance to a restaurant where customers must stand while waiting to get inside can be cold. Radiant heaters can easily be added in the ceiling to prevent the customers from becoming so cold that they leave, Figure 3-16. Remember that radiant heat heats the solid objects, the people, and not the space. This can be very efficient for certain applications. The heat can be manually operated with a switch on the wall (no thermostat is necessary), and an electrician can easily wire these heaters.

Radiant heat can also be built into the home or business. This radiant heat does not glow red-hot but is mounted in panels, usually Sheetrock, in the walls or ceiling, as shown in Figure 3-17. The panels are warm enough to give off heat and can be used for whole-house heating in a residence with a thermostat in each room. Whole-house heating with panel heat can be practical because it provides the home owner with individual room heat control. When some rooms do not require heat, the doors to these rooms can be closed and the heat shut off. Besides being practical, panel heat is even comfortable heat.

Figure 3-16. This radiant heater keeps customers warm until they enter the restaurant.

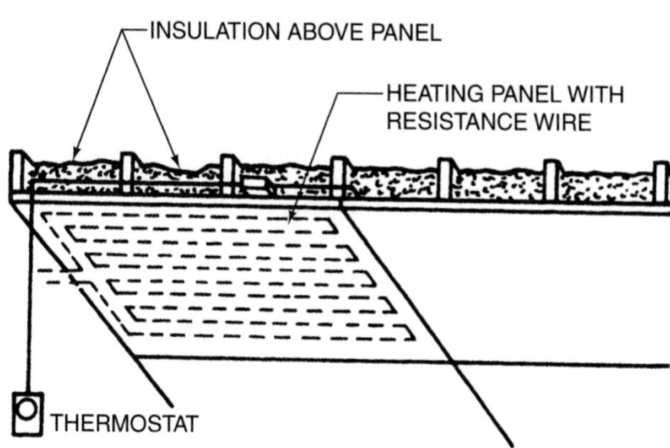

Figure 3-17. Panel radiant heat for home and business.

These radiant heating applications use line-voltage thermostats for controlling the power to the heaters. Panel heaters have a line-voltage thermostat in each room, as shown in Figure 3-18.

Another version of radiant electric heat is the wall-mounted heater, which is recessed in the wall. Figure 3-19 shows a wall-mounted heater. This heater usually has a line-voltage thermostat mounted in the heater.

Caution: These heaters should not be installed close to combustible materials because these heaters radiate a very hot heat.

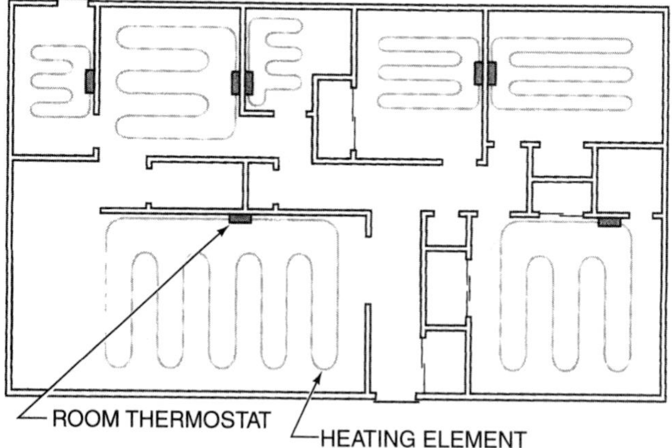

Figure 3-18. A line-voltage thermostat in each room.

Figure 3-19. Wall-mounted radiant electric heater.

3.4 Baseboard Electric Heat

Baseboard electric-heat systems are natural-draft systems. Air next to the heating element is heated by conduction. This air expands and rises. When it rises, new, cooler air takes its place and an air current starts, Figure 3-20. This air current can spread heat through-out the entire room, as shown in Figure 3-21.

The element in a baseboard heater does not glow red-hot; it is operated at a much lower temperature so that it does not set fire to draperies or catch particles on fire that may be dropped into the heater, Figure 3-22. The heating element is enclosed in a tube to isolate the element from the air. Fins are added to the tube to give the element more surface area so that more heat can be added to the air per foot of heater, Figure 3-23.

Baseboard heaters are rated by watts per foot. When more heat is needed in a room, a longer heater is installed. In Figure 3-24 we see that a heating unit is rated at 250 watts per foot and that a 60-inch (5-foot) unit has a capacity of 4,265 Btu per hour, or 1,250 watts at 240 volts. The unit has an amperage draw of 5.2 amperes. Notice that Ohm's law proves the rating of the heater: $P = E \times I = 240 \times 5.2 =$ 1,248 watts, rounded up to 1,250 watts.

Baseboard heat systems can be used for spot-heating or whole-house heating. They provide even, comfortable heat because of the gentle air currents set up by the element, and they are not noisy because there are no fans. Furthermore, line-voltage thermostats are used in the individual rooms, and only those rooms requiring heat need to be heated, Figure 3-25. These thermostats have decorative covers and will not cause electrical shocks because of the type of safety enclosure.

Electricians usually service baseboard heaters. They are mentioned in this text only so that the heating technician can be familiar with them.

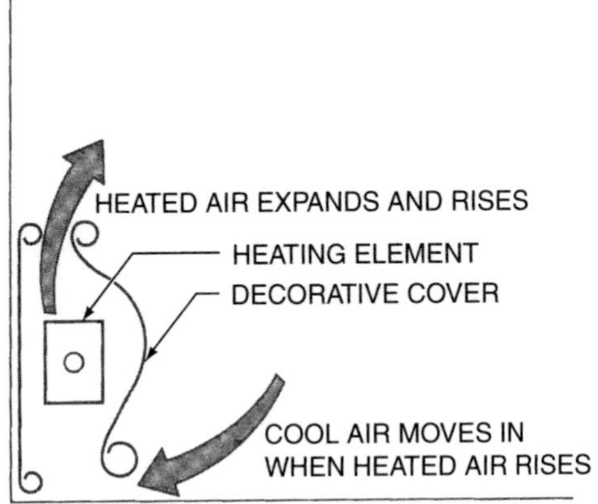

Figure 3-20. Air-flow in a baseboard electric heater.

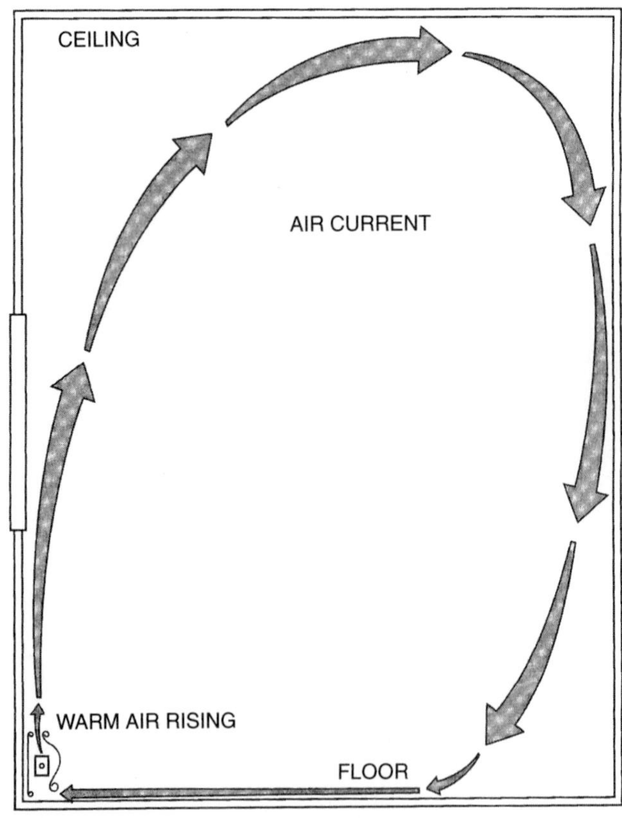

Figure 3-21. With baseboard heat, an air current moves around the entire room.

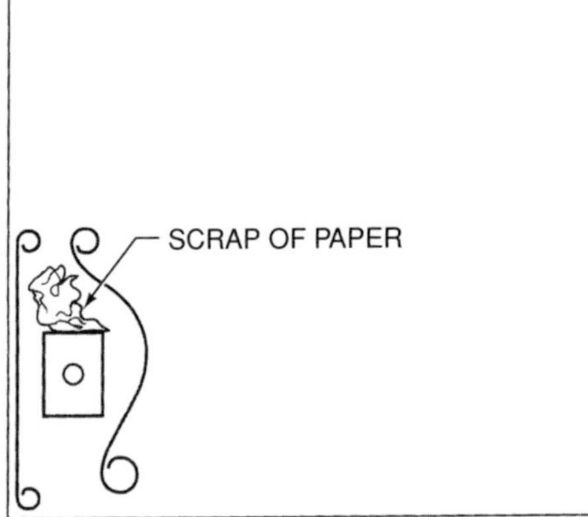

Figure 3-22. Particles can be dropped into the baseboard heaters: as a result, the heating elements do not glow red-hot.

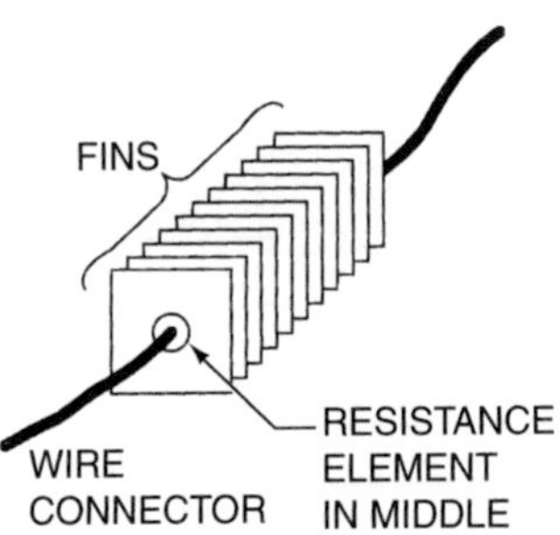

Figure 3-23. Fins are added to the enclosed heating element to increase the surface area of the element for more heat exchange.

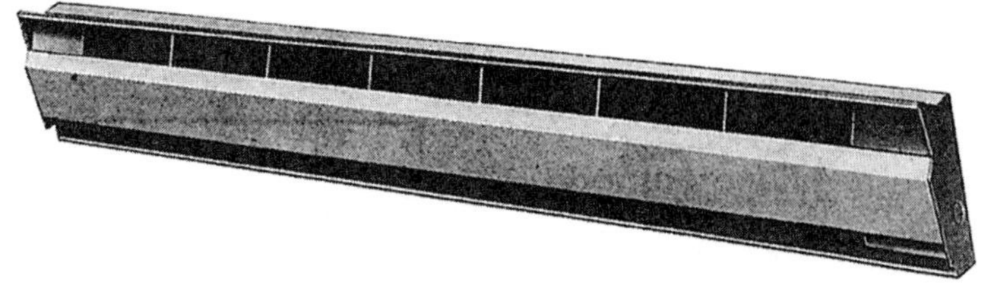

(A)

Length Inches	BtuH	Total* Watts	208V Watts	Watts Per Ft.	Amps
28	1706	500	375	250	2.1
40	2559	750	562	250	3.1
48	3412	1000	750	250	4.2
60	4265	1250	937	250	5.2
72	5118	1500	1125	250	6.3
96	6824	2000	1500	250	8.3
120	8530	2500	1876	250	10.4

(*) Specifications shown are for units operated on 240V. Operation on 208V will result in 25% lower BtuH, total watts, and watts per foot; Amps will be 13% less.

(B)

Figure 3-24. The baseboard electric heater is rated by watts per foot. (A) Baseboard heater. (B) One manufacturer's ratings. *Courtesy W. W. Grainger Company*

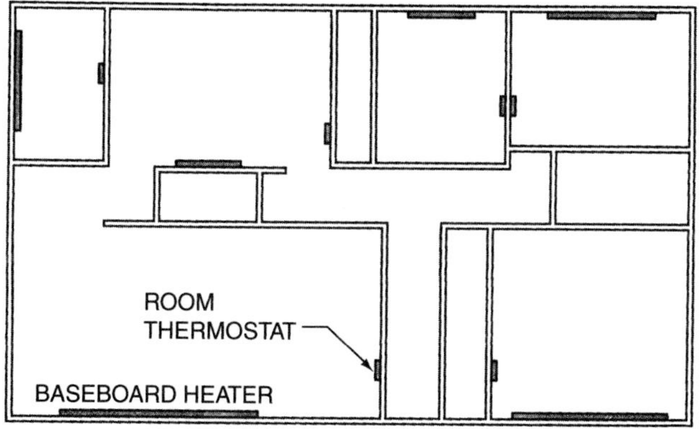

Figure 3-25. With baseboard heat, a line-voltage thermostat is used in each room requiring heat.

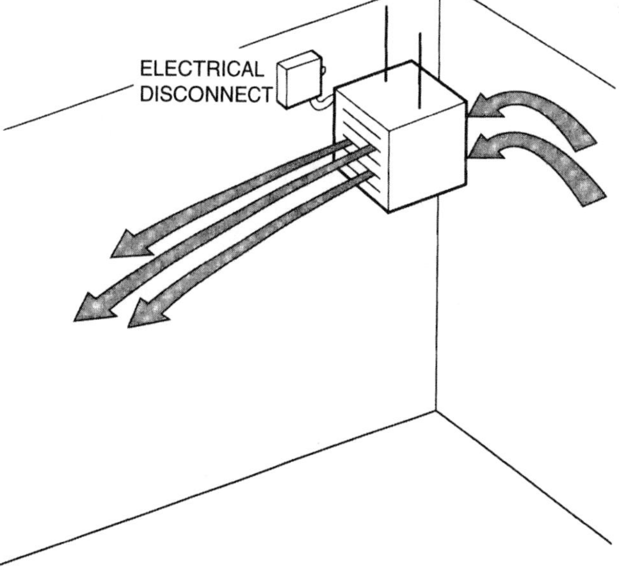

Figure 3-26. The electric unit heater is suspended from the ceiling and often heats large spaces.

3.5 Electric Unit Heaters

Electric unit heaters are suspended from the ceiling and used in installations where large spaces need heat, Figure 3-26. These heaters are compact, as seen in Figure 3-27, and each has a fan to move the heat over the heating element. One advantage of an electric unit heater is the low initial cost of the heater, compared with the cost of a gas or oil unit heater. The electric heater may cost more to operate, but it is easy to install and a gas line or a flue, which costs more money, does not have to be run. The electric power line to operate the heater is costly, however, and must be run by a qualified electrician.

Unit heaters vary in capacity from small (3 kW) to large (50 kW). The larger heaters are supplied with three-phase power to balance the system power consumption.

Unit heaters can be mounted vertically or horizontally. Vertically mounted heaters are used in low-ceiling applications and circulate air in a pattern around the room. These units can move heat from the ceiling level down to the floor. The horizontally mounted unit, such as the one in Figure 3-28 often allows you to operate the fan independently of the heater to remove heat from the ceiling, as shown in Figure 3-29. Heat that gathers at the ceiling increases the ceiling temperature and therefore increases the heat transfer to the outside.

Figure 3-27. Compact unit heater. *Courtesy International Telephone and Telegraph Corporation—Reznor Division*

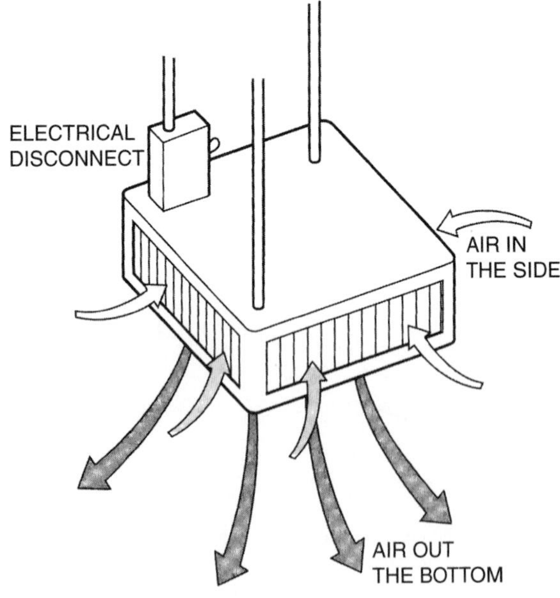

Figure 3-28. This unit heater is designed to mount horizontally, usually from the ceiling.

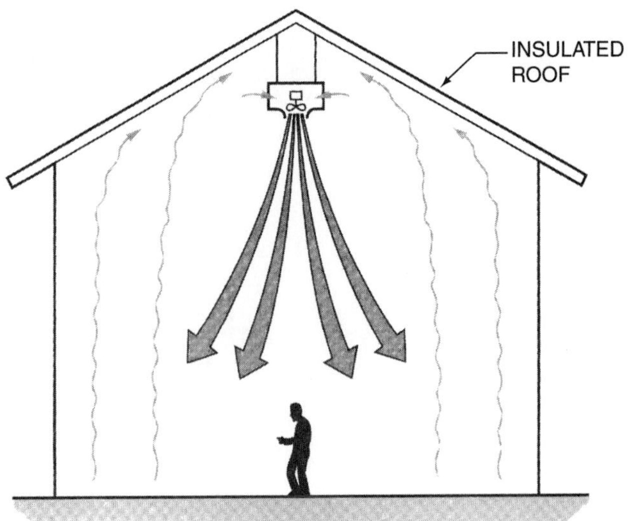

Figure 3-29. When the heater is mounted horizontally and the element is not operating, the fan can blow heat that collects at the ceiling downward.

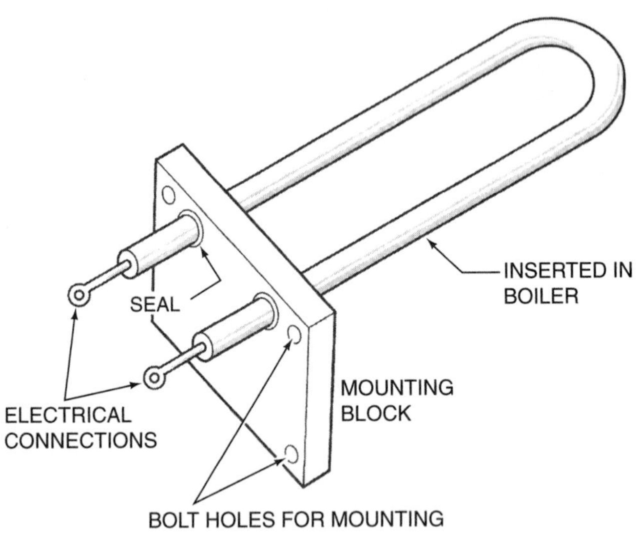

Figure 3-30. Electric heating element.

3.6 Electric Boilers

Electric boilers use the heat energy in electricity to heat water or to make steam. This is accomplished by placing the electric heating element, Figure 3-30, in an enclosure that is electrically insulated from the outside. This enclosure can be inserted into water and used to heat the water. The electrical element is much like the device used to heat water in a coffee cup, Figure 3-31.

The electric boiler is small in comparison with the fossil fuel boiler because the heat exchanger is eliminated and there is no flue gas. These boilers are efficient because most of the heat from the element converts to usable heat for the water. The only heat that may be lost is when the boiler is in an unconditioned space; then, some heat is lost through the jacket of the boiler to the unconditioned space.

Electric boilers are small for spot-heat applications or larger for commercial applications. The only difference between a small and a large boiler is the number of elements. A large boiler has many elements, as shown in Figure 3-32.

Figure 3-31. The electric element is much like a coffee warmer.

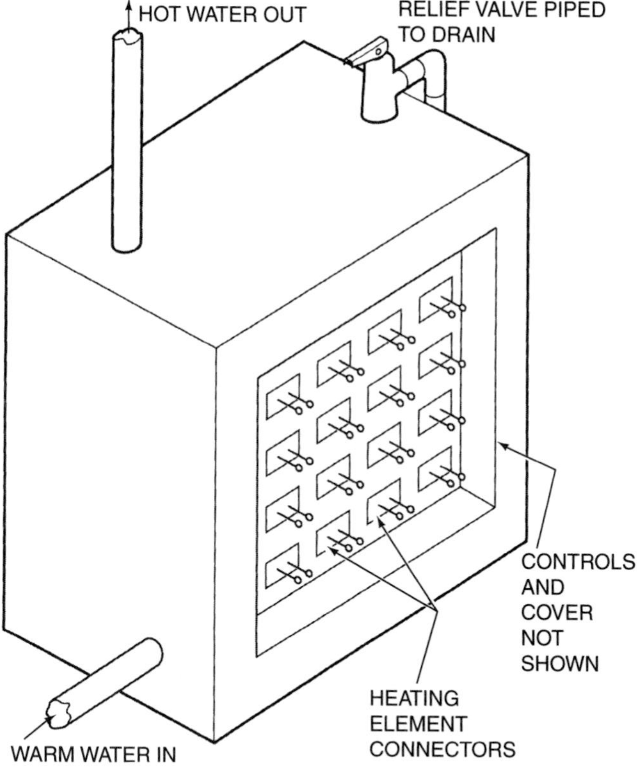

Figure 3-32. A large boiler has many heating elements.

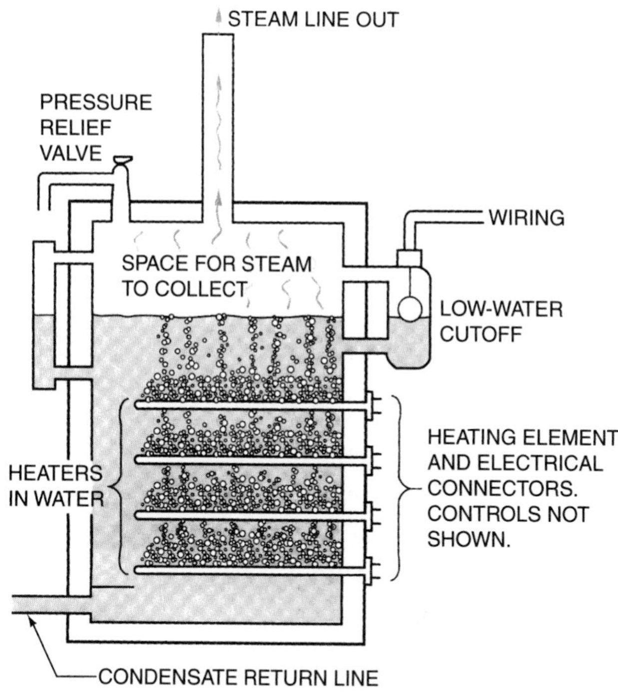

Figure 3-33. The steam boiler is not entirely full of water. Space at the top collects steam.

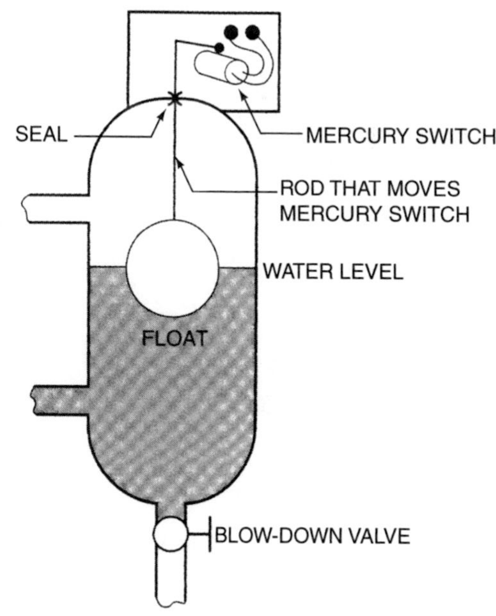

Figure 3-34. The low-water cutoff on a steam boiler.

When the electric boiler is used to make hot water, its thermostat stops and starts the power to the element. A high limit control may be built into the thermostat. The thermostat maintains the temperature, and the limit control stops the heat if the thermostat fails to do so.

The electric steam boiler has a water level, just as the fossil fuel steam boiler does. Figure 3-33 is a schematic of a steam boiler. The boiler must have a low-water cutoff to ensure that the water is at the correct level. Figure 3-34 shows a cross-sectional view of a low-water cutoff. If the water level falls below the electric element, the element will quickly burn out.

3.7 Duct Heaters

Electric heaters are often installed in air ducts to heat the air as it passes over the heaters. Two kinds of elements are used in duct heaters; both use ni–chrome wire. One heater has ni–chrome wire wound in spring-like coils that are suspended in the airstream, as shown in Figure 3-35. The heating element is electrically insulated from the frame or support with ceramic insulators, Figure 3-36. To provide

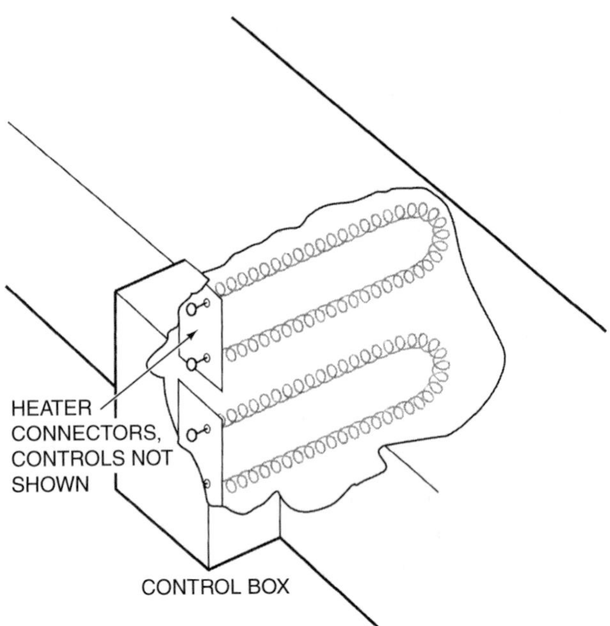

Figure 3-35. Electric duct heater.

Figure 3-36. Ceramic insulators insulate the heating element from the frame. *Photo by Bill Johnson*

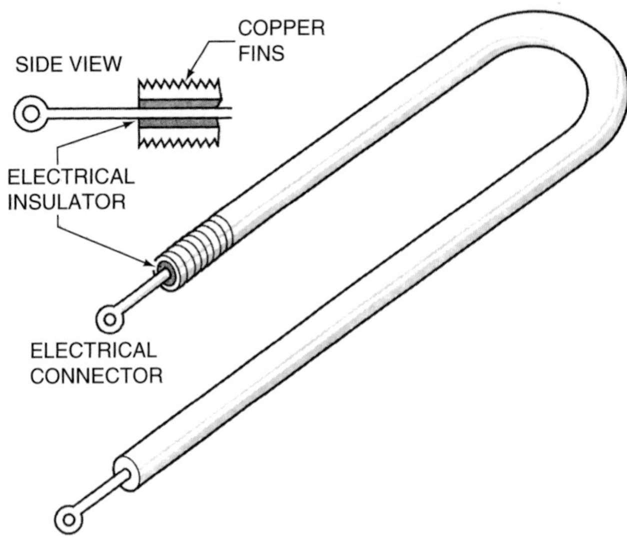

Figure 3-37. Enclosed electric heating element, or finned-tube element, for a duct heater.

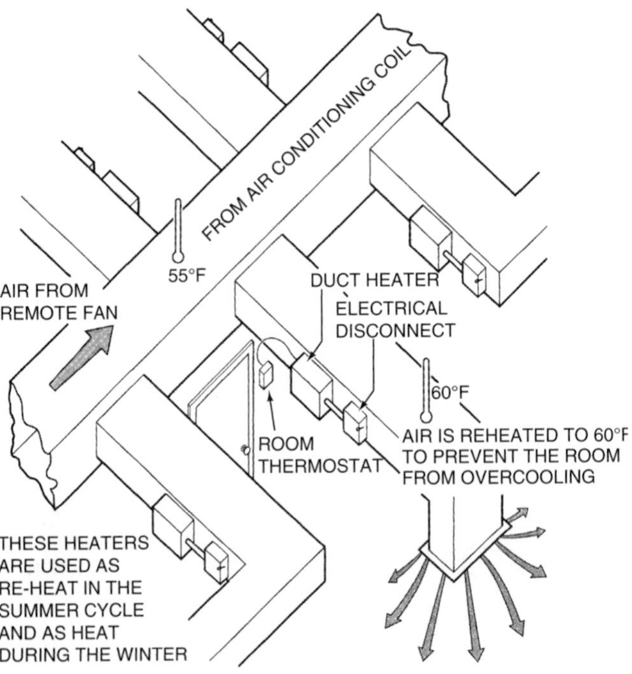

Figure 3-38. Duct heaters used as reheat and terminal heating elements.

the best protection, these insulators must be in good condition and not cracked.

Another type of element has ni–chrome wire enclosed in a tube, called *a finned-tube element*, as shown in Figure 3-37. This element eliminates the hazard of electrical shock in case the technician accidentally touches one of the elements during service procedures. It also protects the heating element against moisture in the air, such as from a humidifier that is upstream of the heater.

Duct heaters can be used as reheat elements or terminal heating elements in a large duct system, Figure 3-38. When used as a reheat element, this heater may be located in a cool air duct to prevent an office from becoming too cool when a large duct system has only one thermostat. The reheat element is installed in the duct, close to the air register. A room thermostat controls the heater at the end of the duct for individual room temperature control. An airflow safety switch must be installed to ensure that the fan is running before the heater is operated. Reheat applications are not efficient and are not usually used for newly designed systems. The heater serves as the heating system in the winter and is called terminal heat.

Each duct heater has its own control package built into the end of the heater, Figure 3-39. All that needs to be

Figure 3-39. The control compartment for a duct heater. *Courtesy Indeeco*

furnished are a power supply for the heater and a control voltage for the thermostat and heater start mechanism.

There must be a 3½-foot clearance at the end of the heater so that it can be serviced. This requirement makes the heater difficult to position in the side of a duct, where ceiling space is crowded, Figure 3-40. Instead, a heater that mounts on the bottom of the duct and has its control panel on the bottom is a better choice for this application. Figure 3-41 shows a bottom-mounted duct heater.

The duct heater must have adequate safety controls to prevent overheating the surroundings or burning out the heating element. It is typical to have two limit (overtemperature) controls for electric heating elements. One is an automatic reset control, and the other is a manual reset control. The automatic reset control, Figure 3-42, is set at a lower value (175°F) than the manual reset control is (250°F). Often, the duct heater has a fuse link or a high-temperature interruptor that opens the line circuit to the heater in case of overtemperature, instead of a manual reset control. When a fuse link is burned in two, the technician may be able to see that the fuse is open with the naked eye. The heater must be removed to see it, Figure 3-43. The other type of temperature interruptor is a solid-state device, and the technician must use an ohmmeter to determine that it is open.

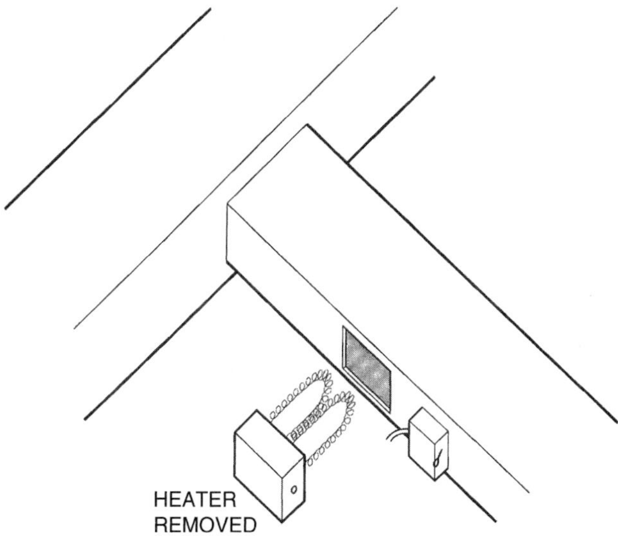

Figure 3-40. The duct heater that mounts in the side of the duct must have the correct clearance so that it can be serviced.

Figure 3-42. Automatic reset control. *Photo by Bill Johnson*

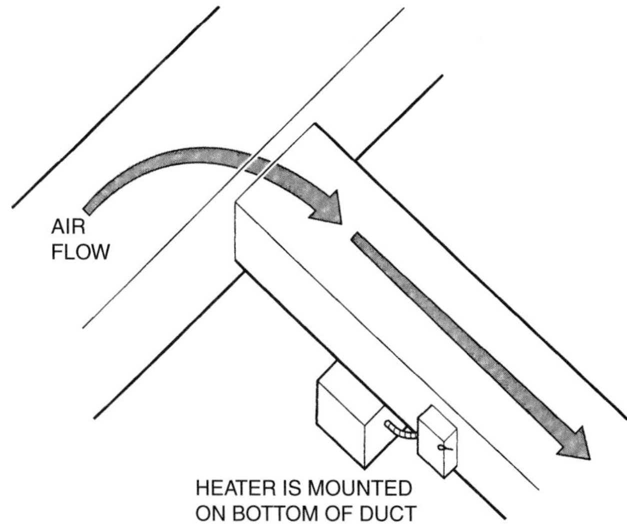

Figure 3-41. Bottom-mounted duct heater.

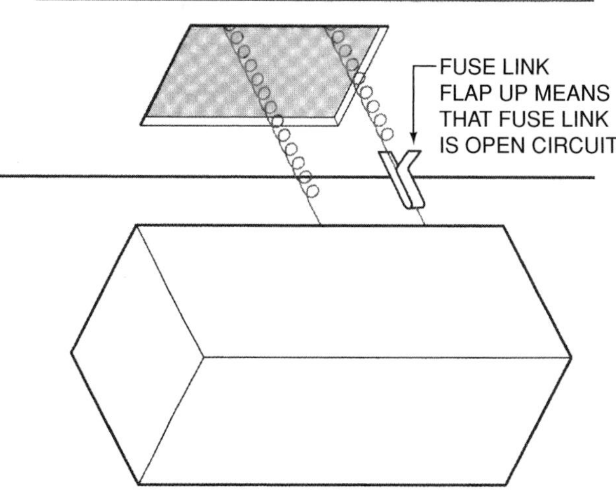

Figure 3-43. Removing a duct heater to examine the fuse link.

3.8 Electric Furnaces

Electric furnaces are similar to gas or oil furnaces, except they have electric heating elements instead of fossil fuel heat exchangers. They also are much smaller because they do not have the heat exchangers and flue systems, Figure 3-44. The electric heating elements are within the furnace housing and are designed for the correct amount of air to flow across them.

Electric forced-air furnaces are usually fastened to duct systems that distribute air throughout the conditioned space. The use of a duct system with the electric furnace allows other features to be added to the system that may not be added to space heating systems or room heaters. Air-conditioning can be added by placing a cooling coil in the airstream, as shown in Figure 3-45, provided that the fan for the electric-heat system has enough capacity. Electronic air filtration and humidification can also be added to the airstream with the proper controls, as shown in Figure 3-46. These features are common in all forced-air systems.

The electric heating elements in a typical residential or light commercial electric furnace consume 4 to 5 kilowatts each. Large heating elements are used only

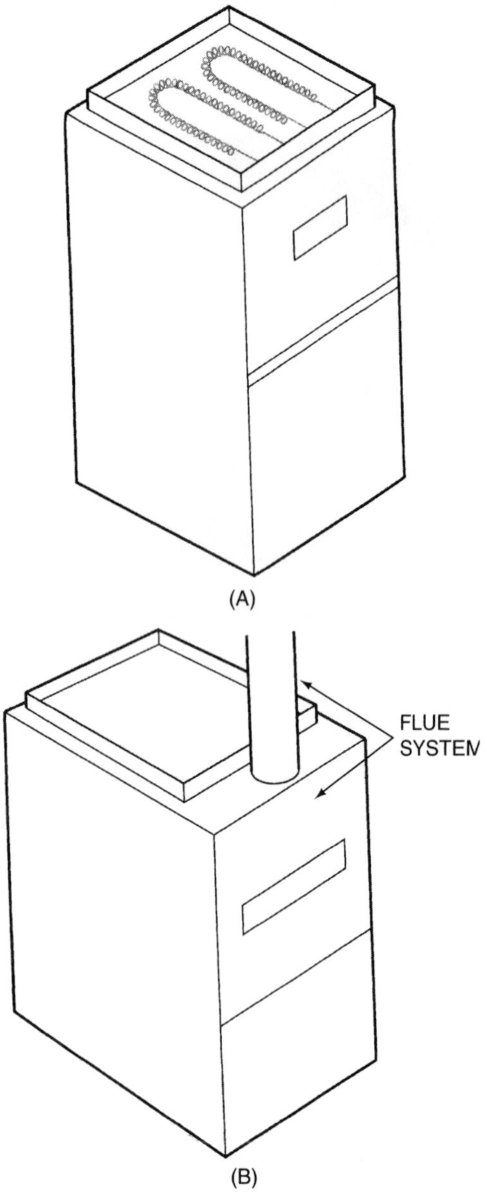

Figure 3-44. The electric furnace does not have the flue system that a fossil fuel furnace has. (A) Electric furnace. (B) Fossil fuel furnace.

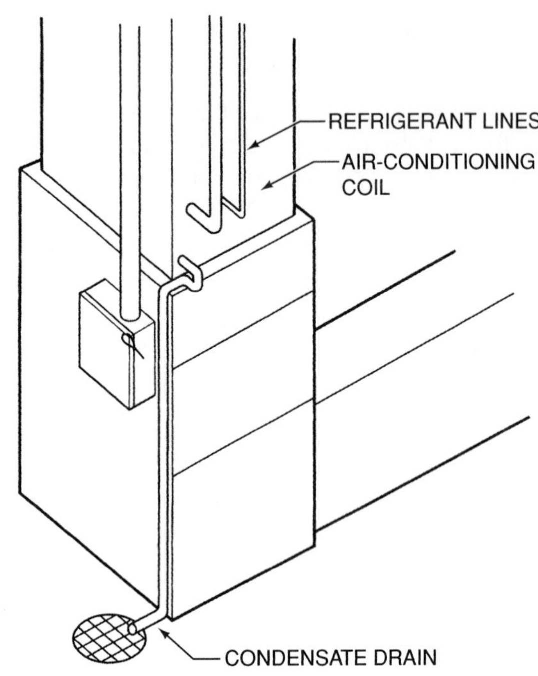

Figure 3-45. Air-conditioning can be added to an electric furnace.

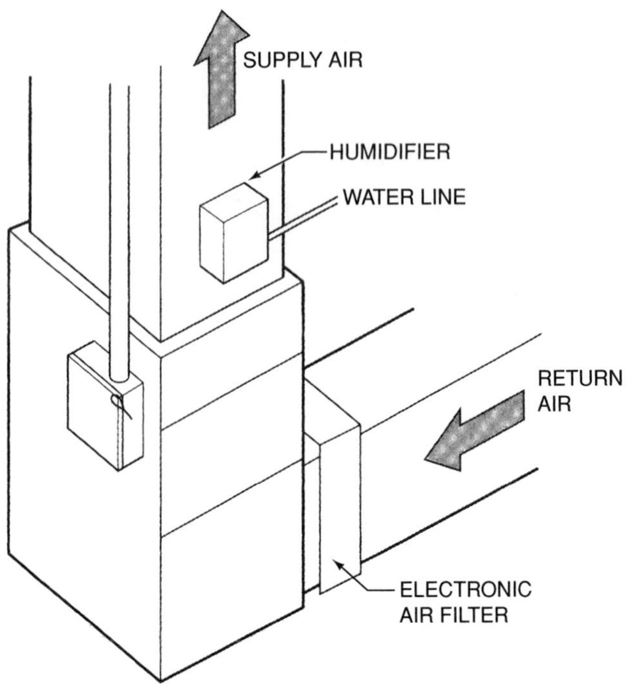

Figure 3-46. Electronic air filtration and humidification are added to an electric furnace.

with much larger furnaces. With the system divided into several smaller elements, multistaging the elements becomes easier. For example, a typical electric furnace may have a capacity of 30 kilowatts at 240 volts. This is a heating capacity of 102,390 Btu (30 kW × 3,413 Btu/kw = 102,390 Btu). When the heaters are energized for 1 hour, 30 kilowatt-hours of heat, or 102,390 Btu per hour of heat, are furnished to the conditioned space. All this heat is released into the airstream in a small area of the furnace.

By using time delay in the control circuit, the manufacturer can stage the furnace for the elements to come on one at a time. Staging is advantageous because if all the elements were energized at once, the lights in the house would dim for a few seconds as a result of the large current draw. The running current draw for the 30-kilowatt (30,000-W) heater rated at 240 volts is 125 amperes ($I = P/E = 30{,}000/240 = 125$ A). This current can be broken into 5-kilowatt increments with time-delay controls so that only 20.8 amperes are energized at a time until the entire heater is operating ($I = P/E = 5{,}000/240 = 20.8$ A). The time-delay control for this type of furnace is discussed in more detail in Unit 4.

An electric furnace housing is made of sheet metal that is insulated and painted. The elements and their controls are accessed through a removable panel, as shown in Figure 3-47. A flange at the furnace outlet provides a method for attaching the duct work to the furnace in a manner that minimizes leaks. Figure 3-48 shows this flange. The outlet to the furnace is in a positive pressure and must be attached tightly to the duct so that heated air is not lost through a poor connection. Heated air that is lost is expensive.

Electric furnaces are designed for upflow, downflow, and horizontal airflow applications, as shown in Figure 3-49. Some manufacturers design furnaces that work in either

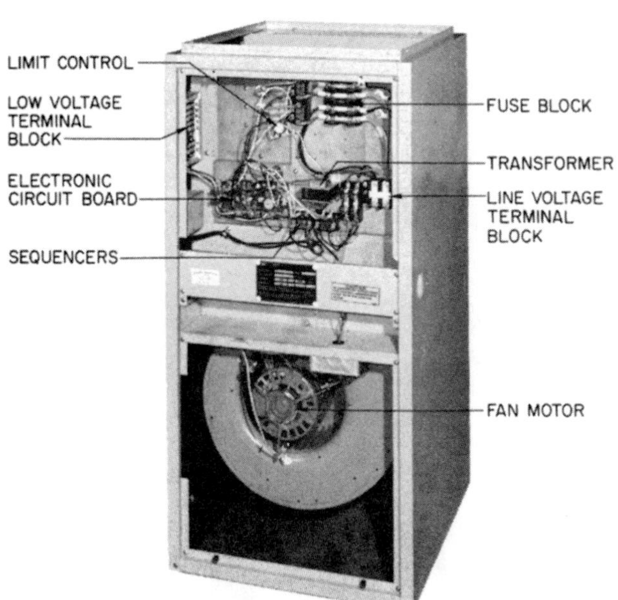

Figure 3-47. The elements and controls are serviced through a removable control panel. *Courtesy The Williamson Company*

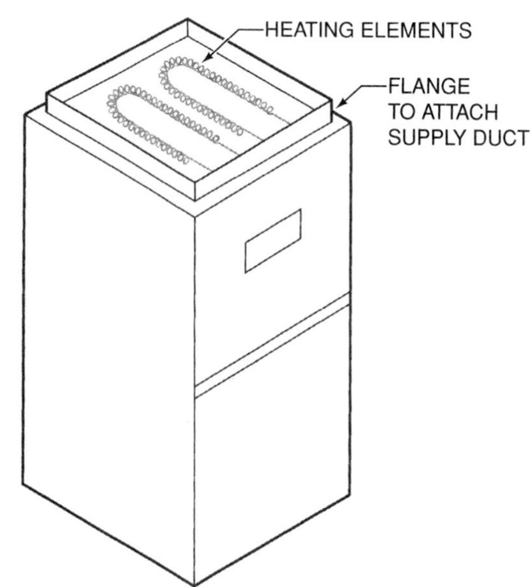

Figure 3-48. Furnace outlet flange.

the upflow or the downflow position (the furnace is turned upside down for the downflow application).

The furnace is insulated on the inside to minimize heat flow to the surroundings because the manufacturer does not know where the furnace will be installed. Some furnaces are installed in attics or a crawl spaces under houses. Heat loss from the attic installation will rise and be lost, as shown in Figure 3-50. Heat loss from a crawl space installation may rise up under the floor, and some of it may be used, Figure 3-51. Some of the heat is lost due to cross ventilation to the outside air.

The return air duct connection may be designed for air to enter either side or the bottom of the furnace, as shown in Figure 3-52. The return air filter may be located in the furnace, or the contractor may need to furnish a filter rack external to the furnace, as shown in Figure 3-53. A tight return air connection is important when the unit is in an unconditioned space. When air must be heated from the crawl space temperature and mixed with room air temperature, extra heat must be added. This extra heat costs money, Figure 3-54.

Note: More attention is usually paid to gaining or losing air from an electric-heat system because of the higher cost of electrical energy.

The controls for electric furnaces are under a control panel in the furnace, usually at the junction of the electric heating elements and the power wiring. This portion of the furnace contains the stop-and-start mechanism for the electric heat and the fan motor, as well as the fuses or breakers used to protect the system. Also, this control panel usually contains the control transformer, Figure 3-55. Because the control panel is next to the electric heating elements, a certain amount of heat is conducted from the heating compartment into the control compartment. If for some reason the fan does not run for a period of time and remove the correct amount

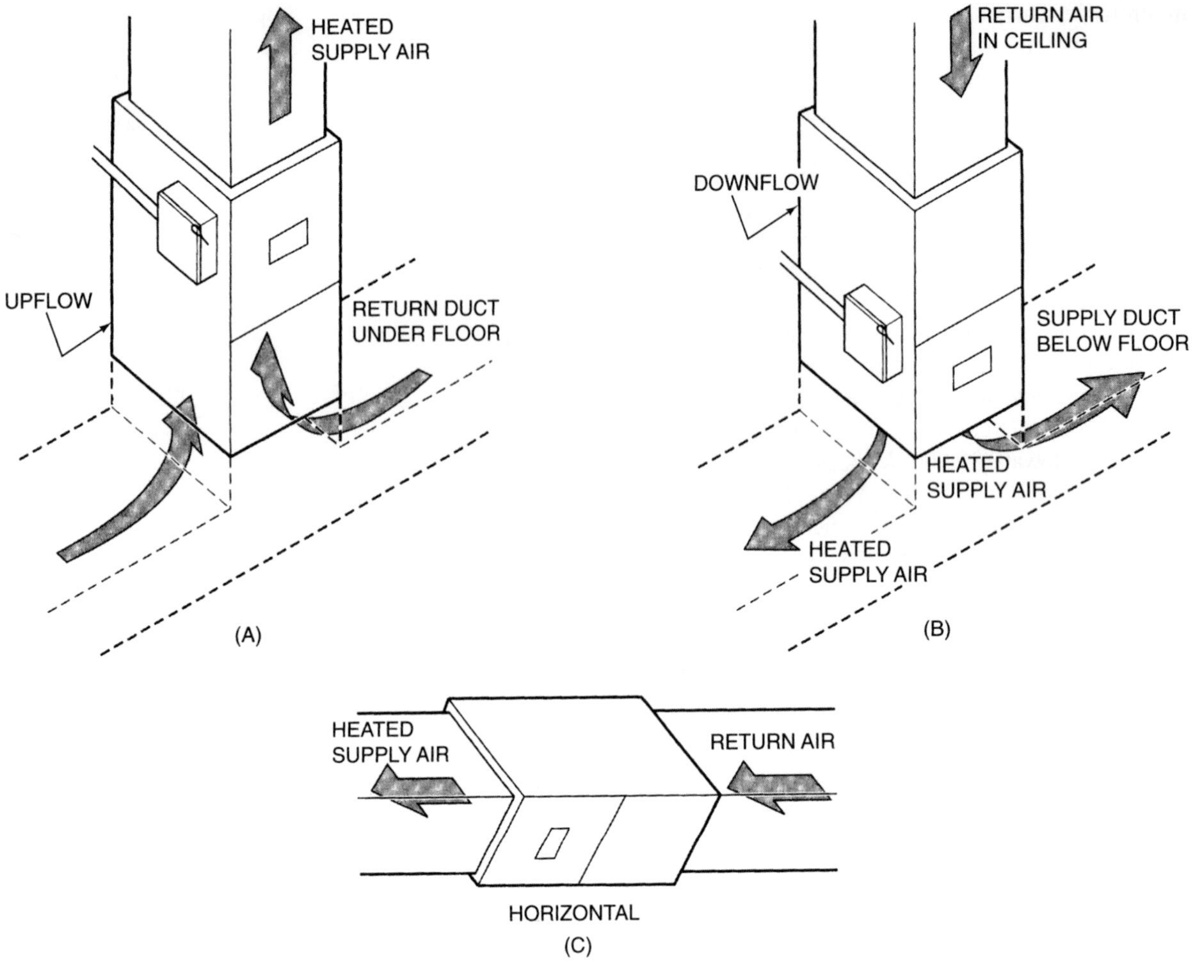

Figure 3-49. (A) Up-flow furnace. (B) Down-flow furnace. (C) Horizontal furnace.

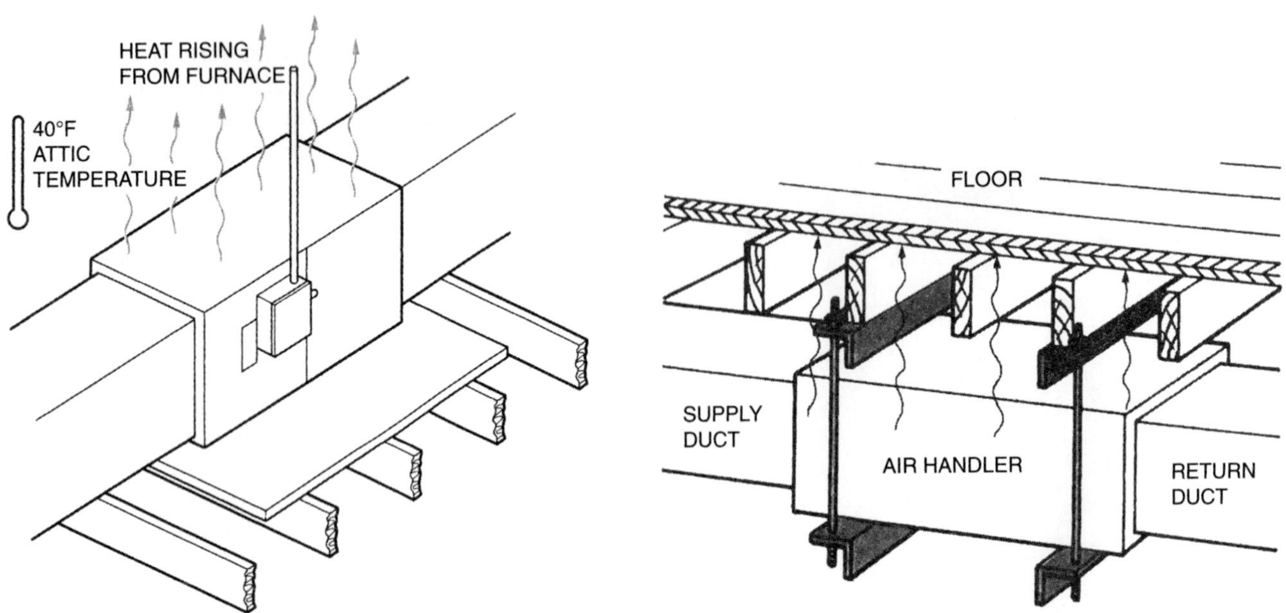

Figure 3-50. Heat loss in the attic.

Figure 3-51. Heat loss in a crawl space.

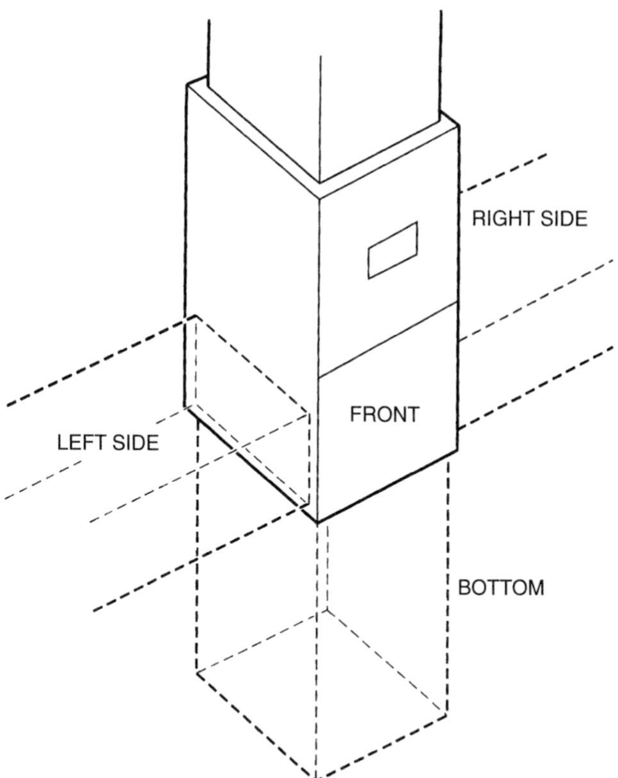

Figure 3-52. The return air duct connection can be on either side or the bottom of the furnace.

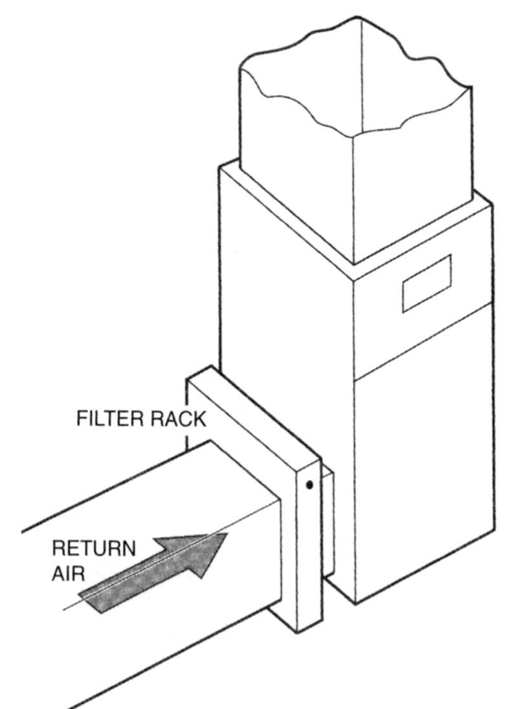

Figure 3-53. Filter rack outside the furnace.

of heat from the elements, a large amount of heat builds up in the control compartment. Much electrical energy is also available in this compartment.

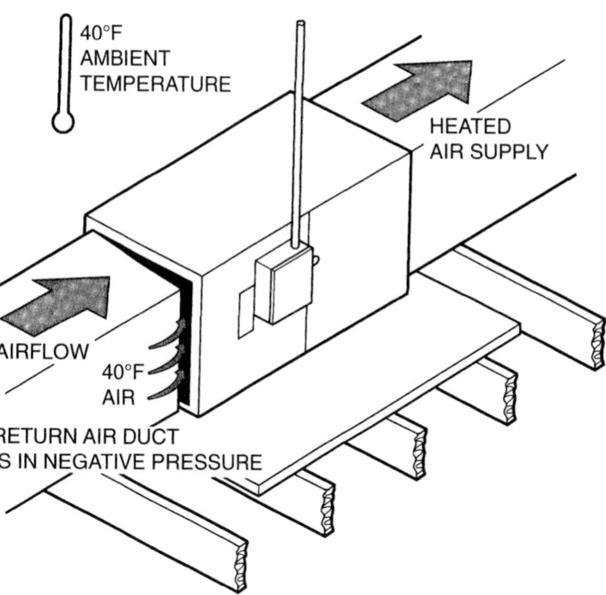

Figure 3-54. Loose return air connection is expensive.

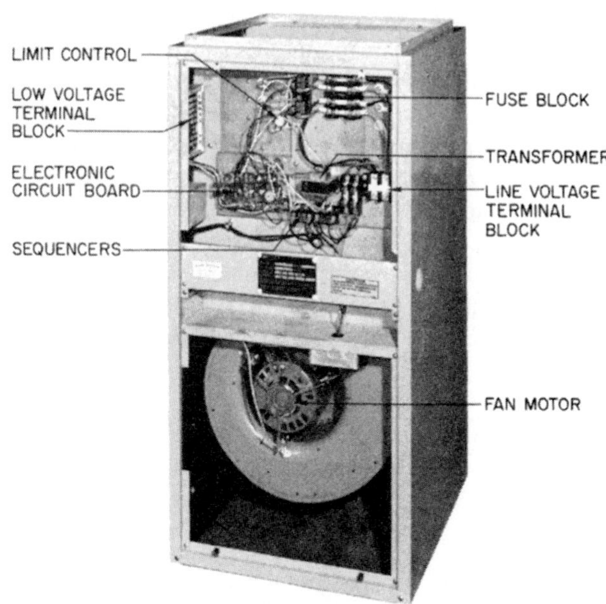

Figure 3-55. Control transformer location in panel. *Courtesy The Williamson Company*

Warning: Use caution when opening this compartment. Turn off the power before trying to service it.

The correct airflow across an electric furnace is important. If the airflow is insufficient, the elements will overheat and stop and start because of limit control operation. If the airflow is excessive, the air leaving the unit will be too cool, and customers will complain about drafts in the conditioned space. Electric-furnace manufacturers usually make one furnace for several ranges of heat. For example, one furnace shell may provide as little as 5 kilowatts and as much as 30 kilowatts of heat. The manufacturer does not know how a furnace will be applied, so the airflow is an unknown to the manufacturer. Electric-furnace manufacturers also do not print the minimum and maximum air temperature rises on the nameplate of the furnace as the fossil fuel furnace manufacturers do. The technician is responsible for calculating the airflow and arriving at the correct volume of air.

The correct air temperature leaving an electric furnace should be no more than approximately 130°F, or the heating elements will become hot enough to cause the limit control to cycle the furnace off, Figure 3-56. If temperatures higher than 130°F are used, and the furnace air filter becomes slightly clogged, the furnace will operate above its correct temperature range.

If the leaving air temperature is less than approximately 100°F, the system will have drafts, unless it is well designed. The reason for this is that the human body temperature is 98.6°F in the trunk of the body. The temperature of the extremities, such as the hands and the feet, is approximately 90°F. When air cooler than 100°F blows, it heats the room to 70°F or 75°F but feels cool to the body. When the air leaves the air distribution register at 100°F, it mixes with some of the room air before going farther into a room. This mixture of supply air and room air lowers the final air temperature to well below body temperature, as shown in Figure 3-57. The farther the mixture is from the register, the lower the temperature of the mixed air and the less velocity it will have. The system must be designed so that air does not blow on a person in the room if the air temperature is less than approximately 100°F.

When measuring the temperature difference across an electric furnace, the technician should take the entering air temperature at the return air inlet. The outlet air temperature should be taken as close to the furnace outlet as possible, but radiant heat from the heating element must not be allowed to affect the temperature reading, Figure 3-58. The difference between the two probe readings is the temperature difference.

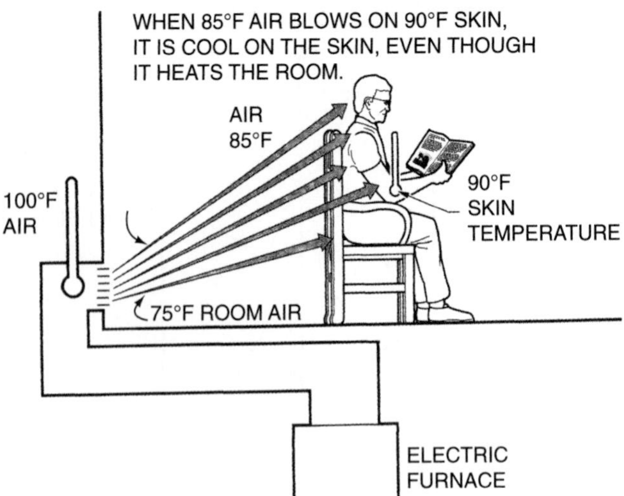

Figure 3-57. A person feels a draft when the supply air temperature is cool.

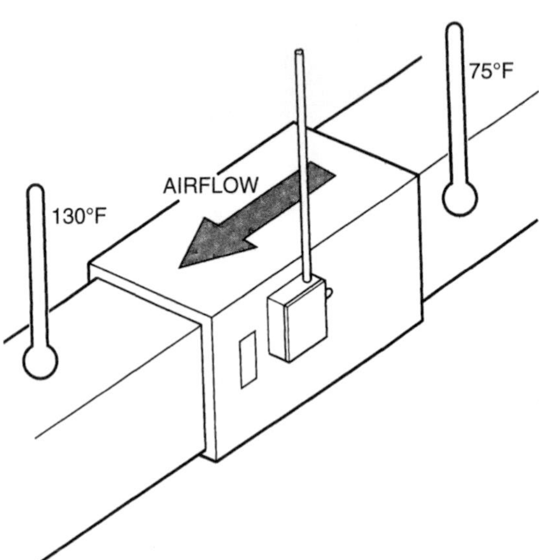

Figure 3-56. Maximum air temperature (130°F) is achieved with the minimum acceptable airflow.

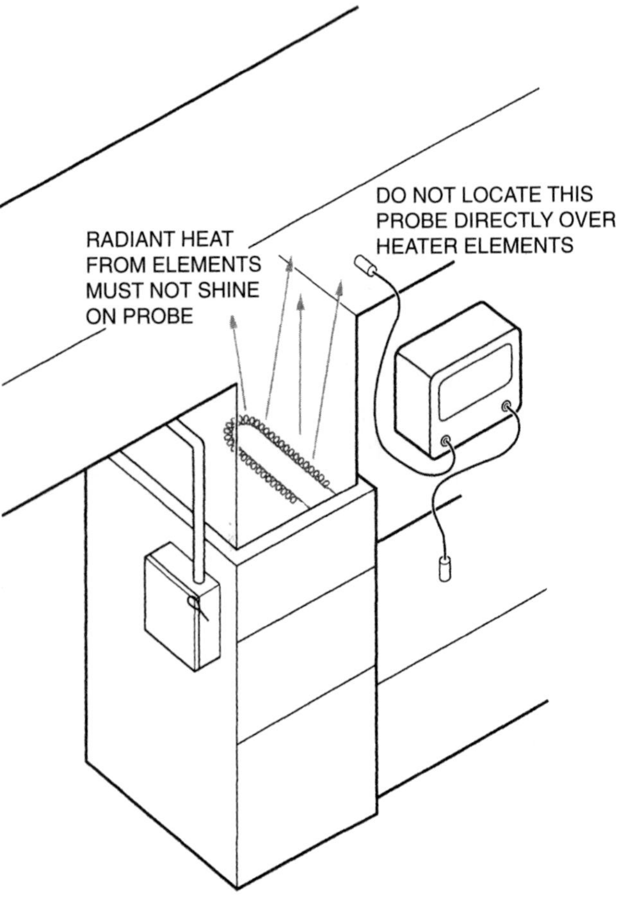

Figure 3-58. Do not allow radiant heat from the heater to affect the temperature-sensing element.

The sensible heat formula can be used to calculate the airflow for an electric furnace. This formula yields the amount of airflow, but not necessarily the correct airflow.

$$q = 1.08 \times \text{cfm} \times \text{TD}$$

where q = Amount of heat added to the airstream in Btu/h.

1.08 = A constant.

cfm = Cubic feet (of air) per minute.

TD = Temperature difference.

The formula can be rearranged to solve for cfm;

$$\text{cfm} = \frac{q}{1.08 \times \text{TD}}$$

Note that the formula calls for the amount of heat transferred to the airstream in Btu per hour and the input to the furnace in kilowatts. Consequently, the technician must measure not only the temperature difference, but also the voltage and amperage in order to determine the actual input to the furnace in kilowatts. This input must then be converted to Btu per hour. The furnace heat input can be measured by using a volt-meter to find the operating voltage and an ammeter to measure the ampere draw of the operating furnace.

Referring to Figure 3-59, let us look at an example. Suppose that a technician measures the return air temperature, which is 72°F, and the leaving air temperature, which is 128°F. The temperature difference is 56°F. The voltage to the unit is measured at the unit disconnect when the amperes are measured. They are recorded as 226 volts and 66 amperes, with all the elements and the fan operating. Note that the fan releases heat to the airstream just as the heaters do. Using the power formula from Ohm's law to calculate the total wattage of heat is accurate for electric heaters. Because the fan portion of the calculation is small, only approximately 5 of 66 amperes, little error is introduced by using the same formula for the fan as for the resistance heating elements.

Using the power formula, we find that the unit out-put is 14.916 kilowatts ($P = I \times E = 66 \times 226 = 14{,}916$ W/ $1{,}000 = 14.916$ kW).

The kilowatts are converted to Btu for use in the sensible heat formula by multiplying kilowatts by 3,413 Btu per kilowatt. The unit output is 50,908.3 Btu (14.916 kW × 3,413 Btu/kW = 50,908.3 Btu).

We can now use the sensible heat formula to find the airflow by solving for cfm:

$$\text{cfm} = \frac{q}{1.08 \times \text{TD}}$$

$$= \frac{50{,}908.3}{1.08 \times 56} = 841.74 \text{ cfm}$$

The technician or the estimator often needs this calculation to determine the airflow. By knowing the airflow, the technician or estimator can then determine whether adding air-conditioning or more air ducts to the system is feasible, particularly when rooms are added to a structure.

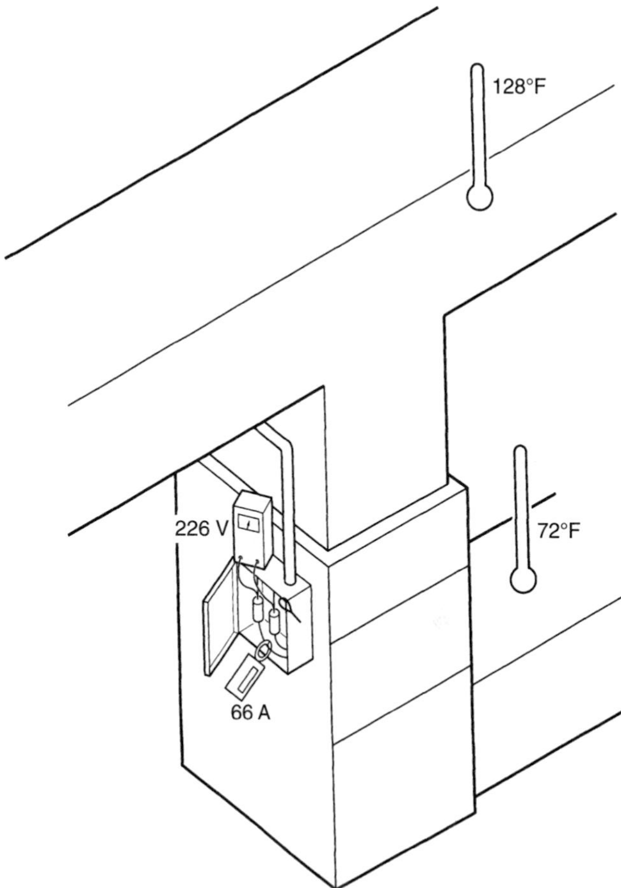

Figure 3-59. Measuring the heat input for an electric furnace.

Summary

- Whole-house electric heat became popular in the early 1960s, when houses were first insulated better.
- Electric heat is 100% efficient in most applications because all the energy can be used. The electric furnace has no flue.
- The cost of electric heat as a fuel is usually more expensive than the cost of fossil fuels.
- The initial cost of the electric-heat appliance is less than that of the fossil fuel appliance.
- The goal of heating with electricity is to convert electrical energy to heat.
- The power formula expresses electric heat as follows: $P = E \times I$, where P = power in watts, I = current flow in amperes, and E = electromotive force in volts.
- Power is expressed in watts (W) and kilowatts (kW). One thousand watts equal one kilowatt.
- One watt of power equals 3.413 Btu; one kilowatt equals 3,413 Btu.

- The power company charges for kilowatt-hours; a 1,000-watt heater operated for 1 hour is a kilowatt-hour (kWh).
- Ohm's law can be used to estimate the output of a heater: $I = E/R$. I the current, can then be converted to watts by using the voltage in the power formula.
- Conductors are sized to concentrate the heat energy at the electric heater.
- Conductors are usually made of copper, which has little resistance.
- The heating element in electric heaters is made of ni–chrome wire.
- Heat from the heating element is transferred by conduction or radiation to the heated medium, air, or objects in the conditioned space.
- Portable space heaters are used for spot-heating.
- Portable heaters are plug-in devices that are purchased at retail stores.
- Some portable heaters have fans to circulate air, and some are radiant heaters.
- A portable heater may have a tilt switch to shut off the heater if it falls over.
- Portable heaters are often equipped with thermostats and with selector switches so that different levels of heat can be selected.
- Portable heaters should be operated at the lowest wattage setting that will heat the space. A hot power cord indicates overheating.
- Radiant electric heat heats solid objects and not the space among the objects.
- Radiant heat can be used for spot-heating.
- Some wall heaters are mounted in the wall or ceiling.
- Baseboard electric heaters heat the air in the heater. This air rises, then more air enters the heater and is heated. This process creates an air current that is not noticeable.
- The baseboard heater element is enclosed and operates at a lower temperature than the element in an electric space heater. Baseboard heaters are often below windows, where drapes may touch the heaters.
- Baseboard heaters are rated by watts per foot. The longer the heater, the more heat.
- Baseboard heaters may be used for spot-heat or whole-house heat. They can be furnished as plug-in appliances.
- Electric unit heaters are ceiling mounted and typically use line-voltage thermostats. When the heater is located at a remote distance, it may have a low-voltage thermostat.
- Unit heaters may vary in capacity from small (3 kW) to large (50 kW).
- Unit heaters are usually mounted vertically; however, some are designed to be mounted horizontally and used to pull the heat from the ceiling downward to reduce heating bills.
- Electric boilers are either hot water or steam.
- The only boiler heat loss is the heat that may escape through the skin to the boiler room.
- Boilers can be small or large. Large boilers have many elements staged to start one at a time to prevent a voltage drop.
- The water boiler has a thermostat to maintain the water temperature and may have a backup high-temperature shutoff.
- Duct heaters are resistance heaters with ni–chrome elements that are mounted in duct work. These heaters are the same as electric furnaces, except they have a different fan control method.
- Some duct heater elements are exposed to air, and some are enclosed in a tube with fins.
- Each duct heater has its own control package, usually including the fan control.
- The safety controls may include an automatic reset and a manual reset limit control. The automatic reset control opens at approximately 175°F, and the manual reset opens at approximately 250°F.
- Electric furnaces use elements and controls similar to those of duct heaters, but these elements and controls are in a furnace housing.
- Electric furnaces are typically fastened to air ducts.
- Heating elements in electric furnaces typically consume 5 kilowatts each. The elements are staged to start one at a time.
- Electric furnaces are designed for upflow, downflow, and horizontal applications.
- The furnace housing is insulated inside to minimize heat transfer to the surroundings.
- The supply duct is fastened to a flange on the furnace outlet. The return air duct is often fastened to the bottom or the sides, with a filter rack between. These duct connections should be as airtight as possible to prevent either losing heated air or heating air that may leak in from a crawl space.
- The controls for an electric furnace are typically located in the compartment, where the power wiring enters and is connected to the heating elements. This compartment contains the power transformer, the controls to start and stop the current flow to the heaters, and the safety controls. The power should be turned off while the technician services this compartment.
- The correct airflow is important across an electric furnace. Insufficient airflow causes the elements to over-heat, and excessive airflow cools the air leaving the furnace and causes drafts.
- The airflow can be determined by using formulas and air temperature measuring devices. Typically, the air temperature leaving an electric furnace should not be less than 100°F or more than 130°F.
- If the airflow is reduced to an outlet temperature of more than 130°F and the filters become restricted, overheating will occur.
- Measure the return air and the supply air at the furnace, being careful not to let the radiant heat affect the temperature lead.
- An ammeter and a voltmeter are used to measure the furnace input.

Review Questions

1. Why is electric heat 100% efficient?
2. How does the cost of electric energy compare with the cost of fossil fuels, more or less?
3. When an electric heating element has an operating voltage of 208 volts and the current draw is 18 amperes, how much heat will the unit produce?
4. One kilowatt is equal to how many watts?
5. One kilowatt is equal to how many Btu?
6. A 25-kilowatt furnace produces how many Btu?
7. The power company uses what unit of measure to calculate its rate?
8. An electric heating element has a resistance of 12 ohms and an operating voltage of 206 volts. What is the amperage of the heater?
9. How are electric baseboard heaters rated?
10. The element in electric heaters is made of what material?
11. Why are time-delay controls used to start electric heating elements in heaters and boilers?

4 Controls and Wiring for Electric Heat

Objectives

Upon completion of this unit, you should be able to

- **describe three types of electric-heat systems.**
- **describe the sequence of events in the control circuit for sequencer control and time-delay electric-heat systems, including electric boilers.**
- **explain the fan sequence of operation for forced-air heating systems.**
- **recognize airflow problems in a forced-air system.**
- **estimate the airflow for a forced-air system.**
- **perform simple service tasks on the basis of the wiring diagrams in the text.**

4.1 Automatic Control for Electric Heat

Room Heaters

The electric-heat room heater is typically controlled by a line-voltage thermostat that is built into the heater. A knob is used to control the temperature, Figure 4-1. Because the thermostat is in the heater, accurate room temperature control cannot be expected. This type of heater also often has an internal tilt switch that shuts off the heater if it falls over, Figure 4-2.

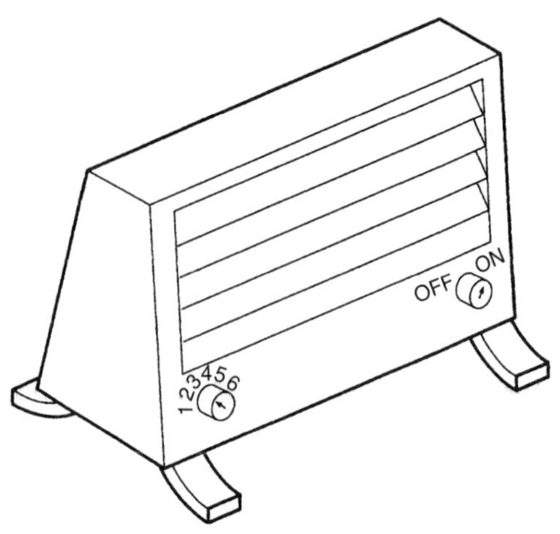

Figure 4-1. Controls for electric heater.

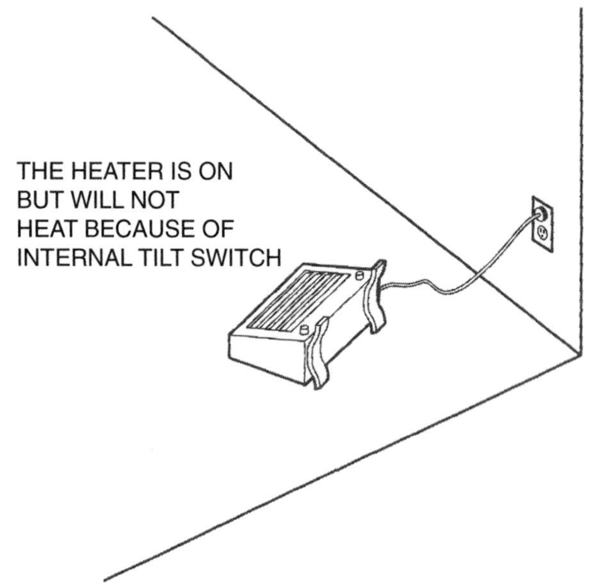

Figure 4-2. A tilt switch shuts off the room heater if it falls over.

Panel Heat

The panel electric-heat system has a line-voltage thermostat in each room. Figure 4-3 shows a typical line-voltage thermostat. This type of thermostat is reliable and causes little trouble. Nevertheless, the service technician may be called to determine why a room is not heating. In such a case, the circuit must be traced. Line voltage is usually furnished to a junction box at the panel heater with a pair of wires routed to the room thermostat from the junction box, as shown in Figure 4-4.

Baseboard Heat

Baseboard heaters are room heaters that plug into a wall receptacle, or they can be hard wired from the house wiring, as shown in Figure 4-5. When a base-board heater is a plug-in appliance, the line-voltage thermostat is built into the heater, Figure 4-6. When wired to the house wiring, the line-voltage thermostat is wired much as the panel heaters are, with a junction box at the heater and a pair of wires routed to the room thermostat, as illustrated in Figure 4-5.

Figure 4-3. Line-voltage thermostat for panel heat. *Courtesy W.W. Grainger, Inc.*

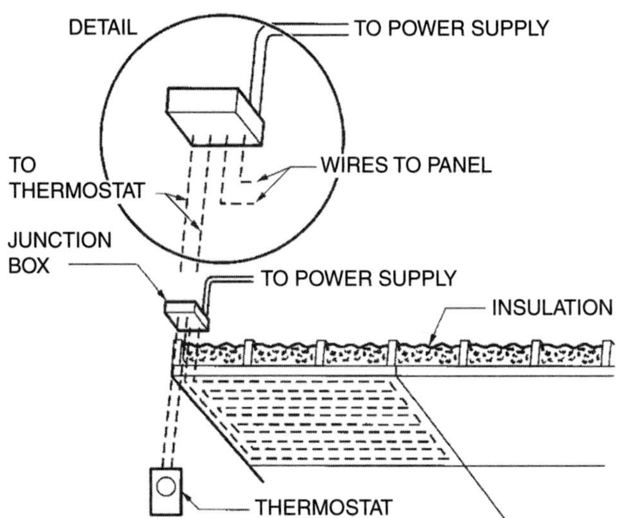

Figure 4-4. Line-voltage wiring for electric panel heat.

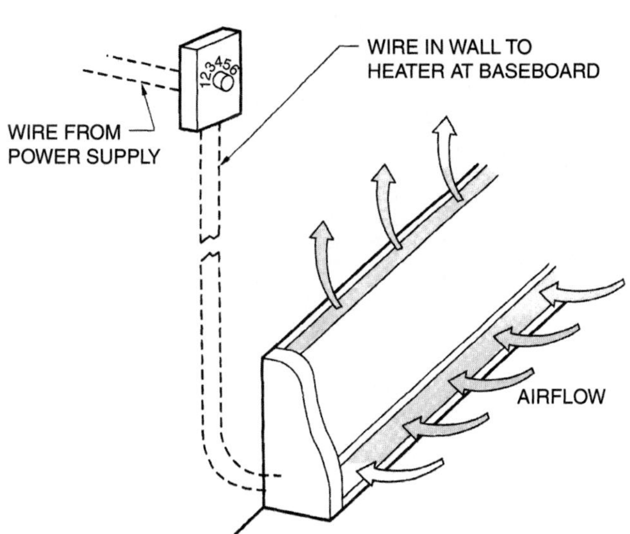

Figure 4-5. Hard-wired baseboard heat system.

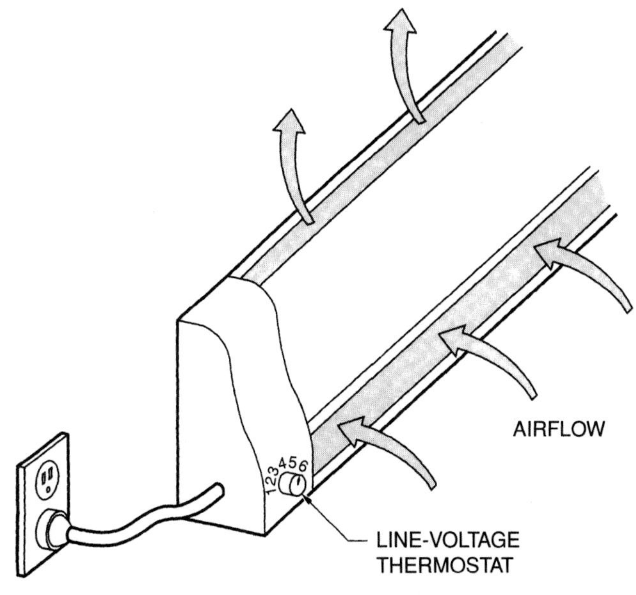

Figure 4-6. Plug-in baseboard heater.

Unit Heaters

The electric unit heater usually has a line-voltage thermostat mounted on the wall to control the room temperature. The main power is routed to the heater, and a pair of wires is routed to the room thermostat, as shown in Figure 4-7. If the thermostat is far from the heater or if a line-voltage thermostat is not wanted, a low-voltage thermostat can be used and the heater controlled with a relay.

The unit heater has a fan to circulate air, so the fan must also be started with the heater. Some installations have unit heaters mounted horizontally so that the air blows straight down. An energy-saving feature on the heater allows you to operate the fan independently of the heater to push the heat from the ceiling to the floor.

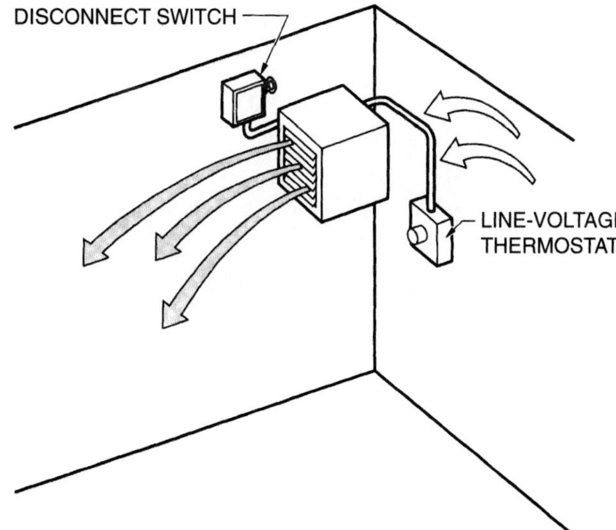

Figure 4-7. Unit heater line-voltage thermostat wiring.

Electric Boilers

The electric water boiler usually requires multiple controls for the heating elements so that they do not start all at once. If all the elements started at once, the voltage would drop and cause the lights to dim for a few moments. Furthermore, if a computer were on the circuit, problems would likely occur because of the voltage drop.

The boiler thermostat is mounted in the water portion of the boiler and senses the water temperature. It starts the sequence of controls. An additional room thermostat can be mounted in the conditioned space in some applications to maintain the room conditions. This room thermostat can be wired in one of three ways: to control the boiler and the pump, to control the pump only, or to control the boiler only. If it controls the boiler and the pump, when the room temperature is satisfied, the boiler and the pump stop. When the thermostat is wired to control the boiler only, the pump continues to run and dissipates the heat left in the boiler after the boiler is turned off. Different wiring sequences are used for different applications. The important thing to remember is that a single control stops and starts the boiler elements sequentially so that all of them do not start or stop at once.

Multiple control for boilers is usually accomplished with step controllers or time-delay relays. Figure 4-8 shows a typical step controller. It has a timing mechanism with multiple switches. A call for heat starts the timing. As time passes the switches are made, one at a time. The switches are low in amperage capacity and are used to stop and start the contactors for the electric heaters. This timer stages the electric elements.

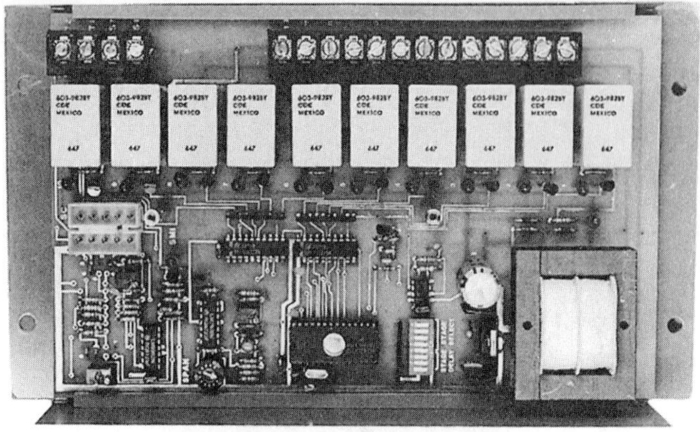

Figure 4-8. Multiple (or step) controller for a boiler. *Courtesy Indeeco*

If the power fails, the timing sequence starts over when power is resumed. This feature prevents the elements from starting all at once after a power outage.

Time-delay relays are used to start the contactors in a boiler one at a time. When the first contactor is energized, a set of auxiliary contacts energizes a time-delay relay, which in turn energizes the second contactor. As many contactors and time-delay relays as needed may be used for a large boiler, Figure 4-9. Each depends on the next and if one relay or contactor fails, the boiler will have no heat from the elements that are wired into the sequence after the failed component.

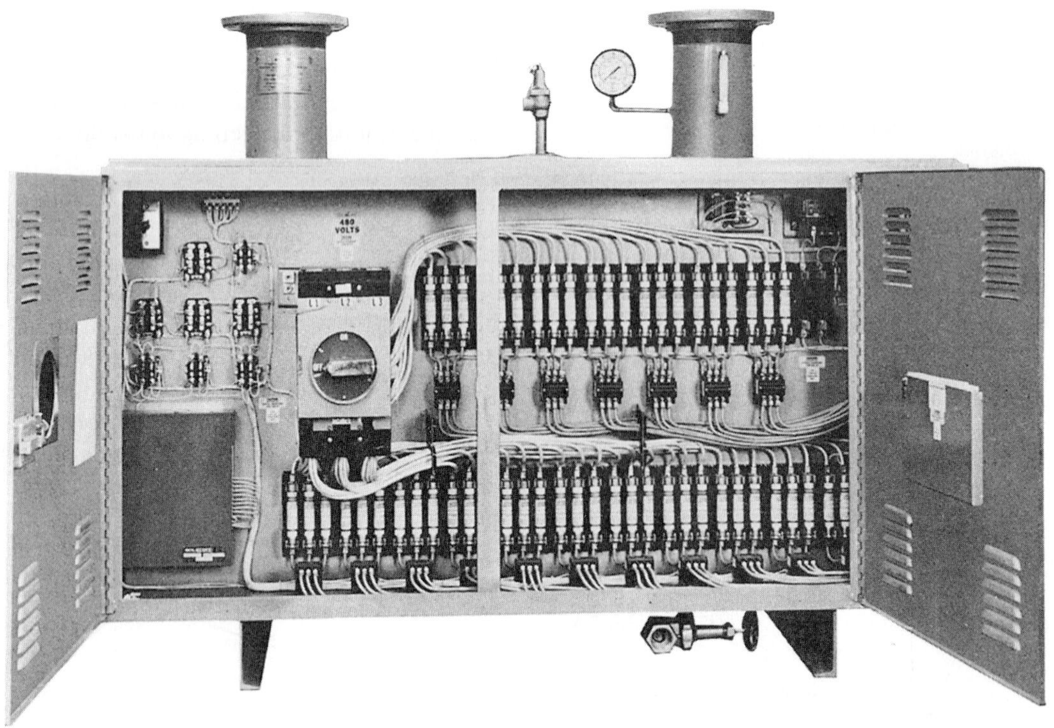

Figure 4-9. High-capacity boiler. *Courtesy Indeeco*

Duct Heaters and Electric Furnaces

Electric duct heaters and electric furnaces are wired similarly, except the electric furnace has a separate fan circuit. Like boilers, they often have many elements to produce a large amount of heat. The duct heaters are mounted in duct work, which is usually metal in large systems. These metal ducts amplify any sound made by the heating system controls and transmit it to the conditioned space, so the controls must be as quiet as possible. Because low noise level is important, these heaters usually use silent controls. A contactor in the duct heater may be too noisy, unless special care is taken to mount it so that it is silent.

Some duct heaters use mercury contactors that operate silently. Each contactor has a magnetic sleeve that moves downward when the coil is energized. When the sleeve moves down into the mercury, it displaces the mercury, which rises and touches the electrode. When the mercury touches the electrode, it completes the circuit with a silent connection. Figure 4-10 shows a typical mercury contactor.

Heating-element control can also be accomplished with sequencer controls. Each control has a bimetal strip that warps when the room thermostat calls for heat. The coil in the sequencer is a resistance heater instead of a magnetic coil. This type of control has a built-in time delay that is in effect while the control element is heating. A sequencer usually requires 15 to 30 seconds to become hot enough to make the contacts. You should also know and remember that the contacts of a sequencer take as long to open at the end of the cycle as they do to close at the beginning of the cycle. There is a time delay to start the heat and a time delay to stop the heat. This feature is important when you incorporate the fan into the circuit. There are two basic types of sequencers: the individual sequencer and the multiple-circuit (package) sequencer. Figure 4-11 shows each.

Figure 4-10. Mercury contactor. *Courtesy Mercury Displacement Industries*

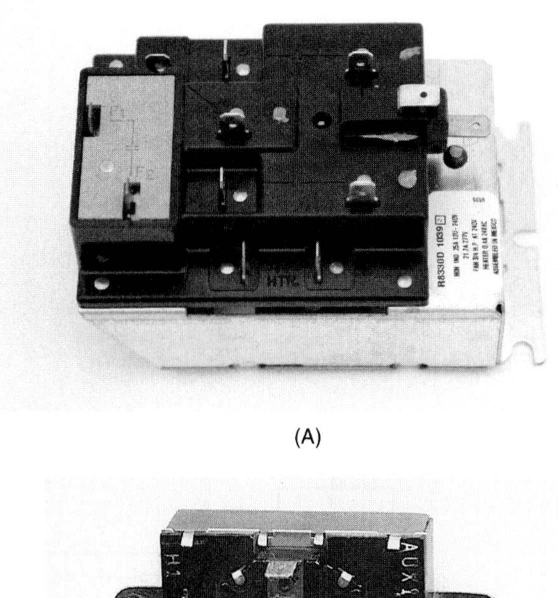

(A)

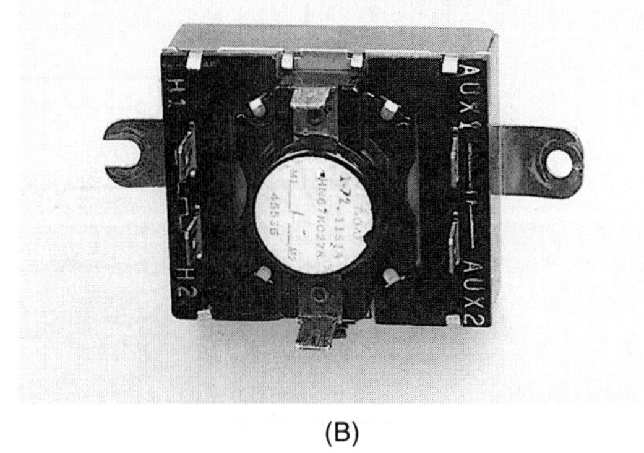

(B)

Figure 4-11. Two types of electrical sequencers. (A) Multiple sequencer. (B) Individual sequencer. *Photos by Bill Johnson*

The individual sequencer has a coil, a line-voltage contact to start the heating element, and a low-amperage auxiliary contact to start another sequencer, as shown in Figure 4-12. The line-voltage contact must have an amperage rating large enough to start the heater, typically 20 amperes at 240 volts. This capacity can operate a 4.8-kilowatt heater, which is typical. The individual sequencer may also have another contact to start the fan motor, as shown in Figure 4-13. When all this is put together to start four stages of electric heat, the panel has many controls and wires. Note that if one of the sequencers fails, none of the following sequencers will energize because they each depend on the previous sequencer.

The control circuit requires a large control transformer to handle all the sequencers. Two methods are used to wire individual sequencers together. One method wires all the sequencers through the room thermostat, as shown in Figure 4-14. You can see that the room thermostat must be able to carry all the amperage. For four sequencers that each have an amperage rating for the bimetal coil of 0.45 ampere, the transformer must be able to have an output of 1.8 amperes, and the thermostat must be rated to handle the load. A 43.2 volt–ampere (VA) transformer is required ($1.8\text{ A} \times 24\text{ V} = 43.2\text{ VA}$).

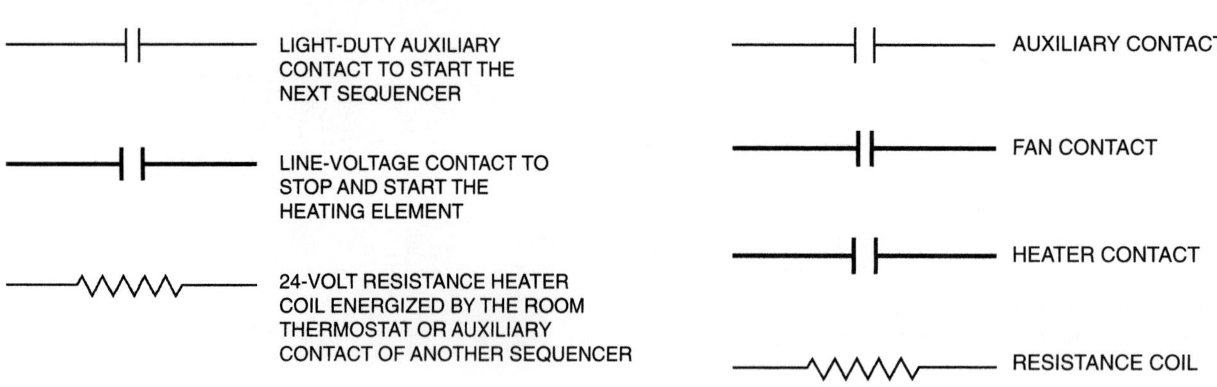

Figure 4-12. Individual sequencer components.

Figure 4-13. Individual sequencer components with a fan contact.

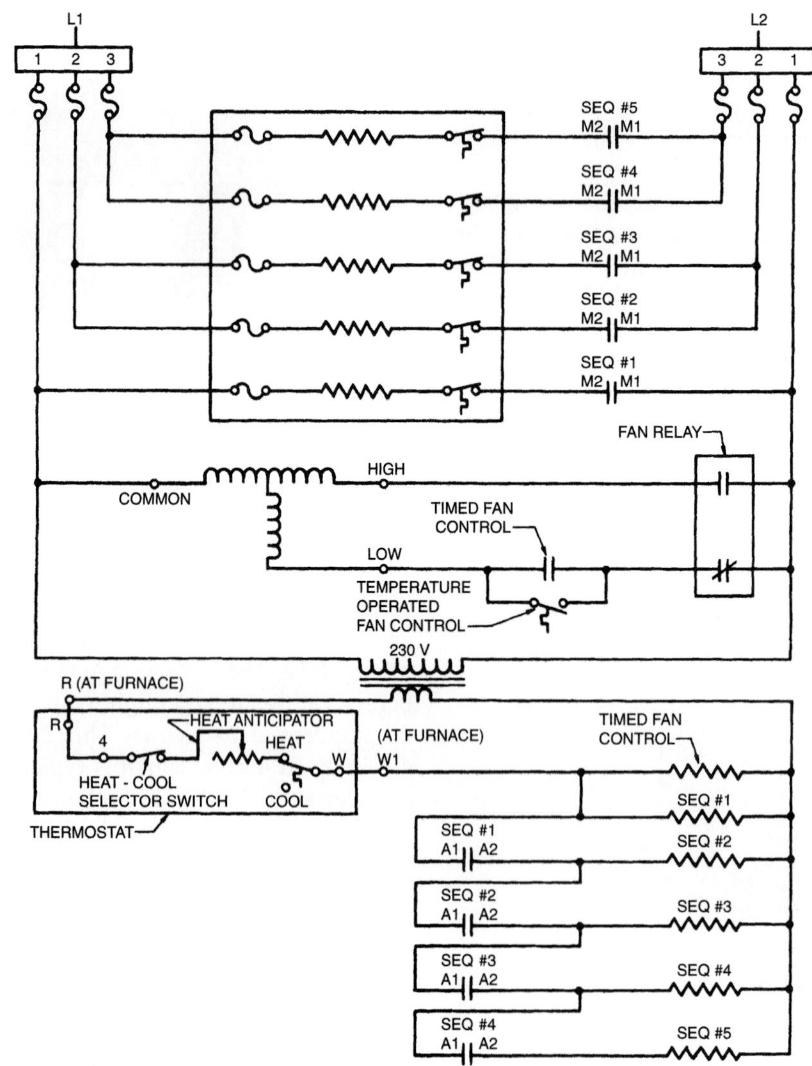

Figure 4-14. All sequencers are wired through the room thermostat.

The other method wires the first sequencer through the thermostat and the others before the room thermostat, as shown in Figure 4-15. This method relieves some of the load on the thermostat, but the transformer must have the same capacity.

In the multiple-circuit, or package, sequencer, several circuits are controlled by one bimetal strip. The package sequencer shown in Figure 4-16 has three heater circuits, a fan circuit, and an auxiliary circuit to start another sequencer. This unit takes up much less space

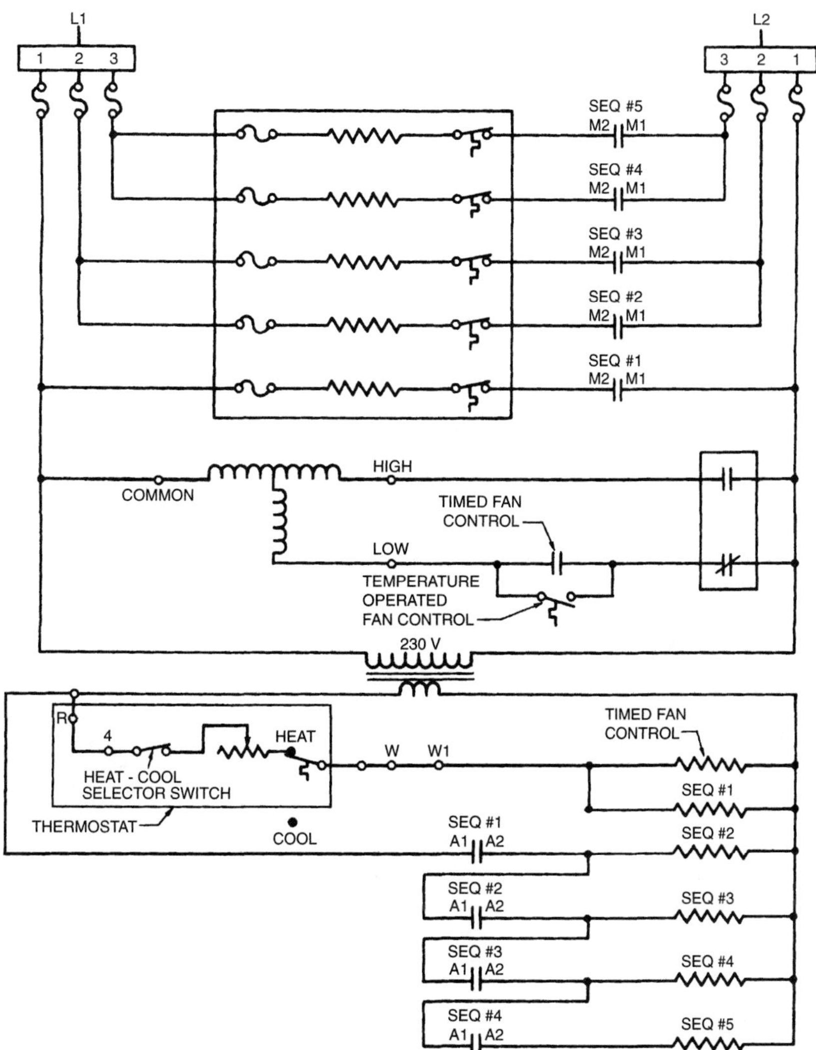

Figure 4-15. These individual sequencers are wired to reduce the load on the thermostat.

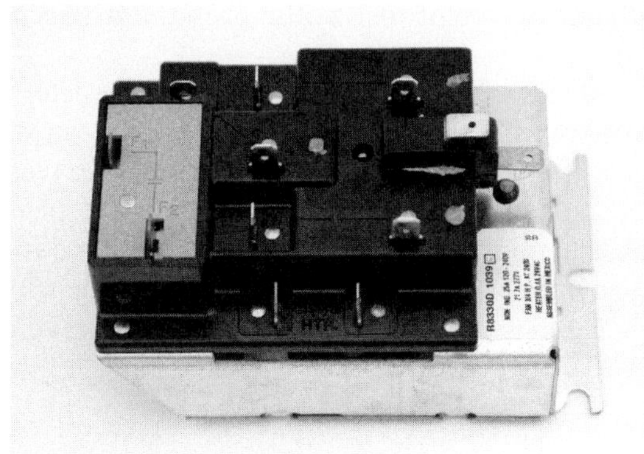

Figure 4-16. Multiple-circuit sequencer for three heaters. *Photo by Bill Johnson*

than several individual sequencers, and all the wires for the three elements, the fan, and the auxiliary circuit, in addition to the control circuit wiring, are fastened to it. Figure 4-17 shows the wiring diagram for a package sequencer.

FAN CONTROL. The fan control for duct heaters is different from that for electric furnaces. In both cases, however, the fan must be operating and air must be flowing over all electric heating elements while they are energized (or at least within a few seconds of energizing), or the heaters will become too hot. When the heating elements become too hot, they can sag or burn in two, as shown in Figure 4-18.

Fan Control for Duct Heaters A remote fan can serve several duct heaters in different ducts, as shown in Figure 4-19. Each heater must have a method of determining that air is flowing over it before it is energized, so the fan control is located with the heater. A pressure differential switch is often used as a means of determining airflow. This switch measures the pressure on each side of the duct heater, and when airflow is adequate, the switch closes and furnishes power to the control circuit. Figure 4-20 shows a typical pressure differential switch. Often a duct heater has an air diffuser just prior to the heater to distribute the air over the heater. This air diffuser causes the required pressure drop for an air switch to function, Figure 4-21.

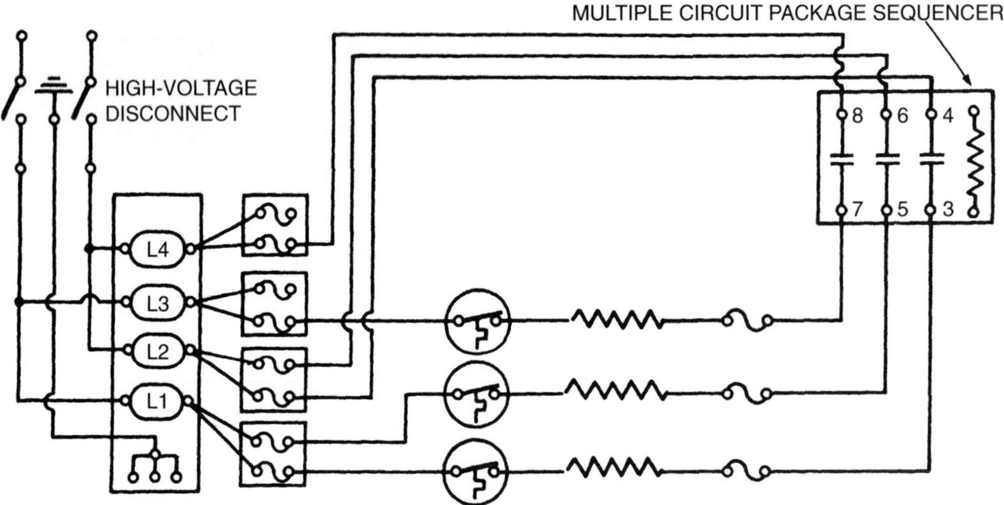

Figure 4-17. Wiring diagram for a package, or multiple-circuit, sequencer.

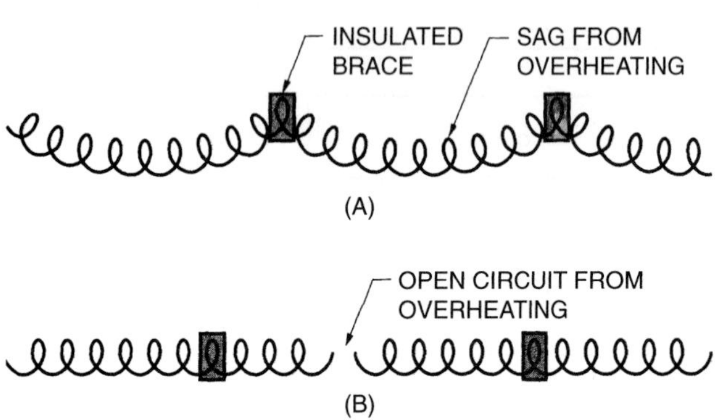

Figure 4-18. A heating element that becomes too hot may (A) sag or (B) burn in two.

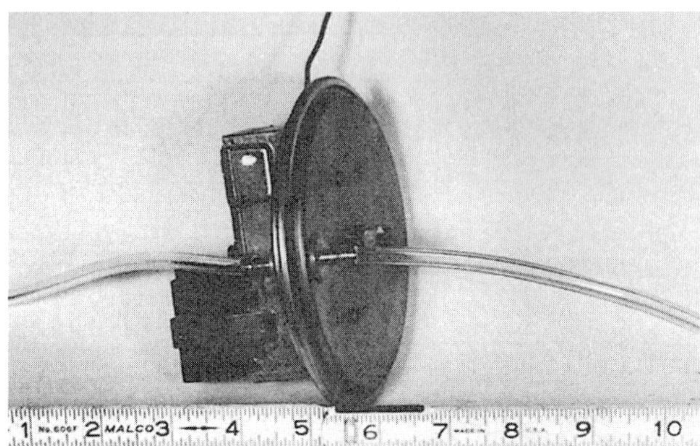

Figure 4-20. Pressure differential switch. *Photo by Bill Johnson*

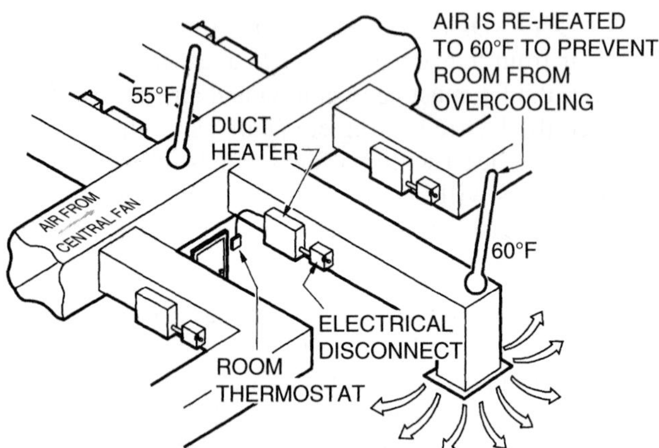

Figure 4-19. Duct heater installation with remote central fan.

The sail switch, which looks like a sail on a boat, is another airflow sensor. It catches the airflow and moves. This movement makes a set of contacts to verify airflow. Figure 4-22 shows a sail switch. This switch can be wired into the control circuit just as the pressure differential switch can.

Both switches also serve a second purpose: to detect reduced airflow while in operation, such as if the fan stops or slows or if the airflow is reduced for some other reason. Filters become clogged, a duct liner can fall into the duct work, obstructions are placed in front of the air outlets, or the occupant of the structure may shut off the supply registers. All these situations can cause airflow reduction, Figure 4-23.

Fan Control for Electric Furnaces The electric furnace has a built-in fan and usually does not use airflow-proving devices. The element is in a controlled location, and there is not as much danger of fire if the fan fails. The fan must be started at the beginning of the cycle and stopped at the end of the cycle. It can be started with the first heating element at the beginning of the cycle, but it should remain on at the end of the cycle until the last heating element is deenergized and should remain on to remove the heat from the furnace after the heaters are shut off.

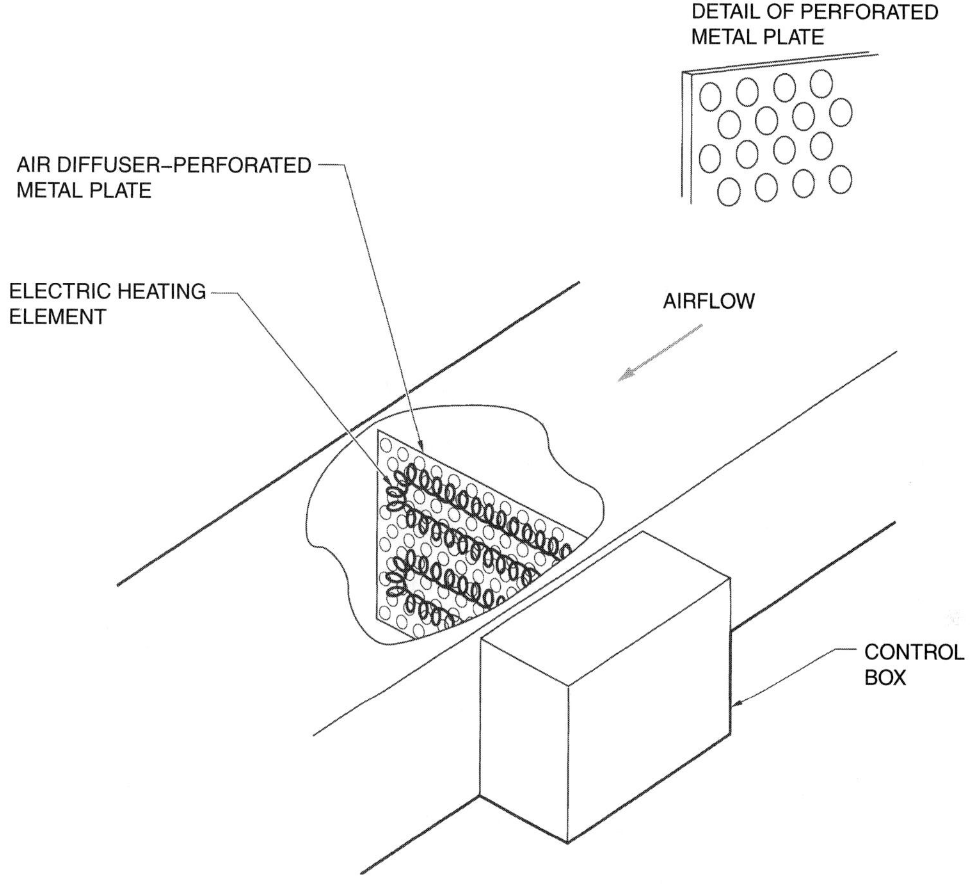

Figure 4-21. Air diffuser creates a differential pressure.

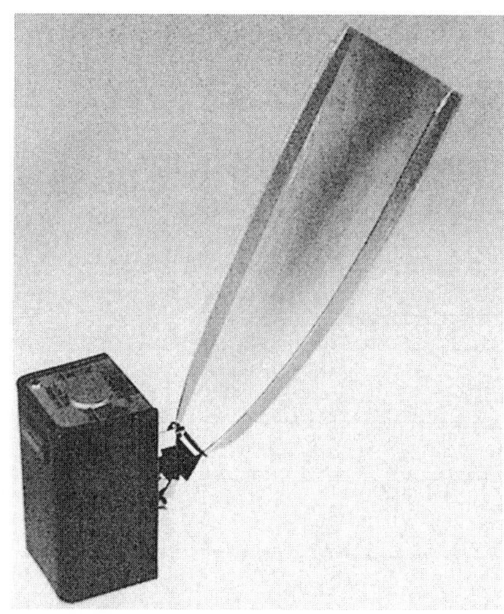

Figure 4-22. Sail switch. *Photo by Bill Johnson*

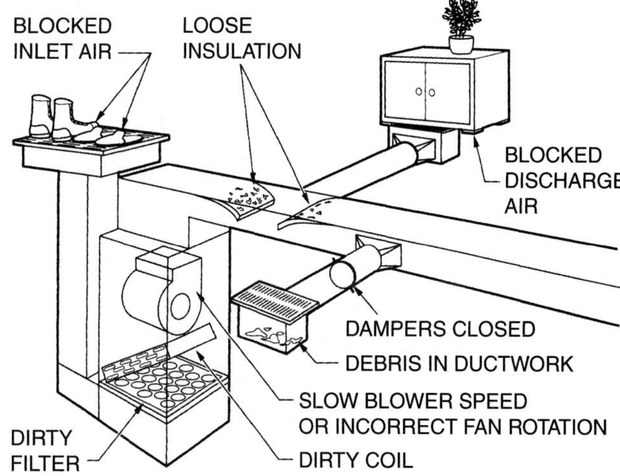

Figure 4-23. Airflow can be reduced for many reasons. *Courtesy Carrier Corporation*

The electric furnace often uses a relay to start the fan when the first heating element is energized, Figure 4-24. The relay can also be used for starting the fan when the system is in the cooling mode. When a relay is used to start the fan, the system blows cold air for the first part of the heating cycle, until at least one of the heating elements is energized and warms up. The cold air may blow for 15 to 30 seconds and may be noticeable in the conditioned space, so another method of energizing the first element may be more practical.

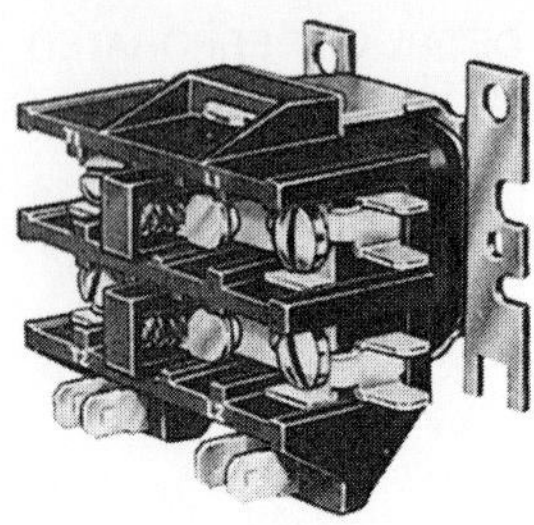

Figure 4-24. Large relay to start electric heating elements and fan. *Courtesy Honeywell, Inc., Residential Division*

The individual sequencer with a fan circuit is one option; the fan starts at approximately the same time as the first heating element. The same is true of the package sequencer.

Remember that when sequencers are deenergized, they cool and break the circuits in about the same amount of time as it takes them to make the circuits from a cold start. The element in the sequencer must cool and the bimetal strip must cool and warp back to the cold position. This keeps the fan on until the last heating element is deenergized, but it does not remove the heat left in the heaters and the furnace. A temperature-activated control can be wired in parallel with the fan contacts on the sequencer to turn off the fan at the end of the cycle, Figure 4-25. This control also starts the fan if the fan contacts in the sequencer fail and prevent the fan from starting.

SAFETY CONTROLS. A large amount of heat is available in the area of a duct heater or an electric furnace. The furnace and the surroundings must be protected in case of reduced airflow or in case one heating element remains energized when no air is flowing. Note from the wiring diagrams that only one side of the electrical line is broken by the sequencer. The other side is wired directly to the heaters. If a heater became too hot and melted in two, one wire could drop to the frame, and a circuit would be established to ground, as shown in Figure 4-26.

Warning: It is always a good practice before touching an appliance to touch one lead of an electrical meter to ground and

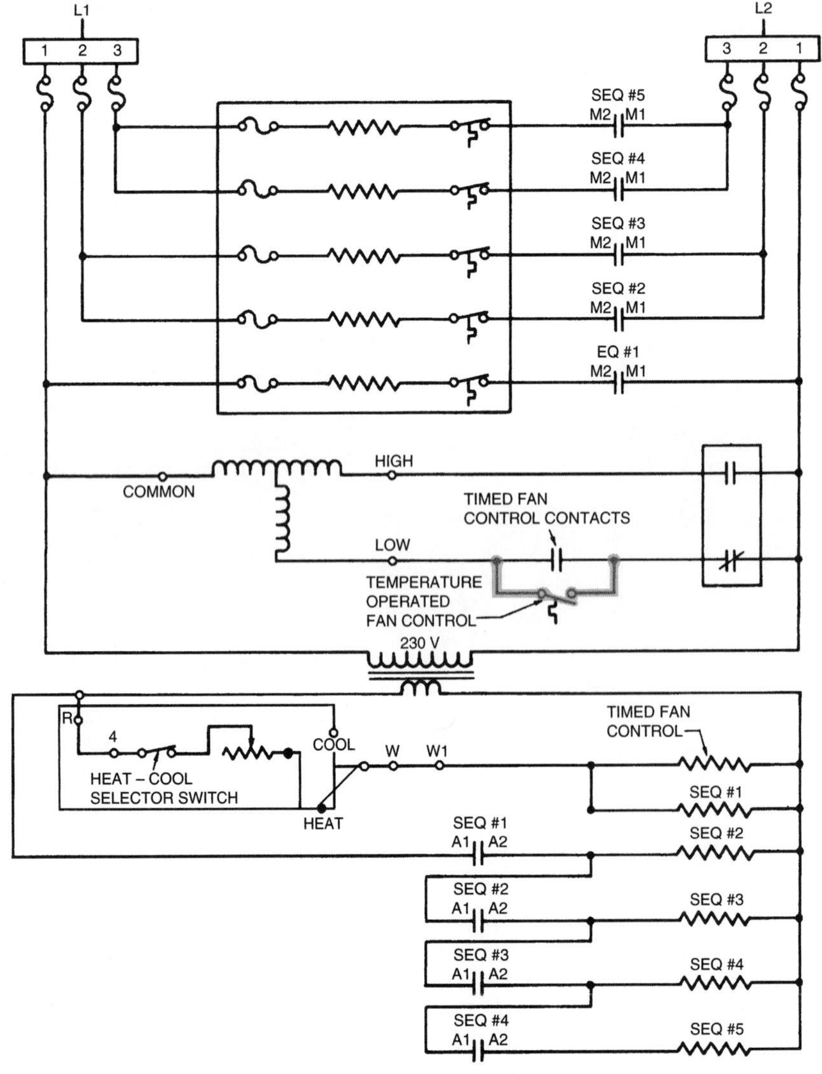

Figure 4-25. Temperature-activated fan control working in parallel with the timed fan control.

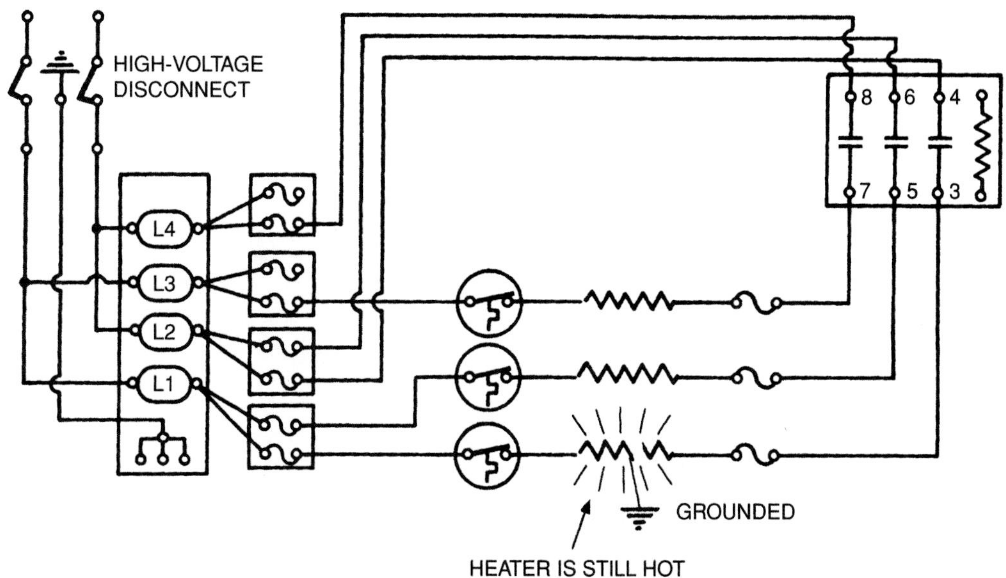

Figure 4-26. Heater with one element grounded to the frame.

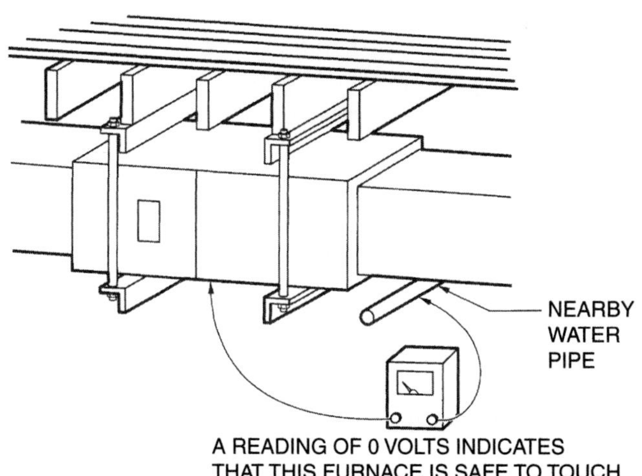

Figure 4-27. A meter is used to check the safety of touching the frame of an electric furnace.

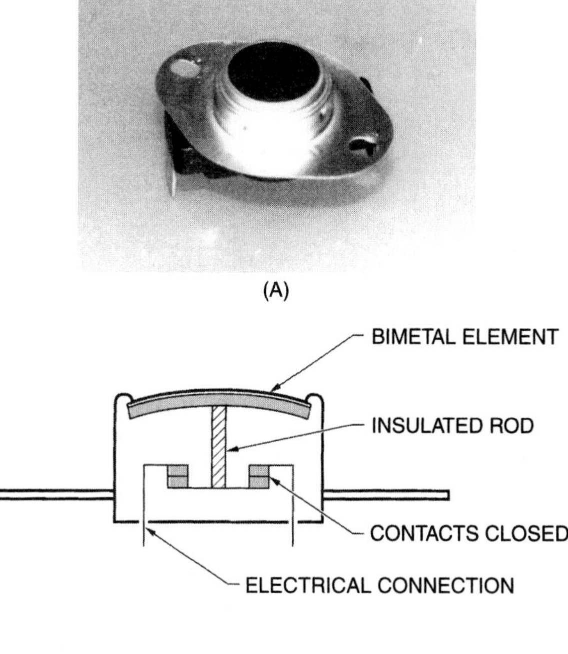

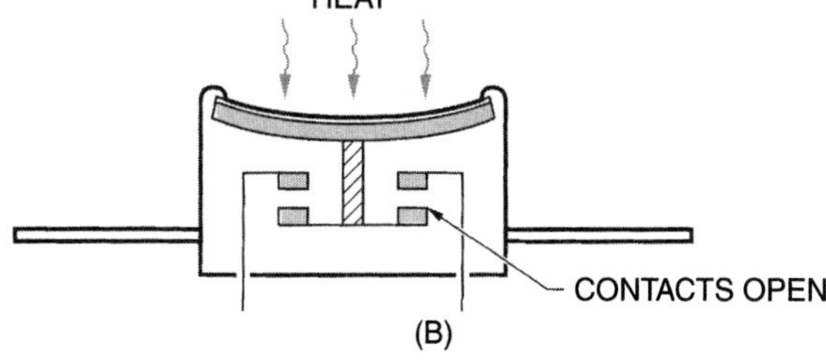

Figure 4-28. (A) Temperature-operated limit control. *Photo by Bill Johnson* (B) Simplified diagram of limit control operation.

the other to the case of the electrical appliance in case of an internal ground. The appliance may not have a good ground circuit, and you may become the circuit, Figure 4-27.

The safety controls for overtemperature protection are furnished in two types: the automatic reset and either the manual reset or the replaceable fuse link. The automatic reset control is set at the lowest practical temperature to prevent the manual reset unit from being tripped because of a momentary problem. The automatic reset control may be in the control circuit or it may be in the line-voltage circuit, but it must be located where it can sense the heat from the heaters, Figure 4-28. Automatic reset controls are usually set to trip at approximately 175°F.

The manual reset device can be in the same location as the automatic reset device, but it has a reset button. It is usually set to trip at approximately 250°F. Figure 4-29 shows the manual reset button. Because someone must push this button to reset the control, this control calls attention to a possible problem.

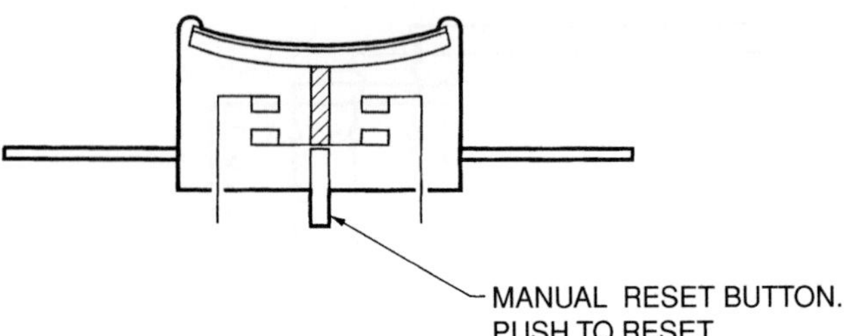

Figure 4-29. Simplified diagram of manual reset control. This high limit control trips at approximately 250°F.

The replaceable fuse link overtemperature device is available in one of two styles, Figure 4-30. One style is in the heater circuit inside the heater. The heater must be removed to service this fuse link. This fuse link is soldered with a metal that has a low melting point. The link is spring-loaded. When the solder reaches its melting point, usually approximately 250°F, the link springs open, as shown in Figure 4-31. Some service technicians reset these used fuse links by soldering them with the solder remaining on the connection. Soldering them with new solder would change the value of the melting point. Reusing this type of fuse link is not recommended. An exact replacement is the best fix. This type of fuse link is outdated and is probably not used in new equipment.

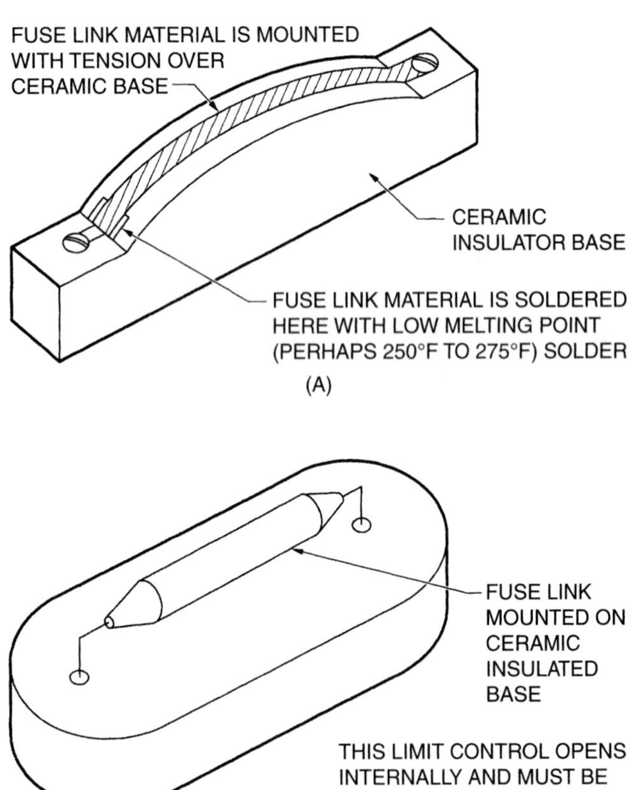

Figure 4-30. Two types of fuse links. (A) Older style that flaps open. (B) Newer style that opens internally.

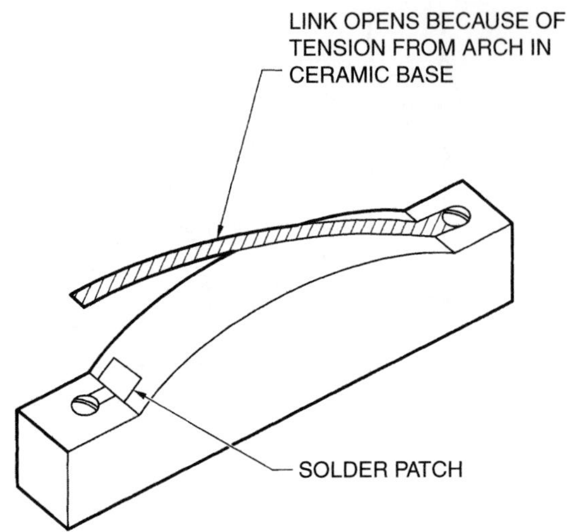

Figure 4-31. This fuse link shows a visible open circuit.

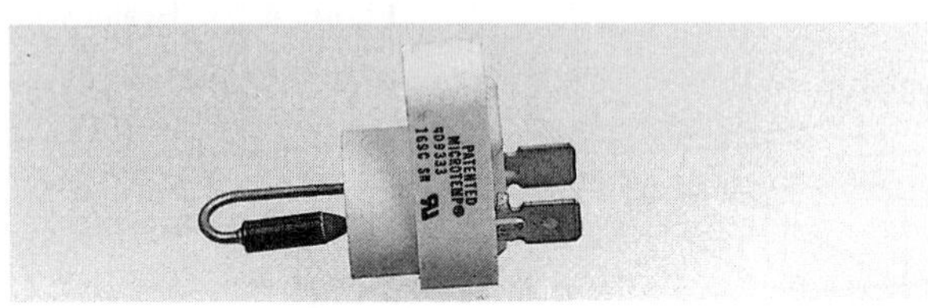

Figure 4-32. The technician can remove this fuse link without removing the heater. *Photo by Bill Johnson*

The newer fuse link, or overtemperature device, is much smaller and often mounted on a ceramic base and inserted into the heating element space through a hole in the heater housing, Figure 4-32. This fuse link is more convenient to replace because the heater need not be removed to be serviced.

LOW-VOLTAGE CONTROLS. The low-voltage controls consist of the power supply, the room thermostat, the interconnecting wiring, and the devices used to start the electric heating elements—sequencers or contactors. These devices are simple, with either resistance coils or magnetic coils as the operators.

As mentioned previously, these controls can be wired so that the total amperage of the control circuit goes through the thermostat or, in the case of some multiple sequencer operations, the power from only one of the sequencers passes through the thermostat. The technician should be aware of both methods, Figure 4-33.

Whichever method is used, the technician should be able to measure the amperage through the circuit and set the heat anticipator in the room thermostat to the correct amperage. If the heat anticipator is not set correctly, noticeable temperature fluctuations will occur in the space temperature.

The technician can measure the amperage at the room thermostat by removing the thermostat from the subbase and applying a jumper with 10 coils of wire (10-wrap wire)

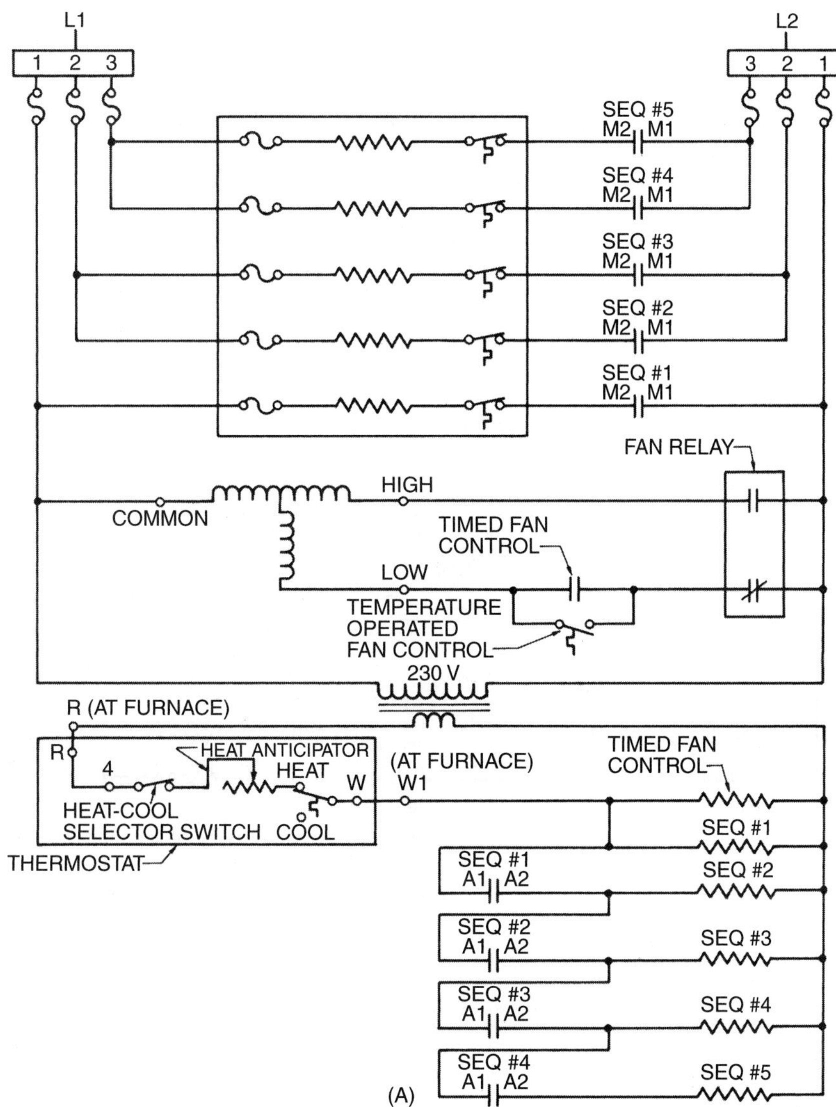

Figure 4-33. Two methods of wiring the sequencer control circuit. (A) All of the amperage passes through the thermostat. (B) Power from only one sequencer passes through the thermostat.

wrapped around the clamp-on ammeter jaws, Figure 4-34, or by using a small, specially designed ammeter. The jumper should be applied from the R terminal to the heating circuit wire; typically this is the W terminal.

The correct amperage for the heat anticipator setting is the total amperage in the circuit that passes through the room thermostat. Suppose that a system has four sequencers that each draw 0.12 ampere. The setting on the room thermostat heat anticipator should be 0.48 ampere (0.12 A × 4 sequencers = 0.48 A, as shown in Figure 4-35).

The technician can also check the control amperage at the control panel in the unit, provided that the control diagram is understood and that the proper checkpoints are used. Often the technician prefers this method because the thermostat need not be removed from the subbase. Just remember this: All the amperage that flows through the thermostat must be measured.

If the system has contactors, the same method can be followed.

Remember, too, that the transformer must be sized to be able to support all the controls in the circuit. Transformers are rated in volt–amperes, or VA. A 60-VA transformer rated at 24 volts can operate at 2.5 amperes (60 VA/24 V = 2.5 A).

4.2 Wiring Diagrams

Electric heaters and boilers can have extensive wiring diagrams because of the numerous elements and controls used. It is important that the technician have a good understanding of what the manufacturer planned before troubleshooting begins. It is a good practice for the technician to study each system encountered before servicing anything, but this is impossible. Some systems will not have diagrams, and a diagram will need to be made for troubleshooting purposes.

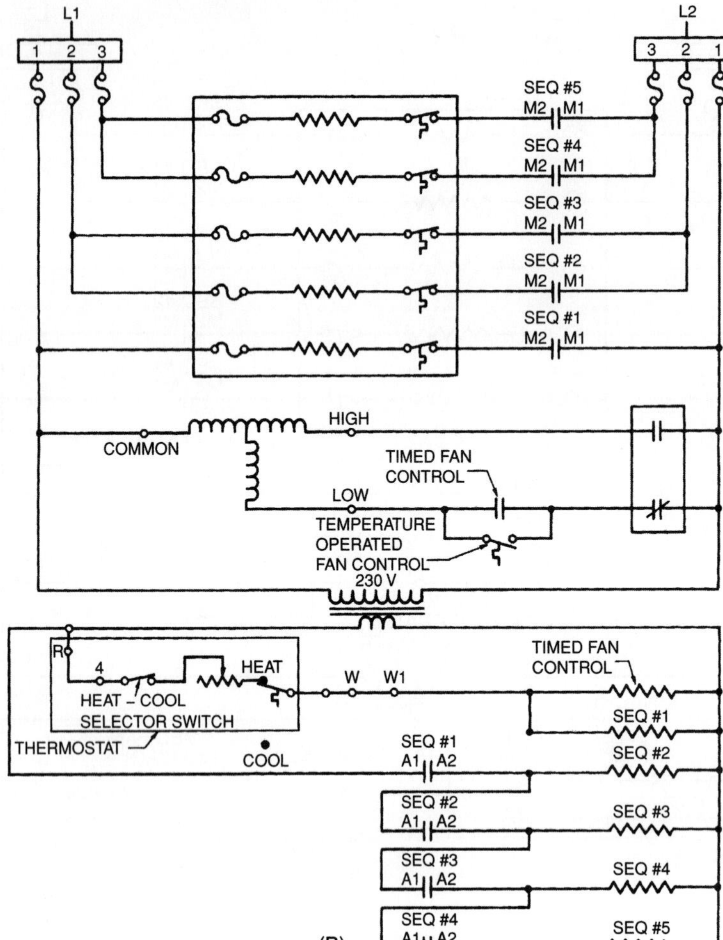

Figure 4-33. *(Continued)*

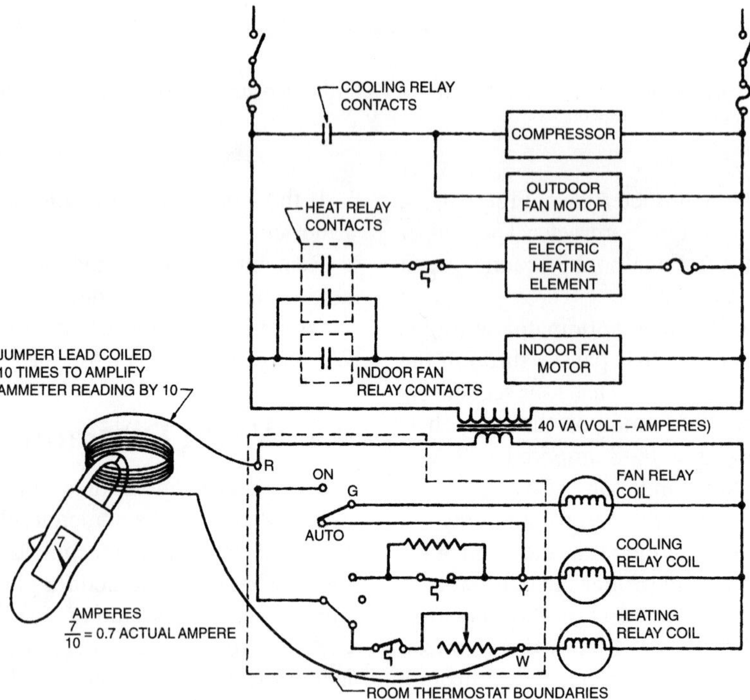

Figure 4-34. Measuring the amperage in the heat anticipator circuit. Notice, the measurement is taken in the sub-base with the thermostat removed.

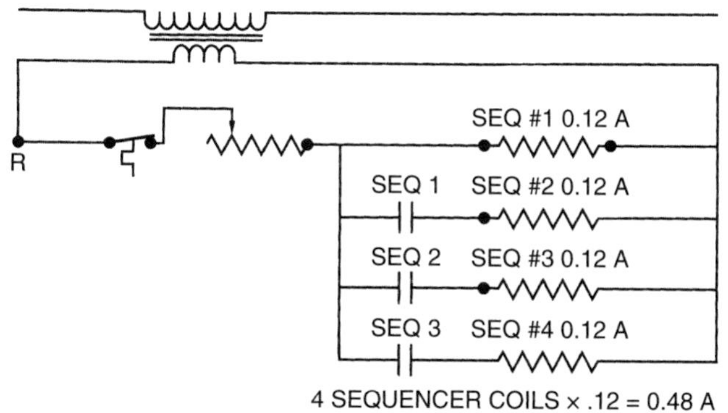

Figure 4-35. Calculating the correct amperage for a heat anticipator circuit.

There are two types of wiring diagrams: the pictorial and the line, or ladder. The pictorial diagram shows the technician where each component is located in relation to the other components. For example, the diagram in Figure 4-36 shows the control transformer in the upper left corner of the control panel and the sequencer in the lower middle of the panel. This diagram helps the technician locate all the major parts of the system.

On the other hand, the line diagram, often called the ladder diagram, is handy for following the route of power to the various components in the circuit. Figure 4-37 shows a ladder diagram for the circuit shown in Figure 4-36. Notice that all the power-consuming devices are located between the Line 1 (L1) and Line 2 (L2) power legs. It is important for the technician to grasp the concept of basic current flow and the fact that for a circuit to function, there must be a power supply, a conductor on which the power can flow, and a device to consume the power, as shown in Figure 4-38. It is also important for the technician to understand which devices are power-consuming devices and which are power passing in a circuit, Figure 4-39.

Use the diagram in Figure 4-40 to follow the route of power to each of five electric heaters and a fan in a typical electric furnace. This diagram will become very complicated if you try to look at it all at once. All technicians should realize that each power-consuming device must have an individual circuit. That circuit may be mixed with other circuits, but it can be singled out with a little selective reasoning. It is suggested that you take this diagram to a copy machine and make an enlarged copy. Then use colored pencils to trace the individual circuits and their interactions with other circuits. You can trace the circuits in the pictorial diagram after locating them in the ladder diagram. We establish power to each side of each power-consuming device to make the circuit function.

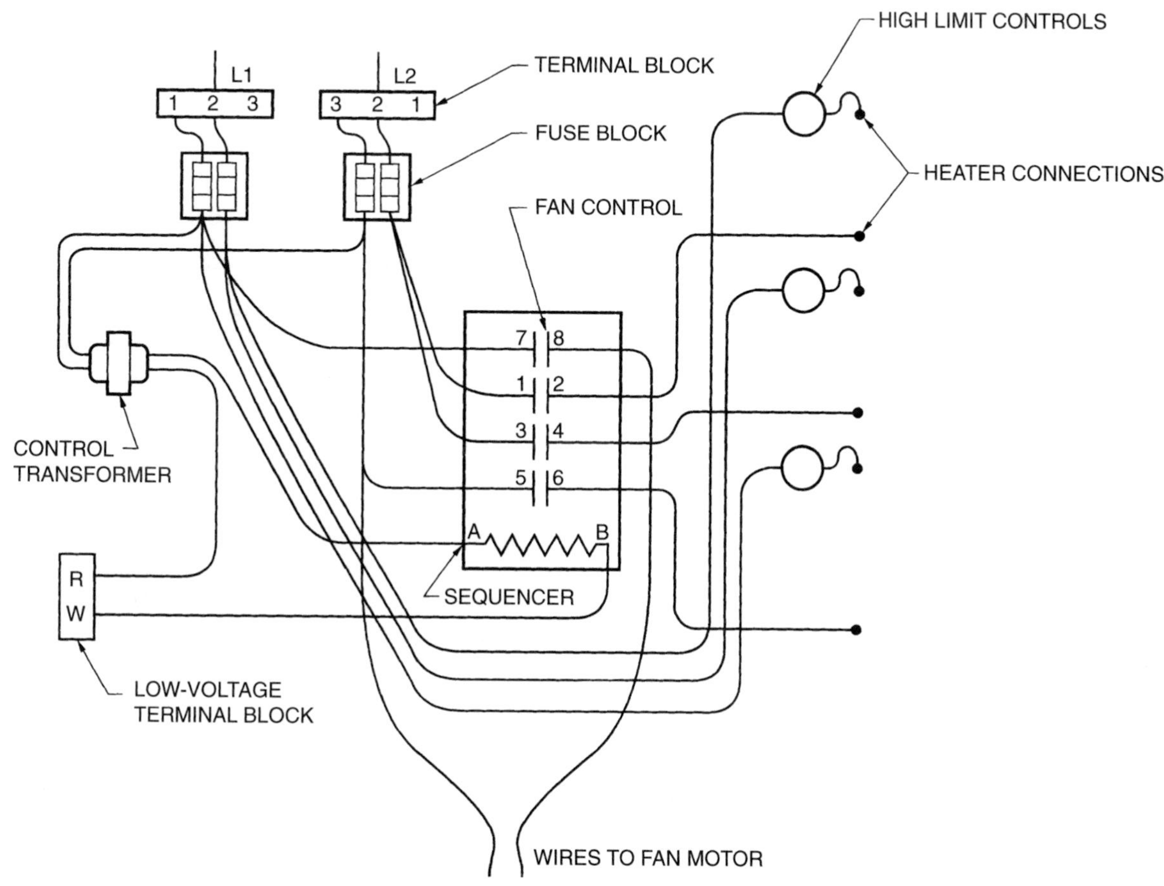

Figure 4-36. Pictorial wiring diagram.

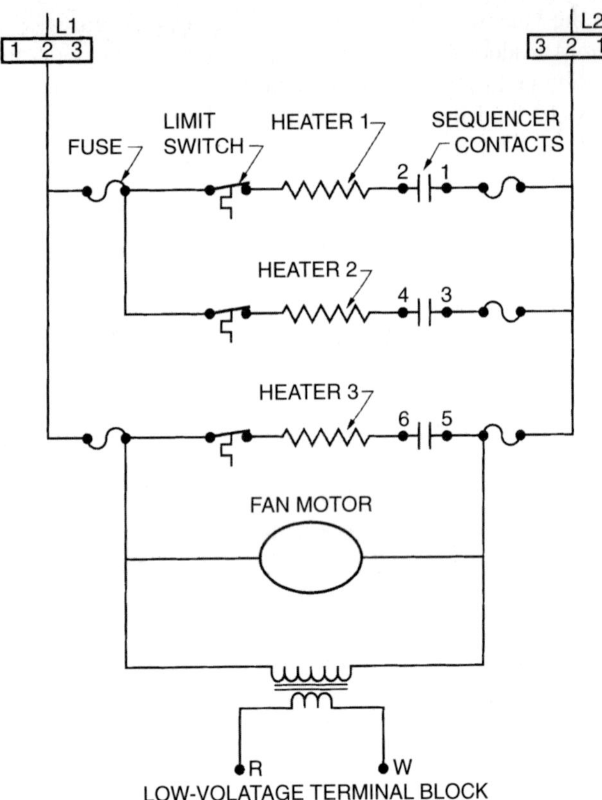

Figure 4-37. Ladder diagram, also called a line diagram.

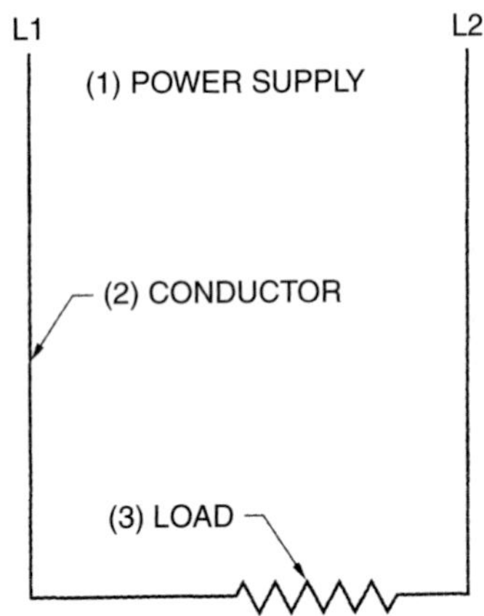

Figure 4-38. The three parts of a complete circuit.

1. Power is supplied to L1, where it is divided into three circuits, labeled 1, 2, and 3.
2. Heating element number 1 is the lower element in the diagram, and the power comes from L1 through the fuse (the fuse is a circuit protector and is discussed further in Section 4.3), then through the fuse link (located at the element) to the left side of the heater. The Ll side of the circuit is satisfied to the power-consuming device.
3. Power is furnished from L2 through the Ml terminal, to the sequencer contact, to the M2 terminal (the sequencer

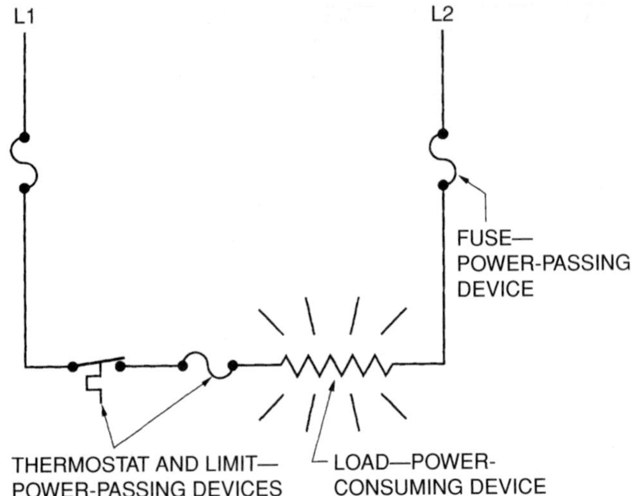

Figure 4-39. Power-consuming and power passing devices in a circuit.

is the operating control for the heating element; its contacts are closed by the low voltage from the room thermostat). Power is then furnished to the automatic reset limit and then to the right side of the number 1 heating element. This satisfies the L2 portion of the circuit. With power to both sides of the element, the element will function if it has a measurable resistance.

Power to the low-voltage sequencer starts the sequence of events. This circuit is mentioned after the heating element because it appears after the heating element in the circuit. The low-voltage circuit follows the same rules as the line-voltage circuit.

1. Power is supplied to the control transformer from circuit 1, from Ll and L2. This energizes the transformer. With these circuits energized, the transformer supplies low voltage, provided that the transformer is functional.
2. The transformer feeds 24-volt power to all the power-consuming devices on the right side of the diagram including the No. 1 sequencer coil. This side of the line is called the *common side* because it is common to all power-consuming devices.
3. Power to the other side of the No. 1 sequencer coil is routed through the room thermostat, entering at terminal R, then through terminal 4, through the heat anticipator circuit, through the Heat terminal to the W terminal, and finally to the W1 terminal, where it leaves the thermostat to energize the No. 1 sequencer. When this sequencer coil is energized, the contacts above Ml and M2 close, and heating element No. 1 provides heat.

This seems like a lot of procedure to energize one element, but it is necessary in order to prevent all the elements from starting at once.

Heating element No. 2 is energized in the same way as No. 1 on the line-voltage portion: from L1 through the fuse, then the fuse link, to the heater. The other side is energized

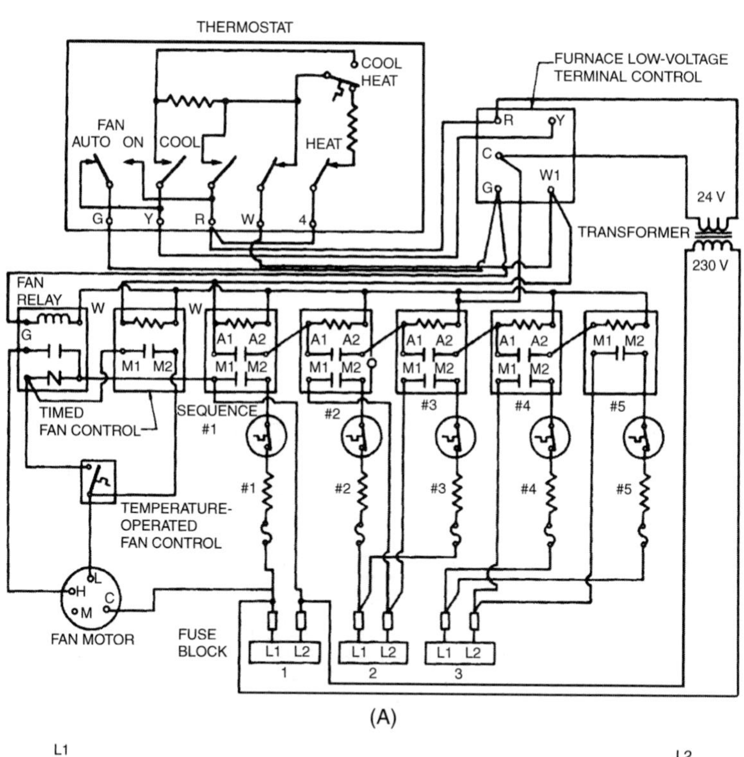

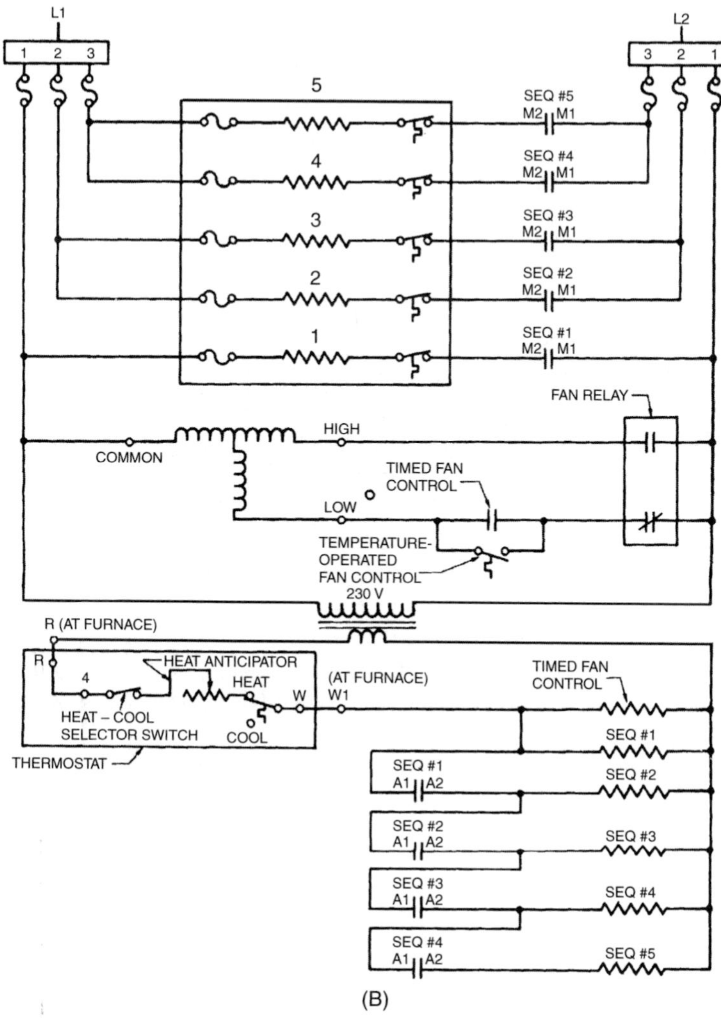

Figure 4-40. A five-element heat circuit for an electric furnace. (A) Pictorial diagram. (B) Ladder diagram.

from L2 through the fuse, then through the Ml and M2 contacts on sequencer No. 2, then to the automatic reset limit to the element. The only difference is how the No. 2 sequencer coil is energized to close the Ml and M2 contacts. This is accomplished in the low-voltage circuit in the following manner:

1. Power is supplied to the right side of the sequencer coil from the common 24-volt line.
2. Power to the other side of the sequencer is supplied from the Al to A2 contacts in the No. 1 sequencer. They are closed by the coil in the No. 1 sequencer. When they are closed, the No. 2 sequencer coil should then be energized, and the No. 2 heating element should then function.

The entire process just described is then used to start all the heaters, one at a time, with a time delay between each start.

The fan is another power-consuming device that must be started when the first heating element is energized. This is accomplished in the following manner:

1. Power is supplied to the common terminal for the fan motor from L1.
2. Power is supplied to the low-speed fan operation from L2, through the NC (normally closed) contacts of the fan relay. Power then moves to the terminals of the two fan operating controls, the timed control and the temperature-operated control. The timed control should close first and start the fan, and the temperature control should heat and close the contacts being used to stop the fan at the end of the cycle.
3. The timed fan control contacts are energized by the timed fan control coil, located just above sequencer No. 1 and energized at the same time as sequencer No. 1. The fan should start at the same time as the No.1 heating element.
4. At the end of the cycle, the timed fan control will try to shut off the fan at the same time as the sequencers, but there is still heat in the furnace that should be dissipated by the fan. This is accomplished with the thermostatic fan switch. It keeps the fan on until the heat is dissipated. It also acts as a backup to the timed fan control. If the timed fan control malfunctions, the thermostatic fan switch starts the fan.

The purpose of the fan relay is to have two-speed fan operation in case the appliance has a cooling system. In a cooling system, the fan relay would be energized by the room thermostat upon a call for cooling. The relay would change

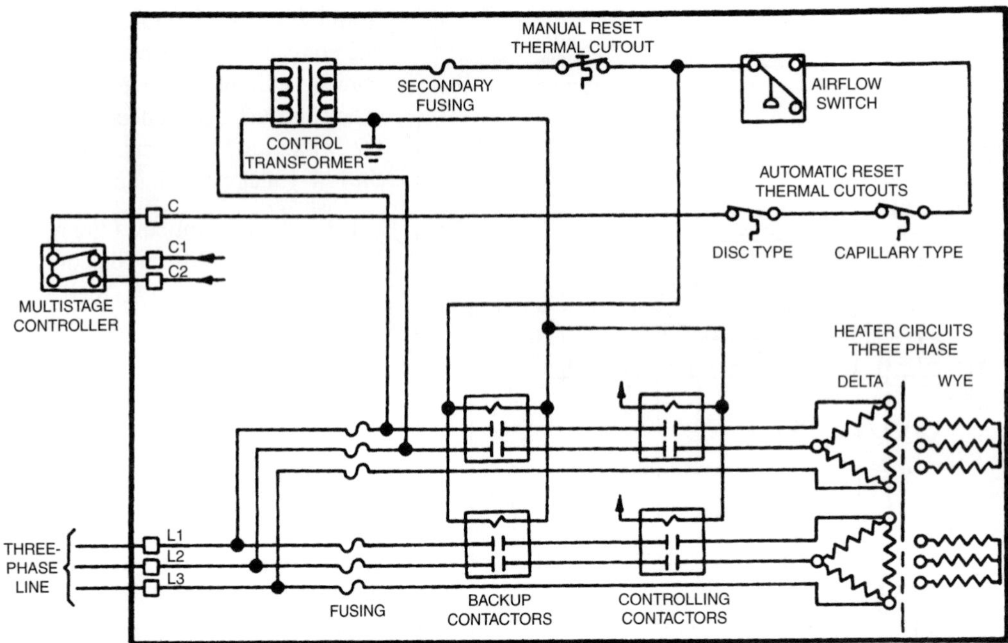

Figure 4-41. Three-phase electric-heat circuit. *Courtesy Indeeco*

positions and start the fan in high speed. The customer can also use the fan-on switch on the room thermostat to circulate air if needed. The relay prevents the fan from trying to run in high speed and low speed at the same time, which would ruin the fan motor.

The diagram in Figure 4-40 is typical of many electric-heat diagrams. The controls may be different because the application is different, but the concept will be the same. It is important to have a working diagram for these complicated applications. In some cases, you may have to develop the diagrams yourself.

The diagram in Figure 4-40 is for a single-phase electric-heat circuit. The diagram in Figure 4-41 is for a three-phase system. The only difference is that the heaters are applied to all three phases of the power supply to balance the load. The control circuit is still across only two of the phases. However, the thermal overloads that will stop the electric heat system are all in the low voltage circuit. Placing controls in all of the circuits can be prevented by placing an automatic reset and a manual reset in the control wiring. If either of them open, it shuts off all control voltage to the contactors. Notice that there are two sets of contactors. This is in case one of them is welded closed, the other one will act as a backup for protection. If one contactor fails in either heater, for any reason, the system will not provide heat.

4.3 Circuit Protection

The circuits in electric-heat systems carry much more current than those in the fossil fuel systems, so circuit protection becomes more complicated. The wiring carries

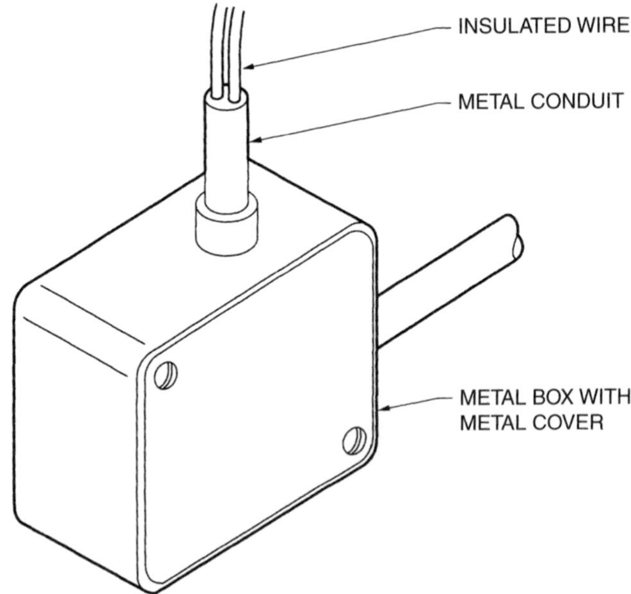

Figure 4-42. Electrical conduit for protection of the wires and connections in a commercial installation.

heavy electrical loads. If the circuits are not purposely protected, they may become part of the load and become hot. In a home, the wires are routed through wood stud walls, and the wire has only plastic for protection. A fire could easily start without proper circuit protection. Commercial installations require that electrical wiring be routed in metal conduit, Figure 4-42, for fire protection and physical protection from damage. The fire protection from the conduit does not excuse poor circuit protection, so these circuits are also carefully sized and protected. Circuit protection is the job of the designer and the installing contractor. It is a broad subject that cannot be covered completely in this text. However, for practical purposes there are certain portions of the subject the technician must be aware of. General principles are discussed so that you have an understanding of the concepts of protection. The technician must be able to recognize undersized wire and circuit protectors to adequately troubleshoot electric-heat applications. The heating technician must often provide proper wire routing, sizing, and connector workmanship.

Proper circuit protection begins with selecting the correct conductor for the application. Conductor selection involves selecting not only the correct cover (insulation) for the wire, but also the correct wire size for the application. The cover for the wire may be different for different applications. For example, some wire covers withstand more heat than others and are used where the wire is routed through extremely hot locations, such as through a boiler room. Figure 4-43 can be used to select the wire size for a typical application. Note that the wire cover (insulation) is categorized according to the allowable temperature rise of the wire.

Table 310-16. Allowable Ampacities of Insulated Conductors Rated 0-2000 Volts, 60° to 90°C (140° to 194°F) Not More Than Three Conductors In Raceway or Cable or Earth (Directly Buried), Based on Ambient Temperature of 30°C (86°F)

Size	Temperature Rating of Conductor. See Table 310–13.						Size
	60°C (140°F)	75°C (167°F)	90°C (194°F)	60°C (140°F)	75°C (167°F)	90°C (194°F)	
AWG kcmil	TYPES TW†, UF†	TYPES FEPW†, RH†, RHW†, THHW†, THW†, THWN†, XHHW†, USE†, ZW†	TYPES TA, TBS, SA SIS, FEP†, FEPB†, MI RHH†, RHW-2, THHN†, THHW†, THW-2, THWN-2, USE-2, XHH, XHHW†, XHHW-2, ZW-2	TYPES TW†, UF†	TYPES RH†, RHW†, THHW†, THW†, THWN†, XHHW†, USE†	TYPES TA, TBS, SA, SIS, THHN†, THHW†, THW-2, THWN-2, RHH†, RHW-2 USE-2 XHH, XHHW XHHW-2, ZW-2	AWG kcmil
	COPPER			ALUMINUM OR COPPER-CLAD ALUMINUM			
18	....	....	14	....	....	....	....
16	....	....	18	....	....	....	....
14	20†	20†	25†	....	....	....	....
12	25†	25†	30†	20†	20†	25†	12
10	30	35†	40†	25	30†	35†	10
8	40	50	55	30	40	45	8
6	55	65	75	40	50	60	6
4	70	85	95	55	65	75	4
3	85	100	110	65	75	85	3
2	95	115	130	75	90	100	2
1	110	130	150	85	100	115	1
1/0	125	150	170	100	120	135	1/0
2/0	145	175	195	115	135	150	2/0
3/0	165	200	225	130	155	175	3/0
4/0	195	230	260	150	180	205	4/0
250	215	255	290	170	205	230	250
300	240	285	320	190	230	255	300
350	260	310	350	210	250	280	350
400	280	335	380	225	270	305	400
500	320	380	430	260	310	350	500
600	355	420	475	285	340	385	600
700	385	460	520	310	375	420	700
750	400	475	535	320	385	435	750
800	410	490	555	330	395	450	800
900	435	520	585	355	425	480	900
1000	455	545	615	375	445	500	1000
1250	495	590	665	405	485	545	1250
1500	520	625	705	435	520	585	1500
1750	545	650	735	455	545	615	1750
2000	560	665	750	470	560	630	2000

CORRECTION FACTORS

Ambient Temp. °C	For ambient temperatures other than 30°C (86°F), multiply the allowable ampacities shown above by the appropriate factor shown below.						Ambient Temp. °F
21-25	1.08	1.05	1.04	1.08	1.05	1.04	70-77
26-30	1.00	1.00	1.00	1.00	1.00	1.00	78-86
31-35	.91	.94	.96	.91	.94	.96	87-95
36-40	.82	.88	.91	.82	.88	.91	96-104
41-45	.71	.82	.87	.71	.82	.87	105-113
46-50	.58	.75	.82	.58	.75	.82	114-122
51-55	.41	.67	.76	.41	.67	.76	123-131
56-60	.....	.58	.71	.....	.58	.71	132-140
61-70	.....	.33	.58	.....	.33	.58	141-158
71-80	.....	.....	.41	.....	.....	.41	159-176

†Unless otherwise specifically permitted elsewhere in this Code, the overcurrent protection for conductor types marked with an obelisk (†) shall not exceed 15 amperes for No. 14, 20 amperes for No. 12, and 30 amperes for No. 10 copper; or 15 amperes for No. 12 and 25 amperes for No. 10 aluminum and copperclad aluminum after any correction factors for ambient temperature and number of conductors have been applied.

Figure 4-43. Wire-sizing table. *Data from the National Electric Code Book*

To ensure that the conductor has enough current-carrying capacity, the NEC (National Electrical Code) calls for wire to carry no more than 75% of its rated capacity, so wire must be oversized by at least 25%. This practice must be followed because it is a national code. There are variations to this 25% oversizing, but they are not covered here because they do not apply to electric-heat systems.

Let us consider an example of sizing the wire for a circuit: Suppose that a wire size for the furnace circuit in Figure 4-44 is needed. The furnace has five electric heating elements that draw 4.8 kilowatts each and a fan motor that draws 6 amperes. The furnace is rated at 240 volts. The total current draw for the furnace is 106 amperes ($I = P/E = (4{,}800 \text{ W} \times 5)/240 \text{ volts} = 100$ A for the heaters + 6 A for the fan motor = 106 A). According to the NEC, we must now add 25% to the current, for a total of 132.5 amperes ($106 \times 1.25 = 132.5$ A), and use this number in the wire sizing table, Figure 4-43. This number is often called the *ampacity* of the circuit; ampacity is the amperage capacity needed for the conductors.

The wire size for this circuit, if THWN wire is used, is No. 1/0. Note that size No. 1 wire carries 130 ampacity. We must round up to the next value, size No. 1/0 wire, which has an ampacity of 150. The extra ampacity is further protection for the circuit.

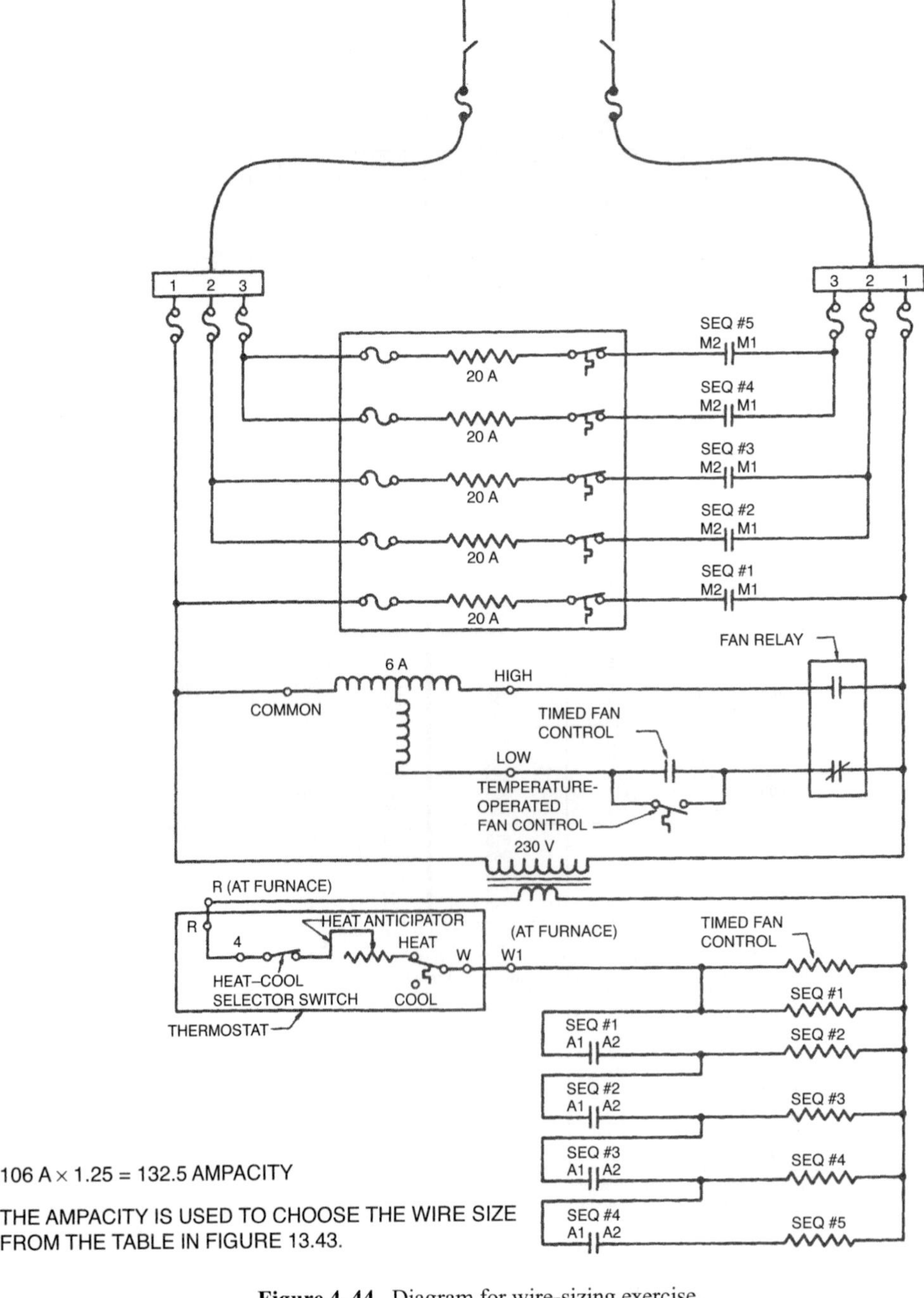

20 A
20 A
20 A
20 A
20 A
6 A
106 A

106 A × 1.25 = 132.5 AMPACITY

THE AMPACITY IS USED TO CHOOSE THE WIRE SIZE FROM THE TABLE IN FIGURE 13.43.

Figure 4-44. Diagram for wire-sizing exercise.

You will also encounter wire that the manufacturers installed in electric-heat appliances. This wire may seem undersized for the application. However, manufacturers are permitted to use smaller wire for two reasons: They have tested their equipment, and the wire runs within the appliance are shorter than long field circuits. They may also use wire rated at higher temperatures than you would typically encounter in field wiring.

The technician should learn the proper methods used to connect wire in these high-amperage circuits. Loose connections can be serious problems in electrical circuits. The technician can learn these techniques from electricians or from electrical supply house personnel, along with practice.

Properly sizing the circuit protector—fuse or breaker—further ensures that the circuit is protected. The general rule for sizing the circuit protection is to size it at 25% more than the circuit amperage. For example, the circuit in Figure 4-44 was estimated to draw 106 amperes. The circuit protector should be sized at 132.5 amperes ($106\ \text{A} \times 1.25 = 132.5\ \text{A}$). It is permissible to round up to the next size protector, which, in this case, would probably be 135 amperes for a fuse or a circuit breaker. The circuit protector is sized at approximately the current-carrying capacity of the actual conductor.

The procedures discussed in this section are general, but when they are used they will prevent the technician from practicing outside the National Electrical Code bounds.

Summary

- A line-voltage thermostat built into the room heater provides automatic control for the heater.
- Panel heat and the baseboard heater are controlled with a line-voltage thermostat in each room.
- The unit heater can be controlled with either a low-voltage or a line-voltage thermostat in the conditioned space.
- The electric boiler is controlled with a room thermostat or a boiler temperature control.
- Multiple-element control for boilers is usually accomplished with step controllers or time-delay relays.
- The duct heater and the electric furnace are controlled similarly, except for the fan. Each typically has two limit controls and a fuse link to prevent overheating. A low-voltage thermostat controls the space temperature.
- Multiple electric heating elements must be step controlled to prevent power fluctuations. Sequencers (individual or package), time-delay relays, and mercury contactors are typically used.
- Electric heating elements are typically 5 kilowatts or less in capacity.
- The fan control for a duct heater is often a sail switch or a pressure differential switch to ensure that the air is moving before the heat is energized.
- The fan control for an electric furnace can be timed or temperature operated, or a combination.
- The room thermostat has a heat anticipator that must be set to the value of the operating control, sequencer or contactor.
- The technician can check the amperage for setting the heat anticipator by applying a 10-wrap wire around the ammeter jaws or by using a special ammeter.
- A wiring diagram is either pictorial or ladder. The technician uses the pictorial diagram to locate the various controls and the ladder diagram to understand the circuit.
- Circuit protection starts with choosing the correct cover, or insulation, for the conductor.
- The size of the conductor is chosen by allowing it to carry only 75% of its actual capacity. This is accomplished by multiplying the estimated current in the circuit by 1.25 and using this value to choose the wire size. Uneven wire sizes are rounded up to the next size.
- The circuit is protected by a fuse or breaker that is rated at 125% of the circuit ampacity (ampere capacity), then rounded up to the next value protector. For example, when a 32-ampere protector is needed, a 35-ampere protector, the next available size, is used.

Review Questions

1. What is the approximate temperature at which the automatic reset limit switch trips? at which the manual reset limit switch trips?
2. Are all high limit controls either automatic or manual reset?
3. A furnace has a return air temperature of 71°F, a supply air temperature of 128°F, an operating voltage of 225 volts, and a current draw of 97 amperes. How many cfm of air are moving?
4. What is the lowest air temperature that should be used for an electric forced-air system? the highest temperature?
5. Why should the supply air temperature not be measured directly over the heating elements in a forced-air system?
6. What type of control can be used to ensure airflow for a duct heater?
7. Why are both time and temperature often used to control the fan on an electric furnace?
8. When a control transformer has a rating of 75 VA and is rated at 24 volts, how many amperes can the transformer handle?
9. What device inside the room thermostat must be set according to the amperage draws of the operating coils for the sequencers or contactors?

5 Troubleshooting and Maintenance for Electric-Heat Equipment

Objectives

Upon completion of this unit, you should be able to

- **determine when a plug-in heater has a problem.**
- **know how to inspect an electric heating system for loose connections.**
- **perform filter maintenance on a forced air system.**

5.1 Preventive Maintenance

Maintenance of electric-heat equipment can involve small appliances or central heating systems. Electric-heat appliances, units or systems, all consume large amounts of power. This power is routed to the unit or system through wires, wall plugs (appliance heaters), and connectors.

The power cord at the plug-in connector is usually the first place where trouble occurs in a heating appliance. The heater should be taken out of service at the first sign of excess heat at the wall plug. The plug may be warm to the touch. This is not a problem, but if it is too hot to hold comfortably, shut off the appliance. If the situation is not corrected, it will become worse every time the heater is operated, and fire may start in the vicinity of the plug. When corrective action is taken to the plug-in cord, the wall receptacle should be checked at the same time. An overheated plug often damages the receptacle beyond repair, so receptacle replacement is often necessary.

Warning: An electric-appliance heater should be observed at all times to ensure safe operation. The heater must not be too close to combustible materials. Operate the heater only in the location for which it was designed. When the line cord for a plug in appliance is too short and an extension cord is used, follow the manufacturers' instructions. Make sure the extension cord is sized correctly and as short as possible. DO NOT run the cord under the floor carpeting.

Large electric-heat systems are either panel, base-board, or forced-air systems. Panel and baseboard systems do not require much maintenance when properly installed. Baseboard heaters have more air circulation over the elements than panel heaters do, so the units must be kept free from lint and dust. Some areas will have more dust, and units in these areas will require more frequent cleaning. The covers should be taken off and the lint and dust removed with a vacuum cleaner.

Panel and baseboard systems both use line-voltage thermostats. These thermostats contain the only moving parts and usually are among the first components of the system to fail. They also collect dust and must be cleaned inside at some point. Be sure to turn off the power before servicing the thermostats.

A central forced-air system has a fan and consequently requires filter maintenance.

Filters should be checked every 30 operating days unless you determine that longer periods are allowed. Use filters that fit the filter rack and be sure to follow the arrows on the filter for the correct airflow direction. When the filter system becomes clogged in a forced-air system, the airflow is reduced and the elements overheat. If this occurs, the automatic reset limit switches should trip and then reset. If the condition continues, these switches may fail and the fuse link in the heater may open the circuit. When this happens, the link must be replaced by a qualified technician. This type of service call can be prevented with proper filter maintenance.

Some fan bearings require lubrication at regular intervals. Motors and fan shafts with sleeve bearings must be kept in good working order because these bearings will not take much abuse. They do not always make excess noise when worn, so they should always be examined. Lift the motor or fan shaft and look for movement. Do not confuse shaft end play with bearing wear. Many motors have as much as $\frac{1}{8}$-inch of end play.

Some systems have belt drives that require belt-tension adjustment or belt replacement. Frayed or broken belts must be replaced to ensure safety and trouble-free operation. The only way to correctly tighten a belt is to use a belt-tension measuring device. Belts should be checked at least every 3 months when a system has many operating hours per day. It is extremely dangerous to allow electric-heat systems to operate with little or no airflow.

The electric heating elements are usually controlled by low-voltage thermostats that open or close the contacts on the sequencers or contactors. The contacts in the sequencers carry large electrical loads that probably receive the most stress. Sequencers or contactors are among the first components to fail in a properly maintained system. You can visually check the contacts on a contactor, but the contacts on sequencers are concealed. When the contacts on a contactor become pitted, change the contacts or the contactor. If you do not, the condition will only become worse and cause other problems, such as terminal and wire damage to the circuit. Often the contacts will weld themselves closed. This can cause the system to be heating all the time. If only one element is still heating during the cooling season, it may not be noticed, but heating and cooling at the same time is inefficient.

The only method you can use to check a sequencer is to visually check the wires for discoloration and to check the current flow through each circuit with the sequencer energized. If power does not pass through the sequencer circuit, the sequencer may be defective. Do not forget that the heater must have a path through which electrical energy can flow. Do not make the common mistake of declaring that a sequencer is defective only to find later that a fuse link in the heater or the heater itself has an open circuit.

Line-voltage electrical connections in the heater junction box should be examined closely. If there is any discoloration or connectors that appear to have been hot, these components must be replaced. If the wire is discolored and is long enough, you can cut off the damaged wire. If it is not long enough, a proper splice can be made or the wire can be replaced.

A fuse holder may lose its tension and not hold the fuse tightly if it has been overheated. This will cause fuses to blow without any apparent reason. When this happens, the fuse holder must be changed for the system to operate correctly.

5.2 Technician Service Calls

Service Call 1

A customer with a 20-kilowatt electric furnace calls. There is no heat. The customer says that the fan comes on, but the heat does not work. This system has individual sequencers with a fan-starting sequencer. The first-stage sequencer has a burned-out coil and will not close its contacts. The first stage starts the rest of the heat, so there is no heat.

The technician is familiar with this system and had started it up after installation. The technician realizes that because the fan starts, the 24-volt power supply is working. Upon arriving at the job, the thermostat is set to call for heat. The technician goes under the house with the spare sequencer, a volt–ohmmeter, and a clamp-on ammeter.

Upon approaching the electric furnace, the technician can hear the fan running. The panel covering the electric heating elements and sequencer is removed. The ammeter is used to determine whether any of the heaters are using current; none are. The technician observes all electrical safety precautions. The voltage is checked at the coils of the sequencers, as shown in Figure 5-1. It is found that 24 volts are present at the timed fan control and the first-stage sequencer, but the

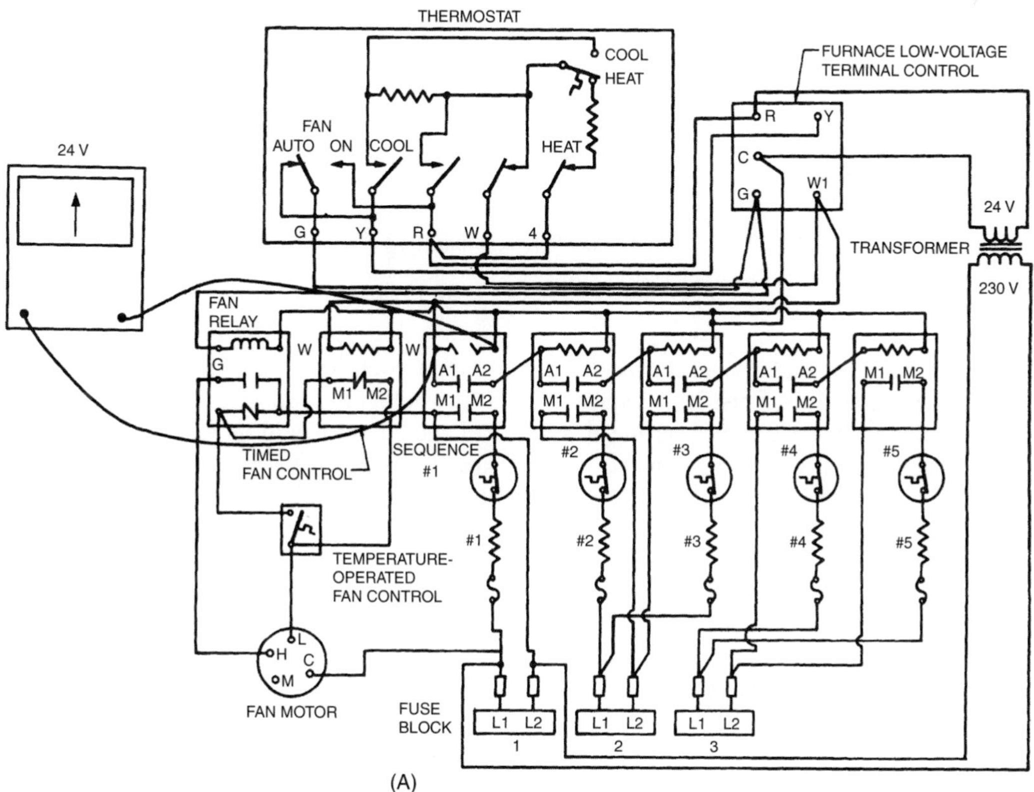

Figure 5-1. Checking the circuit with a meter.

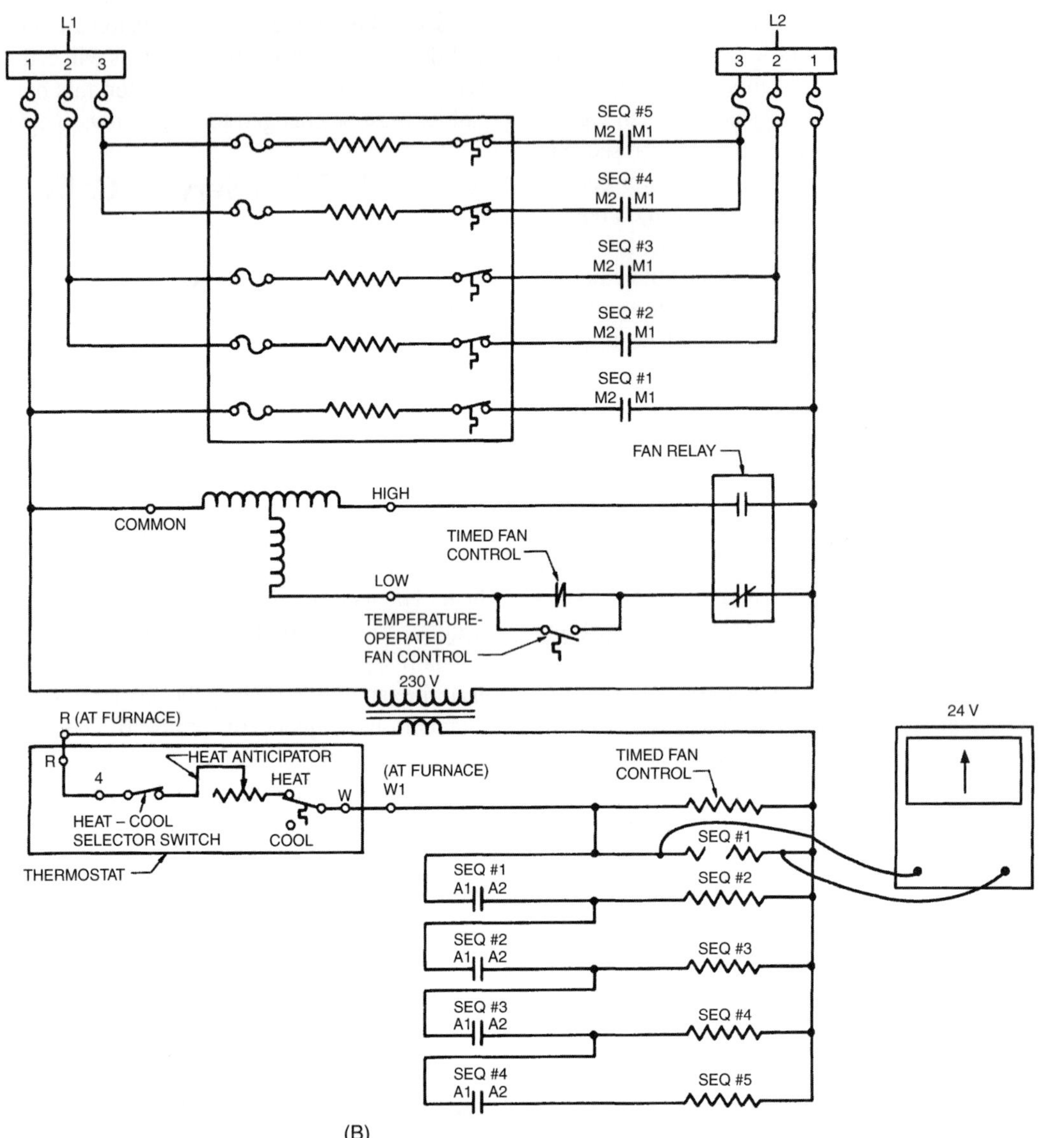

Figure 5-1. (*Continued*)

contacts are not closed. The electrical disconnect is opened, and the continuity is checked across the first-stage sequencer coil. The circuit is open. When the sequencer is changed and power is turned on, the electric heat begins to operate after the sequencer begins operation.

Service Call 2

A residential customer calls. There is no heat, but the customer smells smoke. The company dispatcher advises the customer to turn off the system until the technician arrives. The fan motor has an open circuit and will not run. The smoke smell is resulting from the heating unit cycling on the limit control.

The technician arrives, goes to the room thermostat, and turns the system to heat. This is a system that does not have cooling, or the technician would turn the thermostat to the fan-on position to see whether the fan would start. If it would, there would be power to the unit and the low-voltage circuit would be functional. When the technician goes to the electric furnace in the hall closet, it is noticed that a smoke smell is present. When an ammeter is used to check the current at the electric heater, the technician finds that the heater is pulling 40 amperes, but the fan is not running. The heater is then interrupted by the limit control. The technician observes all electrical safety precautions.

The technician looks at the wiring diagram and discovers that the fan should start with the first-stage heat from a set of contacts on the sequencer. The voltage to the fan motor is checked; the motor is getting voltage but is not running. The power is turned off and a continuity check of the motor proves that the motor winding is open. The technician changes the motor, and the system operates normally.

Service Call 3

The manager of a small business calls. Her heating system is not putting out enough heat. This is the first day of bitterly cold weather, and the space temperature is heating only to 65°F with the thermostat set at 72°F. The system is a package air-conditioning unit with 30 kilowatts (six stages of 5 kilowatts each) of strip heat located on the roof. The fuse links in two of the heaters are open because of dirty air filters.

The technician arrives, goes to the room thermostat, checks the setting of the thermostat, and finds that the space temperature is 7°F lower than the setting. The technician goes to the roof with a volt–ohmmeter and an ammeter. After removing the panels on the side of the unit where the strip heat is located, the technician carefully checks the amperage. Two of the six stages are not pulling any current. There appears to be a sequencer problem. However, a voltage check of the individual sequencer coils shows that all the sequencers have closed contacts; there are 24 volts at each coil.

A voltage check at each heater terminal shows that stages 4 and 6 have voltage but are not pulling any current. The technician shuts off the power, pulls the stage 4 heater element from the unit for examination, and finds that the fuse link is open. The same is true of the stage 6 heater. The fuse links are replaced, and the heater restarted. Stages 4 and 6 now draw current.

The technician knows that the heaters have been too hot for some reason. A check of the air filters shows that they are clean, too clean. The technician looks in the trash can and finds the old filters, which are almost completely clogged. The business manager is informed about changing filters on a regular basis whether they appear to need it or not.

Service Call 4

The service technician is on a routine service contract call and inspection at a store. The terminals on the electric heating units have been hot. The insulation on the wire is burned at the terminals.

When removing the panels on the electric heat panel, the technician sees the burned insulation on the wires and shows it to the store manager. It is suggested that these wires and connectors be replaced. The store manager asks what the consequences of waiting for them to fail would be. The technician explains that they may fail on a cold weekend and allow the building water to freeze. If the overhead sprinkler freezes and thaws, all the merchandise will get wet; the store manager does not want to take this chance. The technician turns off the power. The old wire is cut back to clean new-looking wire on some of the heater connectors. Some of the wire must be replaced with new wire. The technician prepares new wire of the same gauge and insulation quality as the rest of the wire in the heater box and makes the connections using the correct terminals. The cover is replaced and the heater is then started.

Service Call 5

A customer calls at the end of the heating season and explains that the fan is starting and running for every few minutes. One of the four heating elements is burned in two and one leg is touching the frame of the furnace. The heater is heating all the time and creating enough heat to start the fan on the fan temperature control.

The technician arrives and observes the fan starting and stopping without any apparent reason. After a look at the wiring diagram, the technician decides that the fan must be starting from the fan temperature switch but does not know why.

The technician takes some hand tools, a volt–ohmmeter, and a clamp-on ammeter to the furnace, in the attic. The control box cover is removed and it is warm to the touch. The technician clamps the ammeter on the line leading to the heaters, one at a time, and discovers that heater No. 3 is pulling 10 amperes of current. It should draw approximately 20 amperes in normal operation. Power is checked to the heater, and line voltage, 230 volts, is not available, but the heater is still drawing 10 amperes. The wiring diagram shows that power is available at one of the heater terminals at all times, so the heater must be grounded to the frame in order to draw current, Figure 5-2.

The technician pulls out the heater and finds that one leg is grounded. A new heating element is installed and the problem is solved. The fan was responding to a call from temperature at the heater, starting and stopping according to the temperature-operated fan control.

Service Call 6

A customer calls on an extremely cold winter day from a small office building and complains that the electric-heat system is not heating the building. The problem is that the offices were just painted and the room thermostat was damaged. The company maintenance may replace the thermostat with an exact replacement and did not set the heat anticipator. This system has five stages of heat with individual sequencers. The amperage draw in the heat anticipator circuit is causing the thermostat to satisfy too early.

The technician arrives at the job and finds the room thermostat set at 75°F, but the indicator on the thermostat shows the room temperature to be 60°F. The technician takes some hand tools, a volt–ohmmeter, and a clamp-on ammeter to the closet where the electric furnace is located. The control panel is removed. Each circuit is checked with the clamp-on ammeter. The first three heaters are drawing current, suddenly they all shut off. The limit switch appears to be shutting them off, so further investigation is prompted. The technician then uses the volt–ohmmeter, set at 50 volts, and checks the voltage at the first sequencer coil. It does not have 24 volts, so the thermostat is not calling. Note that the limit is in the line-voltage circuit. The sequencer must be energized for the system to function.

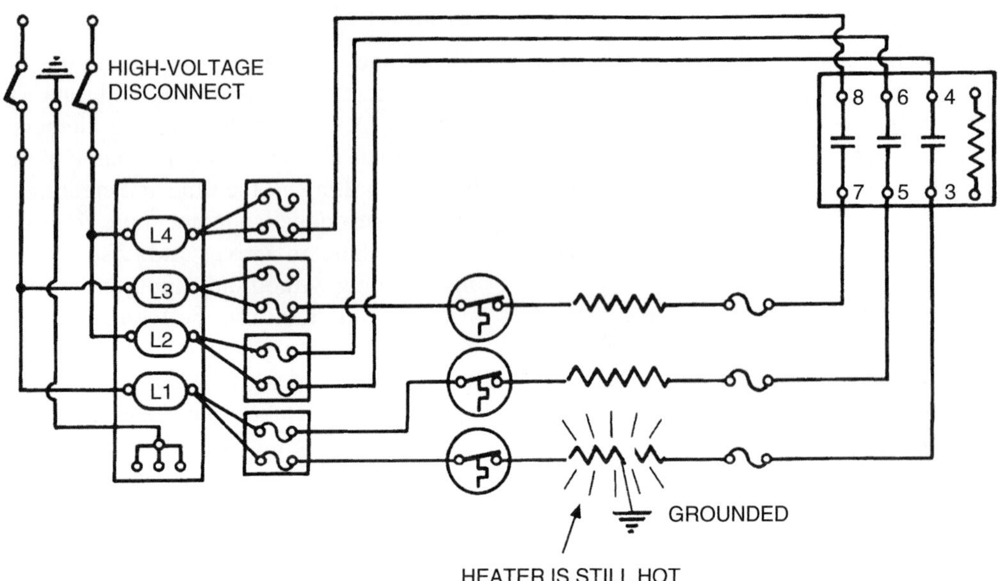

Figure 5-2. Electrical shock is possible on grounded equipment. Follow the safety precautions carefully when working with electrical equipment.

The technician is puzzled and proceeds to the room thermostat. The cover is removed. The technician notices that the bulbs in the thermostat are calling for heat and wonders why there is no power at the sequencer. It is first thought that the thermostat must be defective. While the technician is looking at the room thermostat, it satisfies and the bulbs move to the open position. The temperature indicator is still pointing to 75°F, and the thermostat still indicates 60°F. In a moment, the thermostat bulb swings over for a call for heat. Either this thermostat is defective or the heat anticipator is set incorrectly. The technician decides that the heat anticipator must be set incorrectly.

The technician goes to the furnace and totals the amperage draws for all the sequencers. In this appliance, the fan starts from a set of contacts in the first sequencer, so there are five sequencers. Each sequencer calls for an ampere draw of 0.2 ampere. The total amperage should be 1 ampere (0.2 A × 5 = 1 A).

The technician then shuts off the power and adds an ammeter (10 wraps around the clamp-on ammeter jaws; see Unit 4, Figure 4-34) in the 24-volt circuit that is common to all the sequencers. Power is resumed and the amperage is taken and found to be 1 ampere.

The technician then goes to the room thermostat and sets the heat anticipator to 1 ampere. The technician observes the system, using the ammeter to check for the amperage draw at the heater circuits, and finds the system to be operating correctly.

The technician informs the building manager of the problem and explains that the building will not be warm for several hours because the system must heat the space back to temperature. A call later in the day from the technician to the manager confirms that the system is working correctly.

Service Call 7

A customer at a small building calls and says that there is little heat in the building. The thermostat is set at 72°F, and the temperature indicator shows that the temperature is 62°F. The building temperature is uncomfortable. The problem is that one of the time-delay relays in the boiler is defective and half the heat is not working.

The technician arrives and goes to the boiler room. The boiler is warm but not hot. The leaving water should be 190°F, but it is only 140°F. The weather is cold and 140°F water will not heat the building. The boiler has six heating elements.

The technician removes the control panel and, using an ammeter, checks all the circuits for current draw. The only elements drawing current are 1, 2, and 3. This unit uses contactors and time-delay relays between the contactors to prevent all the elements from energizing at once. The technician uses a voltmeter to trace the circuit and determines that the time-delay relay for circuit No. 4 has power but is not energized. The coil is defective.

The power is turned off and the time-delay relay is replaced. When the power is turned on, all the circuits are energized, one at a time. The boiler repair is complete.

Summary

- The power cord on a plug-in heater must be carefully watched.
- An electric plug-in heater should be operated at the lowest load that will heat the space to reduce the load on the power cord.

- The room thermostat may be the first system component to require maintenance in any system that has a line-voltage thermostat, such as panel or baseboard heat.
- Baseboard heaters may need the covers removed and the dirt and dust vacuumed from time to time, depending on the amount of dust in the area.
- Central forced-air systems move a large amount of air and require regular filter maintenance, once a month until longer periods can be established.
- A central forced-air system also has a fan that requires belt or bearing maintenance.
- The proper method for adjusting the tension on a belt is to use a belt-tension measuring device.
- The wire leading to all heating elements must be inspected from time to time. All wires that are discolored must be changed with a like size wire and insulation value. Contactor contacts may be inspected and replaced when they become pitted or burned.
- Heating elements can be checked with a clamp-on ammeter to make sure that the heater is pulling current.

Review Questions

1. What is the weak point in a room space heater?
2. Why is the room thermostat the most likely part to fail on a panel heating system?
3. What maintenance may be performed on a line voltage thermostat?
4. What maintenance may be performed on baseboard heat units?
5. What is the best method for adjusting belt tension on a belt drive motor?
6. What can be done when loose burned connections are discovered on an electric heater?
7. What is the best method for determining which electric heating elements are not functioning?

6 Theory of Gas and Combustion

Objectives

Upon completion of this unit, you should be able to

- **discuss the origins of natural gas and LP gas.**
- **describe how gas is distributed to the customer.**
- **describe pressure.**
- **express units of pressure.**
- **explain the properties of natural, LP, and LP–air gases.**
- **have a working knowledge of gas combustion and burners.**
- **describe the condensing by-products of combustion.**

6.1 History of Natural Gas

History shows many experiences with natural gas before it was recognized as a fuel for heat and light. Natural gas from deposits deep within the earth sometimes seeped to the surface, and were noticed as places on the earth's surface that would burn. In ancient times, people worshipped these "eternal flames." Later, George Washington purportedly bought some land in Virginia that had a "burning spring," a natural gas deposit that burned when it was ignited. This type of gas deposit—when the gas reaches the surface of the earth—is rare because the gas is usually deep within the earth.

Around 1821, some children in upstate New York ignited an aboveground natural gas leak and reported it to their parents. William Hart from Fredonia, New York, drilled a hole in the ground, and the first natural gas well was established. Soon the town used natural gas for lighting.

Since then, new methods of discovering natural gas have been developed. In addition, pipeline technology has allowed the natural gas, which is cleaned, dried, and slightly refined, to be piped across the nation. Long-distance lines with pumping stations along the way keep the pressure high enough to pump gas to most large communities. Research through the years has also improved the efficiency and safety of natural gas appliances.

Two relatively new sources of small amounts of natural gas are landfills and farms. In landfills, Figure 6-1, the garbage, which contains much animal and vegetable matter, rots. The rotting process generates methane (natural gas), which rises off the garbage, Figure 6-2. Some cities are currently exploring the possibility of using this gas for heating.

Waste from cattle has also been used to generate small amounts of methane in some cases. The waste is dumped into a tank, where it decays, and methane gas is generated,

Figure 6-1. Garbage is dumped in landfills.

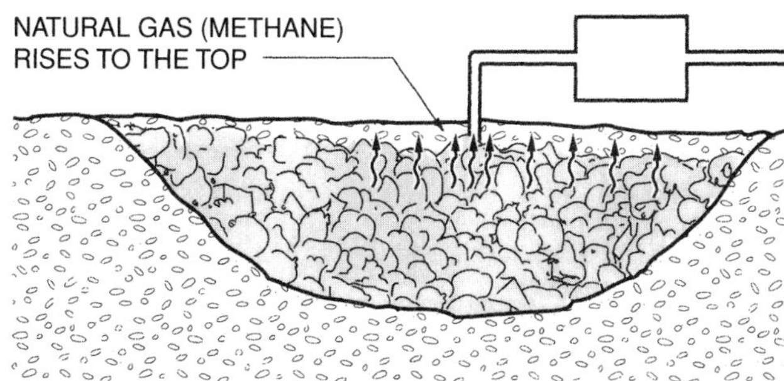

Figure 6-2. The garbage in the landfills contains animal and vegetable matter that decomposes into methane deposits like the natural gas deposits of the past.

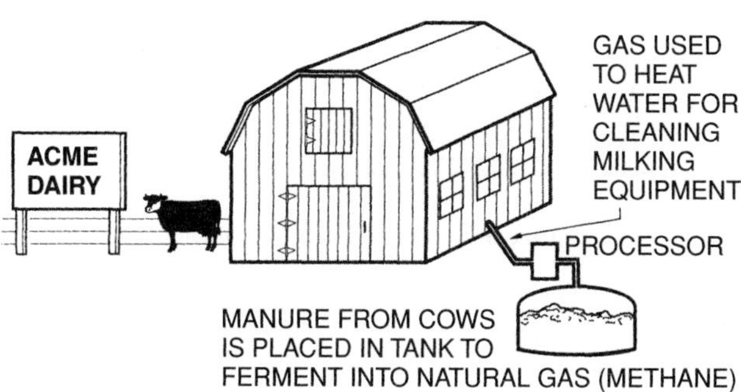

Figure 6-3. Methane can also be generated from farm waste.

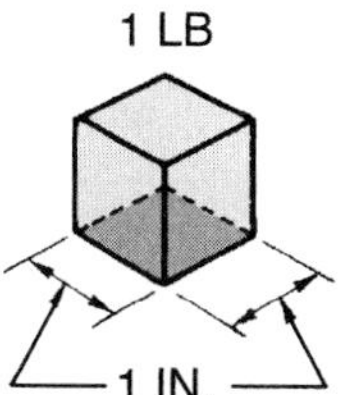

Figure 6-4. This weight is 1 lb and it is resting on a surface of 1 square inch, this represents a pressure of 1 lb per square inch.

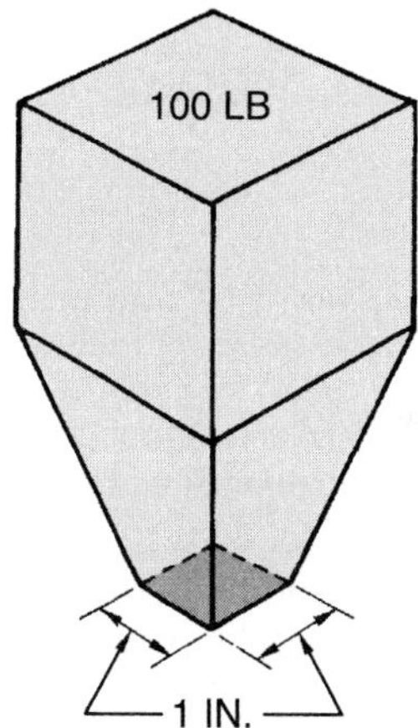

Figure 6-5. This 100 lb weight is resting on a surface area of 1 square inch and represents a pressure of 100 lb per square inch.

Figure 6-3. The farmer must be extremely careful with the tanks because the gas is highly explosive. Specially designed systems such as the one shown in Figure 6-3 use this farm waste.

Gas is used for many applications. In our field, it is used to heat homes and other buildings and to fire the furnaces of industry. Large reserves of natural gas are still in the earth, so it will be one of the fuels used for heating for many more years.

6.2 Pressure

The text in the previous section referred to pressure and how it must be maintained to move gas cross country. Pressure is defined as force per unit of area. The terms most often used are pounds for force and square inch for area. This is commonly abbreviated to psi (pounds per square inch). It is important that the technician understand pressure terminology.

See Figure 6-4 for an example of calculating psi in which a 1-pound weight is resting on a 1-square-inch area. Another example is shown in Figure 6-5: a 100-pound weight is resting on a 1-square-inch surface, resulting in 100 psi.

The previous example seems to imply that pressure always has a downward force, but this is not the only direction that pressure pushes. In a water pipe to a home, the water pressure is also expressed in psig (pounds per square inch gage; because a gage is used to measure the pressure). The typical water pressure at a home should be about 50 psig. That means that the pressure of the water pushing outward is 50 pounds per square inch of surface area of the pipe. It is the same amount of force as used in the example of a weight pressing down, only the weight (pressure) is pushing outward. Knowledge of these forces is used to design and build vessels and piping that hold all fluids under pressure. The water pipe in a residence is typically designed for 125 psig.

You may have noticed that when you go swimming and go to the bottom of the pool, you feel pressure on your body. The water exerts a downward pressure due to its weight. Water weighs 62.4 pounds per cubic foot, so it exerts a downward pressure of 62.4 pounds per square foot. There are 144 square inches on the bottom of the cubic foot, Figure 6-6. This means that a pressure of 0.433 psi ($62.4/144 = 0.433$) is exerted downward for any column of water that is 1 foot high. Remember, this is 0.433 pounds per square inch for a column of water 1 foot high. A column of water that is 10 feet high would exert a pressure of 4.33 pounds per square inch. This number will be important to you for use in this chapter and future chapters.

Air has weight also, and we need to learn how it fits into the pressure application.

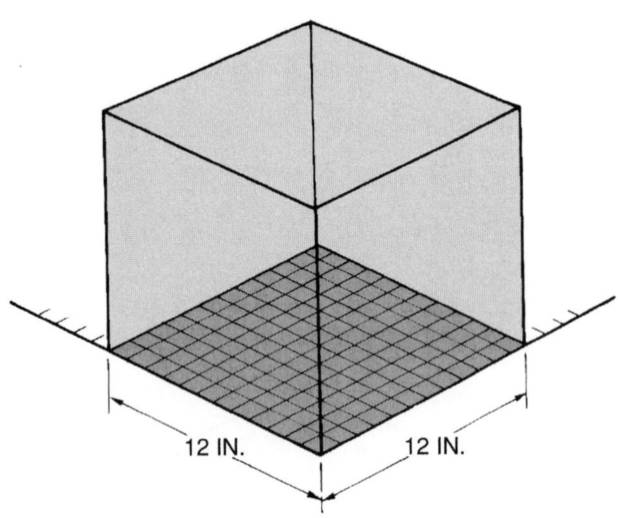

Figure 6-6. One cubic foot (ft^3) of water exerts a pressure downward of 62.4 lb/ft^2 on the bottom surface of a cube.

We live at the bottom of an ocean of air called the atmosphere. We are not aware of the weight of this air until we fly in a plane and rise through the atmosphere. The atmospheric air has a weight of 14.696 psi at standard conditions at sea level. As we rise up in the atmosphere, the air has less weight, and less pressure is pushing down on us. When flying in a plane or climbing a mountain in a car, our ears are the first to feel the pressure difference. We can feel our ears "pop" as we climb. Yawning or chewing gum helps equalize the pressure. The pressure on the inside of our bodies must be equalized to the outside as the pressure is reduced.

The atmosphere's pressure of 14.696 psi can be converted to other units of measure. A column of standing mercury (Hg) is often used to measure pressure. The atmosphere can support a column of mercury (Hg) 29.92 inches high, Figure 6-7. The instrument shown in Figure 6-7 is called a mercury barometer. The weather reporter often refers to atmospheric pressure in terms of inches mercury. The metric term is *millibar* (14.696 psi is equal to 1013.25 millibars or 1.01325 bars).

When air, water, or other fluids are pressurized above atmospheric, they are considered high pressure, even though the pressure may not be much above atmospheric. Natural gas is piped across the country at high pressures to the final point of utilization, where the pressure is reduced to what is known as working pressure. This working pressure is just above atmospheric and is expressed in inches of water column. When the pressure will support a column of water 1 inch high, it is known as 1 inch of water column (in wc). Inches of water column is used to measure very small pressures and pressure differences. A typical natural gas burner operates at a pressure of 7 in wc.

A gauge called a Bourdon tube gauge is a common pressure gauge used by technicians. This gauge is an enclosed flexible brass tube that distorts when pressure is

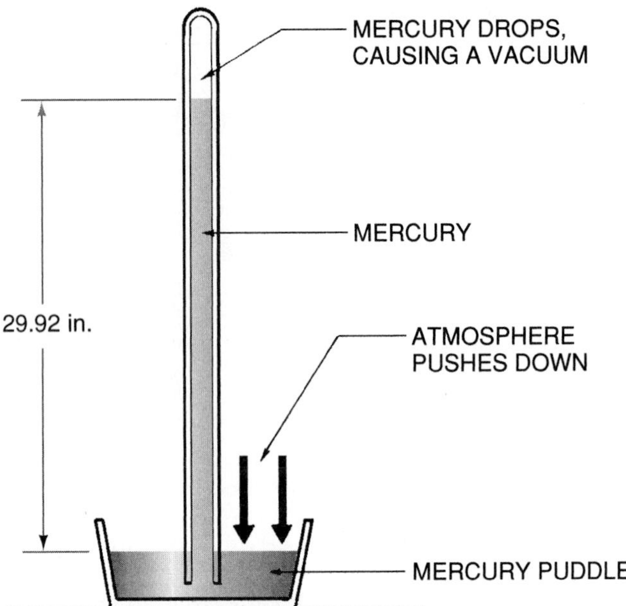

Figure 6-7. Mercury (Hg) barometer.

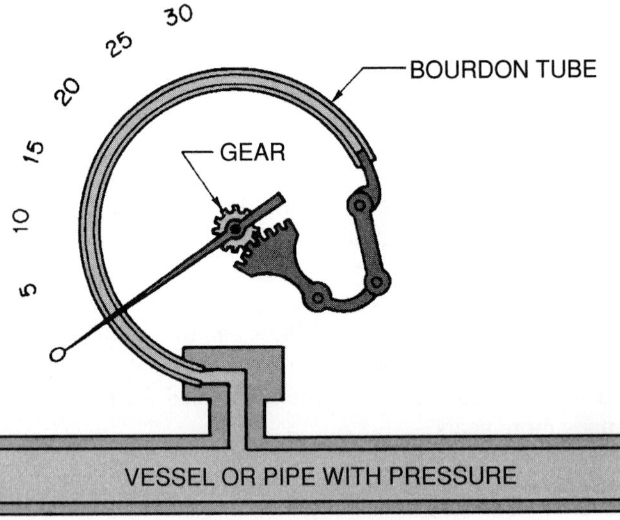

Figure 6-8. The Bourdon tube is made of a thin substance such as brass. It is closed on one end, and the other end is fastened to the pressure being checked. When pressure increases, the tube tends to straighten out. When attached to a needle linkage, pressure changes are indicated with the needle.

added. This distortion is passed through gears, where it is registered with a needle, Figure 6-8.

The pressures that we have talked about up to now are pressures above atmospheric. The pressure from absolutely no pressure to atmospheric pressure is called absolute pressure. Engineering calculations are all done in absolute pressure. These calculations start at 0 pounds per square inch absolute (psia). Atmospheric pressure that we talk about is 0 psig, or 14.696 psia. The technician may use a common pressure gauge to record a pressure. This gauge reads 0 at atmospheric pressure. To change that pressure reading on the gauge to absolute, the technician must add 14.696 to the gauge reading. For example, when a gauge reads 30 psig, the reading can be adjusted to absolute by adding atmospheric pressure to the gauge reading, Figure 6-9. The absolute pressure is 44.696 psia.

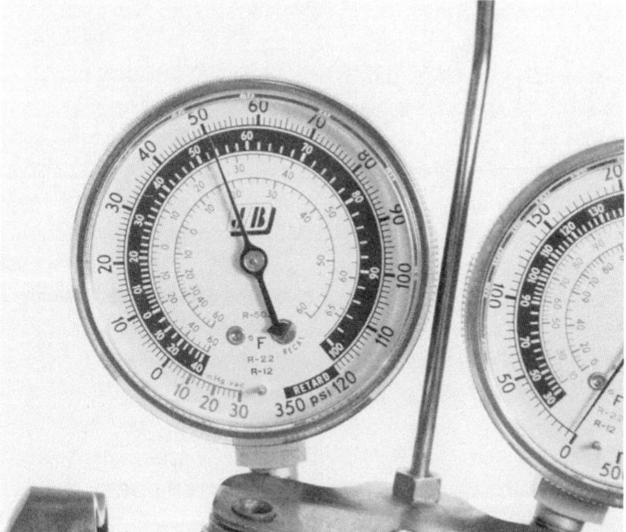

Figure 6-9. The gage reads 50 pounds per square inch (psig). To convert this gage reading to pounds per square inch absolute (psia), add 50 psig to 15 (rounded off atmospheric pressure) for a sum of 65 psia. *Photo by Bill Johnson*

Figure 6-10. A draft gage. Used to measure very small pressures in chimneys or flues. *Courtesy Bacharach, Inc. Pittsburg, Pa*

One must often measure pressure that is just below atmospheric. This is called a vacuum. Vacuum in the combustion industry is measured in inches of water vacuum. It is measured with a meter called a draft meter, Figure 6-10. This instrument reads in hundredths of an inch of water column.

6.3 Properties of Gas

The two gases commonly used for heating are natural gas and LP (liquefied petroleum) gas.

Natural Gas

Natural gas, or methane, is made of carbon and hydrogen, and its chemical formula is CH_4, Figure 6-11. Different gases have different chemical formulas, but all the fuel gases contain carbon and hydrogen. These atoms of carbon and hydrogen vary from gas to gas; the more atoms in the gas, the more the gas weighs.

The heat content varies slightly from one locality to another for methane. Some gas companies say that the heat content of their gas is 1,075 Btu per cubic foot and some say 950 Btu. The difference between the two is slight. Even the heat content of gas from the same company may vary slightly. Therefore, the measure most accepted for calculating gas consumption is 1,000 Btu per cubic foot at the typical burner manifold pressure of 3.5 inches of water column (wc). The quality of the gas cannot be more precise than this without greater expense.

The typical gas appliance is 80% efficient, which means that 20% of the heat from the gas escapes with the by-products of combustion. If the gas has a heat content of 1,000 Btu per cubic foot, 800 Btu of heat are usable in a typical application and 200 Btu escape with the by-products of combustion, Figure 6-12. This loss of heat is necessary in a typical appliance because this heat is needed to push the by-products of combustion from the furnace to the outside, where they are vented. Some of this heat can be saved with special appliances called *condensing furnaces*.

Natural gas is in a vapor state when it is moved across the country in pipelines and when it is burned at the point

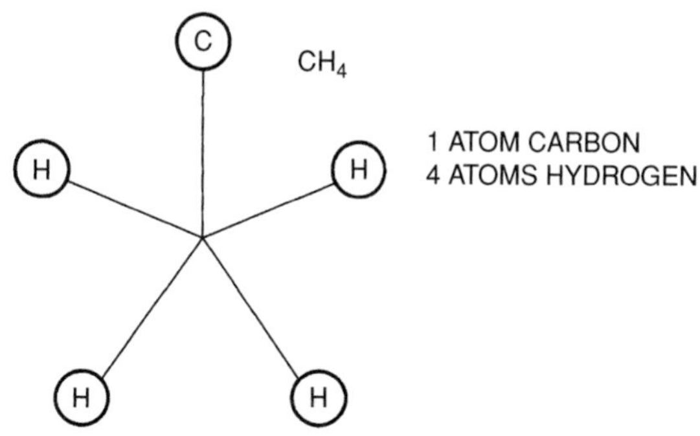

Figure 6-11. Methane consists of carbon and hydrogen.

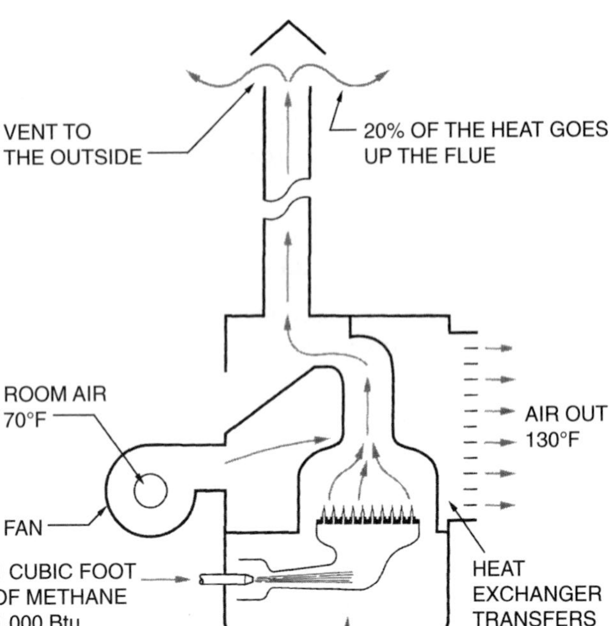

Figure 6-12. The typical gas appliance is 80% efficient. Twenty percent of the heat pushes the by-products of combustion up the flue and outside.

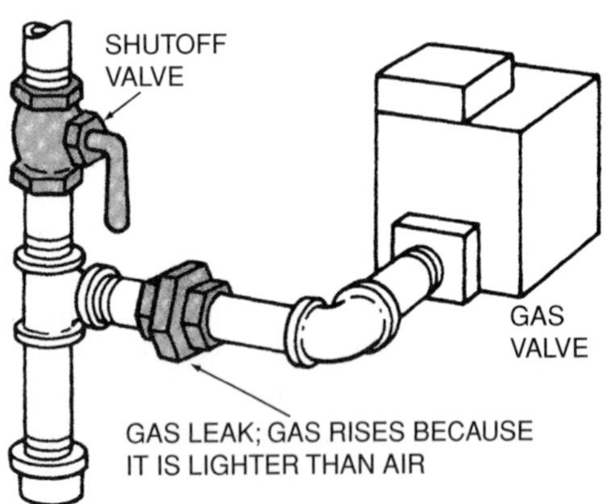

Figure 6-13. Methane is lighter than air.

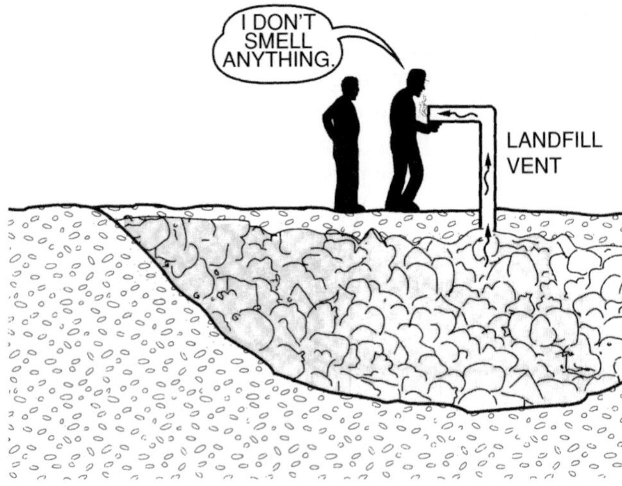

Figure 6-14. Pure natural gas is colorless, odorless, and tasteless.

Figure 6-15. An odorant that smells like rotten eggs is added to methane so that leaks can be detected.

of consumption. The specific gravity of natural gas is 0.6. Specific gravity is a comparison of the weight of a substance to another substance. In this case, it is the comparison of natural gas to the weight of an equal volume of dry air. Natural gas weighs 0.6 as much, or 60% times the amount of air. This is significant because if a natural gas leak occurs, the gas rises into the air, Figure 6-13. Whether a particular gas rises or falls determines the type of safety device that will be used to protect the consumer from explosions in the event of a gas leak.

Natural gas is colorless, odorless, and tasteless in its natural form. If a leak occurred, you could not detect it, Figure 6-14. Consequently, for the consumer's safety, the gas is made detectable. The gas company adds a strong odorant, which smells much like rotten eggs, to the gas so that even a small leak can be detected, Figure 6-15. The consumer should quickly call an expert when a gas leak is suspected.

LP Gas

LP gas can be found in two forms: propane and butane. Both are refined from crude oil and for many years were burned at the refinery as useless by-products. Today, both are valuable gases. LP gas is shipped in a liquid state in pipelines or trucks to the point of consumption. Vapor is removed from the top of the storage tank for the purpose of burning, Figure 6-16. As the vapor is drawn off, more of the liquid boils to a vapor to replace the withdrawn vapor and to keep the pressure high enough to move vapor to the burner section.

PROPANE. Propane is the most popular heating fuel in rural colder climates because the vapor pressure there is higher. It has a vapor pressure of 64 psig (pounds per square inch gauge) when the liquid is 40°F. Propane maintains a satisfactory vapor pressure in temperatures as low as approximately –30°F (5.7 psig). This is significant because LP gas is stored outside in a tank. When the outdoor temperature falls, so does the pressure inside the tank, Figure 6-17. Propane maintains a vapor pressure in temperatures as low as –43.7°F; below this temperature, the tank goes into a vacuum, Figure 6-18. The vapor pressure may be kept high by putting the tank underground, where the temperature is more constant and is warmer in the winter, Figure 6-19.

The heat content of propane differs from that of natural gas. Propane normally burns at a manifold pressure of 11 inches of wc and has a heat content of 2,500 Btu per cubic foot, Figure 6-20. It is also heavier than natural gas or air. Its chemical formula is C_3H_8, and its specific gravity is 1.5, so it is 1.5 times heavier than air. Propane also is colorless, odorless, and tasteless in its pure state.

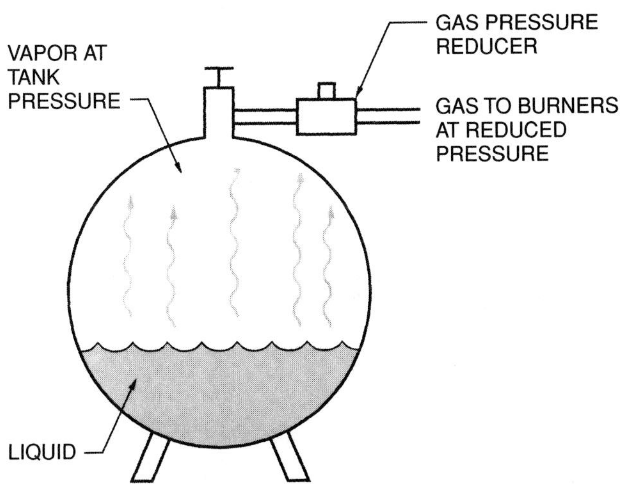

Figure 6-16. Vapor is removed from the top of liquid propane. The liquid then boils and creates more vapor.

Figure 6-17. When the outside temperature is low, the pressure of the tank containing LP gas is also low.

Figure 6-18. When the temperature of the tank falls below –43.7°F, the tank pressure is reduced to a vacuum. No gas can leave the tank when it is in a vacuum.

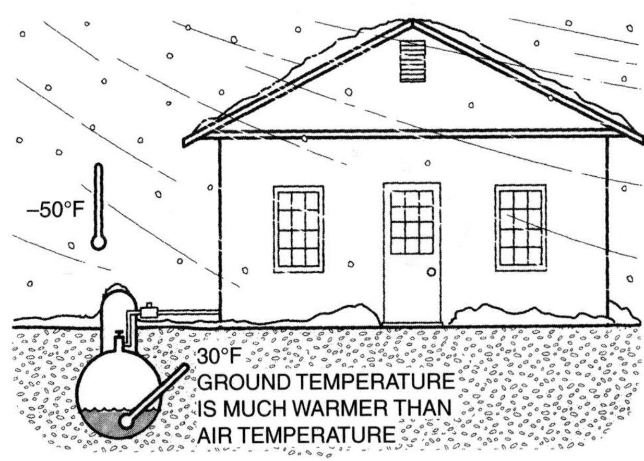

Figure 6-19. Placing the tank underground helps keep the temperature of the tank warmer.

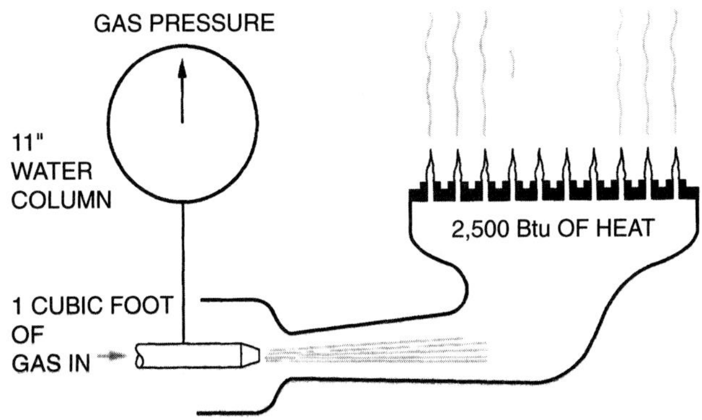

Figure 6-20. Propane heat content.

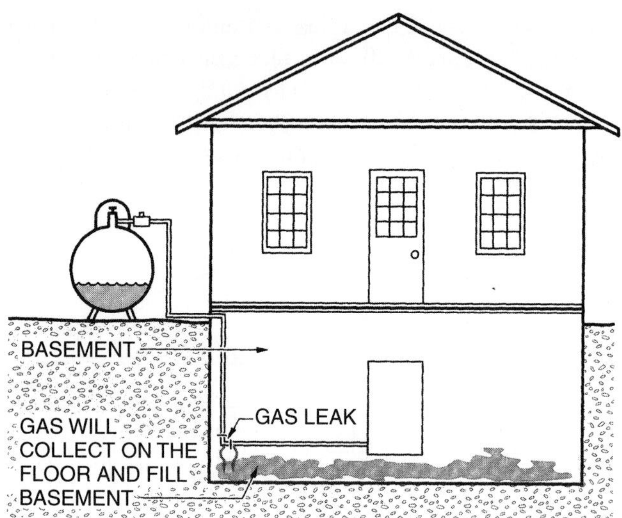

Figure 6-21. Because it is heavier than air, propane tends to fall when a leak occurs.

An odorant that smells much like wild onions is added to it and is evident when even a small leak is present. When a leak occurs, propane tends to stay close to the ground and to accumulate in low places, Figure 6-21.

BUTANE. Butane is similar to propane in that it is a liquefied petroleum gas, but it has some different characteristics. It has a vapor pressure of 2.9 psig at 40°F. When the temperature drops, the pressure also drops. When the temperature reaches 31°F, the pressure goes to 0 psig; below 31°F, the storage container goes into a vacuum. Even in the milder climates where heat is required, butane vapor pressure is too low to pressurize a gas line without some assistance. In some areas, a heater is applied to the bottom of the tank to keep the vapor pressure high, Figure 6-22. Tanks holding butane can also be stored underground, Figure 6-23.

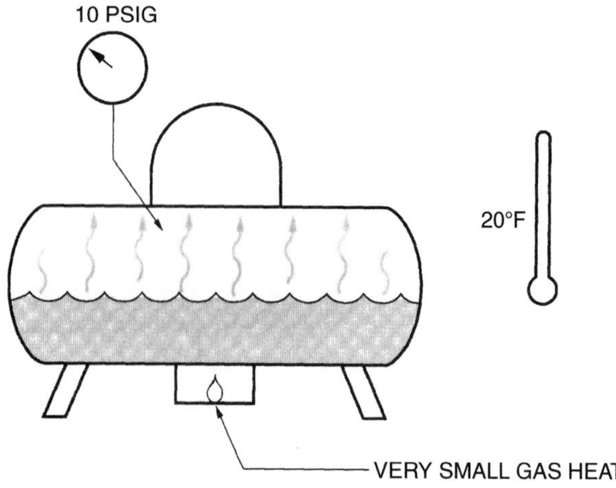

Figure 6-22. This butane tank is outfitted with a heater to keep the pressure of the tank up.

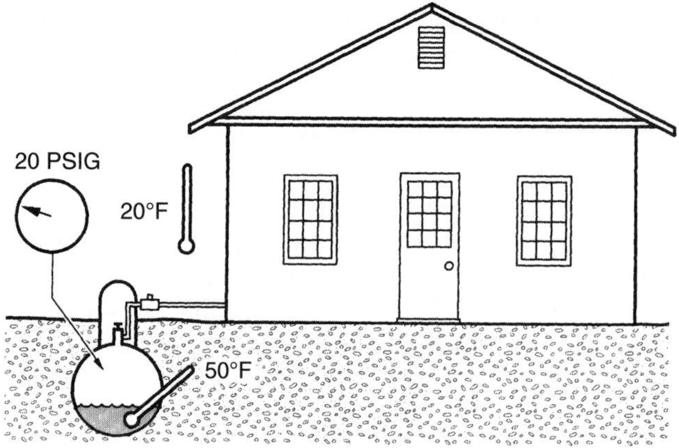

Figure 6-23. Butane tanks can also be placed underground.

The chemical formula for butane is C_4H_{10}. Butane has a heat content of 3,200 Btu per cubic foot at a manifold pressure of 11 inches of wc. This is more heat content than propane, but the disadvantage of butane is that its vapor pressure is too low in cold climates. Butane also is a colorless, odorless, and tasteless gas in its pure state, so the same odorant that is used for propane is used to make butane detectable. The specific gravity of butane is 2.5. This means that it is 2.5 times heavier than air.

MIXTURES. Propane can be mixed with butane to produce heat and pressure values between those of the two gases. For example, a mixture of 50% propane and 50% butane yields a higher fuel heat value than that of propane and a higher vapor pressure than that of butane. The fuel heat value of this mixture is approximately 2,850 Btu per cubic foot ([2,500 × .50 = 1,250 Btu/cu ft] + [3,200 × .50 = 1,600 Btu/cu ft] = 2,850 Btu/cu ft). This is a considerable increase in fuel content per cubic foot of gas. The vapor pressure for a 50/50 mixture is 33.45 psig at 40°F ([2.9 psig for butane × .50 = 1.45 psig] + [64 psig for propane × .50 = 32 psig] = 33.45 psig total pressure). The gas company creating this mixture must ensure that the mixture does not exceed the working pressure for a tank designed for butane.

Propane or butane can also be mixed with air to form either *propane–air gas* or *butane–air gas*. The correct combination of air and LP gas produces a mixture that is similar to natural gas. Natural gas companies often use this mixture as a standby gas in extremely cold weather when the natural gas pressure in the gas mains may drop because of demand or extra use. These companies may have a tank farm of propane and a mixing plant to mix propane–air gas and inject it into the gas main during the peak season, winter. The consumer cannot tell the difference between natural and propane–air gas.

Propane vapor can be mixed with 45% air to create the propane–air mixture, which yields a heat content of approximately 1,400 Btu per cubic foot. The specific gravity of propane–air gas is 1.28, which means that it is 1.28 times heavier than air.

The specific gravities of the various gases and their mixtures determine how the gases flow in pipes and the sizes of the metering devices to be used at the burners. (This is discussed in more detail next.)

6.4 The Combustion Process

Combustion, also called *rapid oxidation*, is the burning of a substance. In this case, the substance is a fuel gas. Combustion is a chemical reaction between the gas and oxygen, which supports the combustion. Oxygen by itself does not burn; it requires a combustible gas for combustion. Consequently, three things must be present for combustion to take place: fuel, oxygen, and heat. If any one is missing, combustion does not occur, Figure 6-24. Methane (natural gas) is used in the following discussion of combustion.

For perfect combustion to occur with methane gas, the correct mixture of oxygen and methane must be present. To burn 1 cubic foot of methane with perfect combustion, 2 cubic feet of oxygen are required. Pure oxygen is rarely used for combustion of methane; air is used. As shown in Figure 6-25, air is 21% oxygen and 79% nitrogen. For every 5 cubic feet of air, there is 1 cubic foot of oxygen, Figure 6-26. To have 2 cubic feet of oxygen for perfect combustion of 1 cubic foot of methane, we must have 10 cubic feet of air, Figure 6-27. The nitrogen in the air passes through the combustion process as a by-product, Figure 6-28.

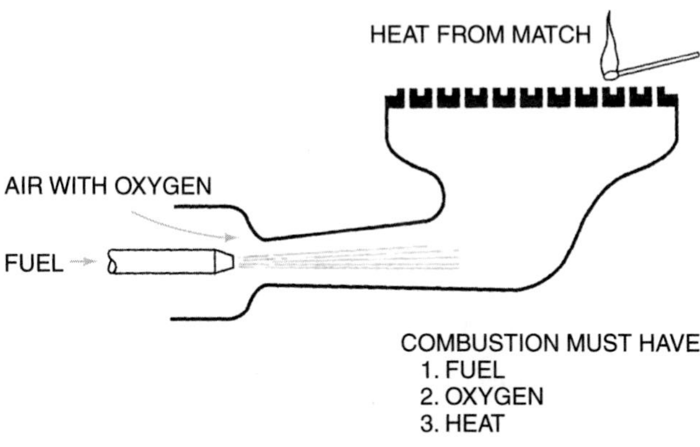

Figure 6-24. Fuel, oxygen, and heat are required for combustion to occur.

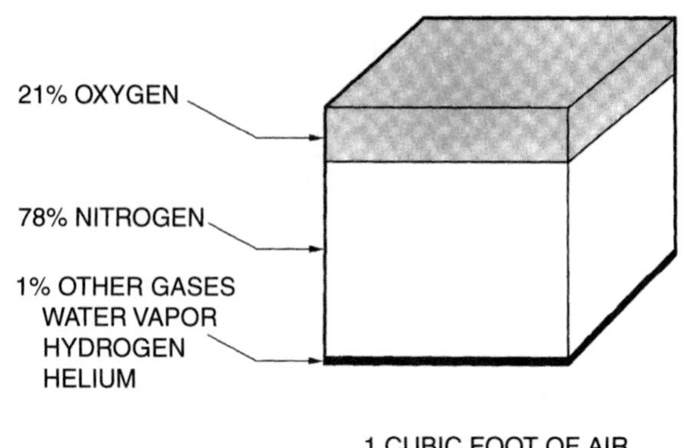

Figure 6-25. The contents of air.

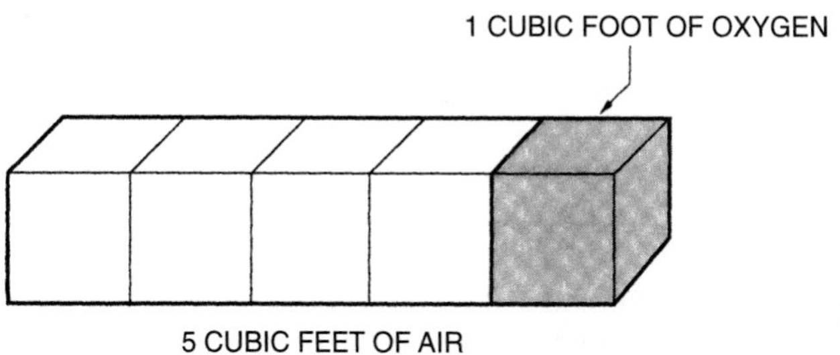

Figure 6-26. There is 1 cubic foot of oxygen for every 5 cubic feet of air.

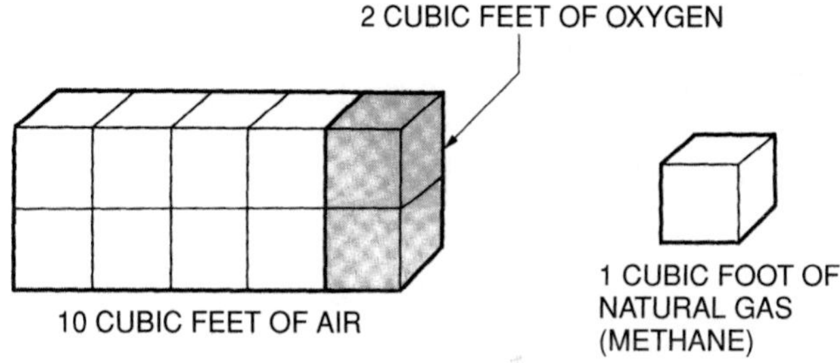

Figure 6-27. For perfect combustion, 2 cubic feet of oxygen are required per 1 cubic foot of methane.

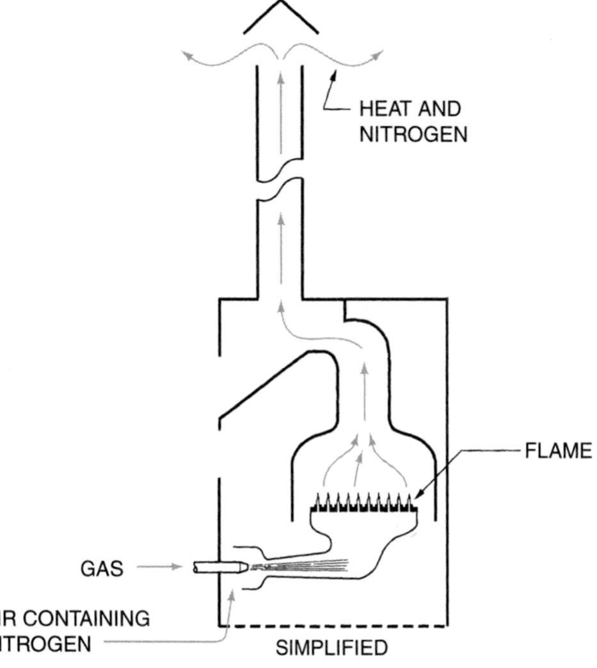

Figure 6-28. The nitrogen in air passes through the combustion process.

Perfect combustion is not practical because of critical air adjustment. If you obtained perfect combustion with the exact amount of air, and the air supply were reduced for some reason, you would then have imperfect combustion and a hazard. The problem of critical air adjustment is solved with an excess of air. For practical combustion, most manufacturers use 50% excess air. In actual practice, 15 cubic feet of air are supplied to the combustion process of methane for each cubic foot of gas, Figure 6-29.

Perfect combustion by-products for natural gas would be 8 cubic feet of nitrogen, 1 cubic foot of carbon dioxide (CO_2), 2 cubic feet of water vapor, and heat, Figure 6-30. The other gases would have the same by-products as natural gas but in different quantities.

Propane and butane gas both require more air for burning than methane does, Figure 6-31. For perfect combustion, propane requires 23.5 cubic feet of air per cubic foot of gas, and butane requires 30 cubic feet of air per cubic foot of gas. When excess air is added, these gases require considerably more (50%) air, but they also yield more heat. Propane—air gas requires 13 cubic feet of air per cubic foot of gas, and butane—air gas requires 11 cubic feet of air per cubic foot of gas. The air quantity for the propane—air and butane—air gases is closer to the air quantity for natural gas.

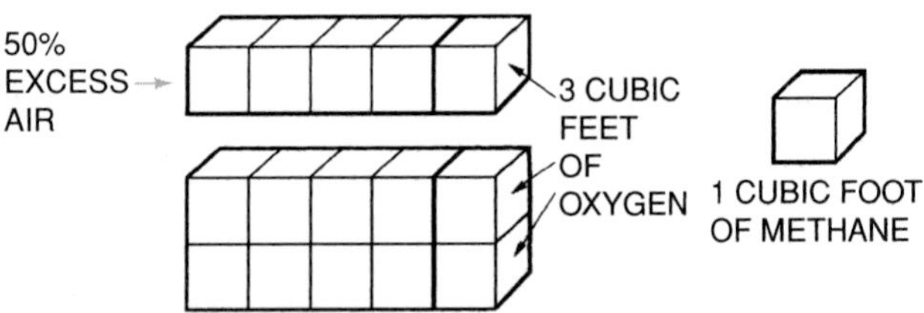

Figure 6-29. For practical combustion, 50% excess air is used, so 15 cubic feet of air are used with 3 cubic feet of oxygen; 1 cubic foot of oxygen is not burned.

Other characteristics of the gases are important. The ignition temperature of the gas–air mixture is an example. This temperature determines which type of device can be used to light the flame for the gas, and it differs for each of the three gases discussed in this text. The ignition temperature for methane is 1,170°F. Propane and butane both have lower ignition temperatures: 900°F for propane and 825°F for butane.

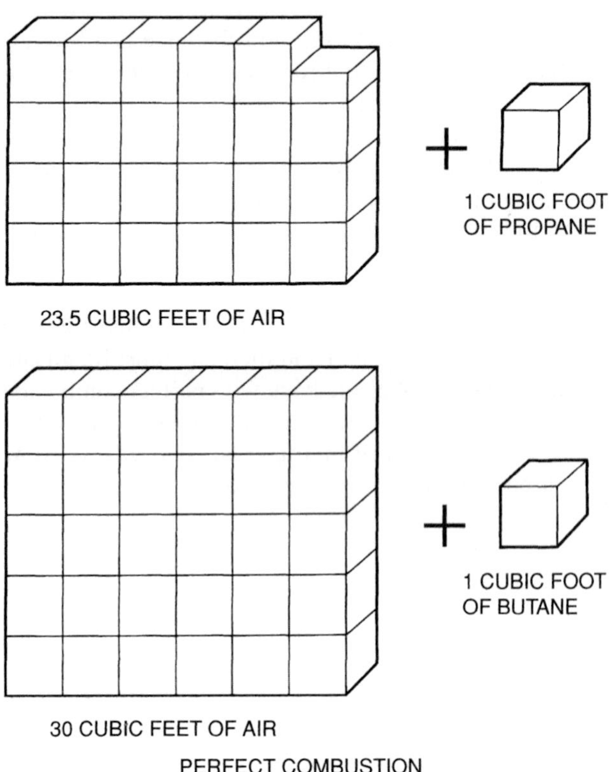

Figure 6-31. Propane and butane require more air for perfect combustion than methane does. (For practical combustion, they require approximately 50% more air than they require for perfect combustion.)

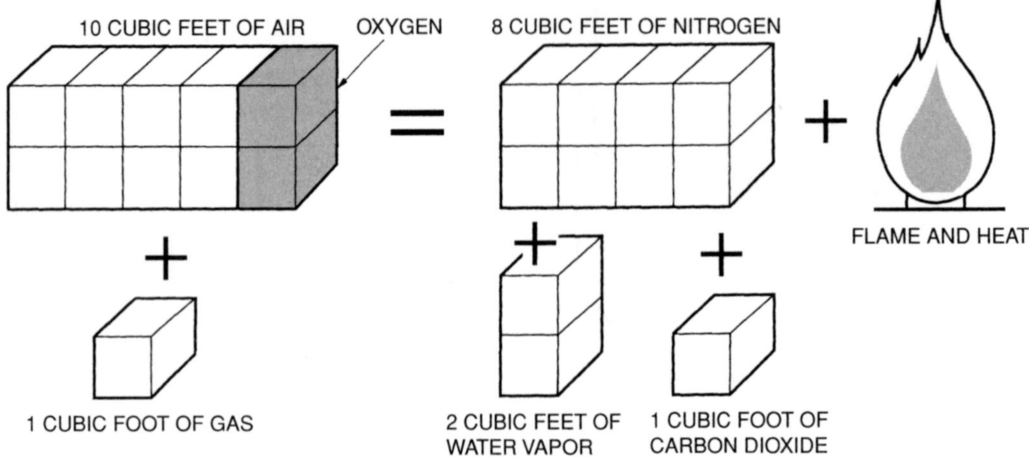

Figure 6-30. The by-products for perfect combustion.

The limits of flammability of the gas describe the lower and upper mixture percentages of gas to oxygen that are required for the mixture to actually ignite, or burn. When the mixture is too lean, has too little gas, it does not burn. When the mixture is too rich, has too much gas, again it does not burn. The limits of flammability are different for the different gases, Figure 6-32.

The rate at which the flame travels is also a consideration in the design of the burners. The flame travel is 25 inches per second for methane, 32 inches per second for propane, and 33 inches per second for butane, Figure 6-33. The flame travel is influenced by the gas–air mixture; thus, if the mixture is not correct, initial ignition of the flame may be delayed. Hydrogen has the fastest flame rate of any gas, so the hydrogen content of the gas determines the flame rate at the correct air–gas mixture. The more hydrogen atoms in the gas, the faster the flame rate.

Combustion is accomplished with burners that combine the gas with the needed air. Two types of burners are used, the atmospheric burner and the power burner, which are shown in Figure 6-34. The atmospheric burner is the most common type used for systems that require millions of Btu per hour and is discussed in detail in this book. The power burner uses a fan to force air into the burner and is used in large applications.

The atmospheric gas burner, Figure 6-35 operates at atmospheric pressure with the air induced into the burner by the gas stream. The burner is made up of a gas orifice (often called *a spud*); a primary air intake; a venturi, or tube, that uses the gas velocity to increase the air velocity; a mixing chamber; and burner ports. The orifice meters the gas into the venturi; the gas induces air into the burner

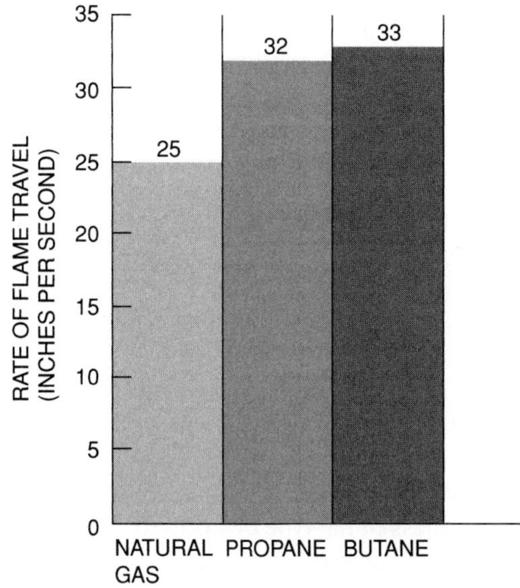

Figure 6-33. Rate of flame travel at perfect combustion.

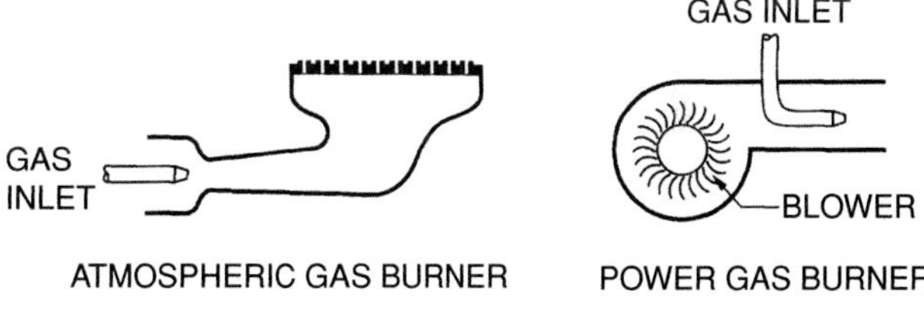

Figure 6-34. The two types of gas burners.

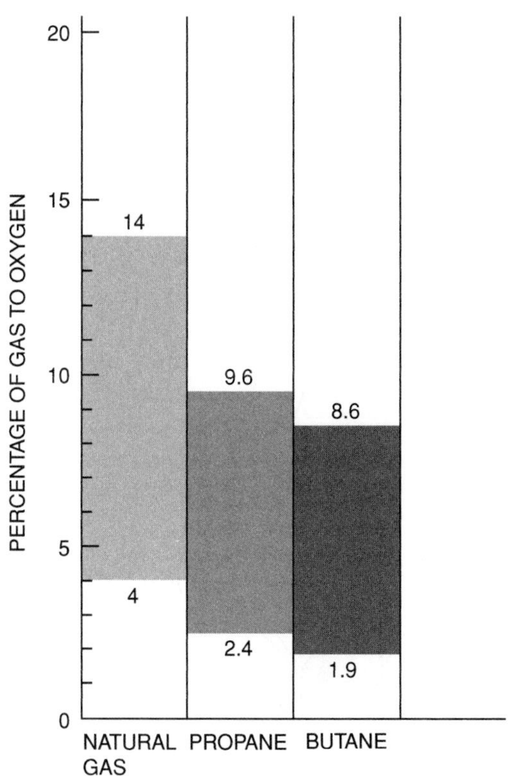

Figure 6-32. Limits of flammability for the fuel gases.

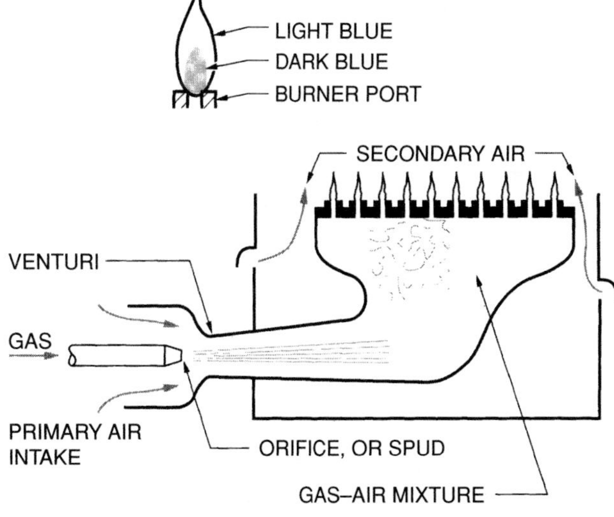

Figure 6-35. The components of an atmospheric gas burner.

at atmospheric pressure through the primary air shutter at the orifice. The gas and air are mixed in the mixing chamber, and the velocity of the gas moves the mixture to the burner ports.

Most burners have a primary air adjustment as the only adjustment. The orifice is a fixed size, so the primary air can be adjusted by one of several methods. The air shutter simply acts as an adjustable valve for the air supply, Figure 6-36. Another adjustment is an air spoiler for reducing the velocity of the gas in the mixing tube, therefore reducing the amount of primary air that is induced with the gas stream, Figure 6-37.

When the gas and air mixture reaches the burner ports, the mixture is ignited. Usually ignition is started with a pilot burner. Once ignition takes place, the flame continues to burn as long as air, gas, and heat are present. The burner port has a secondary air supply that is in the vicinity of the burner, Figure 6-38.

The flame at the burner port should have some distinctive characteristics when combustion is correct. The flame should be firm, not wavering and should sit tight to the burner port. The flame should also have shades of blue, and orange streaks, regardless of the type of gas used, Figure 6-39. The orange streaks are caused by dust particles in the primary air, Figure 6-40. Each shade of blue represents a different stage of the flame. The inner core is light blue, whereas the outer mantle is a darker blue, Figure 6-41. A flame with characteristics different from these is an indication of a poor combustion problem.

Two types of poor combustion, incomplete combustion and combustion with excess air, have different flame characteristics. Incomplete combustion is a result of lack of air, or oxygen starvation. The flame is yellow and lazy looking, and it is not firm, Figure 6-42. This type of flame emits an oxygen-starved by-product, carbon monoxide (CO), instead of carbon dioxide (CO_2), which is emitted in good combustion. Carbon monoxide is a poisonous gas that is colorless, odorless, and tasteless. Yellow flames must be corrected. In addition to the carbon monoxide produced, these flames sometimes smoke and cause soot to form in the appliance. When incomplete combustion occurs, the reason for

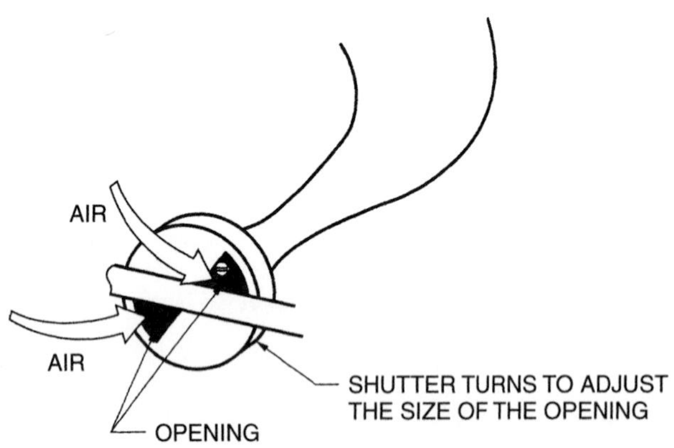

Figure 6-36. Primary air shutter adjustment.

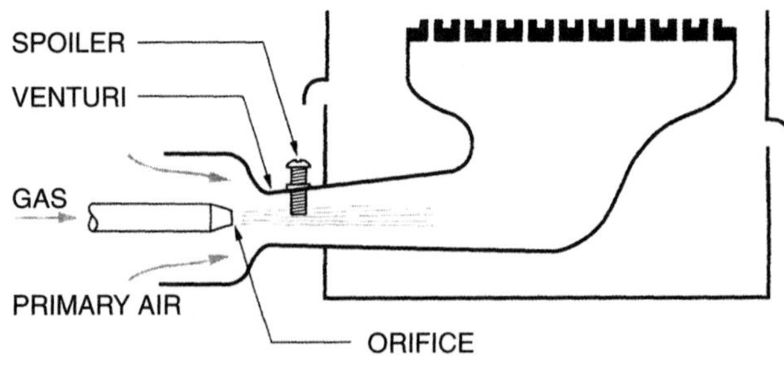

Figure 6-37. Primary air spoiler adjustment.

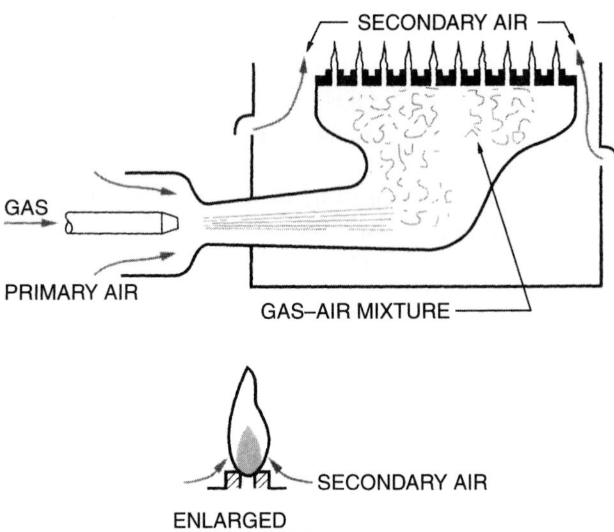

Figure 6-38. Secondary air is furnished at the burner port.

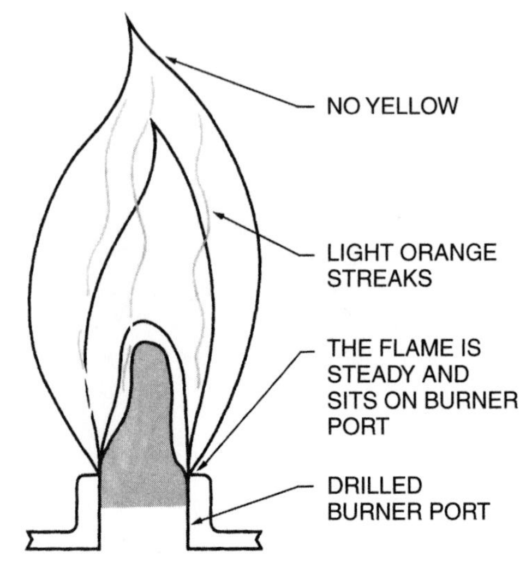

Figure 6-39. Proper flame characteristics.

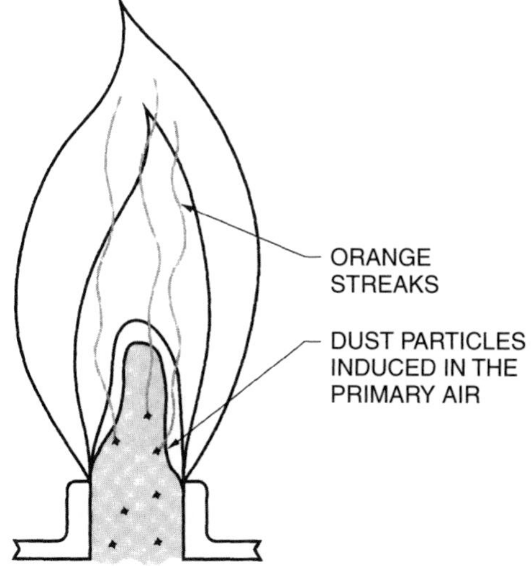

Figure 6-40. Orange steaks caused by dust particles. The dust originates from the primary air in the room.

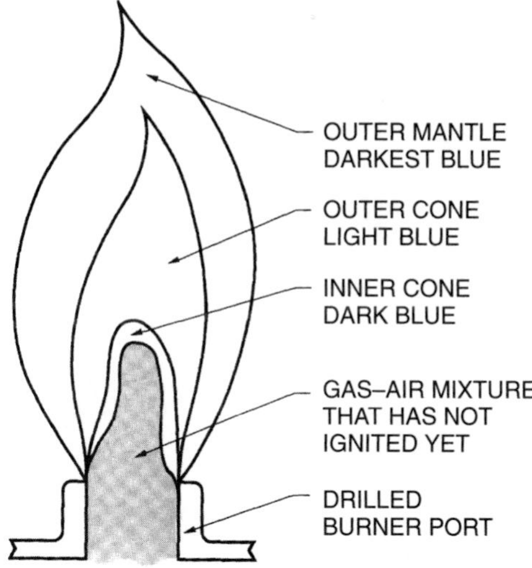

Figure 6-41. The correct flame is several shades of blue.

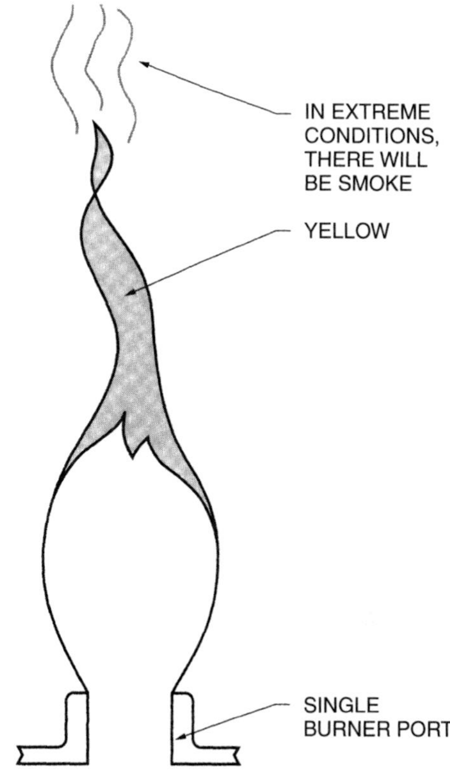

Figure 6-42. A flame with incomplete combustion is lazy and yellow.

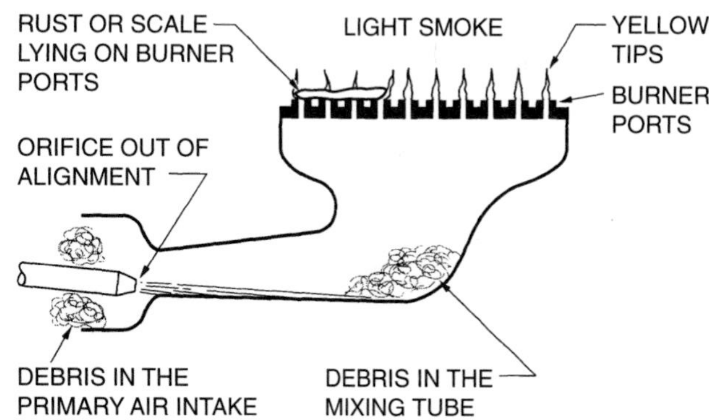

Figure 6-43. Causes of incomplete combustion.

insufficient air must be found. The problem can be either primary or secondary air starvation, Figure 6-43. Fortunately, when a gas flame burns in an oxygen-starved condition, other by-products in the gas, called *aldehydes*, give off an offensive odor smelling like formaldahyde. These aldehydes result from the impurities in the gas that burn when the gas burns, Figure 6-44. Note that if a gas has no aldehydes, there is no smell when carbon monoxide is present.

When the flame has too much air, it tends to lift off the burner port and make a noise like that of a torch, Figure 6-45. The burner should be allowed to burn for a few minutes and get up to temperature before excess air is diagnosed. This condition is usually solved by adjusting the primary air.

Combustion analysis for gas-burning appliances using atmospheric gas burners is not critical in modern factory-built gas appliances when methane gas is used. Many years ago, the manufacturers simplified the burners to the point that not much adjustment is available. The only adjustment is the primary air adjustment. If you were to turn the adjustment from one end of the scale to the other, only a few percentage points of efficiency would change. The key to a well-burning gas appliance is the flame. It must meet the standards of no yellow, firm, and not lifting off the burner port. When these standards are met, the flame is burning efficiently. The primary air adjustment for the LP gases is a little more critical. It requires a finer adjustment because the gas has a greater specific gravity and requires more

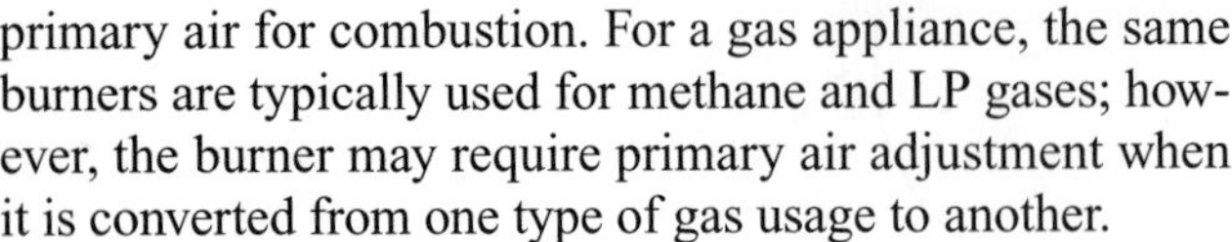

primary air for combustion. For a gas appliance, the same burners are typically used for methane and LP gases; however, the burner may require primary air adjustment when it is converted from one type of gas usage to another.

6.5 Gas Burners

Several types of atmospheric gas burners are used for different reasons. The first and most heavy-duty burner is the cast-iron, drilled-port burner. This burner is heavy and the most expensive to manufacture. It is available in as many designs as are needed for different flame patterns, Figure 6-46. Some cast-iron burners are slotted for ease in manufacturing. The weight of this burner adds to the overall weight of the appliance, which also causes more expensive shipping costs.

The stamped steel burner has taken the place of the cast-iron burner in many applications and is available in two designs, as shown in Figure 6-47. It is much cheaper

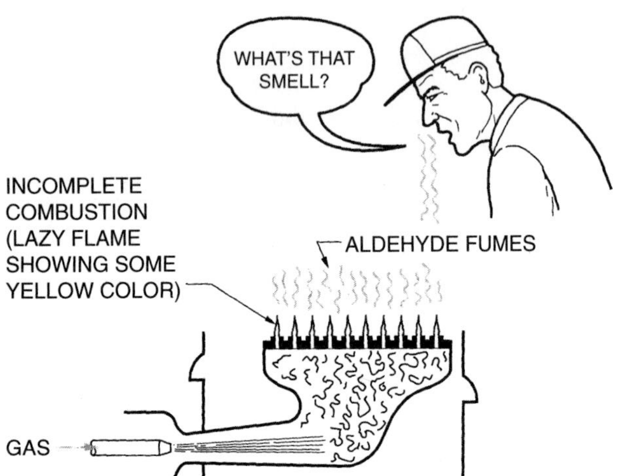

Figure 6-44. The aldehydes in some gases emit an odor with incomplete combustion.

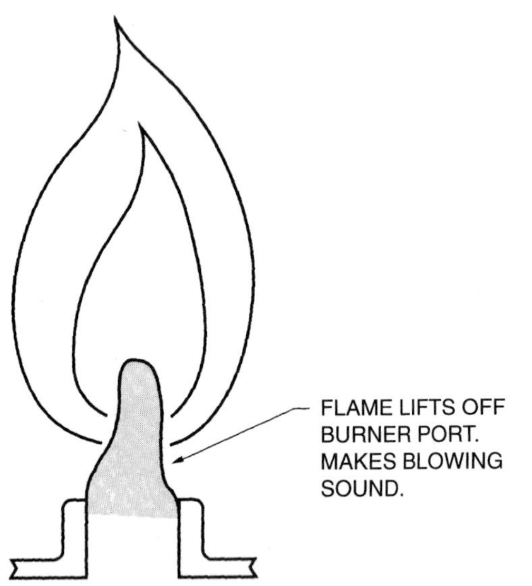

Figure 6-45. Too much air causes the flame of a burner to lift off the port.

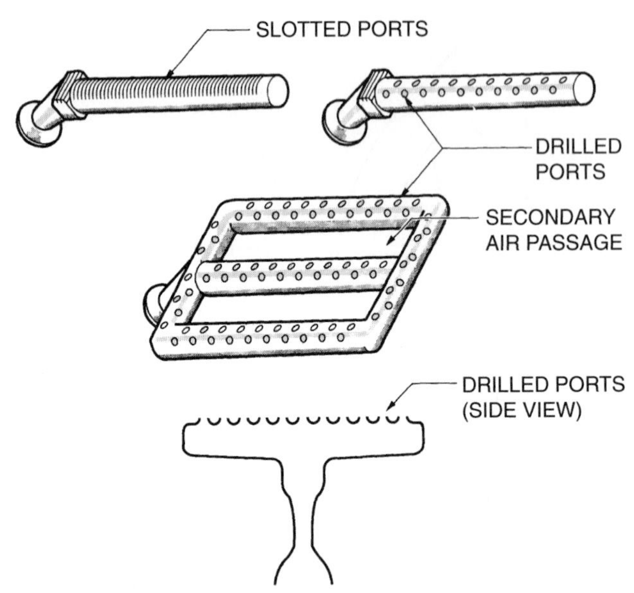

Figure 6-46. Cast-iron burners are available in several designs.

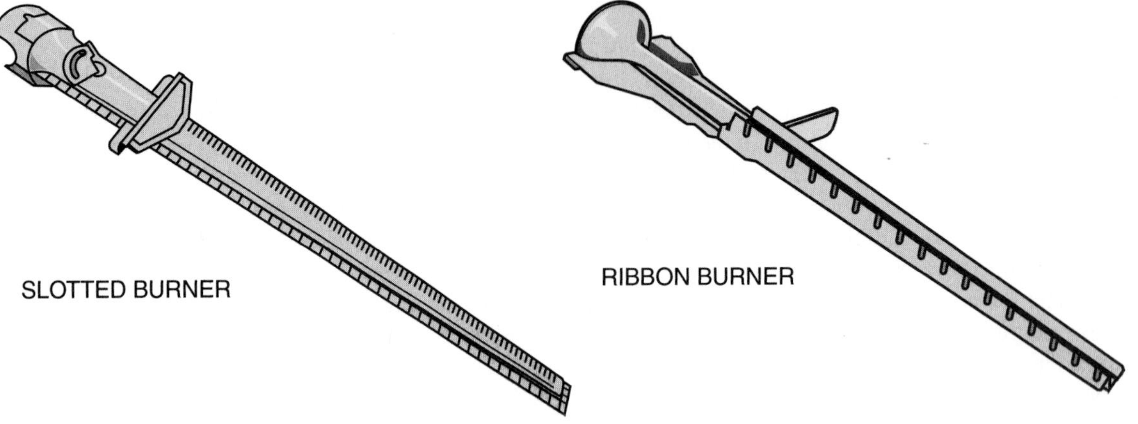

Figure 6-47. Two stamped steel burner designs.

to manufacture and lighter in weight. The top of the burner is often made of stainless steel, which can withstand the heat for long time periods.

The inshot and upshot burners are similar; they are known as single-port burners, Figure 6-48. The flame is at the end of the burner. This burner is often used when direct spark ignition is used. Direct spark ignition is explained later, but for now, when gas is allowed to enter the burner, it is ignited with a spark. A time limit of approximately 3 seconds is given for ignition; if it does not occur, the gas automatically shuts off. This burner is used most commonly for direct spark ignition because all of the flame is at the end of the burner instead of spread down the length of the burner, as it is in the stamped steel burner.

Pilot burners operate on the same principle as the main burners; they are also atmospheric-type burners. There are three types of common pilot burners: aerated, nonaerated, and target. The aerated-type pilot burner has a primary air port, Figure 6-49, and also uses secondary air. The size of the primary air port is extremely small, so this pilot should not be used when many particles are in the air, or the primary air port will become restricted, Figure 6-50.

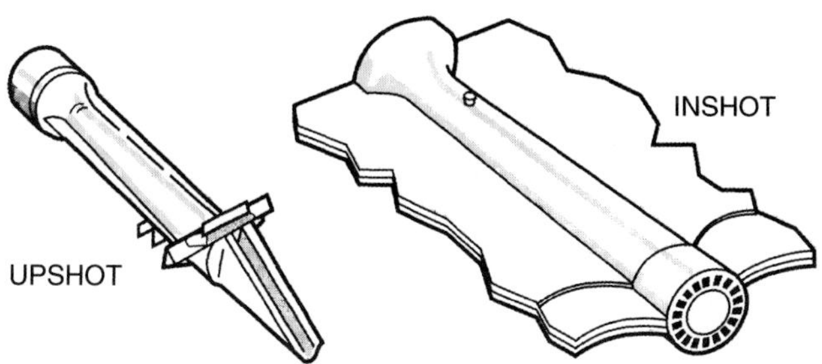

Figure 6-48. Single-port, or endshot, burners. *Reproduced courtesy Carrier Corporation*

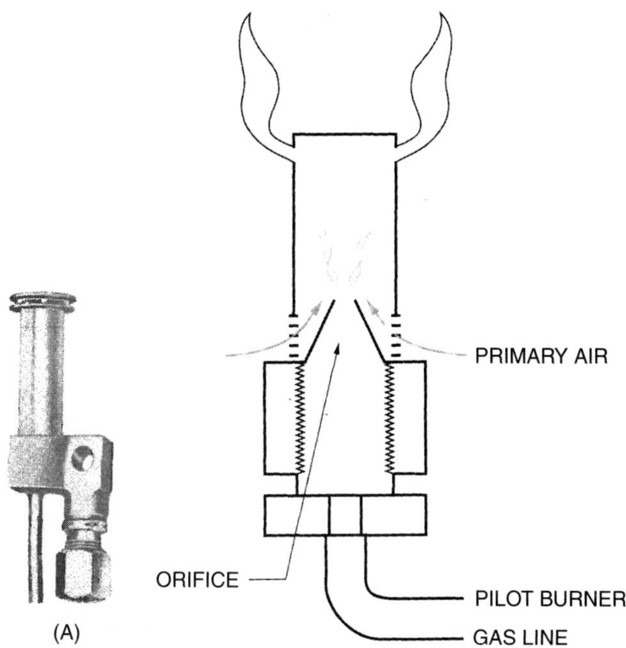

Figure 6-49. (A) Aerated type of pilot burner. *Courtesy Robertshaw Controls Company.* (B) Cross section.

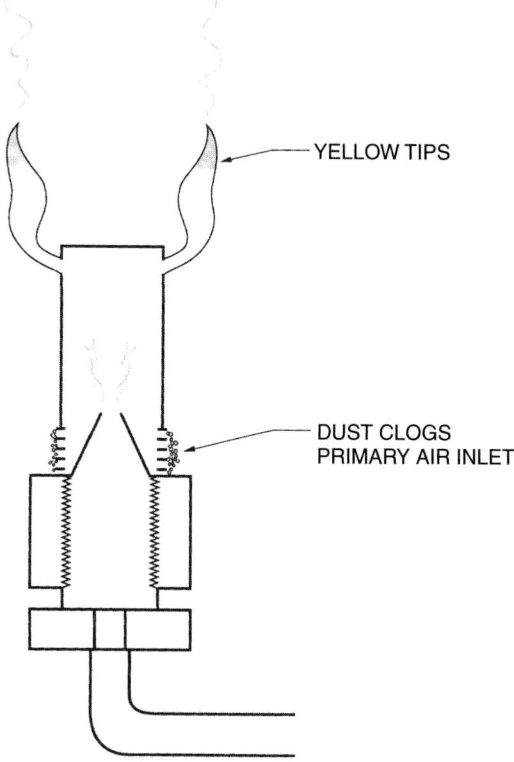

Figure 6-50. The aerated-type pilot burner should not be used when many particles are in the air.

The nonaerated pilot uses only secondary air and is not subject to obstruction by airborne particles, Figure 6-51. The nonaerated pilot flame is not as hot as the aerated flame. This difference may need to be considered in choosing the pilot safety device because some safety devices may require maximum heat.

The target type of pilot can be aerated or nonaerated. The word *target* describes the fact that the flame impinges on an object. The object glows red-hot to help in ignition and stabilizes the flame, Figure 6-52.

For the pilot to light a large burner, it must be able to spread the flame from one burner to another. Crossover ports are used for this purpose, Figure 6-53.

6.6 Condensing the By-Products of Combustion

As explained earlier, in a typical application, 20% of the heat from the gas escapes with the by-products of combustion. There are two kinds of heat: sensible and latent. Sensible heat is easy to understand. The flue

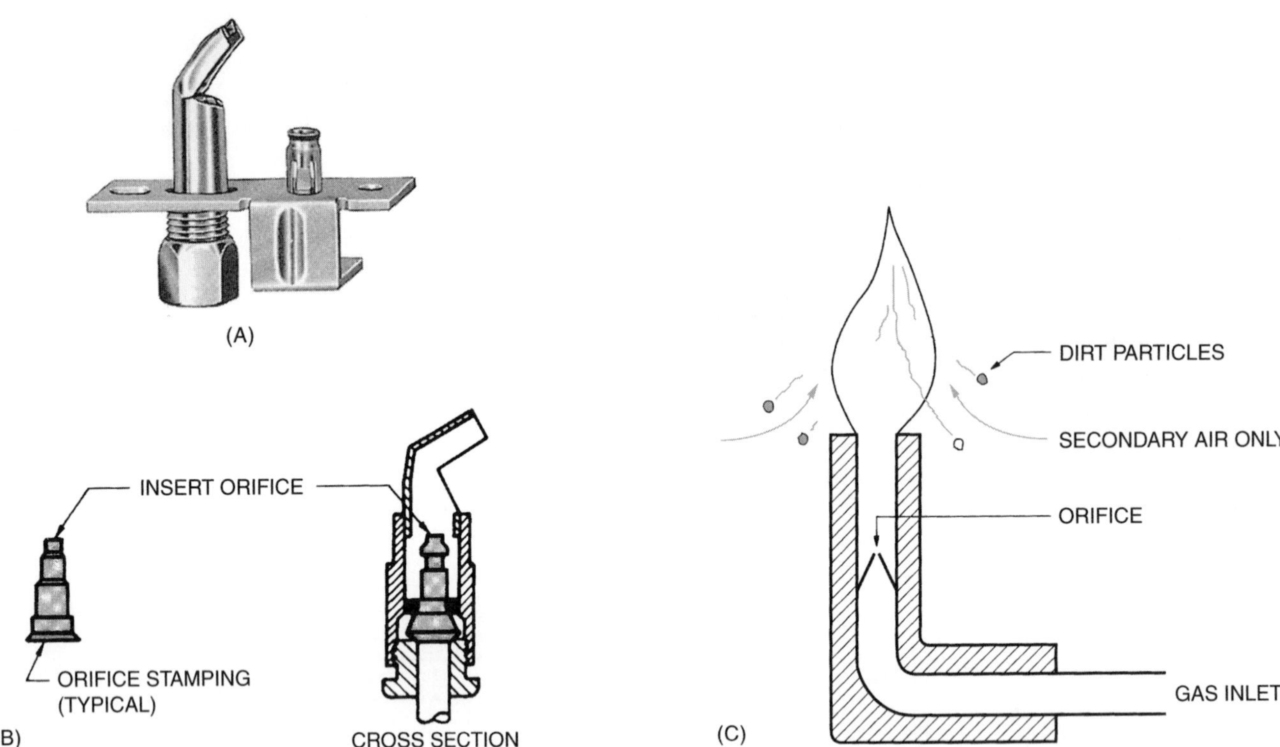

Figure 6-51. (A) The nonaerated pilot. *Courtesy Robertshaw Controls Company* (B and C) Cross sections show how this pilot will not clog with airborne particles.

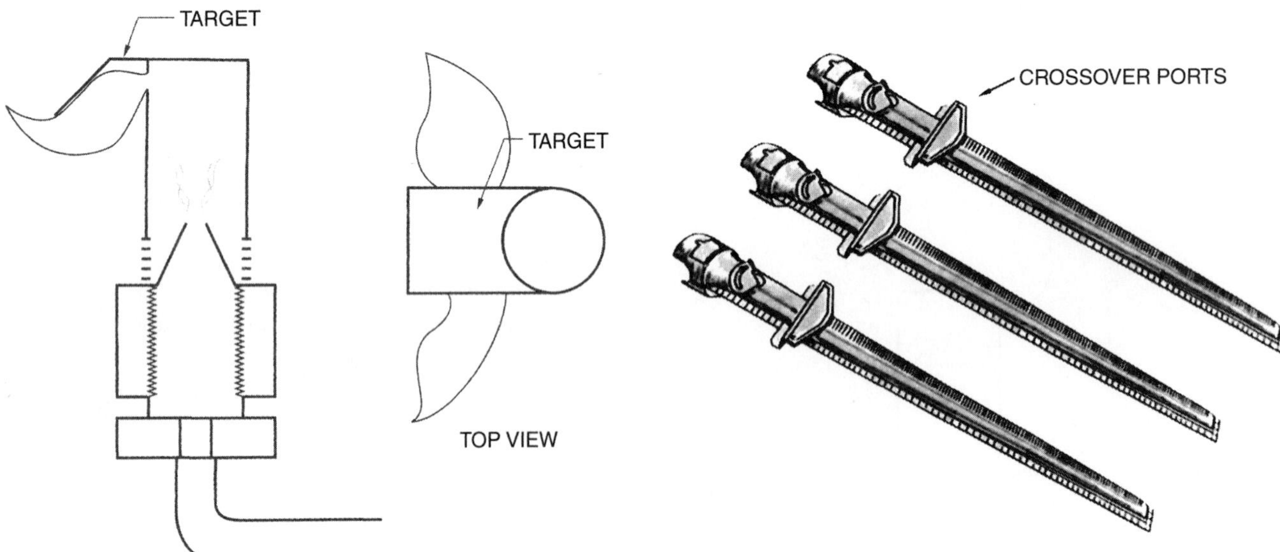

Figure 6-52. Target-type pilot burner, aerated.

Figure 6-53. Crossover ports carry the flame from one burner to another.

pipe is hot, usually nearly 350°F, Figure 6-54. If this heat could be removed from the flue gases and used as heat for the appliance, efficiency would improve and customers would get more for their money.

The other heat, latent heat, is contained in the water vapor. Approximately 2 cubic feet of water vapor are exhausted for each cubic foot of gas burned. This water vapor contains heat that is removed and condensed to a liquid. For this heat to be removed, the flue gases must be reduced to approximately 150°F, at which point the water vapor turns to liquid. This water has a slight acid content, so it must be handled carefully. Figure 6-55 shows

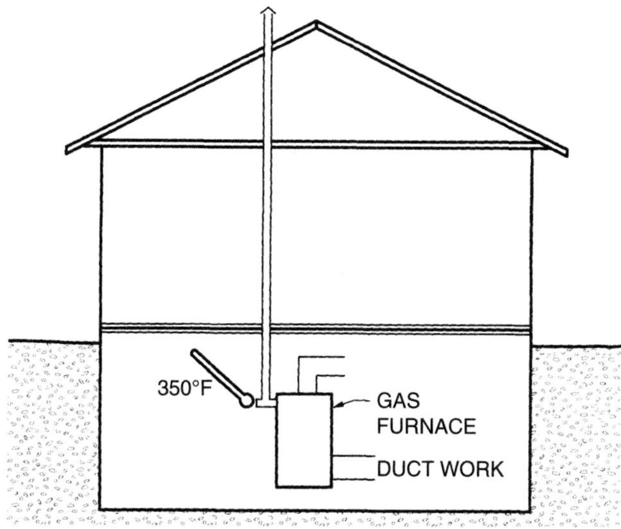

Figure 6-54. The flue of a typical furnace is approximately 350°F when the furnace is in operation.

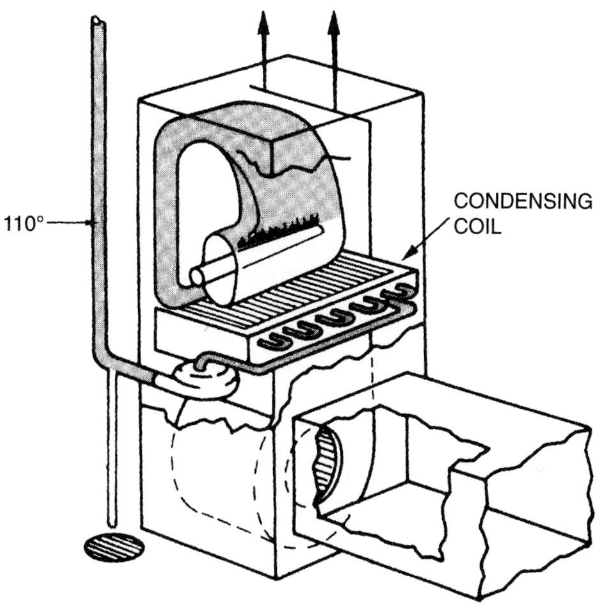

Figure 6-56. Acid must be drained from condensing furnaces.

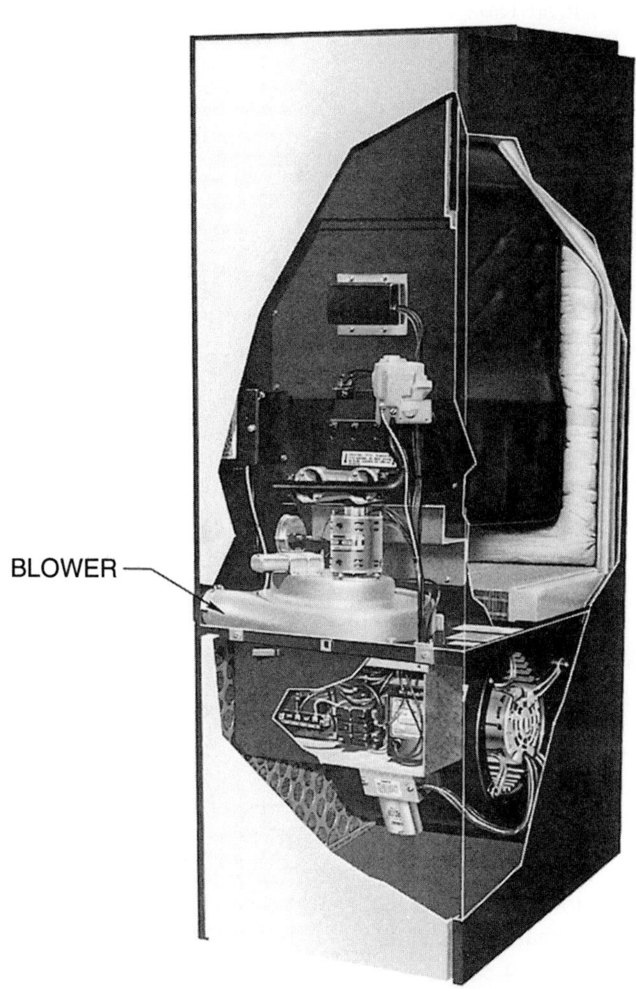

Figure 6-55. A condensing furnace is highly efficient.

a condensing furnace that can be used in a home. These furnaces are nearly 95% efficient, depending on the manufacturer. The flue gas temperature leaving the appliance is approximately 110°F. The furnace must have a drain to route the condensed water, Figure 6-56.

Understanding the combustion process and how burners work is the first step in the technology of using gas for heating.

Summary

- Natural gas, or methane, was first discovered seeping out of the ground because it would burn.
- Natural gas comes from animal and vegetable deposits deep within the ground that are many years old.
- Pressure is defined as force per unit of area, pounds per square inch (psi).
- Water or gas pressure is expressed in pounds per square inch gage (psig) and is indicated on a pressure gage.
- The atmosphere is made up of air and it has weight.
- The atmosphere's pressure is called atmospheric pressure and has a pressure of 14.696 psi, or 29.92 inches of mercury (Hg).
- Gage pressures start at atmospheric pressure, so a gage that reads 0 psig starts at atmospheric pressure.
- Another pressure that is commonly used in engineering measurements is pounds per square inch absolute (psia).
- To convert a gage pressure to absolute, you must add the atmosphere's pressure to the gage reading.
- Natural gas is cleaned and dried, and odorant is added to make it detectable before it is used as fuel for heat and light.

- There are three common gases: methane, propane, and butane. Propane and butane are commonly called LP (liquefied petroleum) gases.
- The LP gases are refined from petroleum products.
- All fuel gases consist of carbon and hydrogen, in different quantities.
- The chemical formula for methane is CH_4. Methane has a heat content of approximately 1,000 Btu per cubic foot at a burner manifold pressure of 3.5 inches of wc.
- The chemical formula for propane is C_3H_8. Propane has a heat content of 2,500 Btu per cubic foot at a burner manifold pressure of 11 inches of wc.
- The chemical formula for butane is C_4H_{10}. Butane has a heat content of 3,200 Btu per cubic foot at a burner manifold pressure of 11 inches of wc.
- Each gas has a different specific gravity: methane, 0.6; propane, 1.5; and butane, 2.5.
- The specific gravity describes the weight of the gas in relation to an equal volume of air.
- Propane and butane can be mixed with air to produce propane–air gas and butane–air gas.
- Perfect combustion for natural gas requires 2 cubic feet of oxygen to burn 1 cubic foot of gas.
- Perfect combustion requires 10 cubic feet of air to furnish 2 cubic feet of oxygen to burn 1 cubic foot of methane.
- Perfect combustion is not practical because if the air supply is slightly restricted when a burner is set up for perfect combustion, a problem occurs. Fifty percent excess air is commonly used to prevent this problem.
- Different amounts of air must be furnished for perfect combustion for each cubic foot of gas to be burned. Methane requires 10 cubic feet, propane requires 23.5 cubic feet, and butane requires 30 cubic feet of air.
- Manufacturers have simplified the primary air adjustment for gas burners to the point that from one end of the adjustment scale to the other only a few percentage points of change occur.
- Proper combustion is indicated by the flame. After the burner warms up, the flame should be blue, with orange streaks, firm, and not lifting off the burner port.
- The ignition temperatures for the various gases are methane, 1,170°F; propane, 900°F; and butane, 825°F.
- The limits of flammability for the gas describe the upper and lower percentages of gas to air that are required for the mixture to burn. These limits are different for each gas.
- The flame travel rates for the various gases are methane, 25 inches per second; propane, 32 inches per second; and butane, 33 inches per second.
- Atmospheric burners operate at atmospheric pressure.
- In atmospheric burners, the gas velocity is used to induce the air into the burner.
- The venturi is used to increase the gas–air mixture into the mixing tube in atmospheric burners.
- A small burner called a *pilot burner* is usually used to ignite the gas.
- There are three types of pilot burners: aerated; nonaerated; and target, which may be aerated or nonaerated.
- By-products of perfect combustion for 1 cubic foot of methane are 8 cubic feet of nitrogen, 1 cubic foot of carbon dioxide, 2 cubic feet of water vapor, and heat.
- There are two types of incomplete combustion: lack of air (air starved) and too much air.
- Lack of air yields a lazy yellow flame.
- The by-product of combustion for a starved-air flame is carbon monoxide.
- A flame with too much air tends to lift off the burner port.
- A flame burning correctly should be blue, with orange streaks from burning dust, and should be firm and steady.
- The only adjustment of gas appliances is the primary air adjustment. The orifice is a constant and the gas pressure is supposed to be a constant for each gas.
- A burner should be allowed to burn for a few minutes and get up to temperature before adjustment is made.
- Combustion analysis for modern gas-burning appliances is not critical. Correct flame characteristics will yield proper combustion. Some adjustment may be necessary when a burner is converted from using one fuel to using another.
- There are two types of gas burners: power and atmospheric.
- Power burners force air into the burner.
- With atmospheric burners, air is induced into the burner with the gas stream.
- There are several types of atmospheric gas burners: cast iron, both drilled and slotted; stamped steel; and single port.
- The atmospheric burner consists of an orifice, a primary air shutter, a venturi, a mixing chamber, and burner ports.
- Pilot burners are used to light the main burners and are atmospheric-type burners also.
- The typical gas-burning appliance is 80% efficient; 20% of the gas is used to produce heat that pushes other combustion by-products up the flue and outside.
- The flue gases can be cooled to remove the sensible heat and to condense the water vapor to a liquid. The condensed water vapor is slightly acidic and must be handled as such.

Review Questions

1. From what three places can methane gas originate?
2. What is the source of LP gas?
3. Define pressure.
4. The atmosphere creates a pressure of __________ psi, or __________ inches of Hg.
5. Convert 33 psig to psia __________.
6. How is natural gas transported across the country?
7. Methane is burned in which of the following states?
   a. Solid
   b. Liquid
   c. Gas

8. What is the difference between methane and natural gas?
9. What are the two LP gases?
10. How are LP gases transported to the customer?
11. Which gas or gases are heavier than air?
12. What is the typical burner pressure for natural gas?
13. What is the typical burner pressure for LP gases?
14. Why is propane preferred to butane?
15. Why is butane better than propane?
16. Is propane or butane most commonly used in colder climates?
17. How many cubic feet of air are required for perfect combustion of natural gas?
18. Why does the industry not use perfect combustion?
19. What are the by-products of complete combustion?
20. What is the efficiency of a typical gas-burning appliance?
21. What is the by-product of combustion that is toxic and how does it occur?
22. How many air adjustments does a gas appliance have?
23. How can more efficiency be accomplished with a gas-burning appliance?

7 Gas Heat: Natural and LP

Objectives

Upon completion of this unit, you should be able to

- **describe the major component of a gas furnace.**
- **list three types burned in gas furnaces and describe characteristics of each.**
- **describe the heat exchange for air and water for gas-burning appliances.**
- **list the types of safety systems used to prove gas combustion.**
- **name three types of ignition systems for gas heat.**

7.1 Introduction to Gas Heat

You should be fully familiar with Unit 6 before starting this unit. Unit 6 covers the combustion process and the characteristics of natural and LP gases. How these gases can be turned into usable heat and how the heat can be used are the topics of this unit.

The energy in the fuel gases can be turned into usable heat in several ways. Three common methods are heat transferred to air for circulation (*forced-air systems*), heat transferred to water for circulation (*hot-water* or *steam systems*), and radiant heat. Each of these three methods has a popular application and is discussed in this unit. Figure 7-1 shows a forced-air system and a hot-water system. These two systems are also called *convection heat.*

Note: The intent of this text is to provide the technician with a good working knowledge of the basics of heating. It is hoped that once the technician has become proficient in this area that they will continue their education and training to advance to more complex equipment. Therefore the scope of this text is limited to light commercial and residential systems only. It should be noted that although larger commercial system can be more complicated in their wiring and controls the basic configuration of the equipment and the equipment itself in most cases is similar to that covered in this text.

Gas heat is a popular method of heating a structure because it is reliable, clean, and readily available. Currently there are numerous appliances that are set up to consume gas (cook stoves, water heaters, clothes dryers) as well as heating equipment (furnaces and boilers).

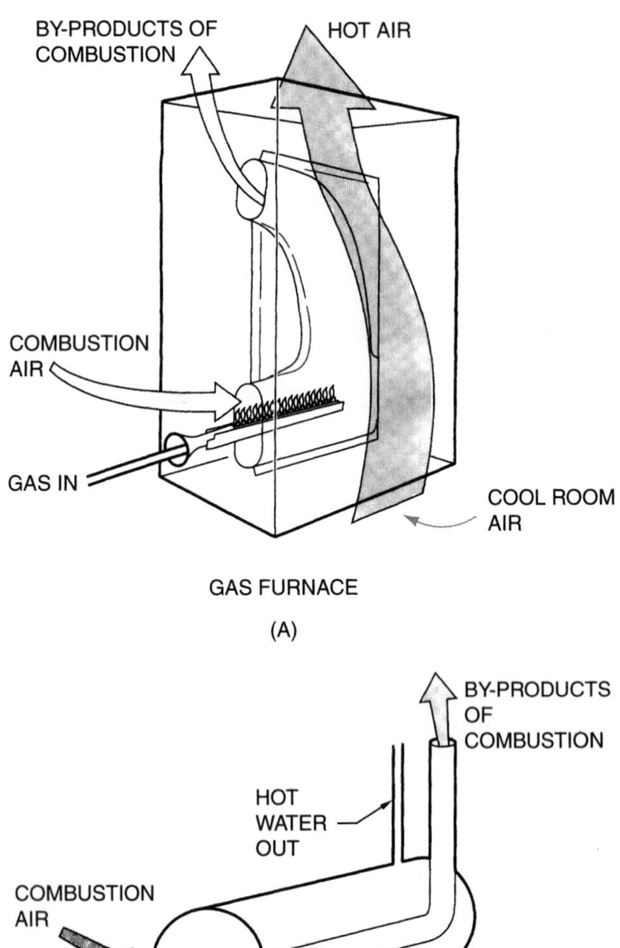

Figure 7-1. The heat energy in gas fuels is turned into usable heat energy by means of a heat exchanger for air and water. (A) A forced-air system. (B) A hot-water system.

Natural Gas

Natural gas is a colorless, odorless gas that is available in most cities and heavily populated areas. It is a mixture of approximately 60–90% methane and 5–40% other hydrocarbons such as ethane, propane, and butane. However, butane is not as common as the other hydrocarbons. Because natural gas is lighter than air, it has a tendency to rise when released into the atmosphere. This is because natural gas has a specific gravity of 0.6, and air has a specific heat of 0.1. Natural gas weighs only 60% as much as air. The heating capacity of natural gas is approximately 1050 Btu/ft^3. Although this varies depending on the chemical makeup of the natural gas being used, the accepted practice is to assume the heating capacity is 1050 Btu/ft^3.

Manufactured Gas

Manufactured gas is made by humans and is produced by combining hydrocarbon gases, coal, and oil. Manufactured gas is not as efficient as natural gas and therefore has a heating capacity of approximately 500–600 Btu/ft^3.

Liquefied Petroleum

Liquefied petroleum is typically not a single product but instead can be propane, butane, or a combination of the two in a liquefied state. Because liquefied propane boils at a temperature of −40°F (at atmospheric pressure), it is commonly used for space heating. The heat capacity of liquefied propane is much higher than natural gas and is rated at approximately 2500 Btu/ft^3, whereas the heating capacity of liquefied butane is even higher than liquefied propane. Its heating capacity is approximately 3200 Btu/ft^3.

It should be noted that liquefied propane and liquefied butane are both heavier than air when in the vapor state. Therefore, when they are released into the atmosphere, they will sink into the lowest possible location. This makes the liquefied products extremely dangerous, and absolute care should be exercised when working around appliances that operate using either of these gases. This is not to say that the other fuels previously mentioned are any less dangerous, and therefore they should be handled with care. Each fuel has its own set of cautions.

7.2 Heat Exchangers

When gas is burned, the intention is to remove as much heat from the by-products of combustion as practical. Two types of systems can accomplish this: those that vent the by-products of combustion to the atmosphere, such as a typical gas furnace, and those that allow the by-products to dissipate into the heated space. The latter type requires special consideration because, as Unit 6 discusses, one of the by-products of combustion is a considerable amount of water vapor, and incomplete combustion can yield CO (carbon monoxide, a toxic gas). The different types of equipment and their venting systems are discussed next.

Note: Be aware of local codes for any installation by contacting the local authority.

Forced-Air Systems

All forced-air furnaces are vented and use atmospheric gas burners for combustion, as shown in Figure 7-2. The by-products of combustion move through a heat exchanger that removes the allotted amount of heat from them.

A typical heat exchanger for a standard-efficiency furnace is shown in Figure 7-3. This type of heat exchanger removes approximately 80% of the heat from the by-products of combustion. The remaining by-products and heat rise up the flue, as shown in Figure 7-4. This is a steady-state efficiency that does not account for the on–off cycling of the furnace. Remember, the flue gases must be warm enough to rise and clear the structure. The temperature must not drop enough for the water vapor to condense.

Expanded-surface heat exchangers were developed to remove more heat from the combustion by-products. The expanded surface is composed of baffles and longer passages so that the combustion by-products maintain more contact with the heat exchanger, as shown in Figure 7-5. These furnaces improve combustion efficiency only slightly, approximately 5% (from 80–85%). Not much more heat can be removed in a standard furnace with-out some assistance,

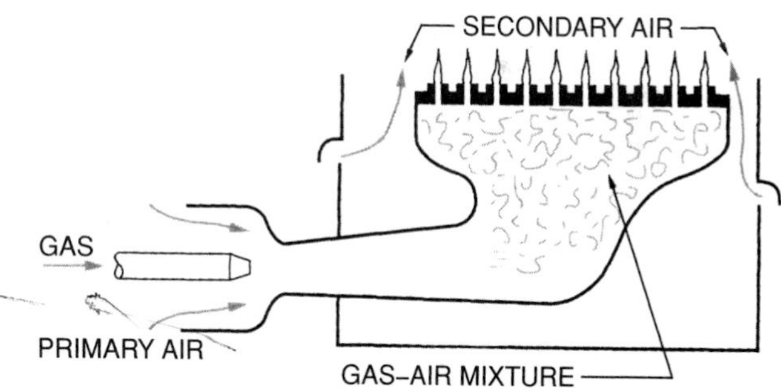

Figure 7-2. Forced-air furnaces use atmospheric gas burners.

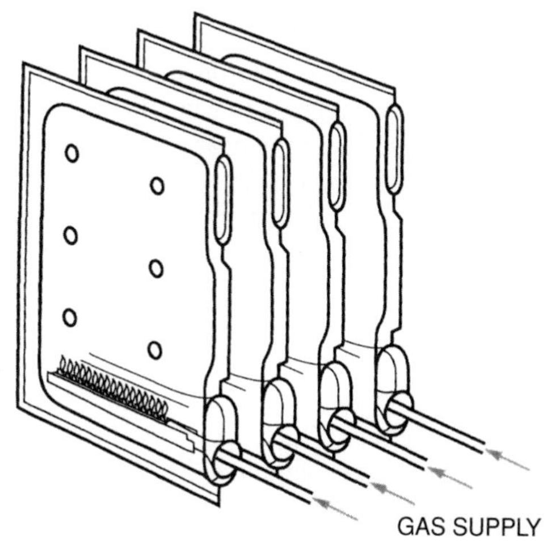

Figure 7-3. A typical heat exchanger for a forced-air gas furnace.

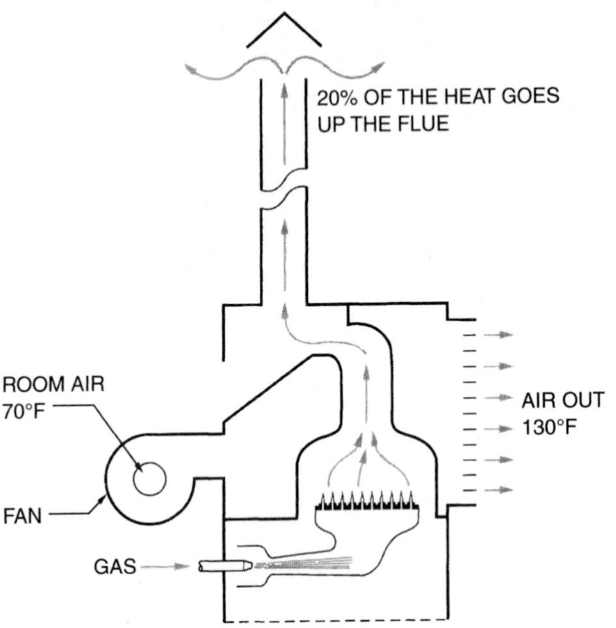

Figure 7-4. The typical gas furnace is 80% efficient.

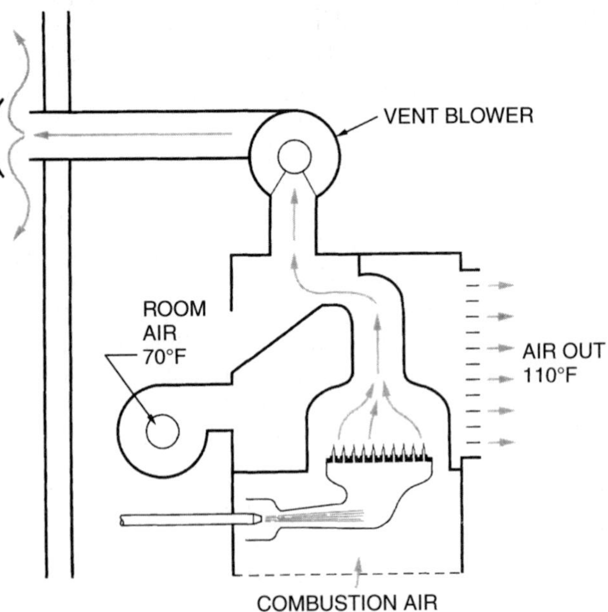

Figure 7-6. If too much heat is removed from the by-products of combustion, a blower must be used to assist in venting.

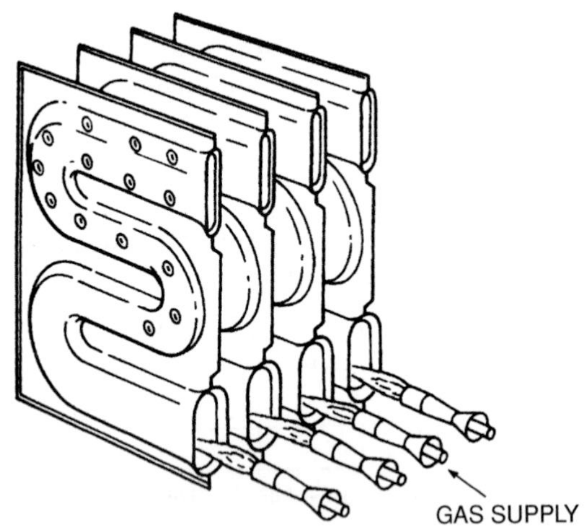

Figure 7-5. Expanded-surface heat exchangers offer more efficiency.

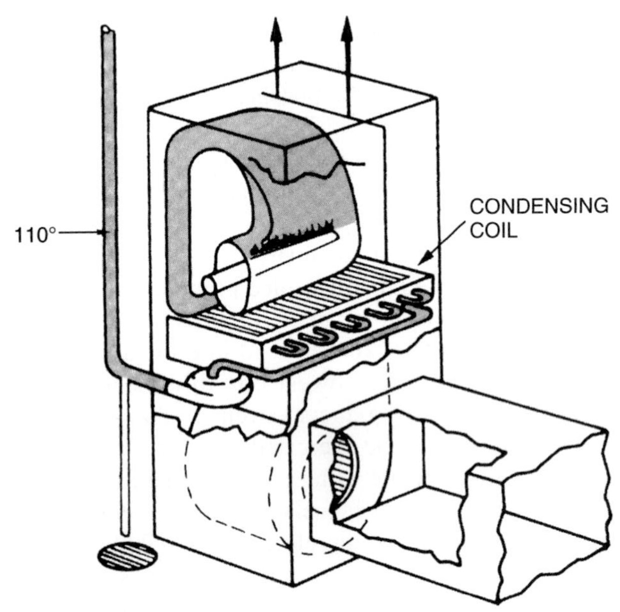

Figure 7-7. Flue gases are reduced to approximately 110°F in high-efficiency furnaces, for nearly 98% efficiency.

such as a combustion blower, to push the by-products out the flue, as shown in Figure 7-6. Manufacturers could remove almost all the heat in the heat exchanger of a typical furnace, but the water vapor in the by-products would condense, corrode the furnace, and make a water mess.

High-efficiency furnaces with efficiencies of nearly 98% condense the water vapor from the flue gases and reduce the temperature to approximately 110°F. These are called *condensing furnaces*. This type of furnace has a standard combustion chamber for the first part of the process and a condensing coil through which the by-products of combustion can pass, as shown in Figure 7-7. This condensing coil must be made of a metal, such as stainless steel, that will not corrode from the slightly acidic water that passes through it.

Regardless of the furnace type, the heat exchanger serves two purposes: to exchange heat between the by-products of combustion and the room air and to prevent the by-products of combustion from mixing with the room air. The room air must be on one side of the heat exchanger and the by-products of combustion on the other. The heat exchanger discussed here is the heat exchanger directly over the fire (the primary heat exchanger in a condensing furnace), not the secondary condensing heat exchanger.

Heat-Exchanger Damage

Heat exchangers should last forever, but they do not. They deteriorate from rust, flame impingement, exposure to

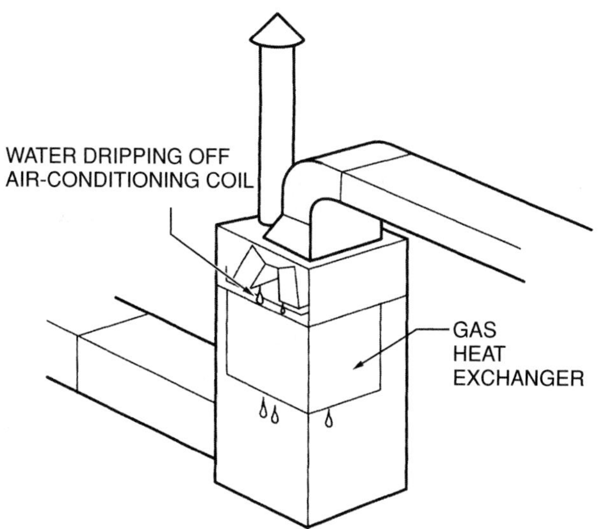

Figure 7-8. Water from the air-conditioning system may fall on the heat exchanger and cause rust.

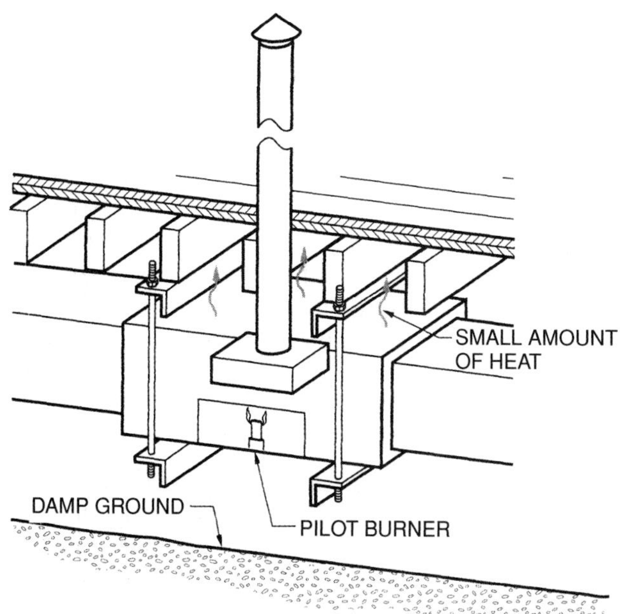

Figure 7-9. For crawl-space installations, a standing pilot burner can keep the heat exchanger dry in the summer.

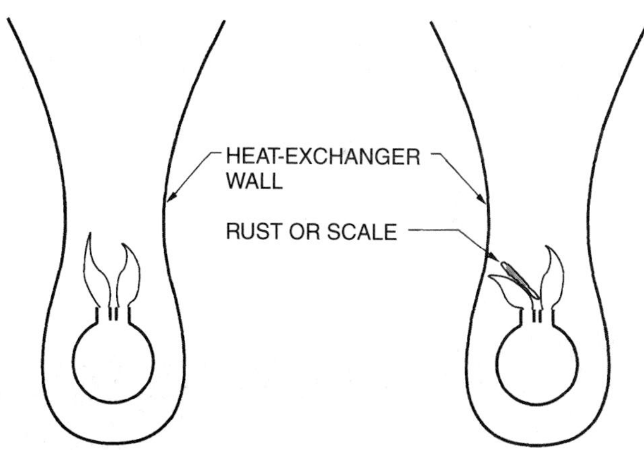

Figure 7-10. Flame impingement on a heat exchanger.

corrosive products in the air, and metal fatigue that results from expansion and contraction.

Rust is caused by moisture that may collect on the surface during the off-season. This may happen when water falls on the heat exchanger from an air-conditioning coil above it (see Figure 7-8) or when it is in a moist location, such as under a house. Many furnaces are located in crawl spaces under houses, where it is damp or wet. Moisture collects on the metal surfaces, and rust starts to form. The rust itself attracts moisture, which accelerates the rust process.

Furnaces have either standing pilot burners or intermittent ignition (pilot burners that are ignited upon a call for heat) for lighting the main gas burner. For years, many service technicians recommended that the pilot burner (for standing pilot furnaces) be left on even during summer to help keep the heat exchanger dry, see Figure 7-9. This solution lengthens the life of the heat exchanger when it is in a damp location, at the cost of a small amount of heat.

Flame impingement on a heat exchanger can be caused by burner misalignment or burner problems, such as a burner that is rusted through and is allowing gas to leak out to the side. In this situation, the burner has an undesirable flame pattern. Flame impingement can also be caused by rust or scale on the burner head, which causes the flame to burn to the side instead of upward, see Figure 7-10. No matter what the cause of flame impingement, the heat exchanger will weaken where the flame touches the steel. Flame impingement must not be allowed.

When gas-burning equipment is located in an installation with a corrosive atmosphere, an uncoated steel heat exchanger will deteriorate early. Installations with large amounts of freshly dyed cloth, such as carpet, drapery, and furniture stores, have slight concentrations of the dye products in the air. Beauty shops and dry-cleaning establishments may have corrosive chemical concentrations in the air. These installations can experience early heat-exchanger failure because when the chemicals burn, they produce corrosive by-products that come in contact with the steel heat exchanger. When a heat exchanger must be located in one of these or other corrosive locations, the installer should require a heat exchanger that will meet the standards for the location, such as a specially coated steel or stainless steel heat exchanger.

A more common reason for heat exchanger failure in most gas furnaces is metal fatigue. This is typically caused by the expansion and contraction of heat exchanger metal. Even though a heat exchanger is subjected to no other stresses than heat, the expansion and contraction of its metal does produce thermal stress so that over time, the characteristics of the metal are changed. It is this changing of the heat exchanger's properties as well as the expansion and contraction of the metal that finally causes the equipment to fail.

When a heat exchanger fails due to thermal expansion and contraction, it typically fails along either a seam or a bend. This is because as the metal is heated and cooled, the stresses associated with that action usually concentrate in these areas. Although in most cases the amount of expansion that occurs in a heat exchanger is less than 1/16 inch, over time the effect is detrimental, Figure 7-11.

Properly sized equipment lasts longer because it has longer running times. If a furnace is oversized, it will cycle on and off more often. Each cycle causes an expansion and contraction. The heat exchanger only expands and contracts a limited number of times before metal fatigue will crack the metal.

Note: If it is determined that the heat exchanger does indeed have either a hole or a crack, that furnace should be condemned and taken out of service until either the heat exchanger or the furnace itself is replaced. A damaged heat exchanger has the potential to leak dangerous gases into the habitable space, which could cause serious injury and even death.

Heat-Exchanger Construction: Forced-Air Furnaces

Most primary heat exchangers are made of stamped steel. They are often called *clam-shell heat exchangers* because two pieces of steel are stamped and then fastened at the seams. These clam-shell sections can be manifolded together to increase the capacity of a furnace (see Figure 7-12).

The burners are inserted into the bottom of each heat exchanger and are aligned so that the flame does not touch the steel sides of the heat exchanger (see Figure 7-13). Air for combustion is drawn from the room where the furnace is located and is pulled into the primary air ports of the burner through the primary air adjustment (see Figure 7-14). Room air must come from outside the building and is discussed in the paragraph on gas vent sizing and practices Unit 9.

Another type of heat exchanger is manufactured by welding the steel pieces together. This is a more expensive heat exchanger, but it does not have the stress applied to the steel that

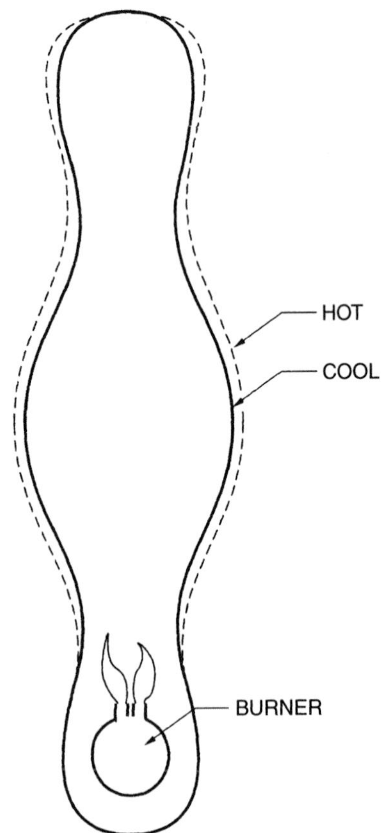

Figure 7-11. Expansion and contraction cause metal fatigue.

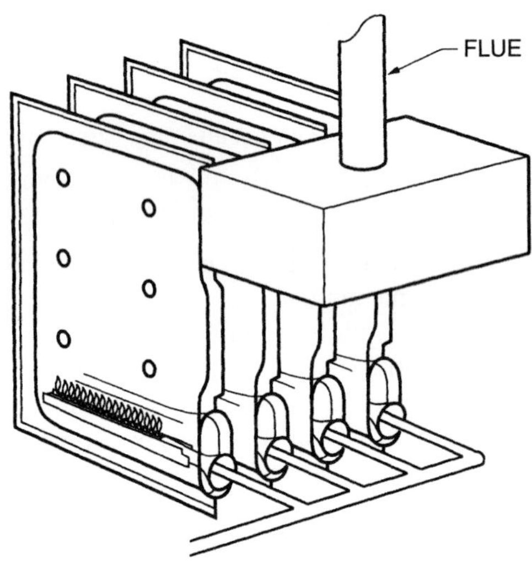

Figure 7-12. Clam-shell heat exchangers can be manifolded together for more capacity.

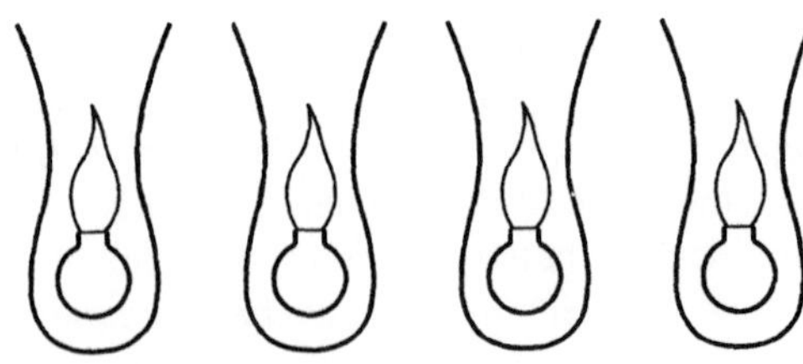

Figure 7-13. Burners are aligned in the bottom of the heat exchanger.

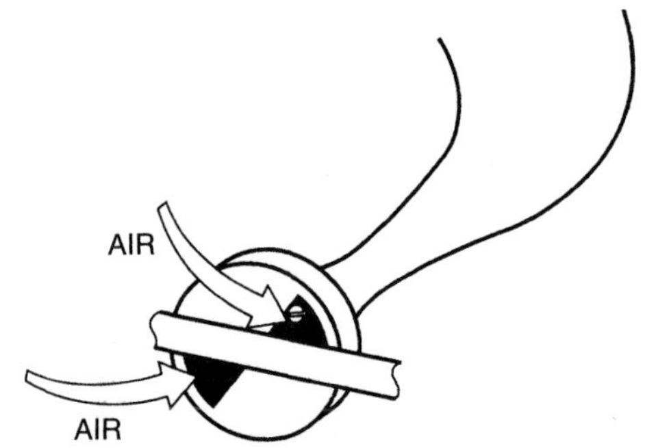

Figure 7-14. Air for combustion enters the burner through the primary air adjustment.

a stamped heat exchanger has during manufacturing. These furnaces are more heavy duty, as shown in Figure 7-15.

The condensing furnace uses a primary and a secondary heat exchanger. The primary heat exchanger is similar to the heat exchangers just described. The secondary heat exchanger is usually made of stainless steel because it exchanges heat between the flue gases and the return air. The temperature is much lower than at the primary heat exchanger, and the moisture is condensed from the by-products of combustion. This moisture is slightly corrosive. The secondary heat exchanger will probably never fail from metal fatigue when it is made of stainless steel, but it may become plugged with dust from combustion or with soot when improper combustion occurs. These heat exchangers may resemble water coils with small passages, as shown in Figure 7-16. Manufacturers usually provide a clean-out access for the secondary heat exchanger in case a restriction occurs.

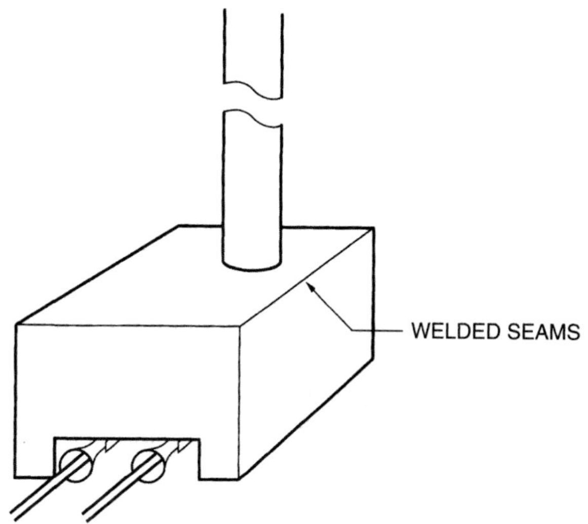

Figure 7-15. Welded-steel heat exchangers are heavy duty.

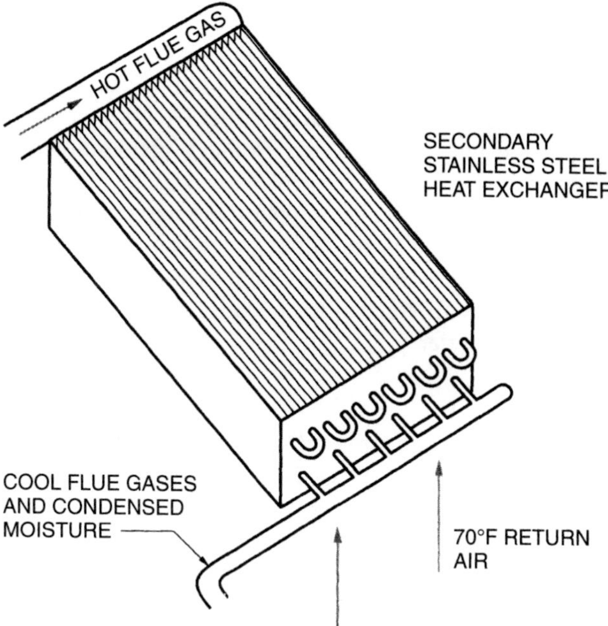

Figure 7-16. A secondary heat exchanger resembles a hot-water coil.

Hot-Water Systems

Water-type heat exchangers transfer heat to water boilers. The purpose of these heat exchangers is much the same as that of air heat exchangers: to transfer as much heat as possible and still maintain enough heat to push the by-products of combustion up the flue.

These heat exchangers are typically made of cast iron, steel, or copper. Cast-iron boilers are manufactured for small applications as well as for midsized commercial use. They are manufactured in sections that can be combined for larger applications. A special end section is used at each end, with center sections between, as shown in Figure 7-17. Center sections can be added to extend the size of the boiler, and only the skin needs to be extended to make a larger boiler. When the boiler becomes too long as judged by the manufacturer, a boiler with larger sections can be used. The fire is below the cast-iron sections, so the flue gases rise through them, as shown in Figure 7-18. Fins or knobs can be used to extend the surface of the cast iron for more efficient heat exchange, as shown in Figure 7-19. The heat exchange between the flue gases and the water is extremely efficient, more than between the flue gases and the air in a forced-air furnace. The cast iron in the heat exchanger is thick and heavy, so it takes longer for the boiler to get hot and to cool. These heat exchangers are designed and manufactured to operate with a water pressure of 15 psig on the water side of the system and are known as *low-pressure boilers.* High-pressure boilers are used for larger applications and when more water pressure is needed, such as in high-rise buildings.

Some boilers have steel tube sheets at the ends of the shell, and water moves through the tubes between the sheets; these

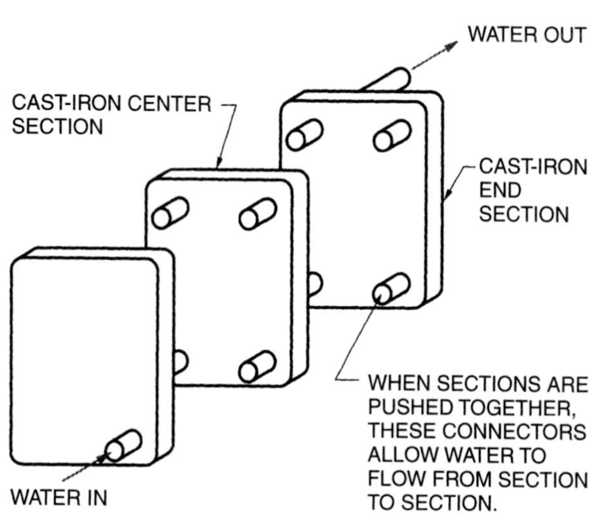

Figure 7-17. Sectional boiler parts.

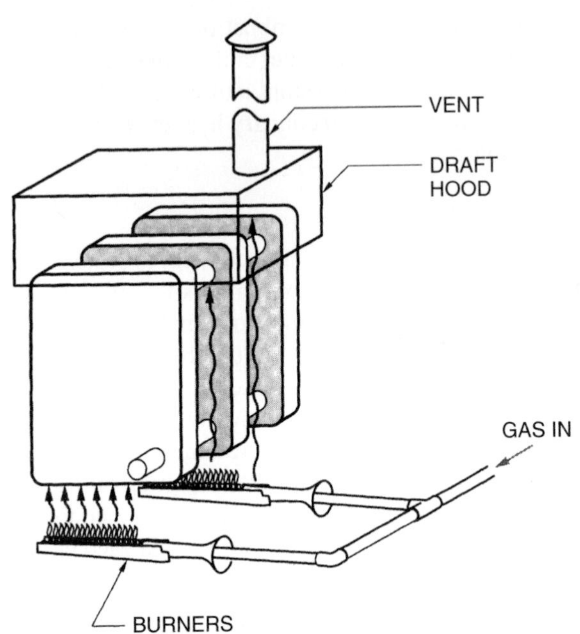

Figure 7-18. The fire is in the lower portion of a sectional boiler.

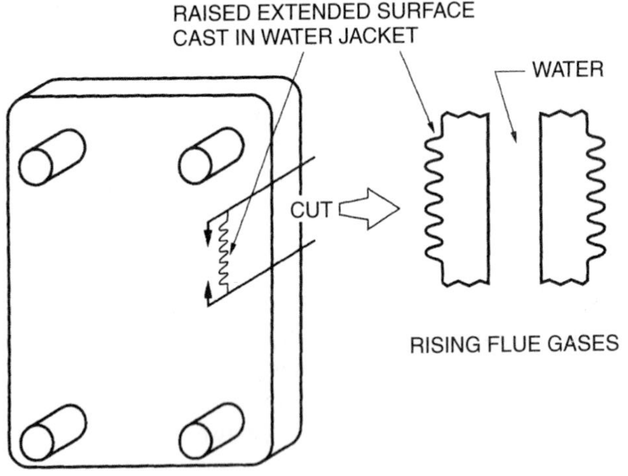

Figure 7-19. Extended surfaces are used to improve the heat exchange of a cast-iron boiler.

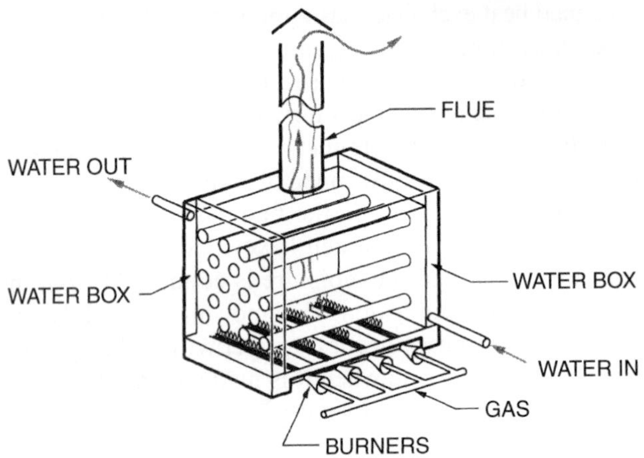

Figure 7-20. The by-products of combustion work their way up the tubes of a water-tube boiler.

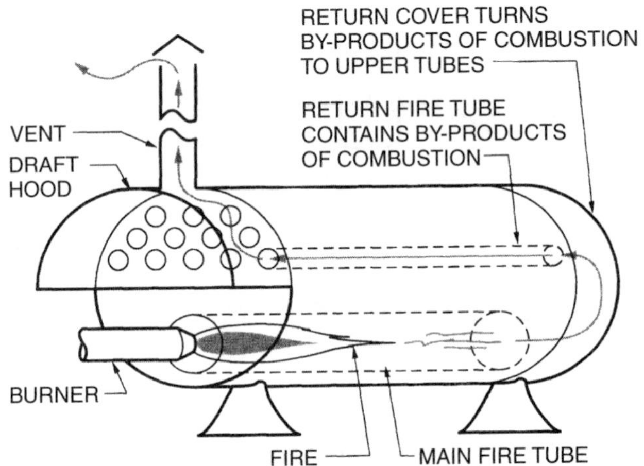

Figure 7-21. The fire-tube boiler has one main tube for the fire.

are known as *water-tube boilers,* as shown in Figure 7-20. The fire is below the tubes and the by-products of combustion work their way up through the tubes to the vent at the top. This is an efficient method of heat transfer because the steel tubes are not thick and the heat transfers quickly.

Other boilers in the large sizes are known as *fire-tube boilers* because the fire is in the tubes. These boilers typically have a large, long tube (called the *burner tube*) that the burner fires within, as shown in Figure 7-21. The fire burns out before it reaches the end of the burner tube, and flue gases leave the tube. The flue gases turn upward, enter a row of smaller tubes, and travel toward the burner end of the boiler and then either exhaust up the flue on the burner end or return to the other end and exhaust up the flue. These boilers are often dual-fuel boilers; they can be manufactured to burn gas or fuel oil. They are compact for their capacity and are often used in larger commercial buildings.

Some manufacturers use copper tubes and cast-iron end sheets. The end sheets are full of water and also exchange heat. One of these boilers may be much smaller than the same boiler with steel tubes because the copper tubes transfer heat better.

Radiant Heat

Radiant heat exchangers exchange heat directly with the heated space. These heaters are either room heaters or commercial heaters that are vented or unvented, depending on the application. According to the *National Fuel Gas Code* (pamphlet 54, paragraphs 6.23.1 and 6.23.2 published by the national fire protection association), unvented radiant room heaters may be used in residential applications other than bathrooms or bedrooms. They may not be used in institutions, such as homes for the aged.

The *National Fuel Gas Code* may have been amended in your community. Get in touch with your local authorities before making any decisions about using radiant heaters.

A room heater, shown in Figure 7-22, has a burner below a set of elements called *radiants*. Radiants for unvented heaters are made of ceramic material and glow red-hot. The heat from the red-hot elements radiates into the room, as shown in Figure 7-23. This heater is efficient and is used for spot-heating in many businesses and homes. It has the advantage of allowing you to heat only the area where the heater is located as shown in Figure 7-24. Room heaters that are not vented release the by-products of combustion into the heated space and are 100% efficient.

Warning: Care must be taken to ensure that unvented heaters are burning correctly, or they can release dangerous carbon monoxide into the room air.

The structure of the heater must also be able to absorb the moisture given off from the by-products of combustion without damage. Usually these structures have a lot of *infiltration air* (outside air that seeps into the room through cracks in the structure and around doors and windows).

Figure 7-22. Radiant room heater. *Courtesy Empire Comfort Systems*

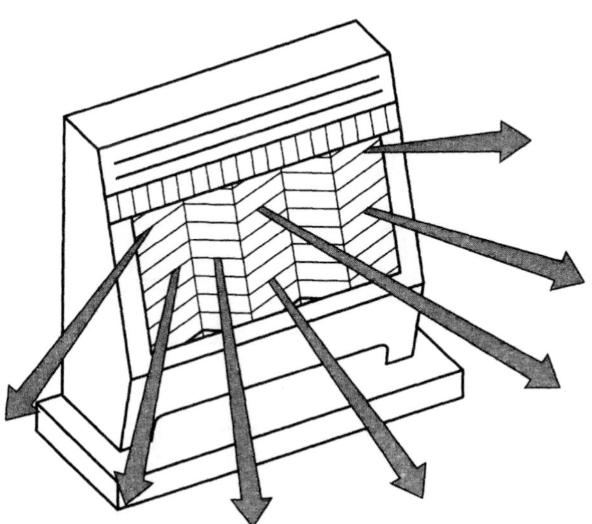

Figure 7-23. Heat radiates from the heater.

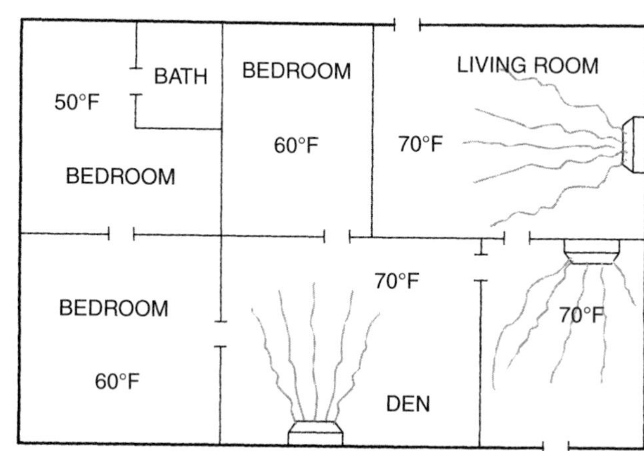

Figure 7-24. A room heater provides zone room temperature control.

Vented room heaters are also common and often have circulating blowers, as shown in Figure 7-25. They also have radiants, that give off the radiant heat to the surroundings, as well as the circulated air. These heaters are approximately 80% efficient; the remaining heat rises up the flue. The vent carries the by-products of combustion out of the building.

Another type of radiant heater is used for commercial and industrial use. These heaters can be vented or unvented. They are usually hung from the ceiling, and their heat energy is focused toward the area to be heated. These heaters do not heat the air but instead heat the solid objects upon which they are focused. They provide efficient heat for applications such as garages or assembly areas in factories, where the whole space need not be heated for a worker to be comfortable. Once again, the by-products of combustion are released in the structure or vented outside the building, depending on the application.

Figure 7-25. Vented room heater with a circulating fan. *Courtesy Empire Comfort Systems*

7.3 Gas Regulators

Natural gas pressure in the supply line does not remain constant and is always at a much higher pressure than required at the manifold. The *gas regulator* drops the pressure to the proper level (in. WC) and maintains this constant pressure at the outlet where the gas is fed to the gas valve. Many regulators can be adjusted over a pressure range of several inches of water column, Figure 7-26. The pressure is increased when the adjusting screw is turned clockwise; it is decreased when the screw is turned counterclockwise. Some regulators have limited adjustment capabilities; others have no adjustment. Such regulators are either fixed permanently or sealed so that an adjustment cannot be made in the field. The natural gas utility company should be contacted to determine the proper setting of the regulator. In most modern furnaces, the regulator is built into the gas valve, and the manifold pressure is factory set at the most common pressure: 3 1/2 in. WC for natural gas.

LP gas regulators are located at the supply tank. These regulators are furnished by the gas supplier. Check with the supplier to determine the proper pressure at the outlet from the regulator. The outlet pressure range is normally 10 to 11 in. WC. The gas distributor would normally adjust this regulator. In some localities a higher pressure is supplied by the distributor. The installer then provides a regulator at the appliance and sets the water column pressure to the manufacturer's specifications, usually 11 in. WC. LP gas installations normally *do not* use gas valves with built-in regulators.

Combination gas valves have built-in regulators in their bodies. Manufacturers usually will provide some sort of retrofit kit for converting from natural gas to LP gas. Many times a higher tension spring will have to be installed in the regulator at the gas valve to hold the valve open all the way. Some gas valves incorporate a push-pull knob. Pushing the knob down into the regulator will open the regulator all the way, which allows the regulator at the tank or the regulator at the side of the house to take control and regulate at 10 1/2–11 in. WC. Gas valves sold as LP valves already may have had this done to their regulators in some fashion. Always consult the manufacturer's specifications in setting regulators for any type of gas.

Converting from natural gas to LP gas involves many steps that may differ from manufacturer to manufacturer because of design differences of gas valves. For example, converting from natural gas to LP gas will include changing the regulator pressure from 3.5 in. WC by some method, installing smaller main burner orifices, and installing a new pilot orifice. This process also may include changing the ignition/control module to a 100% shutoff module because LP gas is heavier than air.

SAFETY PRECAUTION: Only experienced technicians should adjust gas pressures.

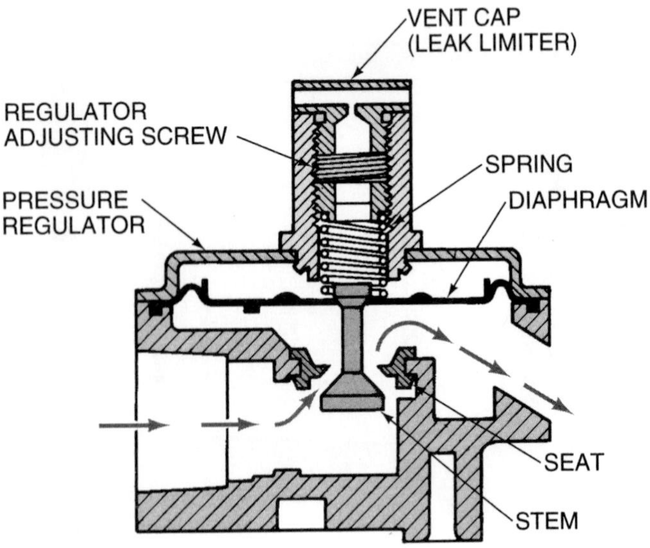

Figure 7-26. The diagram of a standard gas pressure regulator.

7.4 Gas Valve

From the regulator the gas is piped to the *gas valve* at the manifold, Figure 7-27. Several types of gas valves exist. Many are combined with pilot valves and called *combination* gas valves, Figure 7-28. We will first consider the gas valve separately and then in combination. Valves are generally classified as solenoid, diaphragm, and heat motor.

7.5 Solenoid Valve

The gas-type *solenoid* valve is an NC valve, Figure 7-29. The plunger in the solenoid is attached to the valve or is in the valve. When an electric current is applied to the coil, the plunger is pulled into the coil. This opens the valve. The plunger is spring loaded so that when the current is turned off the spring forces the plunger to its NC position, shutting off the gas, Figure 7-30.

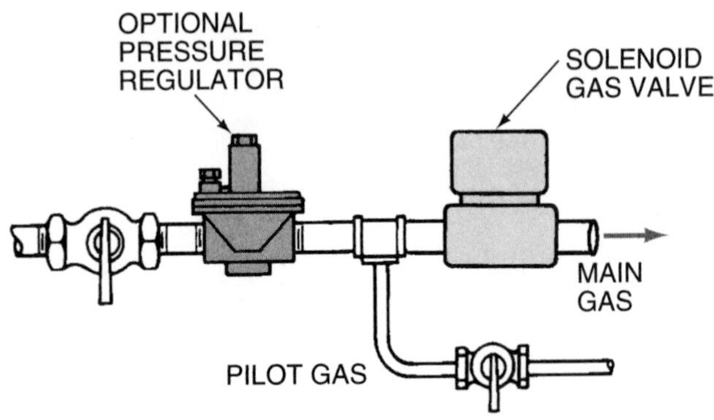

Figure 7-27. Natural gas installation where the gas passes through a separate regulator to the gas valve and then to the manifold.

(A)

(B)

Figure 7-28 (A)–(B) Two gas valves with pressure regulator combined. *(A) Courtesy Honeywell, Inc., Residential Division, (B) Courtesy Robertshaw Controls Company*

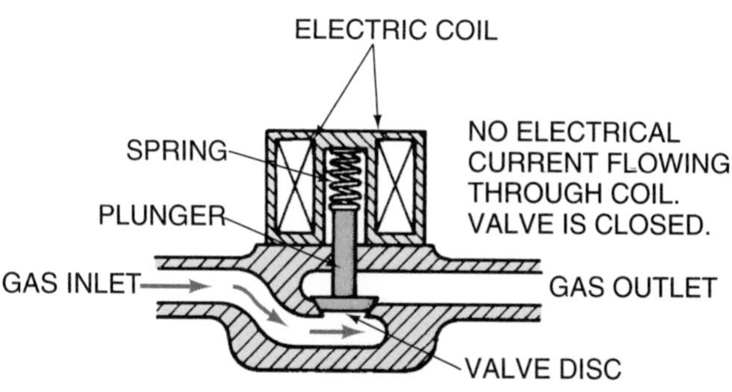

Figure 7-29. A solenoid gas valve in its normally closed position.

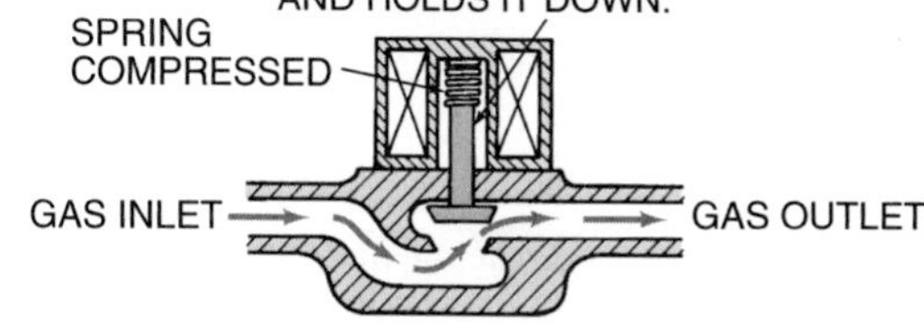

Figure 7-30. A solenoid valve in the open position.

7.6 Diaphragm Valve

The *diaphragm* valve uses gas pressure on one side of the diaphragm to open the valve. When there is gas pressure above the diaphragm and atmospheric pressure below it, the diaphragm will be pushed down, and the main valve port will be closed, Figure 7-31. When the gas is removed from above the diaphragm, the pressure from below will push the diaphragm up and open the main valve port, Figure 7-31. This is done by a very small valve, called a *pilot operated* valve because of its small size. It has two ports—one open while the other is closed. When the port to the upper chamber is closed and not allowing gas into the chamber above the diaphragm, the port to the atmosphere is opened. The gas already in this chamber is vented or bled to the pilot

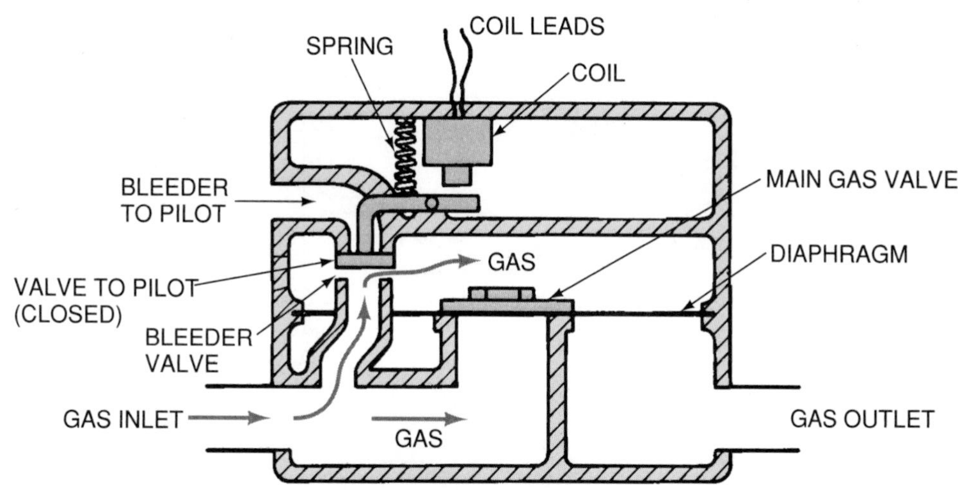

Figure 7-31. An electrically operated magnetic diaphragm valve.

where it is burned. The valve controlling the gas into this upper chamber is operated electrically by a small magnetic coil, Figure 7-32.

When the thermostat calls for heat in the thermally operated valve, a bimetal strip is heated, which causes it to warp. A small heater is attached to the strip, or a resistance wire is wound around it, Figure 7-33. When the strip warps, it closes the valve to the upper chamber and opens the bleed valve. The gas in the upper chamber is bled to the pilot where it is burned, reducing the pressure above the diaphragm. The gas pressure below the diaphragm pushes the valve open, Figure 7-34.

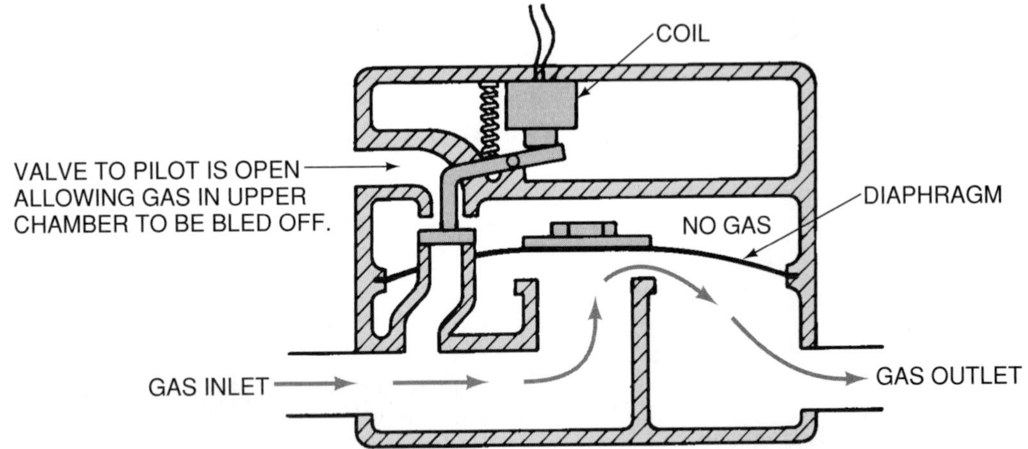

Figure 7-32. When an electric current is applied to the coil, the valve to the upper chamber is closed as the lever is attracted to the coil. The gas in the upper chamber bleeds off to the pilot, reducing the pressure in this chamber. The gas pressure from below the diaphragm pushes the valve open.

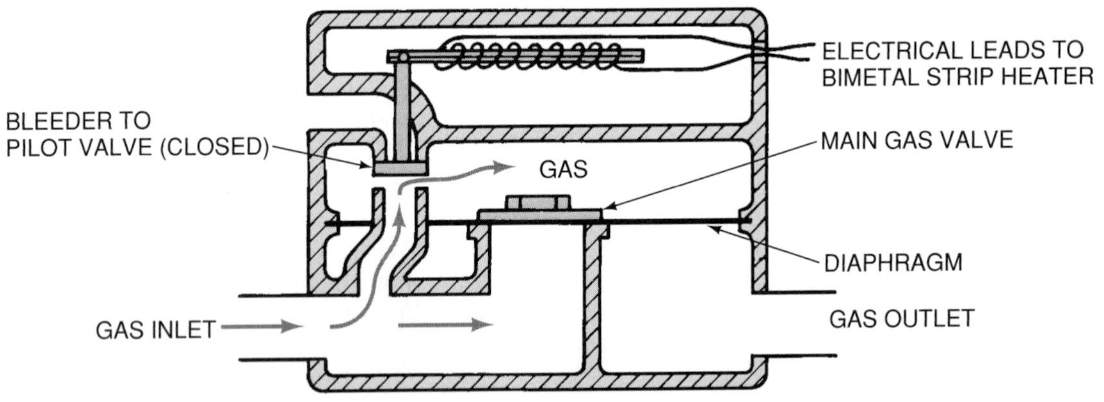

Figure 7-33. The thermally operated diaphragm gas valve.

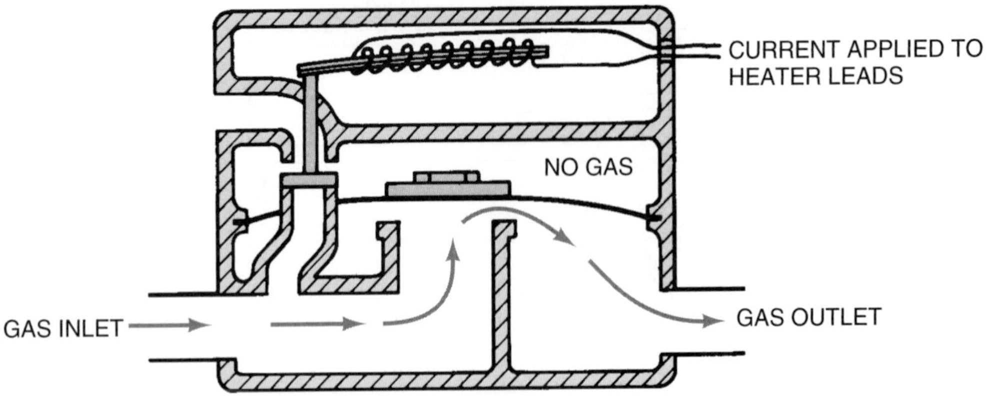

Figure 7-34. When an electric current is applied to the leads of the bimetal strip heater, the bimetal warps, closing the valve to the upper chamber, and opening the valve to bleed the gas from the upper chamber. The gas pressure is then greater below the diaphragm, pushing the valve open.

7.7 Heat Motor-Controlled Valve

In a heat motor-controlled valve an electric heating element or resistance wire is wound around a rod attached to the valve, Figure 7-35. When the thermostat calls for heat, this heating coil or wire is energized and produces heat, which expands, or elongates, the rod. When elongated, the rod opens the valve, allowing the gas to flow. As long as heat is applied to the rod, the valve remains open. When the heating coil is de-energized by the thermostat, the rod contracts. A spring will close the valve.

It takes time for the rod to elongate and then contract. This varies with the particular model but the average time is 20 sec to open the valve and 40 sec to close it.

SAFETY PRECAUTION: Be careful while working with heat motor gas valves because of the time delay. Because there is no audible click, you cannot determine the valve's position. Gas may be escaping without your knowledge.

7.8 Automatic Combination Gas Valve

Many modern furnaces designed for residential and light commercial installations use an automatic combination gas valve (ACGV), Figure 7-36. These valves incorporate a manual control, the gas supply for the pilot, the adjustment and safety shutoff features for the pilot, the pressure regulator, and the controls to operate the main gas valve. They often have dual shutoff seats for extra safety protection.

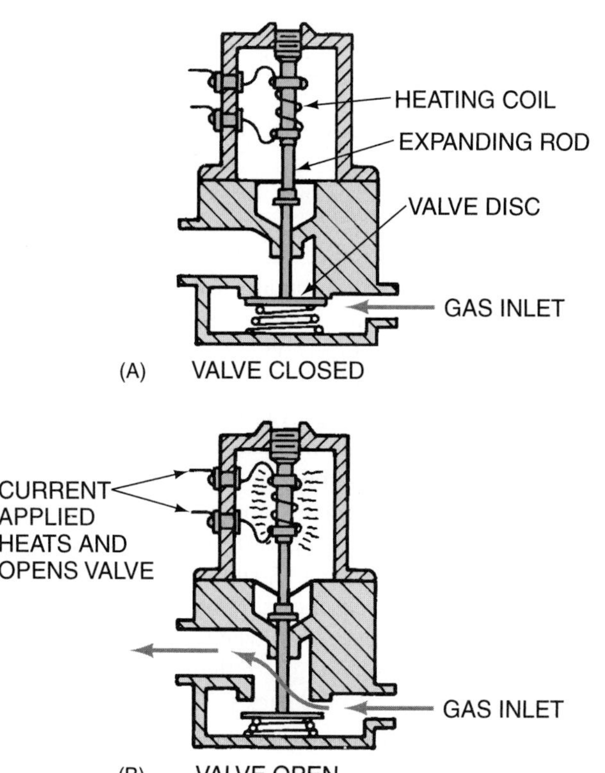

Figure 7-35. A heat motor-operated valve, (A) closed and (B) open.

Figure 7-36. An automatic combination gas valve. *Courtesy Honeywell, Inc., Residential Division*

This is also called the redundant gas valve. These valves also combine the features described earlier relating to the control and safety shutoff of the gas.

Modern combination gas valves also may include programmed safe lighting features, servo pressure regulation, choices of different valve operators, and installation aids.

Standing Pilot Automatic Gas Valves

The internals of a typical combination gas valve for a standing pilot system are shown in Figure 7-37. A standing pilot system has a pilot burner that is lit all the time. The operation of the valve is as follows.

The gas inlet is on the left of the valve. A screen or filter is usually installed at the factory to keep the valve free from dirt and miscellaneous debris. The gas then encounters the safety shutoff valve. Pushing the red reset button can manually open this valve. In fact, this is how the pilot burner is lit. Pushing the red reset button allows gas to flow to the pilot gas outlet. The pilot gas burner is then manually lit. The pilot flame engulfs the thermocouple, which starts generating a direct current (DC) voltage. The voltage generation of a single standard thermocouple is usually 24 to 30 millivolts DC. A direct current is then generated through the thermocouple and the power unit. Thermocouples will be covered in more detail later in this unit.

The power unit consists of a low-resistance coil and an iron core. When the power unit is energized, it has enough power to hold the safety shutoff valve open against its own spring pressure. If the pilot flame is ever lost, the power unit will be de-energized because the thermocouple will

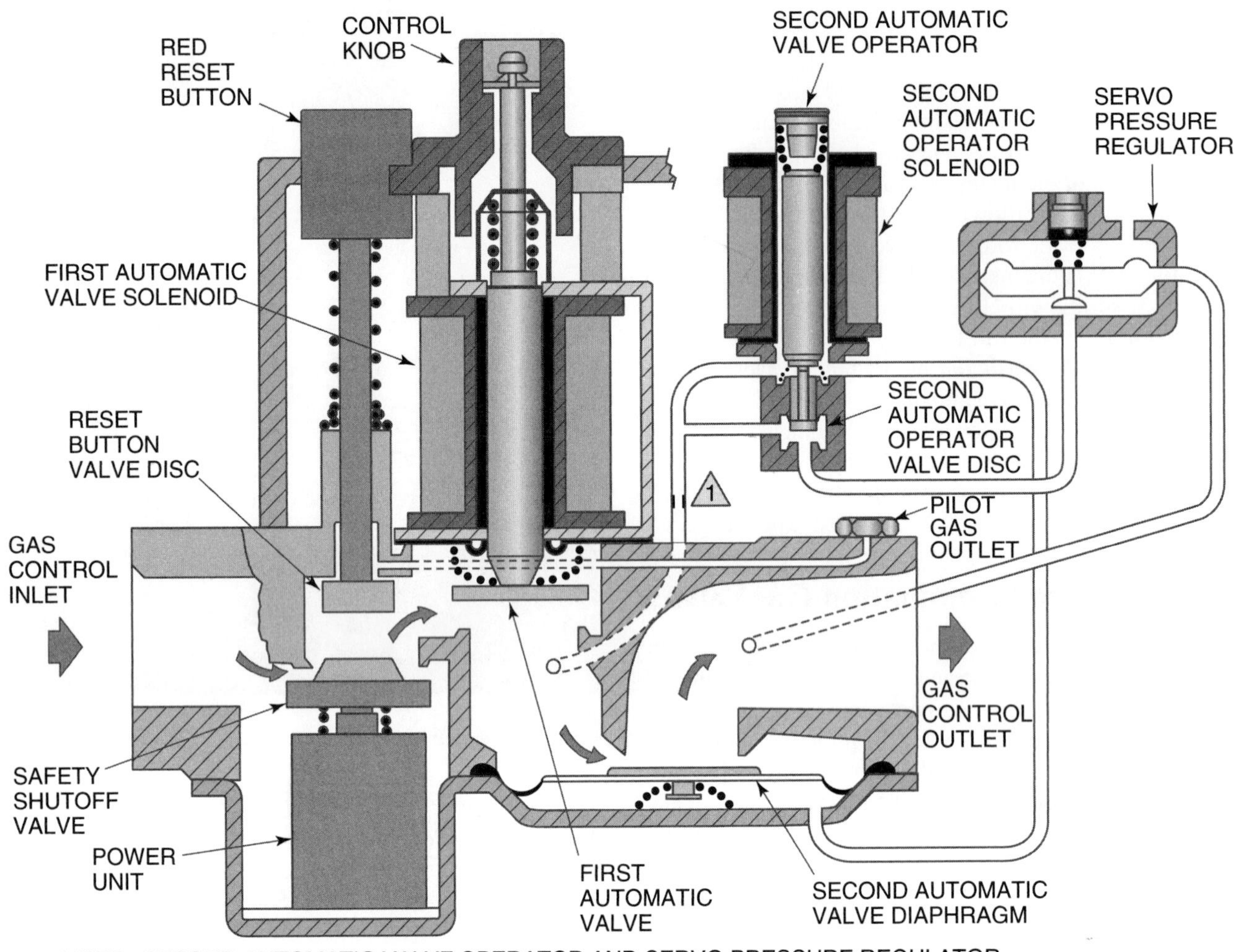

Figure 7-37. A standing pilot combination gas valve with servo pressure regulation. *Courtesy Honeywell, Inc., Residential Division*

quickly stop generating a millivoltage as it cools. This action will stop all gas flow.

The power unit does not have enough power to pull the assembly open once closed because it is simply a holding coil. Once closed, pushing the red reset button on the combination gas valve will manually reset the unit. The red reset button can only be pushed down when it is in the pilot position. Figure 7-38 shows how a safety shutoff mechanism assembly works along with its shutoff circuit.

Once the gas passes the safety shutoff valve, it comes to the first automatic valve. The first automatic valve is controlled by the first automatic valve solenoid coil. The closing of the room thermostat energizes its solenoid. The valve is spring-loaded and opens against flowing gas pressure. When the solenoid is de-energized, the spring closes the valve. This is one of the valve's built-in safeties.

Once the room thermostat calls for heat, the first automatic valve is opened. The closing of the room thermostat also energizes the second automatic valve solenoid, which raises the second automatic operator valve disc. This in turn controls the position of the second automatic valve diaphragm. The second automatic valve diaphragm is a servo-operated valve, which means that its position is controlled by the pressure of gas coming to a chamber beneath its diaphragm. This modulates the valve diaphragm between the open and closed position. Putting more pressure under the valve diaphragm modulates the valve closed, and less pressure modulates it more in the open position.

The servo pressure regulator, an outlet or working pressure regulator, is an integral part of the second automatic gas valve. During the furnace's running cycle, the servo pressure regulator closely monitors the outlet or working pressure of the gas valve, even when inlet pressures and flow rates vary widely. Any gas outlet pressure change is instantly sensed and reflected back to the diaphragm in the servo pressure regulator. The servo pressure regulator then repositions its diaphragm and disc to change the flow rate through the servo pressure regulator valve. This changes the pressure under the second automatic valve diaphragm and thus the position of the diaphragm.

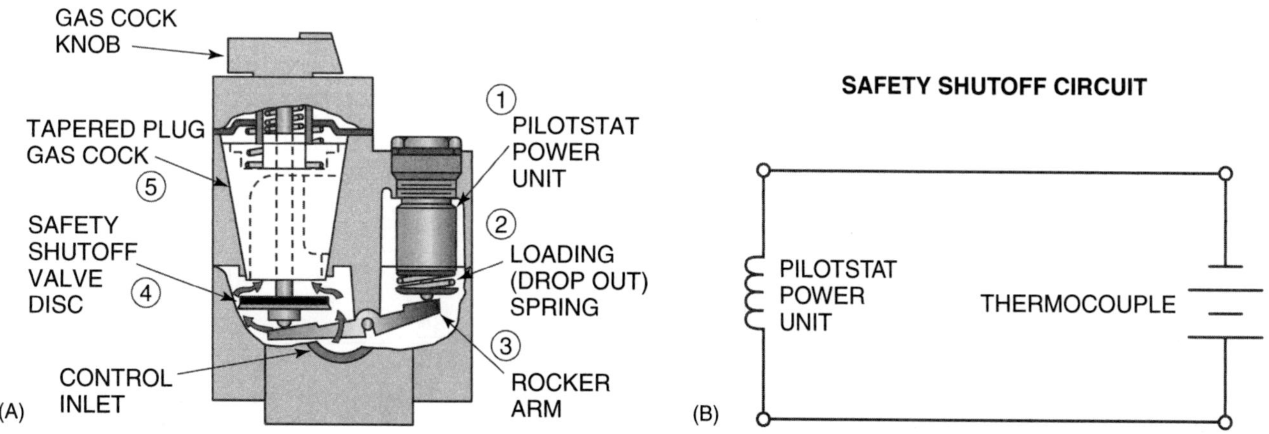

Figure 7-38. (A) A safety shutoff mechanism and (B) a safety shutoff circuit. *Courtesy Honeywell, Inc., Residential Division*

If the working pressure or outlet pressure of the gas valve rises, the servo pressure regulator diaphragm moves slightly higher. This allows less gas to bleed out of the servo pressure regulator to the gas valve outlet. However, it increases the pressure under the second automatic valve diaphragm, causing the valve to close. This action reduces gas valve outlet pressure.

If the working pressure or outlet pressure of the gas valve drops, the servo pressure regulator diaphragm moves slightly lower. This allows more gas to bleed out of the servo pressure regulator to the gas valve outlet. This decreases the pressure under the second automatic valve diaphragm, causing the valve to open. This action increases gas valve outlet pressure.

The servo pressure regulation system is a self-balancing system that works off gas valve outlet or working pressure and is independent of flow rate. It can operate over a wide range of gas flow rates. The gas valve's outlet pressure is adjusted at the top of the servo pressure regulator's body. Putting more or less tension on a spring will change the outlet or working gas valve pressure.

Intermittent Pilot Automatic Gas Valves

In an intermittent pilot (IP) automatic gas valve system, the pilot is lit every time there is a call for heat by the thermostat. Once the pilot is lit and proved, it lights the main burner. When the call for heat ends, the pilot is extinguished and does not relight until the next call for heat. In other words, there is automatic and independent control of the gas flow for both the pilot and the main burner.

Figure 7-39 shows the internals of a typical combination gas valve for an IP system. Note that there is no power unit or reset button. In fact, there is no thermocouple circuit either. There are simply two automatic valves. The first valve has an electric solenoid operator, and the second valve is servo operated. The passage for the pilot gas is located between these two main valves.

The first automatic valve solenoid is energized by an electronic module or integrated furnace controller (IFC) when there is a call for heat from the thermostat. This opens the passage for pilot gas to flow. The electronic module or IFC has about 90 sec to light the pilot gas. Four electronic modules and an IFC are shown in Figure 7-40. Once the pilot is lit and proved, usually through flame rectification, the ignition module or IFC energizes the second automatic valve solenoid. Flame rectification will be covered in detail later in this unit.

Because the second automatic valve is servo operated, the second main valve will open and the main burner will be lit. Its servo operator will operate the same way as described earlier for the standing pilot automatic gas valve. The main burner and the pilot burner will remain lit until the room thermostat opens and the call for heat ends. Both automatic valves will close once the call for heat ends. This action will shut off the main burner and pilot burner until the room thermostat initiates another call for heat.

Direct Burner Automatic Gas Valves

In a direct burner automatic gas valve system, the electronic module or IFC lights the main burner directly without a pilot flame. A spark, a hot surface igniter, or a glow coil usually accomplishes ignition. These types of ignition will be covered in detail later in this unit.

Figure 7-41 shows the internals of a typical combination gas valve for a direct burner system. Notice that there is no power unit or reset button. There are two automatic valves, and one is servo operated. The servo-operated valve will operate the same way as described for both the standing pilot and intermittent pilot combination gas valves.

In this system, when the room thermostat calls for heat, an electronic module or IFC energizes both first and second automatic valve solenoids at the same time along with the ignition source. Remember, there is no pilot in a direct burner system. Gas flows to the main burner and must be ignited within a short period of time, usually 4 to 12 sec. The reason for the short time period is safety because of the large flow of gas to the main burners compared to the small flow of gas to a pilot burner. Once the main burner is lit and proved,

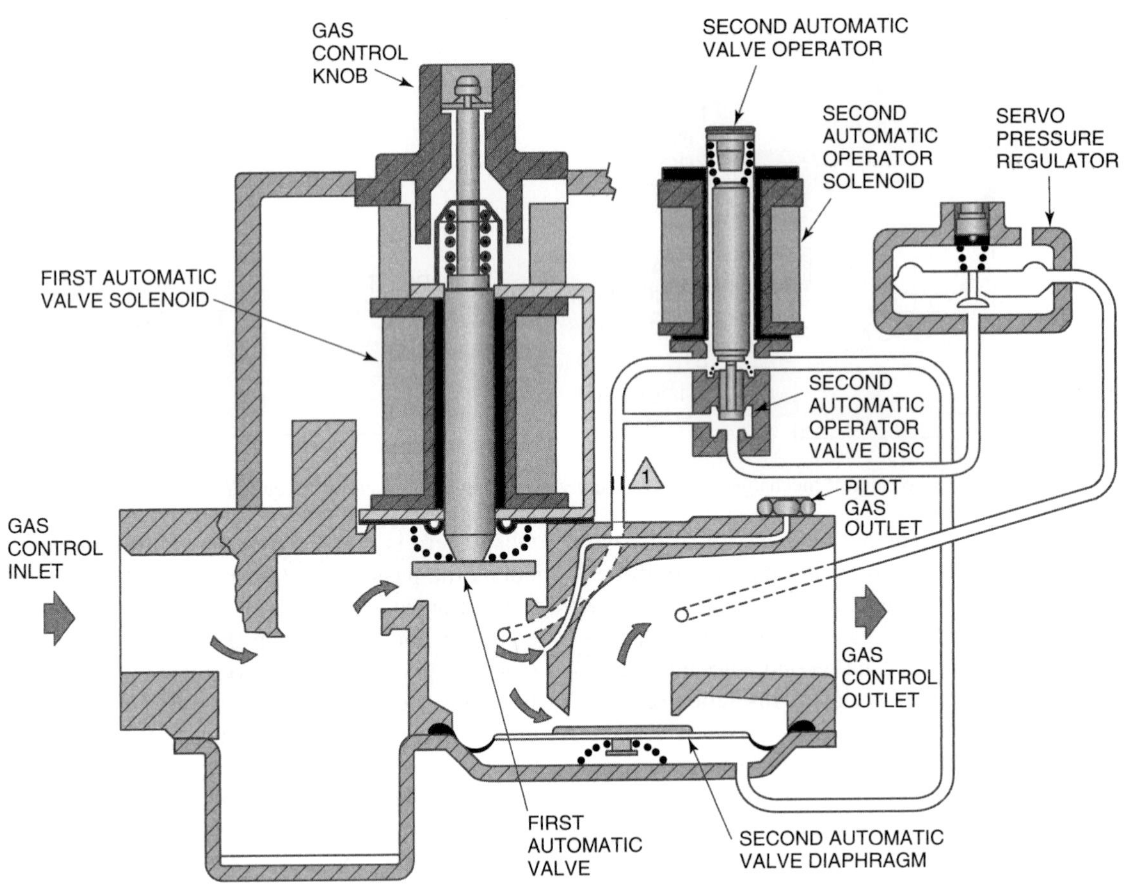

Figure 7-39. An intermittent pilot combination gas valve with servo pressure regulation. *Courtesy Honeywell, Inc., Residential Division*

(A)

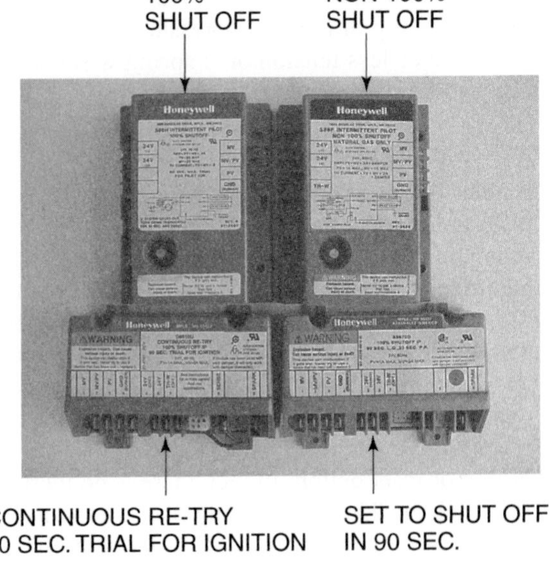

(B)

Figure 7-40. (A) An integrated furnace controller (IFC). (B) Four electronic modules for gas furnaces. *(A) and (B) Courtesy Ferris State University, Photos by John Tomczyk*

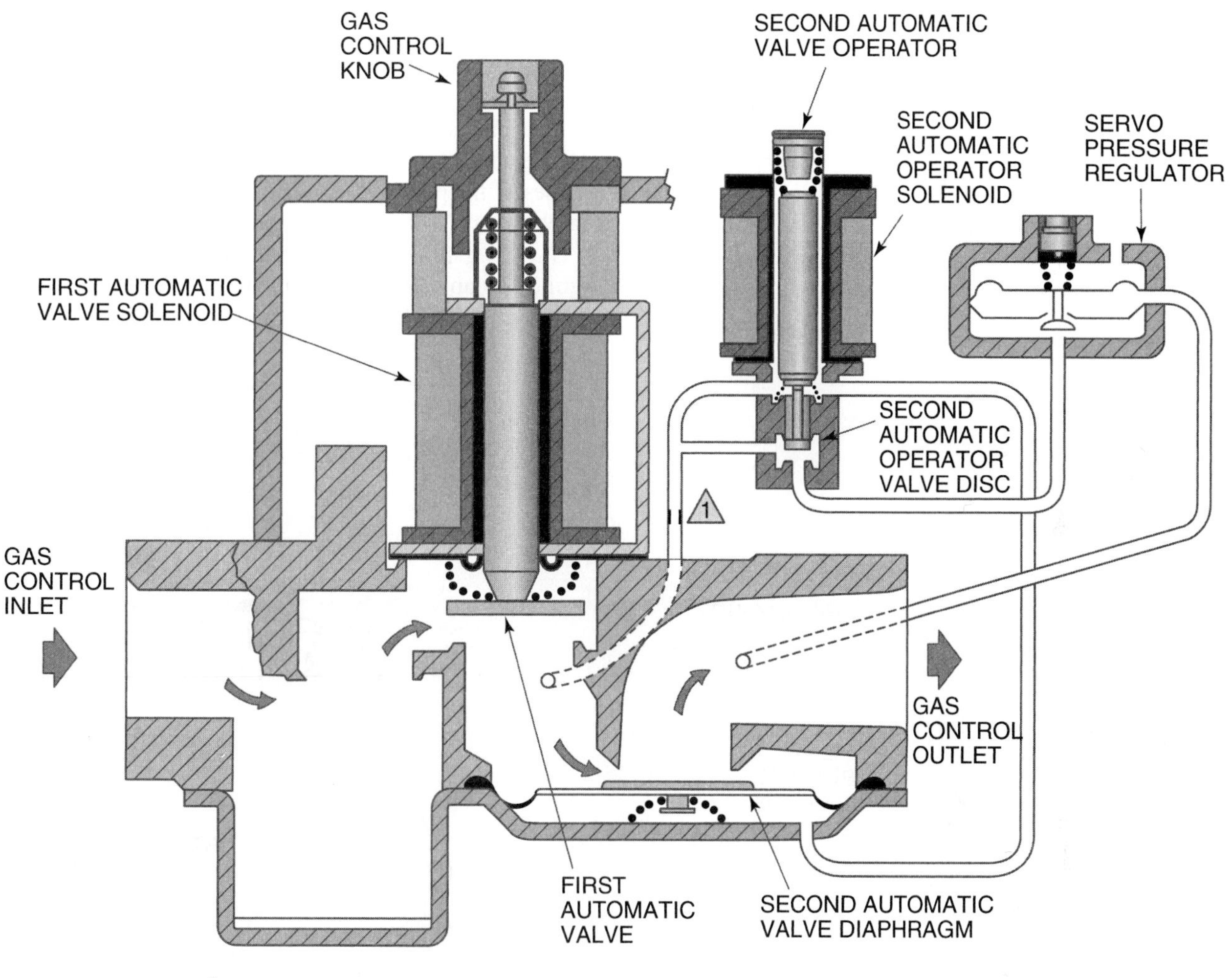

Figure 7-41. A direct burner combination gas valve with servo pressure regulation. *Courtesy Honeywell, Inc., Residential Division*

usually through flame rectification, the furnace will run until the room thermostat opens and the call for heat ends. If the main burner flame is not proved, gas flow will be shut off to the main burner through the electronic module or IFC.

Most modern combination gas valves have a built-in slow opening feature. The second automatic valve opens slowly, letting the flame grow slowly in the main burner. This eliminates noise and concussion in the heat exchanger by giving the main burner a slow and controlled ignition. Slow opening may be accomplished through a restriction in the passage to the second automatic valve operator. This restriction controls and limits the rate at which gas pressure builds in the servo pressure regulator system. The restriction is shown in Figure 7-39.

All three types of gas valves—standing pilot, intermittent pilot, and direct burner—are classified as redundant gas valves. A redundant gas valve has two or three valve operators physically in series but wired in parallel with one another. This built-in safety feature allows any operator (pilot or main) to block gas from getting to the main burner. Figures 7-36, 7-39, and 7-40 are all examples of redundant gas valves. Most gas valves manufactured today are redundant gas valves.

SAFETY PRECAUTION: Most combination gas valves look alike. Their model number may be the only way to differentiate them. The correct combination gas valve must be used in each application, or serious injury can occur. Always contact the furnace or gas valve manufacturer if you are unsure of what gas valve to use.

7.9 Pilots

Pilot flames are used to ignite the gas at the burner on most conventional gas furnaces. Pilot burners can be aerated or nonaerated. In the aerated pilot the air is mixed with the gas before it enters the pilot burner, Figure 7-42.

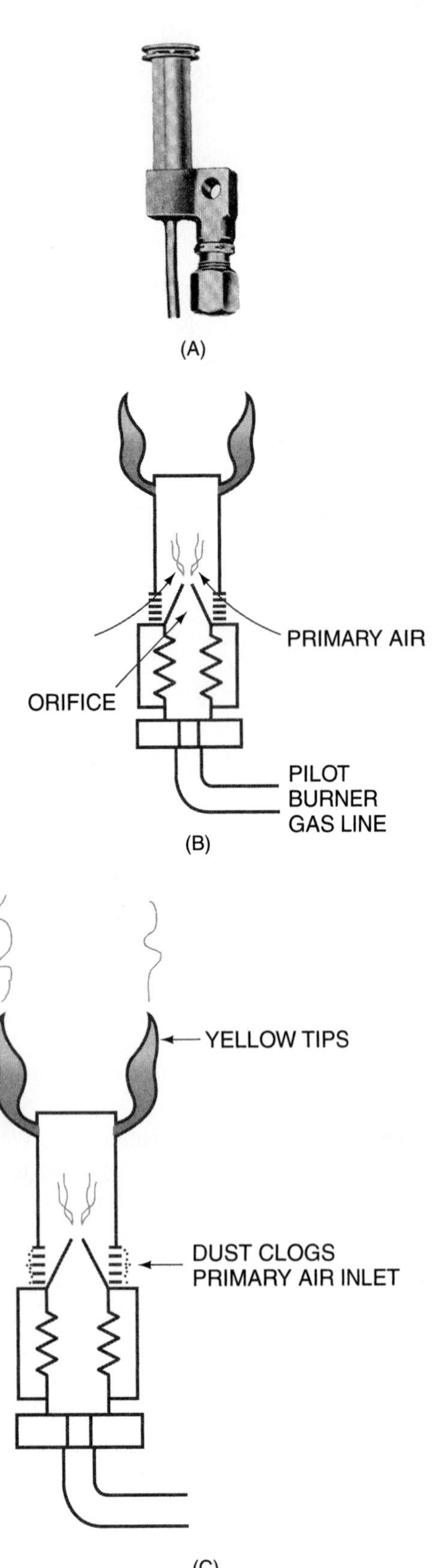

Figure 7-42. (A) The aerated pilot. (B) Cutaway. (C) Clogged, and starved for air. *Courtesy Robertshaw Controls Company*

The air openings, however, often clog and require periodic cleaning if there is dust or lint in the air. Nonaerated pilots use only secondary air at the point where combustion occurs. Little maintenance is needed with these, so most furnaces are equipped with nonaerated pilots, Figure 7-43.

The pilot is actually a small burner, Figure 7-44. It has an orifice, similar to the main burner, through which the gas passes. If the pilot goes out or does not perform properly, a safety device will stop the gas flow.

Standing pilots burn continuously; other pilots are ignited by an electric spark or other ignition device when the thermostat calls for heat. In furnaces without pilots, the ignition system ignites the gas at the burner. In furnaces

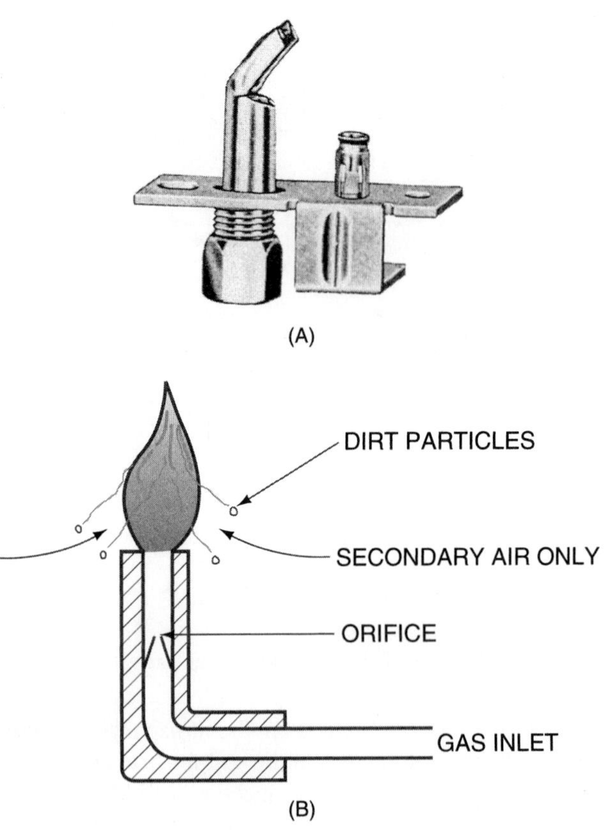

Figure 7-43. (A) A nonaerated pilot. (B) Cutaway. Notice how dust particles burn away. *Courtesy Robertshaw Controls Company*

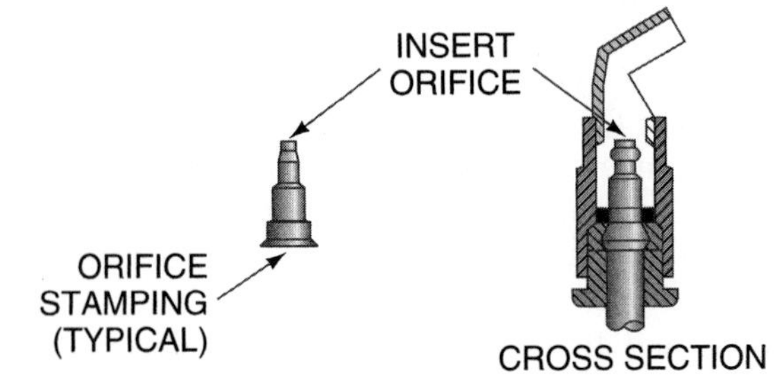

Figure 7-44. A nonaerated pilot burner showing an orifice.

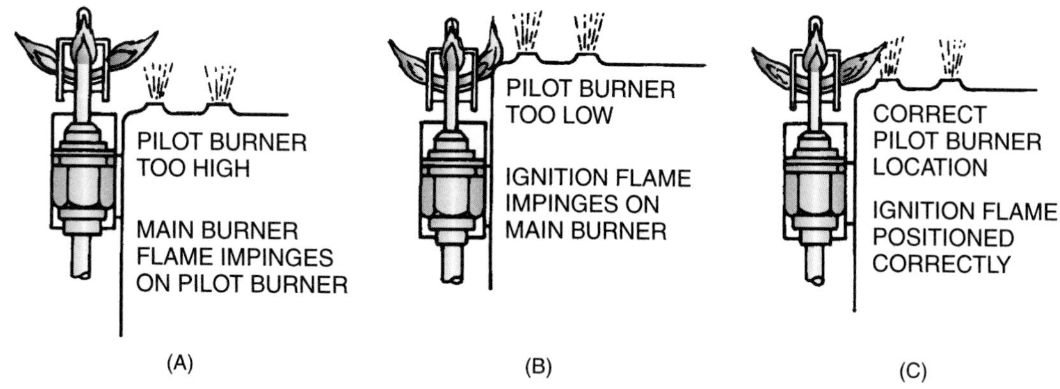

Figure 7-45. (A)–(C) The pilot flame must be directed at the burners and adjusted to the proper height.

with pilots, the pilot must be ignited and burning, and this must be proved before the gas valve to the main burner will open.

The pilot burner must direct the flame so there will be ignition at the main burners, Figure 7-45. The pilot flame also provides heat for the safety device that shuts off the gas flow if the pilot flame goes out.

7.10 Safety Devices at the Standing Pilot

Three main types of safety devices, called *flame-proving* devices, keep the gas from flowing through the main valve if the pilot flame goes out: the thermocouple or thermopile, the bimetallic strip, and the liquid-filled remote bulb.

Thermocouples and Thermopiles

The *thermocouple* consists of two dissimilar metals welded together at one end, Figure 7-46, called the "hot junction." When this junction is heated, it generates a small voltage (approximately 15 mV with load; 30 mV without load) cross the two wires or metals at the other end. The other end is called the "cold junction." The thermocouple is connected to a shutoff valve, Figure 7-47. As long as the electrical current in the thermocouple energizes a coil, the gas can flow. If the flame goes out, the thermocouple will cool off in about 30 to 120 sec, and no current will flow and the gas valve will close. A *thermopile* consists of several thermocouples wired in series to increase the voltage. If a thermopile is used, it performs the same function as the thermocouple, Figure 7-48.

Bimetallic Safety Device

In the *bimetallic* safety device the pilot heats the bimetal strip, which closes electrical contacts wired to the gas safety valve, Figure 7-49(A). As long as the pilot flame heats the bimetal, the gas safety valve remains

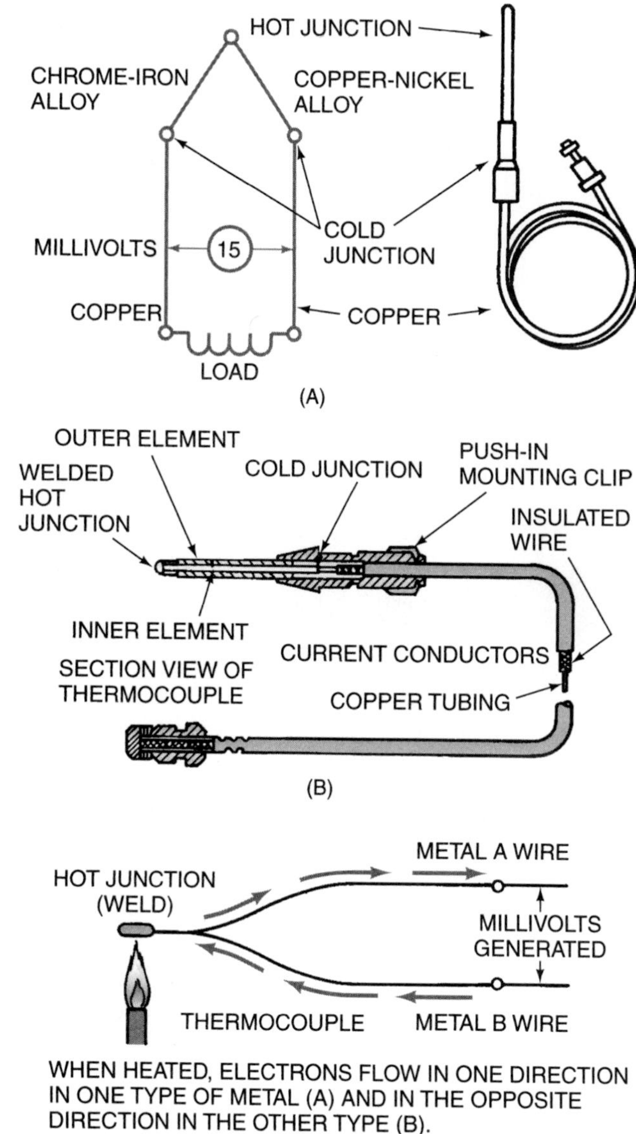

Figure 7-46. (A)–(C) The thermocouple. *Courtesy Robertshaw Controls Company*

energized and gas will flow when called for. When the pilot goes out, the bimetal strip cools within about 30 sec and straightens, opening contacts and causing the valve to close, Figure 7-49(B, C).

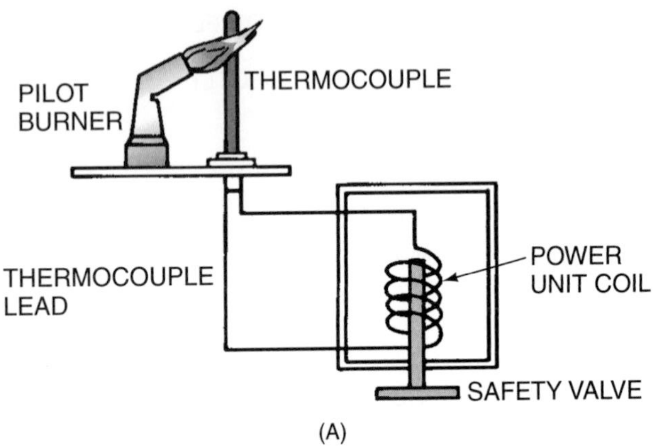

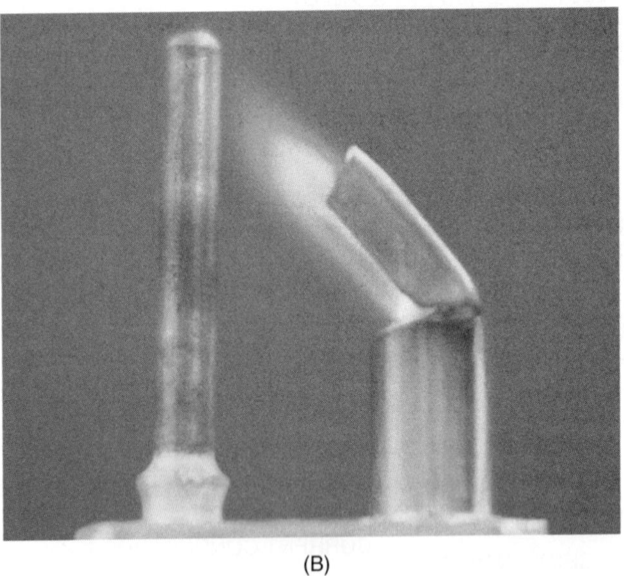

Figure 7-47. The thermocouple generates electrical current when heated by the pilot flame. This induces a magnetic field in the coil of the safety valve holding it open. If flame is not present, the coil will deactivate, closing the valve, and the gas will not flow. *Photo by Bill Johnson*

Liquid-Filled Remote Bulb

The *liquid-filled remote bulb* includes a diaphragm, a tube, and a bulb, all filled with a liquid, usually mercury. The remote bulb is positioned to be heated by the pilot flame, Figure 7-50. The pilot flame heats the liquid at the remote bulb. The liquid expands, causing the diaphragm to expand, which closes contacts wired to the gas safety valve. As long as the pilot flame is on, the liquid is heated and the valve is open, allowing gas to flow. If the pilot flame goes out, the liquid cools in about 30 sec and contracts, opening the electrical contacts and closing the gas safety valve.

7.11 Ignition Systems

Ignition systems can ignite either a pilot or the main burner. They can be divided into three categories:

- **Intermittent pilot (IP)**
- **Direct-spark ignition (DSI)**
- **Hot surface ignition (HSI)**

Intermittent Pilot Ignition

In the intermittent pilot or spark-to-pilot type of gas ignition systems, a spark from the electronic module ignites the pilot, which ignites the main gas burners, Figure 7-51. The pilot burns only when the thermostat calls for heat. This system is popular because fuel is not wasted with the pilot burning when not needed. Two types of control schemes are used with this system. One is used a lot with natural gas and is not considered a 100% shutoff system. If the pilot does not ignite, the pilot valve will remain open, the spark will continue, and the main gas valve will not open until the pilot is lit and proved. The other type is used with LP gas and some natural gas applications and is a 100% shutoff system. If there is no pilot ignition, the pilot gas valve will close and may go into safety lockout after approximately 90 sec, and the spark will stop. Safety lockout systems will be covered in detail later in this unit. This system must be manually reset, usually at the thermostat or power switch.

The 100% shutoff is necessary for any gas fuel heavier than air, because it will accumulate in low places. Even the

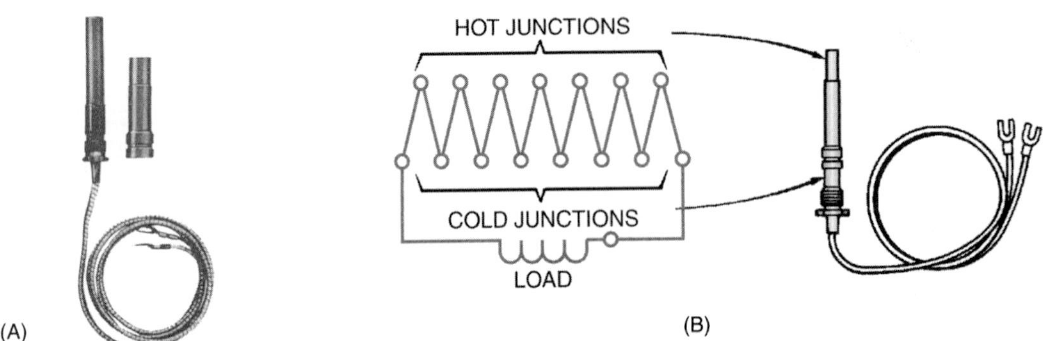

Figure 7-48. (A) A thermopile consists of a series of thermocouples in one (B) housing. *(A) Photo Courtesy Honeywell*

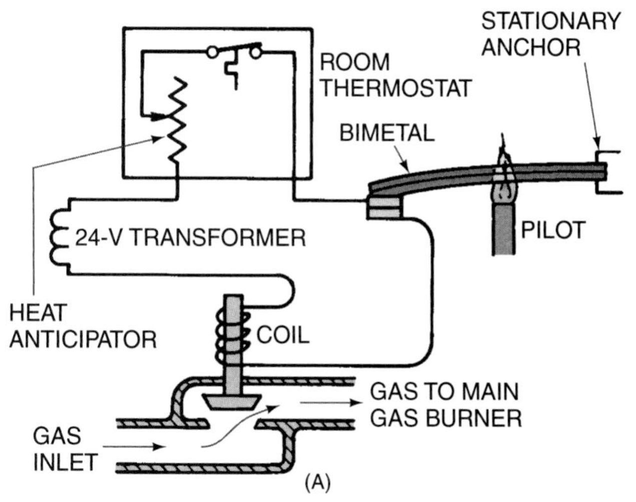

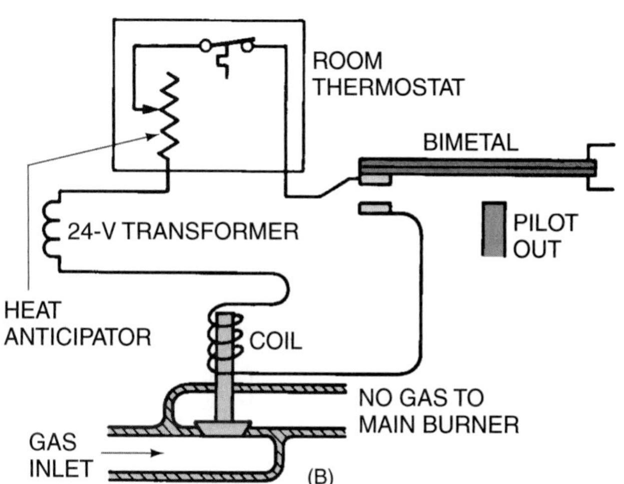

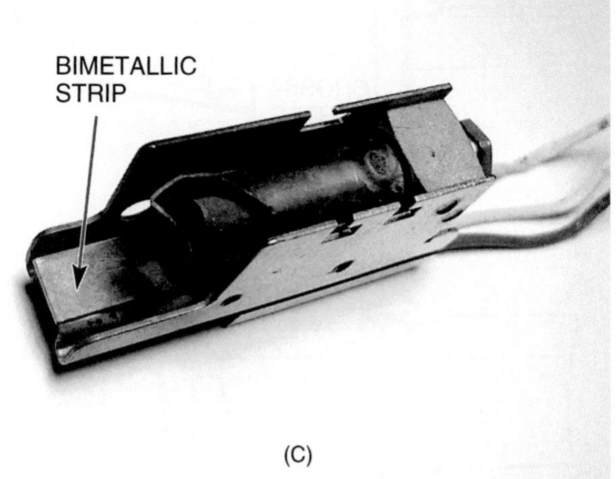

Figure 7-49. (A) When the pilot is lit, the bimetal warps, causing the contacts to close. The coil is energized, pulling the plunger into the coil opening the valve. This allows the gas to flow to the main gas valve. (B) When the pilot is out, the bimetal straightens, opening the contacts. The safety valve closes. No gas flows to the furnace burners. (C) A bimetallic safety device assembly. *Courtesy Ferris State University, Photo by John Tomczyk*

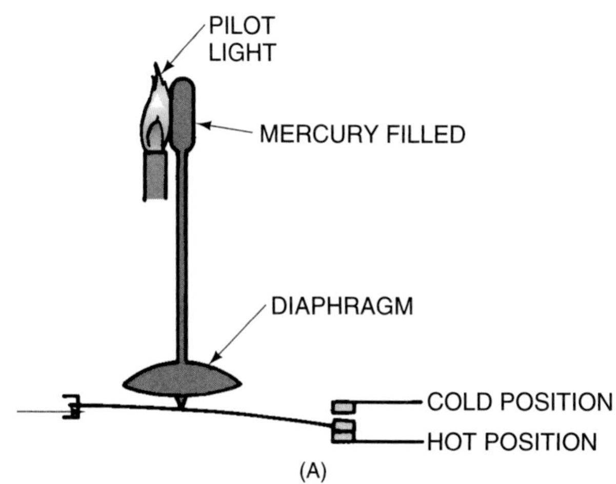

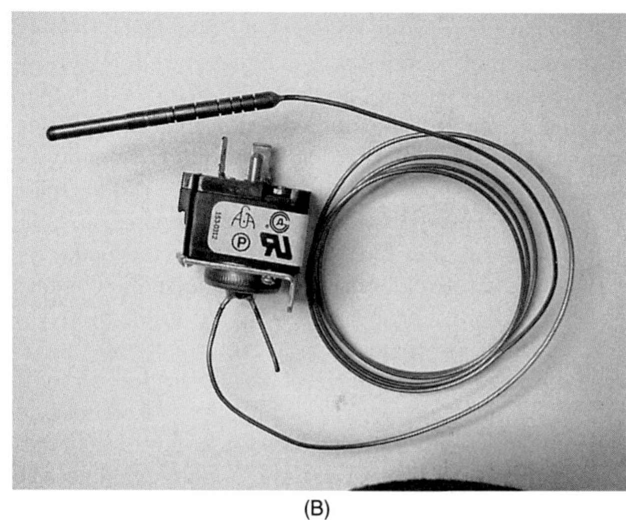

Figure 7-50. (A) A liquid-filled remote bulb. (B) A liquid-filled remote bulb assembly. *Courtesy Ferris State University, Photo by John Tomczyk*

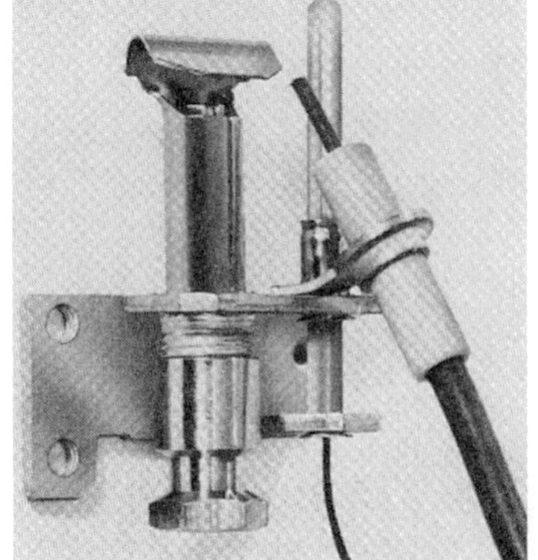

Figure 7-51. A spark-to-pilot ignition system. *Courtesy Robertshaw Controls Company*

small amount of gas from a pilot light could be dangerous over time. Natural gas rises up through the flue system and is not dangerous in small quantities, such as the volume from a pilot light. Even though natural gas is lighter than air, some codes call for 100% shutoff. All technicians that service fuel gas products should be totally aware of the local codes. Both 100% shutoff and less than 100% shutoff systems will be covered in detail later in this unit.

When there is a call for heat in the natural gas system, contacts will close in the thermostat, providing 24 V to the ignition module which will send power to the pilot igniter and to the pilot valve coil. The coil opens the pilot valve, and the spark ignites the pilot.

Once the pilot flame is lit, it must be proved. Flame rectification is the fastest, safest, and most reliable method to prove a flame.

In the *flame rectification* system, the pilot flame changes the normal alternating current to direct current. The electronic components in the system will energize and open the main gas valve only with a direct current measured in microamps. It is important that the pilot flame quality be correct to ensure proper operation. Consequently, the main gas valve will open only when pilot flame is present. Flame rectification will be covered in detail later in this unit.

The spark is intermittent and arcs approximately 100 times per minute. It must be a high-quality arc, or the pilot will not ignite. The arc comes from the control module and is very high voltage. The voltage can reach 10,000 V in some systems but is usually very low amperage. A direct path to earth ground must be provided because the arc actually arcs to ground. A ground strap near the pilot assembly or the pilot hood often acts as ground, Figure 7-52.

Direct-Spark Ignition (DSI)

Many modern furnaces are designed with a spark ignition direct to the main burner or a ground strap, Figure 7-52. No pilot is used in this system. Components in the system are the igniter/sensor assembly and the DSI module. The sensor rod verifies that the furnace has fired and sends a microamp signal through flame rectification to the DSI module confirming this. The furnace will then continue to operate. This system goes into a "safety lockout" if the flame is not established within the "trail for ignition" period (approximately 4 to 11 sec). Gas is being furnished to the main burner, so there cannot be much time delay compared to a 90 sec trial for ignition for IP systems. The system can then only be reset by turning the power off to the system control and waiting 1 min before reapplying the power. This is a "typical" system. The technician should follow the wiring diagram and manufacturer's instructions for the specific furnace being installed or serviced.

Most ignition problems are caused from improperly adjusted spark gap, igniter positioning, and bad grounding, Figure 7-53. The igniter is centered over the left port. Most manufacturers also provide specific troubleshooting instructions. Once the main burner flame is proved by the flame rectification system, the sparking will stop. If the sparking does not stop once the main burner is lit, the electronic DSI module could be defective. The continual sparking may not be harmful to the system, but it is noisy and often can be heard in the living or working area. Read and follow the manufacturer's instructions to repair the system.

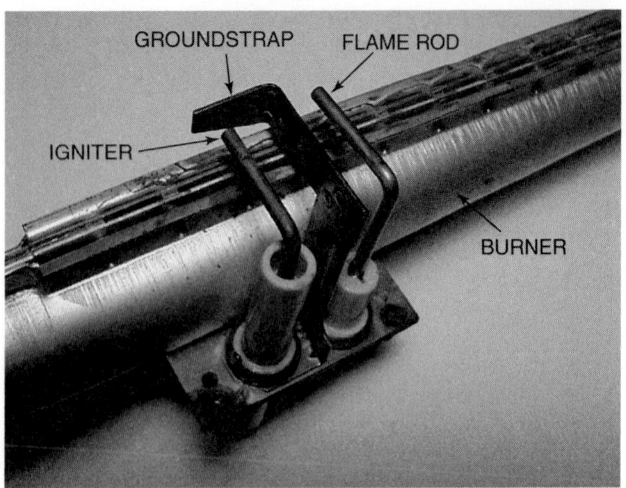

Figure 7-52. A direct-spark ignition assembly near the gas burner. *Courtesy Ferris State University, Photo by John Tomczyk*

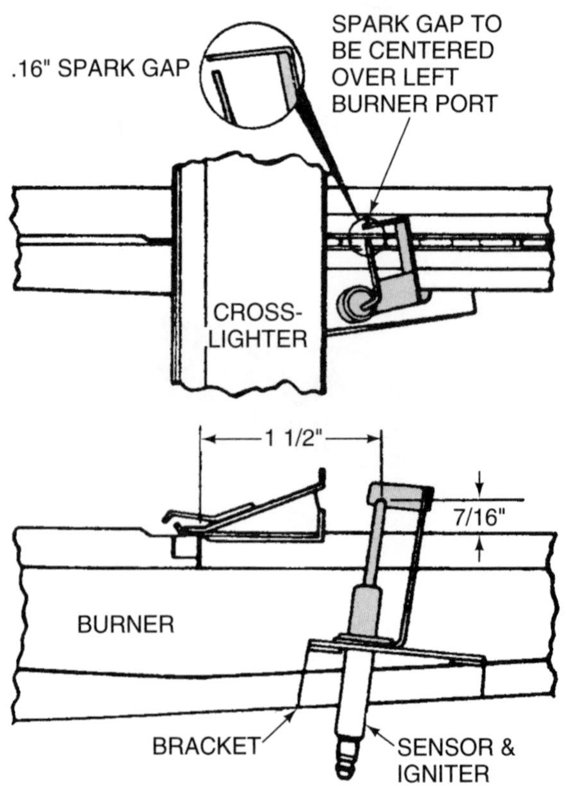

Figure 7-53. The spark gap and igniter position for a DSI system. *Courtesy Heil-Quaker Corporation*

Hot Surface Ignition

The hot surface ignition (HSI) system uses a special product called silicon carbide that offers a high resistance to electrical current flow but it is very tough and will not burn up, like a glow coil. This substance is placed in the gas stream and is allowed to get very hot before the gas is allowed to impinge on the glowing hot surface. Immediate ignition should occur when the gas valve opens.

The hot surface igniter is usually operated off 120 V, the line voltage to the furnace, and draws considerable current when energized. It is energized for only a short period of time during start-up of the furnace so this current is not present for more than a very few minutes per day. Figure 7-54 (A, B, C) shows the hot surface igniter.

Even though the HSI system has been used successfully for many years, some problems do occur. The igniter is very brittle and breaks very easily. If you bump it with a screwdriver, it will break.

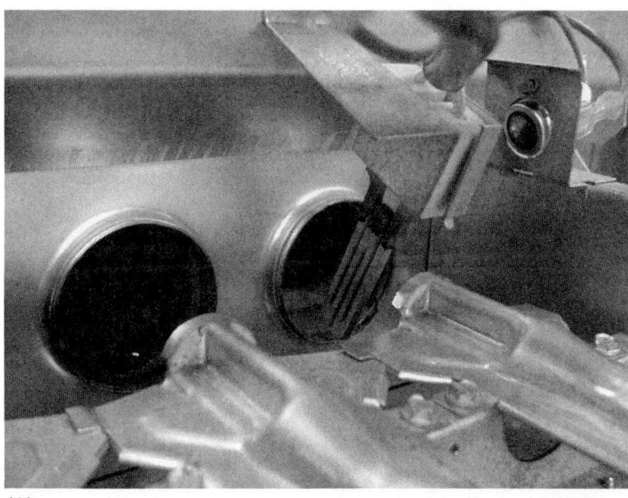
(A)

(B)

(C)

(D)

(E)

Figure 7-54. (A) A hot surface igniter de-energized. (B) The hot surface igniter system energized before main gas is allowed to flow. (C) The hot surface igniter system de-energized during the furnace heating cycle. (D) A 24-V hot surface igniter that lights a pilot flame. (E) The 24-V hot surface igniter energized. *(A)–(E) Courtesy Ferris State University, Photos by John Tomczyk*

If repeated failures occur with an HSI system, look for the following:

- **Higher than normal applied voltage. In excess of 125 V can shorten the igniter's life.**
- **Accumulation of drywall dust, fiberglass insulation, or sealant residue.**
- **Delayed ignition can stress the igniter due to the small explosion.**
- **Overfiring condition.**
- **Furnace short cycling as with a dirty filter may cause the furnace to cycle on high limit.**

The hot surface igniter may be used for lighting the pilot or for direct ignition of the main furnace burner. When it is used as the direct igniter for the burner, it will have very little time, usually 4 to 11 sec. to ignite the burner, then a safety lockout will occur to prevent too much gas from escaping.

Newer hot surface igniters are made of a different material that does not break as easily as older ones. Care should still be taken when working around hot surface igniters because of their fragile nature.

Some HSI systems operate on 24 V. These systems are usually hot surface to pilot, meaning that the 24-V hot surface igniter lights a pilot flame, Figure 7-54(D, E). The pilot flame has been the most reliable ignition source to light the main burner for many decades. Also, by using a low-voltage hot surface igniter, there is less chance for high-voltage electrical shock.

7.12 Flame Rectification

In the flame rectification system, the pilot flame, or main flame, can conduct electricity because if contains ionized combustion gases, which are made up of positively and negatively charged particles. The flame is located between two electrodes of different sizes. The electrodes are fed with an alternating current (AC) signal from the furnace's electronic module. Because of the different size of the electrodes, current will flow better in one direction than in the other, Figure 7-55(A). The flame actually acts as a switch. If there is no flame, the switch is open, and electricity will not flow. When the flame is present, the switch is closed and will conduct electricity.

The two electrodes in a spark-to-pilot (intermittent pilot) ignition system are the pilot hood (earth ground) and a flame rod or flame sensor, Figure 7-56. In a DSI system, the two electrodes can be the main burners (earth ground) and a flame rod or sensor. These types of systems are often referred to as dual-rod or remote sensing systems because one rod is for flame sensing and the other is for the ignition process. Often, the flame rod or sensor can be part of the ignition system and is referred to as a combination sensor and igniter, Figure 7-54. This type of system is referred to as a single-rod or local sensing system because it has a combination sensor and igniter in one rod.

As alternating current switches the polarity of the electrodes 60 times each second, the positive ions and negative electrons move at various speeds to the different size electrodes. This causes a rectified AC signal, which looks much like a pulsating DC signal, because the positively charged particles (ions) are relatively larger, heavier, and slower than the negatively charged particles (electrons), Figure 7-55(B). In fact, the electrons are about 100,000 times lighter than the ions. This DC signal is what the electronics of the furnace recognize. The DC signal is proof that a flame exists, rather than a short from humidity or direct contact between the electrodes. Humidity or direct contact between the electrodes would cause an AC current to flow, and the furnace's electronic module would not recognize the AC signal. The module only recognizes the DC signal. An AC signal would then prevent the main gas valve from opening in a spark-to-pilot system. It would shut down the main gas in a DSI system once the trial for ignition timing has elapsed and may put the furnace in a lockout mode. This DC signal can be measured with a microammeter in series with one of the electrode leads. This signal usually ranges between 1 and 25 microamps. The magnitude of the microamp signal may depend on the quality, size, and stability of the flame and on the electronic module and electrode design. Always consult with the furnace manufacturer for specific information on flame rectification measurements.

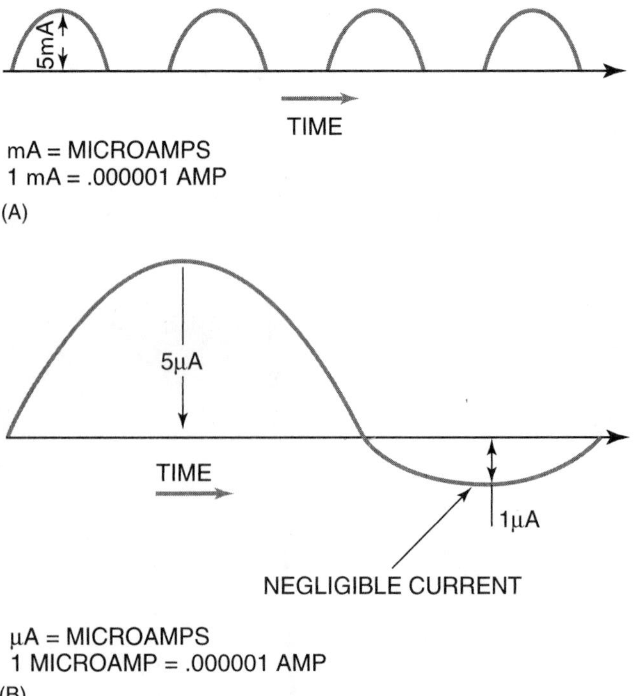

Figure 7-55. (A) The different size of the electrodes causes current to flow better in one direction than the other. (B) A pulsating direct current (DC) signal that can be measured in microamps with a microammeter in series with one of the electrodes.

FLAME RECTIFICATION
PILOT & PROBE
PILOT FLAME
PILOT HOOD–NEGATIVE
PROBE TO GROUND AND
SENSING CIRCUIT
FLAME
SENSOR

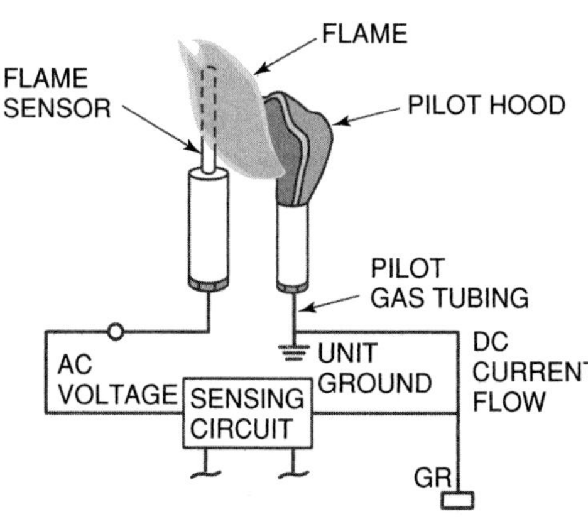

Figure 7-56. An intermittent pilot dual-rod system showing the pilot hood and the flame sensor rod as the two electrodes for flame rectification. Dual-rod systems are often referred to as remote sensing systems. *Courtesy Johnson Controls*

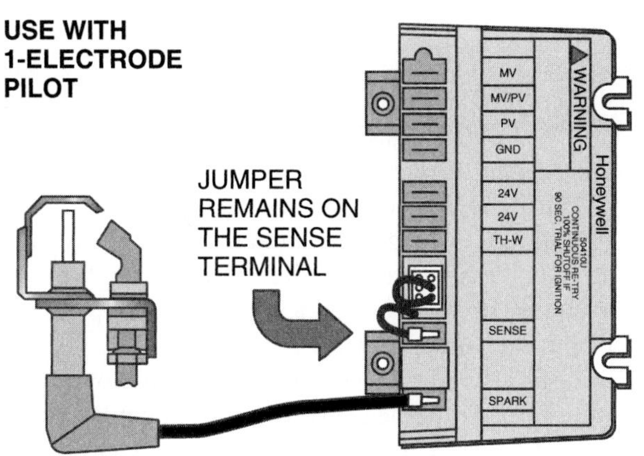

Figure 7-57. A single-rod or local sensing system connected to a furnace electronic module. *Courtesy Honeywell*

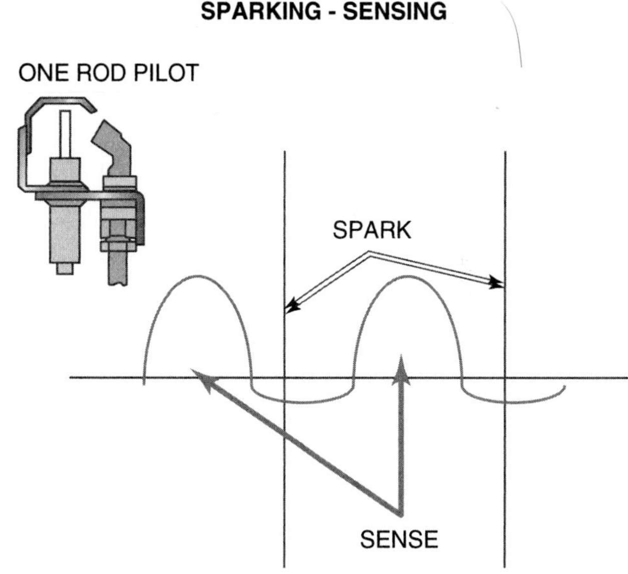

Figure 7-58. Single-rod or local sensing flame rectification senses flame current during one half cycle and the ignition pulse during the other half cycle. *Courtesy Honeywell*

Single-rod or local sensing consists of an igniter and a sensor all in one rod. It is often referred to as a one-rod pilot. Single-rod flame rectification systems use a single rod to accomplish both ignition and sensing. There is only one wire, which is the large ignition wire, running from the pilot assembly to the ignition module, Figure 7-57. This high-voltage lead does both the sparking and the sensing. The same rod used for ignition is used for flame rectification. The ignition pulse is accomplished on one half cycle and the sensing pulse current is accomplished on the other half cycle, Figure 7-58. Signal analyzers within the electronic module can differentiate between the two signals.

Dual-rod or remote sensing uses a separate igniter and flame-sensing rod or electrode, Figure 7-59. The sparking is accomplished with one rod and the sensing is done with the other rod. There are two wires running from the pilot assembly to the module. The spark electrode wire goes to the spark terminal on the module. The flame rod or sensing wire goes to the sense terminal. A jumper wire will have to be cut off on some models, Figure 7-60. The flame sensor works with the grounded burner hood and the ground strap to achieve flame rectification.

The examples in the preceding two paragraphs are only a few of the furnace modules seen in the field today. Many newer furnaces have much more sophisticated IFCs. However, the principles of flame rectification still hold true for these modern furnaces.

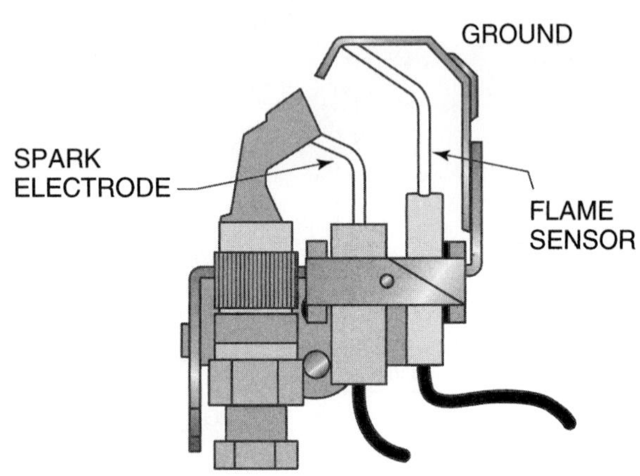

Figure 7-59. A dual-rod or remote sensing flame rectification system showing a separate igniter spark electrode and flame sensing rod. *Courtesy Honeywell*

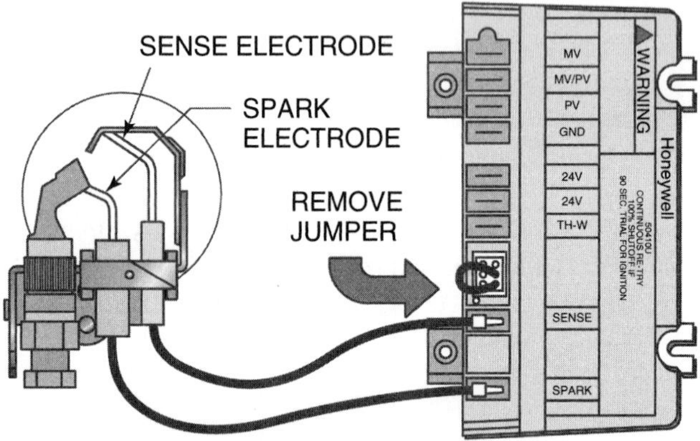

Figure 7-60. A dual-rod or remote sensing flame rectification system showing two separate wires going to the electronic furnace module. *Courtesy Honeywell*

Summary

- Forced-hot air furnaces are normally classified as upflow, downflow (counterflow), horizontal, low-boy, or multipoise.
- Multipoise furnaces can be installed in any position, which adds versatility. Multiple safety controls like bimetallic flame rollout or over-temperature limits must be designed into the furnace for safety.
- Furnace components consist of the cabinet, gas valve, manifold, pilot, burners, heat exchangers, blower, electrical components, and venting system.
- Gas fuels are natural gas, manufactured gas (used in few applications), and LP gas (propane, butane, or a mixture of the two).
- Inches of water column is the term used when determining or setting gas fuel pressures.
- For combustion to occur there must be fuel, oxygen and heat. The fuel is the gas, the oxygen comes from the air, and the heat comes from the pilot flame or other igniter.
- Gas burners use primary and secondary air. Excess air is always supplied to ensure as complete combustion as possible.
- Gas valves control the gas flowing to the burners. The valves are controlled automatically and allow gas to flow only when the pilot is lit or when the ignition device is operable.
- Some common gas valves are classified as solenoid, diaphragm, or heat motor valves.
- A servo pressure regulator is an outlet or working pressure regulator. It monitors gas outlet pressure when gas flow rates and inlet pressures vary widely.
- A redundant gas valve has two or three valve operators physically in series with one another but wired in parallel.
- Ignition at the main burners is caused by heat from the pilot or from an electric spark. There are standing pilots that burn continuously, and intermittent pilots that are ignited by a spark when the thermostat calls for heat. There is also a direct-spark or direct burner ignition, in which the spark or hot surface igniter ignites the gas at the burners.
- The thermocouple, the bimetal, and the liquid-filled remote bulb are three types of safety devices (flame-proving devices) to ensure that gas does not flow unless the pilot is lit.
- Modern gas furnaces use either induced- or forced-draft systems. Induced-draft systems have a combustion blower motor located on the outlet of the heat exchanger. They pull or suck combustion gases through the heat exchanger, usually causing a slight negative pressure in the heat exchanger itself.
- Forced-draft systems have a combustion blower motor on the inlet of the heat exchanger. They push or blow combustion gases through the heat exchanger and cause a positive pressure in the heat exchanger.
- The gas burns in an opening in some heat exchangers. Air passing over the heat exchanger is heated and circulated to the conditioned space.
- A blower circulates this heated air. The blower is turned on and off by a fan switch, which is controlled either by time or by temperature.
- Modern heat exchangers come in many sizes, shapes, and materials. Many are L-shaped and S-shaped and made of aluminized steel. There is actually a serpentine path for combustion gases to travel. Serpentine heat exchangers use inshot burners instead of ribbon or slotted port burners.
- Modern condensing furnaces may have two or even three heat exchangers. The last heat exchanger is usually made of stainless steel or aluminum to resist corrosion. It is designed to intentionally handle condensate and eventually drain it away from the furnace.
- Many modern gas furnaces use electronic modules or integrated furnace controllers (IFCs) to control operations. These controllers often have dual in-line pair (DIP) switches, which can be switched to reprogram the furnace to make it more versatile.

- In a flame rectification system, the pilot flame, or main flame, can conduct electricity because it contains ionized combustion gases that are made up of positively and negatively charged particles. The flame is located between two electrodes of different sizes. The electrodes are fed with an alternating current from the furnace's electronic module or controller. Current will flow in one direction more than the other because of the different size electrodes. The flame acts as a switch. When the flame is present, the switch is closed and will conduct electricity.
- Flame rectification systems can be classified as either single rod (local sensing) or dual rod (remote sensing). Single-rod systems consist of an igniter and a sensor all in one rod. Dual-rod systems use a separate igniter and flame-sensing rod.
- A direct-vented, high-efficiency gas furnace has a sealed combustion chamber. This means that the combustion air is brought in from the outside through an air pipe, which is usually made of PVC plastics. A non-direct-vented, high-efficiency gas furnace uses indoor air for combustion.
- A positive pressure system can be vertically or sidewall vented. Its flue pipe pressure is positive the entire distance to its terminal end.
- The dew point temperature is the temperature at which the condensation process begins in a condensing furnace. The DPT varies depending on the composition of the flue gas and the amount of excess air. As excess air decreases, the DPT increases.
- Excess air consists of combustion air and dilution air. Combustion air is primary air and/or secondary air. Primary air enters the burner before combustion takes place. Secondary air happens after combustion and supports combustion.
- Dilution air is excess air after combustion and usually enters at the end of the heat exchanger.
- Furnace efficiency ratings are determined by the amount of heat that is transferred to the heated medium.
- Electronic ignition modules come in 100% shutoff, non-100% shutoff, continuous retry with 100% shutoff, and many other custom control schemes that are manufacturer dependent. The combustion blower can prepurge, interpurge, or postpurge the heat exchanger depending on the modules' control scheme.
- One of the first steps in systematically troubleshooting a furnace with an integrated furnace controller is to obtain the electrical diagram of the furnace and sequence of operation of the IFC.
- SAFETY PRECAUTION: Venting systems must provide a safe and effective means of moving the flue gases to the outside atmosphere. Flue gases are mixed with other air through the draft hood. Venting may be by natural draft or by forced draft. Flue gases are corrosive.
- A cast iron boiler is manufactured for small applications as well as for midsized commercial use.
- A fire-tube boiler typically has a large; long tube that the burner fires within.

Review Questions

1. The four types of gas furnace air flow patterns are ________, ________, ________ and ________
2. Describe the function of a multipurpose or multipositional furnace.
3. The specific gravity of natural gas is:
   A. .08 C. .42
   B. 1.00 D. .60
4. What is the approximate oxygen content of air?
5. The typical manifold pressure for propane is ________ in WC.
6. The purpose of the gas regulator at a gas burning appliance is to:
   A. Increase the pressure to the burner.
   B. Decrease the pressure to the burner.
   C. Filter out any water vapor.
   D. Cause the flame to burn yellow.
7. True or False: All gas valves snap open and closed.
8. Which of the following features does an automatic combination gas valve have?
   A. Pressure regulator.
   B. Pilot safety shutoff.
   C. Redundant shutoff feature.
   D. All of the above.
9. What is the function of the servo pressure regulator in a modern combination gas valve?
10. What is a redundant gas valve?
11. Briefly describe an intermittent pilot gas valve system.
12. Briefly describe a direct burner gas valve system.
13. What is an integrated furnace controller (IFC)?
14. Describe the two types of pilot lights.
15. True or False: A thermocouple develops direct current and voltage.
16. Describe how a thermocouple flame-proving system functions.
17. Describe how a bimetallic flame-proving system functions.
18. Describe how a liquid-filled flame-proving system functions.
19. Explain why the preceding flame-proving systems are called safety devices.
20. Describe a flame rectification flame-proving system.
21. What is a hot surface ignition system?
22. Describe the difference between a single-rod (local sensing) and a dual-rod (remote sensing) system as they pertain to flame rectification.
23. Flame current is measured in ________.
24. What is a direct-vented furnace?
25. What is meant by a furnace that has a positive pressure venting system?
26. Define the dew point temperature as it applies to a condensing furnace.
27. What are four factors that determine the efficiency of a furnace?
28. What is a 100% shutoff system?
29. A ________ boiler is manufactured for small applications as well as for midsize commercial use.

8 Gas-Burning Equipment

Objectives

Upon completion of this unit, you should be able to

- **state the difference between a natural gas and an LP gas system.**
- **describe several control sequences for gas heat and warm-air furnaces.**
- **describe the difference between a standard-efficiency furnace and a high-efficiency furnace.**

8.1 Forced-Air Furnaces

Gas-fired, forced-air furnaces have a heat-producing system and a heated-air distribution system. The heat-producing system includes the manifold and controls, the burners, the heat exchanger, and the venting system. A forced-air furnace is one of two types: standard efficiency or high efficiency. The standard-efficiency furnace operates at approximately 80–85% combustion efficiency while in a steady state and running all the time. The high-efficiency furnace operates at a combustion efficiency of approximately 90–98%, extracting more heat from the flue gases than the standard-efficiency furnace does.

Regulation of Equipment

Manufacturers do not make equipment without a plan. Equipment of any kind is regulated and approved by various organizations. This equipment must be rated as to capacity so that the public receives what it pays for. These organizations are nonprofit trade organizations that determine the guidelines by which equipment is designed and manufactured. The various organizations work together with the manufacturers to establish reasonable guidelines, then national, state, and local enforcement agencies adopt the guidelines recommended by the organizations. One of the larger organizations is ASHRAE (American Society of Heating, Refrigerating and Air-Conditioning Engineers), which is worldwide. In this book, you will find many tables attributed to ASHRAE.

The AGA (American Gas Association) is an organization that works along with GAMA (Gas Appliance Manufacturers Association) to rate and approve all gas-burning equipment. ANSI (American National Standards Institute) is also instrumental in setting guidelines for gas furnaces. The electrical portion of each appliance is approved by UL (Underwriters Laboratories) for safety of the component.

Figure 8-1. (A) Standard upflow furnace. *Courtesy Tempstar* (B) Airflow.

Note: Before purchasing and installing a gas-burning appliance, you should look on the appliance for the seal of approval of one of these organizations unless your local code recognizes another testing laboratory or procedure.

Airflow

Standard forced-air furnaces are described as upflow, downflow, and horizontal. The flow refers to the direction in which the air moves from the furnace.

UPFLOW FURNACE. The standard upflow furnace is a tall furnace that can be used when a standard 8-foot ceiling is available. The cool room air (called *return air*) enters the bottom or side of the furnace and discharges the hot air out the top, as shown in Figure 8-1.

Another version of the upflow furnace is a low profile. This furnace is short and can be used for low basement applications. It has a low profile because the fan is to the side of the furnace in a separate compartment. The return air enters the top of the side compartment and discharges out the top of the other compartment, as shown in Figure 8-2.

DOWNFLOW FURNACE. The downflow furnace looks much like the standard upflow furnace in that it is tall. This furnace takes in the return air at the top of the furnace and discharges the hot air out the bottom. Ideally, this furnace is located in a hall closet where the air is discharged to the duct system below the house, and the return air duct is located in the attic. Figure 8-3 shows a typical downflow furnace.

HORIZONTAL FURNACE. The horizontal furnace looks much the same as the others, but it lies on its side and can be located under a house or in an attic as shown in Figure 8-4. The return air goes in one end, and the hot discharge air leaves the other end.

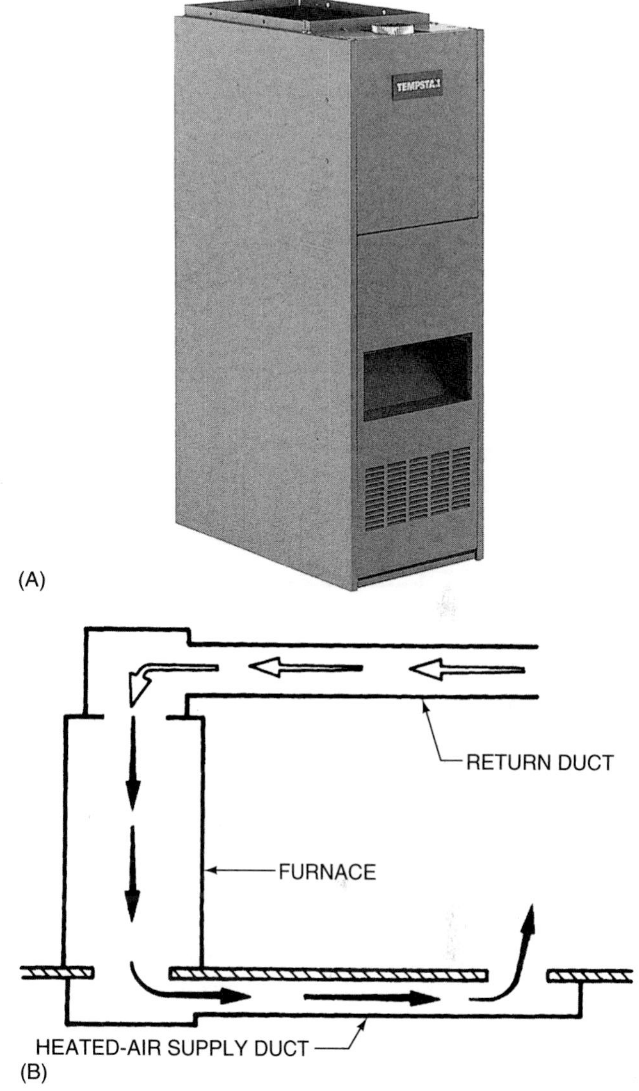

Figure 8-3. (A) Downflow furnace. *Courtesy Tempstar* (B) Airflow.

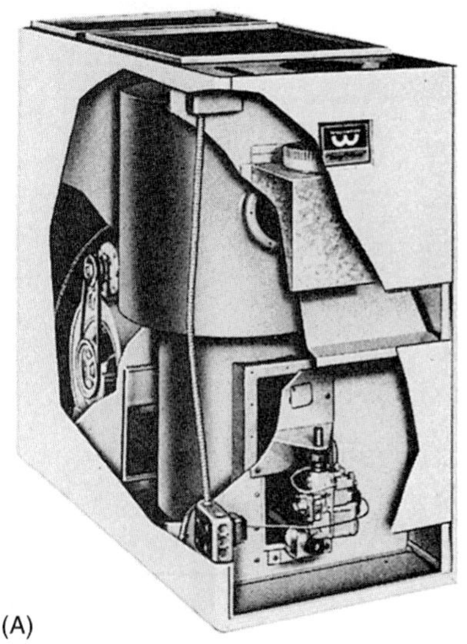

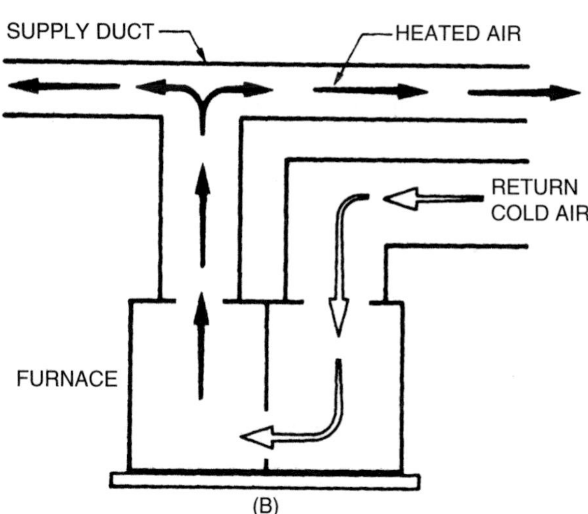

Figure 8-2. (A) Low profile upflow furnace. *Courtesy The Williamson Company* (B) Airflow.

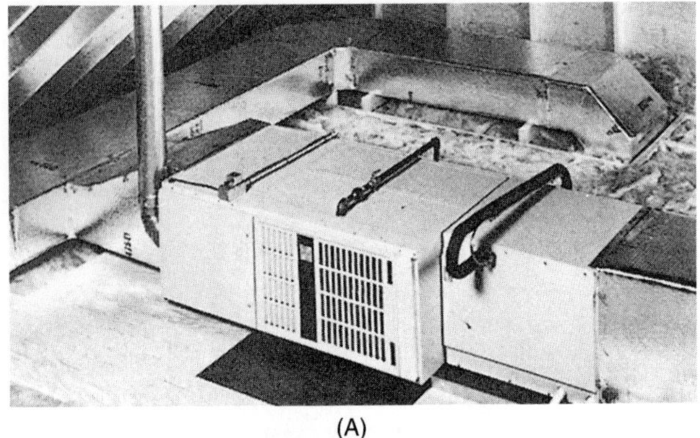

(A)

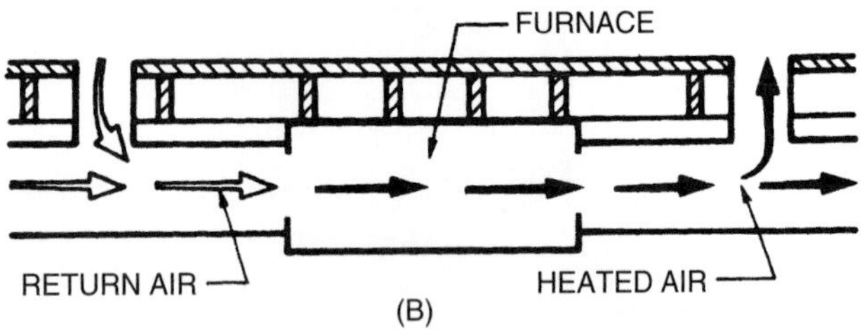

(B)

Figure 8-4. (A) Horizontal furnace. *Courtesy BPD Company* (B) Airflow.

Low-boy Furnace

In applications where space is critical and limited, another type of furnace that can be installed in a residential application is a low-boy furnace. This type of furnace is approximately four feet in height and is typically installed in a basement. In this particular application the connecting duct work is typically located in the first floor.

Multipurpose Furnaces

These furnaces can be installed in any position, thus providing more versatility than other types. This is especially true for installation. Depending upon the configuration, they can be upflow, downflow, and horizontal (left and right). In many cases the preset configuration of these furnaces can be changed in the field with very few modifications. Usually the combustion air pipe and exhaust pipe can be attached to either side of the furnace.

Other Types of Forced-Air Systems

UNIT HEATERS AND DUCT HEATERS. Unit heaters are free-hanging heaters that are used to heat large areas, such as factories or warehouses. They hang from the ceiling and operate just as horizontal furnaces do, except they often use prop-type fans and may not have ducts, as shown in Figure 8-5.

A duct heater looks like a unit heater, except it mounts in the duct. The duct heater does not have its own fan; instead, the system fan moves the air through the heater.

HIGH-EFFICIENCY GAS-FIRED
UNIT HEATER/BLOWER TYPE

STANDARD GAS-FIRED
UNIT HEATER/PROPELLER TYPE

(A)

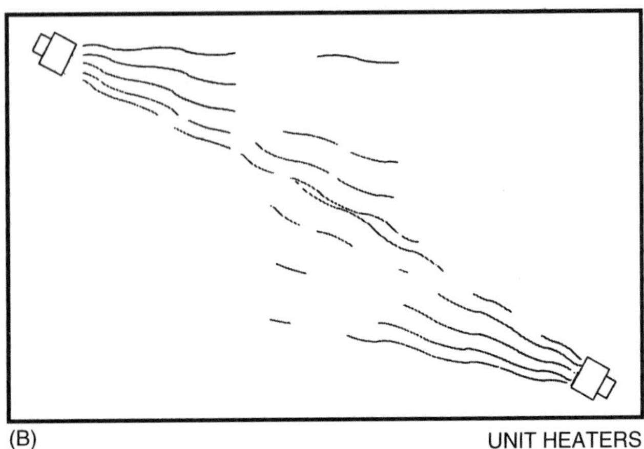

(B) UNIT HEATERS

Figure 8-5. (A) Unit heaters. *Courtesy Modine* (B) Unit heater application.

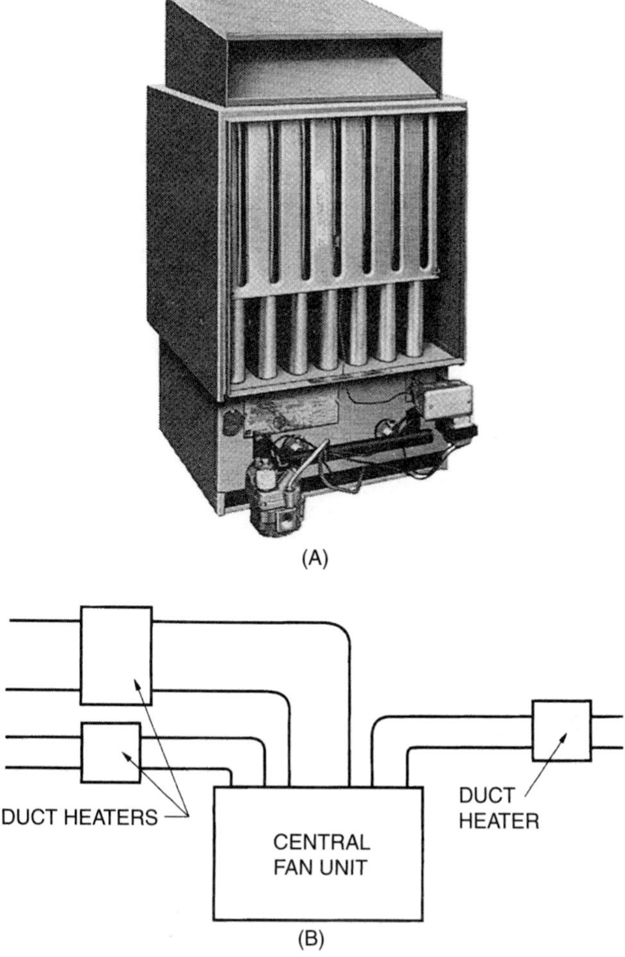

Figure 8-6. (A) Duct heaters. *Courtesy Modine* (B) Duct heater application.

A method must be provided to ensure air-flow from the system fan, as shown in Figure 8-6.

GAS PACKAGE UNITS. The gas package unit, or gas pack, is a special type of forced-air gas heater that is manufactured for outside use. Air-conditioning for the system is built into the unit. One of these units can be located at ground level with the return air entering one side and the supply air leaving the other, as shown in Figure 8-7. It can also be located on the rooftop, where the supply air and the return air are ducted from the unit to the system duct work. Special roof-mount kits are available for downflow applications. Figure 8-8 shows a roof-mounted gas package unit. These units are versatile and can have many features built in for commercial applications, such as fresh-air intake, which allows the unit to cool a business with outside air when the air temperature and humidity are suitable, as shown in Figure 8-9. A gas package unit makes an installation safer because the unit, including the vent and burners, is located outside the structure.

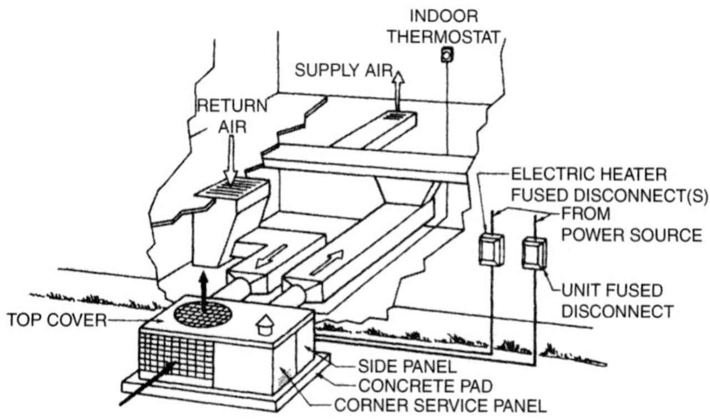

Figure 8-7. Side-by-side duct connections for a gas package unit.

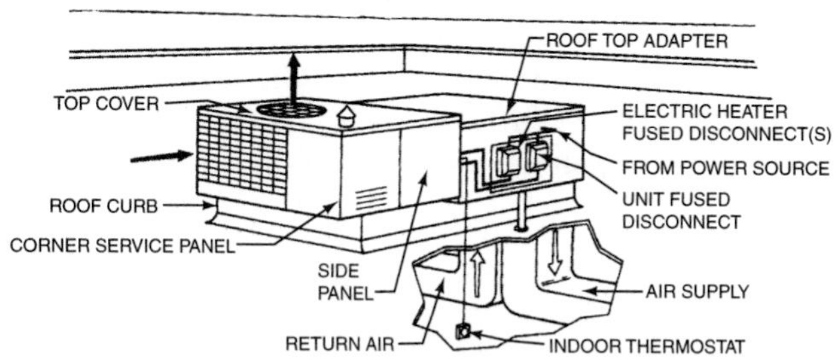

Figure 8-8. Roof-mounted gas package unit.

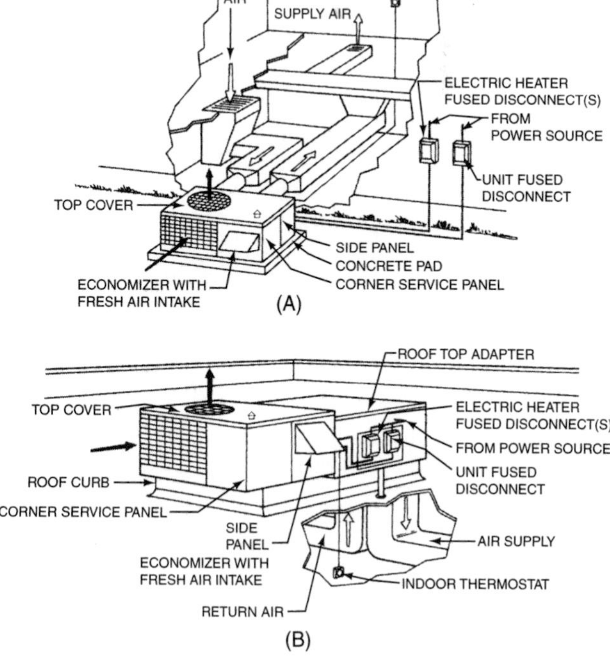

Figure 8-9. Economizer attachment for (A) a ground-level gas package unit and (B) a rooftop gas package unit.

The Air Side of a Furnace

The air side of the forced-air furnace is responsible for moving the air into the furnace inlet, through the heat exchanger, and out into the duct distribution system, as shown in Figure 8-10. The same concept applies to the gas package unit. It consists of a fan and a motor to drive the fan. The fan, or blower, most often used is the *centrifugal*, or *squirrel-cage fan.* This fan type is used because it is efficient, is quiet, and produces the correct pressures for the duct system. (Duct design is discussed in Unit 22.) The centrifugal fan used in furnaces has forward-curved fan blades, as illustrated in Figure 8-11. As the fan wheel turns, the air inside the wheel is forced to the outside of the wheel because of centrifugal force. This causes a low-pressure area in the center of the wheel, where new air takes the place of the air forced to the outside, as shown in Figure 8-12. The air leaving the fan wheel is directed into the fan housing to the fan outlet. Because the fan depends on the centrifugal action of the air, it can overcome only a certain amount of pressure from the inlet to the outlet. When the pressure differential becomes too great, the fan merely spins in the air. Much of the air moves from the inlet to the outlet and spins around there instead of moving down the duct, as depicted in Figure 8-13.

The mechanism that turns the fan is the motor and drive. There are two kinds of fan drives for furnaces: belt drives and direct drives. Belt-drive fans are used on older and some special heavy-duty furnaces. The motor sits to the side with a pulley that turns another pulley on the fan. The motor pulley is called the *drive pulley*, and the pulley on the fan is called the *driven pulley*, as shown in Figure 8-14. The advantage of a belt-drive fan is that different motor pulleys can be used to change the fan speed; if a motor pulley is removed and a larger pulley is put on, the fan will

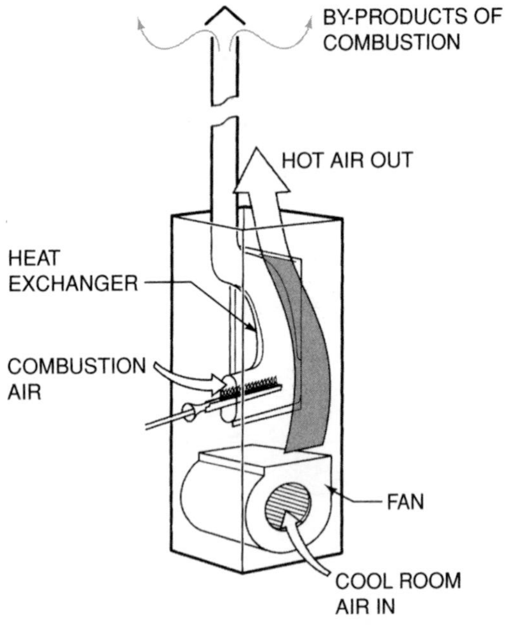

Figure 8-10. The air side of the furnace includes the air inlet connection, the fan, the heat exchanger, and the air outlet connection.

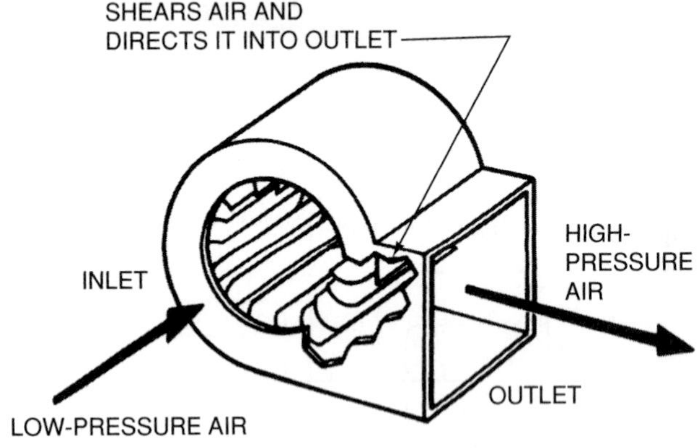

Figure 8-12. Operation of a centrifugal fan.

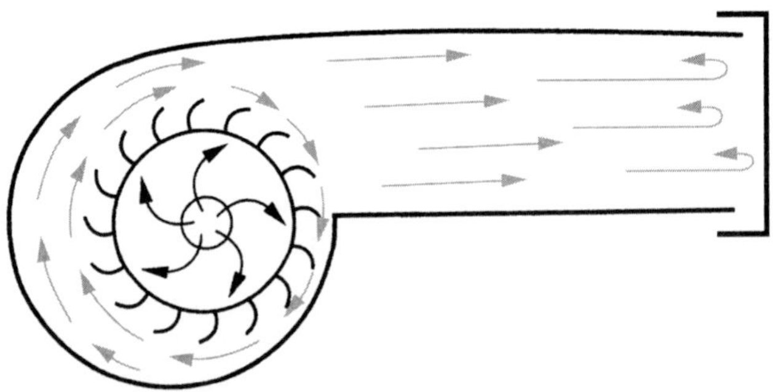

Figure 8-13. When the fan meets too much outlet pressure, the air cannot leave the fan and some of it goes round and round.

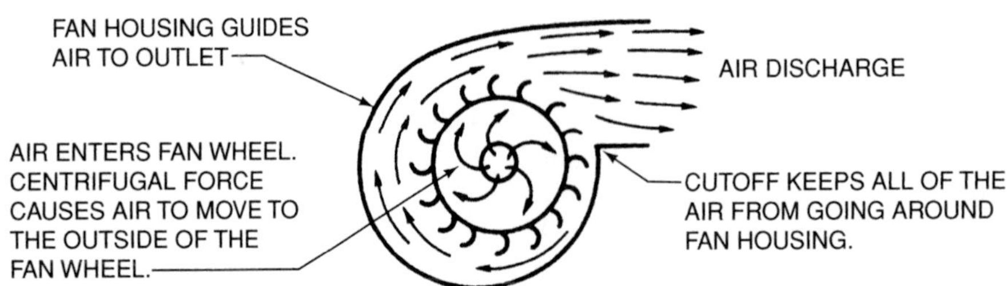

Figure 8-11. Centrifugal fan with forward-curved blades.

turn faster and therefore will move more air, as shown in Figure 8-15. Alternatively, some pulleys are adjustable, as illustrated in Figure 8-16. The pulley is adjusted by turning the outside hub; when the hub is turned inward, the two pulley halves move closer together and the belt rides higher in the pulley. This causes the pulley to become larger in inside diameter and makes the fan turn faster. Figure 8-17 shows what happens when the pulley halves spread apart.

The disadvantage of a belt-drive fan is that the complete assembly requires four bearings, two pulleys, and a belt. The four bearings in Figure 8-18 cause friction. The belt running in the two pulleys also causes friction. Therefore, all these parts can wear out, and the complete mechanism can require more maintenance than the direct-drive system.

The direct-drive system consists of a fan wheel and a motor. The fan wheel is mounted on the motor shaft; therefore, the motor has only two bearings, as shown in Figure 8-19. There are no pulleys or belts to cause friction and wear out. The fan speed can be changed by changing the motor speed when a multiple-speed motor is used. Figure 8-20 shows a multiple-speed motor.

The furnace is usually connected to the duct system at the inlet and the outlet. The outlet of the furnace has a

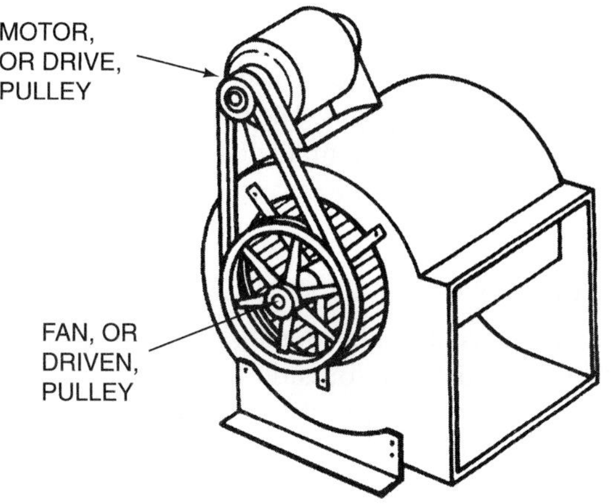

Figure 8-14. A belt-drive fan has a drive pulley (the motor pulley) and a driven pulley (the fan pulley).

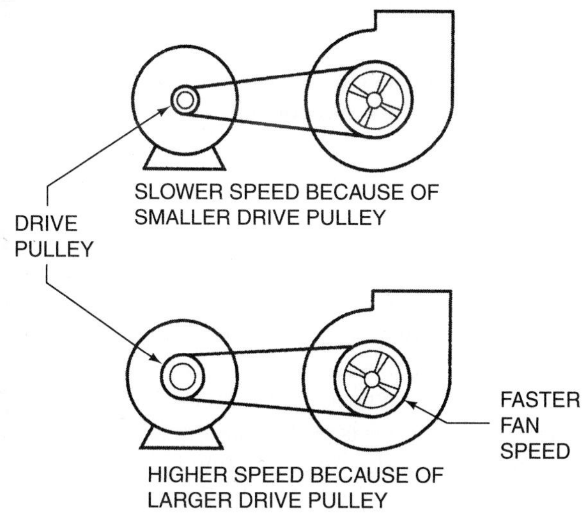

Figure 8-15. A larger drive pulley makes the fan turn faster.

Figure 8-16. Some pulleys are adjustable. *Photo by Bill Johnson*

Figure 8-17. The pulley sides spread apart, which causes the belt to ride lower in the pulley and slows the fan. *Photo by Bill Johnson*

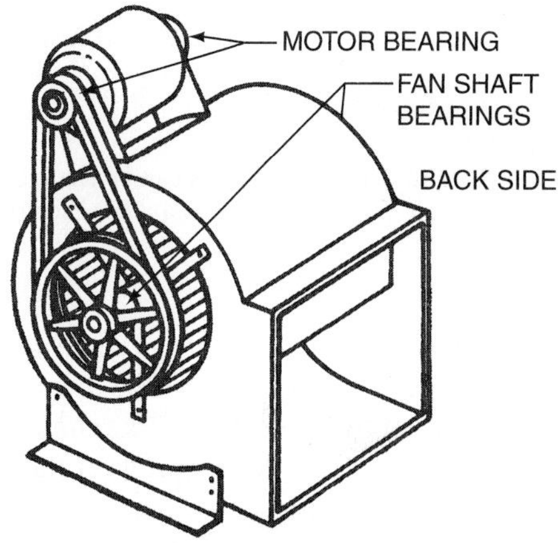

Figure 8-18. Belt-drive fan arrangements have four bearings, two pulleys, and a belt that will wear out.

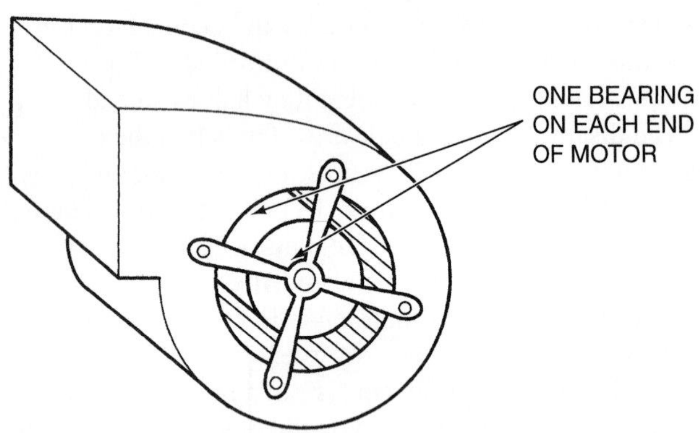

Figure 8-19. Direct-drive fans have only two bearings that will wear.

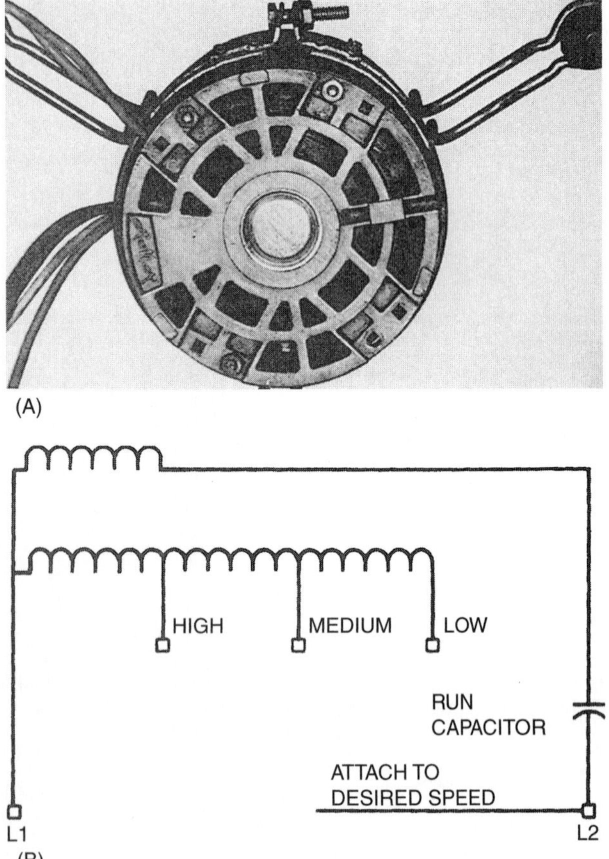

Figure 8-20. (A) Direct-drive fan speeds can be changed by using a multiple-speed motor. *Photo by Bill Johnson* (B) Electrical circuit in motor.

flange that connects to the duct work. The air inlet to the furnace may be one of several types. The low profile furnace has the air inlet at the top, just like the air out-let. The horizontal furnace has the air inlet on the opposite end from the air outlet. The standard upflow and downflow furnaces are versatile and have several options for their air inlets. The upflow or downflow furnace often has knock-out panels for the sides or the back, and the furnace sits on the floor. The upflow furnace may also be set on an air inlet box, called a plenum, to allow the air to enter the bottom. These different combinations allow the installer to fit one of the furnaces to almost any installation.

Whatever the furnace type, some provision should be made for the air filter. This filter is often built into the furnace, and a separate filter rack is furnished for the furnace. The reason for this rack is that the manufacturer does not know how you will use the furnace, particularly in the upflow and downflow designs. The installation may need the air to enter from the bottom, the back, or either side, so if the built-in filter is not where you need it, you can install the rack where needed.

The Vent System

The by-products of combustion for gas-burning appliances must be vented to the outside for forced-air furnaces because of the volume of exhaust. Remember that to burn a cubic foot of natural gas, 15 cubic feet of air are used for practical combustion. More air is required for LP gases. The by-products can be vented to the outside by one of two methods: natural draft or forced draft. According to the *National Fuel Gas Code*, natural-draft furnaces are classified as Category I furnaces. Forced-draft furnaces are either Category III or Category IV furnaces depending on the flue gas temperature. In this portion of the text, we discuss what occurs in the appliance. Venting practices are discussed in Unit 9, Section 9.2.

Natural-draft venting is accomplished with heat. The by-products of combustion must be kept warmer than the surrounding air to prevent them from back drafting (flowing in toward the appliance instead of out of the structure). Also, if they are not hot enough, the moisture in them will condense and run down the flue pipe. As mentioned previously, this moisture is corrosive. The by-products of combustion leave the heat exchanger at the top and are collected. They then rise through the draft diverter and up the flue. The draft diverter is part of the furnace that prevents momentary downdrafts from affecting the burner or pilot flame as illustrated Figure 8-21. The flue pipe is connected to the draft diverter. Another job of the draft diverter is to mix some room air with the gases that rise up the flue. This air is called *dilution air*, as shown in Figure 8-22.

Category III forced-draft systems (also known as *induced-draft systems*) use small blowers to move the flue gases to the outside. These systems are becoming more popular because venting can be controlled more efficiently. Appliances that previously required a vent pipe to the roof of the house can now be vented through the side wall with forced-draft venting. All condensing furnaces use forced-draft venting and are classified as Category IV furnaces.

The forced-draft fan can create either a positive or a negative pressure, or draft, on the heat exchanger, depending on the location of the fan. If the fan is at the inlet of the heat exchanger, the heat exchanger is under a

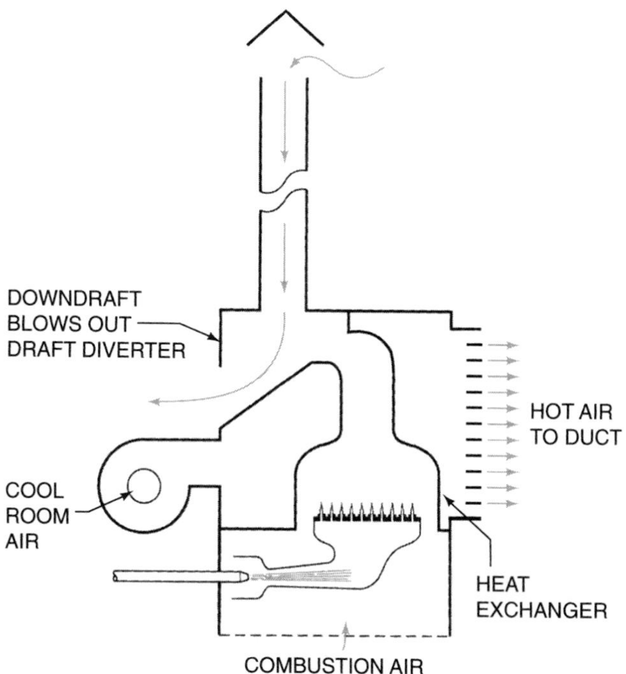

Figure 8-21. The draft diverter prevents momentary down-drafts from affecting burner flame.

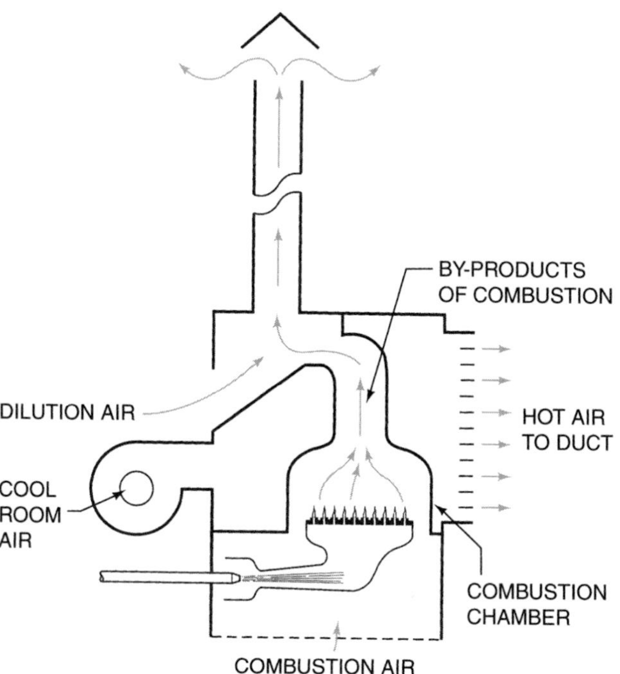

Figure 8-22. Dilution air rises up the draft diverter.

positive pressure (see Figure 8-23). Note that if a crack in the heat exchanger occurs, by-products of combustion may be pushed into the air passing over the heat exchanger and into the structure. If the combustion fan is located at the outlet to the heat exchanger (see Figure 8-24), air is drawn through the heat exchanger. The heat exchanger is then in

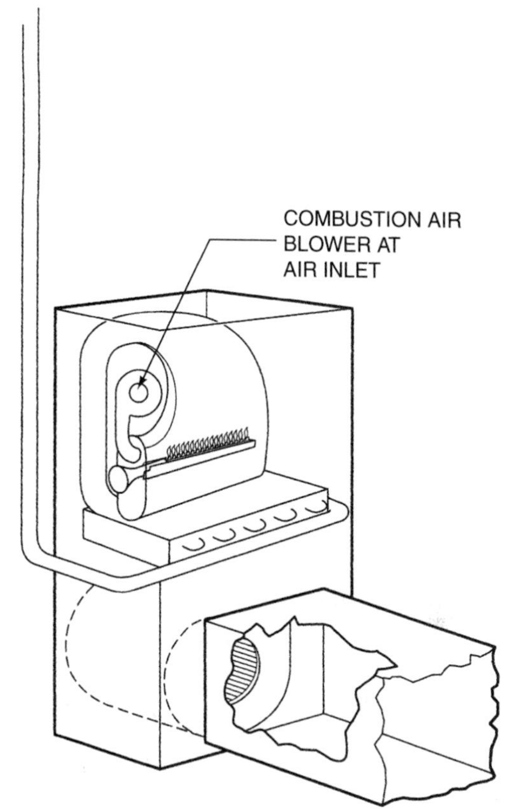

Figure 8-23. Forced-draft furnace with positive pressure on the heat exchanger.

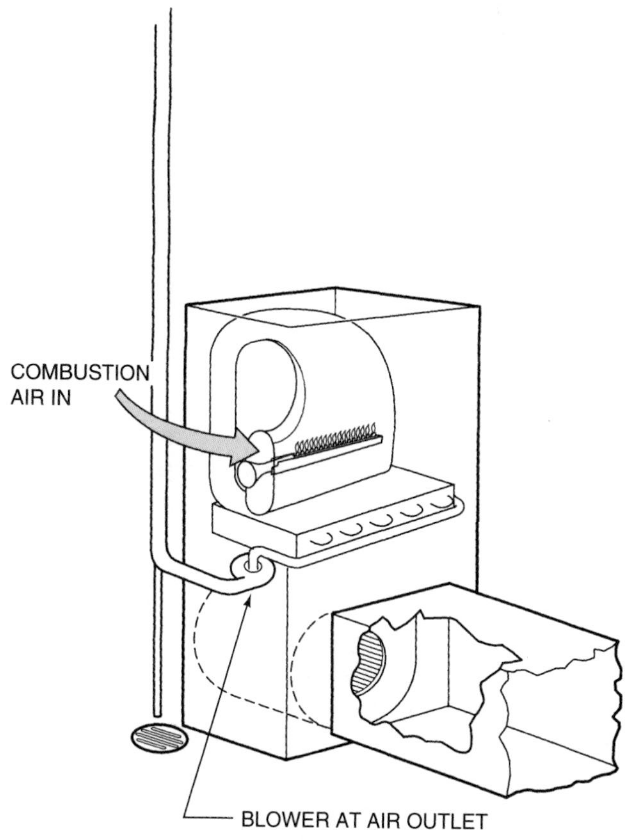

Figure 8-24. Forced-draft furnace with negative pressure.

a negative pressure. If a crack occurs, room air is drawn into the combustion process and no by-products enter the heated structure.

When forced draft is used, the control system is different than with natural draft. The draft system must be proven before the burner is allowed to ignite. This is discussed in Section 8.4.

Draft Diverter

Sometimes high winds can cause a natural draft furnace's pilot light to go out. This is due to the wind entering the flue pipe and blowing out the pilot light. Manufacturers have remedied this problem by installing a draft diverter into the furnace. A draft diverter is designed to re-route any downdraft air back out of the furnace from the point where the dilution air is introduced into the furnace.

Makeup Air

The volume of air that is used during the combustion process in the furnace and the air that goes up the flue pipe or chimney must be replaced or reintroduced to the space surrounding the furnace to prevent creating a partial vacuum in that space. Inadequate air supplied to the furnace can result in a greater air pressure at the chimney than in the structure, thus permitting flue gases to be forced back into the structure in the area surrounding the furnace. If this is permitted to continue, then serious injury and/or death can result.

To prevent this from happening, an adequate amount of air must be supplied to the furnace area. Keep in mind that not all of the air will be used for venting purposes. A portion of it will be used in the combustion process. It takes approximately 30 cubic feet of air for each cubic foot of natural gas that is consumed by a furnace. For example, a furnace rated as 150,000 Btu would require 4286 ft^3 of air. The answer was calculated by:

1. Dividing the furnace rating by 1050. This is the amount of heat that 1 cubic foot of natural gas will produce.

Amount of Natural Gas = 150,000 Btu/h/1050 Btu/h/ft^3

Amount of Natural Gas = 142.85 ft^3

2. Multiplying the answer obtained in step 1 by 30. This is the amount of air per cubic foot of natural gas necessary to sustain combustion and dilution.

Amount of Air Needed = 142.85 ft^3 * 30 ft^3

Amount of Air Needed = 4285.7 ft^3 or 4286 ft^3

As stated earlier, improper air supply to the furnace location can result in the introduction of deadly gases into the air space surrounding the furnace. In most cases, the furnace is located next to habitable space. Therefore, it is recommended to check local building codes and follow manufacturers' recommendations when working with air quantities and venting. In addition, if no additional information is provided by the local building codes or the manufacturer, then refer to the National Fuel Gas Code.

Sizing and Flue Pipe Material

Before installing a vent pipe, always consult the local building codes in your area. This is because different areas of the country allow for the use of different materials to connect the furnace to the chimney. A good rule of thumb is that the size of the flue pipe connecting to the furnace should be the same size as the vent opening on the furnace. If the size of the flue pipe is too small, then a portion of the combustion gases can enter the area surrounding the furnace. Sizing the flue pipe too can result in the flue gases leaving the furnace much slower, causing them to condense and possibly causing deterioration of the flue pipe and/or chimney.

8.2 Boilers

Boilers are like furnaces, except they transfer heat into water rather than into air. The water is either circulated in pipes or changed into steam to heat the structure. There are two types of boilers: hot water and steam. Boilers used in the typical residential and light commercial applications discussed in this book are known as low-pressure boilers because they operate at a maximum working pressure of 15 psig for steam and 30 psig for water. They have relief valves that relieve the pressure if it rises above the working pressure, as shown in Figure 8-25.

Regulation of Equipment

As with gas furnaces, boilers are designed and manufactured according to the guidelines of organizations that regulate the safety and capacity of the equipment.

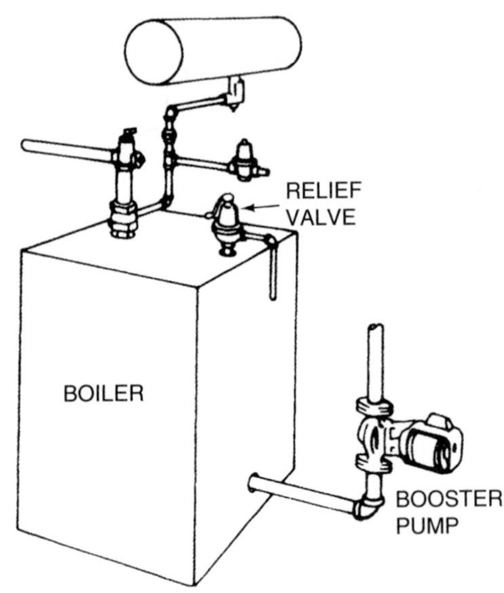

Figure 8-25. The relief valve on a boiler prevents excessive pressure inside the boiler.

ASME (American Society of Mechanical Engineers) determines the safety of the boiler as a pressure vessel. AGA covers gas-fired boilers. UL determines the electrical safety of the components, and NFPA (National Fire Protection Association) helps write the guidelines for boiler safety.

Note: Make sure that any boiler that you buy or install has been approved by the proper organization.

Hot-Water Boilers

Hot-water boiler and piping systems are full of water. The boiler that is most often found in the field has a cast-iron heat exchanger with an insulated jacket around the cast iron. The outer skin is usually painted sheet metal. Cast-iron boilers are manufactured in all sizes. The small boilers (called *packaged boilers*) are for residential use, whereas the larger ones are for larger applications. The large boilers may be sectional and erected on the job because they are too cumbersome to ship and handle in one carton.

The smaller-sized boiler (packaged boiler) is furnished with the accessories: a pressure–temperature gauge, a temperature control, a relief valve, and often the water-circulating pump, as shown in Figure 8-26. The installing contractor must only set the boiler on the required base and run the piping, fuel supply, flue, and wiring to have a complete installation. The base for the boiler must usually be fireproof because the flame for a gas boiler is often on the bottom. Because most of the work involved in installing a packaged boiler is piping, these installations are often completed by plumbing contractors.

Larger sectional boilers are different in that they must be assembled on the job. Doing so requires some-one who can follow the boiler manufacturer's instructions for assembly. The boiler must be assembled on the correct base, often a fireproof base like that of the small boiler, as illustrated in Figure 8-27. However, the piping and controls are much more extensive for larger boilers. The location of a large boiler in a building may be critical to the type of boiler selected. For example, a standing column of water exerts a pressure of 0.433 pound per foot of column.

Figure 8-26. Packaged boiler with all the accessories. *Courtesy Burnham Corporation*

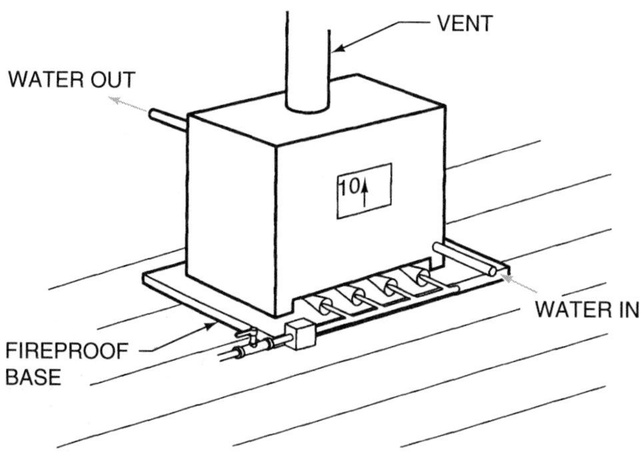

Figure 8-27. A boiler with the fire in the bottom must be set on a fireproof base.

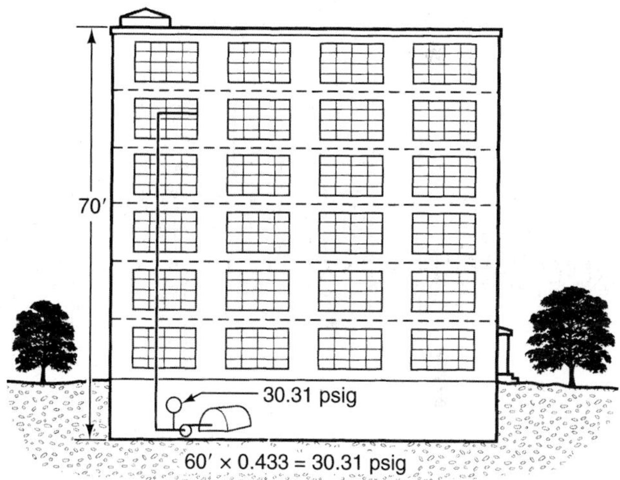

Figure 8-28. A boiler in the basement of a seven-story building must be able to withstand more than 30.31 psig.

A column of water 70 feet high exerts a pressure of 30.31 psig (0.433 × 70 = 30.31). A water boiler in the basement of a seven-story building must be able to withstand a pressure of more than the 30 psig working pressure of a low-pressure water boiler, Figure 8-28. In this case, locating a low-pressure boiler on the top floor is adequate (see Figure 8-29). Many boilers are located on the top floor for this reason. This location makes sense in choosing a boiler, but servicing the boiler becomes more difficult because all tools must be moved to the top floor for service.

Steam Boilers

Steam boilers are much like water boilers; they use water, too, but they are not *full* of water. Instead, the empty space in the top of the boiler collects steam as shown in Figure 8-30. These boilers are also cast iron, just like water boilers. However, the controls of a steam boiler are different than those for a water boiler. A small steam boiler package contains a low-water cutoff, a pressure gauge, a pressure control,

Figure 8-29. The boiler may be located on the top floor to reduce the standing pressure.

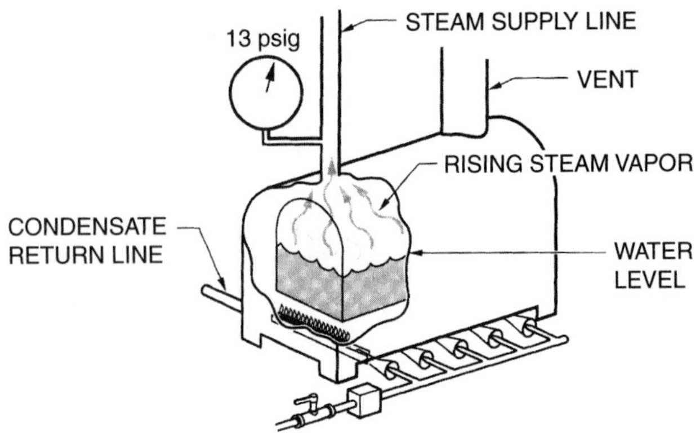

Figure 8-30. The empty space in the top of a steam boiler collects steam.

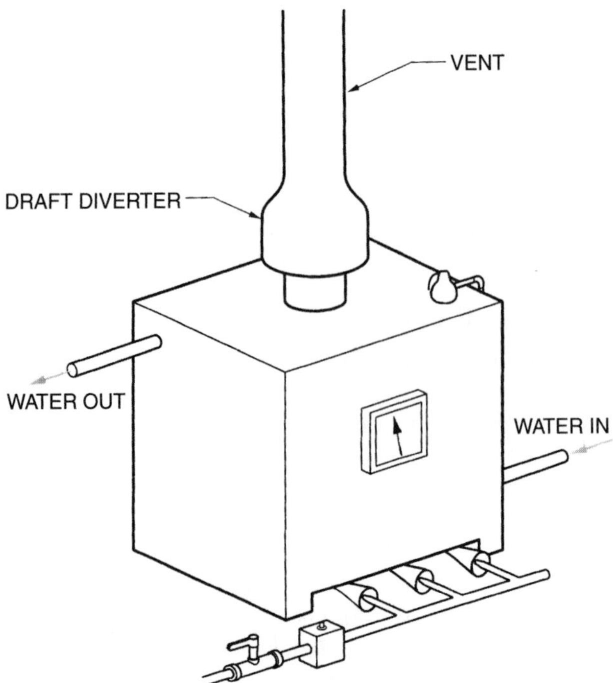

Figure 8-31. The vent system for a boiler is much like the vent system for a gas furnace.

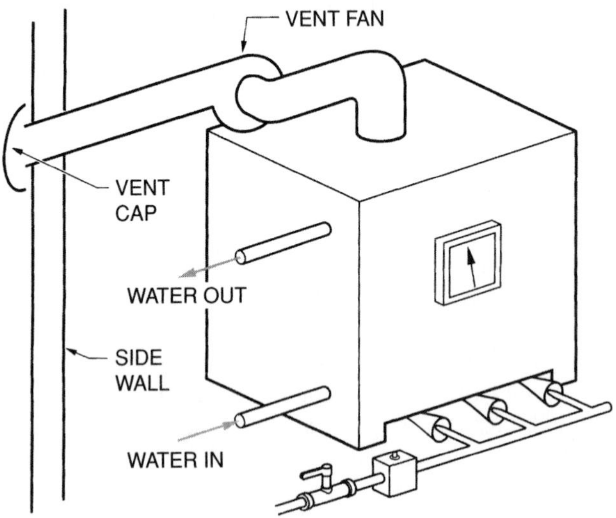

Figure 8-32. Forced-draft boiler.

a pump, a relief valve, and a sight glass for determining the water level in the boiler. With water boilers, the temperature of the water is controlled and maintained up to approximately 220°F. With steam boilers, pressure is controlled. Because the boilers discussed in this text are low pressure, 15 psig is the maximum pressure. Fifteen psig correspond to approximately 250°F, but the boiler cannot be allowed to operate at the full 15 psig. Consequently, steam boilers operate at up to approximately 13 psig and are then shut off by a pressure control.

There is more heat value in a pound of steam than a pound of water. When a pound of steam is condensed, 970 Btu of heat are removed. When a pound of water is reduced from 200°F to 219°F, 1 Btu of heat is removed. The steam is condensed in the system and then returned to the boiler in the liquid state at nearly the saturation temperature.

Other Features of Boilers

The vent system for a boiler is much like that for a forced-air furnace; it consists of a manifold to catch the by-products of combustion and a draft diverter for a natural-draft boiler, as shown in Figure 8-31. Forced- or induced-draft boilers are more efficient and may be vented through side walls. This venting makes them more versatile, as shown in Figure 8-32.

Some small hot-water boilers use pulse ignition and condensing of the flue gases to accomplish high efficiency. This is much like the condensing process described in Unit 6. These boilers are as much as 90% efficient. Boilers do not reach the 98% efficiency that gas furnaces reach because the return water for a boiler is much hotter than the return air in a forced-air system. A forced-air system can expose the exiting flue gases to 70°F air, whereas the boiler return water is not below approximately 120°F. Therefore, not as much heat can be removed from the boiler flue gases.

8.3 Radiant Heaters

Radiant gas heat is accomplished by using the gas flame to heat a surface. The surface itself then radiates heat to the surroundings. Radiant heat appliances for commercial applications are discussed here, not the room heaters that are discussed in Unit 7.

Two types of radiant heat commonly used are hot-tube and ceramic-surface heat. The hot-tube heater fires the gas down a tube to heat the tube's surface, as shown in Figure 8-33. The tube is usually black, which radiates heat better than other colors, and is mounted in a reflector that concentrates the heat toward a specified direction as shown in Figure 8-34. This type of heater can be vented to the outside because the combustion gases can be captured at the far end of the tube, as shown in Figure 8-35. The tube does not glow red-hot in this type of heater.

Other radiant heat appliances use a bed of ceramics to burn the gas, and the ceramics glow red-hot, as illustrated in Figure 8-36. These heaters would be much harder to vent the by-products of combustion to the outside.

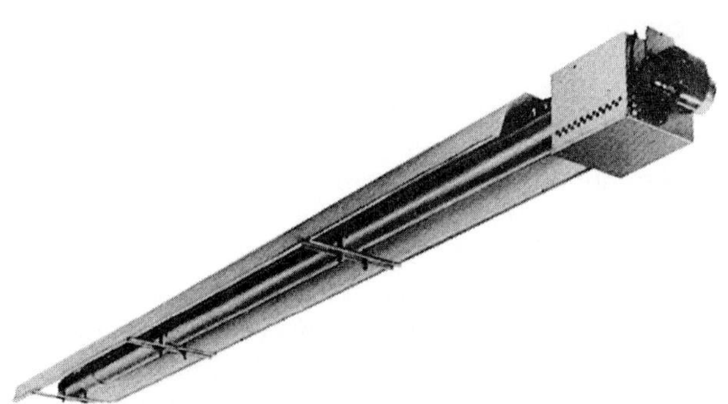

Figure 8-33. Hot-tube radiant heater. *Courtesy Modine*

Figure 8-34. The reflector directs heat downward from the hot tube.

Any radiant heater concentrates the heat in the direction in which the heater is pointed. Typical applications are large room volumes, such as airplane hangars, assembly areas in factories, automotive garages, and warehouse spaces. The heat from the heaters heats only the solid objects and does not heat the air among the objects. In areas where drafts are prevalent, forced-air heat would be swept away by the drafts, whereas radiant heat would heat the surfaces of the work area and the worker.

Rather simple controls are required for radiant heat systems. These systems are available with standing pilot burners or with intermittent ignition pilot systems.

Caution: Systems that are not vented can be used only in areas permitted by the prevailing code because poor ignition can create carbon monoxide. Permissible areas are usually large volume with a lot of fresh air.

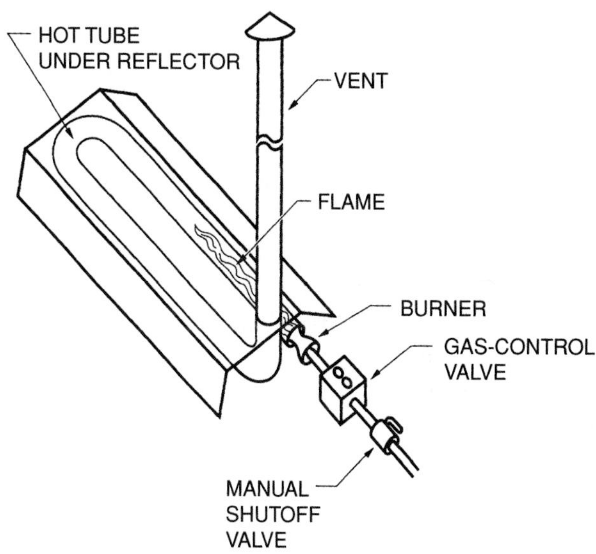

Figure 8-35. A hot-tube heater can have a vent system.

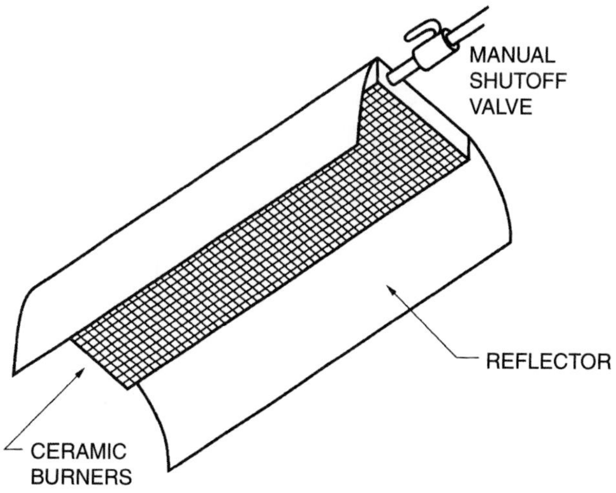

Figure 8-36. This heater has a bed of ceramic burners that glow red hot.

8.4 Automatic Control for Gas Heat

Each gas-burning appliance requires a different type of automatic control. Gas furnaces are either standard efficiency or high efficiency. The gas pack is for rooftop or outdoor ground-level use. Boilers are either hot water or steam. The controls for each type of appliance are discussed as they apply to the appliance only. In the following sections, we apply all the controls discussed up to now to make a working system. Only typical systems are discussed. There are many variations; we attempt only to give you a basic understanding. You can attend specific manufacturers' schools to train for more specific applications.

Standard-Efficiency Furnaces

Standard-efficiency gas furnaces can have standing pilot burners or intermittent pilots. In any case, a method is necessary to prove that the pilot flame is present before power is allowed to energize the main gas valve. Figure 8-37 is a pictorial diagram of a typical gas furnace, and follow the sequence of events for creating heat to the structure.

1. The room thermostat calls for heat by closing its contacts. Note that power will not reach the primary of the transformer unless the limit portion of the fan–limit switch is closed.

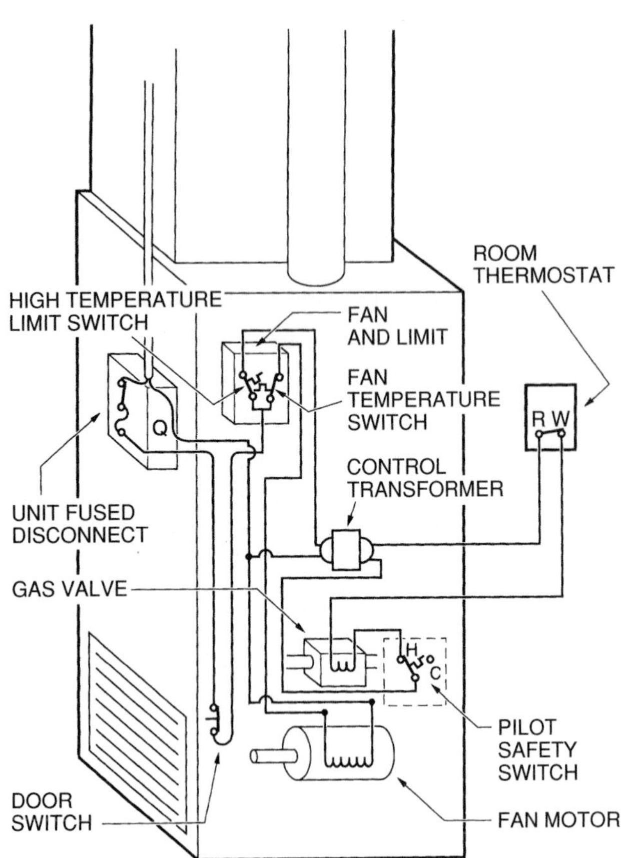

Figure 8-37. Pictorial diagram of a typical gas furnace.

2. Low voltage (24 volts) is applied to the 24-volt coil on the gas valve. Note that two controls are involved; the pilot safety circuit is independent from the 24-volt thermostat circuit. The gas valve coil may be energized, but because of the pilot safety switch the pilot burner must be lit before the main burner can receive gas.
3. Suppose that ignition of the main burner occurs and that the furnace begins to heat. The fan should not be allowed to operate until the heat exchanger becomes hot, or the furnace will blow cold air to the conditioned space. The furnace fan operates on a separate circuit. There are two common types of circuits: timed and temperature controlled. The temperature-controlled circuit is used in this example. A temperature sensor is part of the fan–limit control. When the temperature reaches the fan setting, the fan starts. Note that if the fan does not start, the temperature sensor continues to heat and shuts off the burner when the high limit is reached. Usually, the fan portion of the control is set to start the fan at approximately 50°F to 75°F above the room temperature. If the planned room temperature is 70°F, the setting is 120°F to 145°F (50 + 70 = 120°F, or 75 + 70 = 145°F).
4. The fan and burner are now both on and running. They can continue until the room thermostat is satisfied. When the thermostat is satisfied, the contacts open and shut off the main gas burner. The pilot burner continues to burn.
5. The fan continues to run until the fan portion of the fan–limit control stops the fan. The fan control is usually set to shut off the furnace when the air temperature reaches approximately 90°F. If it is set any lower than this, cold air drafts will be noticed.

The timed fan control can be used for the same application and is better able to ensure fan operation. For example, many manufacturers use the timed fan control so that installation and service people do not have to set the temperature-operated control. People often do not set the correct temperature initially, then fan short cycles can occur at the beginning or the end of the cycle. The timed fan control starts the fan at a predetermined time after the thermostat calls for the burner to be fired. This is accomplished with a 24-volt control similar to an electric heat sequencer, as shown in Figure 8-38. Note that if the burner does not fire, the fan will start anyway because it is on a separate circuit. Timed control is necessary for attic-installed furnaces. Otherwise, the temperature in an attic can become hot enough to cause a temperature-operated control to start the furnace fan in the summer when the burner is off.

Another standard-efficiency application uses a forced-draft, or combustion, fan for positive draft and more efficiency. This is a Category III furnace. It is not considered a high-efficiency furnace, but it is more efficient

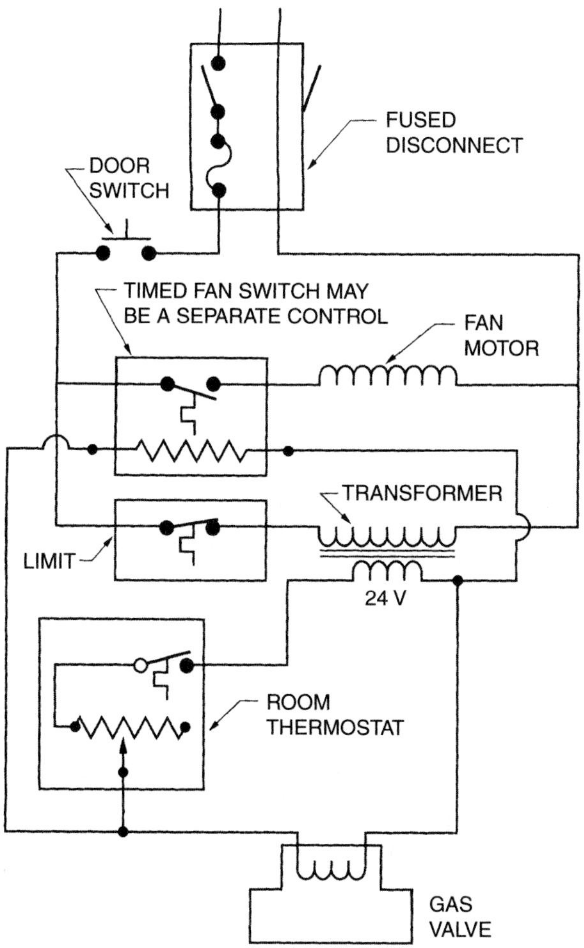

Figure 8-38. Timed fan control.

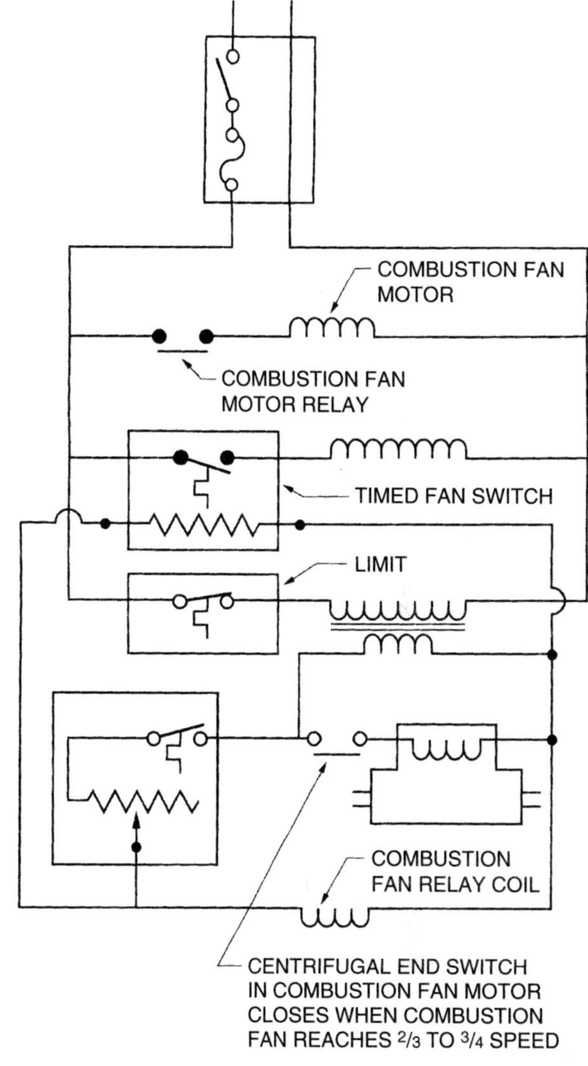

Figure 8-39. This furnace has a positive vent fan. See the text for the sequence of control operation.

than the standard furnace. This furnace operates in much the same way as the standard furnace, with a few exceptions. See Figure 8-39 for an illustration of the following sequence:

1. The thermostat calls for heat and energizes a 24-volt relay that in turn starts the combustion fan. The fan starts to turn. When it reaches approximately ⅔ to ¾ of its rated speed, an end switch in the motor closes. The main burner control circuit is wired through the contacts, so the main burner starts. Note that the transformer primary is wired through the limit switch and there can be no call for heat if the temperature of the furnace is too high.
2. When the burner ignites, the furnace heat exchanger becomes hot, and the fan control starts the fan.
3. When the space temperature thermostat is satisfied, the contacts open and the combustion fan stops. When it starts slowing, the end switch opens and stops the main burner. Note that if the combustion blower fails, the burner shuts off. This blower must be operating for the burner to operate. Other variations of starting the burner include using a sail switch or a pressure differential switch to determine positive blower operation, as shown in Figure 8-40.
4. When the furnace cools, the fan stops because of the fan control.

High-Efficiency Furnaces

High-efficiency furnaces operate much like the standard-efficiency furnaces with combustion blowers, with the addition of automatic pilot relight features.

1. When the thermostat calls for heat, the contacts close and send 24 volts to the furnace control circuit, usually starting the combustion blower with a 24-volt relay. After a positive draft is proven by one of the previously mentioned methods (end switch, sail switch, or pressure differential switch), the 24-volt signal moves to the pilot burner circuit.
2. The signal starts one of the pilot relight sequences mentioned in Unit 7, Section 7.5. When the pilot is

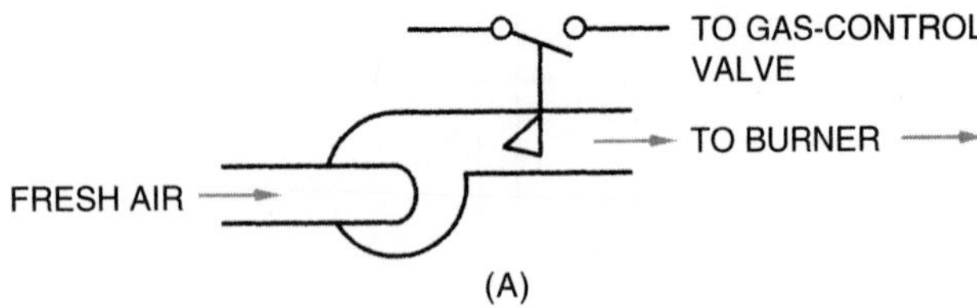

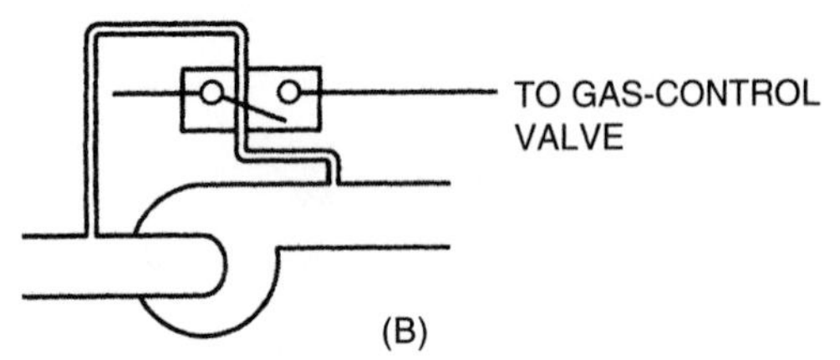

Figure 8-40. (A) A sail switch or (B) a pressure differential switch can be used to start the burner.

lit and proven, only then is power allowed to energize the main gas valve.

3. The main burner ignites and heats the heat exchanger.
4. The fan switch starts the circulating fan.
5. When the space temperature thermostat is satisfied, the blower stops and turns off the main burner through the positive draft switch.
6. The circulating fan continues to run until the furnace cools to the prescribed temperature, then it stops.

Gas Package Units

The gas package unit operates on much the same control sequence as the high-efficiency furnace. It usually has intermittent ignition and a combustion blower. The equipment is located outside the structure with the duct work entering the structure. A control circuit must be wired to the space thermostat location.

1. Upon a call for heat, the contacts in the space temperature thermostat close and send power to the combustion blower through a relay. When the blower is nearly up to speed, the positive draft mechanism (end switch, sail switch, or pressure differential switch) closes, sending power to the intermittent pilot circuit. Note that power to the control transformer is supplied through the limit control, and the burner sequence cannot start if the furnace is hot or the limit control is open for any reason.
2. The pilot lights and proves, then power goes to the main burner gas valve. The main burner ignites.
3. The heat exchanger heats, and the fan starts, distributing air to the conditioned space through the duct work.
4. When the space temperature thermostat is satisfied, the power to the combustion blower is interrupted, and the blower stops. This shuts off the main burner.
5. The fan continues to run until the heat exchanger cools and the fan switch stops the indoor fan.

Boilers

The control of gas flow for a hot-water boiler is much like that for a forced-air furnace. There are two types of controls: the standing pilot and intermittent ignition in which the pilot is lit upon a call for heat. The call for heat may be different from one system to another. Some systems may be gravity hot-water systems, in which the hot water in the system moves on its own on the basis of temperature differences. Some systems operate the boiler from the room thermostat and allow the boiler to cool between calls for heat. Some installations keep the boiler hot all season, and the circulating pump is controlled by the space temperature thermostat. This works successfully in systems that have several zones and several pumps (see Unit 17 for other methods of zone temperature control). Figure 8-41 along with the following sequence, depicts a system that cycles the boiler with the space temperature thermostat.

1. When the space temperature thermostat calls for heat, the contacts close and send a 24-volt signal to the boiler control circuit. This system may have a circulating pump that runs all season, or the pump may cycle with the boiler.
2. The 24-volt signal powers the gas valve on a standing pilot boiler and causes gas to flow to the main burner, provided that the pilot burner is lit and proved. If the

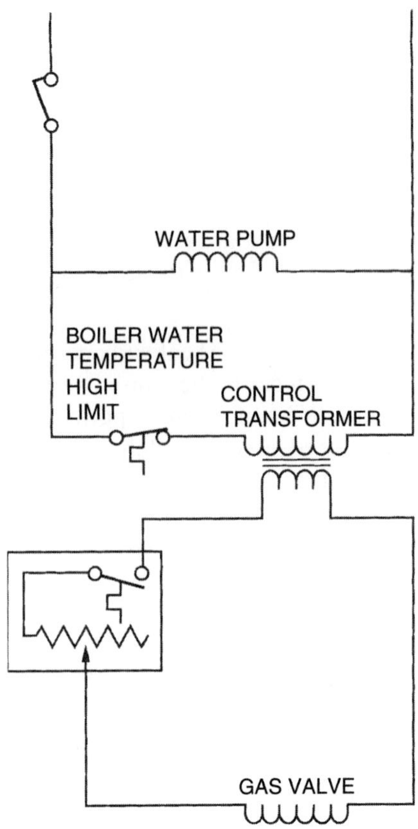

Figure 8-41. This boiler is cycled with the space temperature thermostat.

circulating pump is running, water is circulating. The mass of the water in the system and the mass of the boiler take a while to heat, so instant response is not available if this is the first time the boiler is fired for the season. If the boiler has been operating, the water in the lines and in the boiler will be warm, and a gradual warming will be noticed.

3. The boiler continues to operate until the space temperature thermostat stops the boiler. Then the water and the boiler cool gradually. The heat anticipator in the room thermostat helps to take care of some of the time lag while heating and cooling take place.

A hot-water boiler with intermittent pilot ignition operates similarly to a high-efficiency furnace:

1. Upon a call for heat from the space temperature thermostat, a 24-volt signal returns to the boiler and starts the pilot burner ignition process. After the pilot burner is established and proved, the main burner is ignited.
2. As with the previous example, the pump may be running and a gradual warming will take place.

When you encounter systems in which the boiler is kept hot all the time, the space temperature thermostat (or thermostats in the event of several pumps), merely stops and starts the pumps. These systems are often used when the boiler is also used to heat the domestic hot water. When the boiler does not heat hot water, an outdoor thermostat can be used to shut off the boiler when the weather is warm.

Summary

- Several rating agencies determine manufacturing standards for equipment from a safety and capacity standpoint.
- Forced-air furnaces are designed to be fastened to duct work and are either upflow, downflow, or horizontal in nature. The flow pattern is the direction of the airflow through the furnace.
- Unit heaters are free-hanging, forced-air heaters that do not attach to duct work.
- The duct heater is mounted in the duct and has no fan.
- The gas package unit is located outside, on the roof or next to the house, with the duct work routed to the structure.
- The air side of a forced-air furnace consists of the inlet connection, which may contain the filter rack, the blower, and the furnace outlet connection. The connections are the locations where the ducts fasten to the furnace.
- The centrifugal fan is called a squirrel-cage fan and has forward-curved blades.
- The centrifugal fan may be either belt drive or direct drive.
- The vent system moves the by-products of combustion from the furnace to outside the structure. A Category I furnace has a natural-draft vent. Category III furnaces use forced-draft vents to vent the by-products but do not condense the water vapor. Category IV furnaces are condensing furnaces with forced-draft vents.
- Most residential and light commercial boilers are low pressure, either hot water or steam, and operate at no more than 15 psig for steam and 30 psig for water. Each boiler consists of a heat exchanger covered with insulation and an outer skin of sheet metal. Most small boilers are packaged, but some are disassembled.
- Hot-water boiler heat exchangers are full of water, and the water is piped throughout the building where heat is removed and added to the structure. Steam boilers are partially filled with water and have a dome of steam on top. The steam is piped throughout the building, where heat is removed and added to the structure. The steam is condensed and returns as liquid condensate (water) to the boiler for reheating.
- The vent systems for boilers are much like those for forced-air systems.
- Radiant heaters concentrate their heat on the surfaces to be heated without heating the air among the surfaces and are used for open areas, such as automotive garages and airplane hangars.
- Automatic control of the gas supply for gas furnaces is accomplished with low voltage (24 V). The fan is controlled from a separate circuit and is high voltage.
- When forced, or induced, draft is used, the room thermostat typically energizes a small 24-volt fan relay to start the draft fan. The gas valve is opened by proving that there is a draft. This can be accomplished with an end switch in the fan motor or with a pressure switch or a sail switch.
- High-efficiency condensing furnaces are controlled in much the same manner as standard-efficiency furnaces with combustion blowers.
- A water boiler is controlled by the water temperature in the boiler.
- Steam boilers are controlled by steam pressure.
- Hot-water and steam systems should have low-water cutoffs; in the hot-water system, the cutoff is located in the piping. In the steam system, it is located in the boiler.

Review Questions

1. What are the three types of airflow for gas furnaces?
2. What are the two types of fan motors used in gas furnaces?
3. What is the working pressure of most residential and light commercial steam and water boilers?
4. What type of control is used in hot-water boilers? in steam boilers?

9 Gas Piping, Venting, and Appliance Efficiency

Objectives

Upon completion of this unit, you should be able to

- **state the difference between low-pressure and 2-pound piping systems.**
- **size the pipe for a simple low-pressure system and a 2-pound piping system.**
- **describe several good piping practices for gas piping.**
- **describe two types of venting systems for gas appliances.**
- **size the vent for a simple gas-appliance system with two appliances using a common flue.**

9.1 Gas Pipe Practices and Sizing

In this unit, we discuss first the practices of piping gas systems. We discuss the piping from the meter to the point of consumption for natural gas and from the tank regulator to the point of consumption for LP gases. Much of the information here is extracted from the *National Fuel Gas Code*, which is typically used in many parts of the United States. However, your area may have local codes that go beyond the national code and should be consulted.

The correct materials must be used in piping gas systems to ensure that the pipe does not corrode (either inside or outside), that the pipe does not contain particles that will loosen and plug gas orifices, and that the pipe protects the gas within. Leakage cannot be tolerated because escaping gas can cause fire and explosion.

Piping materials that are used most often and that are approved for natural and LP gases are steel and wrought iron (of at least schedule 40 in weight), both of which can be either black or hot-dipped and zinc-coated; copper; brass; and aluminum. All these pipes can be threaded and used provided that they do not corrode. Copper and aluminum tubing are also used when various compression or flare connectors are used. The gas company in some localities may use plastic pipe from the gas line to the meter or to the entrance to the house. Again, you must check your local code for the exact materials required in your area. The supply house will help you choose the correct materials for any job in question. Note that the piping systems that you will see most often are from the meter or regulator to the gas-burning appliance and are under a pressure of 7 inches of water column (wc) for natural gas and 11 inches of wc for LP gases.

A new system, called the *2-pound system*, has been developed for natural gas.

Rigid Piping Systems

A system piped with iron pipe that has threaded connections is typically called a *rigid piping system.* This type of system is the most difficult to install because each connection must be an exact length. Rigid piping systems provide the best physical protection for the gas in the piping system and are the safest. They have been used for many years.

The rigid piping system can be run above or below ground. When run above ground, the pipe must be supported to prevent sagging and undue stresses on the piping. Figure 9-1 shows the proper spacing for supporting piping and the different types of pipe and tubing materials that correspond to the pipe or tubing size. This information is pertinent for pipe that is fastened to walls, floors, or ceilings. Proper pipe hangers must be used; they must not corrode the pipe, and they must hold it steady. Figure 9-2 shows a hanger and two types of fasteners. When pipe is run above ground outside, such as across the roof, the pipe

Support of Piping

Steel Pipe, Nominal Size of Pipe (Inches)	Spacing of Supports (Feet)	Nominal Size of Tubing (Inch O.D.)	Spacing of Supports (Feet)
½	6	½	4
¾ or 1	8	⅝ or ¾	6
1¼ or larger (horizontal)	10	⅞ or 1	8
1¼ or larger (vertical)	every floor level		

For SI units: 1 foot = 0.305 m

Figure 9-1. This table from the *National Fuel Code* shows the proper spacing for gas piping supports. *Data from page 45 of the National Fuel Gas Code*

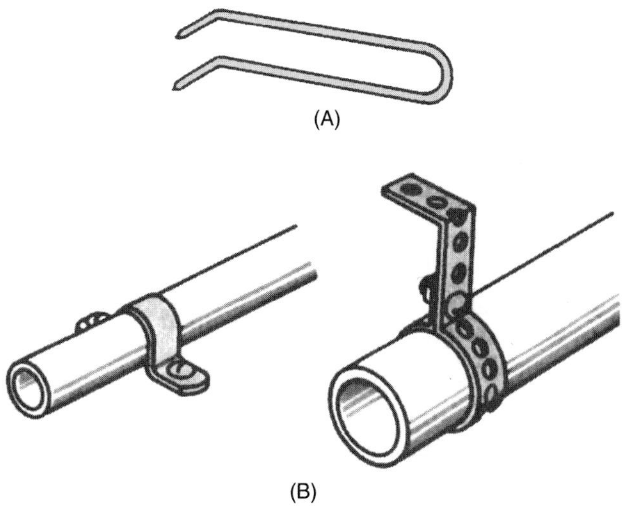

Figure 9-2. (A) A hanger and (B) supports for piping.

must be supported and fastened to protect it from physical damage.

Care should be taken in the design of piping systems that are routed through walls. Certain connections should not be used in walls where piping is concealed. These connections can leak and must be accessible. They include pipe unions, shutoff valves (GAS lock), right and left elbows for swing joints, and bushings, as shown in Figure 9-3 which depicts some examples of these.

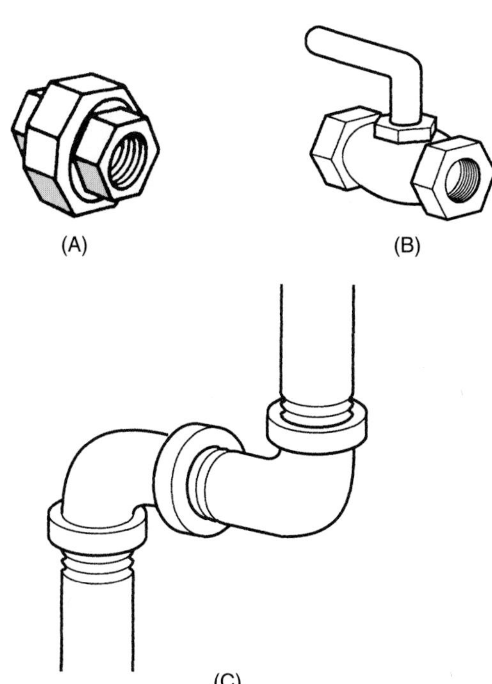

Figure 9-3. These fittings should not be installed in walls and sealed over because they may need serviced. (A) Pipe union. (B) Manual shoutoff valve. (C) Swivel joint.

All piping systems should be installed on a slope of ¼ inch per 15 feet toward the meter or regulator so that any moisture in the system will flow away from the gas-control valves and orifices, as shown in Figure 9-4. Another good practice is to install a drip leg, or drip trap, just before the gas valve (in addition to the slope). With the drip leg, if condensed moisture reaches the equipment, it moves to the bottom of the drip leg and evaporates into a vapor that can move through with the gas vapor. Figure 9-5 shows a drip leg.

Piping run underground should have a protective coating to retard corrosion, see Figure 9-6. The pipe should be laid in the bottom of the ditch, where settling dirt cannot cause stress on the pipe, as illustrated in

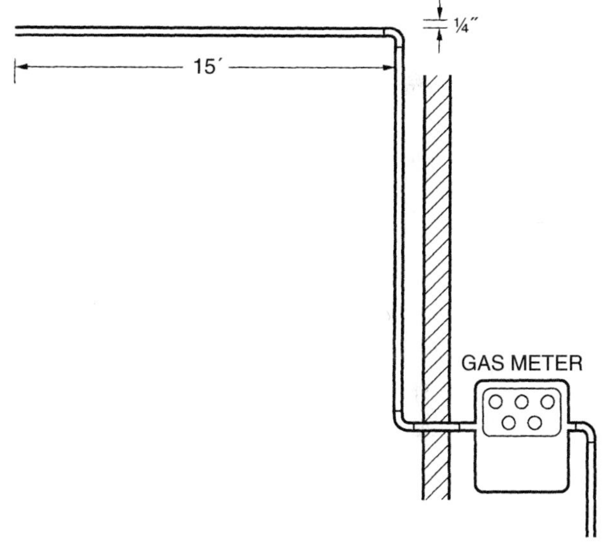

Figure 9-4. Gas piping should slope downward ¼ inch per 15 feet back toward the supply.

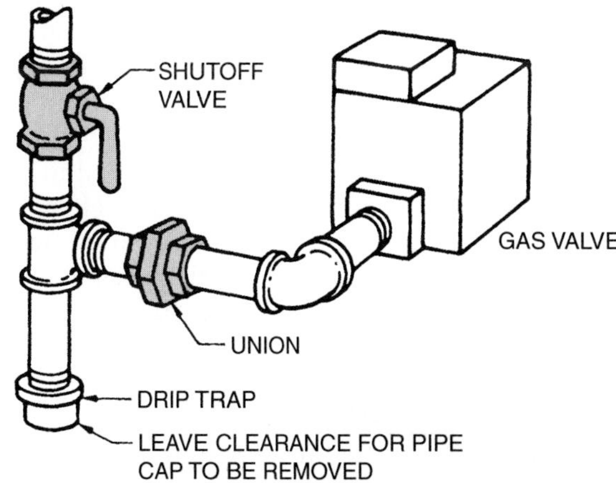

Figure 9-5. The drip leg before a valve prevents moisture from entering the valve. The moisture evaporates from the drip leg with time.

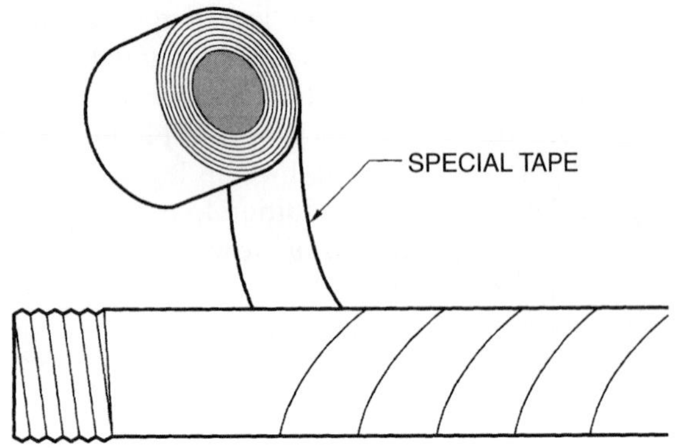

Figure 9-6. This piping, which will be routed underground, has a protective coating.

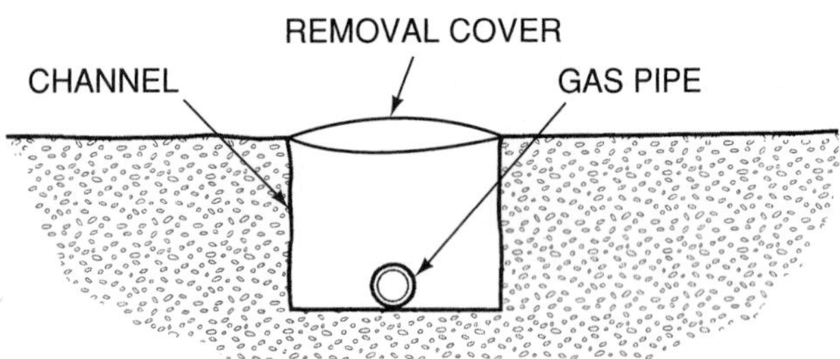

Figure 9-7. Lay the pipe in the bottom of the ditch so that the settling dirt does not stress it.

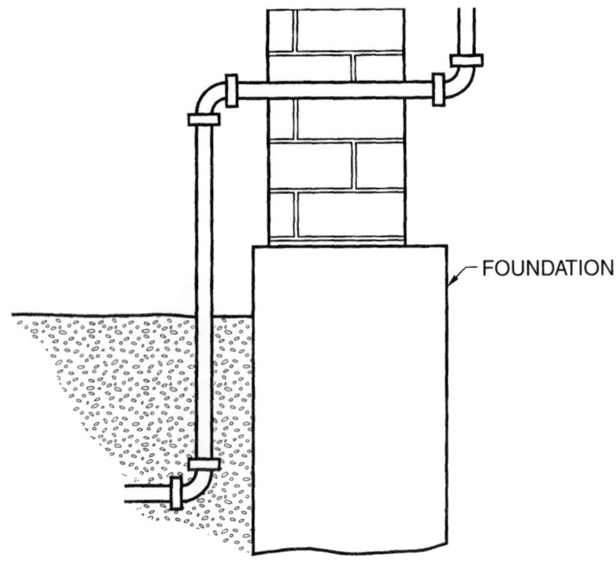

Figure 9-8. Bring the pipe above ground before it enters the structure so that if a leak occurs, the escaping gas will follow the pipe out of the ground.

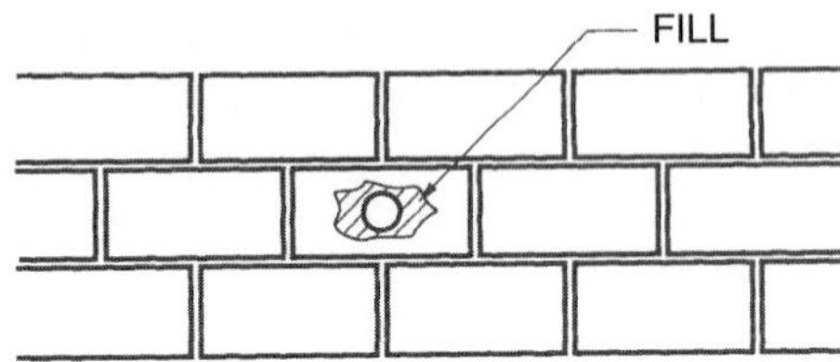

Figure 9-9. Seal the passage around the pipe where it enters the structure.

Figure 9-7. When the underground pipe enters the building (e.g., through a basement wall), it should be brought above ground before it enters the building, as shown in Figure 9-8. If an underground leak occurs, the gas will then follow the pipe and escape to the outside and not collect under the building.

At the point where the pipe enters the building, whether above or below ground, the pipe should be coated to prevent corrosion. The space around the pipe must be sealed to prevent moisture, insects, and rodents from entering, as shown in Figure 9-9.

A rigid piping system can also be installed in concrete floors when proper precautions are taken. The piping should be installed in a channel to allow access to it. In some instances the piping can be routed in the concrete when certain types of noncorrosive concrete are used.

Most of the rigid pipe systems that you will encounter will be assembled with threaded fittings. Figure 9-10 shows a section view of threading and the number of threads each type of pipe should have. If the pipe has too many threads, the fitting will bottom out and may leak, as depicted in Figure 9-11. If the pipe has too few threads, the fitting may not make a tight fit and leak, as shown in

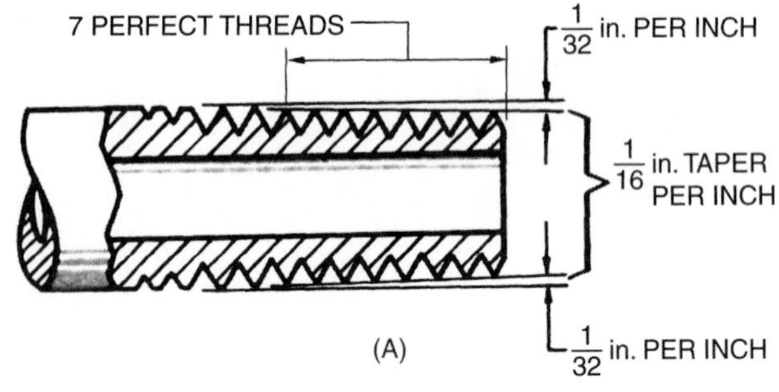

PIPE SIZE (INCHES)	THREADS PER INCH
$\frac{1}{8}$	27
$\frac{1}{4}$, $\frac{3}{8}$	18
$\frac{1}{2}$, $\frac{3}{4}$	14
1 TO 2	$11\frac{1}{2}$
$2\frac{1}{2}$ TO 12	8

(B)

Figure 9-10. Proper thread lengths. (A) Section view. (B) Threads per inch for different pipe sizes.

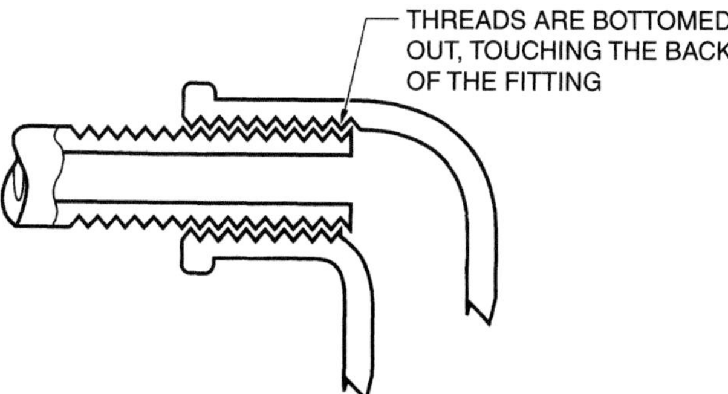

Figure 9-11. Too many threads.

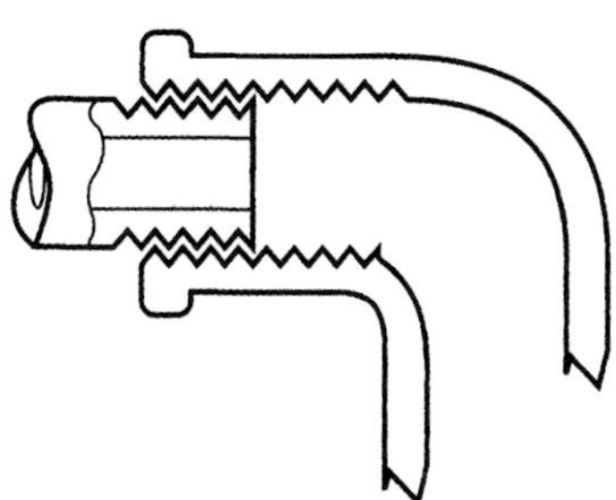

Figure 9-12. Too few threads.

Figure 9-12. All pipe fittings should be wiped clean of cutting oil and burrs after the cut is deburred. A thread seal (commonly called a *pipe dope*) should be used for each threaded connection and should be compatible with gas. Thread seals that are acceptable for natural gas can also be used for LP gases.

Used gas piping and fittings should be reused only after close inspection and a thorough cleaning. No rust or dirt particles should be allowed inside the pipe, or they will eventually loosen and plug the orifices.

Note: Good piping practices are acquired with experience. Work with someone who has experience before practicing on your own.

Gas Cock

Gas cocks are a uniquely designed valve used for gas distribution systems. Because a ball valve design is approved by most codes for isolation of individual gas equipment, a gas cock is used more by preference. Many ball valve designs for gas isolation are manufactured with a unique T-handle design, which differs from a typical WOG ball valve lever handle. A ball valve specifically designed for gas systems is not rated for use with other systems. A gas cock is used more as a means of isolating entire systems, and utility providers commonly use it for isolating gas meters. Many gas cock designs do not have a manual handle such as a lever or wheel handle, but instead require a wrench to open and close the gas cock. The wrench is placed on a raised lug and turned clockwise to close and counterclockwise to open.

Most codes do not allow a ball valve design for isolation of a system because they can be operated without a wrench. Most system isolation valves are located on the exterior of a building near a gas meter. Most gas cocks used in conjunction with a gas meter have an alignment hole in which to place a padlock to secure the gas distribution system when not in use. The unique stem of a gas cock is also the flow channel for the gas. The stem is a round steel rod that has a portion removed to create a flow slot (passageway). The ½-inch and ¾-inch sizes are the most commonly used for residential applications and have female threaded connections.

Tubing Systems for Gas

As mentioned previously, copper or aluminum tubing can be used to run natural and LP gas lines. This tubing is usually used at the end of a rigid piping system to make the last few feet of connection and to align the rigid pipe with the appliance (see Figure 9-13). Copper and aluminum tubing should not be used outside or underground because they do not provide much physical protection for the gas.

The fittings typically used for copper are flare-type fittings with water or gas flare nuts, not refrigeration flare nuts. The water or gas flare nut provides a little more support to the back of the fitting because of its longer skirt, see Figure 9-14. Compression fittings can be used in some applications for aluminum and copper tubing, as illustrated in Figure 9-15.

Copper can be fastened together with a brazed connection that has a melting point higher than 1,000°F if the brazing compound contains no phosphorus. The high

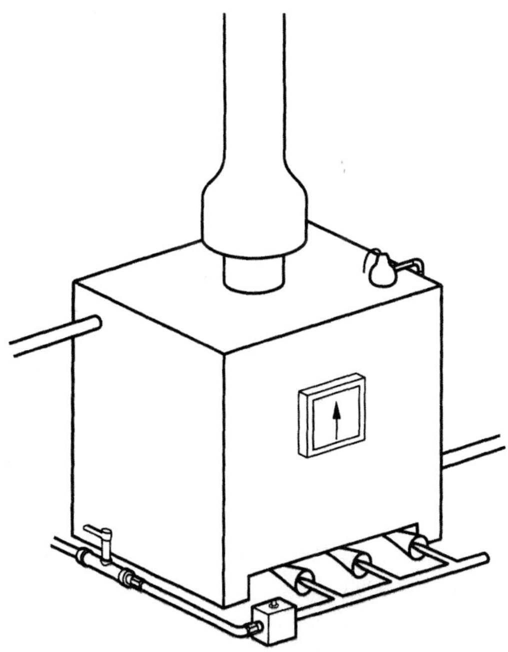

Figure 9-13. Tubing, a copper connector, at the end of the rigid pipe length can be used to make the final connection.

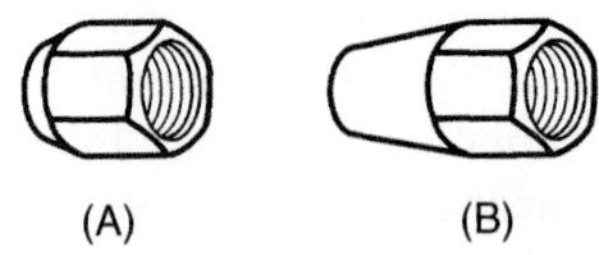

Figure 9-14. (A) Short-skirt refrigeration flare nut. (B) Long-skirt gas flare nut. Gas flare nuts give more support.

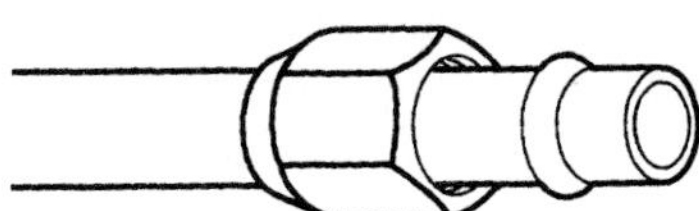

Figure 9-15. Compression fittings are used for some applications.

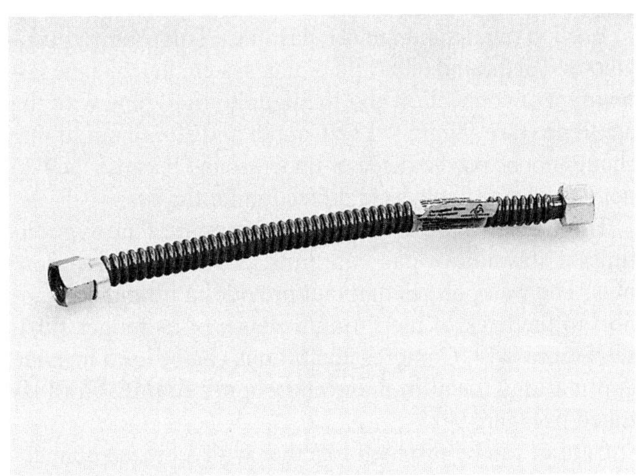

Figure 9-16. Steel tubing connector. *Photo by Bill Johnson*

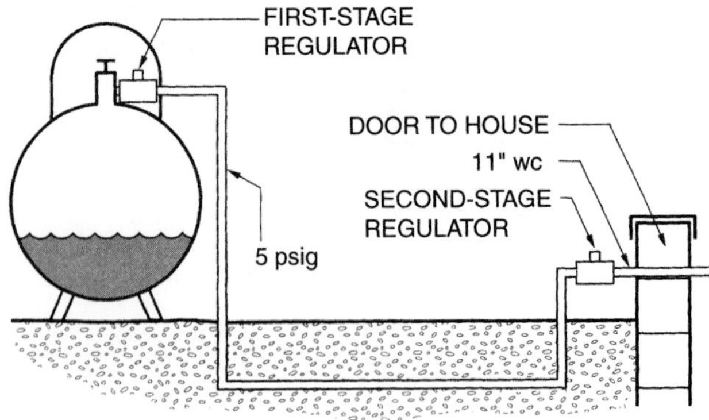

Figure 9-17. This LP system has two stages of regulation.

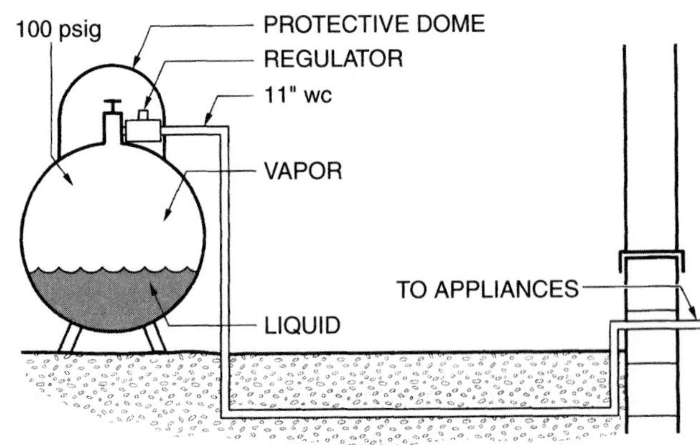

Figure 9-18. LP tank piping installation.

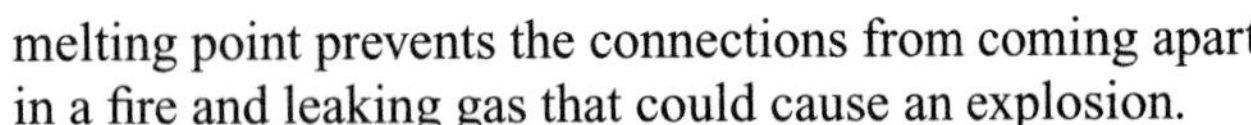

melting point prevents the connections from coming apart in a fire and leaking gas that could cause an explosion.

Another type of tubing that is often used at the end of a rigid piping system is called a *gas connector.* It is made of steel and is very flexible, see Figure 9-16. Its flexibility allows the installer to align the rigid pipe with the appliance.

LP piping systems are piped much as natural gas systems are. However, the tank may be farther from the house than the meter is for natural gas. Sometimes the tank is across the yard or behind the house, and underground piping is necessary. The regulator for the system is always under the tank lid and may reduce the pressure to either 15 psig or 11 inches of wc. If it reduces the pressure to only 15 psig, another regulator must be installed at the house to further reduce the pressure to 11 inches of wc. If only one regulator is used, the pressure is reduced to 11 inches of wc at the tank, and the pipe is run to the house. The advantage of two stages of regulation is a more constant pressure at the appliance. Two-stage regulation is used for systems with long piping runs, as shown in Figure 9-17. The connection between the regulator and the piping to the house can start with copper and switch to iron pipes, or it can be piped entirely with iron pipe, as shown in Figure 9-18.

As with natural gas piping, bring the piping above ground before it enters the structure, and fasten the piping under the house. The final connection between the iron piping and the appliance can be made with a gas connector or with copper tubing. A gas shutoff valve should be located just before each appliance, as shown in Figure 9-19.

Two-Pound Piping System

Any gas-distribution system is responsible for supplying the correct volume of gas at the correct pressure to the gas manifold. One distribution method is to supply the gas to the natural gas installation at the pressure supplied by the gas company after the meter. This pressure is typically 7 inches of wc. (Smaller pipe sizes can be used when greater pressures are used.) This method can be costly, so a less expensive piping system, the 2-pound system, was developed.

The 2-pound system supplies gas to the meter at 2 psig, where it is metered and then distributed to the various appliances, as shown in Figure 9-20. A *pounds-to-inches regulator* at each appliance reduces the pressure to 7 inches of wc;

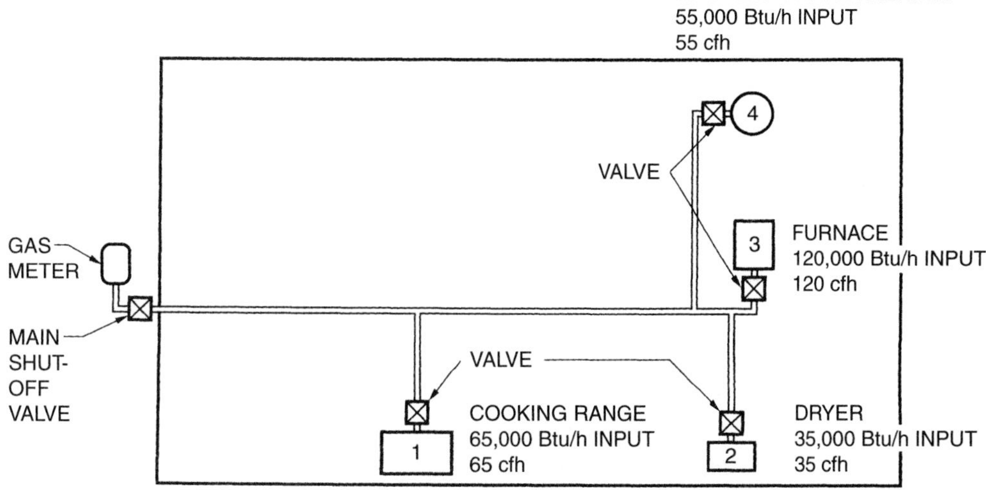

Figure 9-19. A gas shutoff valve should be located before each appliance.

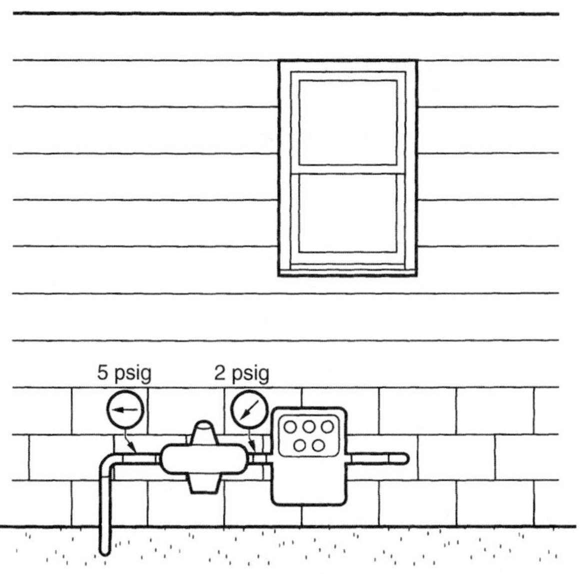

Figure 9-20. This system is metered at 2 psig, and most of the piping is at the 2 psig pressure.

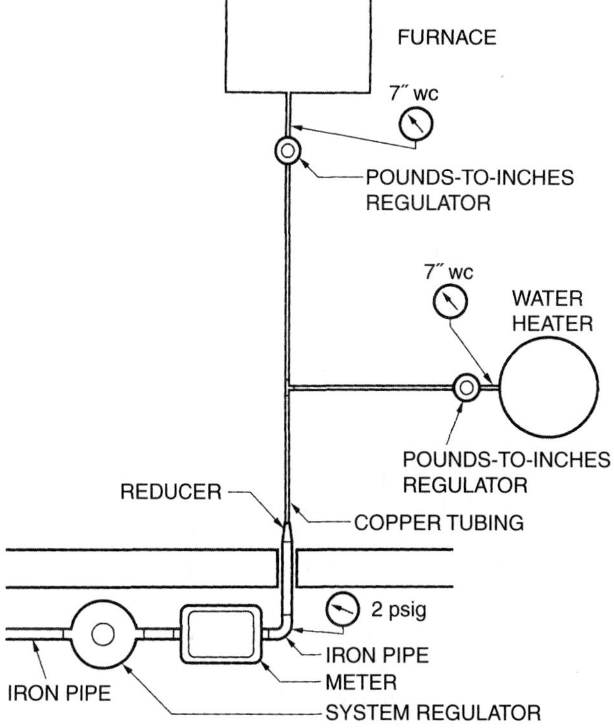

Figure 9-21. A pressure, or pounds-to-inches, regulator reduces the pressure to 7 inches of water column just before the appliance regulator.

the gas is then furnished to each appliance's gas valve for a further reduction to 3.5 inches of wc for the burner manifold, as shown in Figure 9-21. This distribution method requires much smaller pipes, which are less costly. Therefore, many houses can be piped with copper tubing rather than with rigid iron pipe. Most appliances have their own pressure regulators to reduce the pressure from 7 inches of wc to 3.5 inches of wc. If not, one must be supplied.

The 2-pound system must be designed because a prescribed pressure drop is incorporated into the piping system. For example, most 2-pound systems are designed to have a 1.5 psig pressure drop. This furnishes the pounds-to-inches regulator before the appliance with 0.5 psig of pressure instead of 2 psig, see Figure 9-22. The advantages of this procedure are the savings in pipe size and the safety factor. If a pipe breaks at the end, there are less than 2 psig of gas pressure to push gas into the structure. Some statistics show that a broken ¼-inch inner-diameter (ID) copper pipe at 2 psig pressure leaks only approximately ⅓ the amount of gas that a ¾-inch iron pipe does at 6 inches of wc.

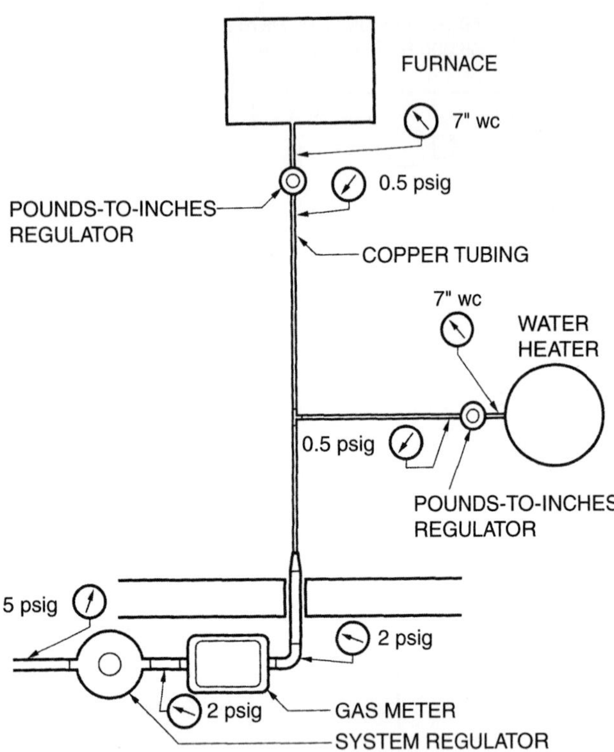

Figure 9-22. A pressure drop in the system with the appliances operating reduces the pressure to approximately 0.5 psig prior to the regulator.

Good piping practices for the 2-pound system include good workmanship with copper tubing. This tubing is run through walls, similarly to electrical wire; however, gas is inside the tubing. If you drive a nail into an electrical wire, sparks may fly and a fuse or a breaker may trip, but if you drive a nail into a copper tube with gas inside, you may not know it until enough gas escapes to cause an explosion. Consequently, you must prevent this from occurring. For example, when the copper tubing is routed through a wall, a metal sleeve, or plate, should surround the copper tubing where it penetrates the lower and upper plates of the wall. This sleeve will prevent a nail from penetrating the copper. Also, the tubing should not be fastened in the wall so that the tubing has room to move if a nail is driven through the wall and makes contact with it, as shown in Figure 9-23. Tubing can be run through floor joists at a right angle by drilling the joists, as shown in Figure 9-24. Tubing run parallel to the floor joist should be run close to water lines, ducts, or drain lines for protection. Tubing should not be run next to the floor because a person may drive a nail or drill the floor and puncture the tubing. The tubing should be supported every 4 feet for horizontal runs. Flared or brazed connections should be used. Compression fittings should not.

All piping must be free of debris because of the marginal size. It should be blown out with air before final assembly, as shown in Figure 9-25.

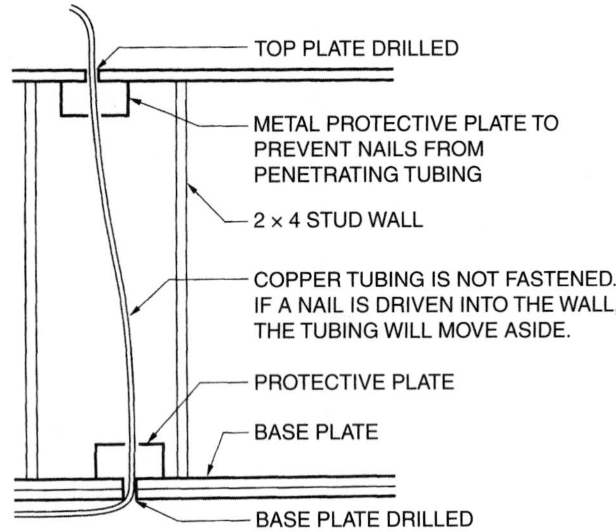

Figure 9-23. The tubing must be protected where it penetrates the lower and upper plates of the structure, and it must not be fastened to the wall.

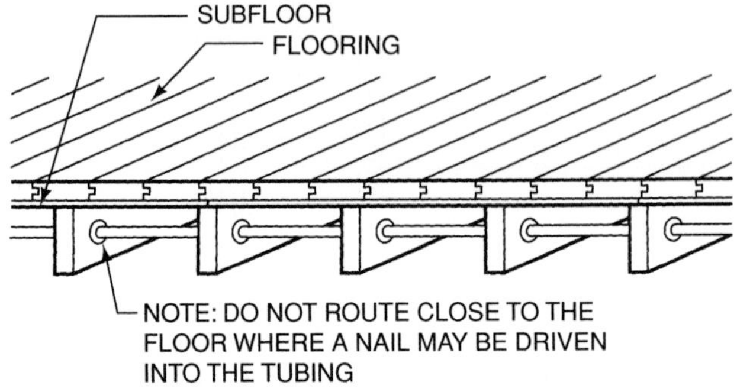

Figure 9-24. Tubing can be routed through the floor joists just as electrical wire can.

Drilling and Notching

Drilling and notching codes vary extensively in each region of the country and often in each state or local area. States that experience different climate conditions may have varying codes within the particular state. Minimum standards are adopted by the state, and local governments adopt ordinances that strengthen state codes. Before drilling and notching any structural member, know the local regulations to avoid expensive replacement of wooden stuctural boards in a building. Drilling though wood boards can weaken the structural strength of the board, and incorrectly positioning or sizing a hole can create unsafe conditions. Climate and regional occurrences such as snowfall, hurricances, tornados and earthquakes are main reasons for variances in drilling and notching codes.

Notching is a cutting process on the edge of a wooden board; the depth of a cut is strictly regulated and varies based on the area of a joist being notched. Hole and notch locations that place the outside diameter of a pipe close to the edge of a stud, joist or plate must be protected against

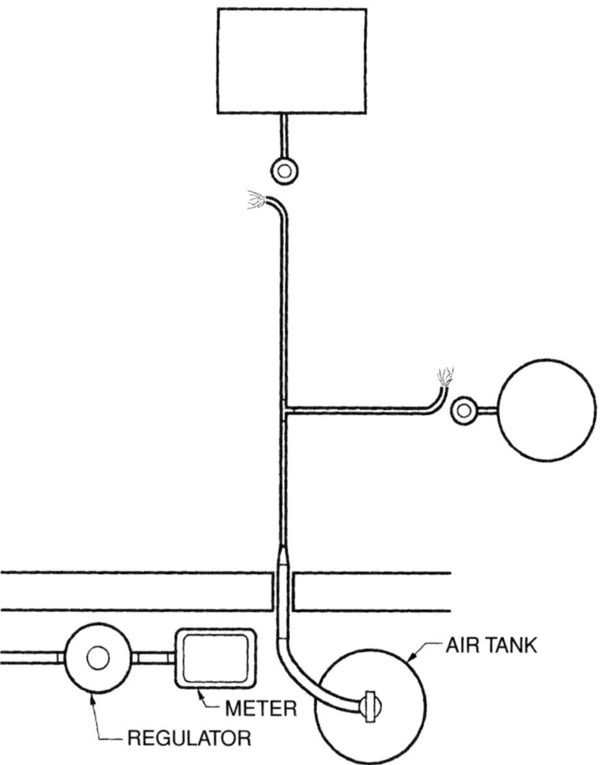

Figure 9-25. Use air to blow out the tubing before making the final connections.

nails and screws. Most codes require that a nail plate (stud guard) be installed in any location where a 1½ inches nail or screw could penetrate the piping. A nail plate is a 1/16 inches thick steel plate that is available in varying lengths and widths. Piping installed horizontally through a wall stud only requires a 1½ inches wide nail plate, but pipes passing vertically through a top or bottom plate must be protected with a 4 inches wall plate. Some codes dictate the specific type of approved nail plates that can be installed based on local codes, and the specific information is listed in a local code book.

Pressure Testing All Systems

All gas systems must be pressure tested to ensure that they are leak free. The *National Fuel Gas Code* states, "The test pressure to be used shall be no less than 1½ times the proposed maximum working pressure, but not less than 3 psig, irrespective of the design pressure." The conversion from inches of wc to psig is 7 inches equals ¼ psig, or 14 inches equals ½ psig. When this is applied to systems that operate at up to 11 inches of wc, a 3-pound test is sufficient. The 2-pound system can also be tested at 3 psig ($2 \times 1.5 = 3$).

The amount of time that the test should take is explained in another paragraph of the code. It covers systems of up to 10 cubic feet of pipe volume in single-family dwellings and states that the test should last for 10 minutes. Air should be used and there should be no pressure drop during the 10 minutes. All appliance shut-off valves must be closed to prevent the test pressure from reaching the automatic gas valve and the pressure regulator if it is not rated at the selected test pressure. Most automatic gas valves are unable to withstand the test pressure.

Note: Your local authority may have additional standards; check your local code.

Sizing Gas Pipe for Low-Pressure Systems

As mentioned previously, the purpose of the gas piping system is to distribute the gas to the gas manifold at the correct pressure. The gas must reach the appliance regulator with a pressure of approximately 6 inches of wc for the regulator to be able to properly control the pressure at the outlet. Any pressure drop below 6 inches of wc will cause fluctuations in the manifold pressure. The size of the gas pipe distribution system determines the pressure drop. If the pipe is sized too small, the pressure drop will be too great; the gas may leave the meter at 7 inches of wc but would be below the 6 inches of wc requirement at the appliance regulator. If the pipe is sized too large, the system will not be economical. The larger the pipe, the higher the cost of the system. A correct pipe size is important.

We use the *National Fuel Gas Code* for gas pipe sizing for low-pressure natural and LP gas. The following factors must be considered in sizing pipe for gas piping systems:

- allowable loss in pressure from point of delivery to equipment
- maximum gas demand
- length of piping and number of fittings
- diversity factor

The allowable loss in pressure is usually 0.3 inches of wc, but pipe can be sized for 0.5 inches of wc pressure loss.

The length of pipe is the distance from the meter to the farthest appliance, in feet of pipe, plus the pressure drop in the pipe fittings, usually expressed in *equivalent length of pipe.* Equivalent length of pipe expresses the pressure drop of a fitting as though it were a length of pipe, as shown in Figure 9-26. For example, suppose that you want to know what the pressure drop is in a ¾-inch, threaded, 90-degree elbow. Refer to Figure 9-26 and find the entry for ¾-inch in the left-hand column; move to the right and find the entry for a 90-degree screwed elbow. The pressure drop is equivalent to 2.06 feet of pipe. This amount can be added to the length of pipe in the system for pipe sizing purposes. Notice that valves are also expressed in equivalent feet of pipe.

The diversity factor in sizing pipes involves the specific gravity of the gas to be moved. The specific gravity of the gas is important if gases other than those in the table are used. In general, the heavier the gas, the less gas will move in a pipe size. In our discussions, we work only with natural gas and LP gases and the tables that apply to them.

		Screwed fittings[2]				90° welding elbows and smooth bends[3]					
		45° ell	90° ell	180° close return bends	Tee	R/d = 1	R/d = 1⅓	R/d = 2	R/d = 4	R/d = 6	R/d = 8
k factor =		0.42	0.90	2.00	1.80	0.48	0.36	0.27	0.21	0.27	0.36
L/d′ ratio[5] n =		14	30	67	60	16	12	9	7	9	12
Nominal pipe size, in.	Inside diam. d, in., Sched. 40[7]	L = equivalent length in feet of Schedule 40 (standard weight) straight pipe[7]									
½	0.622	0.73	1.55	3.47	3.10	0.83	0.62	0.47	0.36	0.47	0.62
¾	0.824	0.96	2.06	4.60	4.12	1.10	0.82	0.62	0.48	0.62	0.82
1	1.049	1.22	2.62	5.82	5.24	1.40	1.05	0.79	0.61	0.79	1.05
1¼	1.380	1.61	3.45	7.66	6.90	1.84	1.38	1.03	0.81	1.03	1.38
1½	1.610	1.88	4.02	8.95	8.04	2.14	1.61	1.21	0.94	1.21	1.61
2	2.067	2.41	5.17	11.5	10.3	2.76	2.07	1.55	1.21	1.55	2.07
2½	2.469	2.88	6.16	13.7	12.3	3.29	2.47	1.85	1.44	1.85	2.47
3	3.068	3.58	7.67	17.1	15.3	4.09	3.07	2.30	1.79	2.30	3.07
4	4.026	4.70	10.1	22.4	20.2	5.37	4.03	3.02	2.35	3.02	4.03
5	5.047	5.88	12.6	28.0	25.2	6.72	5.05	3.78	2.94	3.78	5.05
6	6.065	7.07	15.2	33.8	30.4	8.09	6.07	4.55	3.54	4.55	6.07
8	7.981	9.31	20.0	44.6	40.0	10.6	7.98	5.98	4.65	5.98	7.98
10	10.02	11.7	25.0	55.7	50.0	13.3	10.0	7.51	5.85	7.51	10.0
12	11.94	13.9	29.8	66.3	59.6	15.9	11.9	8.95	6.96	8.95	11.9
14	13.13	15.3	32.8	73.0	65.6	17.5	13.1	9.85	7.65	9.85	13.1
16	15.00	17.5	37.5	83.5	75.0	20.0	15.0	11.2	8.75	11.2	15.0
18	16.88	19.7	42.1	93.8	84.2	22.5	16.9	12.7	9.85	12.7	16.9
20	18.81	22.0	47.0	105	94.0	25.1	18.8	14.1	11.0	14.1	18.8
24	22.63	26.4	56.6	126	113	30.2	22.6	17.0	13.2	17.0	22.6

	Miter elbows[4] (No. of miters)					Welding tees		Valves (screwed, flanged, or welded)			
	1-45°	1-60°	1-90°	2-90°[6]	3-90°[6]	Forged	Miter[4]	Gate	Globe	Angle	Swing Check
k factor =	0.45	0.90	1.80	0.60	0.45	1.35	1.80	0.21	10	5.0	2.5
L/d′ ratio[5] n =	15	30	60	20	15	45	60	7	333	167	83
Nominal pipe size, in.	L = equivalent length in feet of Schedule 40 (standard weight) straight pipe[7]										
½	0.78	1.55	3.10	1.04	0.78	2.33	3.10	0.36	17.3	8.65	4.32
¾	1.03	2.06	4.12	1.37	1.03	3.09	4.12	0.48	22.9	11.4	5.72
1	1.31	2.62	5.24	1.75	1.31	3.93	5.24	0.61	29.1	14.6	7.27
1¼	1.72	3.45	6.90	2.30	1.72	5.17	6.90	0.81	38.3	19.1	9.58
1½	2.01	4.02	8.04	2.68	2.01	6.04	8.04	0.94	44.7	22.4	11.2
2	2.58	5.17	10.3	3.45	2.58	7.75	10.3	1.21	57.4	28.7	14.4
2½	3.08	6.16	12.3	4.11	3.08	9.25	12.3	1.44	68.5	34.3	17.1
3	3.84	7.67	15.3	5.11	3.84	11.5	15.3	1.79	85.2	42.6	21.3
4	5.04	10.1	20.2	6.71	5.04	15.1	20.2	2.35	112	56.0	28.0
5	6.30	12.6	25.2	8.40	6.30	18.9	25.2	2.94	140	70.0	35.0
6	7.58	15.2	30.4	10.1	7.58	22.8	30.4	3.54	168	84.1	42.1
8	9.97	20.0	40.0	13.3	9.97	29.9	40.0	4.65	222	111	55.5
10	12.5	25.0	50.0	16.7	12.5	37.6	50.0	5.85	278	139	69.5
12	14.9	29.8	59.6	19.9	14.9	44.8	59.6	6.96	332	166	83.0
14	16.4	32.8	65.6	21.9	16.4	49.2	65.6	7.65	364	182	91.0
16	18.8	37.5	75.0	25.0	18.8	56.2	75.0	8.75	417	208	104
18	21.1	42.1	84.2	28.1	21.1	63.2	84.2	9.85	469	234	117
20	23.5	47.0	94.0	31.4	23.5	70.6	94.0	11.0	522	261	131
24	28.3	56.6	113	37.8	28.3	85.0	113	13.2	629	314	157

1. Values for welded fittings are for conditions where bore is not obstructed by weld spatter or backing rings. If appreciably obstructed, use values for "Screwed Fittings."
2. Flanged fittings have three-fourths the resistance of screwed elbows and tees.
3. Tabular figures give the extra resistance due to curvature alone to which should be added the full length of travel.
4. Small size socket-welding fittings are equivalent to miter elbows and miter tees.

For SI units: 1 foot = 0.305 m

5. Equivalent resistance in number of diameters of straight pipe computed for a value of f — 0.0075 from the relation n — $k/4f$.
6. For condition of minimum resistance where the centerline length of each miter is between d and 2½d
7. For pipe having other inside diameters, the equivalent resistance may be computed from the above n values.

Figure 9-26. The pressure drop in the pipe fittings is expressed in equivalent length of pipe. *Data from pages 214 and 215 of the National Fuel Gas Code*

Figure 9-27 shows two tables for sizing natural gas piping for nominal iron pipe sizes. One of the tables is for a pressure drop of 0.3 inch of wc and the other is for 0.5 inch of wc. You can use either table; however, perhaps the 0.5 inch of wc table is the most practical because, as you will see, you will need to round up to the capacities of the equipment and will ultimately oversize the pipe. For this reason, you will probably have a final pressure drop of approximately 0.3 inch of wc while sizing for 0.5 inch of wc.

Maximum Capacity of Pipe in Cubic Feet of Gas per Hour for Gas Pressures of 0.5 psig or Less and a Pressure Drop of 0.3 Inch Water Column

(Based on a 0.60 Specific Gravity Gas)

Nominal Iron Pipe Size, Inches	Internal Diameter, Inches	Length of Pipe, Feet													
		10	20	30	40	50	60	70	80	90	100	125	150	175	200
1/4	.364	32	22	18	15	14	12	11	11	10	9	8	8	7	6
3/8	.493	72	49	40	34	30	27	25	23	22	21	18	17	15	14
1/2	.622	132	92	73	63	56	50	46	43	40	38	34	31	28	26
3/4	.824	278	190	152	130	115	105	96	90	84	79	72	64	59	55
1	1.049	520	350	285	245	215	195	180	170	160	150	130	120	110	100
1 1/4	1.380	1,050	730	590	500	440	400	370	350	320	305	275	250	225	210
1 1/2	1.610	1,600	1,100	890	760	670	610	560	530	490	460	410	380	350	320
2	2.067	3,050	2,100	1,650	1,450	1,270	1,150	1,050	990	930	870	780	710	650	610
2 1/2	2.469	4,800	3,300	2,700	2,300	2,000	1,850	1,700	1,600	1,500	1,400	1,250	1,130	1,050	980
3	3.068	8,500	5,900	4,700	4,100	3,600	3,250	3,000	2,800	2,600	2,500	2,200	2,000	1,850	1,700
4	4.026	17,500	12,000	9,700	8,300	7,400	6,800	6,200	5,800	5,400	5,100	4,500	4,100	3,800	3,500

Figure 9-27. Pipe sizing tables for natural gas. *Data from pages 145 and 146 of the National Fuel Gas Code*

Maximum Capacity of Pipe in Cubic Feet of Gas Per Hour for Gas Pressures of 0.5 psig or Less and a Pressure Drop of 0.5 Inch Water Column

(Based on a 0.60 Specific Gravity Gas)

Nominal Iron Pipe Size, Inches	Internal Diameter, Inches	Length of Pipe, Feet													
		10	20	30	40	50	60	70	80	90	100	125	150	175	200
1/4	.364	43	29	24	20	18	16	15	14	13	12	11	10	9	8
3/8	.493	95	65	52	45	40	36	33	31	29	27	24	22	20	19
1/2	.622	175	120	97	82	73	66	61	57	53	50	44	40	37	35
3/4	.824	360	250	200	170	151	138	125	118	110	103	93	84	77	72
1	1.049	680	465	375	320	285	260	240	220	205	195	175	160	145	135
1 1/4	1.380	1,400	950	770	660	580	530	490	460	430	400	360	325	300	280
1 1/2	1.610	2,100	1,460	1,180	990	900	810	750	690	650	620	550	500	460	430
2	2.067	3,950	2,750	2,200	1,900	1,680	1,520	1,400	1,300	1,220	1,150	1,020	950	850	800
2 1/2	2.469	6,300	4,350	3,520	3,000	2,650	2,400	2,250	2,050	1,950	1,850	1,650	1,500	1,370	1,280
3	3.068	11,000	7,700	6,250	5,300	4,750	4,300	3,900	3,700	3,450	3,250	2,950	2,650	2,450	2,280
4	4.026	23,000	15,800	12,800	10,900	9,700	8,800	8,100	7,500	7,200	6,700	6,000	5,500	5,000	4,600

Figure 9-27. *(Continued)*

The following explanation of how to use the gas piping capacity tables is summarized from the *National Fuel Gas Code:*

1. Determine the gas demand of each appliance to be attached to the piping system. When Figure 9-27 is used to select the piping size, calculate the gas demand in terms of cfh (cubic feet per hour) for each piping system outlet. For natural gas, it is sufficient to use the appliance input divided by 1,000 Btu to determine the cfh. For example, suppose that a furnace has an input of 120,000 Btu per hour. The furnace would use 120 cfh (120,000/1,000 = 120). Figure 9-28 shows typical capacities for some gas-burning appliances.
2. Determine the design system pressure, the specific gravity of the gas, and the allowable pressure loss. For example, the design system pressure is usually 7 inches of wc, the specific gravity of natural gas is 0.6, and

Approximate Gas Input for Some Common Appliances

Appliance	Input Btu per hr. (Approx.)
Range, Free Standing, Domestic	65,000
Built-In Oven or Broiler Unit, Domestic	25,000
Built-In Top Unit, Domestic	40,000
Water Heater, Automatic Storage 30 to 40 Gal. Tank	45,000
Water Heater, Automatic Storage 50 Gal. Tank	55,000
Water Heater, Automatic Instantaneous	
Capacity: 2 gal. per minute	142,800
Capacity: 4 gal. per minute	285,000
Capacity: 6 gal. per minute	428,400
Water Heater, Domestic, Circulating or Side-Arm	35,000
Refrigerator	3,000
Clothes Dryer, Type 1 (Domestic)	35,000
Gas Light	2,500
Incinerator, Domestic	35,000

For SI units 1 Btu per hour = .0293 Watts

For specific appliances or appliances not shown above, the input should be determined from the manufacturer's rating.

Figure 9-28. Typical capacities for some gas-burning appliances. Data from NFPA, National Fuel Gas Code. Copyright © 1992, National Fire Protection Association, Quincy, MA 02269. This reprinted material is not the complete and official position of the National Fire Protection Association, on the referenced subject which is represented only by the standard in its entirety.

the allowable pressure loss is either 0.3 inch of wc or 0.5 inch of wc, preferably the latter, as mentioned previously.
3. Measure the length of pipe from the point of delivery to the most remote outlet in the building—from the meter to the farthest point. If the system seems to have more fittings than usual, the chart in Figure 9-26 for pipe fittings can be used to add to the total feet of pipe. A typical system with some elbows and tees and a valve does not require adding the fittings to the pipe length because we round up appliance capacities and pipe sizes.
4. From the appropriate capacity table, select the column showing the measured length, or the next longer length if the table does not give the exact length (as mentioned, we round up to create less of a pressure drop). This is the only length used in determining the size of any section of gas piping.
5. Use this vertical column to locate *all* gas-demand figures for this particular piping system.
6. Starting at the most remote outlet, find the vertical column just selected for the gas demand for that outlet. If the exact figure of demand is not shown, choose the next larger figure in the column.
7. Opposite this demand figure, in the first column at the left, find the correct size of gas piping.
8. Proceed in a similar manner for each outlet and each section of gas piping. For each section of piping, determine the total gas demand supplied by this section (some sections will be gas mains with more than one appliance pipe from them).

Using Figure 9-29 and the following information, size the pipes for the installation shown in the figure. This house has four appliances:

1. *Cooking range:* 65,000 Btu/h input/1,000 = 65 cfh.
2. *Dryer:* 35,000 Btu/h input/1,000 = 35 cfh.
3. *Furnace:* 120,000 Btu/h input/1,000 = 120 cfh.
4. *Water heater (50 gallon):* 55,000 Btu/h input/1,000 = 55 cfh.

We start by measuring the longest run of piping. Because there is no excessive use of fittings in this piping, the fittings are disregarded. The longest run is to the water heater, 65 feet. Because we will likely round up some values, we will use the 0.5 inch of wc table for natural gas (0.6 specific gravity). From Figure 9-27 find the column for 70 feet (this is rounded up from 65 feet).

Note: In the remainder of this problem, the column for 70 feet will be used. Always use the longest run of pipe for the complete problem.

Follow these steps to determine the branch line pipe size from the main gas line supplying all appliances:

1. The cooking range has a capacity of 65,000 Btu/h and consumes 65 cfh. Find the column for 70 feet and move down to 65 cfh, or the next highest value. A value of 61 cfh is shown but is not enough. Round up to 125 cfh, and a branch line pipe size of ¾ inch is found.
2. The dryer uses 35 cfh. Move down the column for 70 feet to 35 or the next highest value, and find 61. Looking to the left, we find the branch line pipe size to be ½ inch.
3. The furnace uses 120 cfh. Move down the column for 70 feet to 120 or the next highest value, and we find 125 cfh. The branch line pipe size is 1 inch.
4. The water heater uses 55 cfh. Move down the column for 70 feet until you find 55 or the next highest number, and we find 61. The branch line pipe size is ½ inch.

There are 3 main lines, that serve more than one appliance. The same pipe length should be used for each, and

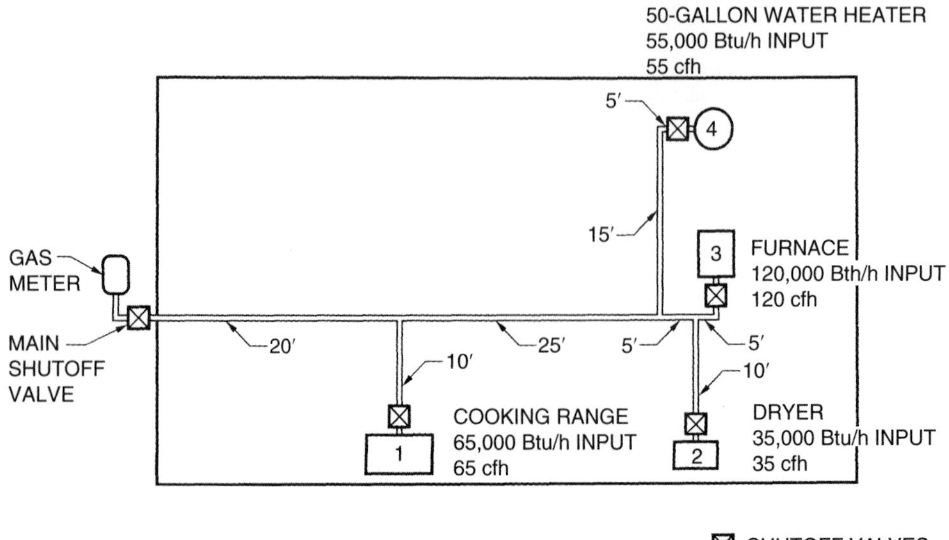

Figure 9-29. Building layout for sizing piping for a low-pressure system.

the total gas carrying capacity for each will be used. Follow these steps:

1. The main serving the furnace and the dryer must carry 155 cfh of gas. Move down the column for 70 feet until you find 155 or the next largest number. This number is 240. Looking left, you should find this line size to be 1 inch.
2. The main that serves outlets 2, 3, and 4 must carry 210 cfh because it must have the capacity to carry gas to all three appliances. Look down the column for 70 feet to 210 or the next highest number. Again, the size is 1 inch.
3. The main that carries gas from the meter to all the appliances must carry 275 cfh. Look down the column for 70 feet and find 275 or the next highest number. This pipe must be 1¼ inches.

At this point, the complete piping system has been sized except for connectors that may be used on some systems between black iron pipes and the appliances. For all practical purposes, these connectors should be the same size as the branch pipe line to the appliance. Notice how many times you needed to round up. We sized the system according to a pressure drop of 0.5 inch of wc, but the pressure drop will be less than 0.5 inch of wc. It may be near the pressure drop in the other table for 0.3 inch of wc.

When pipe is sized for an LP gas system, the same procedures are used, but the table for LP gas is used as shown in Figure 9-30.

Sizing Pipe for the Two-Pound System

As mentioned, the 2-pound system for sizing and routing pipe can be less costly than the 7 inches of wc low-pressure system. The gas can be routed through smaller tubing rather than iron pipe. Smaller tubing costs less and takes less time to install. The system that we discuss next was established by Minnegasco and has been used for more than 15 years. However, as the company points out, there are other approved systems and methods.

Two systems are discussed next: a system in which all the appliances are grouped close to one another and a system in which the appliances are scattered throughout the structure. The gas is metered at 2 psig, and the gas pressure downstream from the meter starts at 2 psig and reduces in pressure, as a result of a pressure drop in the pipe, to a final system pressure regulator, called a *psig-to-inches-of-wc regulator* (this is shortened to pounds-to-inches regulator). Figure 9-31 shows a typical

Figure 9-31. Pounds-to-inches regulator. *Photo by Bill Johnson*

Pipe Sizing
Sizing Between Single or Second Stage (Low Pressure Regulator) and Appliance
Maximum undiluted propane capacities listed are based on 11" W.C. setting and a 0.5" W.C. pressure drop.
Capacities in 1,000 Btu/hr

Pipe Length Feet	Nominal Pipe Size, Schedule 40								
	½" 0.622	¾" 0.824	1" 1.049	1¼" 1.38	1½" 1.61	2" 2.067	3" 3.068	3½" 3.548	4" 4.026
10	291	608	1146	2353	3525	6789	19130	28008	39018
20	200	418	788	1617	2423	4666	13148	19250	26817
30	161	336	632	1299	1946	3747	10558	15458	21535
40	137	287	541	1111	1665	3207	9036	13230	18431
50	122	255	480	985	1476	2842	8009	11726	16335
60	110	231	435	892	1337	2575	7256	10625	14801
80	94	198	372	764	1144	2204	6211	9093	12668
100	84	175	330	677	1014	1954	5504	8059	11227
125	74	155	292	600	899	1731	4878	7143	9950
150	67	141	265	544	815	1569	4420	6472	9016
200	58	120	227	465	697	1343	3783	5539	7716
250	51	107	201	412	618	1190	3353	4909	6839
300	46	97	182	374	560	1078	3038	4448	6196
350	43	89	167	344	515	992	2795	4092	5701
400	40	83	156	320	479	923	2600	3807	5303

Figure 9-30. LP gas pipe sizing table. *Data from page 158 of the National Fuel Gas Code*

pounds-to-inches regulator. Both systems have at least one common feature: the piping from the meter to the system regulator (the pounds-to-inches regulator) is smaller than the piping from the pressure-reducing regulator to the appliance.

When a system is arranged so that the appliances are grouped, only one pounds-to-inches regulator is necessary. The savings is in the long main line from the meter to the regulator. This system is sized like the main gas pipe in the multiple system and is not discussed further.

When the appliances are scattered, the system between the pounds-to-inches regulator and the appliance regulator is sized just as the low-pressure system is. It operates at approximately 0.5 psig, or 14 inches of wc, and is usually short.

Using the same piping system as before, let us size the pipe for 2 psig with individual regulators for each appliance, see Figure 9-32.

First, we size the branch gas lines between the main and the pounds-to-inches regulator, starting with the water heater because it is the farthest away.

1. The water heater has a capacity of 55 cfh (55,000 Btu/h) and is 65 feet from the gas meter, so the entry for 70 feet is used, just as in the previous problem. Note that with the 2-pound system we use the correct pipe lengths for each run instead of the maximum run length only. Use Figure 9-33 to find the entry for 70 feet, then move to the right until you reach 55 cfh. It is the first column. The tubing size is ¼-inch ID (the same as ⅜-inch OD [outer diameter] shown at the bottom of the column).
2. The furnace has a capacity of 120 cfh. Use the pipe length of 55 feet, rounded to 60 feet. Find the entry for 60 feet and move to the right until you find 120 cfh. Still, ¼-inch ID tubing can be used.
3. The dryer has a capacity of 35 cfh, and it is 60 feet from the meter. Find the entry for 60 feet and move to the right to 35 cfh; ¼-inch tubing can be used.
4. The cooking range has a capacity of 65 cfh and is 30 feet from the meter. Find the entry for 30 feet and move to the right to 65 cfh; ¼-inch tubing can be used.

The main lines are sized this way:

1. The main serving the furnace and the dryer must carry 155 cfh of gas. The distance to the farthest appliance is 70 feet. Find the entry for 70 feet and move to the right to 155 cfh; ⅜-inch ID tubing is required.
2. The main serving the water heater, the furnace, and the dryer must carry 210 cfh and will be sized for the longest run, the water heater, 65 feet. Find the entry for 70 feet and move to the right to 210 cfh; ⅜-inch ID tubing can be used.
3. The remaining trunk line carries the total system capacity of 275 cfh. Find the entry for 70 feet and move to the right to 275 cfh. You will find that ⅜-inch ID tubing can be used.

If the pounds-to-inches regulator is far from the appliance, the lines from the regulator to the appliance can be sized with the low-pressure method because they are operating at approximately 7 inches of wc. In some cases the line will size small in comparison with the connection on the appliance. If this occurs, use the connection on the appliance as the guide for the line size.

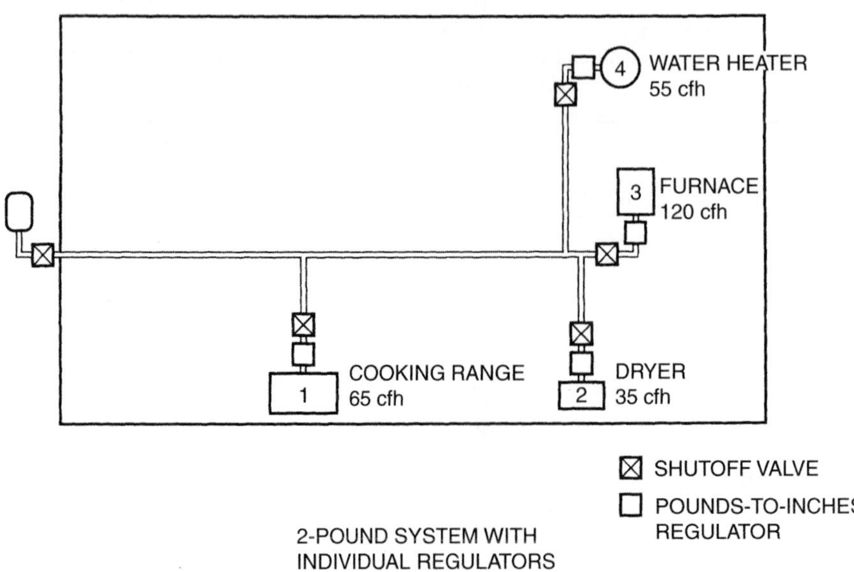

Figure 9-32. Building layout for sizing a 2-pound system.

Chart 3
Sizing Chart For Two Pound Pressure Pipe

Cfh at 1½ Pounds Pressure Drop - .65 Gravity Gas - Nominal Pipe or I.D. Tubing Size

INSIDE DIAMETER OF TUBING

Length in Ft.	1/4	3/8	1/2	5/8	3/4	1	1 1/4	1 1/2	2	2 1/2	3	4
5	540	1260	2400	4150	6500	10500	21000	31000	58000	90000	150M	310M
10	360	850	1630	2780	4350	7600	15000	22000	41000	64000	110M	220M
15	285	670	1280	2150	3450	6200	12000	18000	34000	52000	90000	180M
20	240	570	1080	1860	2950	5400	10500	15000	29000	45000	79000	150M
30	192	450	860	1480	2300	4400	8600	13000	24000	36000	63000	125M
40	163	380	730	1250	2000	3800	7500	11000	20000	32000	55000	110M
50	143	335	645	1100	1750	3350	6700	9800	18000	28000	49000	97000
60	130	300	580	1000	1560	3050	6100	9000	17000	26000	45000	90000
70	118	275	530	910	1430	2800	5600	8200	15000	24000	41000	82000
80	110	255	490	850	1330	2650	5200	7700	14000	22000	38000	77000
90	102	240	460	790	1230	2500	4900	7200	13500	21000	36000	72000
100	96	225	430	740	1160	2350	4700	6800	12500	20000	34500	70000
125	85	198	380	650	1025	2100	4150	6100	11300	18000	31000	62000
150	76	178	340	585	920	1900	3800	5600	10400	16000	28000	56000
175	69	164	315	540	845	1800	3550	5200	9700	15500	26000	53000
200	64	146	290	500	780	1700	3300	4900	9000	14000	24000	49000
250	58	140	255	440	690	1500	2950	4300	8100	12500	22000	44000
300	51	120	230	395	620	1350	2700	4000	7400	11500	19000	40000
	3/8	1/2	5/8	3/4	7/8	1 1/8	1 3/8	1 5/8	2 1/8	2 5/8	3 1/8	4 1/8

OUTSIDE DIAMETER OF TUBING

Use this chart when metering or reduced pressure is 2 PSIG. Minimum inlet pressure to pounds-to-inches regulator will be 1/2 pound or 14 inches.

Figure 9-33. Two-pound pipe sizing chart. *Courtesy Minnegasco*

9.2 Gas Vent Practices and Sizing

The content and the temperature of the by-products of combustion are discussed in previous units. These by-products must be vented outside the structure for safe, efficient operation. There are two different venting systems, natural draft and forced draft. Both have the same purpose; the only difference is the intention of the manufacturer. Natural-draft appliances and equipment rely on heat to cause the by-products to rise up the flue. Therefore, the flue gases must remain hot enough to rise in the vent. On the other hand, forced-draft equipment has a small combustion blower to force the by-products out of the appliance through the vent system to the atmosphere. Most systems require the flue gases to be hot enough to prevent moisture from condensing in the venting system; however, in some equipment, such as the condensing furnace, the flue gases are cooled until moisture condenses from the by-products. The moisture, when condensed, is removed with a piping system as part of the process.

Natural-Draft Venting Practices

The natural-draft furnace must keep the flue gases at the correct temperature for venting. These appliances are classified as Category I appliances according to the *National Fuel Gas Code.*

Many years ago, when the gas-burning appliance replaced the wood- or coal-burning appliance, the installing technician often made the mistake of venting the gas appliance through the old chimney that handled the wood or coal by-products of combustion. These chimneys were coated on the inside with soot. It is easy to imagine that it would be hard to heat a large chimney to the correct temperature with a gas appliance. In many instances, the chimney was not cleaned before the gas appliance was vented. This practice was undesirable because the chimney would allow the flue gases to condense. The condensate mixed with the soot already in the chimney and created an acid that attacked the chimney mortar joints, as shown in Figure 9-34.

Newer, safer methods for using chimneys are now used. For example, the modern chimney has a ceramic lining that heats faster than the chimney itself, as shown in Figure 9-35. The lining is not in contact with the brick-work, so there is an air space between the liner and the brick. This air space acts as an insulator from the mass of the chimney. When an old, unlined chimney is encountered, a flexible liner can be used, as shown in Figure 9-36.

In new constructions, several types of venting material are used. The most common is the single-wall or the

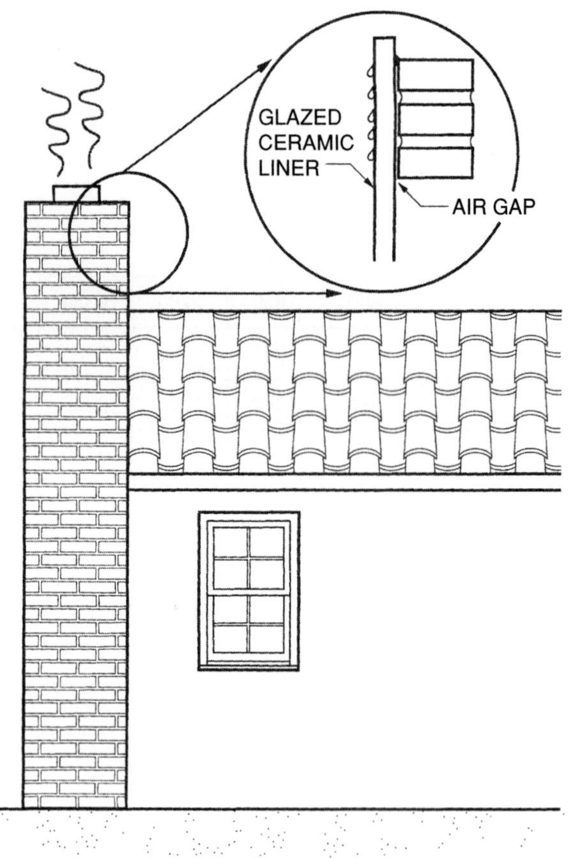

Figure 9-35. This chimney has a glazed ceramic liner much like the glazing on a coffee cup.

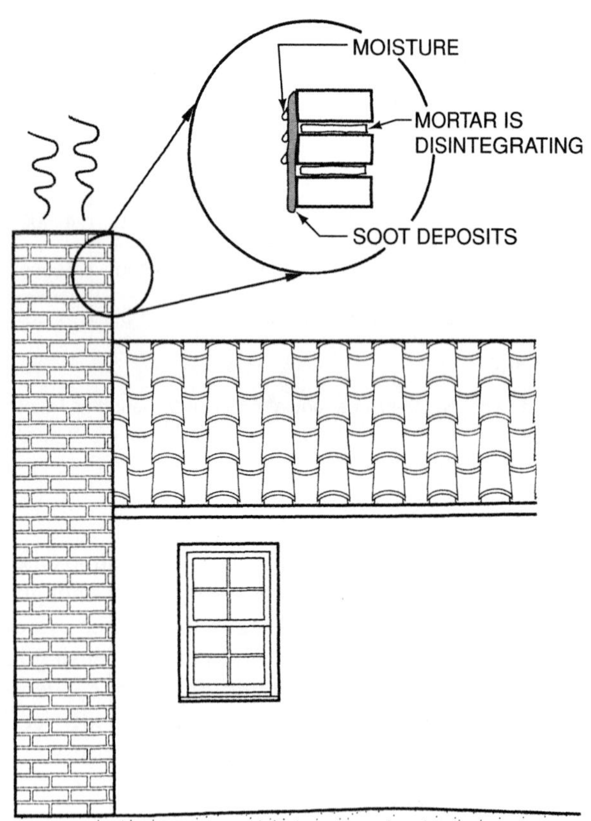

Figure 9-34. The vent gases attack the mortar joints in this old chimney.

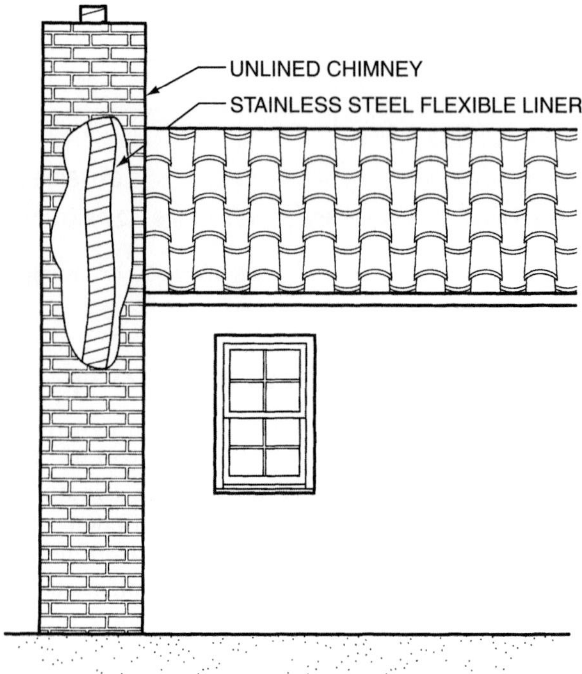

Figure 9-36. Flexible liner for an old chimney.

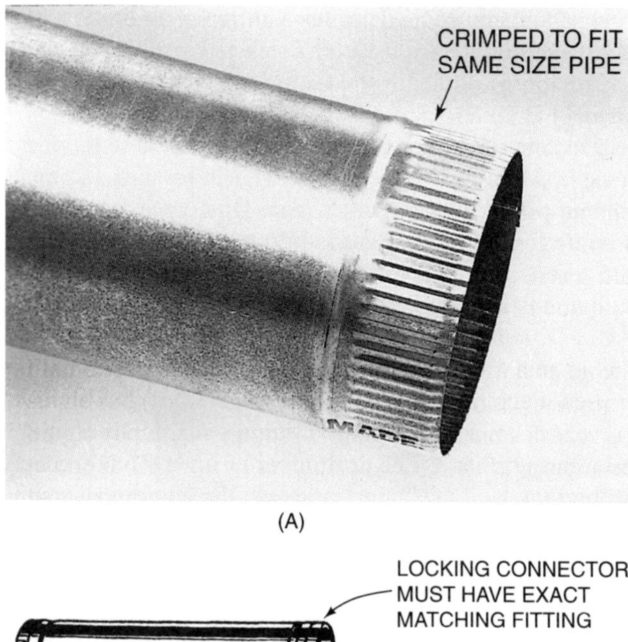

Figure 9-37. (A) Single-wall vent pipe. *Photo by Bill Johnson* (B) Double-wall vent pipe.

double-wall metal vent pipe, as shown in Figure 9-37. These pipes may also be called type B vent materials, which designates the temperature of the flue gases vented in this material. Type B vent material is for low-temperature gas appliances with flue gases below 550°F and with draft diverters.

Single-wall vent pipe is made of metal and is usually used when the flue pipe passes through a conditioned (heated) space, such as when the furnace is located in a heated basement. The single-wall vent pipe cannot keep the flue gases up to temperature when a flue passes through a cool, unheated space, such as an attic in winter.

Double-wall vent pipe is used when the flue passes through a cool (unconditioned) space. Many people think that the double-wall feature prevents the flue from starting a fire, but it instead prevents the flue gases from cooling, see Figure 9-38.

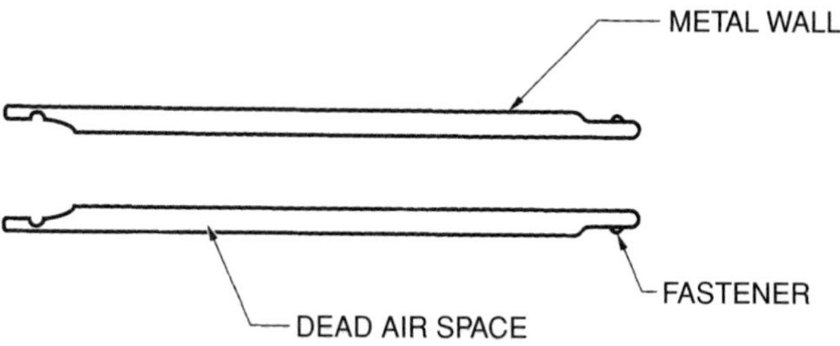

Figure 9-38. A double-wall vent pipe uses the air space between the pipes as an insulator.

Note: Care should be used with double-wall pipe and fittings. The purpose of the double wall is the air space between the inside and the outside of the pipe. This air space acts as insulation and must be preserved. Screws must not be driven into the pipe; it should usually be fitted together with twist-lock or snap-lock fittings. Each manufacturer has its own fittings, which may not be interchangeable with those of other manufacturers.

Forced-Draft Venting Practices

Some Category I furnaces and boilers are manufactured without draft hoods. This type of furnace has a combustion blower to move the flue gases out of the heat exchanger to the flue, see Figure 9-39. The flue gases operate differently with this type of furnace.

One purpose of the draft hood is to reduce the dew-point temperature of the flue gases by mixing them with room air. The dew-point temperature of the flue gases is the temperature at which moisture begins to condense from the flue gases. The higher the dew-point temperature, the more likely condensation will occur. The room

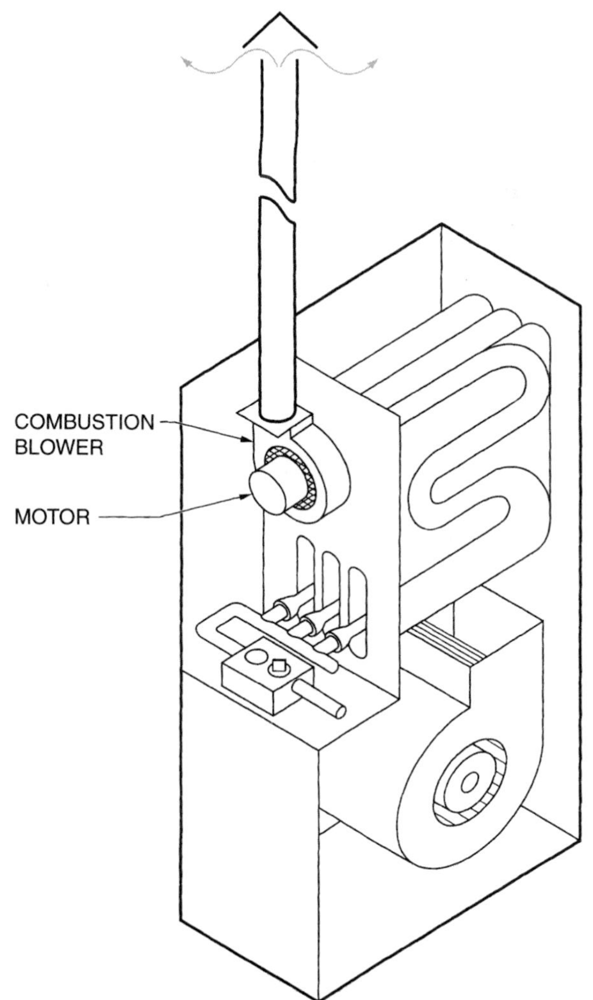

Figure 9-39. Category I furnace with a combustion blower.

air added to reduce the dew-point temperature by dilution causes an inefficiency of the system because this room air must be made up from the cold outside air. (Dilution air is discussed in more detail later in this section.)

When a furnace is manufactured with a fan instead of a draft hood, the flue gases are much hotter, with a higher dew-point temperature. Condensation is more likely to occur. Because of this, the vent practices for these furnaces is more critical. The technician may encounter a problem when an old oversized furnace is replaced with a newer furnace of less capacity that requires a smaller flue. Sometimes the older furnace was vented into a chimney that was nearly oversized for the larger appliance and that is definitely oversized for the newer, smaller furnace. In situations such as this, be sure to consult the manufacturer's literature for vent practices.

In addition to Category I furnaces, there are categories II, III, and IV. The Category II furnace is a condensing furnace with a natural-draft vent system. These systems are rare. The Category III appliance operates at a vent gas temperature of 140°F above the dew-point temperature of the flue gases. An example of a Category III furnace is a gas furnace vented through the side wall of the structure instead of through the roof. This vent system must be airtight. The Category IV vent system is the condensing furnace. The flue gases are below the dew point and condensation occurs.

Air

Regardless of the venting type, the correct amount of air must be supplied to the equipment room for the equipment to vent properly. Remember from Unit 3 that every time 1 cubic foot of natural gas is burned, 15 cubic feet of air are taken into the combustion process. The total amount of combustion products is 16 cubic feet (1 cubic foot of gas + 15 cubic feet of air = 16). This is actually not all the air that moves up the flue when natural-draft venting is used. Each natural-draft appliance has a draft diverter to prevent downdrafts from affecting the main burner or the pilot burner flame. The draft diverter diverts a downdraft around the heat exchanger and into the equipment room, rather than into the heat exchanger and down through the burner, see Figure 9-40.

When the burner is ignited, and a flame is causing flue gases to rise, air is taken in through the draft diverter, see Figure 9-41. Approximately 15 cubic feet of dilution air are moved up the flue for each cubic foot of natural gas burned. This air and the air for combustion must be replenished in the equipment room. When the air for combustion and the air for dilution are added, the total is

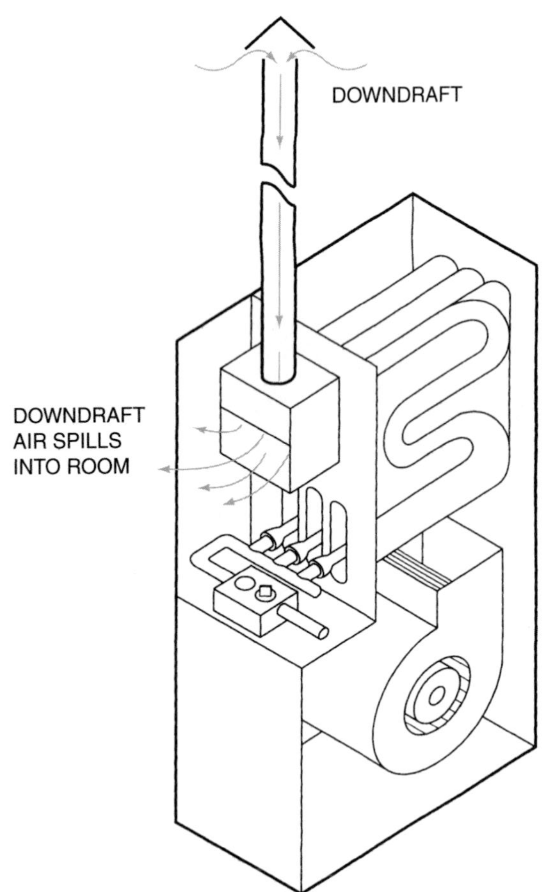

Figure 9-40. Draft diverter action.

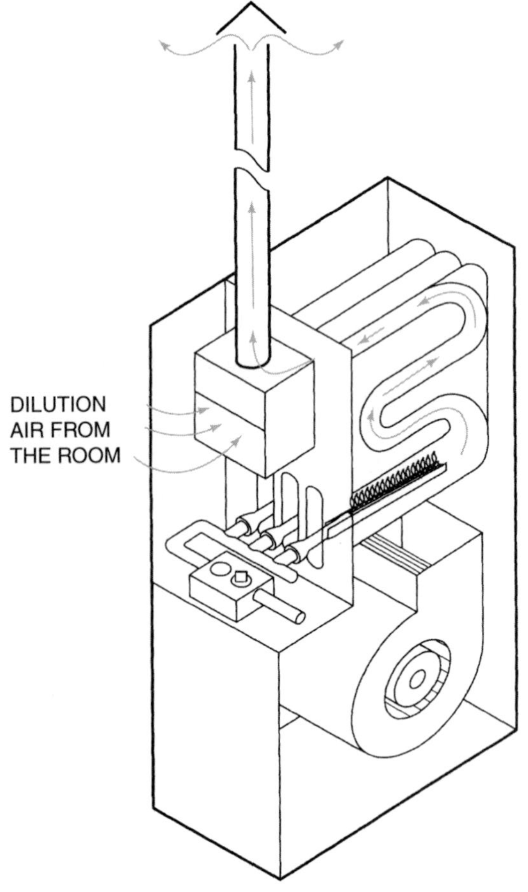

Figure 9-41. Dilution air moving through the draft diverter.

approximately 31 cubic feet of combustion gas per cubic foot of natural gas burned as shown in Figure 9-42. When a furnace has an input of 100,000 Btu/h, it consumes approximately 100 cubic feet of gas an hour and requires 3,100 cubic feet of air to be furnished in the vicinity of the furnace. This includes combustion air and dilution air (31 × 100 = 3,100).

Makeup air for combustion and dilution must come from outside the structure. If it is taken from inside the structure, the structure may become oxygen depleted and the building pressure can move into a vacuum. In the past, when makeup air came from outside, the fresh air was pulled in through the cracks in the building (infiltration) as shown in Figure 9-43. However, today construction is so tight and leak free, that infiltration can not be relied upon. If the building does not have enough cracks or openings, fresh air will spill down the draft diverter and the appliance will vent into the structure, as shown in Figure 9-44. The correct amount of fresh air must be supplied to the equipment. This is easy if the equipment room has easy access to the outside, but an equipment room may be located in the middle of the structure.

All fresh air for combustion and dilution air must have a free path to the equipment. There should be a fresh air supply from low in the room, in the vicinity of the burner, for combustion, and a fresh air supply high in the room for dilution.

Four locations for fresh air are typically encountered, and the *National Fuel Gas Code* has recommendations for each area. These recommendations apply to combustion air and ventilation air for natural and LP gases.

First, when a structure is large and has considerable infiltration or ventilation, the furnace may be located in the center of the structure with an occupied floor above and below the equipment room. It may not be practical

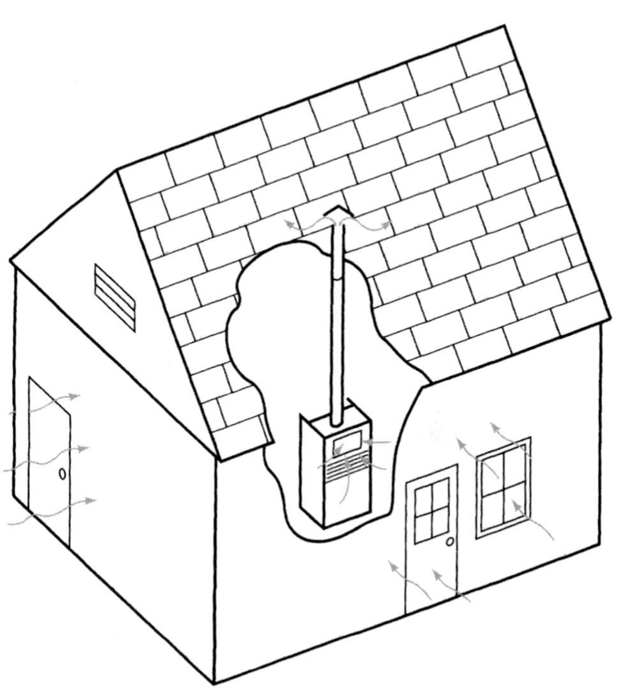

Figure 9-43. Makeup air can be pulled through the cracks in a loose structure.

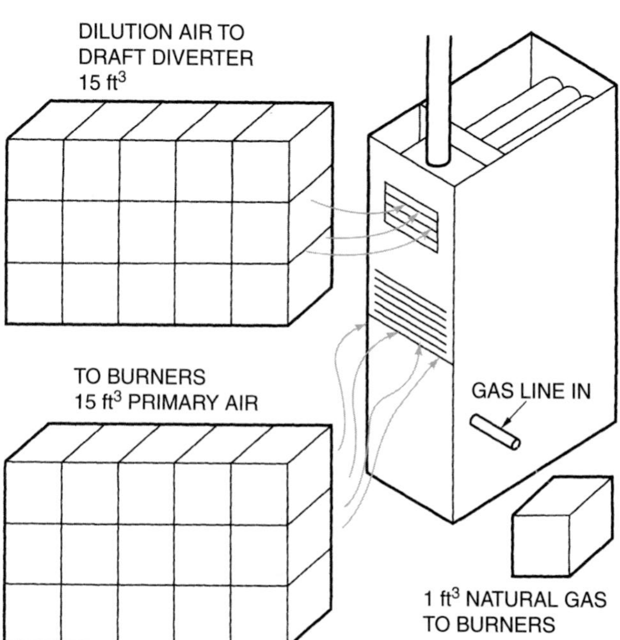

Figure 9-42. Thirty-one cubic feet of combustion gases are used per cubic foot of natural gas burned.

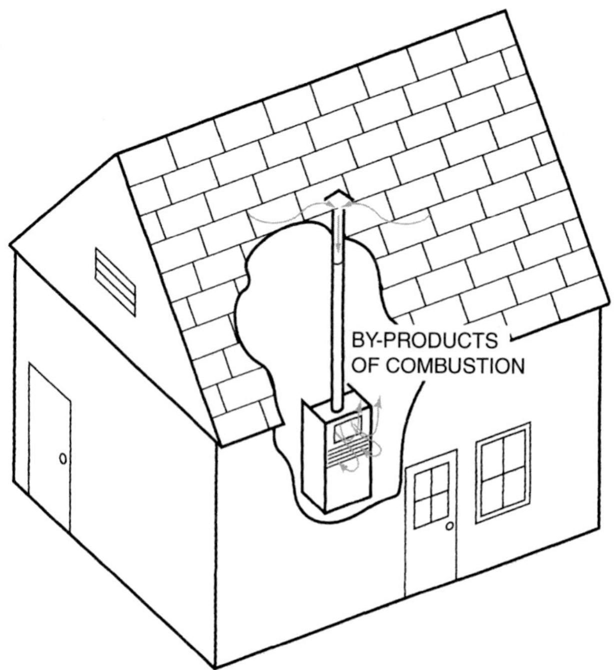

Figure 9-44. If the structure is too tight and no markup air is available, it is pulled down the draft diverter.

to run a duct to the outside for fresh air. In this case, fresh air may be taken from inside the structure. When all the air must come from inside the building, two permanent connections between the furnace room and the additional rooms shall be furnished with a free area of 1 square inch each per 1,000 Btu/h of the total input of all gas-burning equipment, but not less than 100 square inches. One opening shall be within 12 inches of the top and one within 12 inches of the bottom of the enclosure. Figure 9-45 shows how makeup air can be furnished for the system.

Second, an equipment room may be located close enough to an outside wall that ducts can be run to the outside of the structure as shown in Figure 9-46. One opening shall be within 12 inches of the top and one within 12 inches of the bottom of the enclosure. Furthermore, the following apply:

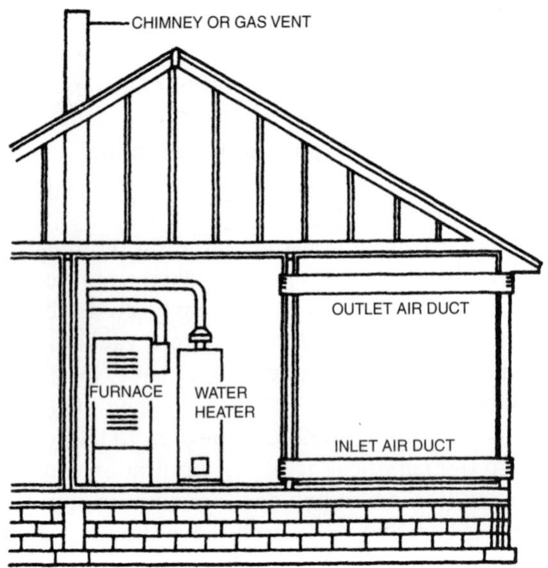

Figure 9-46. Equipment room next to the outside of the structure. *Reprinted by permission from page 69 of the National Fuel Gas Code*

1. When directly communicating with the outdoors (no duct), each opening shall have a minimum free area of 1 square inch per 4,000 Btu/h of the total input rating of all equipment in the enclosure.
2. When communicating with the outdoors through a vertical duct, each opening shall have a minimum free area of 1 square inch per 4,000 Btu/h of the total input rating of all equipment in the enclosure.
3. When communicating with the outdoors through horizontal ducts, each opening shall have a minimum free area of 1 square inch per 2,000 Btu/h of the total input rating of all equipment in the enclosure.
4. When ducts are used, they shall be of the same cross-sectional area as the free area of the openings to which they connect. The minimum dimension of each rectangular air duct shall be not less than 3 inches.

Third, a one-story structure with an attic can furnish all the air for combustion and dilution from the attic space as shown in Figure 9-47. A duct must be run to the floor for proper air circulation.

Fourth, a one-story structure has an attic and a crawl space, where an opening from each can provide fresh air in some instances, as shown in Figure 9-48.

Note: It is important that you follow the rules for the four locations of fresh air. Be sure to check your local code for any additions or changes that apply to your area.

When the combustion and dilution air are taken from a space above or below the structure, this space must be

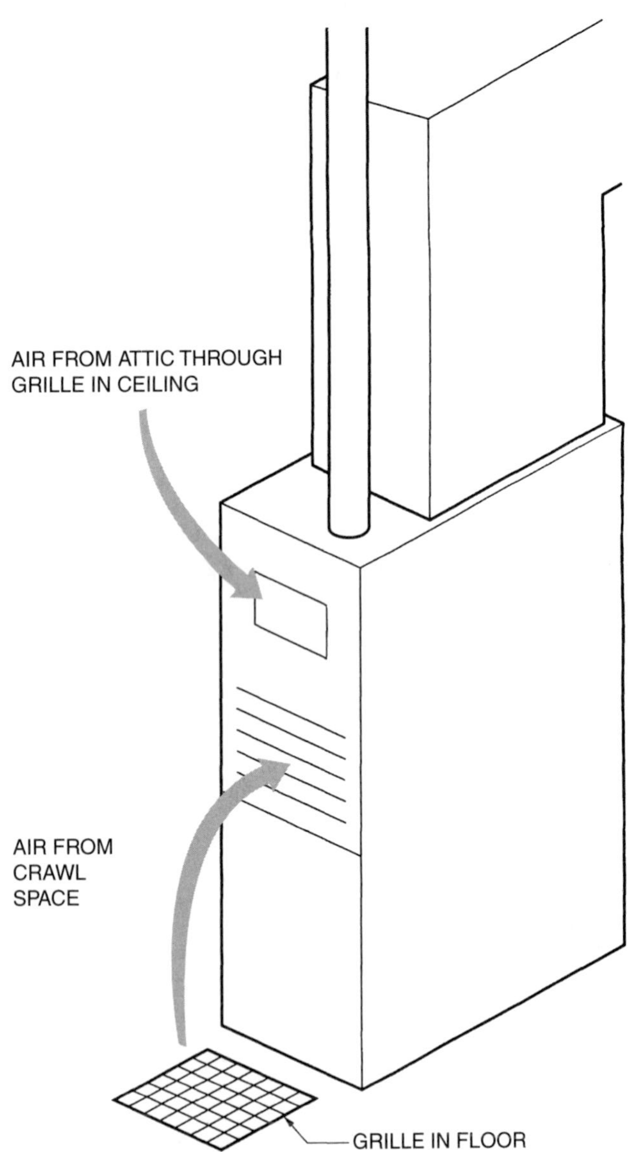

Figure 9-45. Makeup air.

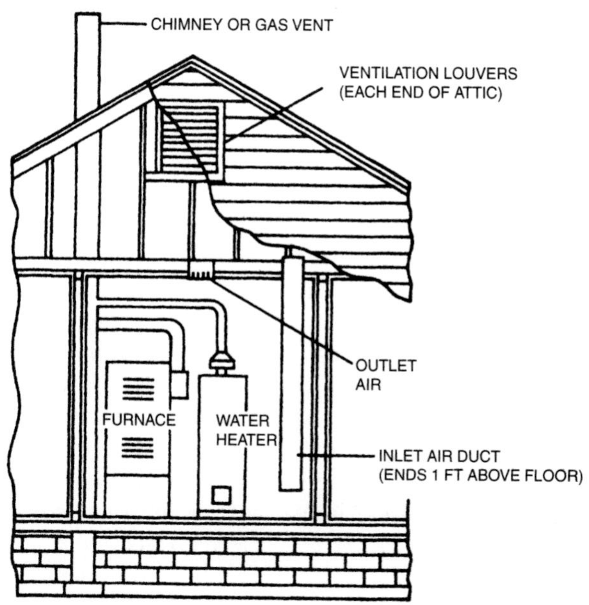

Figure 9-47. One-story structure with the makeup air furnished from the atttic. *Data taken from page 68 of the National Fuel Gas Code*

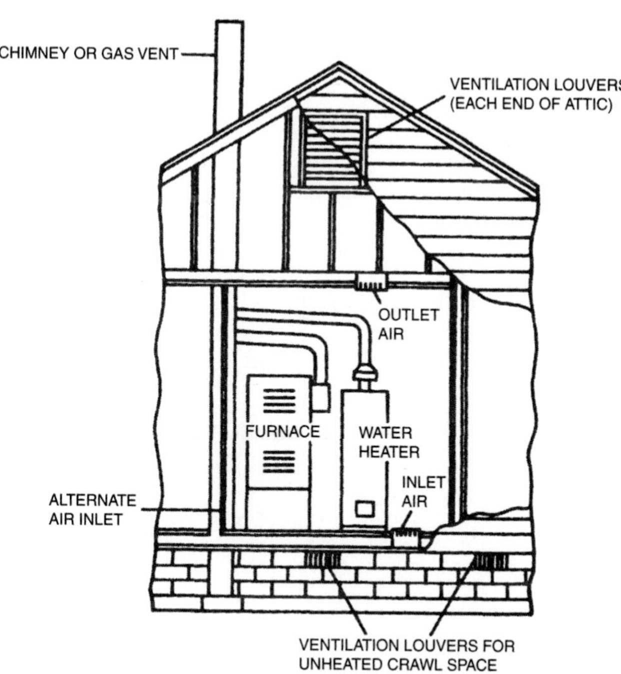

Figure 9-48. This structure has an attic and a crawl space for makeup air. *Data taken from page 67 of the National Fuel Gas Code*

connected to the outside of the house by means of a permanent opening through the foundation and eaves of the house, see Figure 9-49.

All the openings mentioned are for the free area of the grille that covers the opening. The free area is the actual usable area. Louvers or screen wire take up part of the

Figure 9-49. The attic and crawl space must be connected to the outside air with permanent connections if they are to be used for makeup air.

actual opening, so most grilles have a 70% free area. The free area can be calculated by measuring the opening to the grille and multiplying it by .70, see Figure 9-50.

During combustion, dilution air and air that rises through the heat exchanger rise up the flue as long as a negative pressure is created by the temperature difference in the flue gases. The temperature difference from the bottom of the flue to the top causes the flue gases to rise. Flue gases rising from the equipment room during the off cycle create infiltration that is unnecessary and actually an expense. Devices called *vent damper shutoffs* were developed to minimize this rise in flue gases during the off cycle. There are two types: bimetal and motor driven.

The bimetal vent damper shutoff either has a set of bimetal blade dampers that open upon a rise in temperature or has a bimetal element that opens a damper inside

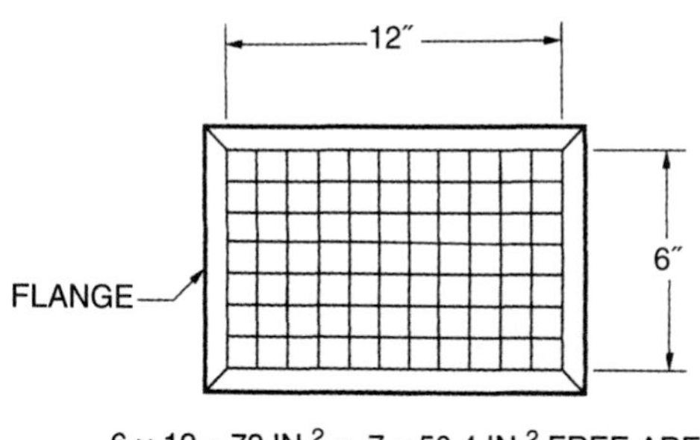

6 × 12 = 72 IN.^{2} × .7 = 50.4 IN.^{2} FREE AREA

Figure 9-50. Calculating the free area of a grille.

the device. This damper is simple and quite reliable. Figure 9-51 shows both types of bimetal vent damper shutoffs. Some of these vent damper shutoff models have a high-temperature thermostat that shuts off the burner in the event of a high-temperature condition.

The motor-driven vent damper shutoff requires a motor to close a damper against a spring tension. This tension is a fail-safe feature and allows the damper to open in case of motor failure. The motor is in locked rotor while it is powered open, but it does not pull enough power (only about 40 watts) to create enough heat to damage itself while in locked rotor. The motorized vent damper shutoff is energized by the room thermostat through a relay and has an end switch that does not allow the furnace burner to start until the damper is open, as illustrated in Figure 9-52. This end switch is a safety feature to ensure that the by-products of combustion leave the structure.

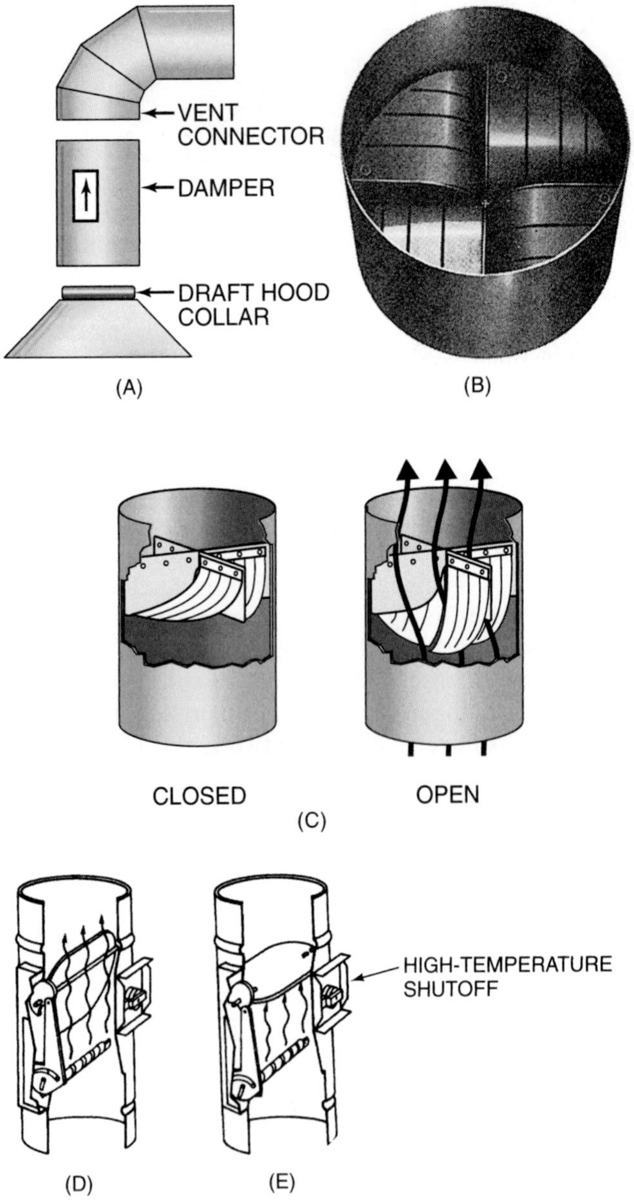

Figure 9-51. Bimetal vent damper shutoffs. (A–C) Bimetal blade dampers. (D–E) Bimetal element opens and closes a damper.

Vent Sizing

The *National Fuel Gas Code* also established vent sizing tables and procedures that are used in many states as part of their codes. Vent sizing and practices are an extremely important part of any gas-burning installation because the vent must be run from the appliance to the roof in the correct manner for natural-draft equipment. The vent takes up space and must be routed in the correct manner with the correct materials. Vent sizing can be complicated for a complex installation, such as an apartment house with many furnaces and water heaters.

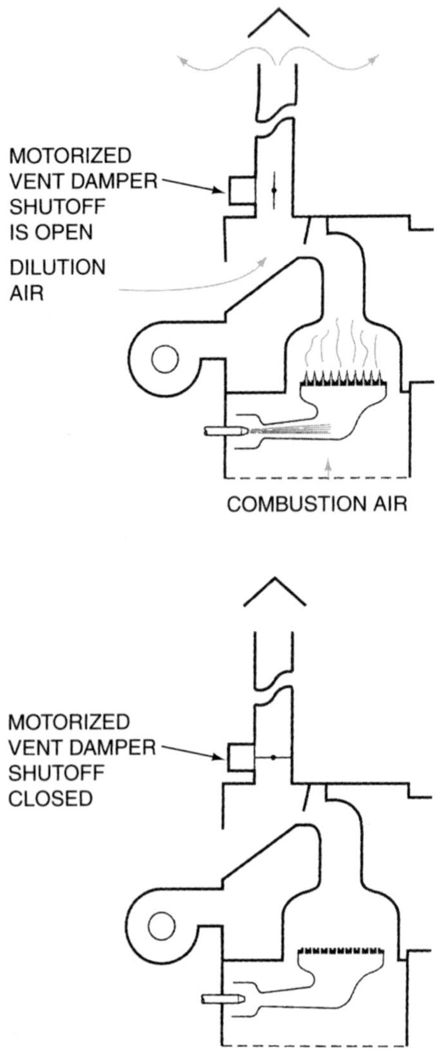

Figure 9-52. Motorized vent damper shutoff.

Note: We provide only the basics of how to size a vent for an installation. You must follow your local code, and it is strongly recommended that you use the *National Fuel Gas Code* as part of your reading material.

Our vent sizing example involves a two-story house with a furnace and a water heater in the basement, as shown in Figure 9-53. Following are the steps and decisions used to size the vent and connectors:

1. Double-wall vent pipe must be used in the basement and in the attic because they are unconditioned. If the basement were conditioned, we probably would use single-wall vent pipe up to where the vent penetrates the first unconditioned space, the attic, because single-wall vent pipe is much less expensive and is easier to install than double-wall vent pipe.
2. Using Table 11–5 from the *National Fuel Gas Code* in Figure 9-54, we start with the water heater. Read down the Total Vent Height H column and find 30 feet. Read to the right across the 1-foot rise (for the connector) and find that a 4-inch connector has a capacity of 59,000 Btu/h; this is adequate. This is found in the NAT Max column Fig 9–54B because this is a natural draft system with a draft diverter.
3. The furnace connector is sized the same way. Read down the H column to 30 feet. Read to the right and find that a 6-inch connector will handle 134,000 Btu/h. This is the size we will use.
4. The common connector is sized from the Common Vent Connector portion of the table. The combined capacity of the appliances is 165,000 Btu/h. Move down the table to 30 feet and move to the right and find that a 5-inch vent is suitable for up to 185,000 Btu/h. We will use a 6-inch vent for the common vent because we have a 6-inch connector on the gas furnace. The common vent must be equal to or larger than the largest connector.

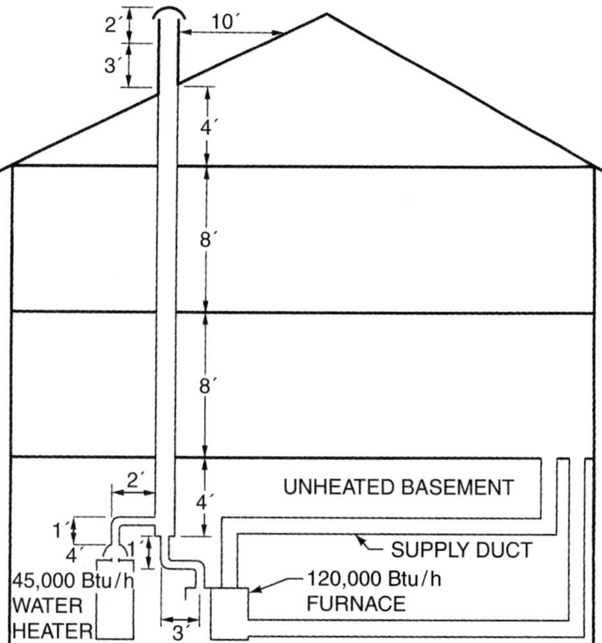

Figure 9-53. Layout for sizing a vent system.

Notice the method used to terminate the vent through the roof, see Figure 9-53. The vent is carried to a height of 2 feet above a portion of the roof that is within 10 feet of the vent. This is part of the code to ensure correct venting and to minimize downdrafts.

Also notice that the water heater was vented in above the furnace in the common connector. The smaller-capacity appliance should vent above the larger-capacity appliance to ensure that the smaller appliance vents properly when it is the only one burning.

Note: There is much more to vent sizing than we present here. You should study the *National Fuel Gas Code* and

Capacity of Single-Wall Metal Pipe or Type B Asbestos Cement Vents Serving a Single Draft Hood Equipped Appliance

Height H (ft)	Lateral L (ft)	Vent Diameter — D: 3"	4"	5"	6"	7"	8"	10"	12"
		Maximum Appliance Input Rating in Thousands of Btu per Hour							
6	0	39	70	116	170	232	312	500	750
	2	31	55	94	141	194	260	415	620
	5	28	51	88	128	177	242	390	600
8	0	42	76	126	185	252	340	542	815
	2	32	61	102	154	210	284	451	680
	5	29	56	95	141	194	264	430	648
	10	24*	49	86	131	180	250	406	625
10	0	45	84	138	202	279	372	606	912
	2	35	67	111	168	233	311	505	760
	5	32	61	104	153	215	289	480	724
	10	27*	54	94	143	200	274	455	700
	15	NR	46*	84	130	186	258	432	666
15	0	49	91	151	223	312	420	684	1040
	2	39	72	122	186	260	350	570	865
	5	35*	67	110	170	240	325	540	825
	10	30*	58*	103	158	223	308	514	795
	15	NR	50*	93*	144	207	291	488	760
	20	NR	NR	82*	132*	195	273	466	726
20	0	53*	101	163	252	342	470	770	1190
	2	42*	80	136	210	286	392	641	990
	5	38*	74*	123	192	264	364	610	945
	10	32*	65*	115*	178	246	345	571	910
	15	NR	55*	104*	163	228	326	550	870
	20	NR	NR	91*	149*	241*	306	525	832
30	0	56*	108*	183	276	384	529	878	1370
	2	44*	84*	148*	230	320	441	730	1140
	5	NR	78*	137*	210	296	410	694	1080
	10	NR	68*	125*	196*	274	388	656	1050
	15	NR	NR	113*	177*	258*	366	625	1000
	20	NR	NR	99*	163*	240*	344	596	960
	30	NR	NR	NR	NR	192*	295*	540	890
50	0	NR	120*	210*	310*	443*	590	980	1550
	2	NR	95*	171*	260*	370*	492	820	1290
	5	NR	NR	159*	234*	342*	474	780	1230
	10	NR	NR	146*	221*	318*	456*	730	1190
	15	NR	NR	NR	200*	292*	407*	705	1130
	20	NR	NR	NR	185*	276*	384*	670*	1080
	30	NR	NR	NR	NR	222*	330*	605*	1010

*See Note 6

Figure 9-54. A & B Vent sizing table from the *National Fuel Gas Code. Data from pages 182, 186, and 187 of the National Fuel Gas Code*

Table 11-6

Capacity of Type B Double-Wall Vents with Type B Double-Wall Connectors Serving Two or More Category I Appliances

Vent Connector Capacity

		Type B Double-Wall Vent and Connector Diameter — D																							
		3"			4"			5"			6"			7"			8"			9"			10"		
Vent Height H (ft)	Connector Rise R (ft)	Appliance Input Rating Limits in Thousands of Btu per Hour																							
		FAN		NAT	FAN		NAT	FAN		NAT	FAN		NAT	FAN		NAT	FAN		NAT	FAN		NAT	FAN		NAT
		Min	Max	Max	Min	Max	Max	Min	Max	Max	Min	Max	Max	Min	Max	Max	Min	Max	Max	Min	Max	Max	Min	Max	Max
6	1	22	37	26	35	66	46	46	106	72	58	164	104	77	225	142	92	296	185	109	376	237	128	466	289
	2	23	41	31	37	75	55	48	121	86	60	183	124	79	253	168	95	333	220	112	424	282	131	526	345
	3	24	44	35	38	81	62	49	132	96	62	199	139	82	275	189	97	363	248	114	463	317	134	575	386
8	1	22	40	27	35	72	48	49	114	76	64	176	109	84	243	148	100	320	194	118	408	248	138	507	303
	2	23	44	32	36	80	57	51	128	90	66	195	129	86	269	175	103	356	230	121	454	294	141	564	358
	3	24	47	36	37	87	64	53	139	101	67	210	145	88	290	198	105	384	258	123	492	330	143	612	402
10	1	22	43	28	34	78	50	49	123	78	65	189	113	89	257	154	106	341	200	125	436	257	146	542	314
	2	23	47	33	36	86	59	51	136	93	67	206	134	91	282	182	109	374	238	128	479	305	149	596	372
	3	24	50	37	37	92	67	52	146	104	69	220	150	94	303	205	111	402	268	131	515	342	152	642	417
15	1	21	50	30	33	89	53	47	142	83	64	220	120	88	298	163	110	389	214	134	493	273	162	609	333
	2	22	53	35	35	96	63	49	153	99	66	235	142	91	320	193	112	419	253	137	532	323	165	658	394
	3	24	55	40	36	102	71	51	163	111	68	248	160	93	339	218	115	445	286	140	565	365	167	700	444
20	1	21	54	31	33	99	56	46	157	87	62	246	125	86	334	171	107	436	224	131	552	285	158	681	347
	2	22	57	37	34	105	66	48	167	104	64	259	149	89	354	202	110	463	265	134	587	339	161	725	414
	3	23	60	42	35	110	74	50	176	116	66	271	168	91	371	228	113	486	300	137	618	383	164	764	466
30	1	20	62	33	31	113	59	45	181	93	60	288	134	83	391	182	103	512	238	125	649	305	151	802	372
	2	21	64	39	33	118	70	47	190	110	62	299	158	85	408	215	105	535	282	129	679	360	155	840	439
	3	22	66	44	34	123	79	48	198	124	64	309	178	88	423	242	108	555	317	132	706	405	158	874	494
50	1	19	71	36	30	133	64	43	216	101	57	349	145	78	477	197	97	627	257	120	797	330	144	984	403
	2	21	73	43	32	137	76	45	223	119	59	358	172	81	490	234	100	645	306	123	820	392	148	1014	478
	3	22	75	48	33	141	86	46	229	134	61	366	194	83	502	263	103	661	343	126	842	441	151	1043	538
100	1	18	82	37	28	158	66	40	262	104	53	442	150	73	611	204	91	810	266	112	1038	341	135	1285	417
	2	19	83	44	30	161	79	42	267	123	55	447	178	75	619	242	94	822	316	115	1054	405	139	1306	494
	3	20	84	50	31	163	89	44	272	138	57	452	200	78	627	272	97	834	355	118	1069	455	142	1327	555

Common Vent Capacity

	Type B Double-Wall Common Vent Diameter — D																				
	4"			5"			6"			7"			8"			9"			10"		
Vent Height H (ft)	Combined Appliance Input rating in Thousands of Btu per Hour																				
	FAN +FAN	FAN +NAT	NAT +NAT	FAN +FAN	FAN +NAT	NAT +NAT	FAN +FAN	FAN +NAT	NAT +NAT	FAN +FAN	FAN +NAT	NAT +NAT	FAN +FAN	FAN +NAT	NAT +NAT	FAN +FAN	FAN +NAT	NAT +NAT	FAN +FAN	FAN +NAT	NAT +NAT
6	92	81	65	140	116	103	204	161	147	309	248	200	404	314	260	547	434	335	672	520	410
8	101	90	73	155	129	114	224	178	163	339	275	223	444	348	290	602	480	378	740	577	465
10	110	97	79	169	141	124	243	194	178	367	299	242	477	377	315	649	522	405	800	627	495
15	125	112	91	195	164	144	283	228	206	427	352	280	556	444	365	753	612	465	924	733	565
20	136	123	102	215	183	160	314	255	229	475	394	310	621	499	405	842	688	523	1035	826	640
30	152	138	118	244	210	185	361	297	266	547	459	360	720	585	470	979	808	605	1209	975	740
50	167	153	134	279	244	214	421	353	310	641	547	423	854	706	550	1164	977	705	1451	1188	860
100	175	163	NR	311	277	NR	489	421	NR	751	658	479	1025	873	625	1408	1215	800	1784	1502	975

Figure 9-54. *(Continued)*

your local code for more specific applications. Venting is one of the most important portions of the installation because of the safety that it provides, and it should not be treated lightly.

9.3 Appliance Efficiency

Determining appliance efficiency can be divided into several parts: the gas input, the combustion efficiency, and the air-side heat exchange.

The Gas Input

The gas input can be checked by using the house meter to measure the gas going into the appliance (if there is a meter; many LP gas installations use tank delivery to determine the amount of gas used). Gas meters have dials calibrated in several increments; the smallest increment can be used to determine the gas flow to an appliance, provided that the appliance is the only one operating. Figure 9-55 shows the dials of a typical gas meter. Notice that this meter has a ½ cubic foot dial and a 2 cubic foot

Figure 9-55. Typical gas meter dials. *Photo by Bill Johnson*

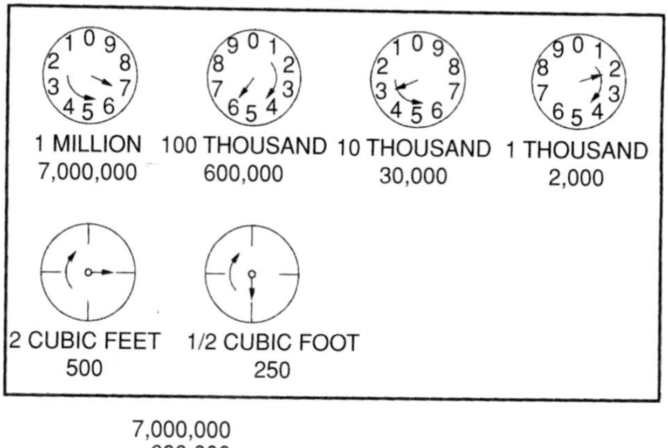

7,000,000
600,000
30,000
2,000
500
250
7,632,750 CUBIC FEET

Figure 9-56. Reading the gas meter dials.

dial. When the ½ cubic foot needle makes one round, ½ of a cubic foot of gas has passed through the meter. The 2 cubic foot dial makes one turn every time 2 cubic feet of gas pass through the meter. When the ½ cubic foot needle makes one round, the 2 cubic foot needle makes ¼ of a round. Suppose that a natural gas furnace is rated at 100,000 Btu/h input, and the gas company rates its gas at 1,050 Btu per cubic foot as shown in Figure 9-56. This furnace should consume 95 cubic feet of gas per hour (100,000/1,050 = 95.23, rounded to 95). Suppose that you want to check the gas consumption of this furnace. You must do the following:

1. Turn off all other appliances, including the pilot burners. (Note that this test can be difficult if several pilot burners must be relit after the test.)
2. Start the furnace burner.
3. Time the ¼ cubic foot hand for 5 minutes, using a stopwatch. You can also use the ½-foot dial to accumulate the reading.
4. Multiply the reading by 12 (there are twelve 5-minute segments in 1 hour) to arrive at the cubic feet of gas consumed in 1 hour.
5. Multiply the cfh by the heat content of the gas to arrive at the furnace capacity.

The furnace in our example has a capacity of 100,000 Btu/h, so the meter should register 7.9 cubic feet in 5 minutes (7.9 × 12 = 94.8, rounded to 95; 95 × 1,050 = 99,750, which is close enough).

Suppose that the meter registered 9.5 cubic feet in 5 minutes. This means that the furnace is firing at 119,700 Btu/h (9.5 × 12 × 1,050 = 119,700). This furnace is overfiring by 19,700 Btu/h and should be corrected. The overfiring may be caused by oversized orifices or too much gas pressure at the manifold.

Measure the pressure at the manifold with a water manometer or other suitable gauge and check the orifice size for the rating of the furnace as shown in Figure 9-57.

The pressure reading should match the manufacturer's recommendation, usually 3.5 inches of wc, unless the application is commercial and the appliance is large. Look on the nameplate of the furnace. Figure 9-58 shows a typical furnace nameplate. If the furnace is an LP gas furnace with no meter, you must rely on the gas pressure and

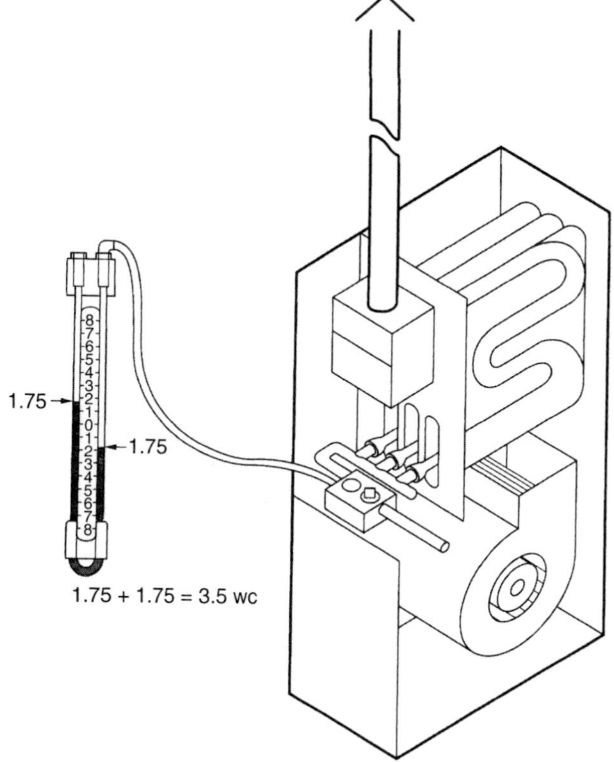

Figure 9-57. Checking the capacity of a furnace by measuring the gas pressure and knowing the orifice sizing.

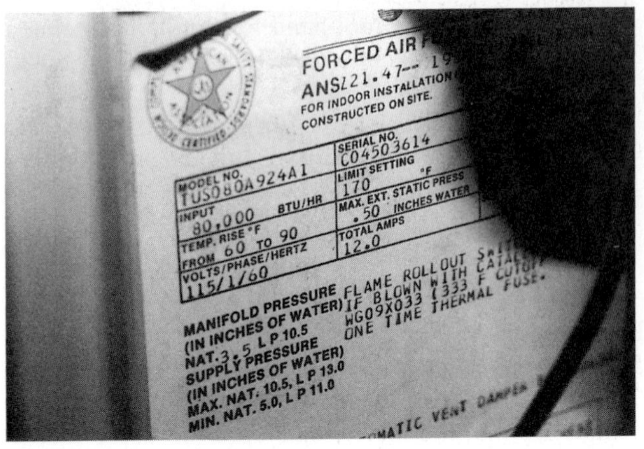

Figure 9-58. Typical furnace nameplate, *Photo by Bill Johnson*

orifice size for the correct firing rate. The pressure for an LP furnace is 11 inches of wc.

Orifice sizing tables for natural gas can be found in Figure 9-59. Notice that these sizes are for altitudes below 2,000 feet. The capacity of the orifice is the same at a higher altitude, but the furnace capacity must be reduced because less oxygen is in the atmosphere. The air is expanded and the heat exchanger and flue passages cannot handle the additional air that would be required for the rated combustion. Figure 9-60 has a table of corrections for altitudes above 2,000 feet. Figure 9-61 is a table for LP gases.

Suppose that a furnace is a natural gas furnace with a rated capacity of 100,000 Btu/h, that the gas has a capacity of 1,000 Btu per cubic foot, that the specific

Flow of Gas Through Fixed Orifices

This Appendix is informative and is not a part of the Code.

Table F-1 Utility Gases

(Cubic feet per hour at sea level)

Specific Gravity = 0.60
Orifice Coefficient = 0.90
For utility gases of another specific gravity, select multiplier from Table F-3.
For altitudes above 2,000 feet, first select the equivalent orifice size at sea level from Table F-4.

Orifice or Drill Size	Pressure at Orifice—Inches Water Column								
	3	3.5	4	5	6	7	8	9	10
80	.48	.52	.55	.63	.69	.73	.79	.83	.88
79	.55	.59	.64	.72	.80	.84	.90	.97	1.01
78	.70	.76	.78	.88	.97	1.04	1.10	1.17	1.24
77	.88	.95	.99	1.11	1.23	1.31	1.38	1.47	1.55
76	1.05	1.13	1.21	1.37	1.52	1.61	1.72	1.83	1.92
75	1.16	1.25	1.34	1.52	1.64	1.79	1.91	2.04	2.14
74	1.33	1.44	1.55	1.74	1.91	2.05	2.18	2.32	2.44
73	1.51	1.63	1.76	1.99	2.17	2.32	2.48	2.64	2.78
72	1.64	1.77	1.90	2.15	2.40	2.52	2.69	2.86	3.00
71	1.82	1.97	2.06	2.33	2.54	2.73	2.91	3.11	3.26
70	2.06	2.22	2.39	2.70	2.97	3.16	3.38	3.59	3.78
69	2.25	2.43	2.61	2.96	3.23	3.47	3.68	3.94	4.14
68	2.52	2.72	2.93	3.26	3.58	3.88	4.14	4.41	4.64
67	2.69	2.91	3.12	3.52	3.87	4.13	4.41	4.69	4.94
66	2.86	3.09	3.32	3.75	4.11	4.39	4.68	4.98	5.24
65	3.14	3.39	3.72	4.28	4.62	4.84	5.16	5.50	5.78
64	3.41	3.68	4.14	4.48	4.91	5.23	5.59	5.95	6.26
63	3.63	3.92	4.19	4.75	5.19	5.55	5.92	6.30	6.63
62	3.78	4.08	4.39	4.96	5.42	5.81	6.20	6.59	6.94
61	4.02	4.34	4.66	5.27	5.77	6.15	6.57	7.00	7.37
60	4.21	4.55	4.89	5.52	5.95	6.47	6.91	7.35	7.74
59	4.41	4.76	5.11	5.78	6.35	6.78	7.25	7.71	8.11
58	4.66	5.03	5.39	6.10	6.68	7.13	7.62	8.11	8.53
57	4.84	5.23	5.63	6.36	6.96	7.44	7.94	8.46	8.90
56	5.68	6.13	6.58	7.35	8.03	8.73	9.32	9.92	10.44
55	7.11	7.68	8.22	9.30	10.18	10.85	11.59	12.34	12.98
54	7.95	8.59	9.23	10.45	11.39	12.25	13.08	13.93	14.65
53	9.30	10.04	10.80	12.20	13.32	14.29	15.27	16.25	17.09
52	10.61	11.46	12.31	13.86	15.26	16.34	17.44	18.57	19.53
51	11.82	12.77	13.69	15.47	16.97	18.16	19.40	20.64	21.71
50	12.89	13.92	14.94	16.86	18.48	19.77	21.12	22.48	23.65
49	14.07	15.20	16.28	18.37	20.20	21.60	23.06	24.56	25.83
48	15.15	16.36	17.62	19.88	21.81	23.31	24.90	26.51	27.89
47	16.22	17.52	18.80	21.27	23.21	24.93	26.62	28.34	29.81
46	17.19	18.57	19.98	22.57	24.72	26.43	28.23	30.05	31.61
45	17.73	19.15	20.52	23.10	25.36	27.18	29.03	30.90	32.51
44	19.45	21.01	22.57	25.57	27.93	29.87	31.89	33.96	35.72
43	20.73	22.39	24.18	27.29	29.87	32.02	34.19	36.41	38.30
42	23.10	24.95	26.50	29.50	32.50	35.24	37.63	40.07	42.14

Table F-1 (Continued)

Orifice or Drill Size	Pressure at Orifice—Inches Water Column								
	3	3.5	4	5	6	7	8	9	10
41	24.06	25.98	28.15	31.69	34.81	37.17	39.70	42.27	44.46
40	25.03	27.03	29.23	33.09	36.20	38.79	41.42	44.10	46.38
39	26.11	28.20	30.20	34.05	37.38	39.97	42.68	45.44	47.80
38	27.08	29.25	31.38	35.46	38.89	41.58	44.40	47.27	49.73
37	28.36	30.63	32.99	37.07	40.83	43.62	46.59	49.60	52.17
36	29.76	32.14	34.59	39.11	42.76	45.77	48.88	52.04	54.74
35	32.36	34.95	36.86	41.68	45.66	48.78	52.10	55.46	58.34
34	32.45	35.05	37.50	42.44	46.52	49.75	53.12	56.55	59.49
33	33.41	36.08	38.79	43.83	48.03	51.46	54.96	58.62	61.55
32	35.46	38.30	40.94	46.52	50.82	54.26	57.95	61.70	64.89
31	37.82	40.85	43.83	49.64	54.36	58.01	61.96	65.97	69.39
30	43.40	46.87	50.39	57.05	62.09	66.72	71.22	75.86	79.80
29	48.45	52.33	56.19	63.61	69.62	74.45	79.52	84.66	89.04
28	51.78	55.92	59.50	67.00	73.50	79.50	84.92	90.39	95.09
27	54.47	58.83	63.17	71.55	78.32	83.59	89.27	95.04	99.97
26	56.73	61.27	65.86	74.57	81.65	87.24	93.17	99.19	104.57
25	58.87	63.58	68.22	77.14	84.67	90.36	96.50	102.74	108.07
24	60.81	65.67	70.58	79.83	87.56	93.47	99.83	106.28	111.79
23	62.10	67.07	72.20	81.65	89.39	94.55	100.98	107.49	113.07
22	64.89	70.08	75.21	85.10	93.25	99.60	106.39	113.24	119.12
21	66.51	71.83	77.14	87.35	95.63	102.29	109.24	116.29	122.33
20	68.22	73.68	79.08	89.49	97.99	104.75	111.87	119.10	125.28
19	72.20	77.98	83.69	94.76	103.89	110.67	118.55	125.82	132.36
18	75.53	81.57	87.56	97.50	108.52	116.03	123.92	131.93	138.78
17	78.54	84.82	91.10	103.14	112.81	120.33	128.52	136.82	143.91
16	82.19	88.77	95.40	107.98	118.18	126.78	135.39	144.15	151.63
15	85.20	92.02	98.84	111.74	122.48	131.07	139.98	149.03	156.77
14	87.10	94.40	100.78	114.21	124.44	133.22	142.28	151.47	159.33
13	89.92	97.11	104.32	118.18	128.93	138.60	148.02	157.58	165.76
12	93.90	101.41	108.52	123.56	135.37	143.97	153.75	163.69	172.13
11	95.94	103.62	111.31	126.02	137.52	147.20	157.20	167.36	176.03
10	98.30	106.16	114.21	129.25	141.82	151.50	161.81	172.26	181.13
9	100.99	109.07	117.11	132.58	145.05	154.71	165.23	175.91	185.03
8	103.89	112.20	120.65	136.44	149.33	160.08	170.96	182.00	191.44
7	105.93	114.40	123.01	139.23	152.56	163.31	174.38	185.68	195.30
6	109.15	117.88	126.78	142.88	156.83	167.51	178.88	190.46	200.36
5	111.08	119.97	128.93	145.79	160.08	170.82	182.48	194.22	204.30
4	114.75	123.93	133.22	150.41	164.36	176.18	188.16	200.25	210.71
3	119.25	128.79	137.52	156.26	170.78	182.64	195.08	207.66	218.44
2	128.48	138.76	148.61	168.64	184.79	197.66	211.05	224.74	235.58
1	136.35	147.26	158.25	179.33	194.63	209.48	223.65	238.16	250.54

For SI units: 1 Btu per hour = 0.293 watts; 1 foot^3 = 0.028 m^3; 1 foot = 0.305 m; 1 inch water column = 249 pa.

Figure 9-59. Natural gas orifice sizing table from the *National Fuel Gas Code. Data from pages 222 and 223 of the National Fuel Gas Code*

Table F-4
Equivalent Orifice Sizes at High Altitudes
(Includes 4% input reduction for each 1,000 feet)

Orifice Size at Sea Level	Orifice Size Required at Other Elevations								
	2000	3000	4000	5000	6000	7000	8000	9000	10000
1	2	2	3	3	4	5	7	8	10
2	3	3	4	5	6	7	9	10	12
3	4	5	7	8	9	10	12	13	15
4	6	7	8	9	11	12	13	14	16
5	7	8	9	10	12	13	14	15	17
6	8	9	10	11	12	13	14	16	17
7	9	10	11	12	13	14	15	16	18
8	10	11	12	13	13	15	16	17	18
9	11	12	12	13	14	16	17	18	19
10	12	13	13	14	15	16	17	18	19
11	13	13	14	15	16	17	18	19	20
12	13	14	15	16	17	17	18	19	20
13	15	15	16	17	18	18	19	20	22
14	16	16	17	18	18	19	20	21	23
15	16	17	17	18	19	20	20	22	24
16	17	18	18	19	19	20	22	23	25
17	18	19	19	20	21	22	23	24	26
18	19	19	20	21	22	23	24	26	27
19	20	20	21	22	23	25	26	27	28
20	22	22	23	24	25	26	27	28	29
21	23	23	24	25	26	27	28	28	29
22	23	24	25	26	27	27	28	29	29
23	25	25	26	27	27	28	29	29	30
24	25	26	27	27	28	28	29	29	30
25	26	27	27	28	28	29	29	30	30
26	27	28	28	28	29	29	30	30	30
27	28	28	29	29	29	30	30	30	31
28	29	29	29	30	30	30	30	31	31
29	29	30	30	30	30	31	31	31	32
30	30	31	31	31	31	32	32	33	35
31	32	32	32	33	34	35	36	37	38
32	33	34	35	35	36	36	37	38	40
33	35	35	36	36	37	38	38	40	41
34	35	36	36	37	37	38	39	40	42
35	36	36	37	37	38	39	40	41	42
36	37	38	38	39	40	41	41	42	43
37	38	39	39	40	41	42	42	43	43
38	39	40	41	41	42	42	43	43	44
39	40	41	41	42	42	43	43	44	44
40	41	42	42	42	43	43	44	44	45

Table F-4 (Continued)

Orifice Size at Sea Level	Orifice Size Required at Other Elevations								
	2000	3000	4000	5000	6000	7000	8000	9000	10000
41	42	42	42	43	43	44	44	45	46
42	42	43	43	43	44	44	45	46	47
43	44	44	44	45	45	46	47	47	48
44	45	45	45	46	47	47	48	48	49
45	46	47	47	47	48	48	49	49	50
46	47	47	47	48	48	49	49	50	50
47	48	48	49	49	49	50	50	51	51
48	49	49	49	50	50	50	51	51	52
49	50	50	50	51	51	51	52	52	52
50	51	51	51	51	52	52	52	53	53
51	51	52	52	52	52	53	53	53	54
52	52	53	53	53	53	53	54	54	54
53	54	54	54	54	54	54	55	55	55
54	54	55	55	55	55	55	56	56	56
55	55	55	55	56	56	56	56	56	57
56	56	56	57	57	57	58	59	59	60
57	58	59	59	60	60	61	62	63	63
58	59	60	60	61	62	62	63	63	64
59	60	61	61	62	62	63	64	64	65
60	61	61	62	63	63	64	64	65	65
61	62	62	63	63	64	65	65	66	66
62	63	63	64	64	65	65	66	66	67
63	64	64	65	65	65	66	66	67	68
64	65	65	65	66	66	66	67	67	68
65	65	66	66	66	67	67	68	68	69
66	67	67	68	68	68	69	69	69	70
67	68	68	68	69	69	69	70	70	70
68	68	69	69	69	70	70	70	71	71
69	70	70	70	70	71	71	71	72	72
70	70	71	71	71	71	72	72	73	73
71	72	72	72	73	73	73	74	74	74
72	73	73	73	73	74	74	74	74	75
73	73	74	74	74	74	75	75	75	76
74	74	75	75	75	75	76	76	76	76
75	75	76	76	76	76	77	77	77	77
76	76	76	77	77	77	77	77	77	77
77	77	77	77	78	78	78	78	78	78
78	78	78	78	79	79	79	79	80	80
79	79	80	80	80	80	.013	.012	.012	.012
80	80	.013	.013	.013	.012	.012	.012	.012	.011

Figure 9-60. Natural gas orifice sizing table for high altitudes. *Data from pages 226 and 227 of the National Fuel Gas Code*

gravity is 0.6 (natural gas), and that the furnace has four burners. What should the orifice size be? Each burner should consume 25,000 Btu/h each. Looking at the table for natural gas, move down the column for 3.5 inches of wc until you find 25; the closest is 24.95. Look left and you will find that the orifices should be size 42. This number should be stamped on the orifice, or you can check the size with an orifice drill set, shown in Figure 9-62. Orifice sizes are keyed to drill sizes.

Caution: Use care when checking an orifice size with a drill set, or you may drill the orifice larger by mistake.

Notice that as the orifice size number becomes larger, the orifice capacity becomes less. For example, using the table for natural gas at a pressure of 3.5 inches of wc, a size 40 orifice has a capacity of 27.03 cfh, whereas a size 39 orifice has a capacity of 28.20 cfh.

If the furnace in our example is in Denver, Colorado, where the altitude is approximately 5,000 feet, the orifice should be size 43. This size is determined by using the table for high altitudes and finding size 42, then moving to the right to the column under 5,000 feet and finding size 43. Notice that size 43 is workable for altitudes as high as 5,000 feet. When the altitude is higher than 5,000 feet (e.g., 6,000 feet), the orifice size changes to 44. As the altitude increases, the orifice size also increases, for less furnace capacity. Furnaces at high altitudes must be derated.

Suppose that the furnace is at sea level (or an altitude as high as 2,000 feet) and needs to be converted to propane gas. Use the table for LP gas and move down the column for propane. Find the nearest value to 25,000 Btu/h, which is 24,630 Btu/h. Look to the left and find a new orifice size of 54, as shown in Figure 9-63. This is a much larger number and a smaller orifice. Propane appliances use smaller orifices than natural gas appliances do because of the higher heat content of propane.

Combustion Efficiency

The combustion efficiency for gas-burning appliances can be checked with a flue gas analysis test kit. There are two basic types of kits: mechanical and electronic, as shown in Figure 9-64. The mechanical test kit is used most because it is simple to operate and maintain and is the least expensive.

Table F-2 LP-Gases
(Btu per hour at sea level)

	Propane	Butane
Btu per Cubic Foot =	2,516	3,280
Specific Gravity =	1.52	2.01
Pressure at Orifice, Inches Water Column =	11	11
Orifice Coefficient =	0.9	0.9

For altitudes above 2,000 feet, first select the equivalent orifice size at sea level from Table F-4.

Orifice or Drill Size	Propane	Butane
.008	519	589
.009	656	744
.010	812	921
.011	981	1,112
.012	1,169	1,326
80	1,480	1,678
79	1,708	1.936
78	2,080	2,358
77	2,629	2,980
76	3,249	3,684
75	3,581	4,059
74	4,119	4,669
73	4,678	5,303
72	5,081	5,760
71	5,495	6,230
70	6,375	7,227
69	6,934	7,860
68	7,813	8,858
67	8,320	9,433
66	8,848	10,031
65	9,955	11,286
64	10,535	11,943
63	11,125	12,612
62	11,735	13,304
61	12,367	14,020
60	13,008	14,747
59	13,660	15,486
58	14,333	16,249
57	15,026	17,035
56	17,572	19,921
55	21,939	24,872
54	24,630	27,922
53	28,769	32,615
52	32,805	37,190
51	36,531	41,414
50	39,842	45,168
49	43,361	49,157
48	46,983	53,263

Table F-2 (Continued)

Orifice or Drill Size	Propane	Butane
47	50,088	56,783
46	53,296	60,420
45	54,641	61,944
44	60,229	68,280
43	64,369	72,973
42	71,095	80,599
41	74,924	84,940
40	78,029	88,459
39	80,513	91,215
38	83,721	94,912
37	87,860	99,605
36	92,207	104,532
35	98,312	111,454
34	100,175	113,566
33	103,797	117,672
32	109,385	124,007
31	117,043	132,689
30	134,119	152,046
29	150,366	170,466
28	160,301	181,728
27	168,580	191,114
26	175,617	199,092
25	181,619	205,896
24	187,828	212,935
23	192,796	218,567
22	200,350	227,131
21	205,525	232,997
20	210,699	238,863
19	223,945	253,880
18	233,466	264,673

Figure 9-61. Orifice sizing table for LP gases. *Data from pages 224 and 225 of the National Fuel Gas Code*

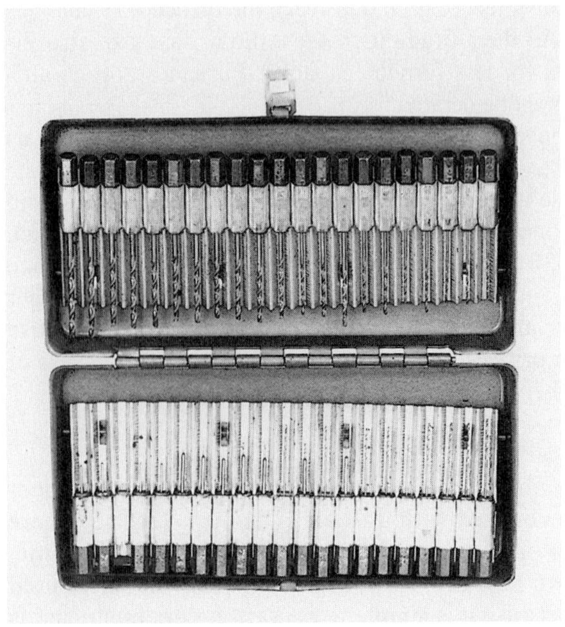

Figure 9-62. Orifice drill set. *Photo by Bill Johnson*

Most modern gas furnaces have a narrow operating efficiency. If you turned the adjustment from one extreme to the other, the percentage of efficiency would change only a few percentage points for natural gas. The flame characteristics would remain about the same. So, typically, natural gas furnaces do not require flue gas analysis. With LP gases, however, the flame characteristics and efficiency can be more sensitive because LP gases require more air. Other gas-burning appliances, such as boilers, may require a flue gas analysis in order to set the efficiency upon initial start-up.

Because the flue by-products have a known chemical makeup, a test for certain quantities of a chemical is the test used for combustion analysis. Combustion analysis kits basically test the flue gas for CO_2 content or O_2 content. From these contents, indicated in percentage of CO_2 or O_2, the combustion efficiency can be determined.

The CO_2 test is the most common and is used here. The kit has a small hand pump that is used to pump a sample of the flue gas into a chamber that contains a fluid that absorbs CO_2. When the liquid absorbs the CO_2, the volume of liquid rises, as shown in Figure 9-65. Following

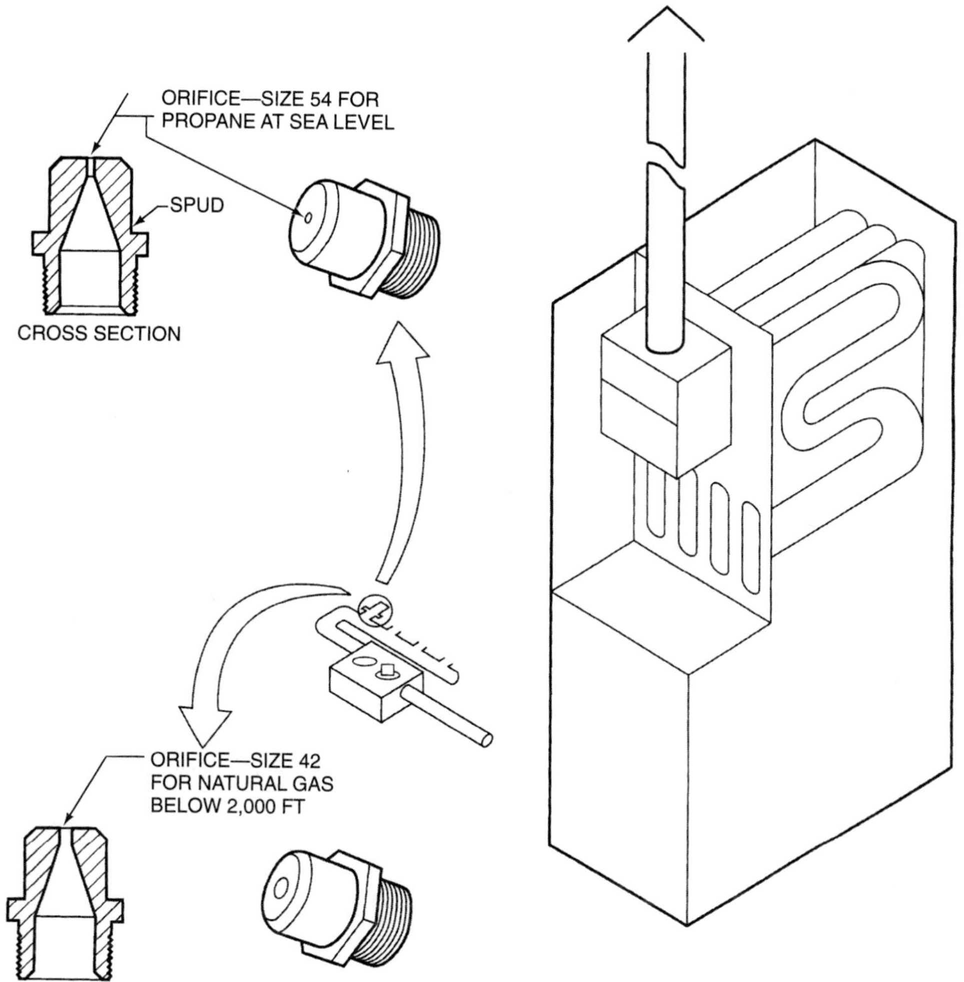

Figure 9-63. Converting a natural gas furnace to propane.

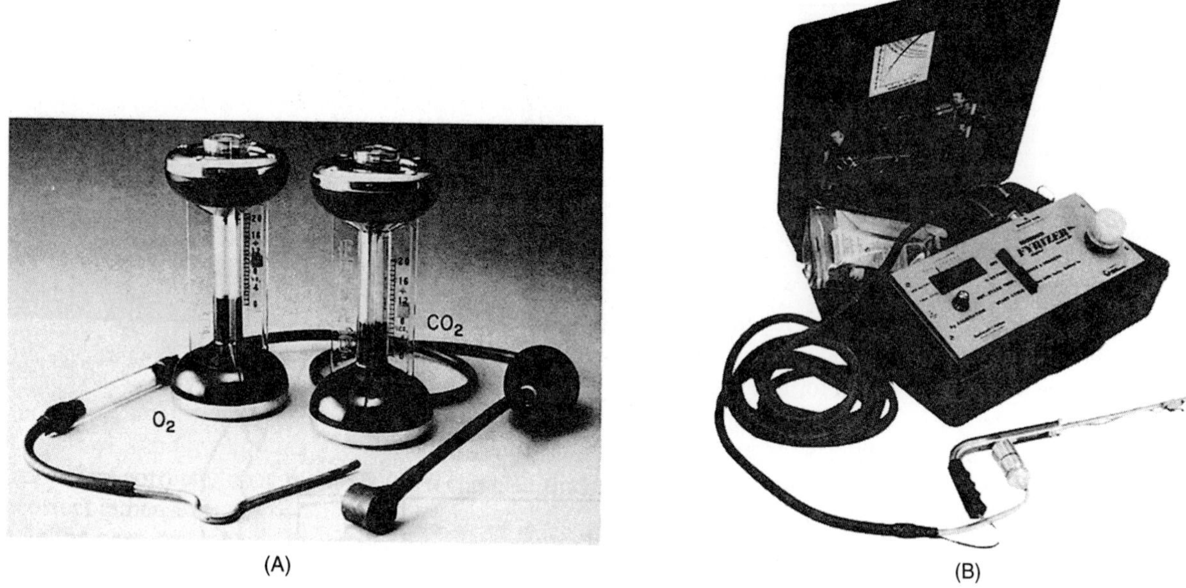

Figure 9-64. Flue gas analyzers. (A) Mechanical. *Courtesy United Technologies Bacharach* (B) Electronic. *Courtesy United Technologies Bacharach*

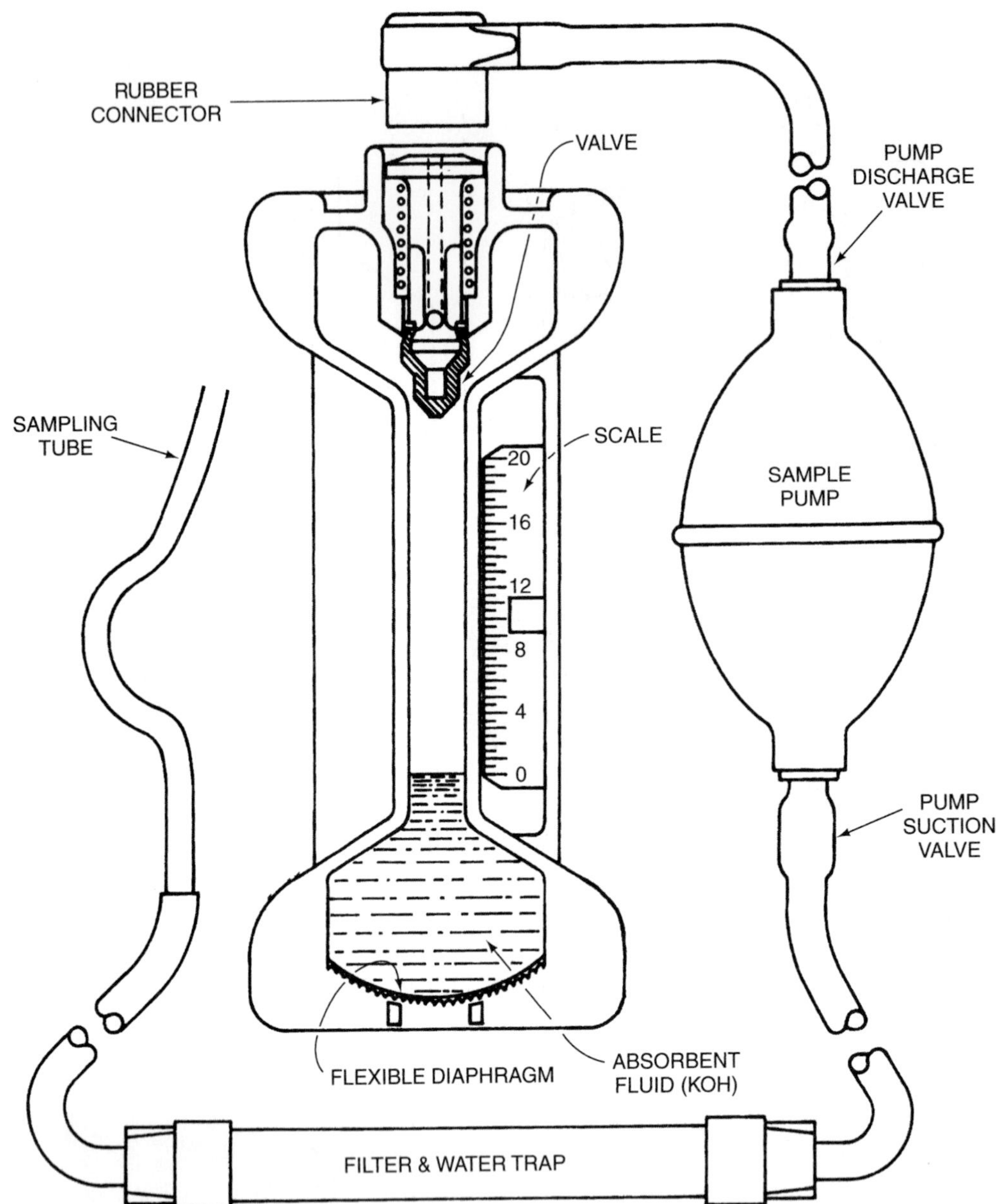

Figure 9-65. The CO_2 is absorbed into the liquid in the analyzer. *Reprinted with permission of R.W. Beckett Corporation*

are the steps for performing a typical CO_2 test by using a mechanical test instrument that has a liquid absorbent:

1. Drill a ¼ inch hole for the thermometer in the flue pipe just above the draft diverter.
2. Start the furnace and allow it to run for approximately 15 minutes, or until the flue gas temperature stops rising.
3. Check the burner flame characteristics and adjust the primary air to each burner if needed. The flame should appear to be firm with a definite cone showing and should not lift off the burner head. No yellow tips should be present. Orange streaks caused by burning dust should appear in the flame. These streaks are normal because dust is always in the air.
4. With fresh fluid in the analyzer container, vent the fluid to the atmosphere to be sure that there is no CO_2 in the analyzer from a previous test. Hold the analyzer level and adjust the zero point on the sliding scale.
5. Record the temperature of the flue gases from the flue gas thermometer, and remove it from the stack.
6. Place the analyzer probe into the sample hole in the flue. Hold the connector tightly down on the analyzer, and using the prescribed number of hand pumps in the order that the tester recommends, pump a sample of CO_2 into the analyzer chamber. (The typical number of pumps is 18. Hold the last pump while removing the connector from the analyzer. Doing so ensures that the last pump sample is not pulled from the sample chamber.)

7. Turn the analyzer upside down and then upright three times, letting the fluid drain to each end each time. This thoroughly mixes the fluid and the CO_2 sample. Then, turn the analyzer upright for a few seconds. To help all the fluid drain to the bottom, hold the analyzer at a 45-degree angle for a few seconds.
8. When all the fluid is in the bottom, read and record the percentage of CO_2.
9. Using a separate thermometer, record the ambient temperature around the furnace so that you know the temperature of the air entering the primary air inlet of the burner.
10. Subtract the ambient temperature from the flue gas temperature.
11. Use the slide calculator furnished in the kit to determine the percentage of efficiency of the furnace.

The efficiency should be approximately 80% for a standard gas furnace. If it is less than 75%, you should suspect that something is wrong with the firing of the furnace. If the flue gas temperature is too high (e.g., more than 500°F), you should suspect that too much heat is rising up the flue. The heat exchanger may be coated with soot and allowing too much heat to rise. If the efficiency is more than 80% for a standard furnace, you may suspect too much airflow across the heat exchanger. The flue gas temperature should then be too low (e.g., below 300°F) and the flue gases may condense.

Another test that is routine and should be performed at start-up of any atmospheric gas-burning appliance is a draft check to determine whether the flue is in a negative pressure. Figure 9-66 shows a draft gauge instrument that can be used. Another method that many technicians use is to hold a lit candle or match beside the draft diverter. An updraft should occur even if the furnace is off. When the furnace is burning, the updraft should be considerable. If the candle flame moves away from the draft diverter when it is moved toward the draft diverter, the furnace is drafting into the room, as shown in Figure 9-67.

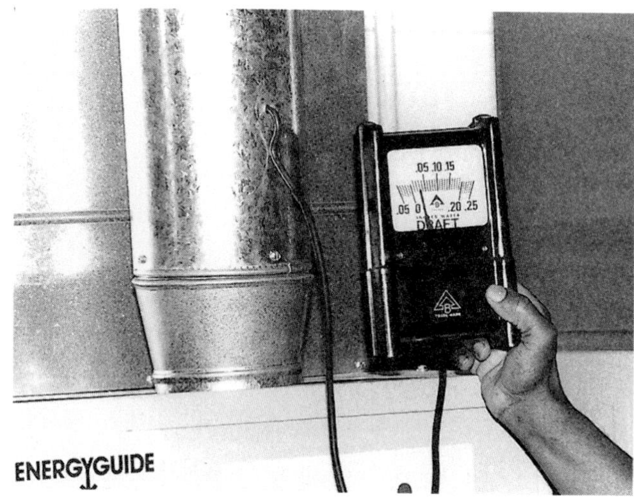

Figure 9-66. Simple draft gauge instrument. *Photo by Bill Johnson*

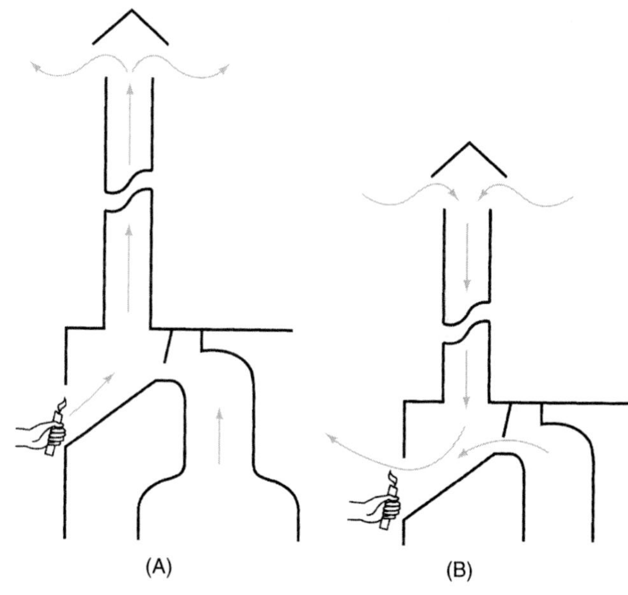

Figure 9-67. (A) Furnace updraft. (B) Furnace downdraft.

Air-Side Heat Exchange

The heat exchanger transfers heat into the room air that passes across the heat exchanger. When the furnace input is known, such as from the tests just described, and the combustion efficiency is correct, the air-side efficiency or capacity can be checked by using a temperature tester and the sensible heat formula. This test tells us the amount of airflow in cfm (cubic feet per minute) moving across the heat exchanger. The airflow across a furnace can be too little, too much, or correct. The furnace nameplate should state the recommended air temperature differences for each specific furnace. Typically for a gas furnace, the temperature difference should be between 45°F and 80°F.

Note: The air temperature rise should be routinely checked across any new installation to see if it is within the range stated on the nameplate.

Figure 9-68 is a photograph of a typical furnace nameplate, which states the manufacturer's recommendations.

If too little air is flowing across the heat exchanger, the air temperature rise across the heat exchanger is more than that recommended by the manufacturer because the same amount of heat is being transferred into less air, as shown in Figure 9-69. The furnace burner is likely to shut down periodically because of high operating temperatures. The high limit switch will cycle the burner. The owner of the structure may not even be aware of a problem when the house temperature is being maintained. High temperatures are damaging to the furnace heat exchanger because it is being stressed through a wider temperature range. When too much air

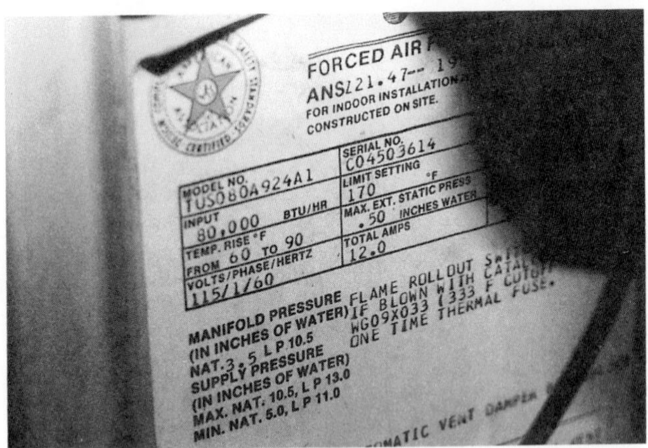

Figure 9-68. Furnace nameplate. *Photo by Bill Johnson*

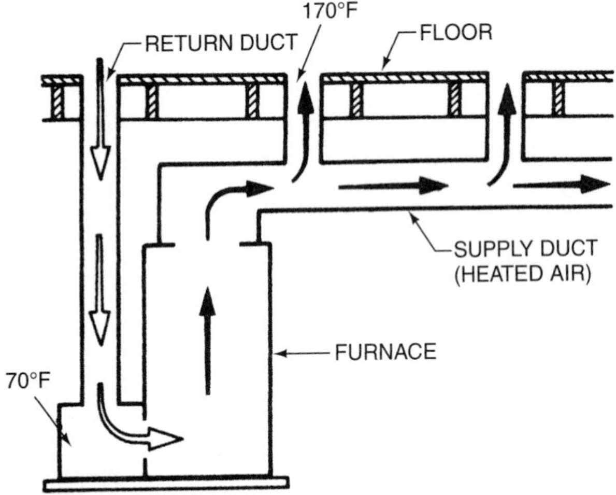

Figure 9-69. This furnace does not have enough air flowing over the heat exchanger.

Figure 9-70. Taking the flue gas temperature. *Photo by Bill Johnson*

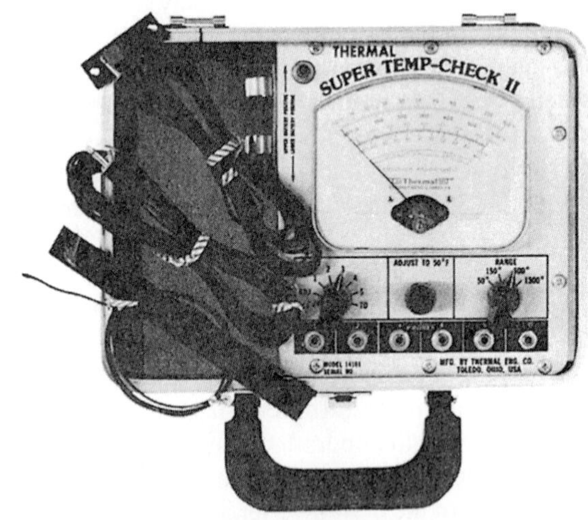

Figure 9-71. An electronic thermometer is a good choice for troubleshooting gas furnaces. *Courtesy Thermal Engineering*

is flowing across the heat exchanger, the flue gases are cooled too much, and the water vapor in the vent system may condense.

To test air-side efficiency, you must take an accurate thermometer reading of the air entering the furnace and the air leaving the furnace after the furnace has warmed to normal operating temperature. Allowing the furnace to operate for 15 minutes before beginning the test will ensure that the furnace is warmed up. Another method of determining the operating temperature is to place a thermometer in the flue. When the mercury stops rising, the furnace is warmed up, as shown in Figure 9-70.

Electronic thermometers are a good choice for taking the required temperature readings. Figure 9-71 shows a typical electronic thermometer. To obtain an accurate reading of the entering and leaving air, do not allow the leaving air thermometer probe to be influenced by the radiant heat from the heat exchanger. The probe must be shielded. The return air probe may be inserted in the return duct or the filter rack.

Let us use our previous furnace example and check the air-side capacity in cfm. Figure 9-72 illustrates the problem. The natural gas (1,000 Btu/cu ft) furnace is rated at an input of 100,000 Btu/h with four burners, each having an orifice size of 42. A size 42 orifice has a capacity of 24.95 cubic feet per hour or an actual capacity of 99,800 Btu/h ($24.95 \times 1{,}000 \times 4$ burners $= 99{,}800$). When the pilot burner is added, the furnace has a capacity of slightly more than 100,000 Btu/h. The furnace output (the amount of heat transferred into the air) is 80% from the previous test, or 80,000 Btu/h ($100{,}000 \times .80 = 80{,}000$). The air temperature rise across the furnace is checked and is 65°F. What is the airflow through the furnace?

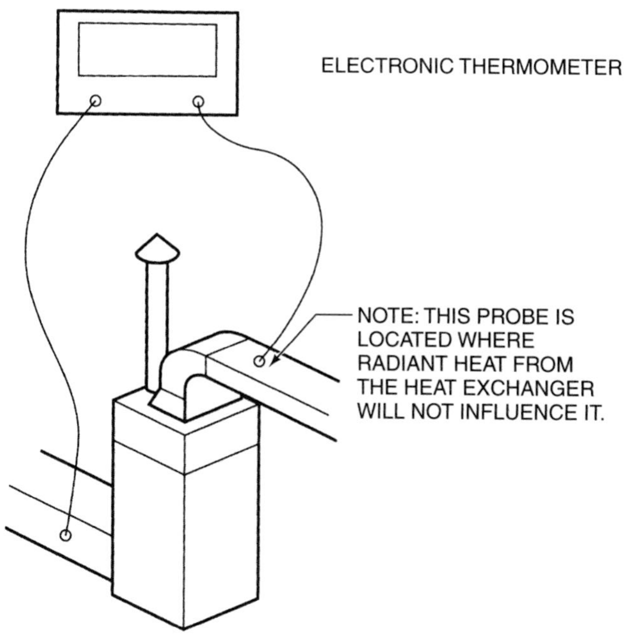

Figure 9-72. Checking the airflow across a gas furnace.

The sensible heat formula is used for determining this:

$$q = 1.08 \times \text{cfm} \times \text{TD}$$

where q = The sensible heat transferred into the air.
1.08 = A constant that converts pounds of air to cfm. It is derived from multiplying the specific heat of air (0.24) × 60 minutes per hour divided by the specific volume of air (13.33) = 1.0803, rounded to 1.08.
cfm = Cubic feet (of air) per minute.
TD = Temperature difference across the furnace in degrees Fahrenheit.

We will solve the formula for cfm, so it must be rearranged to read as follows:

$$\text{cfm} = q/(1.08 \times \text{TD})$$
$$= 80{,}000/(1.08 \times 65)$$
$$= 1{,}140$$

This cfm procedure is often used to prove the airflow for adding air-conditioning systems onto gas furnaces. This procedure can also be used with LP gas systems with equal success. With a little practice, a new technician can use this procedure and quickly determine if a system is performing correctly.

Performance testing for boilers is accomplished in much the same manner as for gas furnaces. First, check the gas input using the gas meter, then check the orifice size and manifold pressure to determine whether they are correct. Adjust the flame if needed, then do a combustion analysis. A performance check of the heat exchanger may also be necessary.

In a boiler installation, the water flow rate should be known. We can use the water flow rate to prove the capacity of the boiler by using the following variation of the sensible heat formula:

$$q = 500 \times \text{gpm} \times \text{TD}$$

where q = Boiler output in Btu/h.
500 = A constant that converts pounds of water to gpm (gallons per minute). It is derived from multiplying the specific heat of water (1 Btu/lb °F) × 60 minutes per hour × the weight of 1 gallon of water (8.33 lb/gal) = 499.8, rounded to 500.
gpm = Gallons per minute.
TD = Temperature difference.

A typical problem follows: Suppose that a system check is required and the output of the boiler to the building must be calculated. A look at the building prints and specifications shows that the water flow through the boiler is 30 gpm. The boiler is rated at 750,000 Btu/h input, which yields 600,000 Btu/h output when the boiler is operating at 80% efficiency (750,000 × .80 = 600,000). The orifice check reveals that the correct sizes are used, and the manifold pressure is correct. A check of the gas meter is not possible because you cannot shut down all the appliances in the building. You take the temperature rise across the boiler after it is up to temperature and discover it to be 40°F. Using the previous formula, we find the following:

$$q = 500 \times \text{gpm} \times \text{TD}$$
$$= 500 \times 30 \times 40$$
$$= 600{,}000 \text{ Btu/h output}$$

This boiler is transferring the correct amount of heat into the water, so it is operating correctly.

The gpm of water flow can also be verified by using water flow meters. Determining water flow is discussed in more detail in Unit 17.

Summary

- Most local gas piping practices follow the *National Fuel Gas Code.*
- Iron pipe, aluminum tubing, and copper tubing are the most common materials used in piping systems.
- LP and natural gas systems are piped similarly. The pipe must be sized correctly by an accepted method so that the correct pressure reaches the appliance.
- The 2-pound piping system is used in residences and businesses to save materials and installation costs. This system meters and furnishes 2 psig of gas pressure to the system up to a regulator just prior to the appliance. The system should have 1.5 psig of pressure drop when the appliances are operating; therefore, only 0.5 psig is actually reaching the regulator.

- The 2-pound system must be carefully sized and a different set of piping practices are used because the system is tubed with copper tubing instead of iron pipe.
- All piping systems should be pressure tested per local requirements.
- The flue vent portion of the installation is as important as the piping system. The proper amount of makeup air must be supplied to the appliance. Remember that there are dilution air for the draft diverter and combustion air for the burner when natural-draft equipment is used. The amount of the makeup air is vital and the vent must be sized correctly according to the *National Fuel Gas Code* and local regulations.
- Two types of vents are in common use for natural-draft equipment; single-wall and double-wall vents. The double-wall vent prevents the vent from cooling and moisture from condensing; it must be used when the vent is routed through an unconditioned space.
- A vent damper shutoff is a device for improving the efficiency of natural-draft gas-burning equipment. It is a device that shuts off the flue gases during the off cycle. There are motor-driven and bimetal types of vent damper shutoffs.
- Natural-draft vents must be routed correctly all the way to the roof and to the vent cap for correct venting.
- Gas-appliance efficiency can be determined by checking the gas input, the combustion efficiency, and the air-side heat exchange.
- The gas input is checked by measuring the gas flow with the house meter or the gas pressure and orifice size method.
- Combustion analysis can be accomplished by using mechanical or electronic instruments. The mechanical instruments are used most often.
- The air-side heat-exchange efficiency is calculated by determining the cfm of air. A known input of heat to an airstream is used and the air temperature rise is measured. These values are entered into a formula, and the cfm is calculated.
- Furnaces must be derated when they are located at high altitudes.
- A gas appliance can be converted from natural gas to LP gas and back by changing the orifices.

Review Questions

1. Which three piping materials are used most often for gas systems?
2. Suppose that a piping system has one appliance with a capacity of 105,000 Btu/h input and is 75 feet from the gas meter. The system pressure is 3.5 inches of wc. What should the pipe size be for natural gas with an allowable pressure drop of 0.5 inch of wc? for propane?
3. What type of test is used to ensure that gas piping is not leaking after the pipe is installed?
4. What are two types of vent damper shutoffs for gas-burning appliances?
5. A gas furnace has three heat exchangers and orifices on the burner manifold. The furnace is rated at 100,000 Btu/h. What size orifice should be used for natural gas with a capacity of 1,000 Btu per cut and a specific gravity of 0.6 with a manifold pressure of 3.5 inches of wc? for propane?
6. What are two methods of determining the combustion efficiency of a gas furnace?

10 Troubleshooting and Maintenance for Gas-Fired Equipment

Objectives

Upon completion of this unit, you should be able to

- **identify the electrical checkpoints for troubleshooting.**
- **describe how to check a thermocouple.**
- **describe how to set a heat anticipator.**

10.1 Troubleshooting Gas-Fired Equipment

Troubleshooting gas-burning equipment involves solving electrical or electronic control problems and mechanical problems. The majority are control problems, which involve the electrical controls and their failures, such as the thermocouple on a standing pilot system or an ignition problem with intermittent ignition. Mechanical problems can be burner or gas-pressure problems. The technician must decide which type of problem a system has. Fortunately, a system usually has only one problem at the time. If more problems exist, often one problem starts a sequence of events that creates the others. This type of problem can be difficult to diagnose, and even if there is no chain reaction, there may be more than one source for a problem. For example, suppose that a pilot burner is not staying lit properly on a standing pilot furnace. Your first thought may be that the thermocouple is defective. This may be the case, but something may be blowing out the pilot or cooling the thermocouple, such as a cracked heat exchanger, Figure 10-1.

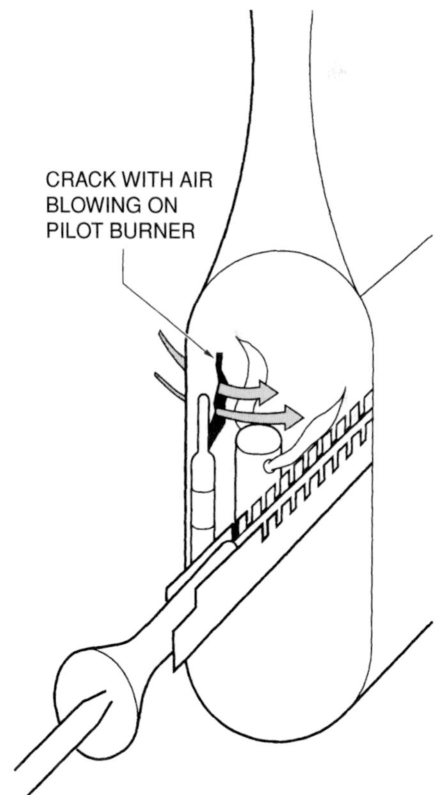

Figure 10-1. The heat exchanger has a crack, which allows air to blow on the thermocouple and put out the pilot burner.

Control Problems

Forced-air furnaces and boilers have much the same control circuits. These circuits are either electrical or electronic. We discuss each type next.

ELECTRICAL CONTROLS. When you suspect an electrical problem, it is always best to check the power supply first unless you have prior knowledge of what may be causing the problem. By starting at the power supply, you can systematically work your way down the circuit to the problem. Figure 10-2 shows a diagram of a basic furnace with a glow coil. Suppose that this furnace has a defective gas valve with an open circuit through the 24-volt coil. We start with the power supply and follow the circuit to the power-consuming device. Following are the steps that can be used to solve the problem. A table outlining some of the common problems associated with gas furnaces is provided at the end of these steps. This is the long method, but you should follow it until you have enough knowledge to take shortcuts.

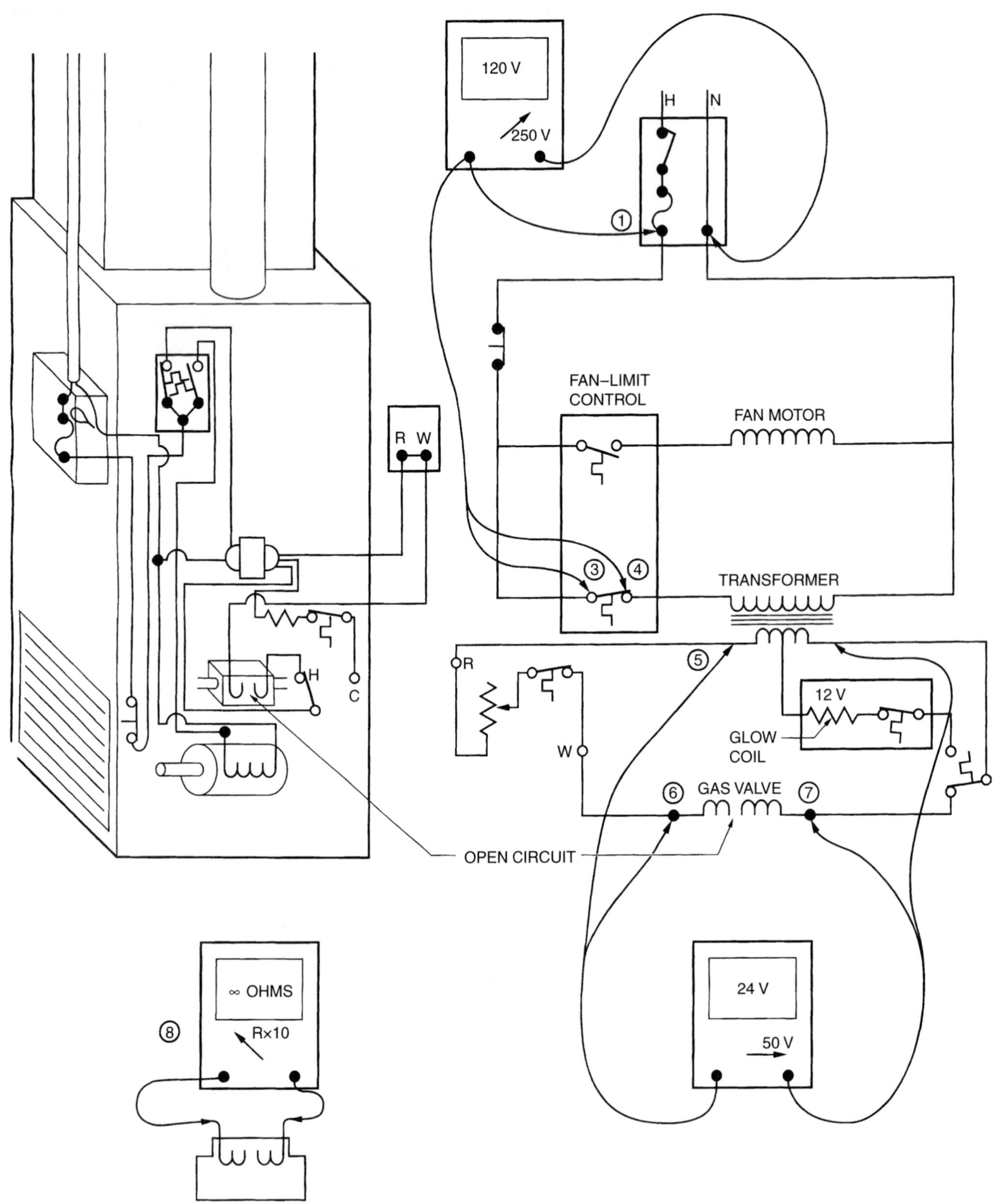

Figure 10-2. Wiring diagram for glow-coil troubleshooting.

1. Using a VOM (Volt–ohm–milliammeter) set on the 250-volt scale, check the voltage at the disconnect on the side of the furnace by placing one lead on the neutral wire and the other on the hot wire. You should read approximately 120 volts.

Warning: Use caution; the circuit is energized.

2. Remove the furnace burner compartment door.
3. Remove the cover to the fan–limit control. Carefully touch the meter lead to the common wire entering the fan–limit control. The meter should indicate 120 volts.
4. Now locate the limit side of the fan–limit control and carefully touch the meter lead to this connection. The meter should indicate 120 volts. It is fairly conclusive at this point that 120 volts are furnished to the control transformer. You could check the primary connections of the transformer, but they are not easily accessible.
5. Next, check secondary connections of the transformer for output. Remove the lead from the neutral connection and place it on one of the secondary terminals on the transformer. Place the other lead on the other secondary terminal of the transformer. The meter should indicate approximately 24 volts.
6. Now place one meter lead on each side of the gas valve terminals; the meter should indicate 24 volts.

Note: Some gas valves have three wires. One of the terminals is called an *accommodation terminal* and is used only to make a junction. The valve should have a symbol, such as that shown in Figure 10-3, to show the coil circuit.

7. The gas valve is being furnished 24 volts, and it is not opening, so the next step is to check for continuity through the gas valve's electrical coil. Shut off the power at the disconnect.

Warning: Set the VOM to the 250-volt scale and verify that the power is off. This may seem unnecessary, but your hand may accidentally touch the hot leads during the next step, so you want to be certain that the power is off.

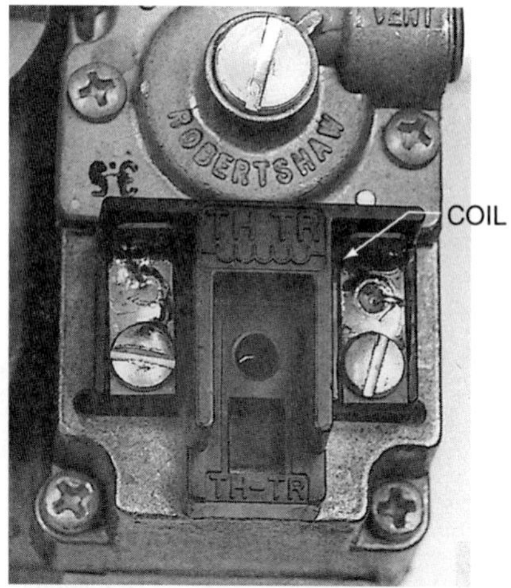

Figure 10-3. The symbol on the valve body shows which terminals apply to the coil circuit. *Photo by Bill Johnson*

8. Remove one of the coil wires (to prevent reading through another circuit), and fasten the meter leads to both sides of the coil. Turn the meter to the R×10 scale. The reading will be 0 ohms because there is no circuit.

The diagram in Figure 10-2 is an example of a *ladder diagram*, and this method of using the voltmeter is called *hopscotching* down the circuit. Following the procedure

Some common problems associated with gas furnaces are outlined in the table below:

Problem	Possible cause	Solution
Furnace won't run		
	1. No power.	1. Check for blown fuses or tripped circuit breakers at main entrance panel, at separate entrance panel, and on or in furnace; restore circuit.
	2. Switch off.	2. Turn on separate power switch on or near furnace.
	3. Motor overload.	3. Wait 30 minutes; press reset button. Repeat it necessary.
	4. Pilot light out.	4. Relight pilot.
	5. No gas.	5. Make sure gas value to furnace is fully open.
Not enough heat		
	1. Thermostat set too low.	1. Raise thermostat setting 5°.
	2. Filter dirty.	2. Clean or replace filter.
	3. Blower clogged.	3. Clean blower assembly.
	4. Registers closed or blocked.	4. Make sure all registers are open; make sure they are not blocked by tugs, drapes or furniture.
	5. System out of balance.	5. Balance system.
	6. Blower belt loose or broken.	6. Adjust or replace belt.

(*Continued*)

Problem	Possible cause	Solution
Pilot won't light		
	1. Pilot opening blocked.	1. Clean pilot opening.
	2. No gas.	2. Make sure pilot light button is fully depressed; make sure gas valve to furnace is fully open.
Pilot won't stay lit		
	1. Loose or faulty thermocouple.	1. Tighten thermocouple nut slightly, if no results, replace thermocouple.
	2. Pilot flame set too low.	2. Adjust pilot so flame is about 2 inches long.
Furnace turns on and off repeatedly		
	1. Filter dirty.	1. Clean or replace filter.
	2. Motor and/or blower needs lubrication.	2. If motor and blower have oil ports, lubricate.
Blower won't stop running		
	1. Blower control set wrong.	1. Reset thermostat from ON to AUTO.
	2. Limit switch set wrong.	2. Reset limit switch for stop-start cycling.
Furnace noisy		
	1. Access panels loose.	1. Mount and fasten access panels correctly.
	2. Belts sticking, worn, or damaged.	2. Spray squeaking drive belts with belt dressing; replace worn or damaged belts.
	3. Blower belts too loose or too tight.	3. Adjust belt.
	4. Motor and/or blower needs lubrication.	4. If motor and blower have oil ports, lubricate.

on the ladder diagram is fairly easy. Next, you should follow the procedure on the second type of diagram, the *pictorial diagram*. The accomplished technician can complete this procedure in a matter of moments because this technician recognizes all the test points on the furnace.

The experienced technician will take a shortcut to analyze the data at hand and make some deductions. The technician will first check whether the pilot burner is lit by touching the flue pipe. If it is warm to the touch, the pilot is lit as shown in Figure 10-4. The technician will then check the power at the disconnect, just as we did. After finding the power to be present, the technician will most likely go straight to the gas valve for a voltage check.

Pilot safety devices also have problems, and the technician must know how to troubleshoot them. The glow-coil system is considered an electrical control, not electronic. The circuit to the glow coil can be followed in much the same manner as the gas valve circuit was followed: Start at the beginning of the circuit and follow it to the power-consuming device. For example, suppose that the glow-coil is open circuit, as shown in Figure 10-5, and there is no fire upon a call for a pilot burner. We find that the pilot is out and will not relight. We must follow these steps to find the problem:

1. Turn the meter selector to the 50-volt scale.
2. Check the voltage across the transformer secondary and find 24 volts.
3. Place one lead of the meter on the 12-volt tap and the other on the common tap of the transformer, and read 12 volts. The transformer is providing the correct voltage to the glow coil.
4. The next step involves using the ohm feature of the meter. The reason for using this feature is that the glow-coil assembly would need to be removed to check the voltage at the actual power-consuming device, so a continuity check must be performed. Turn the power off to the furnace and remove the 12-volt wire from the transformer or the wire from the cold junction of the pilot safety control.
5. Turn the meter selector switch to the R×1 scale and place one lead on one side of the circuit and the other lead on the other side of the circuit. The meter should show an open circuit. The pilot burner and glow-coil assembly can be removed to verify that the glow coil circuit is open.

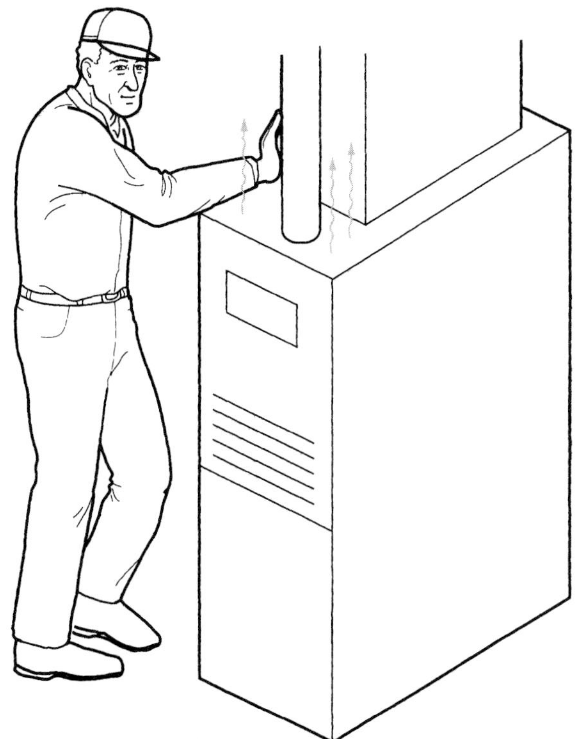

Figure 10-4. We can tell that this furnace pilot burner is lit without removing the front cover of the furnace.

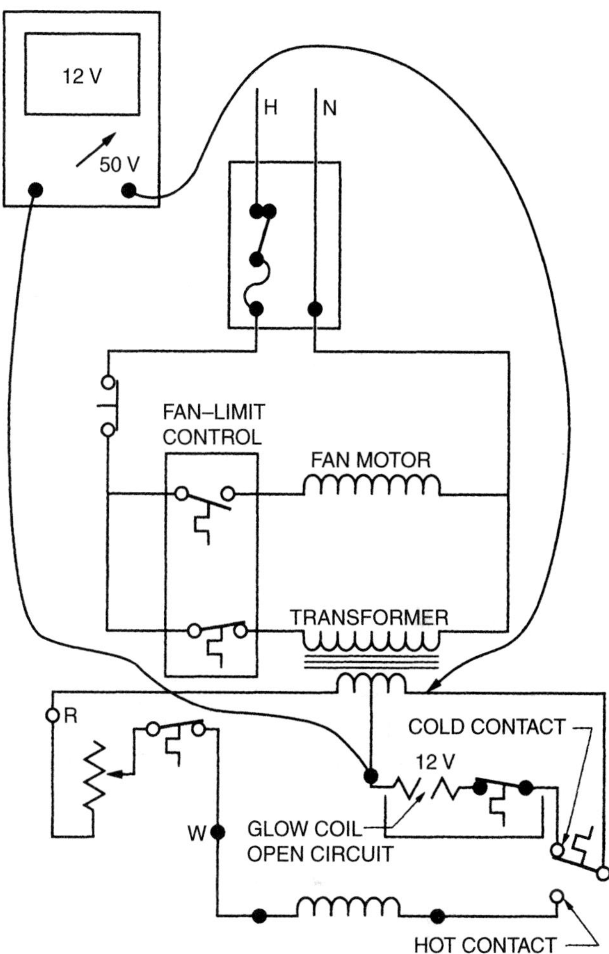

Figure 10-5. Furnace glow coil is open circuit.

If it is defective, the thermocouple pilot safety device can cause a furnace to malfunction. This device is easy to check with the correct instrument, a millivoltmeter, and an adapter. If the pilot burner goes out on a standing pilot furnace, the gas valve cannot open to allow gas to the main burner. If the system has a timed fan switch, the fan may start upon a call for heat, but it will blow cold air. If the furnace has a temperature-activated fan switch, the heat exchanger will not heat, so the fan will not start.

The thermocouple can be checked by disconnecting it from the gas valve and checking the voltage that it generates when it is heated with the pilot flame. This test is easy because the thermocouple does not need to be disconnected from the end of the pilot burner that is in the burner. Follow these steps:

1. Disconnect the thermocouple from the gas valve.
2. Fasten the millivoltmeter leads to the end of the thermocouple as shown in Figure 10-6.

Note: If you have a digital voltmeter that measures millivolts, you can use it instead of a special meter. Most analog meters with needles do not have millivolt scales.

3. Light the pilot burner and allow it to burn. Watch the thermocouple voltage as it rises.

Note: This test is performed with no load on the thermocouple except the meter movement. The thermocouple should generate at least 20 millivolts, or it is defective and should be replaced.

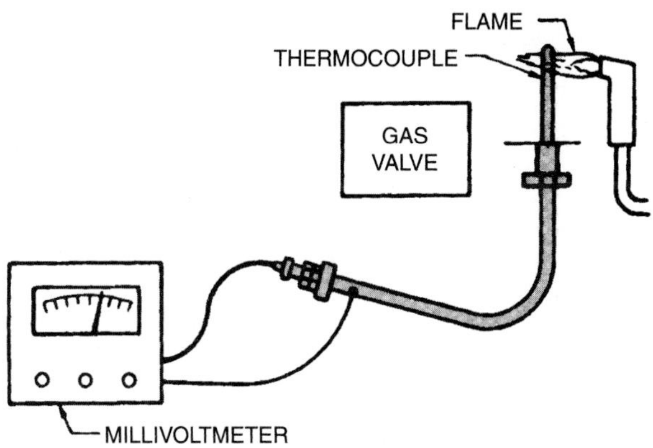

Thermocouple no load test. Operate the pilot flame for at least 5 min with the main burner off. (Hold the gas cock knob in the pilot position and depressed.) Disconnect the thermocouple from the gas valve while still holding the knob down to maintain the flame on the thermocouple. Attach the leads from the millivoltmeter to the thermocouple. Use DC voltage. Any reading below 20 mV indicates a defective thermocouple or poor pilot flame. If pilot flame is adjusted correctly, test thermocouple under load conditions.

Figure 10-6. Thermocouple operating without a load.

Another test should be performed if the thermocouple voltage is more than 20 millivolts. This test puts the thermocouple under its normal working load. To perform this load test, follow these steps:

1. Use the adapter illustrated in Figure 10-7 by fastening it in the thermocouple valve body. The thermocouple can be screwed into the adapter while the adapter is in the gas valve.
2. Light the pilot burner and observe the voltage. The voltage generated by the thermocouple should be at least 9 millivolts under its normal operating load, or the thermocouple is becoming weak and should be replaced.
3. While the voltage is applied to the coil in the valve, try to reset the pilot safety to see if the thermocouple will hold the valve's plunger in place to allow gas to flow through the valve. (The valve may have a button to hold down and/or a control knob to hold in place.) If the thermocouple generates the voltage, but the valve does not allow gas to flow, the valve coil may be defective.

Another electrical problem that can occur with a standard gas furnace involves the fan circuit. The fan circuit controls the fan for two purposes. One is to delay the start of the fan at the beginning of the cycle. This delay gives the heat exchanger time to become warm so that cold air does not blow into the conditioned space. The other is to keep the fan running long enough at the end of the cycle to dissipate the heat in the heat exchanger. Otherwise, this heat moves up the flue during the off cycle.

Two fan circuits are common. One uses the conventional fan and limit control. The other uses the timed fan switch. Figure 10-8 shows both types. Each type functions differently.

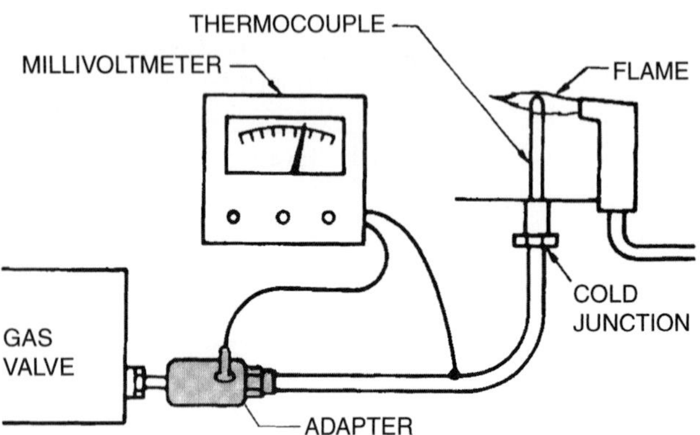

Figure 10-7. Thermocouple operating under a load.

In the conventional fan and limit type, the element for the fan control is part of the limit switch and is mounted next to the heat exchanger. It senses the heat in the heat exchanger area, and the fan runs when this area is within the set points.

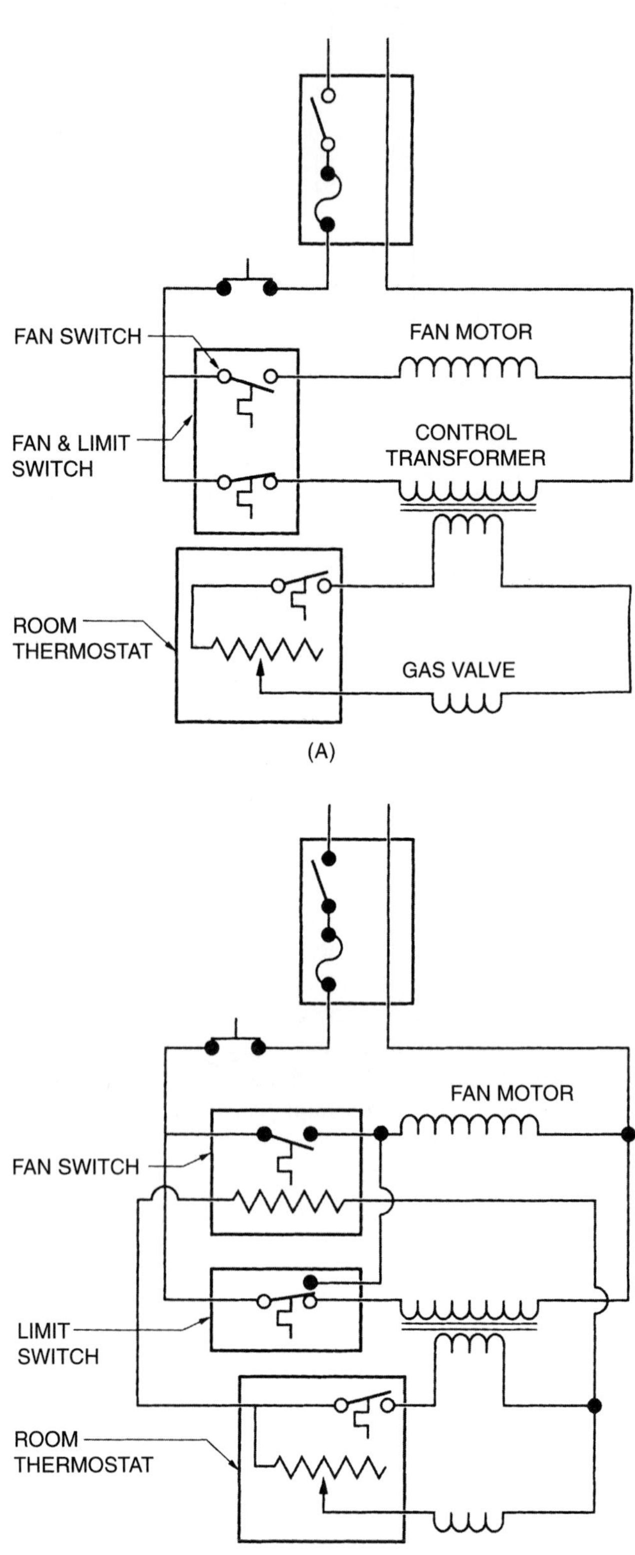

Figure 10-8. Two fan circuits for gas furnaces. (A) Temperature-operated fan switch. (B) Timed fan switch.

The set points start and stop the fan. The fan will start only when the furnace is hot. If the control is set too low, the fan may start in a hot attic. The control is factory set to prevent fan short cycles at the beginning and end of the furnace cycle. One short cycle can occur at the beginning of the cycle when the cool return air hits the thermostatic element in the fan control. Another short cycle can occur at the end of the cycle when residual heat in the heat exchanger rises and reheats the bimetal element, which restarts the fan for a moment. These short cycles do no harm, but they are annoying to the customer, so you should change the fan settings up or down to prevent the short cycles. The voltage can be traced similarly to the method used for finding the defective gas valve. A troubleshooting sequence for this type of problem is discussed later in this section.

Timed fan controls are used in many downflow and horizontal furnaces when the rising heat does not successfully heat a temperature element. The timed fan control has a completely different action. The fan is started from a call for heat by the thermostat. When a call for heat occurs, voltage is supplied to the gas valve to start the flame in the main burner. At the same time, voltage is applied to the heating element in the fan control, and it starts to heat. Approximately 1 minute passes before the control becomes hot enough to make the contacts. This gives the burner time to heat the heat exchanger slightly and prevent cold air from blowing at the beginning of the cycle. At the end of the cycle, the 24-volt circuit of the fan control is deenergized at the same time as the gas valve. The element takes a few minutes to cool, so the fan continues to run. This control does not create short cycles, as the temperature-activated control does; however, if the main burner does not light, the fan starts anyway. When a call is received and the customer explains that the fan is running and blowing cold air, the technician knows that there is a pilot burner or main burner problem.

The thermostat's heat anticipator must be set to the correct value for the timed fan control because it draws more current than a gas valve does. If the heat anticipator is not set correctly, the circuit draws more amperage and cycles the furnace burner off, as shown in Figure 10-9.

You can follow the control sequence on the ladder diagram, using the hopscotch method, to find a defective fan motor with a timed fan switch. See Figure 10-10, and follow these steps:

1. Turn the thermostat to the "heat on" position, and push the temperature lever to call for heat. The burner should light.
2. After a while you will notice that the fan motor will not start.
3. Set the meter to the 250-volt scale and place one lead on the neutral terminal in the disconnect box. There is no need to check the hot lead in this disconnect; the burner's operation is proof of line voltage.
4. Touch the other lead to the power terminal entering the timed fan switch. The meter should read 120 volts.

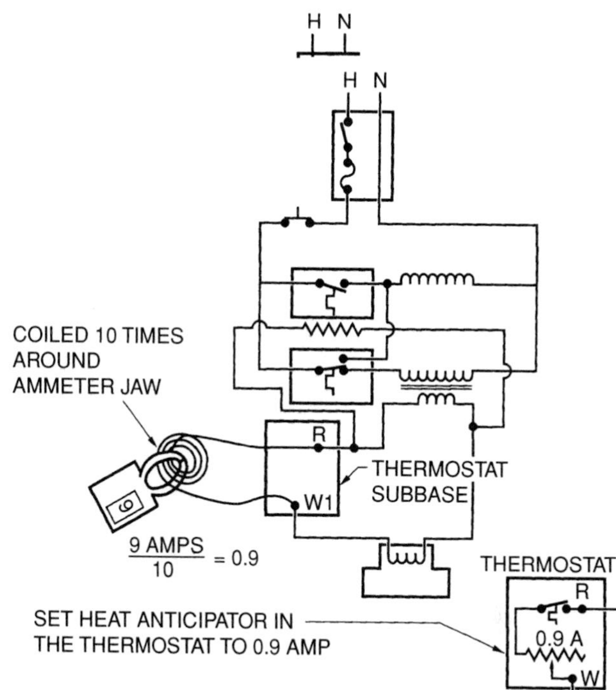

Figure 10-9. Setting the heat anticipator.

5. Now touch the lead to the other side of the fan switch. The meter should read 120 volts. Power is going to the fan motor, which is not turning.
6. The next test is an ohm test of the motor. Turn the power off at the disconnect and use the voltmeter to prove that the power is off.
7. Set the meter to the R×1 scale.
8. Remove the lead wires to the motor and attach one to each lead on the meter. The meter should show that the winding is an open circuit.

ELECTRONIC CONTROLS. Troubleshooting electronic controls usually becomes an electrical troubleshooting job because electronic control circuits are printed on circuit boards. Electrical circuits lead up to and out of the boards. The boards act much like switches to the electrical circuits and are treated as switches when we troubleshoot them. There is no difference between a circuit passing through a switch and then to the power-consuming device and a circuit passing through a circuit board to the power-consuming device, as shown in Figure 10-11. The circuit board can have more than one circuit passing through it, so the board acts much like a thermostat, in which one hot wire enters and power is distributed to the various circuits, as shown in Figure 10-12.

Furnaces with electronic circuit boards are much easier to troubleshoot than conventional furnaces because all of the unit's major circuits pass through the board and can be identified at the board.

There is nothing we can do in the actual electronic circuit to troubleshoot a problem because manufacturers do not publish the internal workings of the boards. This can be

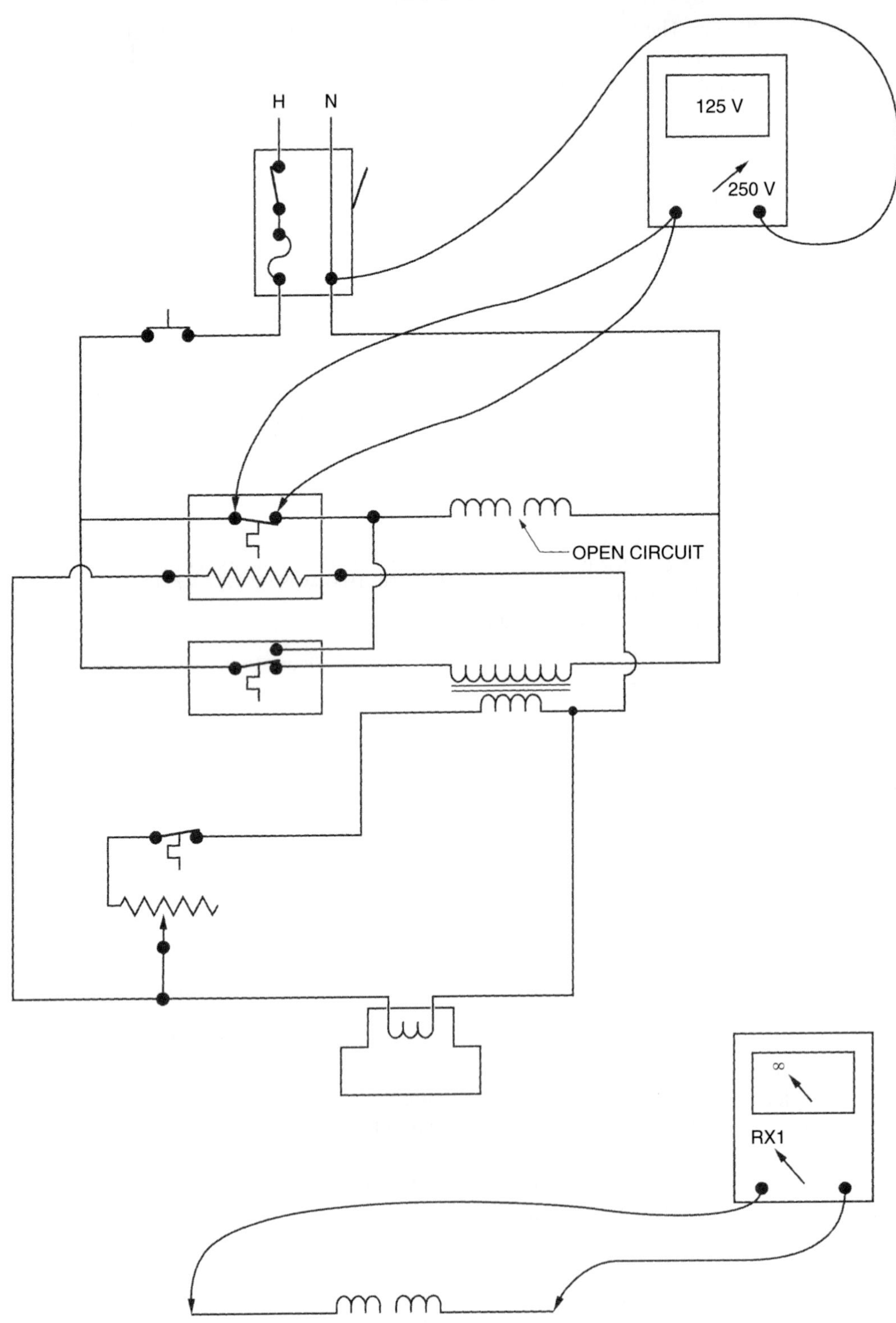

Figure 10-10. Finding a defective fan motor.

fortunate for the heating technician because most of our electronic troubleshooting can be accomplished with the basic VOM. The service technician will have all the tools.

An electronic circuit board for a heating system may look like the one in Figure 10-13. The first thing to know about an electronic circuit board is not to fear it. You do not need electronic experience to service one. Your electrical training and familiarity with the workings of a gas system will go a long way toward your understanding the purpose of a circuit in the circuit board.

One advantage of an electronic circuit board furnace is that auxiliary components can be added to the

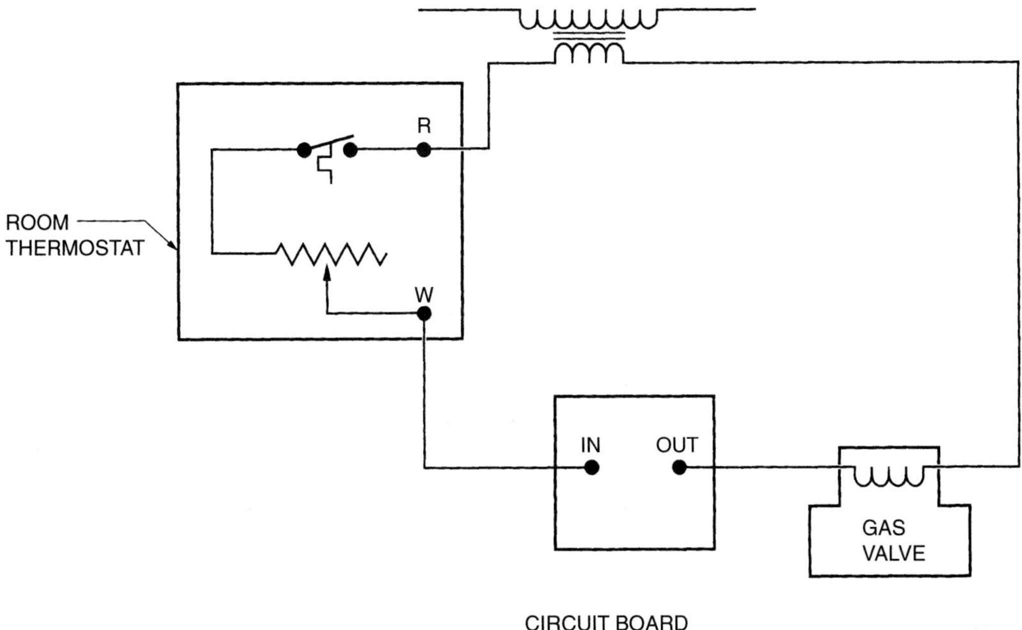

Figure 10-11. A circuit board can often be much like a switch in a circuit, except you cannot see the contacts.

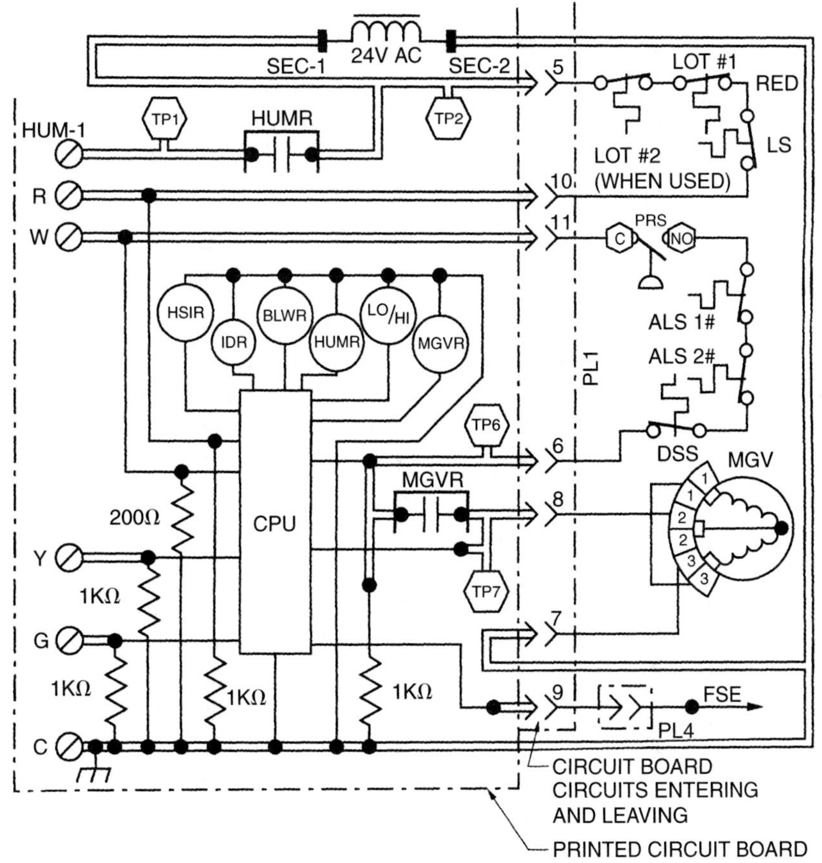

Figure 10-12. Several circuits can pass through a circuit board. *Courtesy Carrier*

system more easily than to a conventional furnace. For example, a humidifier must be interlocked so that it operates only when the system is in the heating mode. The humidifier must not operate while air-conditioning runs if air-conditioning is added to the furnace. The interlock feature can be built into the electronic circuit

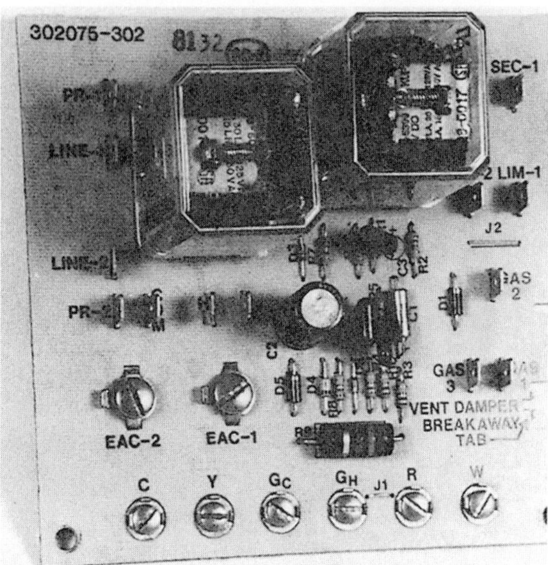

Figure 10-13. Electronic circuit board for a gas furnace. *Courtesy Carrier*

board. An electronic air cleaner can be interlocked on an electronic board so that it operates any time the fan runs. The circuit board can have the capacity for adding the control circuits for central air-conditioning. In this case, the fan needs to run in low speed while in heating and in high speed while in cooling. Figure 10-14 depicts a board that can accommodate all the features just discussed.

Let us use the circuit board in Figure 10-14 to illustrate a troubleshooting procedure. This circuit board is for an upflow gas furnace with a power vent motor and the controls that would be used for it. Take time to study the diagram and its legend. Note the MGVR symbol, which stands for main gas valve relay. When this relay is energized, the MGVR contacts close and low-voltage power is furnished to terminal 2 on the MGV (main gas valve). Power from the other side of the transformer's circuit is furnished to the MGV at terminal 3. When a meter probe is placed on the common terminal of the circuit board and the other lead is placed on TP7 (test point 7), if 24 volts are present, the gas valve must open to supply gas to the main burner, or the valve is defective.

Note: Be sure that the gas valve knob is turned to *on* before condemning the valve.

The circuit board is divided into two sets of test points. Some test points are line voltage (115 V) and some are low voltage (24 V) for the control circuits, as depicted in Figure 10-15.

The technician can use the hopscotch method to troubleshoot the complete board and find where power is furnished and where it is not. Power is not furnished to all points at all times; the sequence of events determines where the circuit is energized.

The sequence of events are as follows for an operating furnace:

1. A circuit from R to W is completed by the room thermostat. This starts the microprocessor through the timed sequence.
2. The induced-draft relay (IDR), which starts the fan and purges any gas fumes from the heat exchanger, is started for 15 seconds.
3. When the draft is proved and purged for 15 seconds, the hot-surface igniter is energized for 17 seconds.
4. At the end of 17 seconds, the hot-surface igniter is deenergized, and the gas valve opens. Within 2 seconds the flame must ignite and be proven through the FSE (flame-sensing electrode). If the flame is not proven, the gas valve will be closed, and the ignition sequence will start over with a 45-second warm-up for the hot-surface igniter. After four tries, the circuit board will lock out the sequence, and it can be restarted only by turning off the power to the microprocessor, waiting a few minutes, and turning it back on. This can be accomplished from the room thermostat or the unit.

When troubleshooting a circuit board, the technician should first test the CPU (central processing unit), which is the microprocessor, the brains used to distribute the power. This test can be performed by applying a jumper wire across test terminals 1 and 2, found in the lower right-hand corner of the diagram (when installed in the furnace, the board is turned 90 degrees clockwise). The following sequence of events occurs when the test port is jumped:

1. The inducer fan runs for 10 seconds.
2. The hot-surface igniter is energized for 15 seconds.
3. The fan runs in high speed for 10 seconds.
4. The fan runs in low speed for 10 seconds.

If the microprocessor does not perform this test, check power to the board at transformer PR-1 to PR-2 for line voltage. Then check the low voltage at the secondary of the transformer at SEC-1 to SEC-2. All these connections are in the middle of the board and are easily accessible. If power is supplied to the board, and it does not perform the test, the board is defective and must be replaced.

Testing a furnace for a defective component requires three main steps. First, memorize the sequence of events; not all components are energized at the same time. Second, use the hopscotch method to test each component when it is to be energized. Start with the line-voltage portion of the board as shown in Figure 10-16 and follow these steps:

1. Set the VOM to a scale to measure 115 volts, and fasten one probe on PR-2.

Note: The PR-2 terminal is common to all line-voltage power-consuming devices.

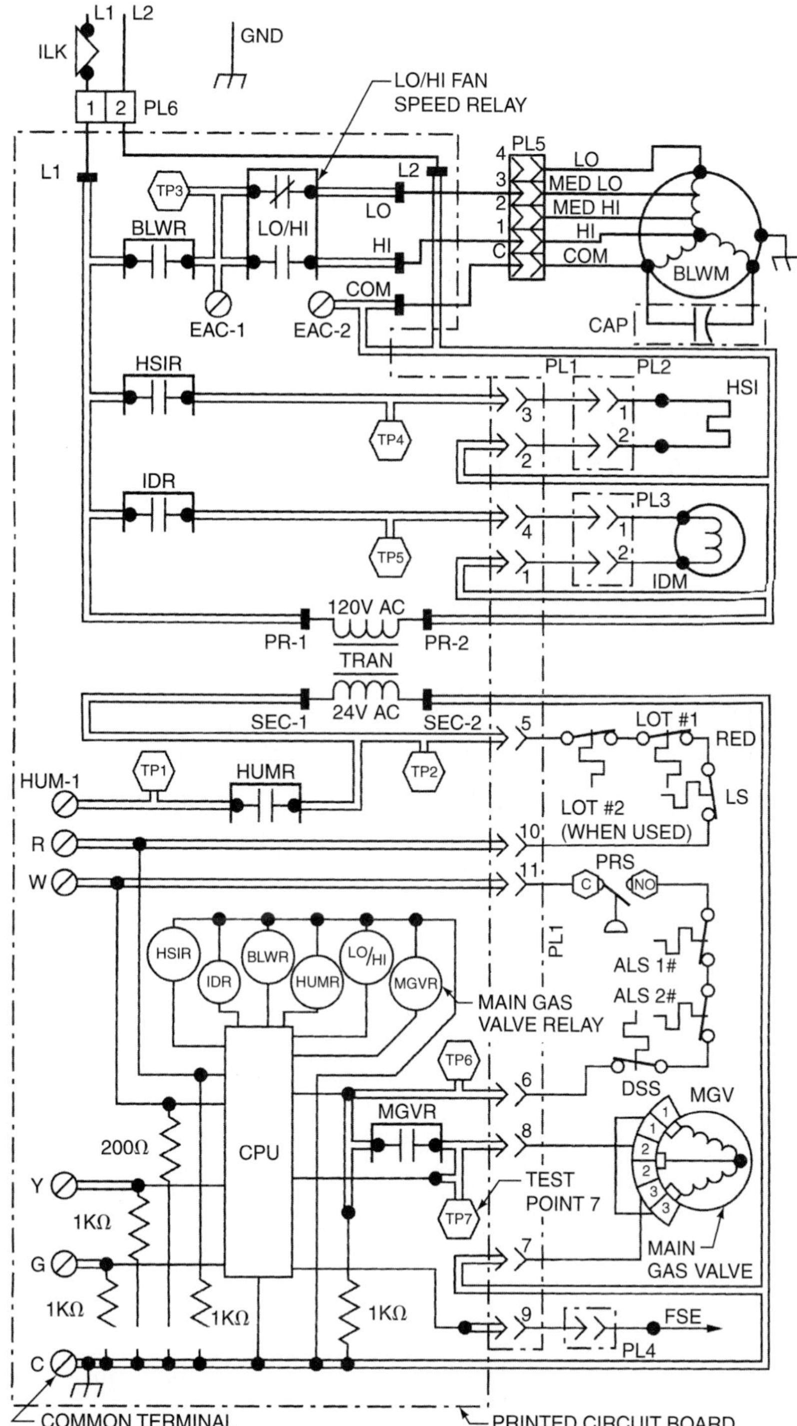

NOTES: (SEE FIGURE 10-16B FOR SYMBOLS LEGEND)

1. COMMON SIDE (SEC-2 AND C) OF 24V AC TRANSFORMER CONNECTED TO GROUND THROUGH THIS MOUNTING SCREW.
2. IF ANY OF THE ORIGINAL EQUIPMENT WIRE IS REPLACED USE WIRE RATED FOR 105 C, OR EQUIVALENT.
3. INDUCER AND BLOWER MOTORS CONTAIN INTERNAL AUTO-RESET THERMAL OVERLOAD SWITCHES.
4. BLOWER MOTOR SPEED SELECTIONS ARE FOR AVERAGE CONDITIONS. SEE INSTALLATION INSTRUCTIONS FOR DETAILS ON OPTIMUM SPEED SELECTION.
5. USE COPPER WIRE ONLY BETWEEN THE DISCONNECT SWITCH AND THE FURNACE JUNCTION BOX.

Figure 10-14. Circuit board description. *Courtesy Carrier*

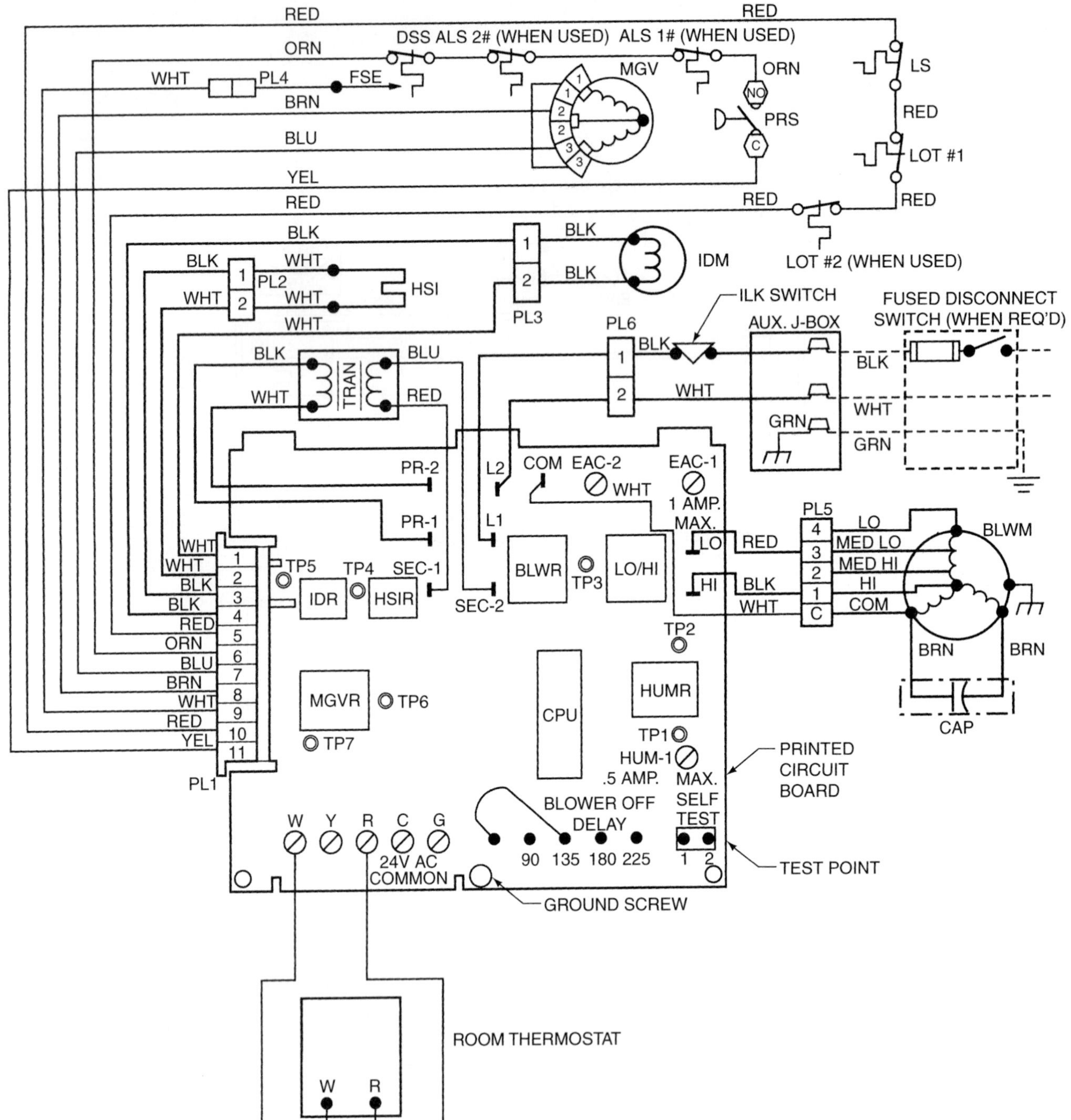

Figure 10-15. Line-voltage and low-voltage test points. *Courtesy Carrier*

2. Touch probe 2 to PR-1. The meter should read 115 volts. This verifies the power supply.
3. When the fan is supposed to be operating, touch probe 2 to TP3. The meter should read 115 volts. This is the main blower relay. To check power further at either high or low speed, touch the probe to either the *lo* or *hi* terminal, depending on which should be calling.
4. When the hot-surface igniter is supposed to be energized during ignition, touch probe 2 to TP4. The meter should read 115 volts.
5. When the induced-draft motor is supposed to be operating, touch probe 2 to TP5. The meter should read 115 volts.

Third, to test the low-voltage portion of the circuit, follow these steps:

1. Set the VOM to a scale to measure the low-voltage (24-V) portion of the circuit, and fasten one probe to the SEC-2 terminal.

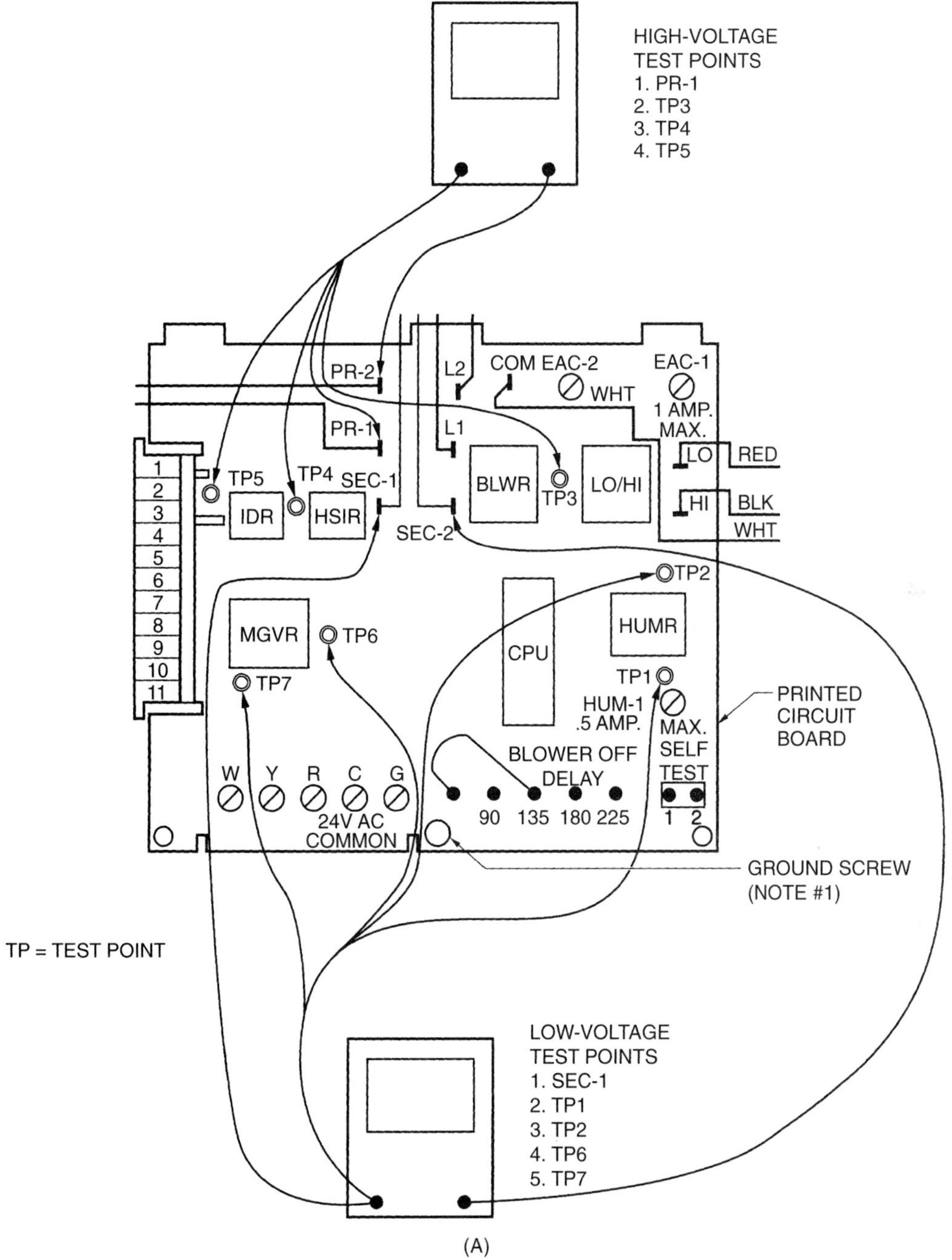

Figure 10-16. (A) Checking the line-voltage and low-voltage test points. (B) The circuit board. *Courtesy Carrier*

Note: The SEC-2 terminal is common to all low-voltage power-consuming devices.

2. Touch probe 2 to SEC-1. The meter should read 24 volts. This verifies the power supply.
3. When the humidifier is supposed to operate, touch probe 2 to TP1. The meter should read 24 volts.
4. Touch probe 2 to TP2. The meter should read 24 volts at all times because this is the same as the transformer secondary terminal.
5. When all the components in the safety circuit are satisfied, touch probe 2 to TP6. The meter should read 24 volts. The only component that usually opens the circuit is the pressure switch that proves that the induced-draft fan motor is creating a pressure for venting purposes.
6. When power is supposed to be furnished to the main gas valve, touch probe 2 to TP7. The meter should read 24 volts.

There are many variations of electronic controls and there is no substitute for knowing the manufacturer's intentions when the circuit was designed. Often, however, this information is not pasted to the furnace or left with the

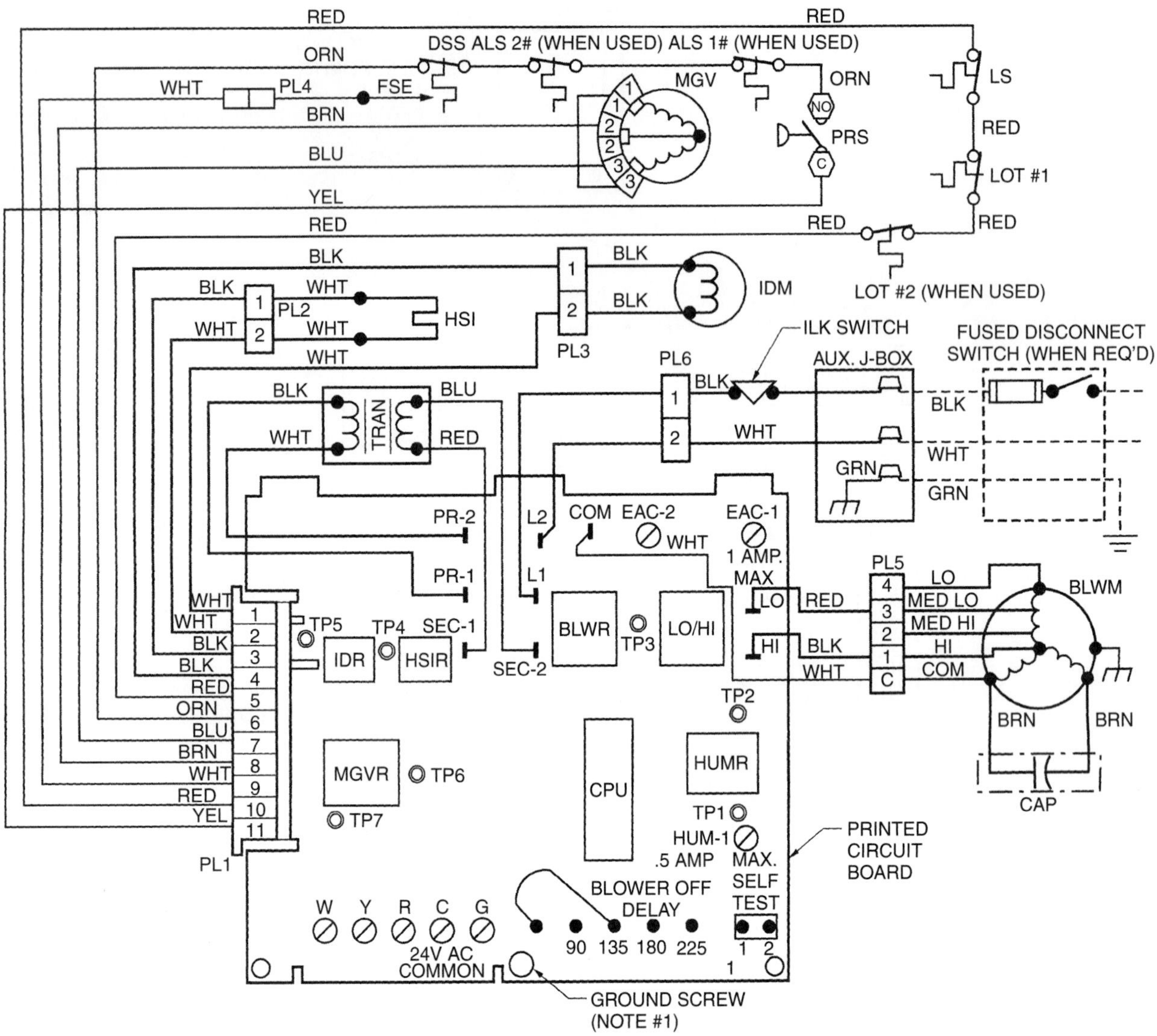

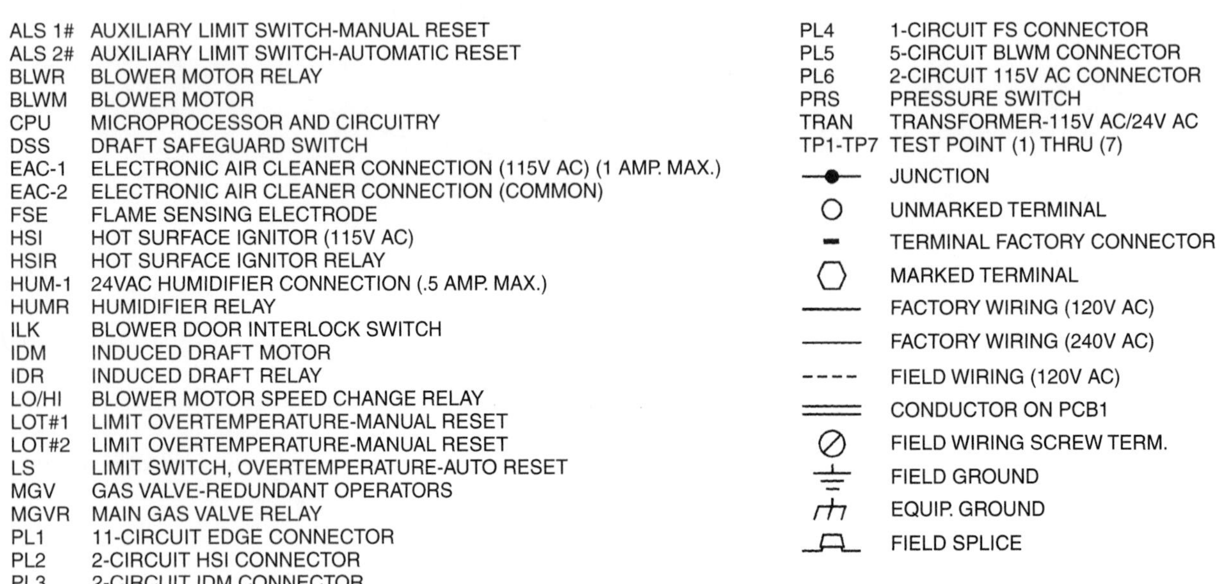

ALS 1#	AUXILIARY LIMIT SWITCH-MANUAL RESET
ALS 2#	AUXILIARY LIMIT SWITCH-AUTOMATIC RESET
BLWR	BLOWER MOTOR RELAY
BLWM	BLOWER MOTOR
CPU	MICROPROCESSOR AND CIRCUITRY
DSS	DRAFT SAFEGUARD SWITCH
EAC-1	ELECTRONIC AIR CLEANER CONNECTION (115V AC) (1 AMP. MAX.)
EAC-2	ELECTRONIC AIR CLEANER CONNECTION (COMMON)
FSE	FLAME SENSING ELECTRODE
HSI	HOT SURFACE IGNITOR (115V AC)
HSIR	HOT SURFACE IGNITOR RELAY
HUM-1	24VAC HUMIDIFIER CONNECTION (.5 AMP. MAX.)
HUMR	HUMIDIFIER RELAY
ILK	BLOWER DOOR INTERLOCK SWITCH
IDM	INDUCED DRAFT MOTOR
IDR	INDUCED DRAFT RELAY
LO/HI	BLOWER MOTOR SPEED CHANGE RELAY
LOT#1	LIMIT OVERTEMPERATURE-MANUAL RESET
LOT#2	LIMIT OVERTEMPERATURE-MANUAL RESET
LS	LIMIT SWITCH, OVERTEMPERATURE-AUTO RESET
MGV	GAS VALVE-REDUNDANT OPERATORS
MGVR	MAIN GAS VALVE RELAY
PL1	11-CIRCUIT EDGE CONNECTOR
PL2	2-CIRCUIT HSI CONNECTOR
PL3	2-CIRCUIT IDM CONNECTOR
PL4	1-CIRCUIT FS CONNECTOR
PL5	5-CIRCUIT BLWM CONNECTOR
PL6	2-CIRCUIT 115V AC CONNECTOR
PRS	PRESSURE SWITCH
TRAN	TRANSFORMER-115V AC/24V AC
TP1-TP7	TEST POINT (1) THRU (7)

JUNCTION
UNMARKED TERMINAL
TERMINAL FACTORY CONNECTOR
MARKED TERMINAL
FACTORY WIRING (120V AC)
FACTORY WIRING (240V AC)
FIELD WIRING (120V AC)
CONDUCTOR ON PCB1
FIELD WIRING SCREW TERM.
FIELD GROUND
EQUIP. GROUND
FIELD SPLICE

(B)

Figure 10-16. *(Continued)*

customer. Instead, you must locate the information. Go to factory schools and any training programs that are available. Manufacturers want you to be able to service their equipment. If you cannot find information locally, call or write the company, and its representatives should be eager to supply you with any information that you need.

Mechanical Problems

Mechanical problems are problems with the prime mover of the air or, in the case of a boiler, the water; problems with the gas-supply components, such as pressure regulators or water pressure regulators; and combustion problems.

Pressure regulators have a few moving parts that eventually wear. These regulators rarely cause any trouble, so adjustment is seldom necessary. Usually when a problem does occur, the entire regulator is replaced. One problem you may encounter with a regulator is that the last person who serviced the system and adjusted the pressure did not have a manometer.

A regulator has a vent tube above the diaphragm that must be kept clear of obstacles so that it can function correctly. When a regulator is located indoors, either the vent has an orifice restrictor or the vent is routed to the outside. Both are safety features in the event that the diaphragm were to rupture and allow gas in large quantities to push to the upper side of the diaphragm.

Caution: Do not terminate the vent where water can run into it or insects can build nests in it. A regulator that is located outside should have a screen covering over the vent to protect it from insects that may otherwise try to build nests in the vent hole. If water covers the regulator vent and freezes, the regulator diaphragm may be pushed up, which will allow too much gas to flow to an appliance and overfire the appliance.

Water-pressure regulators are used on boilers to limit the water pressure entering the boiler. These regulators are discussed in Unit 17.

Combustion problems can have several sources. Some of the problems can be delayed ignition, burning at the orifice, slow crossover ignition, poor pilot burner positioning, sooting of the main burner or the pilot burner, flame impingement, flame roll out, orifice misalignment, and venting problems.

Delayed ignition can be caused from too much primary air. The burners can experience slow ignition and "pop" when the gas valve opens. This problem may not be harmful, but it may scare someone. It can be remedied with a proper air adjustment.

If there is not enough primary air or if the orifice is misaligned, the gas may be slow in reaching the burner ports and ignition may occur in the mixing tube under the burner ports as shown in Figure 10-17. This situation causes a blowing sound, like that of a blow torch. The problem must be corrected, or poor combustion will cause soot to build up in the burners and the heat exchanger.

The crossover pipes or slots on the burner are intended to ignite the next burner. These are provided so that a pilot burner does not have to be located at each main burner as shown in Figure 10-18. If these pipes or slots are misaligned, slow ignition will occur with a bang.

The primary air to the main burner or the pilot burner may become obstructed with dust or dirt. When this occurs, the flame burns yellow, and, with the main burner, you cannot use the primary air adjustment to correct the flame, as shown in Figure 10-19. The aerated pilot burner may need to be disassembled and cleaned (because there is no adjustment and the air passages are tiny). When a burner becomes dirty inside from lint or other obstructions, you should remove the burner from the furnace and, using compressed air, blow down through the burner ports and back toward the primary air opening, as shown in Figure 10-20.

Warning: Be sure to wear eye protection when using compressed air to blow out burners because flying particles can cause eye injury.

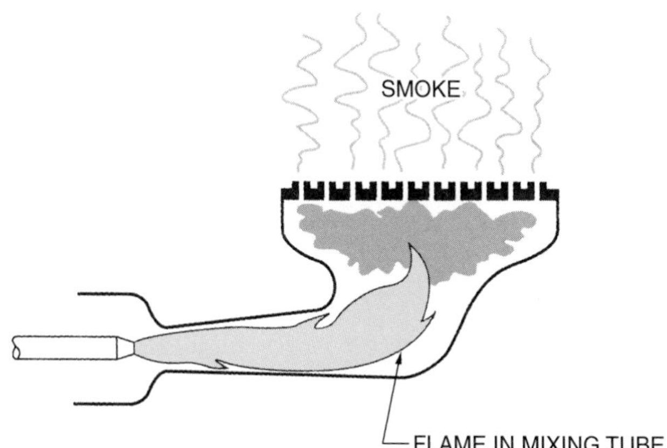

Figure 10-17. Gas burns in the mixing tube of the burner.

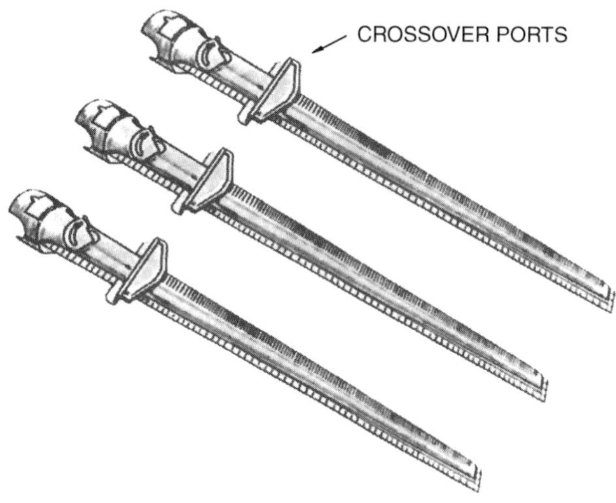

Figure 10-18. You must be sure that the burner crossover tubes are aligned.

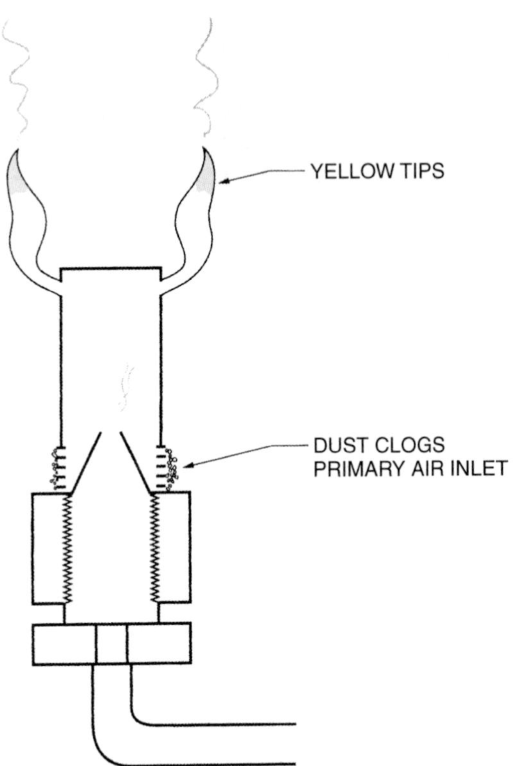

Figure 10-19. When the primary ports of an aerated pilot burner become clogged, the pilot flame burns yellow.

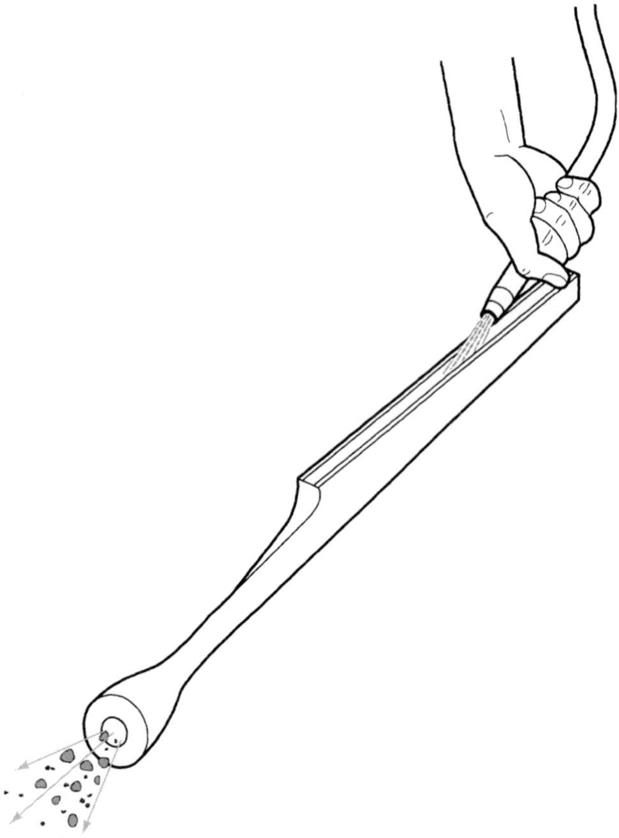

Figure 10-20. Cleaning a burner by blowing it out with compressed air.

Flame impingement can happen to either the pilot burner or the main burner. When the pilot burner orifice is oversized, the flame may be too large. This happens when an appliance is converted from natural gas usage to LP gas usage and the orifice for natural gas is mistakenly left in the pilot burner. Misaligned burners can also cause flame impingement, as illustrated in Figure 10-21. Regardless of the reason for flame impingement, it must be corrected or a hole will burn in the heat exchanger.

Flame roll out can occur when soot is in the heat exchanger or when a heat exchanger is cracked. If a burner has operated for any time with a yellow flame, a sooted heat exchanger should be suspected. Soot accumulates quickly. LP gas is more prone to build soot than natural gas because it requires more air. This soot prevents the flue gases from rising through the heat exchanger, so they roll out the front of the burner compartment. When soot is suspected, the furnace draft diverter and burners should be removed and the soot removed from above and below. Round brushes or wire from coat hangers, and a good vacuum cleaner, are the best tools for this job as shown in Figure 10-22. A badly cracked heat exchanger can also cause flame roll out.

Warning: Furnaces with cracked heat exchangers should not be allowed to operate because of the danger of CO (carbon monoxide) poisoning. When a cracked heat exchanger is found, it should be shut down immediately, and either the heat exchanger or the furnace should be replaced.

The pilot burner must be positioned to ignite the gas as it pours out the burner ports. Natural gas rises, so the pilot flame may need to be slightly above the burner port. When LP gases are used, the flame may need to be located a little lower because LP gases are heavier than air and fall as shown in Figure 10-23. Consequently, the pilot burner may need to be repositioned after a furnace is converted from one gas usage to another.

Venting problems can involve oversized or undersized vents, cold vents, downdrafts, negative pressures in the structure, and vent obstructions. Oversized vents and cold

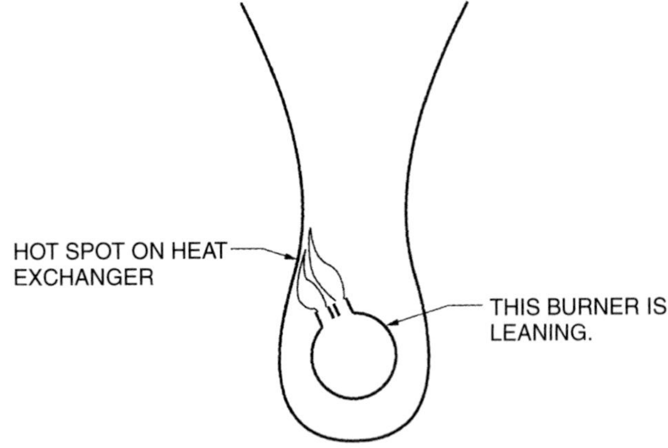

Figure 10-21. Misaligned burners cause flame impingement.

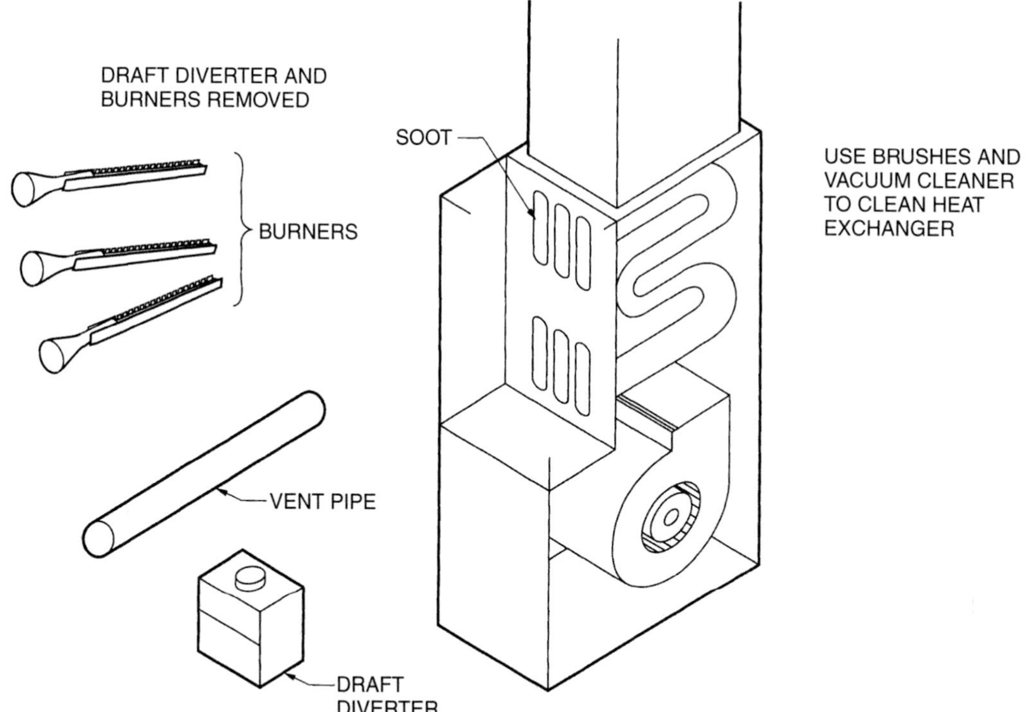

Figure 10-22. Use a vacuum cleaner and brushes to clean a heat exchanger from above and below.

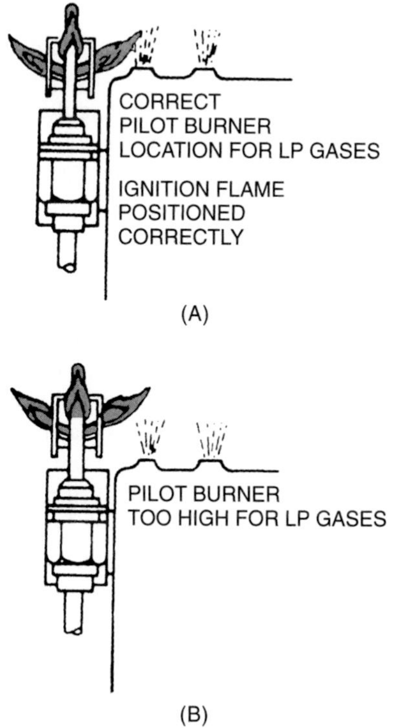

Figure 10-23. Positioning a pilot burner (A) for natural gas and (B) for LP gases.

vents may have the same identifying characteristic: moisture condensing and running down the draft diverter. A cold vent can occur when a single-wall vent is installed instead of a double-wall vent and the space where the vent is routed is not conditioned. Undersized vents cause the by-products of combustion to spill out the draft diverter. This can be checked by holding a match or a candle next to the draft diverter. If the flame pulls away from the diverter, there is a problem. Down-drafts can occur when the horizontal length-to-height relationship of the vent is wrong. The horizontal run must be no more than 75% of the height. Downdrafts can also be caused by the roof configuration around the vent termination. The *National Fuel Gas Code* says that the vent must terminate at least 2 feet above any portion of the roof within 10 feet of the termination point. A prevailing wind moving over a rooftop on top of a terminated vent can cause downdrafts.

10.2 Preventive Maintenance

Gas equipment maintenance consists of servicing the air-side components; filter, fan, belt and drive, and burner section.

Central forced-air systems have a fan to move the air, and therefore they require filter maintenance. Filters should be changed or cleaned regularly, depending on the rate of dust accumulation.

The bearings of some fans require lubrication at regular intervals. Motor bearings and fan shaft bearings must be kept in good working order because sleeve bearings do not take much abuse. They do not always make excess noise to indicate a problem, so be sure to examine them. Lift the motor or fan shaft and look for movement. Do not confuse shaft end play for bearing wear; many motors have as much as $\frac{1}{8}$ inch of end play.

Some systems may have belt drives and require belt tension adjustment or belt replacement. Frayed or broken belts must be replaced to ensure safety and trouble-free operation. Belts should be checked at least once a year. It is dangerous to operate an electric heating system with no airflow or low airflow, so it must be maintained.

Filters should be checked every 30 days of operation unless you determine that longer periods can be allowed. Use the correct filter that fits the holder and be sure to follow the arrows on the filter that indicate the correct airflow direction. When the filter system becomes clogged in a forced-air system, the airflow is reduced and causes the heat exchanger to overheat. Automatic reset limit switches should be tripped and then reset. If the condition continues, these switches may fail, and the unit will be off the line until the switch is replaced. Replacing the switch requires a service call that can be prevented if the filters are kept unclogged.

The burner section consists of the actual burner, the heat exchanger, and the venting system for the by-products of combustion. The burner portion of the modern gas furnace does not need adjustment from season to season. It burns efficiently and reliably if it is kept clean. Rust, dust, or scale must be prevented from accumulating on top of the burner where the secondary air supports combustion or in the actual burner tube where primary air is induced into the burner. When a burner is in a clean atmosphere (no heavy particles are in the air as they are in a manufacturing area), most of the dust particles in the air burn and rise up the venting system. The burner may not need cleaning for many years. A vacuum cleaner with a crevice tool can be used to remove small amounts of scale or rust deposits.

The burner and manifold can be removed for more extensive cleaning when deposits are inside the burner tube. Each burner should be removed and tapped lightly to loosen the scale. An air hose should be used to blow down through the burner ports while the burner is out of the system. The particles will blow back out the primary air shutter and out of the system. Do not attempt to clean the burners this way while they are in the furnace.

Burner alignment should be checked to make sure that no flame impingement to the heat exchanger is occurring. The burners should set straight and should feel secure while in place.

A combustion analysis is not necessary on the modern gas furnace. The primary air adjustment is the only adjustment on the burners, and it cannot be out of adjustment enough to create much inefficiency when natural gas is the fuel. However, LP fuel burners may need adjustment for best performance. All gas burners should burn with a clear blue flame that has orange streaks. The flame should never be yellow or have yellow tips. If yellow tips cannot be adjusted from the flame, the burner contains dirt or trash, and it should be removed for cleaning.

Observe the burner for correct ignition as it is lit. Often the crossover tubes that carry the burner flame from one burner to the next become misaligned, and the burner lights with an irregular pattern. Sometimes it will "puff" when one of the burners has slow ignition.

The pilot burner flame should be observed for proper characteristics. It should not have yellow tips, and it should not impinge on the burner. It impinges on the thermocouple only. If the flame characteristics are not correct, the pilot assembly may have to be removed and cleaned. Compressed air will usually clear any obstructions from the pilot ports.

Caution: Do not use a needle to make the orifice holes larger. If they are too small, obtain the correct size from a supply house.

The venting system should be examined for obstructions. Birds may nest in the flue pipe, or the vent cap may become damaged. Visually inspect the vents first, then strike a match in the vicinity of the draft diverter and look for the flame to pull into the diverter. If it does, the furnace is drafting correctly.

Observe the burner flame while the furnace fan runs. A flame that blows while the fan is running indicates a crack in the heat exchanger and warrants further heat-exchanger examination. A small crack is difficult to find in a heat exchanger and may only be seen by taking the heat exchanger out of the furnace housing. Extensive tests have been developed for proving whether a gas furnace heat exchanger has a crack. These tests may be beyond what a typical technician would use because they require expensive equipment and experience in using them.

High-efficiency condensing furnaces require maintenance of the forced-draft blower motor if it is designed to be lubricated. The condensate drain system should be checked to ensure that it is free of obstructions and will drain correctly. Manufacturers of these furnaces sometimes design the condensing portion to be in the return air section where all air must move through the coil. This coil is finned and collects dirt if the filters are not cleaned. Routine inspection of the condensing coil for obstruction is recommended.

Another routine procedure at each installation is to set the room thermostat to call for heat and to make sure that the proper sequence of events takes place. For example, when you check a standard furnace, the pilot may be lit constantly and the burner should ignite, then the fan should start. A furnace may have intermittent pilot ignition. If so, the pilot lights first, the burner lights next, then the fan starts. With a condensing furnace, the power vent motor may start first, the pilot burner next, then the burner, and finally the fan motor.

Some technicians also disconnect the circulating fan motor and allow the furnace to operate until it shuts off because of the high limit control. This is a good practice because it proves that the safety device will shut off the furnace.

10.3 Technician Service Calls

Service Call 1

A residential customer calls. The furnace stopped in the middle of the night. The furnace is old, and the thermocouple is defective. The furnace is in an upstairs closet.

The technician arrives and goes directly to the thermostat. The thermostat is set correctly for heating. The thermometer in the house shows a temperature 10°F below the thermostat's setting. This unit has air-conditioning, so the technician turns the fan switch to on to see if the indoor fan will start. It does. This proves that the system's low-voltage power supply (transformer) is working.

The technician goes upstairs to the furnace and sees that the pilot burner is not burning. This system has 100% shutoff, so there is no gas to the pilot unless the thermocouple holds the pilot valve solenoid open. The technician positions the main gas valve to the pilot position and presses the red button to allow gas to the pilot burner. The pilot burns when a lit match (an igniter may be used) is placed next to it. The technician then holds down the red button for 30 seconds and slowly releases it. The pilot burner goes out. This is a sure sign that either the thermocouple or the pilot solenoid coil is defective.

The technician uses an adapter to check the thermocouple voltage output and determines that it will generate only 2 millivolts when connected to the coil (this is its load). This can mean either that the coil is pulling too much current because of a short or that the thermocouple is defective. When the thermocouple is disconnected from the coil, it still generates only 7 millivolts. It should easily generate 20 millivolts when disconnected from the load, so the thermocouple is defective. The technician replaces the thermocouple and relights the pilot burner. The furnace then operates as normal. The technician changes the air filter, tightens the belt on the fan, and oils the fan and motor bearings as added value to the customer.

Service Call 2

A residential customer calls. The furnace is located in the basement and is very hot, but no heat is moving into the house. The dispatcher tells the customer to turn off the furnace until a service technician can get to the site. The furnace fan motor is defective and will not run, so the burner is cycling on and off because of the limit control.

From the customer's description of the problem, the technician suspects that the fan motor is bad. The technician knows the furnace type from previous service calls, so a fan motor is taken along to the residence. When the technician arrives, the room thermostat is set to call for heat. From the service request it is obvious that the low-voltage circuit is working because the customer says that the burner is heating the furnace. The technician then goes to the basement and hears the burner operating. When the furnace has enough time to warm up and the fan has not started, the technician takes the front control panel off the furnace and notices that the temperature-operated fan switch dial (this is a circular dial-type control) has rotated as if it were sensing heat. The technician carefully checks the voltage entering and leaving the fan switch. (This is done by placing one meter lead on the neutral wire and the other lead first on the wire going into the fan switch and then on the wire leaving the fan switch that goes to the fan motor.) There is a voltage reading of 122 volts going to the fan motor, but the motor is not turning. The technician then shuts off the power and removes the panel to the fan motor. The motor housing is cool to the touch; its coolness means that it has not been trying to run. If the motor was locked up or had a bad start capacitor, it would be cycling on overload as a result of high current and therefore would be warm or hot to the touch. The technician replaces the motor and the panels then starts the furnace. At the proper time, the fan motor starts. The technician checks the fan motor amperage at the disconnect to verify that it is not pulling too much amperage. The air filters are changed while changing the motor as an added value to the customer.

Service Call 3

A customer in a retail store calls. There is no heat. The furnace is an upflow gas furnace with a standing pilot and air-conditioning. The low-voltage transformer is burned out because the gas valve coil is shorted. This short causes excess current to the transformer. The furnace is located in the stockroom.

The technician arrives at the job, goes to the room thermostat, and turns the fan switch to on to see if the indoor fan will start. It will not start, so the technician suspects that the low-voltage power supply is not operable. The thermostat is set to call for heat. The technician goes to the stockroom, where the furnace is located. The voltage is checked at the transformer secondary and found to be 0 volts. The power is turned off and the ohmmeter is used to check continuity of the transformer. The secondary (low-voltage) coil has an open circuit, so the technician changes the transformer. Before connecting the secondary wires, the technician checks the continuity of the low-voltage circuit. Only 2 ohms of resistance are in the gas valve circuit. (The resistance should be at least 20 ohms.) The technician goes to the truck and checks the continuity of another gas valve, finds it to be 50 ohms, and replaces the gas valve in the furnace. If the gas valve was not changed and the furnace started up with only a new transformer, the new transformer would likely burn in a matter of minutes.

The system is started with the new gas valve and transformer. A current check of the gas valve circuit shows that it is pulling only 0.5 amp, which does not overload the circuit; this is normal.

The technician changes the air filter and oils the fan motor, then calls the store manager to the stockroom for a conference before leaving. The store manager is informed that the boxes of inventory must be kept away from the furnace because they present a fire hazard.

Service Call 4

A customer in a residence calls. Fumes can be smelled and are probably coming from the furnace in the hall closet. The furnace has not been operated for the past 2 weeks because the weather has been mild. Because the fumes might be harmful, the dispatcher tells the customer to shut off the furnace until a technician arrives. A roofer who had been making repairs laid a shingle on top of the flue liner, so the flue is blocked.

The technician arrives, starts the furnace, and holds a match at the draft diverter to see if the flue has a negative pressure. The flue gas fumes are not rising up the flue because the flame blows away from the draft diverter. The technician turns off the burner and examines it and the heat exchanger area with a flashlight. There is no soot to indicate that the burner had been burning incorrectly. The technician goes to the roof to check the flue, finds a shingle on top of the chimney, and lifts off the shingle. Heat rises from the chimney.

The technician goes back to the basement, restarts the furnace and holds a match at the draft diverter. The flame draws toward the flue. The furnace is operating correctly. The technician changes the furnace filter and oils the fan motor before leaving the job.

Service Call 5

A residential customer calls. There is no heat. The furnace is in the basement, emitting a noise that sounds like a clock ticking. The electrode in the pilot has shifted positions and the electronic ignition is arcing and trying to light. It cannot light, but the arc can be heard.

The service technician arrives, goes to the basement, and removes the furnace door. The technician hears the arcing sound. The shield in front of the burner is removed. The technician sees the arc and lights a match near the pilot to see if gas is at the pilot and if the pilot will light. It does. The arcing stops, as it should. The burner lights after the proper time delay. Because there is a gas stream in the vicinity of the arc, the problem must be the electrode.

The technician turns off the power and removes the pilot burner assembly and finds that the screw holding the electrode is loose and that the electrode is slightly out of alignment. The technician looks at the installation start-up manual to find the correct alignment tolerance.

The electrode assembly is aligned and replaced in the furnace, and power is restored. The electronic ignition arcs five times and lights. The technician starts and stops the ignition sequence several times with good results, then changes the filter and oils the motor before leaving the job. This is an added value to the service call for the customer.

Service Call 6

A customer calls and wants an efficiency check on a furnace. This customer thinks that the gas bill is too high for the conditions and wants the entire system checked.

The technician arrives and meets the customer, who wants to watch the complete procedure of the service call. The technician will run an efficiency test on the furnace burners at the beginning and at the end so that the customer can see the difference.

The furnace is in the basement and easily accessible. The technician turns the thermostat to 10°F above the room temperature setting to ensure that the furnace will not shut off during the test. The technician then goes to the truck and gets the flue gas analyzer kit. Being careful not to touch the vent pipe, the technician drills a hole in the flue pipe and inserts a thermometer in the hole. The flue gas temperature is taken and recorded, then a flue gas sample is taken from the same hole after the flue gas temperature stops rising. Room temperature is taken and subtracted from the flue gas temperature to yield the net flue gas temperature.

The flue gas analysis shows that the furnace is operating at 80% efficiency. This is normal for a standard-efficiency furnace. After the test is completed and recorded, the technician checks the temperature across the furnace to see if the air temperature rise is within the amount specified on the furnace nameplate. It is a little high, so the technician asks the home owner if any air registers are shut off. The home owner says that the house has ten rooms and four of the rooms are not heated because the registers are shut off. The technician advises the owner to open the vents in at least two of the ten rooms. The owner does this, and the air temperature rise across the furnace lowers to a more reasonable value.

Note: If the allowable rise is 70°F for a particular furnace and the technician finds a 70°F rise, this rise is too high because it does not allow for the filter to become slightly plugged. A 60°F rise is preferable to be safe.

The technician shuts off the gas to the burner and allows the fan to run and cool the furnace. When the furnace has cooled enough to allow the burners to be handled, the technician takes them out. The burners are easily removed in this furnace by removing the burner shield and pushing the burners forward one at a time while raising the back. They then clear the gas manifold and can be removed. There is a small amount of rust on the burners, which is normal for a furnace in a basement. The draft diverter and flue pipe are also removed so that the technician and customer can see the top of the heat exchanger. All is normal; there is no rust or scale at the top of the heat exchanger.

With a vacuum cleaner, the technician removes the small amount of dirt from the bottom side of the heat exchanger and the burner area. The burners are taken outside, where they are blown out with a compressed air tank at the truck.

After the cleaning is complete, the technician assembles the furnace and tells the customer that his furnace is in good condition and that no difference in efficiency will be seen. Modern gas burners do not plug as badly as the older ones, and the air adjustments will not allow the burner to get out of adjustment more than 2% to 3% at the most.

The technician oils the fan motor and changes the filter, then starts the system and allows it to get up to normal operating temperature. While the furnace is warming up, the technician checks the fan current and finds it to be running at full load. The fan is doing all the work that it can. The efficiency check is run again, and the furnace is still running at approximately 80%. It is clean, and the customer has peace of mind. The technician sets the thermostat before leaving.

Service Call 7

A customer calls. The pilot burner will not stay lit. It goes out a few minutes after it is lit. The heat exchanger has a hole in it close to the pilot burner. The pilot burner will light, but when the fan starts, it blows out the pilot burner.

The technician arrives, checks the room thermostat to see that it is set above the room temperature, then goes to the basement. The standing pilot burner is not lit, so the technician lights it and holds the button down until the thermocouple keeps the pilot burner lit. When the gas valve is turned to on, the main burner lights. Everything is normal until the fan starts, then the flame starts to waver and the pilot burner goes out. The thermocouple cools and the gas to the main burner goes out. The technician shows the customer the hole in the heat exchanger, turns the gas valve off, and explains to the customer that the furnace cannot be allowed to operate in this condition because of the potential danger of gas fumes.

The technician explains to the customer that this furnace is 18 years old and recommends that the customer consider getting a new furnace. The heat exchanger can be replaced in this furnace, but doing so requires much labor plus the price of the new heat exchanger. A more modern furnace is more efficient and a better choice. The customer decides to buy a new furnace.

Service Call 8

A customer calls and reports that the high-efficiency furnace is not heating. The problem is that the power vent fan motor is open circuit and will not run.

The technician arrives, sets the room thermostat to call for heat, then goes to the basement, where the furnace is located, and sees that the power vent fan is not running. The technician removes the burner compartment cover. A check at the low-voltage terminal block shows that there is low voltage. A check from common to W (the terminal for heat) shows that there is a call for heat. Next, the technician checks the voltage across the power vent fan. It should be 120 volts, and it is. Power is going to the fan, but the fan is not running. The technician replaces the vent fan motor. The system is then started and operates correctly. The technician then changes the air filter and oils the furnace fan motor.

Service Call 9

A customer calls and says that the furnace quit operating in the middle of the night last night. The main gas valve has an open circuit. This furnace is a high-efficiency furnace with an electronic circuit board, as shown in Figure 10-16a.

The technician arrives, finds the house cold, and proceeds to the basement, where the furnace is located. The front of the furnace is removed so that the circuit board can be seen. The technician checks the voltage at the thermostat secondary terminals by attaching one meter lead to the SEC-2 and the other to the SEC-1 terminal. The meter reads 24 volts. Next, the technician leaves one lead of the meter attached to the common terminal of the 24-volt circuit (SEC-2 terminal), touches the other meter lead to the gas valve terminal (TP7), and the meter reads 24 volts. The gas valve is energized with 24 volts and will not open. The technician turns off the power, removes the wires from the gas valve, performs a continuity check of the gas valve coil, and finds an open circuit. This is accomplished by using the R×1 scale. The gas valve is defective, and the technician replaces it.

The gas is turned on, the valve is leak checked on the inlet side with soap. The power is turned on again, and the pilot burner circuit energizes and lights the pilot burner. Then the main burner lights. The system seems to be working correctly. The technician then leak checks the main gas valve outlet and finds it leak free. The technician wipes the excess soap from the connections, then shuts off the furnace, changes the air filter, and oils the fan.

Summary

- Troubleshooting gas-burning equipment involves electrical or electronic control problems and mechanical problems.
- Forced-air furnaces and boilers have many of the same types of controls.
- When an electrical problem is encountered, you should start at the power supply and work toward the power-consuming device until you become experienced. You can use the method called *hopscotching*.
- The pilot safety portion of the circuit may be a separate circuit, as with the thermocouple. With electronic ignition, the pilot safety is part of the low-voltage circuit.
- There are two types of fan circuits: temperature controlled and timed. Each operates differently. The temperature-controlled circuit must be set properly or fan short cycles will occur.
- The heat anticipator in the room thermostat must be set correctly for proper furnace operation.
- Electronic controls are mounted on circuit boards and act much like switches in the circuit. You must know the sequence of events and use the hopscotch method of determining which components are defective. The manufacturer will furnish you with the sequence of events and a control checkout procedure if you request them.

Questions

1. How many millivolts should a thermocouple generate when not under a load?
2. Why is a thermal switch often included on the control circuit for a glow-coil pilot burner ignition?
3. What could cause a pilot burner to blow out when the fan starts on a furnace?
4. Name the two types of fan starting device mentioned in the text for gas furnaces.
5. What is the purpose of the burner crossover tubes on a gas burner?
6. What would be the result of misaligned crossover tubes?
7. What must be done when a burner flame burns yellow and it can not be cleared up with the air adjustment?
8. Why is burner alignment important?
9. What is the recommended method for cleaning a pilot burner?
10. Where should the technician start when a power problem is suspected?

11 Theory of Oil Heat and Combustion

Objectives

Upon completion of this unit, you should be able to

- **discuss the origin of fuel oil.**
- **describe the types of fuel oil.**
- **describe some characteristics of No. 2 fuel oil.**
- **describe the process of combustion.**
- **explain the by-products of combustion.**
- **describe a condensing oil furnace.**

11.1 History of Oil Heat

Years ago, animal fats were reduced to oils that were burned in lamps for light and used for heating and cooking. Whale oil, shown in Figure 11-1, was one of the favorite oils because it is clean. However, there was not an abundance of animal fat oils, so other sources of fuel were sought and found. When crude oil was discovered, it became the major source of oils for fuel.

Fuel oils are refined from crude oil and therefore originate at the refineries, much as LP gases do. They are transported across the United States by pipeline and large tank trucks to the various distribution points. The local distributor maintains a large storage of fuel oils that can then be transferred to the local dealer by tank trucks, as shown in Figure 11-2. The local dealer then delivers the fuel oil to the consumer's tank, which may be above or below ground as seen in Figure 11-3. Unlike the LP gases, the oil is moved and stored at atmospheric pressure. The consumer's equipment can then burn the fuel oil, as shown in Figure 11-4. Fuel oil is easy to transport and can be readily converted to heat in a modern oil burner. It is the fuel of choice for many people who live away from the gas mains that usually serve only large-population areas.

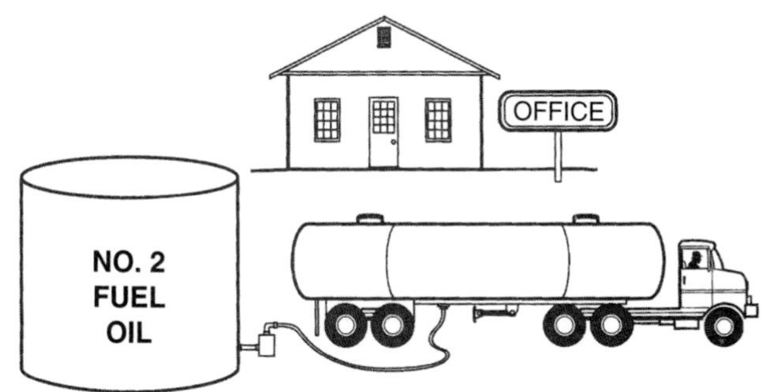

Figure 11-2. The fuel oil distributor maintains a large storage of fuel oil and delivers it to the local dealer in tank trucks.

Figure 11-1. Whale oil was used for heat and light many years ago.

Figure 11-3. The local dealer delivers the fuel oil to the home owner in small trucks to either aboveground or underground storage tanks.

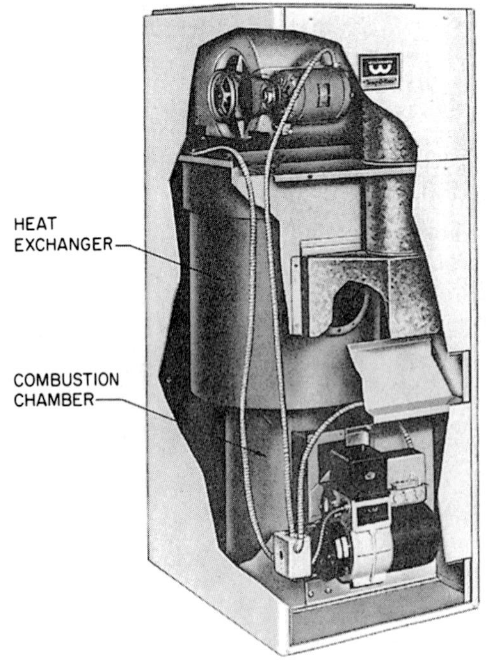

Figure 11-4. The consumer has equipment to burn the fuel oil. *Courtesy The Williamson Company*

11.2 Properties of Fuel Oil

Fuel oil is the only liquid fuel used for residential and light commercial heating. It is refined from petroleum and is called a *distillate* because it is distilled from crude oil. Many other liquid fuels exist, but the others have qualities that make them undesirable as fuel for residences and businesses. For example, gasoline is a liquid fuel, but it is too dangerous to be a heating fuel.

Six fuel oils are in common use: Nos. 1, 2, 4, 5L (light), 5H (heavy), and 6. These fuel oils vary greatly among one another, and only one is used for residential and light commercial heating: No. 2 fuel oil. You need to be aware of the other oils, but you need not have a working knowledge of them for most residential and light commercial work.

The oil is furnished from the refinery, and we accept it as is. The refinery has researched the properties of fuel oil and the best combination of oils for us to use, and it maintains a standard for the fuel oils that is clean and efficient. The properties of oil determine the grade and the reason a particular fuel oil is chosen for a specific application. The *Fundamental Handbook* by the American Society of Heating, Refrigerating and Air-Conditioning Engineers (ASHRAE) covers these properties in detail. They are heating value, carbon and hydrogen content, flash point, viscosity, carbon residue, water and sediment content, pour point, ash content, and distillation quality.

Heating Value

The heating value of the oils varies from approximately 135,000 Btu per gallon for No. 1 oil to approximately 153,000 Btu per gallon for No. 6 fuel oil. No. 2 fuel oil has a heating range of 137,000 to 141,800 Btu per gallon. Most people round this off to 140,000 Btu per gallon for calculating the heat value of this oil, as depicted in Figure 11-5. All of this heating value is not usable, as discussed later in this unit in Section 11.4.

Carbon–Hydrogen Content

Like the fuel gases, fuel oils are made of carbon and hydrogen. With the fuel gases, a chemical formula tells us the carbon–hydrogen relationship. Because fuel oils vary so much, formulas are not furnished. No. 2 fuel oil contains approximately 85% carbon and 15% hydrogen. The more carbon the oil contains, the more heating value it contains. The higher-number oils, such as No. 6, may contain 88% carbon. These oils are also called *heavy oils* because they weigh more than the other oils.

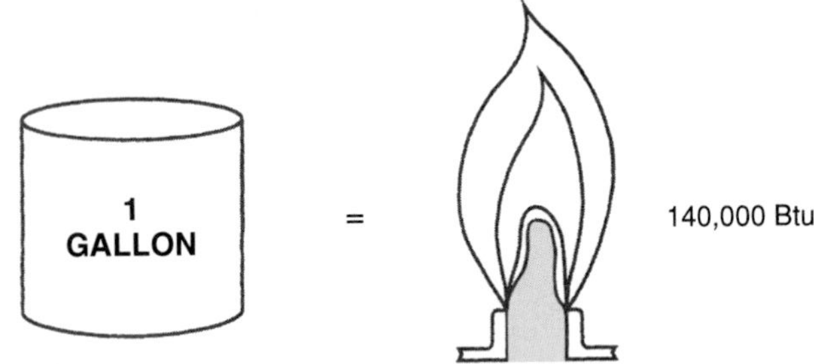

Figure 11-5. One gallon of No. 2 fuel oil has a heating capacity of 140,000 Btu.

For example, No. 2 fuel oil weighs from 6.960 to 7.296 pounds per gallon, and No. 6 oil weighs 8.053 to 8.448 pounds per gallon, as shown in Figure 11-6.

Flash Point

The flash point involves the storage and handling of the fuel oil. All liquid fuels have a maximum safe-storage-and-handling temperature. The *flash point* is the lowest temperature at which vapors above the fuel oil ignite in air when exposed to a flame. The lighter the oil, the lower the flash point, and the lower the maximum storage temperature must be. The storage of heavy oils is not as critical as that of the lighter oils because the flash point of the heavy oils is much higher.

Viscosity

Viscosity is the thickness of the oil under normal temperatures. This thickness determines the size of the metering devices used to govern the amount of fuel flowing to an oil burner. The heavier oils are thicker and require larger metering devices at the same temperatures as the lighter oils. Viscosity is expressed in Saybolt Universal Viscosity seconds (SUS), which describe how much oil will drip through a calibrated hole at a certain temperature. As the temperature is decreased for a particular oil, it becomes thicker, and less oil will drip through the calibrated orifice, as depicted in Figure 11-7. The viscosities of different oils are shown in Figure 11-8.

Carbon Residue

A carbon residue test can be performed on an oil sample by vaporizing a known weight of the oil in a process in which oxygen is absent. The remaining carbon residue is measured. Properly burned fuel oil does not leave any appreciable carbon residue.

Water and Sediment Content

The fuel oil should be refined and delivered with a minimum of water and sediment. Too much sediment plugs filters and strainers. Water in the oil causes poor flame characteristics and burning problems. It also corrodes the inside of the fuel containers and pipelines that are made of ferrous metals (iron or steel). This corrosion loosens and lodges in the filters and strainers, and in colder climates, water turns into ice and blocks the piping. The refinery, the distributor, and the dealer are responsible for caring for and delivering clean, dry fuel oil. As discussed in Unit 13, the service technician can also help protect the system from water after it is installed.

Pour Point

Pour point is the lowest temperature at which the fuel can be stored and handled. No. 2 fuel oil is one of the lower pour-point fuels, meaning that it can be used at lower temperatures and stored outdoors. In the northern parts of the United States, where temperatures get bitterly cold, the storage container can be located inside the structure or underground. The heavier fuel oils are more sensitive to cold and become thick when exposed to low temperatures. Some of them nearly turn into the consistency of grease and require heat in order to flow.

Ash Content

The ash content of fuel oil indicates the amount of noncombustible materials in the fuel oil. These materials pass through the flame without burning and are contaminants. They also can be abrasive and wear down burner components. The refinery is responsible for keeping the ash content within the required tolerances. If there is any doubt about the ash content of an oil, an independent laboratory can analyze the oil if you send a sample to be checked.

Distillation Quality

Oil must be turned into a vapor before it can be burned. The distillation quality describes the ability of the oil to be vaporized.

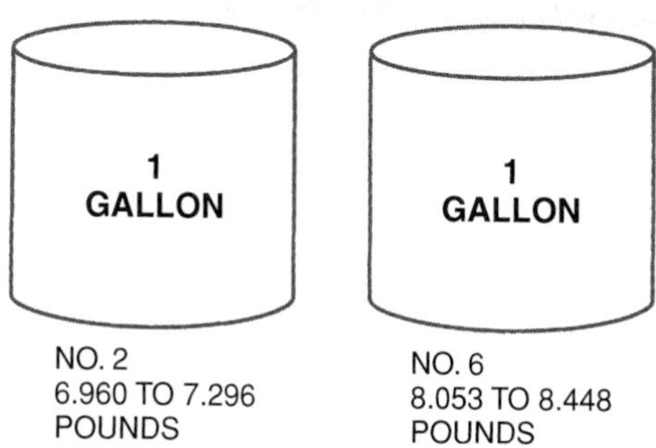

Figure 11-6. Comparison of weights of different fuel oils.

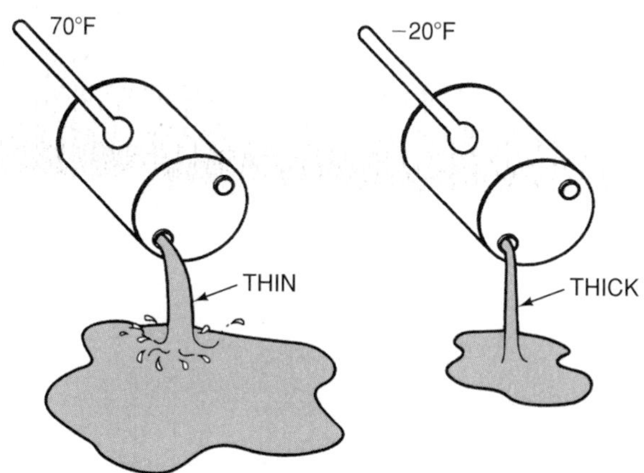

Figure 11-7. As the temperature decreases, oil becomes thicker.

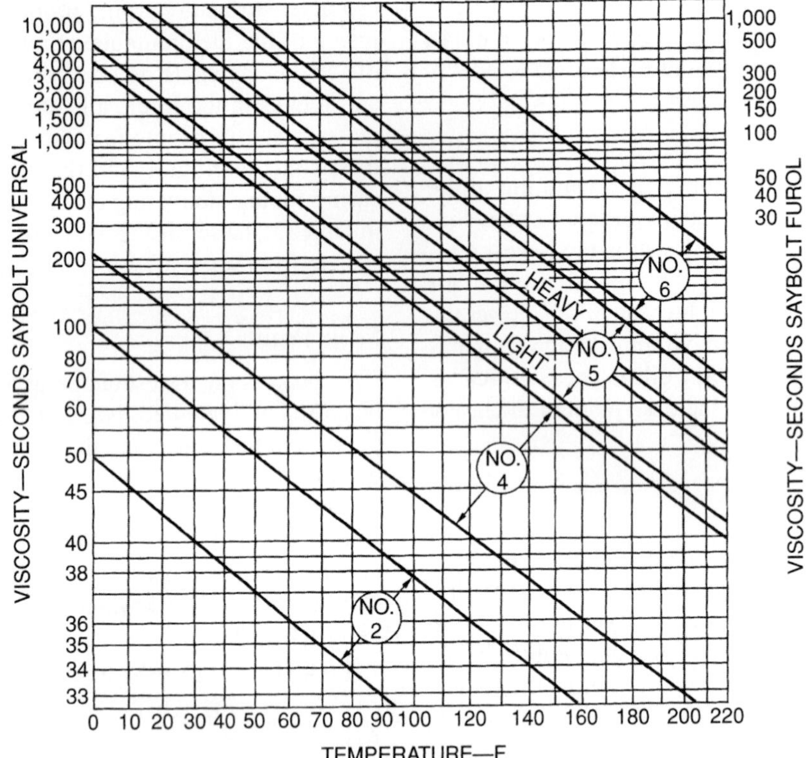

Figure 11-8. This graph shows the different viscosities of oils. *Courtesy American Society of Heating, Refrigerating and Air-Conditioning Engineers*

11.3 The Combustion Process

Combustion is the burning of a substance. It is sometimes called *rapid oxidation*, which is combining fuel with oxygen while ignition occurs. Oil combustion is difficult because liquid does not burn. Therefore, liquid oil must be converted into a vapor before it can burn. Fuel oils are constantly vaporizing at the surface of the liquid, as shown in Figure 11-9. Lighter oils vaporize faster than heavy oils, as shown in Figure 11-10.

Pot Burners

The early oil burners used lighter oil and burned the vapors from the top of a small puddle of oil. This type of burner was called a *pot burner*. The flame on the puddle heated the oil in the puddle to vaporize more oil, and a large flame was produced above the puddle, as shown in Figure 11-11. This type of oil heater used a float mechanism called a *carburetor* to meter the oil to the burner.

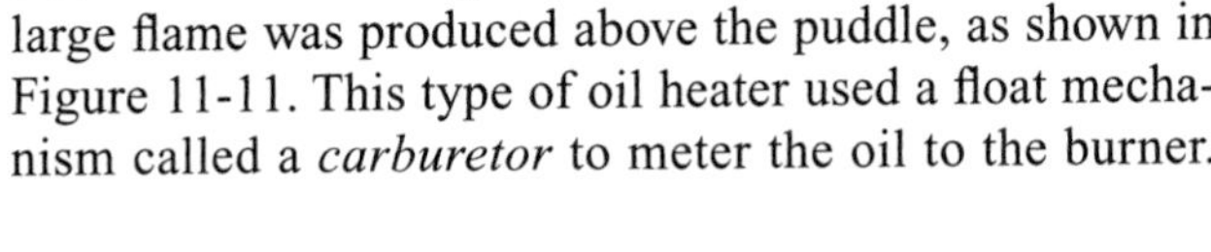

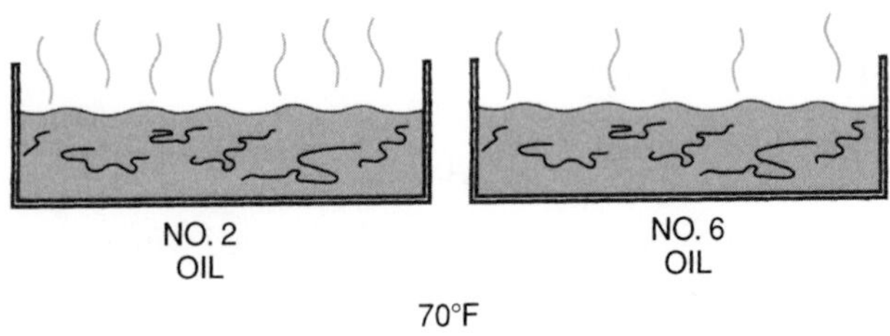

Figure 11-10. Lighter oils vaporize faster than heavy oils.

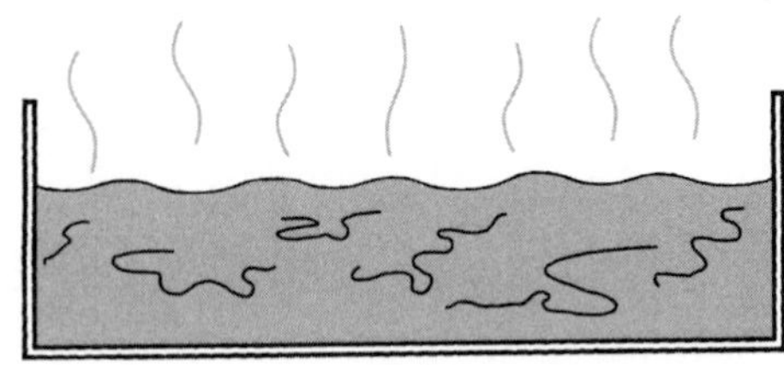

Figure 11-9. Fuel oil constantly changes into a vapor at the surface of the oil.

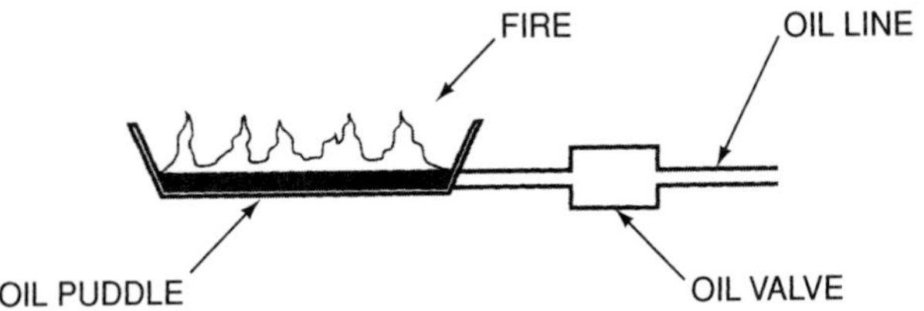

Figure 11-11. The flame in a pot burner burns above the oil puddle.

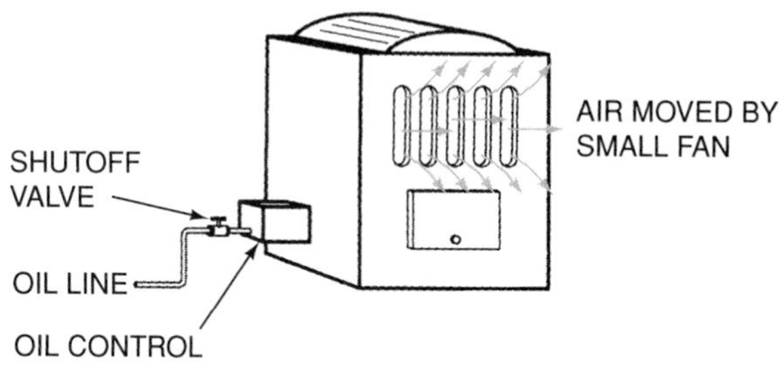

Figure 11-12. A room oil heater with a pot burner.

This method is still used for some room oil heaters, as seen in Figure 11-12. However, the pot burner is limited in size to the area of the puddle of oil.

Gun Burners

More modern types of oil burners use power to break the oil droplets into smaller particles. This process is called *atomizing* the oil and is similar to a high-pressure water hose with a nozzle at the end that lets the water out through a small hole. A fog is formed as the water is atomized. Figure 11-13 shows a typical oil nozzle. Because oil is transferred and stored at atmospheric pressure, power is required to force the oil out the nozzle. This power is supplied by an electric motor, shown in Figure 11-14, that turns a high-pressure pump.

Figure 11-14. The power to atomize oil is supplied by an electric motor. *Courtesy W.W. Grainger, Inc.*

Oxygen must be mixed with the oil for burning, so the pump motor also drives a small fan and air is mixed with the droplets of oil for a better mixture of oil and gas, as depicted in Figure 11-15. This type of oil burner is called a *gun burner* and is the most common burner used today. Figure 11-16 shows a typical gun burner.

There are two types of gun burners: low pressure and high pressure. The low-pressure burner was used for many years and is not as popular now as the high-pressure burner is.

(A)

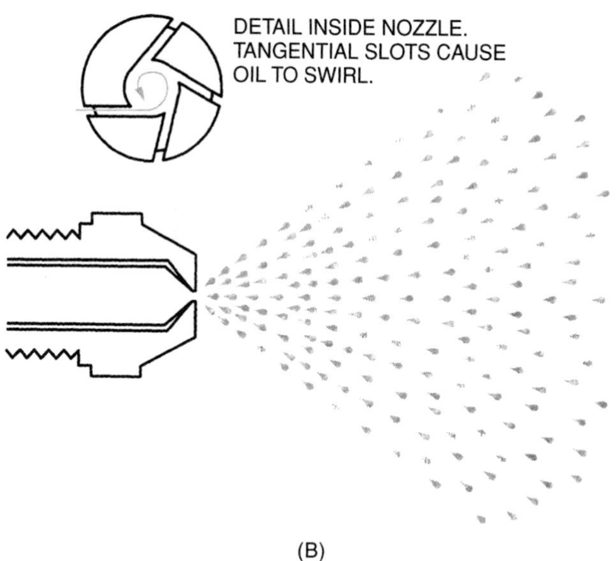

(B)

Figure 11-13. (A) The oil is atomized by the oil nozzle into a fog. *Courtesy Delavan Corporation* (B) Detail of the nozzle.

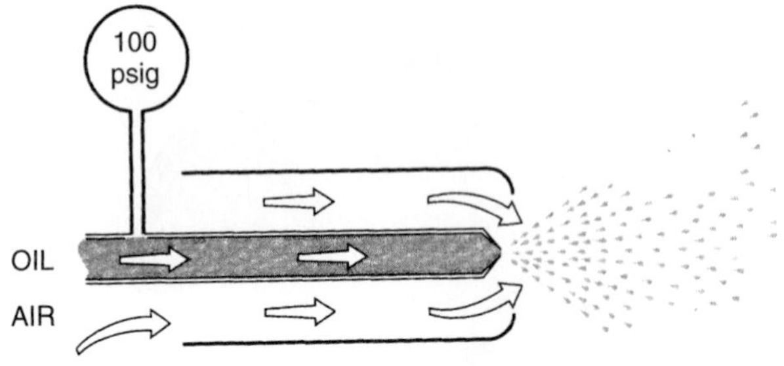

Figure 11-15. Air is mixed with the oil droplets for burning.

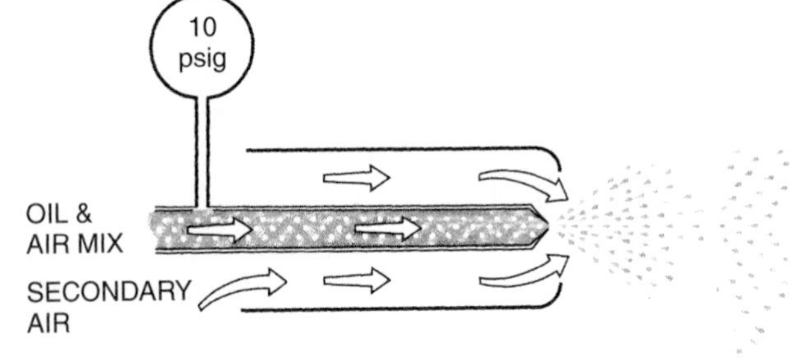

Figure 11-17. Low-pressure oil burner.

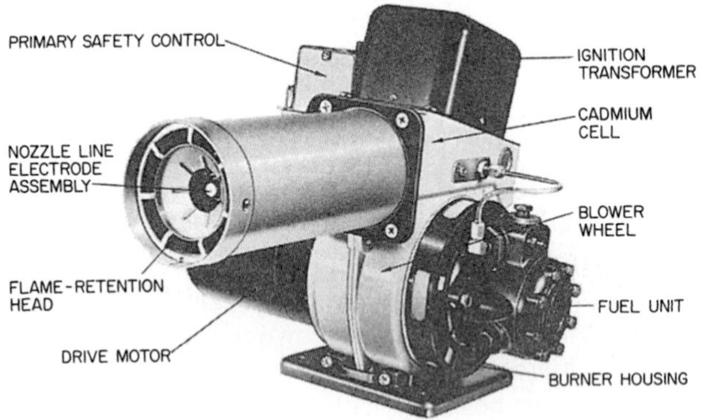

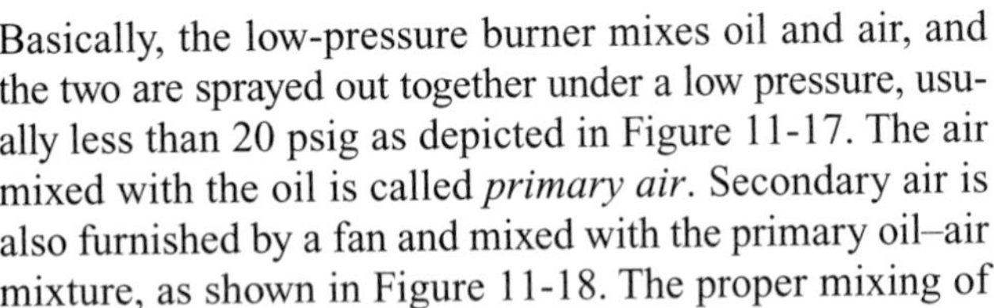

Figure 11-16. Typical gun burner. *Courtesy R.W. Beckett Corporation*

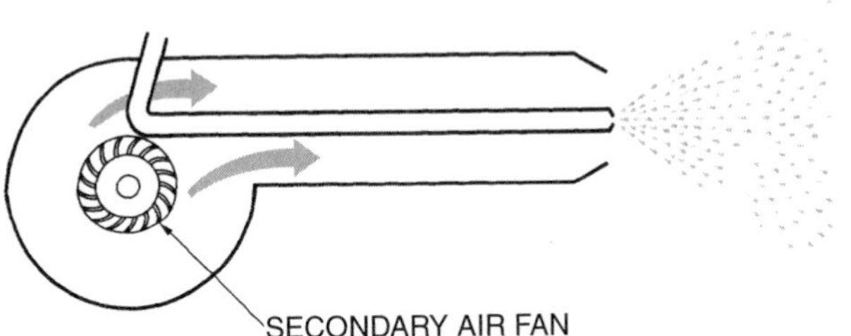

Figure 11-18. Secondary air is furnished by a fan to be mixed with the primary oil–air mixture.

Basically, the low-pressure burner mixes oil and air, and the two are sprayed out together under a low pressure, usually less than 20 psig as depicted in Figure 11-17. The air mixed with the oil is called *primary air*. Secondary air is also furnished by a fan and mixed with the primary oil–air mixture, as shown in Figure 11-18. The proper mixing of air and oil in the low-pressure burner makes it a complicated application.

The high-pressure gun burner is by far the most used burner today and is less complicated than the low-pressure gun burner. Figure 11-19 shows an exploded view of this type of gun burner and its components. Each component is discussed next.

The small, fractional-horsepower *motor* turns the fan and fuel pump. It is mounted on the burner by an end

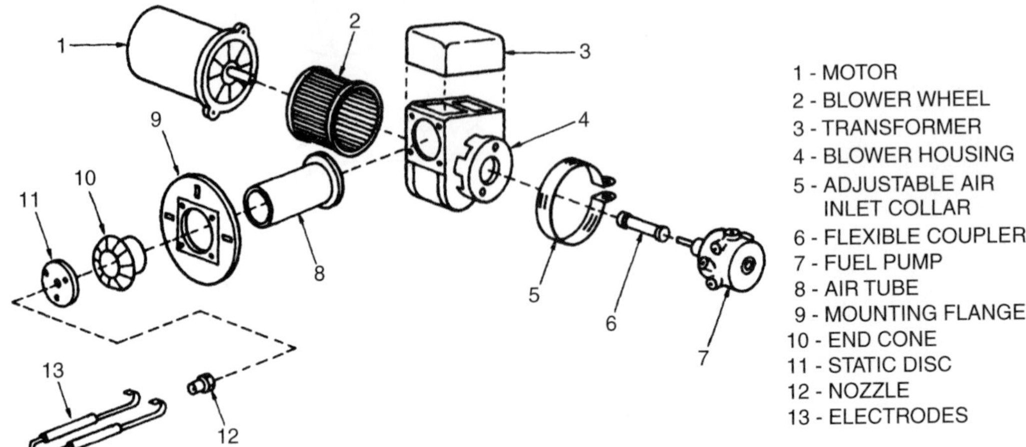

Figure 11-19. Exploded view of a high-pressure burner.

flange, as depicted in Figure 11-20. The shaft protrudes into the burner assembly through the blower wheel and has a flexible coupler (or coupling) on the end. The unloaded speed of this motor is either 1,800 or 3,600 revolutions per minute (RPM). When turning the fan and pump, it slows slightly to approximately 1,750 or 3,450 RPM.

The *fan wheel* is fastened to the motor shaft as shown in Figure 11-21 and turns at the same speed as the motor. This fan wheel is called a *squirrel-cage fan wheel* because it resembles the treadmill device used in a squirrel cage. This fan can create pressure and force air into the air tube of the burner for combustion.

The *flexible coupler* extends the motor shaft to the shaft of the fuel pump and turns the pump as shown in Figure 11-22. The shaft is not always in perfect alignment, therefore the flexible coupler helps prevent vibration.

The *fuel pump* is a rotary pump that can pump high pressure, demonstrated in Figure 11-23. This pressure is normally limited to 100 psig. An internal spring-loaded bypass allows some oil to circumvent to the pump inlet in order to limit the pressure to the nozzle. Several types of these oil pumps, including single-stage and two-stage pumps, are used. They are discussed in more detail in Unit 12.

The high pressure of 100 psig forces the oil through a calibrated *oil nozzle*, as shown in Figure 11-24. The nozzle prepares the oil to be mixed with the air by breaking the oil into tiny droplets that vaporize easily. It also meters the oil in the correct amount and, as the oil leaves the nozzle, it is set into a swirling pattern to help it mix with the air. This pattern can be adjusted to fit the particular application. Figure 11-25 shows a typical nozzle and its swirl chamber. Oil nozzles are also discussed in more detail in Unit 12.

The oil must be ignited to burn. A high-voltage *transformer* supplies approximately 10,000 volts at a very low amperage to furnish an arc for combustion. Although the voltage is high, it is not dangerous because of the low amperage. It is much like the voltage at the spark plug on an automobile; it will give you an electrical shock, but it will not injure you. The transformer is mounted on the gun burner as part of the assembly, as shown in Figure 11-26.

The voltage from the transformer is supplied to *electrodes* that perform much the same function as the spark plug in an automobile. They control the spark and ignite the oil. Figure 11-27 shows two electrodes.

The *blower housing* holds the blower wheel and directs the airflow into the air tube. It is mounted as part of the gun burner assembly as shown in Figure 11-28.

The *adjustable air inlet collar*, shown in Figure 11-29, is used to regulate the air to the blower. This is the only air adjustment for combustion and is important for correct combustion.

The *air tube* surrounds the oil-nozzle assembly. Air from the blower is forced down the air tube under a slight pressure from the fan wheel as depicted in Figure 11-30.

The *static disc* in the air tube creates a resistance that backs up the air. The backed up air creates static pressure

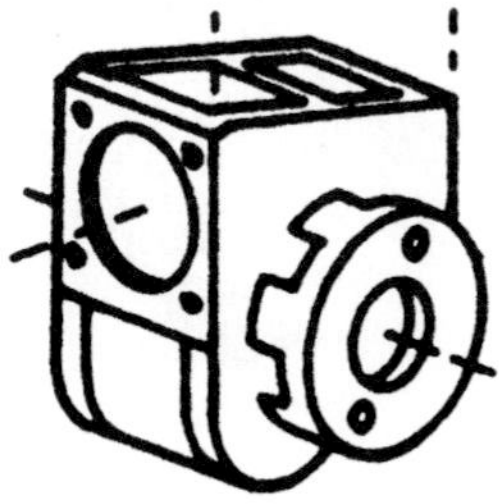

Figure 11-20. The motor is mounted by an end flange to the burner housing.

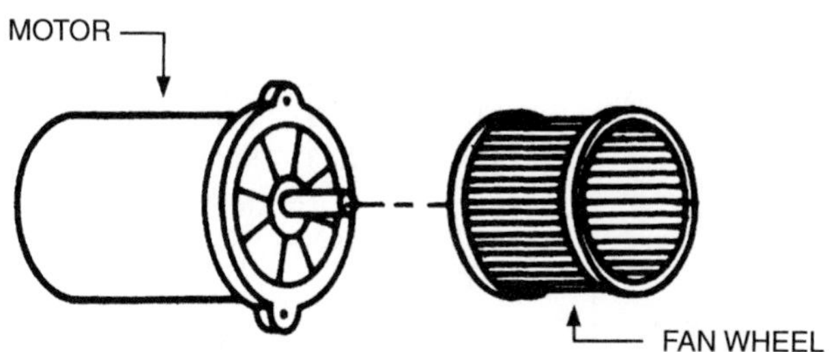

Figure 11-21. The fan wheel is fastened to the motor shaft and turns at the same speed as the motor.

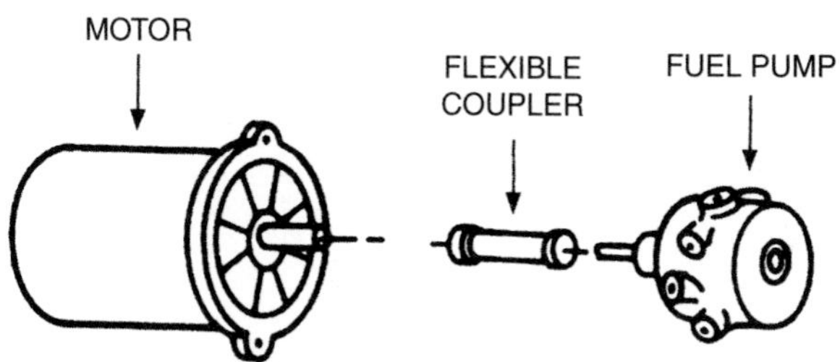

Figure 11-22. Flexible coupler.

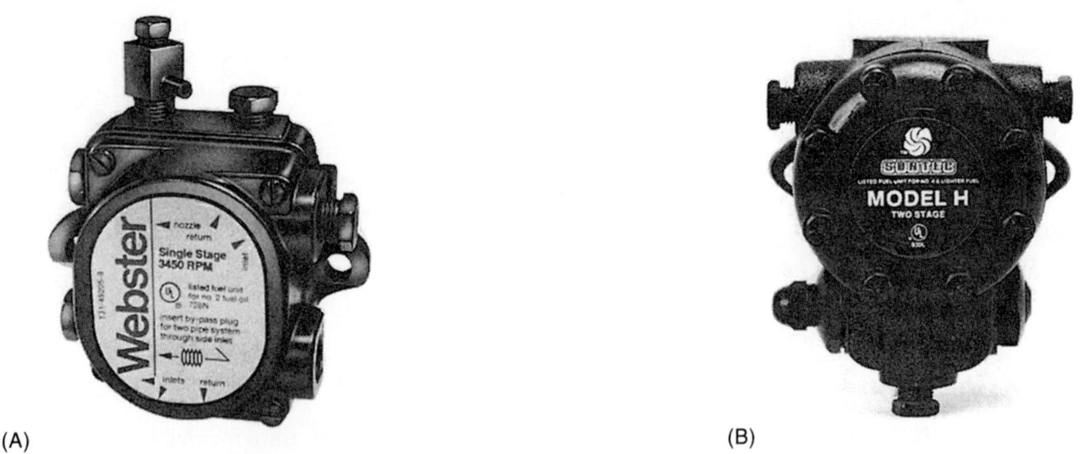

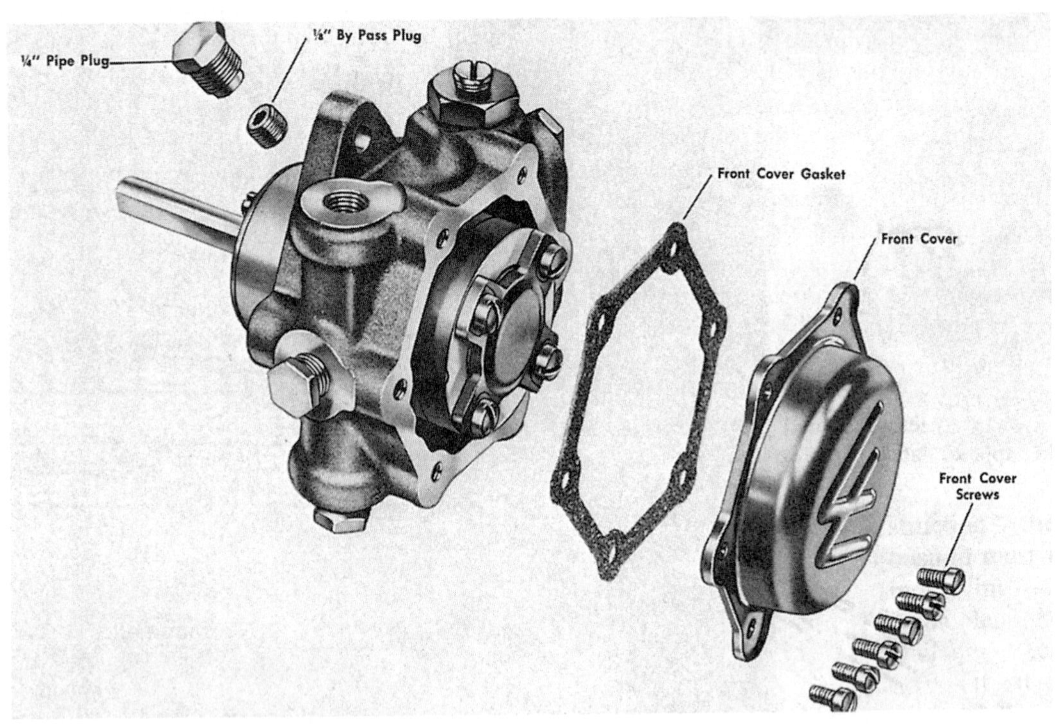

Figure 11-23. (A) Single-stage high-pressure fuel pump. *Courtesy Webster Electric Company* (B) Two-stage high-pressure fuel pump. *Courtesy Suntec Industries Incorporated, Rockford, Illinois* (C) Fuel pump parts. *Courtesy Webster Electric Company*

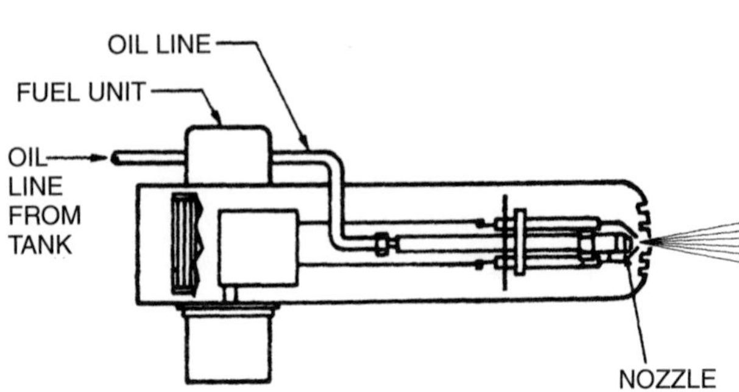

Figure 11-24. High pressure forces the oil through the nozzle. *Courtesy Honeywell, Inc., Residential Division*

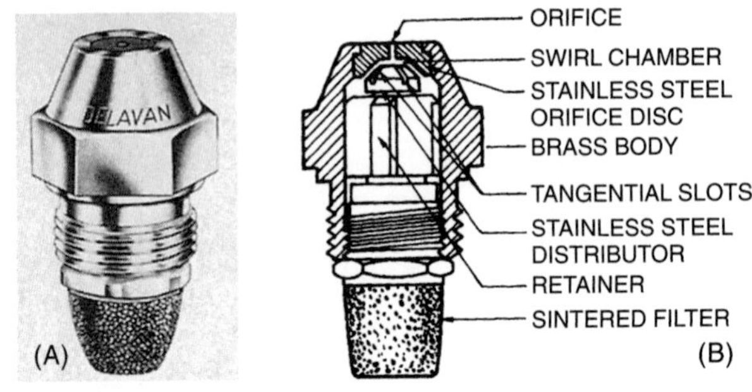

Figure 11-25. The nozzle furnishes the oil in a swirling pattern to mix with the air. (A) A typical nozzle. *Courtesy Delavan Corporation* (B) Cross section of a nozzle.

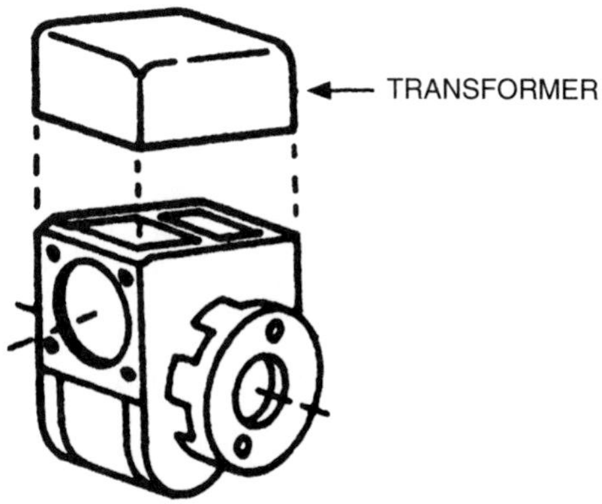

Figure 11-26. The transformer is mounted on the burner assembly and furnishes approximately 10,000 volts for ignition.

Figure 11-27. The electrodes control the spark for the fire.

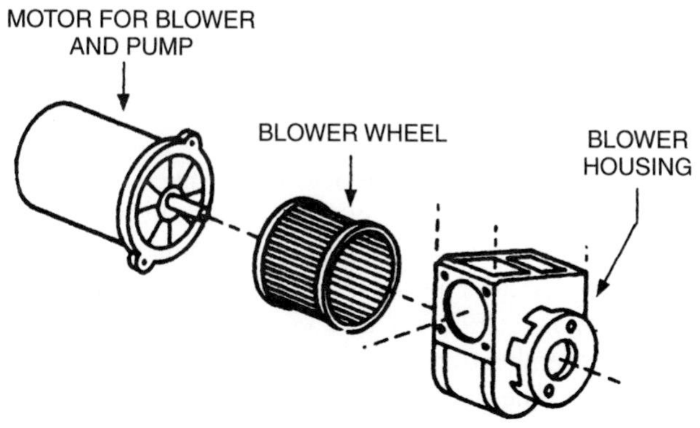

Figure 11-28. The blower housing holds the blower wheel and gathers the air to force it down the air tube.

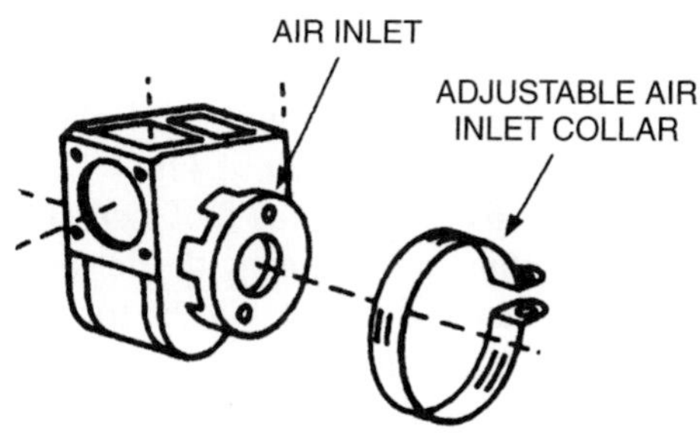

Figure 11-29. The adjustable air inlet collar is used to regulate the air to the blower.

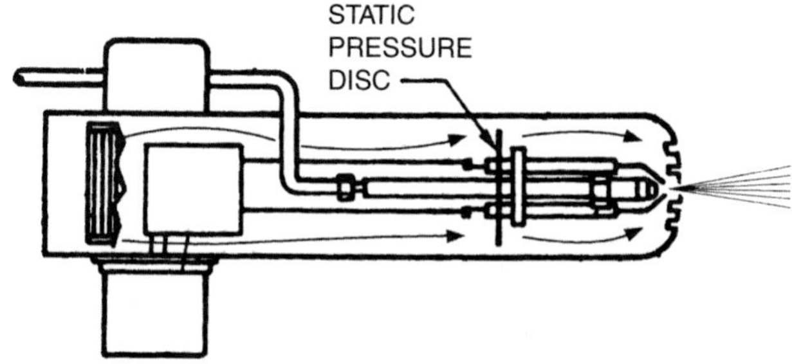

Figure 11-30. The air tube surrounds the oil-nozzle assembly and feeds the air to the end cone.

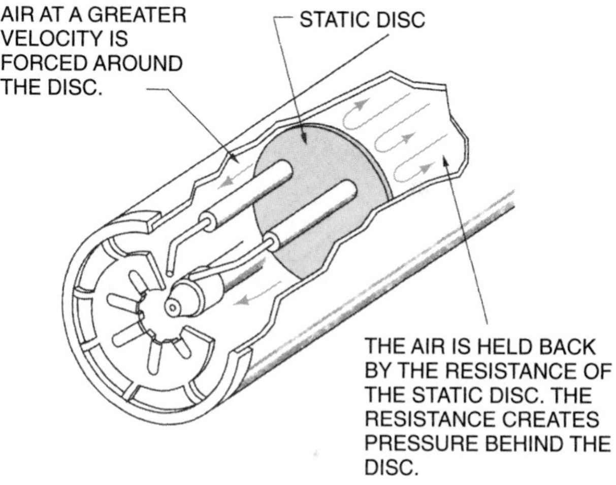

Figure 11-31. For good combustion, the static disc holds the air back and turns the velocity of the air to static pressure for mixing with the oil.

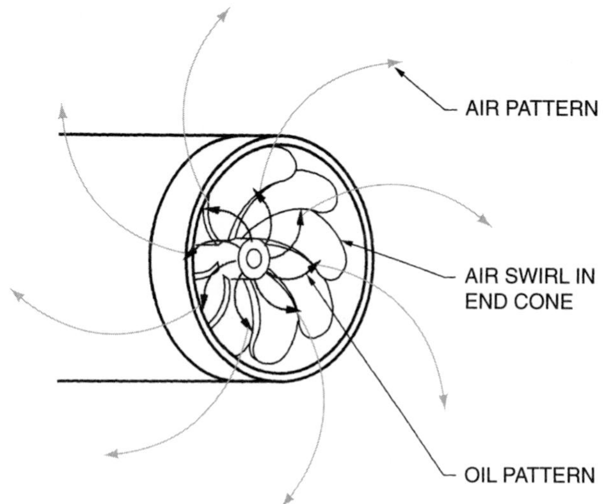

Figure 11-32. The end cone swirls the air into the swirling oil for the best oil–air mixture.

and increases the air velocity for mixing the air with the atomized oil. Figure 11-31 shows how a static disc operates.

The *end cone* is the last component before the actual fire. It swirls the air from the air tube to mix it with the atomized oil droplets, as shown in Figure 11-32. The more activity at the end of the burner, the better the oil droplets will be aerated and the better they will burn. This activity must be controlled in order to hold the flame in a correct pattern.

The *mounting flange* attaches the burner assembly to the oil-burning appliance. It is designed to allow the service technician to remove the complete gun burner assembly from the appliance for servicing as shown in Figure 11-33.

The Combustion Chamber

Combustion for an oil appliance must take place in a controlled compartment called a *combustion chamber.* This chamber controls the shape of the fire and ensures that the maximum benefit is derived from the fuel oil. Combustion chambers are made of *refractory materials*, such as fire brick and stainless steel. The combustion chamber receives the atomized oil and air mixture as it swirls out of the burner head and keeps the fire contained until nothing is left but the hot by-products of combustion. The flame must not impinge on any of the surfaces in the combustion chamber. If the flame strikes or touches any surface, it cools, some of the oil condenses, and smoke occurs. Smoke is a result of unburned fuel and causes soot deposits in the system. Figure 11-34 shows a combustion chamber. The refractory material gets red-hot and helps keep the oil and air mixture ignited by reflecting heat into the mixture, as depicted in Figure 11-35.

Flame Color

The flame from a gun burner is usually bright yellow-orange and has distinctive tips. The flame should appear to hold together and not appear to be lazy, as shown in Figure 11-36. This flame can be viewed through a port in case it needs adjustment. However, adjusting a burner using only flame characteristics is not recommended. Professional technicians also conduct a lengthy analysis of the combustion by-products and perform a smoke density test. These tests are discussed in Unit 14.

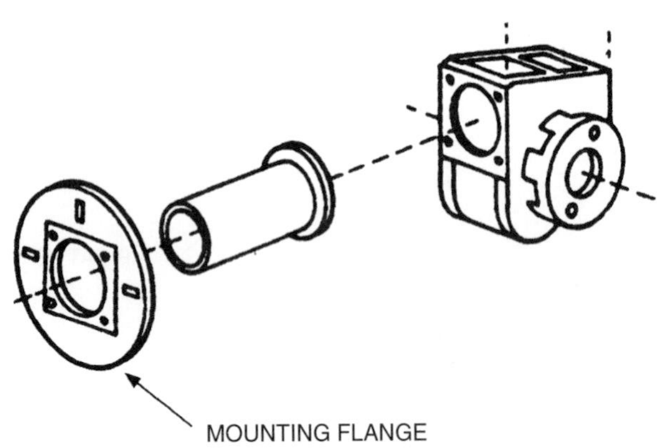

Figure 11-33. The mounting flange holds the gun burner to the appliance for easy removal and servicing.

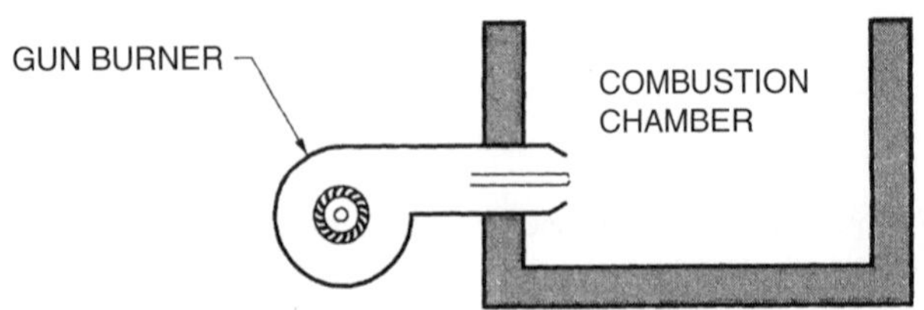

Figure 11-34. The combustion chamber burns the fuel efficiently and does not allow the flame to touch a cool surface.

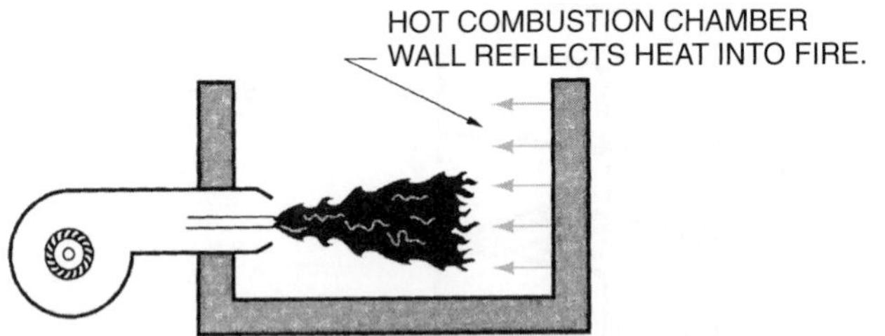

Figure 11-35. The combustion chamber reflects heat into the burning oil to help heat the mixture.

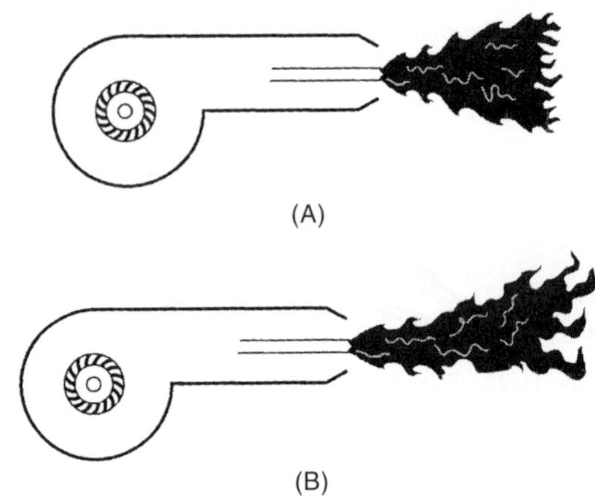

Figure 11-36. The flame should be (A) firm and not (B) lazy.

11.4 By-Products of Combustion

The ratio of fuel oil to air must be correct for efficient combustion. The fuel to be burned is made of carbon and hydrogen, just as the fuel gases are. When 1 pound of fuel oil is burned, it must be mixed with 14.4 pounds of air. This is nearly 192 cubic feet of air (14.4 lb × 13.33 cu ft/lb at standard conditions = 191.95 cu ft of air). This amount is for perfect combustion and, as with the fuel gases, it would be dangerous to set up a burner for exactly the correct amount of air. If the air were restricted for some reason, incomplete combustion would occur. Practical combustion requires excess air to ensure that there is enough oxygen to come in contact with all the carbon and hydrogen particles for complete combustion. Again, as with the gas burner, the air supplied to an oil burner is normally in excess of the exact amount needed. Most oil burners are set up to burn with approximately 50% excess air.

When 1 pound of oil with 50% excess air is burned, 21.6 pounds, or 288 cubic feet, of air, must be furnished with the 1 pound of oil (21.6 lb × 13.33 cu ft/lb of standard air = 287.9 cu ft of air). One gallon of No. 2 fuel oil weighs approximately 7 pounds, so 1 pound of oil is one-seventh of a gallon; this is .88 of a pint, or nearly the contents of a soda can. Figure 11-37 depicts complete combustion.

When the correct amounts of oil and air are mixed and the mixture is ignited (with electrodes and a transformer), the result is a fire that is contained in the combustion chamber. Heat is given off to the appliance for use in heating. The typical oil-burning appliance uses nearly 75% of the heat energy, with approximately 25% escaping with the flue gases. When a gallon of fuel oil contains 140,000 Btu of heat, 105,000 Btu are used for heating and the remaining 35,000 Btu push the by-products of combustion up the flue (140,000 × .75 = 105,000, and 140,000 × .25 = 35,000). See Figure 11-38.

The by-products of combustion result from the chemical reaction in which the stored energy in the fuel oil is released as heat. As with the gas burners, the nitrogen in the air passes through the combustion process. Of the 288 cubic feet of air that are used with 1 pound of fuel oil, 227.5 cubic feet are nitrogen (288 × .79 = 227.5 cu ft of nitrogen).

The excess air contains oxygen that also passes through the combustion process and is not combined with the carbon and hydrogen particles of the fuel oil and burned. Of the 288 cubic feet of air, the combustion process uses only 192 cubic feet, so 96 cubic feet pass through and are only heated. This air contains 21%, or 20 cubic feet, of oxygen (96 cu ft × .21 = 20 cu ft).

As with the fuel gases, the other by-products of combustion are CO_2 and water vapor. The flue gases contain approximately 27 cubic feet of CO_2 and more than a pound of water vapor per pound of fuel. The flue gases must remain hot enough to prevent the water vapor from condensing and hot enough to rise up the flue to the outside. Figure 11-39 shows a detailed example of combustion.

The flue gases from fuel oil furnaces contain the by-products just mentioned when *complete* combustion occurs. If incomplete combustion occurs, carbon monoxide (CO) and smoke form. Carbon monoxide is toxic even in small amounts. Fortunately, when an oil burner is not burning correctly, the smoke will probably enter the living space and alert you to the presence of CO.

The preceding discussion is an overview of combustion. Combustion analysis for oil appliances is discussed in detail in Unit 14.

Condensing the By-Products of Combustion

The flue gas temperature of a fuel oil appliance can easily be 500°F. The flue gases contain approximately 288 cubic feet of 500°F combustion by-products and a pound of water vapor for each pound of fuel burned. If the heat from these by-products is removed, condensation will occur, and this heat can be used to increase the efficiency of the appliance. Most oil burners approximately 75% efficient, so any improvement is desirable. Some manufacturers designed and built a condensing oil furnace that is approximately 96% efficient. It has a heat exchanger in the return air of the furnace that lowers the flue gas temperatures and allows condensation to occur. The water is then piped down the drain. Figure 11-40 shows this high-efficiency furnace. Reducing the flue gas temperature simplifies venting the appliance and provides a more efficient installation.

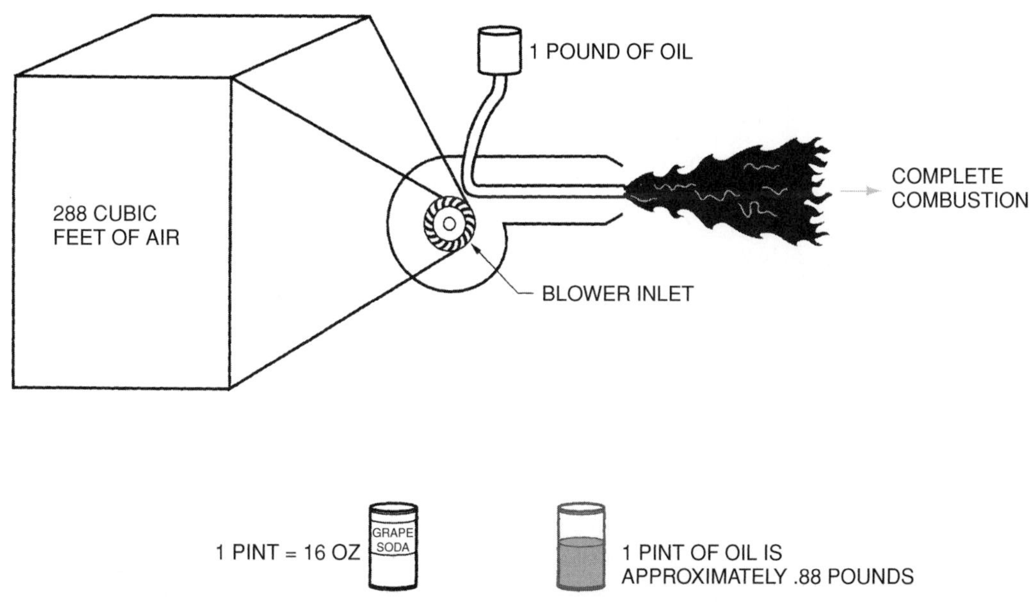

Figure 11-37. One pound of fuel oil combined with approximately 288 cubic feet of air will ensure complete combustion.

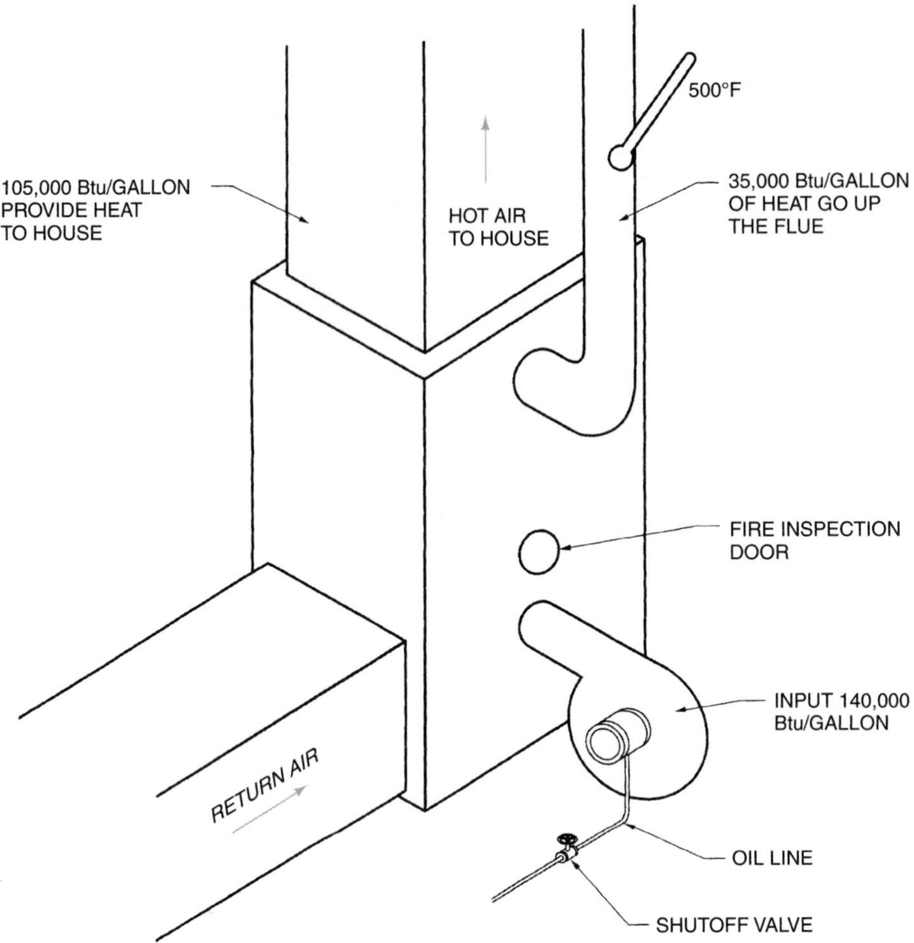

Figure 11-38. In the typical fuel oil appliance, 105,000 Btu heat the structure and 35,000 Btu escape with the by-products of combustion for each gallon of oil burned.

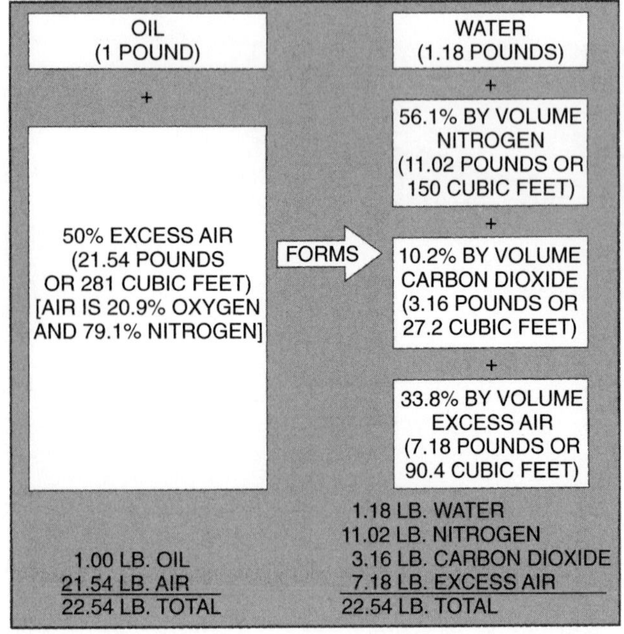

Figure 11-39. Combustion and its by-products. *Courtesy R.W. Beckett Corporation*

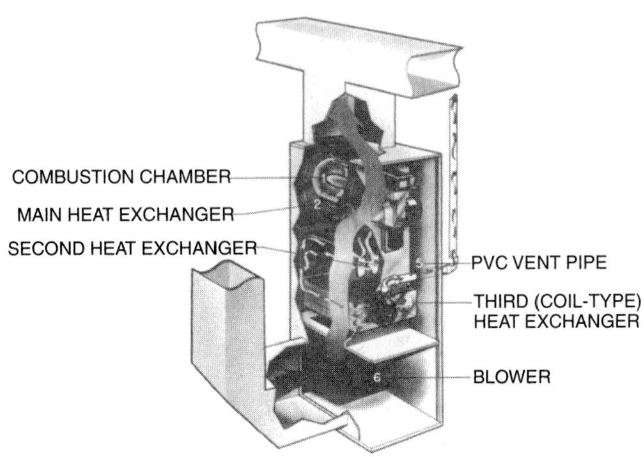

Figure 11-40. Condensing oil furnace. *Courtesy Yukon*

Summary

- Animal fats were the first oils used for heat and light.
- There was not an abundance of animal fats, so other oils were needed.
- Fuel oils as we use them today are refined from crude oil, so they originate at the refinery.
- Fuel oil is moved across the United States through pipelines and in large tank trucks, then is delivered to homes in small delivery trucks.
- Fuel oil is stored and transported at atmospheric pressure, unlike the pressurized LP gases.
- Fuel oil is stored in aboveground or underground tanks. In extremely cold climates, it is sometimes stored inside the structure.
- Fuel oil is easy to transport and store and is the fuel of choice for many people who live away from the gas mains.
- Fuel oil is called a distillate fuel because it is distilled from crude oil.
- The six fuel oils are Nos. 1, 2, 4, 5L (light), 5H (heavy), and 6.
- The higher the number of the oil, the greater the heating value because of a higher carbon content. For example, No. 2 oil has a heat content of approximately 140,000 Btu per gallon, whereas No. 6 has 153,000 Btu per gallon.
- The carbon content is approximately 85% for No. 2 oil and approximately 88% for No. 6 oil.
- The flash point concerns the temperatures for safe storage and handling.
- The viscosity describes how thick the oil is at different temperatures and is expressed in Saybolt Universal Viscosity seconds (SUS).
- When a portion of oil is distilled without oxygen, the amount of carbon residue can be determined. This residue has nothing to do with soot deposits because a properly burned oil sample has no appreciable carbon in the by-products of combustion.
- The refinery should keep the water and sediment content of oil low. Water causes corrosion in the system's ferrous metal parts (iron and steel). This corrosion loosens and becomes sediment, which plugs the strainers and filters. Water also freezes in the winter and blocks the pipes.
- The pour point is the lowest temperature at which the fuel can be stored and handled. All oils become thicker as the temperature becomes colder. Some of the heavier oils must be heated in order to use them in the winter.
- The ash content describes the amount of noncombustible materials in the oil. These materials pass through the fire unburned and become air pollutants. They are also abrasive and can wear out burner parts.
- The distillation quality of the oil describes the ease with which the oil can be vaporized.
- Combustion, also called rapid oxidation, is burning a substance by combining it with oxygen and igniting it.
- Fuel oil is a liquid, and liquids do not burn. It must be vaporized before it can burn.
- Vaporization of oil requires heat or power.
- The first type of burner ignited the vapors above a puddle of oil. As the vapors burned, they heated the puddle and warmed the oil in the puddle, which created more vapors. A float maintained the level of oil in the puddle.
- These first burners were called pot burners and were limited to the size of the oil puddles.
- More efficient gun burners were later developed.
- In the low-pressure gun burner, the fuel and primary air are mixed at the nozzle at less than 20 psig; this mixture is further mixed with secondary air.
- High-pressure gun burners furnish the oil at the nozzle at 100 psig, where the oil is atomized and swirled and combined with a swirling air mixture. The swirling permits the oil droplets to be better aerated.
- The oil is atomized by means of an oil pump that is driven by a small, fractional-horsepower motor that turns at 3,450 or 1,750 RPM.
- The motor also turns the fan for furnishing the air.
- The air is captured in the fan housing, where it is directed down the air tube. The static disc in the air tube backs up the air and creates static pressure. This pressure increases the velocity of the swirling air.
- A flexible coupler is fastened between the motor and the pump shafts to help in case alignment is not perfect.
- For ignition, the oil–air mixture is passed by the electrode tips, where a high-voltage arc (10,000 volts) is furnished by the transformer.
- The end cone of the burner swirls the air into the swirling oil at the nozzle.
- The mounting flange fastens the burner to the appliance so that the burner can be easily removed for servicing.
- The burning of the oil, or combustion, takes place in the combustion chamber.
- The combustion chamber contains the flame in a small area and ensures that all the oil is burned without touching any surface. If the oil touches a surface, the flame cools and soot results.
- An oil flame should be bright yellow-orange with a firm, not lazy, pattern.
- Oil flames should not be set and adjusted by flame characteristics alone. A detailed combustion analysis is necessary for professional results.
- The fuel oil and air must be in the correct proportions for efficient combustion.
- For perfect combustion of one pound of oil, 192 cubic feet of air are required. Perfect combustion is not practical.
- Practical combustion requires excess air. When approximately 50% of excess air for one pound of fuel is furnished, 288 cubic feet of air are used.
- The by-products of complete combustion are 227.5 cubic feet of nitrogen, 20 cubic feet of oxygen, 27 cubic feet of CO_2, and approximately 1 pound of water vapor.

- Heat pushes the flue gases out of the appliance and keeps the water in a vapor state.
- Most oil-burning appliances are approximately 75% efficient; therefore, when one gallon of oil is burned (140,000 Btu/gal), 105,000 Btu are used as heat and 35,000 Btu go up the flue with the by-products of combustion.
- Flue gas temperature can easily be 500°F with a volume of 288 cubic feet of by-products for one pound of oil (one-seventh of a gallon). If any of this heat could be saved, it would add to the efficiency of the appliance.
- The efficiency of an appliance can be improved to nearly 96% by cooling the flue gases and condensing the water vapor.

Review Questions

1. What elements compose No. 2 fuel oil, and what are their percentages?
2. Will liquid fuel oil burn? If not, what must happen so that it can burn?
3. What is the approximate weight of a gallon of No. 2 fuel oil?
4. How does flash point affect the choice of fuel oil?
5. Viscosity is measured in what unit?
6. In what two ways does water in the oil affect the system?
7. How does pour point affect the choice of fuel oil?
8. What does excess ash content do to the system and the atmosphere?
9. From where does excess ash content originate?
10. Who controls the quality of fuel oil?
11. In what two ways can oil be vaporized for burning?
12. At what pressure does the modern gun burner operate?
13. What is combustion?
14. What are four by-products of combustion?
15. What is the voltage of a typical oil-burner transformer?
16. What is used to ignite the oil?
17. In which component does the fire burn in an oil-burning appliance?
18. The oil nozzle serves what three purposes?
19. What is the color of a typical oil flame?
20. What is the efficiency of a typical oil-burning appliance?

12 Oil Heat Components

Objectives

Upon completion of this unit, you should be able to

- **state how oil is used for fuel.**
- **describe the components of a gun burner.**

12.1 Oil Heat

You should have a complete understanding of Unit 11, Theory of Oil Heat and Combustion, before studying this unit.

Oil heat is popular in many areas because it is reliable, simple, and available in locations away from gas mains. Like LP gases, it may be more expensive than natural gas where natural gas is available, but it is much less expensive than electric heat. Heat pumps are a choice in many locations away from gas mains, but these systems are usually applied only where cooling is needed in the summer. Therefore, oil is often the fuel of choice, particularly in the North.

Concentrated in one gallon of No. 2 oil are 140,000 Btu of heat. As with gas heat, the goal is to remove as much heat as practical from the oil and to transfer it into air or water to be circulated throughout a structure. We discuss forced-air furnaces and boilers in this unit.

12.2 Combustion and Heat Exchange

In Unit 11, combustion is described as rapid oxidation, or burning, when fuel and oxygen are combined and a flame is lit. This process occurs in the combustion chamber of an oil appliance. The combustion chamber contains the fire and aids in combustion by reflecting the heat into the fire to help ignite unburned oil. Because of this, the combustion chamber must quickly become hot for the burner to start up smoothly. With proper combustion, the chamber glows red-hot. The burner pump pressurizes the liquid oil into tiny droplets, and these droplets must remain airborne until they burn completely. If any of them touch a combustion chamber wall, they cool and cause incomplete combustion and smoke.

The shape of the combustion chamber and the shape of the flame must be compatible for best combustion. The shape of the flame is discussed in Unit 11. Combustion chambers can be made from different materials and are available in several shapes. They can be square, round, triangular, or rectangular, as shown in Figure 12-1. When the chamber does not fit the flame, such as in a rectangular chamber, the corners should be filled to prevent currents (called *eddy currents*) from forming in them. The oil droplets are not likely to reach these corners and create the heat required to keep the corners hot enough to reflect heat.

The combustion chamber must be, within reason, the correct size. If it is too small, as shown in Figure 12-2,

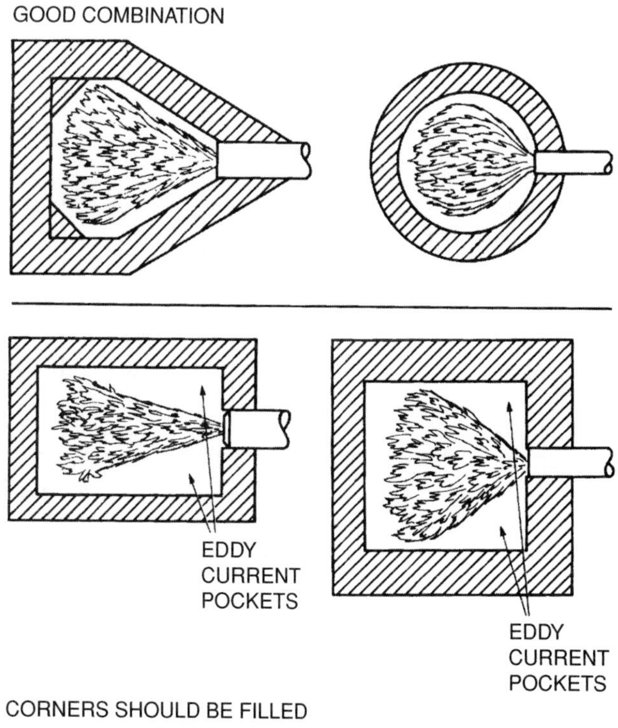

Figure 12-1. Combustion chambers can be square, round, triangular, or rectangular. *Courtesy R.W. Beckett Corporation*

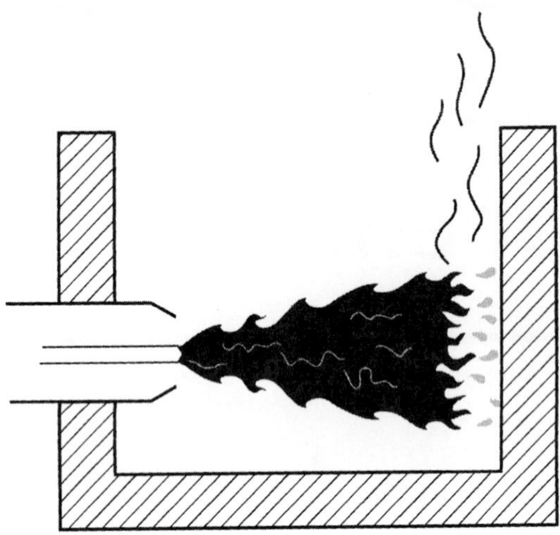

Figure 12-2. If unburned oil droplets hit the combustion chamber walls, they cool and may not ignite. They can then turn into smoke.

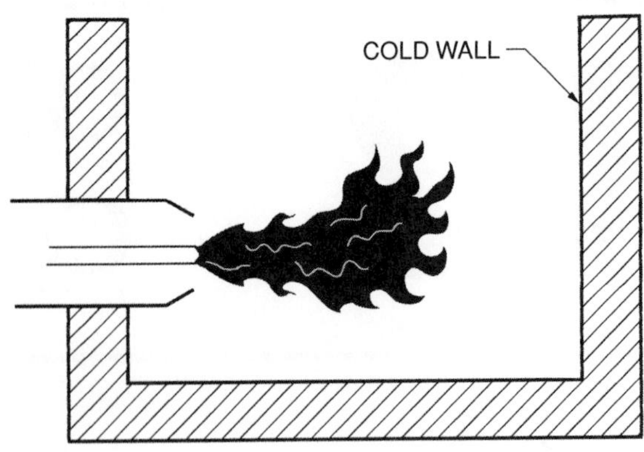

Figure 12-3. If the combustion chamber is too large, it does not reflect heat into the flame.

unburned oil droplets touch the sides, cool, and cause smoke. If the chamber is too large, as shown in Figure 12-3, the surfaces are too far away to reflect heat into the flame, and complete combustion is less likely. Most combustion chambers have 80 to 90 square inches of base surface area per gallon of oil to be burned. Figure 12-4 shows the relative sizes of combustion chambers. The height of the combustion chamber is also important because it should be tall enough to aid in complete combustion but not too tall. Figure 12-5 shows some combustion chamber dimensions that are helpful to the technician.

1 Firing Rate (GPH)	2 Length (L)	3 Width (W)	4 Dimension (C)	5 Suggested Height (H)	6 Minimum Dia. Vertical Cyl.
0.50	8	7	4	8	8
0.65	8	7	4.5	9	8
0.75	9	8	4.5	9	9
0.85	9	8	4.5	9	9
1.00	10	9	5	10	10
1.10	10	9	5	10	10
1.25	11	10	5	10	11
1.35	12	10	5	10	11
1.50	12	11	5.5	11	12
1.65	12	11	5.5	11	13
1.75	14	11	5.5	11	13
2.00	15	12	5.5	11	14
2.25	16	12	6	12	15
2.50	17	13	6	12	16
2.75	18	14	6	12	18

NOTES:

1. Flame lengths are approximately as shown in column (2). Often, tested boilers or furnaces will operate well with chambers shorter than the lengths shown in column (2).
2. As a general practice any of these dimensions can be exceeded without much effect on combustion.
3. Chambers in the form of horizontal cylinders should be at least as large in diameter as the dimension in column (3). Horizontal stainless steel cylindrical chambers should be 1 to 4 inches larger in diameter than the figures in column (3) and should be used only on wet base boilers with non-retention burners.
4. Wing walls are not recommended. Corbels are not necessary although they might be of benefit to good heat distribution in certain boiler or furnace designs, especially with non-retention burners.

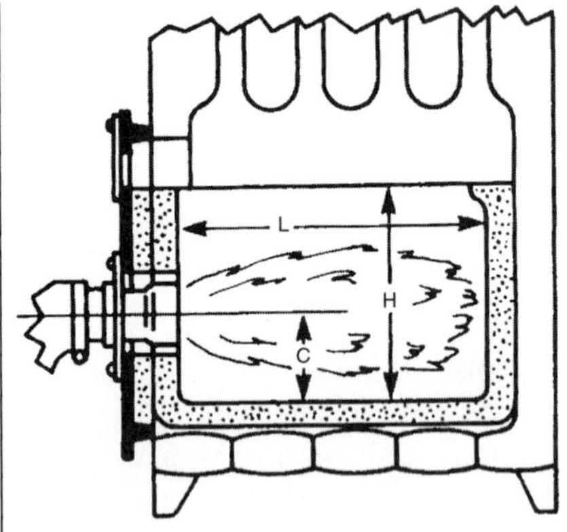

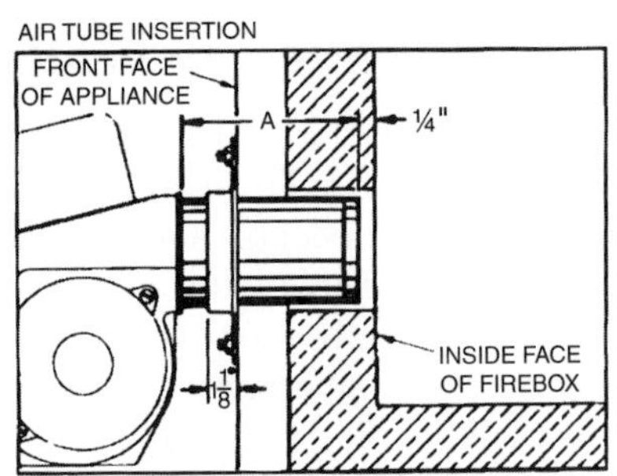

"A" = Usable air tube length.
The burner head should be ¼" back from the inside wall of the combustion chamber. Under no circumstances should the burner head extend into the combustion chamber. If chamber opening is in excess of 4⅜", additional set back may be required.

Figure 12-4. This chart shows the relative sizes of combustion chambers and burner positioning. *Courtesy R.W. Beckett Corporation*

	Oil Consumption G.P.H.	Sq. Inch Area Combustion Chamber	Square Combustion Chamber Inches	Dia. Round Combustion Chamber Inches	Rectangular Combustion Chamber Inches	HEIGHT FROM NOZZLE TO FLOOR INCHES			
						Conventional Burner Wth. x Lgth.	Conventional Burner Single Nozzle	Sunflower Flame Burner Single Nozzle	Sunflower Flame Burner Twin Nozzle
80 Sq. In. Per Gal.	.75	60	8 x 8	9	.....	5	x	5	x
	.85	68	8.5 x 8.5	9	.....	5	x	5	x
	1.00	80	9 x 9	10⅛	.....	5	x	5	x
	1.25	100	10 x 10	11¼	.....	5	x	5	x
	1.35	108	10½ x 10½	11¾	.....	5	x	5	x
	1.50	120	11 x 11	12⅜	10 x 12	5	x	6	x
	1.65	132	11½ x 11½	13	10 x 13	5	x	6	x
	2.00	160	12⅝ x 12⅝	14¼	6	x	7	x	
	2.50	200	14¼ x 14¼	16	12 x 16½	6.5	x	7.5	x
	3.00	240	15½ x 15½	17½	13 x 18½	7	5	8	6.5
90 Sq. In. Per Gal.	3.50	315	17¾ x 17¾	20	15 x 21	7.5	6	8.5	7
	4.00	360	19 x 19	21½	16 x 22½	8	6	9	7
	4.50	405	20 x 20		17 x 23½	8.5	6.5	9.5	7.5
	5.00	450	21¼ x 21¼		18 x 25	9	6.5	10	8
100 Sq. In. Per Gal.	5.50	550	23½ x 23½	Round Combustion Chambers usually not used in these sizes	20 x 27½	9.5	7	10.5	8
	6.00	600	24½ x 24½		21 x 28½	10	7	11	8.5
	6.50	650	25½ x 25½		22 x 29½	10.5	7.5	11.5	9
	7.00	700	26½ x 26½		23 x 30½	11	7.5	12	9.5
	7.50	750	27¼ x 27¼		24 x 31	11.5	7.5	12.5	10
	8.00	800	28¼ x 28¼		25 x 32	12	8	13	10
	8.50	850	29¼ x 29¼		25 x 34	12.5	8.5	13.5	10.5
	9.00	900	30 x 30		25 x 36	13	8.5	14	11
	9.50	950	31 x 31		26 x 36½	13.5	9	14.5	11.5
	10.00	1000	31¾ x 31¾		26 x 38½	14	9	15	12
	11.00	1100	33¼ x 33¼		28 x 29½	14.5	9.5	15.5	12.5
	12.00	1200	34½ x 34½		28 x 43	15	10	16	13
	13.00	1300	36 x 36		29 x 45	15.5	10.5	16.5	14
	14.00	1400	37½ x 37½		31 x 45	16	11	17	14.5
	15.00	1500	38¾ x 38¾		32 x 47	16.5	11.5	17.5	15
	16.00	1600	40 x 40		33 x 48½	17	12	18	15
	17.00	1700	41¼ x 41¼		34 x 50	17.5	12.5	18.5	15.5
	18.00	1800	42½ x 42½		35 x 51½	18	13	19	16

Figure 12-5. These combustion chamber dimensions are helpful to the technician. *Courtesy R.W. Beckett Corporation*

The manufacturer establishes the dimensions of a furnace or a boiler, and they are not usually changed. However, the technician needs to know where to find the tables of dimensions in order to troubleshoot a particular piece of equipment. For example, suppose that a technician is having trouble with ignition on a furnace or a boiler. Suppose that it is noisy on start-up for a few minutes, and the customer objects to this noise. The technician can use the tables as a guideline to find the problem. Manufacturers do not always design heat exchangers correctly, so sometimes the technician finds incorrect dimensions.

In the past, combustion chambers were made of fire brick or a similar hard refractory material, as shown in Figure 12-6.

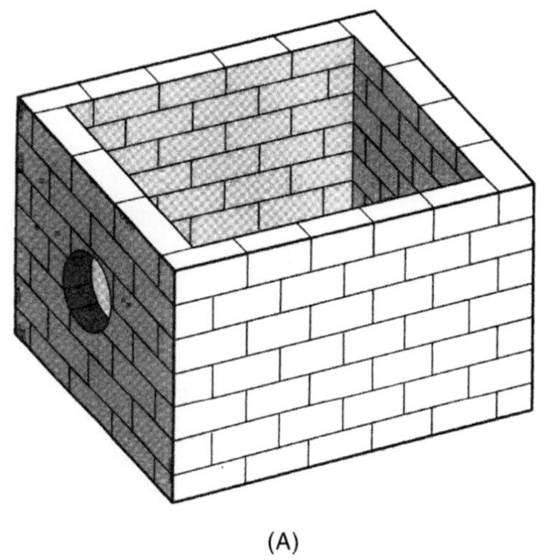

(A)

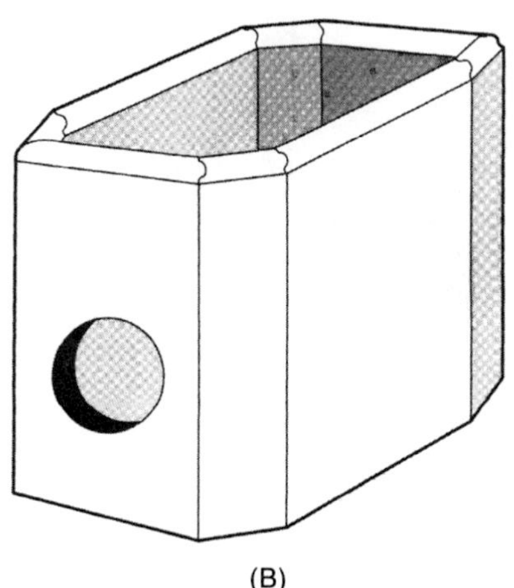

(B)

Figure 12-6. Common combustion chamber materials in the past were (A) fire brick and (B) hard refractory material.

This material worked well for combustion chambers, but it was heavy and fragile when shipped. Stainless steel combustion chambers are now used in some applications. Combustion chambers made of soft fiber refractory material are also used. Figure 12-7 shows an example of each. These materials are much lighter, so the chamber is less expensive for the manufacturer to ship and handle.

Often a combustion chamber must be repaired. A soft, wet blanket-type material can be used to repair many combustion chambers. This material looks like a wet blanket; however, it is not wet with water. The blanket can be installed through the burner opening and laid over the existing heat-exchanger outline for shape. The furnace is then fired and the blanket becomes strong, as shown in Figure 12-8. The refractory material of this blanket quickly becomes hot, so it is a good repair material.

The heat exchanger, which surrounds the combustion chamber and contains the flue gases as they move toward the flue, is responsible for exchanging heat between the flue gases and the medium to be heated, either air or water. Two kinds of heat transfer into the heat exchanger; heat by conduction and radiant heat from the yellow-orange flame to the walls of the heat exchanger, as shown in Figure 12-9.

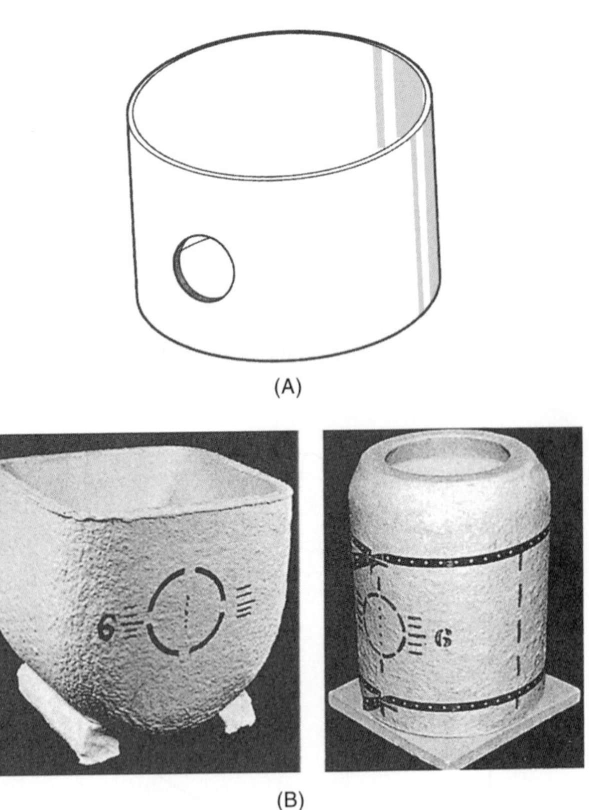

Figure 12-7. (A) Stainless steel combustion chamber. (B) Combustion chambers made of soft refractory material. *Courtesy R.W. Beckett Corporation*

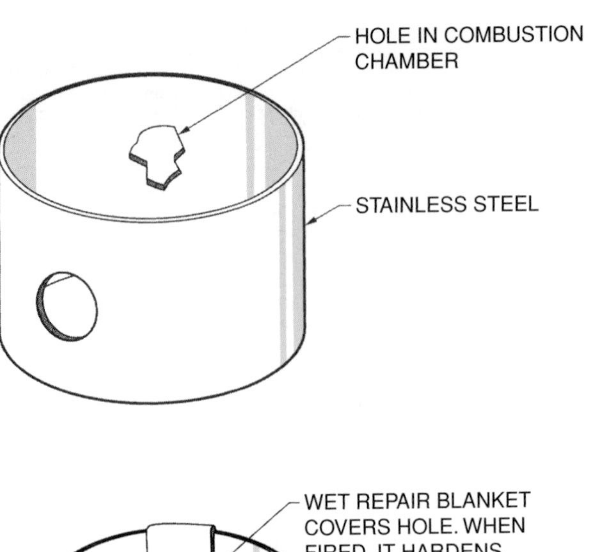

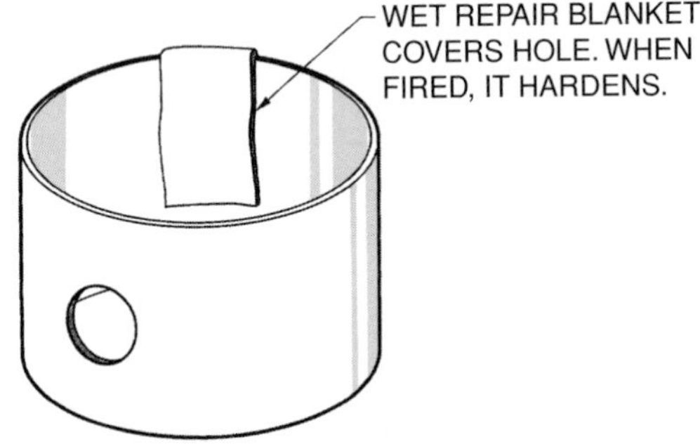

Figure 12-8. Blanket type of repair for combustion chambers. Blanket is molded into place and then heated to harden it.

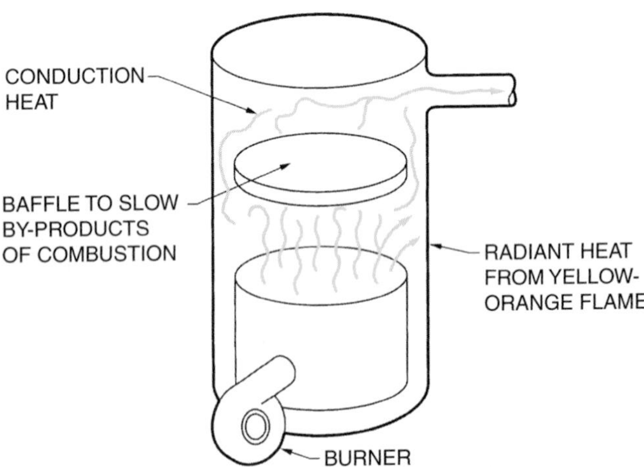

Figure 12-9. Two kinds of heat transfer into the heat exchanger: radiant and conduction.

Heat exchangers for forced-air furnaces are steel. They are welded or folded at the seams to prevent the by-products of combustion from mixing with the air circulating around the heat exchanger. Figure 12-10 shows two styles of heat exchangers. Regardless of the construction of the heat exchanger, the by-products of combustion

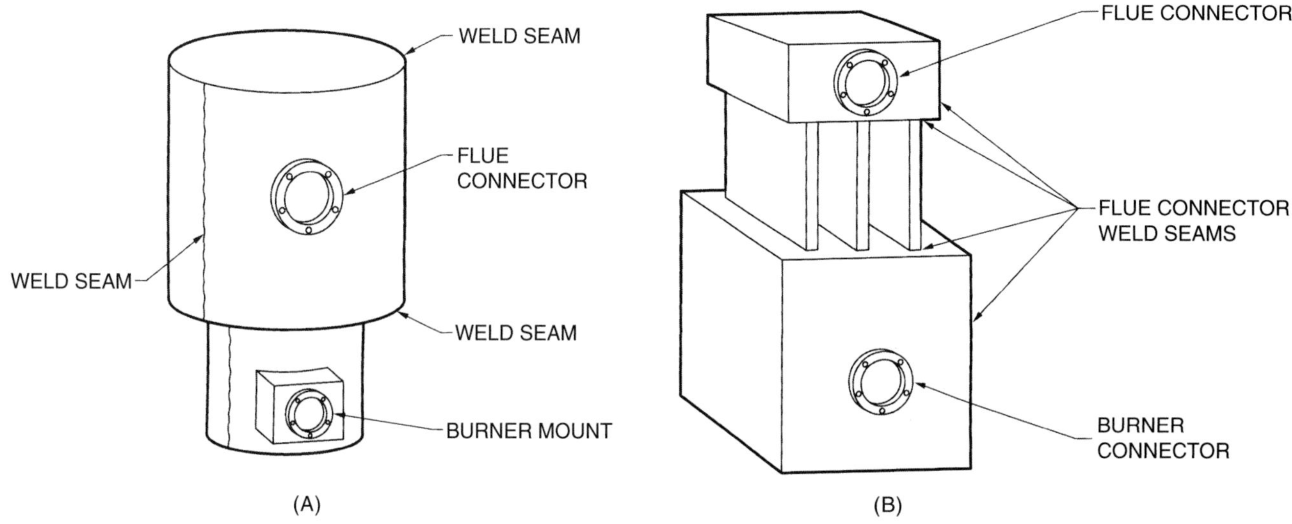

Figure 12-10. Steel heat exchangers are formed and welded. (A) Round heat exchanger. (B) Square heat exchanger.

must not be allowed to mix with the air that moves into the conditioned space. If combustion is incomplete, the by-products contain carbon monoxide, which is toxic. Unlike gas-appliance flue gases, which have a mild odor, flue by-products from oil-burning appliances have a stronger odor. They smell like burning fuel oil and may contain smoke.

The heat exchanger is usually in a positive pressure when the system fan is running because the fan is located before the heat exchanger, as seen in Figure 12-11.

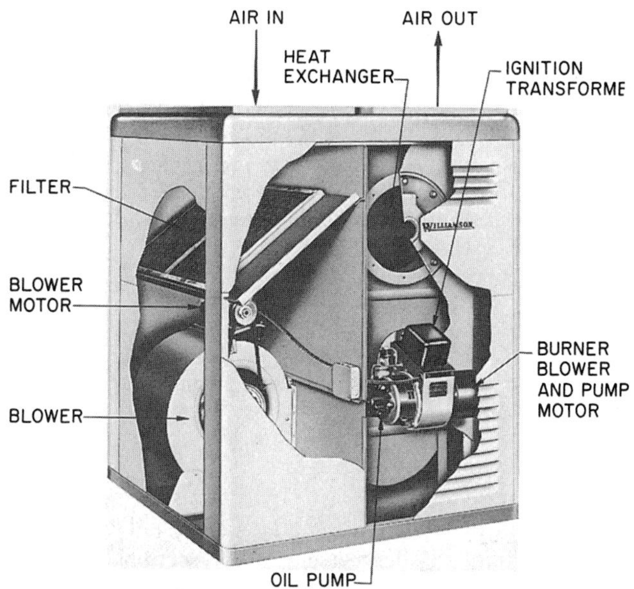

Figure 12-11. The fan side of the typical heat exchanger is in a positive pressure when the fan is running. *Courtesy The Williamson Company*

Because of this positive pressure, when a heat exchanger is cracked, fumes from the by-products of combustion are unlikely while the fan is running. More likely, air will enter the combustion chamber and possibly disturb combustion. At the beginning of the cycle, when the burner is operating without the fan, fumes from combustion can exit through a crack to the air side of the heat exchanger. Then, after the fan is started, these fumes are blown into the heated space and may be noticed by the occupants of the structures as shown in Figure 12-12. The complaint may be that the customer smells oil for a few minutes after the furnace starts. The occupants usually notice when the fan starts and think that the furnace is starting because they often cannot hear only the burner. An oil smell at start-up can be a sign of a cracked heat exchanger, but the smell can also result from a furnace that is not drafting correctly. In this case, the furnace puffs smoke out the inspection door and sucks it into the return air as shown in Figure 12-13.

As with gas furnaces, oil heat exchangers are responsible for extracting as much heat as practical from the by-products of combustion. The flue gases must be warm enough when they leave the furnace to rise up the flue and not condense in the flue pipe. These by-products of combustion are hotter for oil furnaces than for gas furnaces.

Inside the heat exchanger, the by-products of combustion must be kept in close contact with the heat exchanger surface for as long as possible so that the correct heat exchange takes place. The large volume of the heat exchanger over the top of the combustion chamber is reduced in size to smaller tubes to keep the by-products in contact longer. Figure 12-14 shows these smaller passages, which are likely places for smoke deposits, or *soot.* Soot insulates the heat exchanger from the by-products of

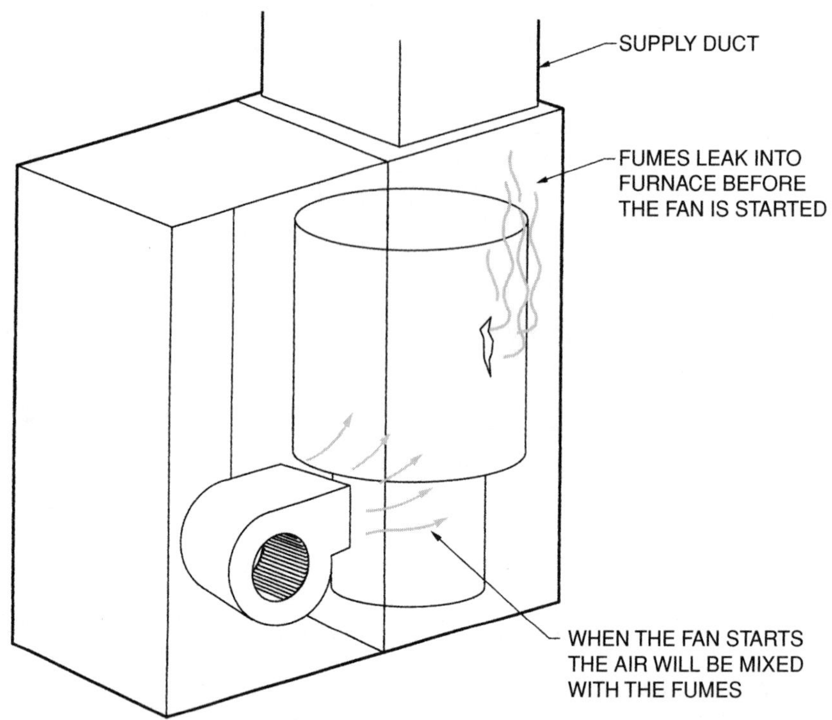

Figure 12-12. If the heat exchanger is cracked, fumes may leak to the air side and be blown into the heated space when the fan starts.

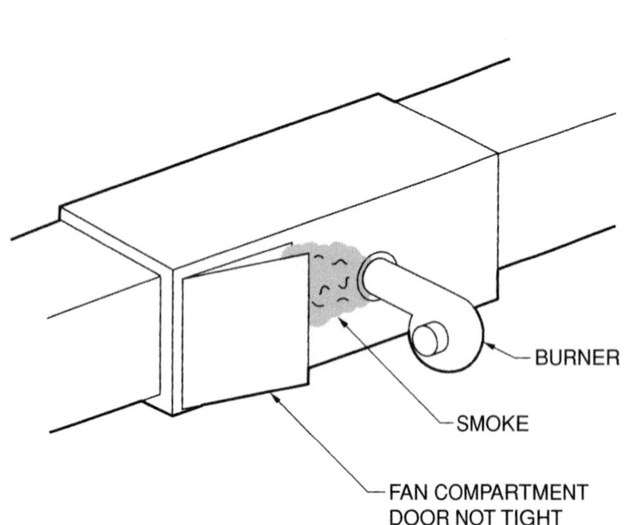

Figure 12-13. An oil smell on start-up can result from improper venting. The flue gases are drawn into the fan compartment on start-up.

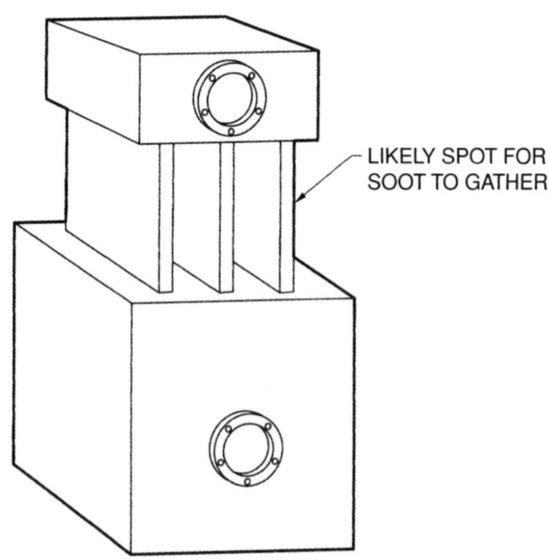

Figure 12-14. The volume of the heat exchanger is reduced into smaller passages to extract as much heat from the by-products as practical.

combustion, so it must be kept to a minimum. It is controlled by burner performance and is discussed in more detail in Unit 14.

Boilers have heat exchangers that extract the heat from the by-products of combustion and heat the water to make hot water or steam, which is circulated throughout a building. These heat exchangers are usually steel or cast iron. Smaller boilers have cast-iron sectional heat exchangers, much like those of the gas boilers mentioned in Unit 2, Section 2.2. Figure 12-15 shows a sectional heat exchanger for an oil boiler, and Figure 12-16 shows the extended surface of the exchanger. Larger oil boilers are

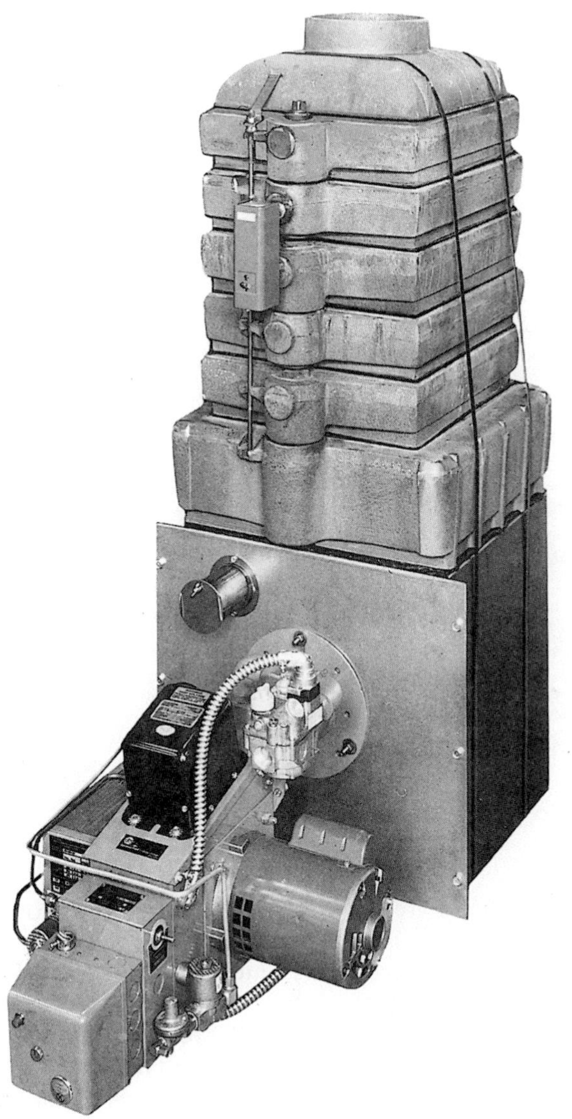

Figure 12-15. The small oil boiler can have a sectional heat exchanger. *Courtesy HydroTherm*

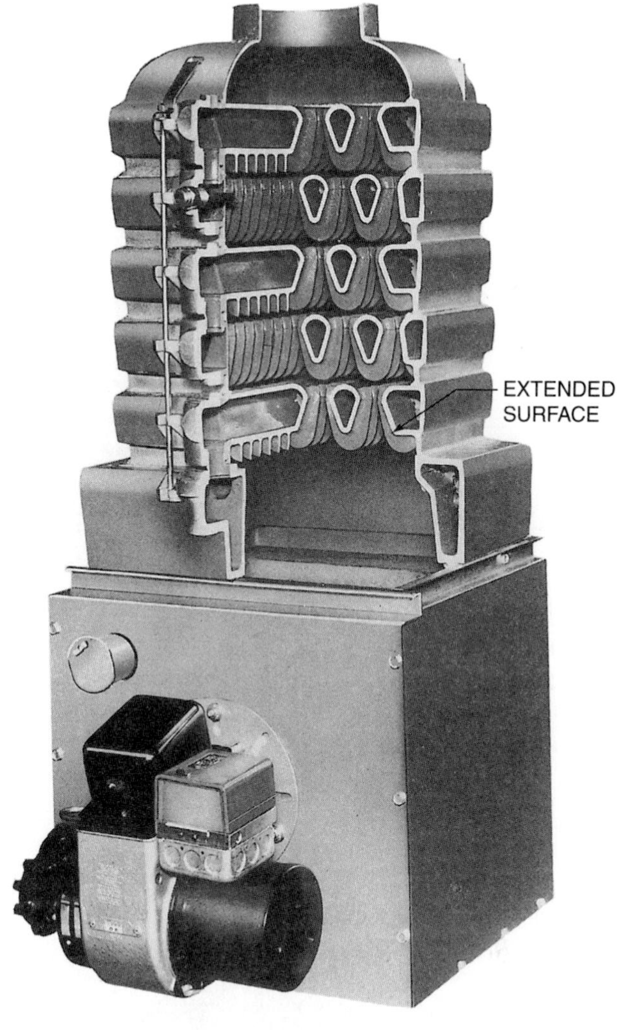

Figure 12-16. The cast-iron heat exchanger can have an extended surface for better heat exchange. *Courtesy HydroTherm*

either water-tube or fire-tube boilers. Figure 12-17 shows a fire-tube boiler.

12.3 Oil Burner Components

All oil burners provided for central heating systems manufactured today are gun burners, which are described in Unit 11. More detail is provided in this unit for each burner component.

For a small application such as a residential furnace or a boiler, the burner assembly is designed to mount on the front of the appliance as depicted in Figure 12-18. The burner for a larger appliance is mounted on a floor pedestal in front of the appliance as shown in Figure 12-19. Burners for both applications have many of the same components. Each component is mounted on the blower housing, as depicted in Figure 12-20.

Burner Motor

The burner motor is mounted on the housing with flange bolts and can be removed for servicing by turning off the power to the unit and loosening the bolts. Mounted on the motor shaft is the fan wheel. It is usually fastened with a hexagonal-head or a square-head set screw. Figure 12-21 shows this assembly. The end of the motor shaft protrudes through the blower housing to the other side, where a flexible coupler connects the motor shaft to the oil pump as shown in Figure 12-22. The flexible coupler may have set screws that must be removed before the motor can be removed. These set screws can be accessed through the transformer housing.

Warning: Before attempting to remove set screws through the transformer housing, make sure the power is off.

Figure 12-17. Larger oil boilers are either fire-tube or water-tube boilers. This is a fire-tube boiler. *Courtesy Cleaver Brooks*

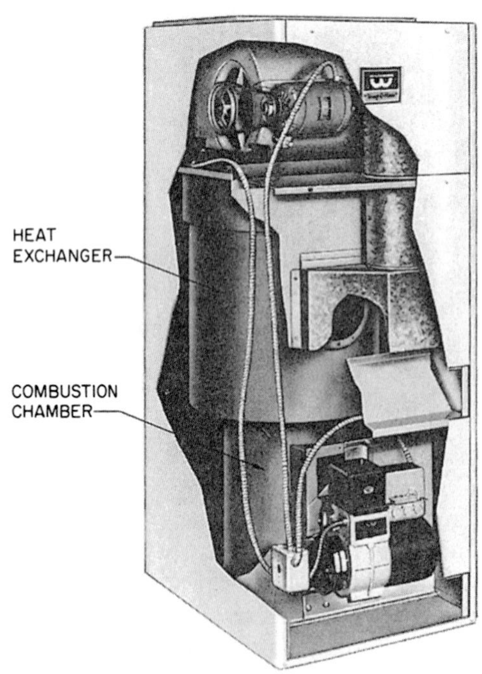

Figure 12-18. The burner is mounted on the front of this furnace. *Courtesy Ducane Corporation*

The set screw for the oil pump can be accessed through the air adjustment shutter, as shown in Figure 12-23.

The motor for the small oil burner is usually a fractional-horsepower, split-phase motor that operates at either 1750 or 3450 RPM (revolutions per minute). This type of motor usually has a manual reset overload button, as depicted in Figure 12-24. This button should trip only when the motor is under excess load, which can be caused by either the oil pump or the motor. The oil pump can become hard to turn or the motor bearings can fail and cause the motor to become hard to turn. The technician will not know which is the problem. When this happens, the technician should remove the motor and determine which component is hard to turn as demonstrated in Figure 12-25.

The fan mounted on the motor shaft must remain clean so that it can push the required amount of air into the combustion process. The air to the fan is controlled by the adjustable air inlet collar and is not filtered. Therefore, dust or lint in the air will clog the blower wheel, as shown in Figure 12-26. The wheel can be cleaned by removing the motor from the housing and using a screwdriver blade to scrape the dirt off the wheel, as shown in Figure 12-27. The fan wheel itself requires little care other than dirt removal.

Ignition Circuit

The high-voltage transformer ignition circuit is composed of the transformer, the insulators, the electrodes, and the interconnecting high-voltage wiring. Figure 12-28 shows this circuit. The transformer has an output voltage of approximately 10,000 volts and cannot be checked with a conventional voltmeter. Service technicians use several methods to check the transformer; one is the voltage

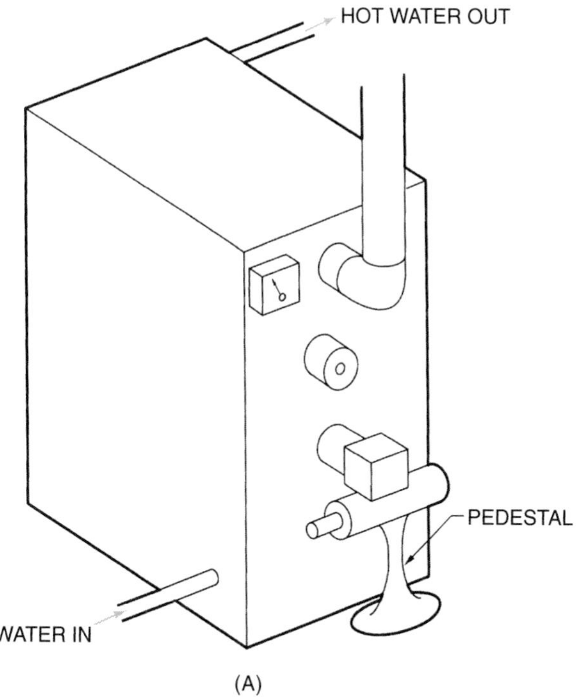

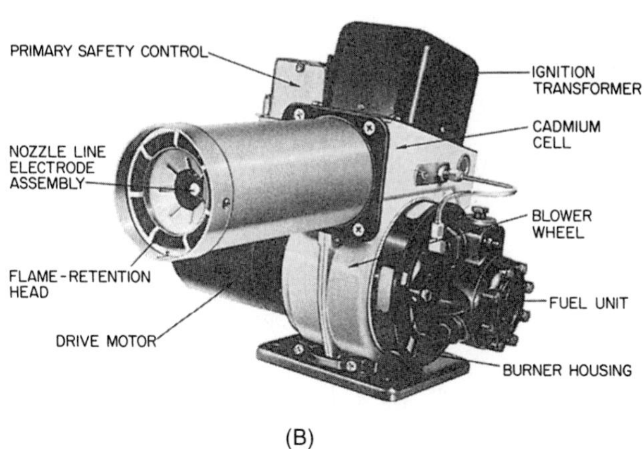

Figure 12-19. The burner is standing on a pedestal on this boiler. *Courtesy R.W. Beckett Corporation*

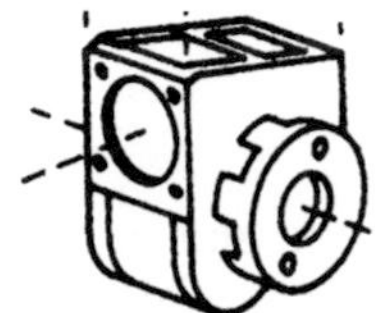

Figure 12-20. Blower housing.

arc test. With this test, an insulated screw-driver handle is used to create an arc across the terminals of the transformer, as shown in Figure 12-29.

Warning: Use care while performing the voltage arc test because of the high voltage. This voltage will not injure you because of the low amperage, but it can give you an unpleasant electrical shock.

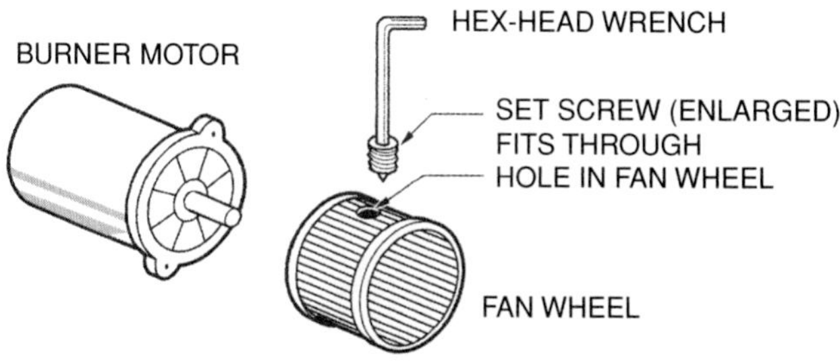

Figure 12-21. The fan wheel is mounted on the motor shaft with a set screw.

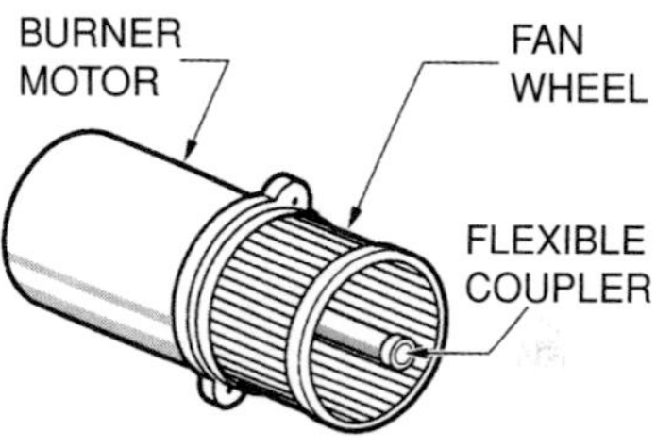

Figure 12-22. The motor shaft extends through the fan wheel into the flexible coupler, which connects the shaft to the oil pump.

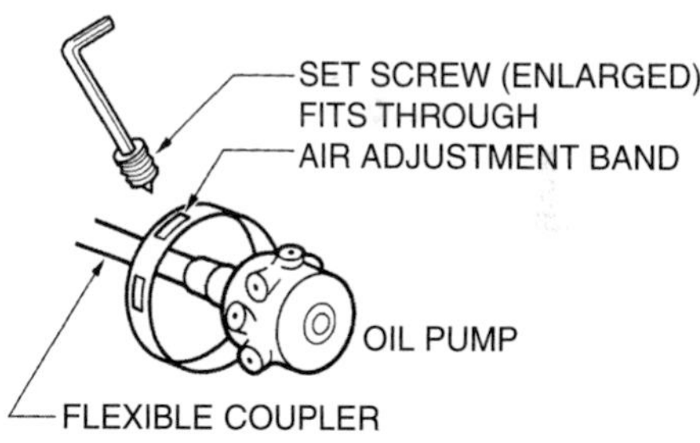

Figure 12-23. The set screw for the oil pump can be accessed through the air shutter adjustment hole.

A good transformer should be able to hold an arc for approximately ½ inch as the screwdriver is moved away from the terminals. The arc should be blue and strong. The arc test shows that a transformer will create an arc and hold it for a ½-inch gap, but it does not indicate the voltage of the arc and whether the transformer will perform under load for long time periods. A transformer tester, as shown in Figure 12-30, is a better choice for serious troubleshooting.

The transformer makes contact with each high-voltage conductor, or lead, through a spring when the transformer is fastened to the housing as depicted in Figure 12-31. This contact surface should be clean, and a good contact should be made. The leads are fastened to the electrodes either by direct connection (a brass band) or through a short run of high-voltage electrical wire, much like the spark plug wire on an automobile engine, as shown in Figure 12-32. Each lead connects to an electrode, which

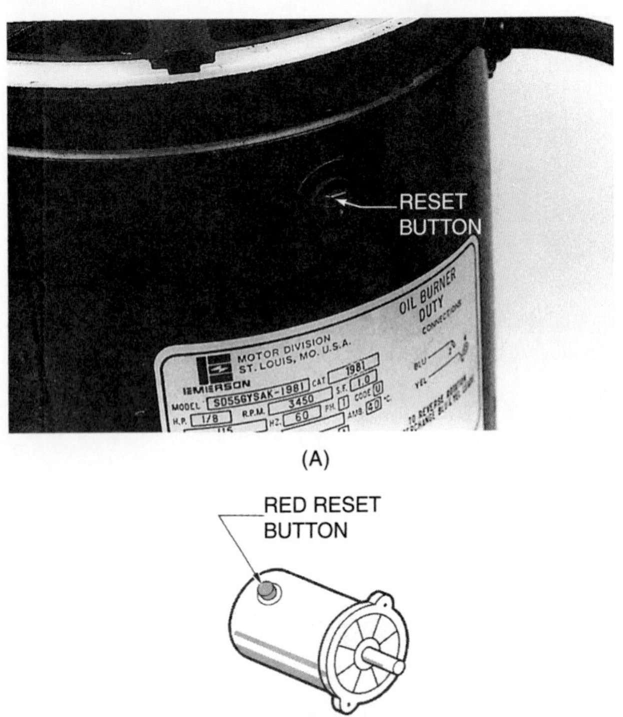

Figure 12-24. (A) Motor with manual reset overload button. *Photo by Bill Johnson* (B) Sketch of motor with reset button.

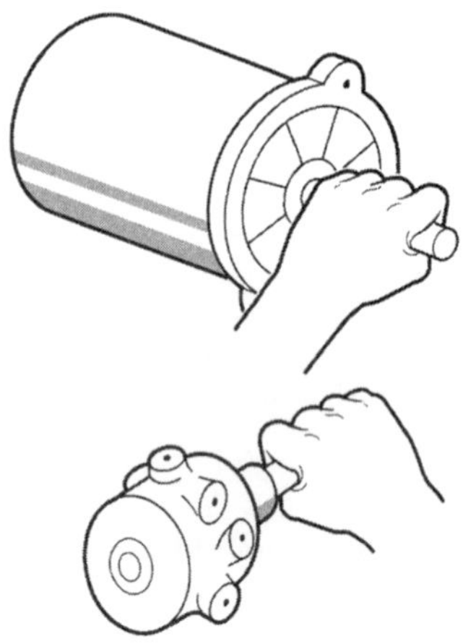

Figure 12-25. This technician removes the motor to determine whether the motor or the pump is hard to turn.

is in a ceramic insulator, or tube as seen in Figure 12-33. This tube is held to the burner assembly by a bracket that allows the tube to be moved back and forth for electrode adjustment. The tube must be clean and without cracks, or the high-voltage arc will leak and short to ground. A short will weaken or stop the arc at the electrodes and can prevent the burner from firing. Carbon can be cleaned from the ceramic tubes with a good degreasing solvent. Many technicians use fuel oil because it is available at every job. Cracked tubes must be replaced.

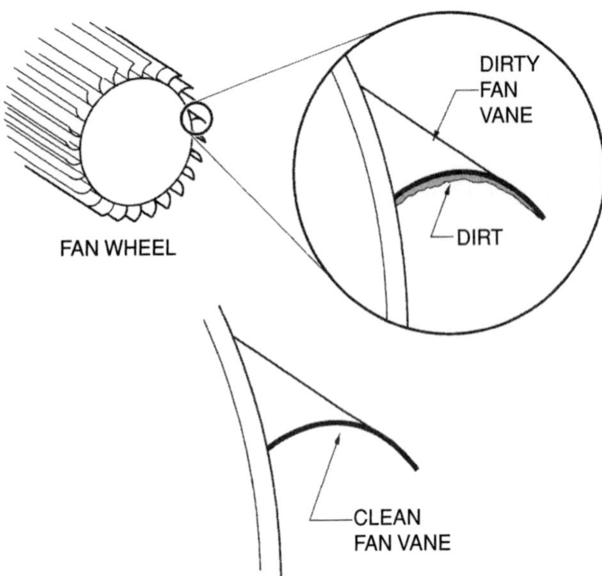

Figure 12-26. The fan, or blower, wheel must be free of dirt.

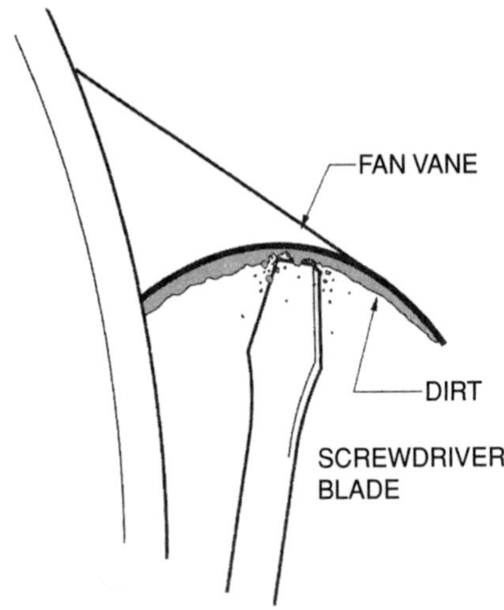

Figure 12-27. The blower wheel can be cleaned by removing it and scraping the dirt from the blower blades.

The electrodes at the end of the assembly must be adjusted for correct firing of the oil burner. If the electrodes are not adjusted correctly, the burner may not fire every time or it may be slow to fire. If the burner does not fire, it must be reset. This is an inconvenience to the customer. If ignition is delayed, the appliance will be noisy on start-up. This adjustment is delicate, and proper practices must be followed. The correct method is to use an electrode-adjustment tool. Printed on this tool is a list of many combinations of burner and nozzle types. All the technician must do is turn the tool to the correct combination and slip it over the nozzle. The electrodes

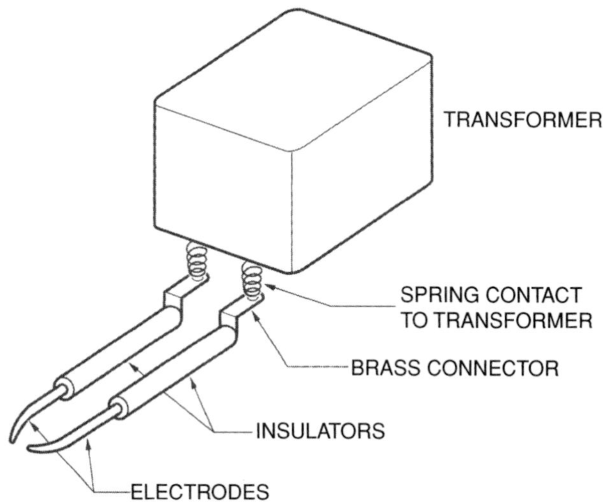

Figure 12-28. The transformer and ignition circuit.

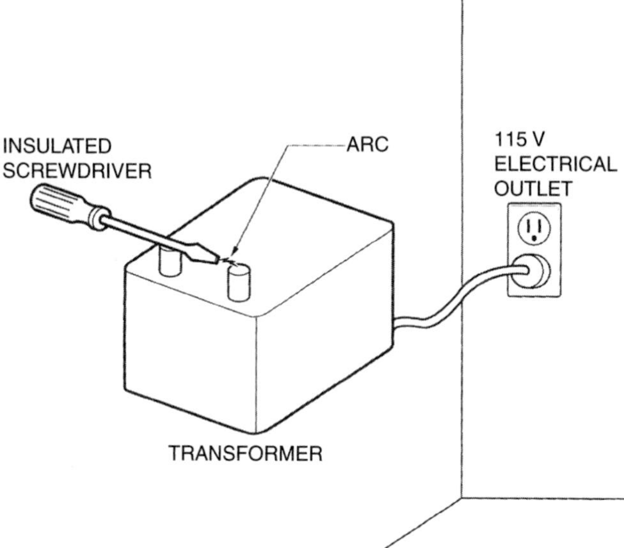

Figure 12-29. The transformer should be able to hold an electrical arc for approximately ½ inch as the screwdriver is moved away from the terminals.

are then adjusted to the recommended gap, as shown in Figure 12-34. The general position of the electrodes can be seen in Figure 12-35.

Caution: Be sure to keep the electrode-adjustment tool clean. Do not allow dirt and oil to build up on it, or you can clog the nozzle.

Note that the electrodes are not in the flame; the arc is. The electrodes are positioned just outside the oil stream as shown in Figure 12-36. The fan blowing the air into the burner tube blows the spark into the oil stream, where ignition occurs.

Figure 12-30. Transformer tester. *Photo by Bill Johnson*

Burner Air-Tube Assembly

The electrode assembly is mounted in the burner air tube that carries the air alongside the oil supply line and the electrodes. The static pressure disc in the air tube creates the correct air pressure before the air mixes with the oil for combustion. The burner air tube is designed by the manufacturer and requires little care except for proper placement in the combustion chamber. The air tube is adjustable in the mounting collar, and this adjustment must be correct. The end of the burner air tube should be recessed in the combustion chamber wall by approximately ¼ inch to prevent it from overheating. If the opening around the end of the air tube is large, it should be packed with insulation to prevent the end from overheating. Figure 12-37 shows a diagram for proper placement of the air tube.

The end of the burner air tube contains the air swirl cone, called the *combustion head.* It turns the air in the correct pattern to meet the oil that is exiting the nozzle under pressure. The air channels of the cone must not become coated with soot, or the air pattern will become distorted. The cone also must not become too hot, or it will melt and deteriorate. The nozzle and the electrodes are also in the end of the tube, recessed slightly. Figure 12-38 shows these components. Their positions relative to the swirling air must be correct.

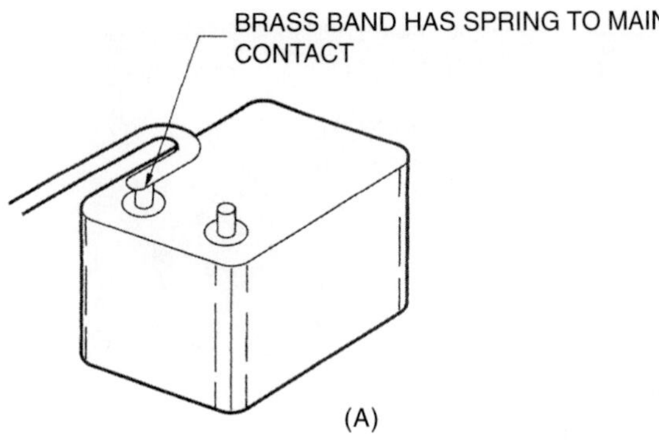

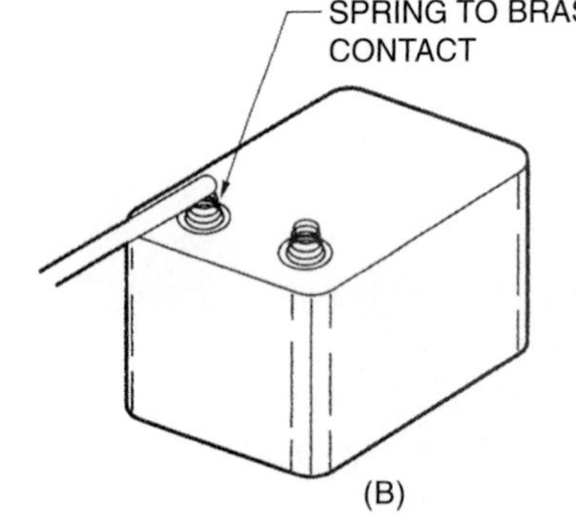

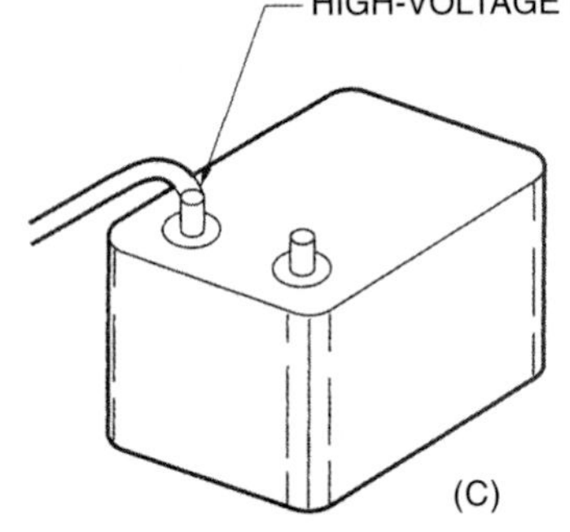

Figure 12-31. Contact between the transformer lead and the electrode terminal. (A) Lead band has spring. (B) Transformer has spring. (C) Lead wire attaches to transformer.

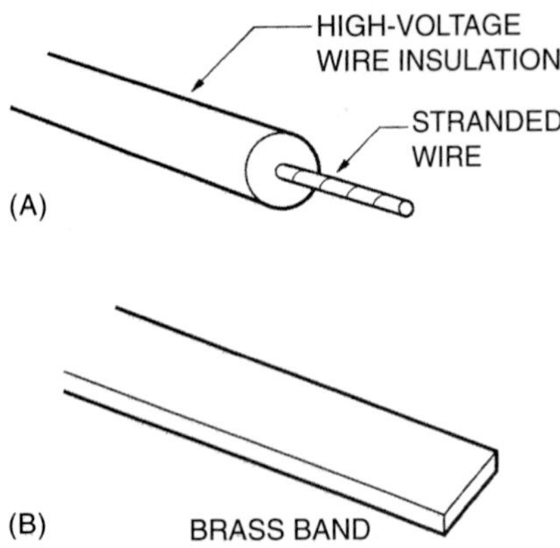

Figure 12-32. The high-voltage lead can be (A) wire (like spark plug wire) or (B) metal band.

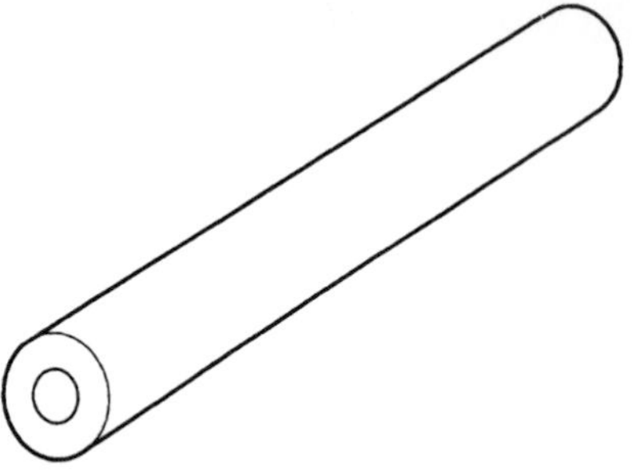

Figure 12-33. Ceramic electrode tube.

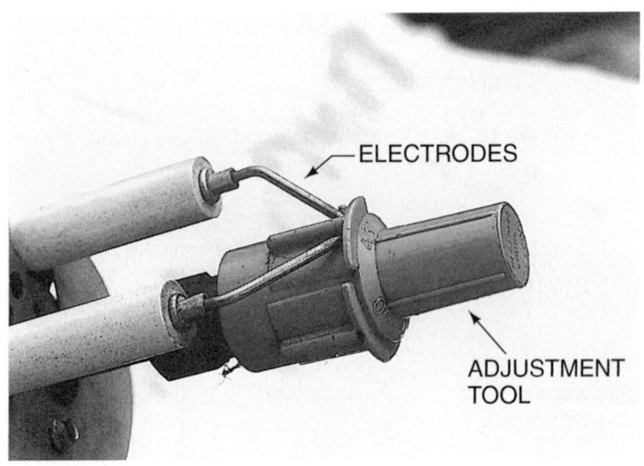

Figure 12-34. Adjusting the electrodes to the recommended gap. *Photo by Bill Johnson*

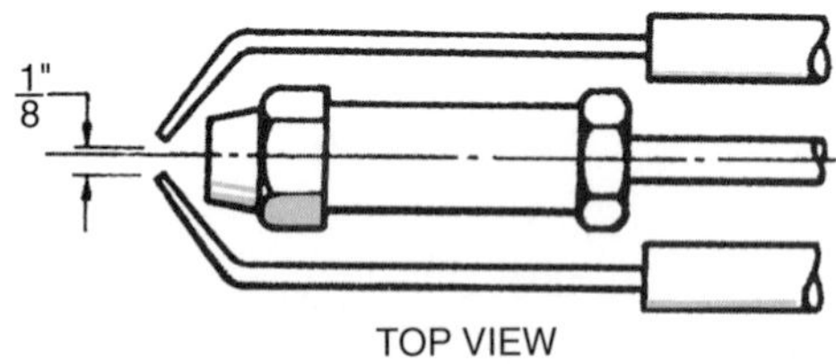

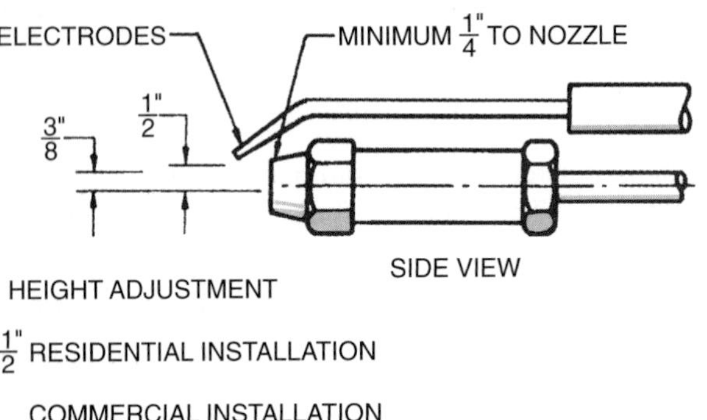

Figure 12-35. General position of electrodes.

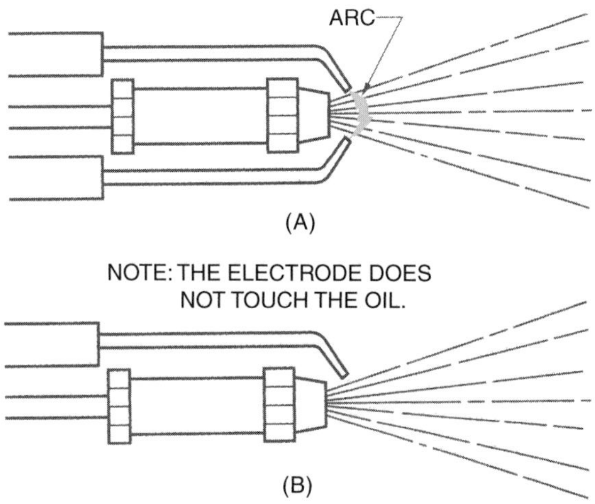

Figure 12-36. The fan blows the spark into the path of the oil droplets. The electrodes are well back from the oil. (A) Top view. (B) Side view.

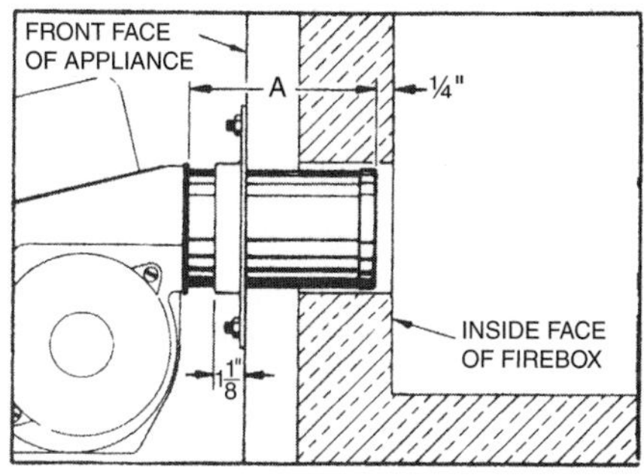

"A" = Usable air tube length.
The burner head should be ¼" back from the inside wall of the combustion chamber. Under no circumstances should the burner head extend into the combustion chamber. If chamber opening is in excess of 4⅜", additional set back may be required.

Figure 12-37. Proper placement of the air tube. *Courtesy R.W. Beckett Corporation*

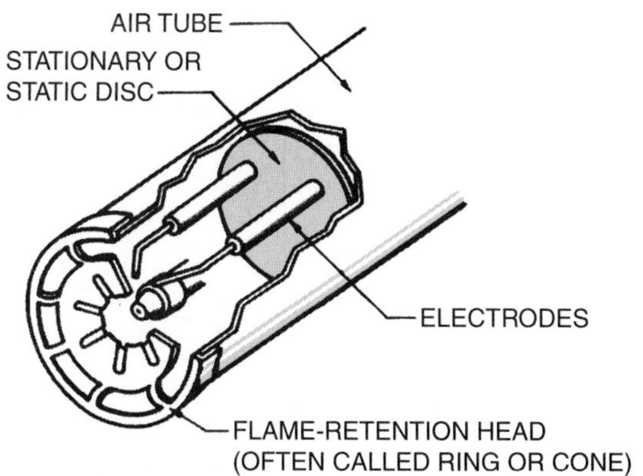

Figure 12-38. The nozzle and electrodes are slightly recessed in the air tube.

Note: It is always a good practice when you are disassembling a furnace or boiler burner to mark all relative positions of the parts. Many a technician has taken apart a working burner for cleaning and then could not get it to fire because the burner parts were not replaced in the same positions.

There are two types of air cones for oil burners; the older combustion head that merely swirls the air and the newer flame-retention combustion head. The older type is not as efficient as the newer type and has a much lazier flame. The flame-retention head has a much tighter flame shape, as shown in Figure 12-39.

Air is blown into the combustion chamber through the air tube. Within this tube, the straight or lateral movement of the air is changed to a circular motion that is opposite the circular motion of the fuel oil. The swirling air mixes with the swirling oil in the combustion chamber.

The amount of air required for the fuel oil preignition treatment is greater than the amount required for combustion. Prior to ignition approximately 2,000 cubic feet of air are needed per gallon of No. 2 fuel oil. Combustion requires 1,540 cubic feet of air per gallon. The excess air ensures correct combustion. If the amount of air exceeds 2,000 cubic feet per gallon, the flame can become too long and can impinge on the far side of the combustion chamber, as shown in Figure 12-40.

Oil Pumps

The modern oil pump is a positive-displacement pump that can pump high pressures. It must have an outlet for the pressure, or damage will occur. The pump has a pumping capacity of approximately 2.5 gallons per hour (gph). This is an input capacity of 350,000 Btu/h and covers all furnaces and many boiler sizes. The pump can be used on a furnace with an oil nozzle of 0.5 gph but will be oversized for the application. The pump is oversized so that it can fit many appliances. The unused portion of the oil is bypassed to the pump inlet or the tank, depending on the application, shown in Figure 12-41.

The pump is driven by an external motor. The shaft protrudes out of the pump casing. A pump shaft seal prevents oil from leaking around the shaft, as shown in Figure 12-42.

Several types of pumps are used in gun burners, as shown in Figure 12-43. A single-stage pump is used when the fuel oil storage tank is above the burner as seen in Figure 12-44. The fuel oil flows to the pump inlet by gravity, and the pump provides oil pressure to the nozzle.

A one-pipe or a two-pipe system can be used with a single-stage pump. A one-pipe system has one pipe from the tank to the burner. In a normal operation a surplus of oil is pumped to the nozzle. The nozzle cannot handle this surplus, so the excess fuel is returned to the low-pressure, or inlet, side of the pump, as shown in Figure 12-45.

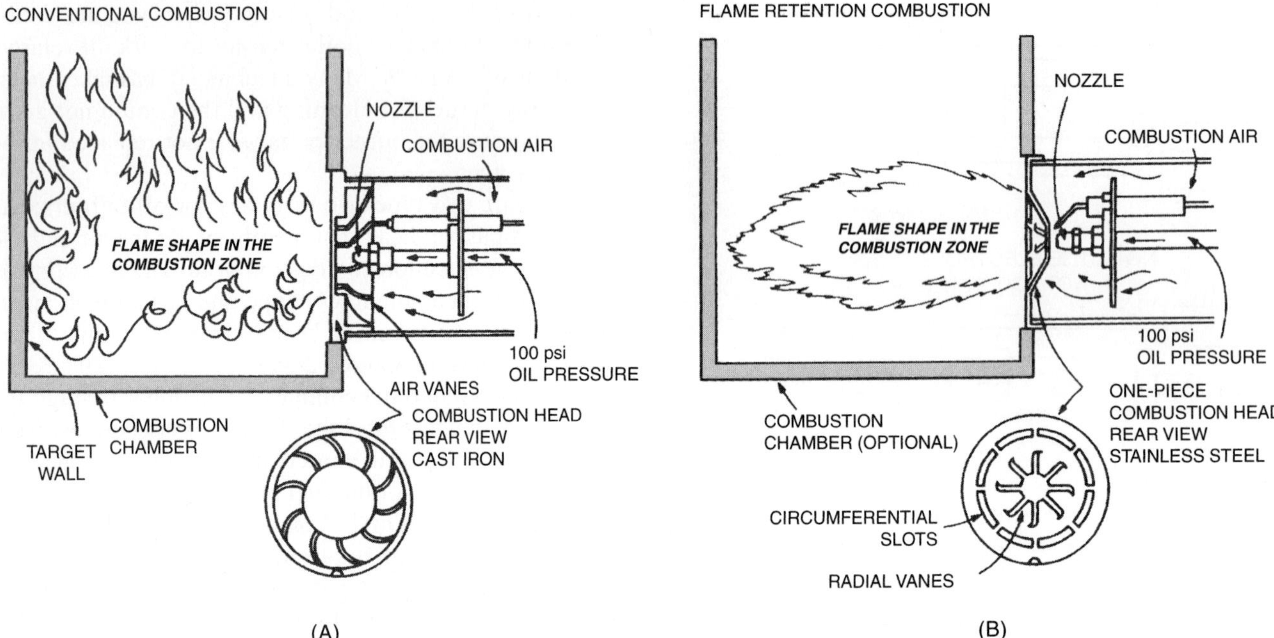

Figure 12-39. The flame-retention combustion head produces a much tighter flame than the older combustion head. (A) Conventional combustion. (B) Flame-retention combustion. *Courtesy R.W. Beckett Corporation*

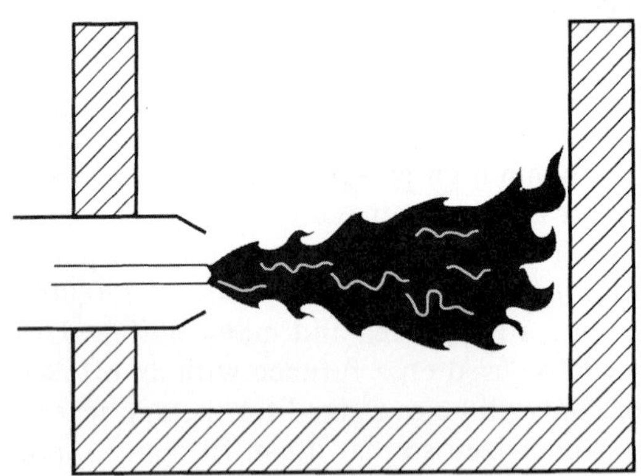

Figure 12-40. Too much air causes the flame to impinge on the far side of the combustion chamber and cool.

Caution: When installing a one-pipe system, make sure that the bypass plug is not in place so that the surplus oil can return to the inlet side of the pump, as shown in Figure 12-46.

If the fuel oil becomes too low in the tank or if the system piping is opened for any reason in a one-pipe system, the air that enters must be bled from the system at the pump bleed port, as seen in Figure 12-47. One-pipe systems are not self-priming. (*Priming* means removing the air from the oil line with fresh oil.) A two-pipe system must have the bypass plug in place; it returns the oil to the tank rather than to the pump inlet, as shown in Figure 12-48. The two-pipe system is self-priming, so it need not be bled.

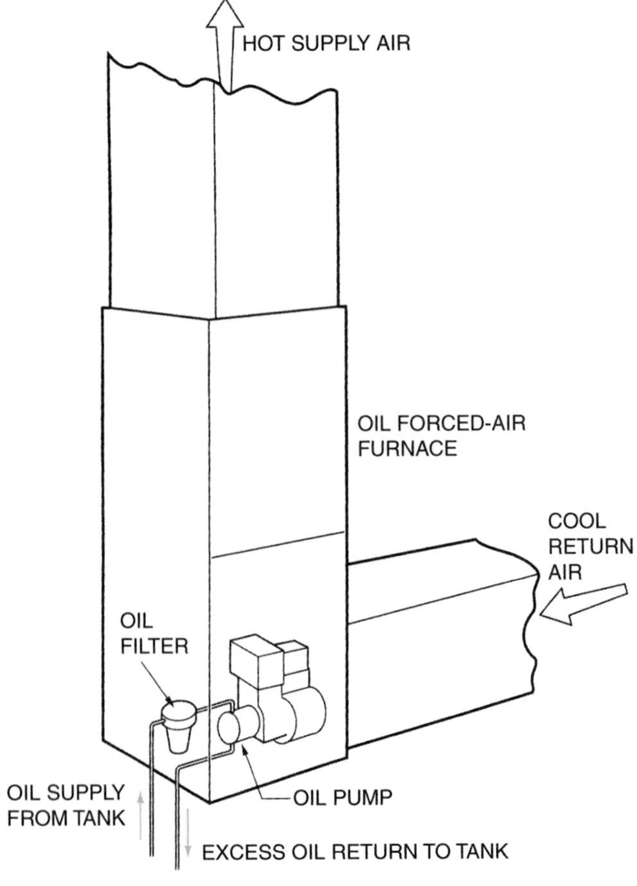

Figure 12-41. The unused oil is routed back to the pump inlet or the tank, depending on the application.

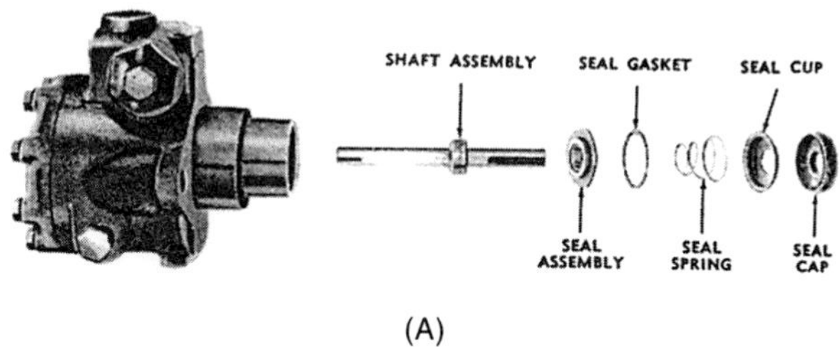

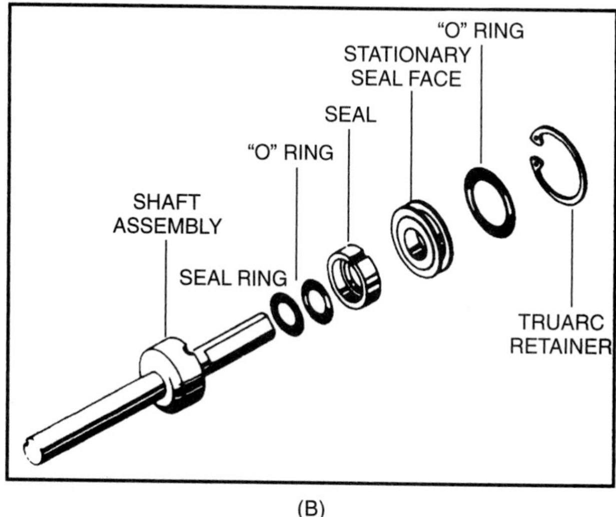

Figure 12-42. The pump must have a shaft seal to prevent oil from escaping down the shaft. (A) The pump and its components. (B) Schematic of the components. *Courtesy Suntec Industries Incorporated, Rockford, Illinois*

A dual-, or two-stage, pump is used when the oil is stored below the burner. One stage of the pump lifts the oil to the inlet of the second stage, and the other provides the pressure for the oil nozzle. The two-stage pump does the same job as the single-stage pump except the first stage provides the oil to the second stage. When the tank is below the oil pump, the bypass plug is inserted in the pump, and an oil return line to the tank is provided for the excess oil. It is necessary

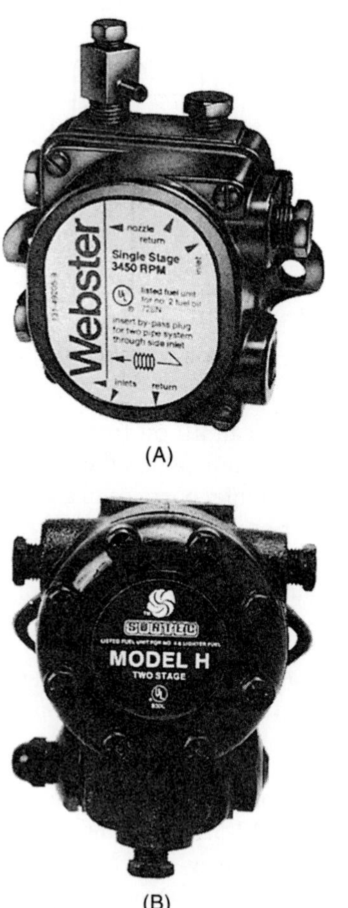

Figure 12-43. A selection of oil pumps. *Photo by Bill Johnson*

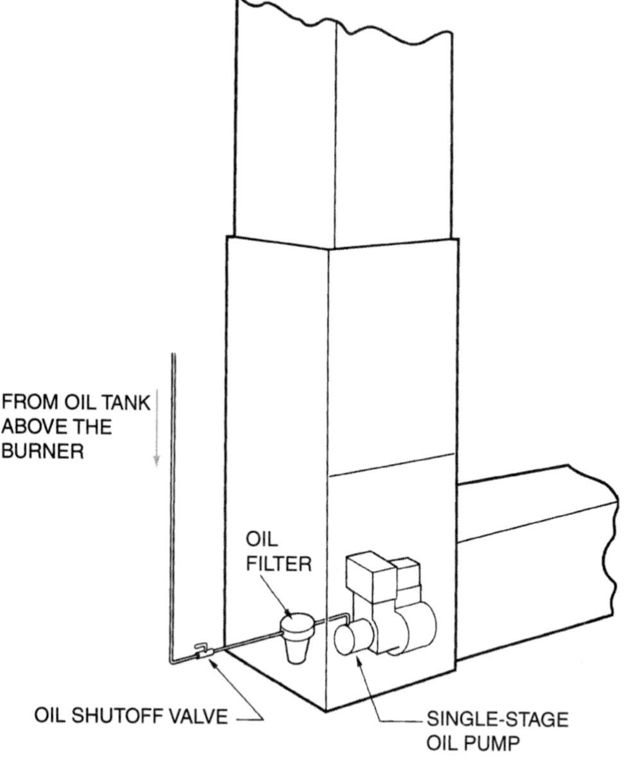

Figure 12-44. Single-stage pump with fuel flowing to the pump inlet by gravity.

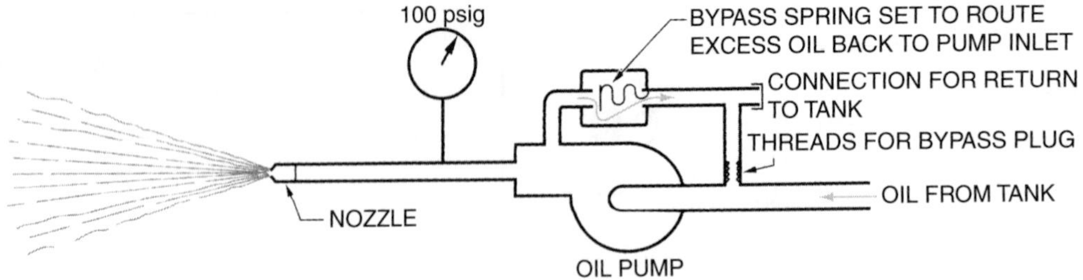

Figure 12-45. One-pipe system with the oil returning to the pump inlet.

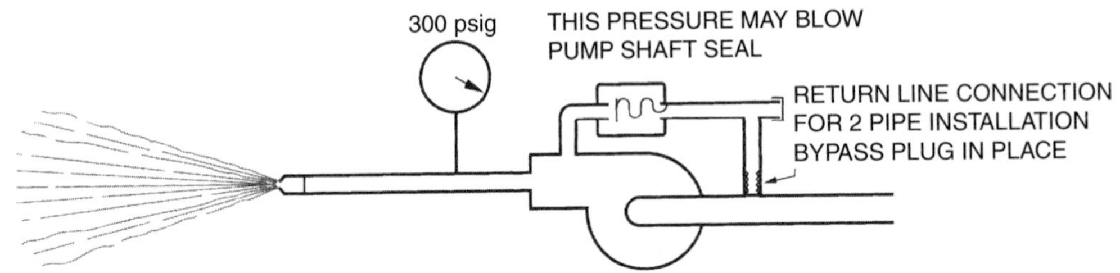

Figure 12-46. The bypass plug must not be in place when the pump is used in a one-pipe system.

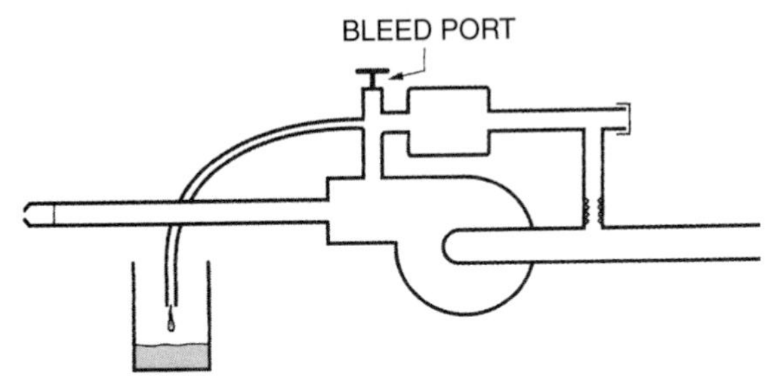

Figure 12-47. When the system runs out of oil with a one-pipe system, the air must be bled from the pump on start-up.

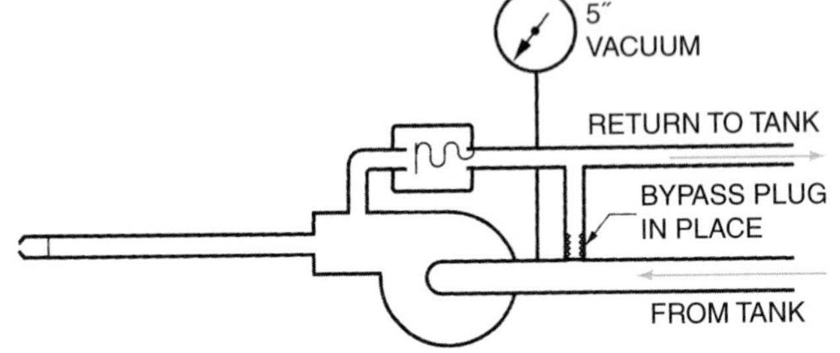

Figure 12-49. When the oil is returned to the tank, there is a vaccum at the pump inlet to move the oil from the tank to the pump inlet.

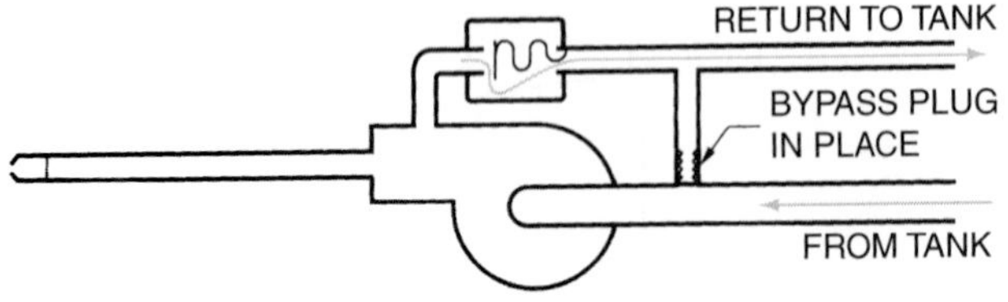

Figure 12-48. The bypass plug is in place for a two-pipe system, and the oil returns to the tank in the second pipe.

to return the oil to the tank rather than to the pump inlet in order to create the vacuum required to pull the oil from the lower tank as demonstrated in Figure 12-49.

The pump used in a high-pressure gun burner is a rotary pump using either a cam system, gears, or a combination to provide the pressure. The pump itself should not be repaired by the technician. When defective, pumps are usually replaced with new or rebuilt ones furnished by oil heat parts supply companies. The cost of a new or rebuilt pump is much less than the cost of having a technician rebuild the old pump in the field.

These oil pumps can lift oil from tanks approximately 15 feet below the pump. When the oil tank is below the pump, the oil pump inlet operates in a vacuum of approximately 1 inch for each foot that the oil level is below the pump. If the distance is more than 15 feet, a secondary pump should be used to lift the oil to the inlet of the oil pump as shown in Figure 12-50.

Note: Do not forget that when the oil tank becomes low, the oil must be lifted from the *bottom* of the tank. See Figure 12-51.

Built into each oil pump is a pressure-regulating valve, as shown in Figure 12-52. The pump can provide pressures in excess of the 100 psig used by the nozzle. The pressure-regulating valve can be set so that the fuel oil being delivered to the nozzle is at the set pressure of 100 psig. Again, the excess oil is diverted to the tank or the pump inlet, depending on the system.

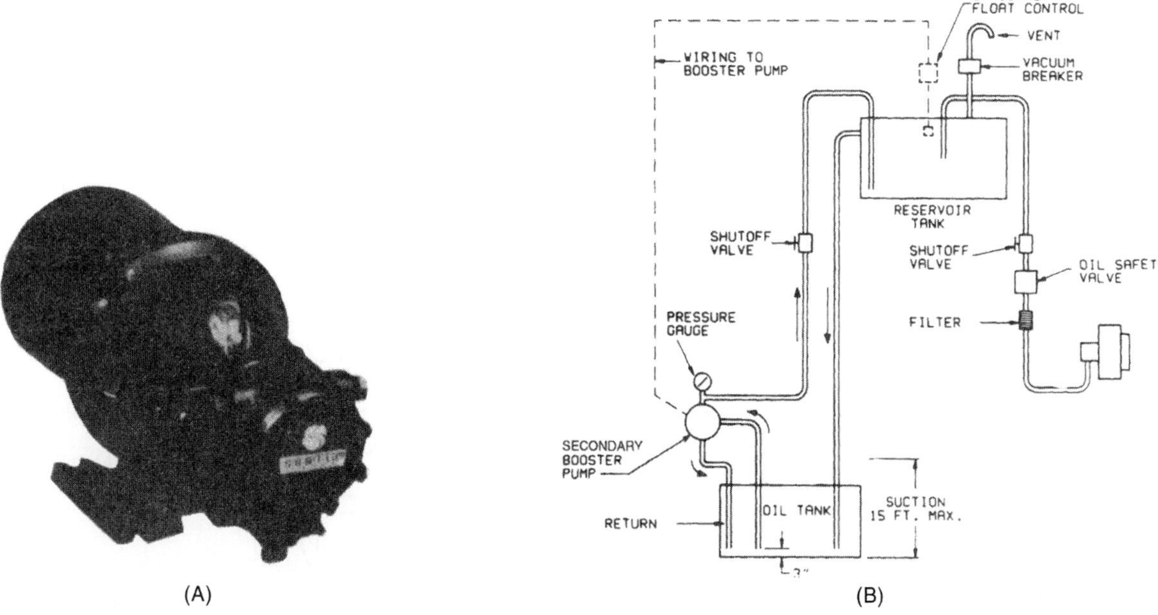

Figure 12-50. If the vertical distance is more than 15 feet, a secondary oil pump must be used. (A) Secondary pump. *Courtesy Suntec Industries Incorporated, Rockford, Illinois* (B) A pumping system for a gun burner.

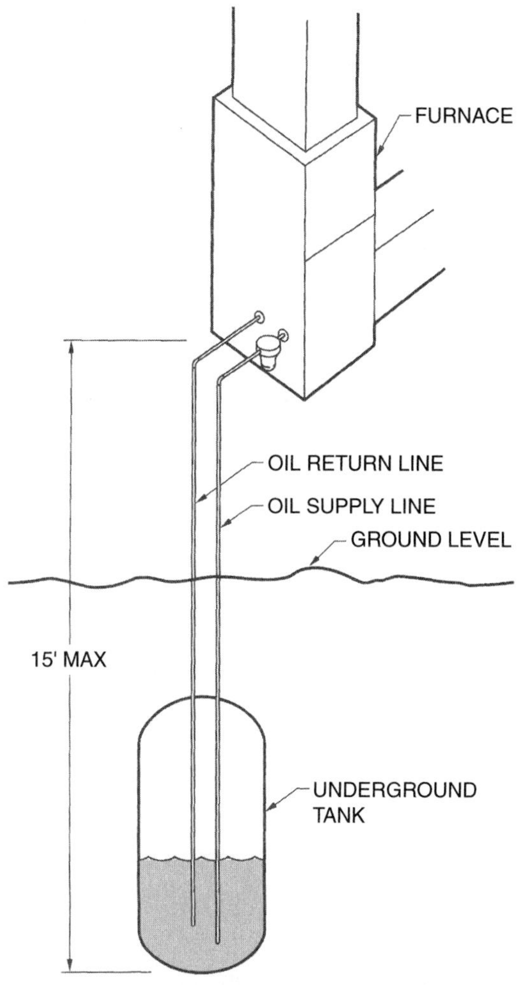

Figure 12-51. The vertical distance must be measured from the bottom of the tank to pull out oil when the oil level is low.

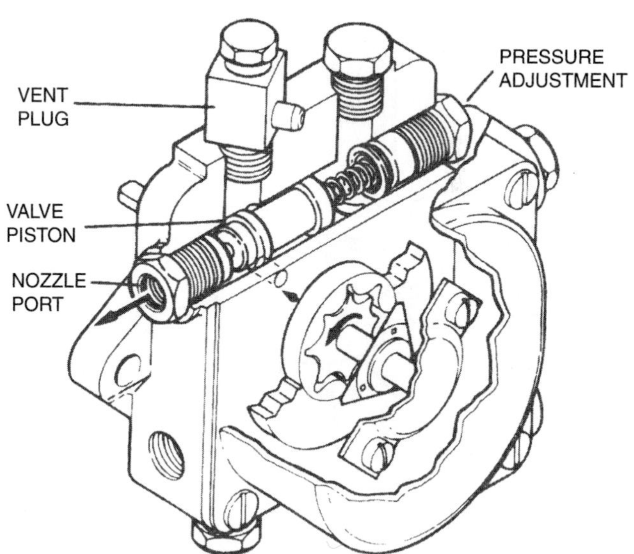

Figure 12-52. The oil pump has a pressure-regulating valve. *Courtesy Webster Electric Company*

Also built into the pump is a fuel shutoff valve. This valve prevents the fuel oil from burning at very low pressures at the end of the cycle. For example, when the burner is operating at 100 psig and then stops, the pressure reduces as the pump slows. When the pressure reaches approximately 85 psig, the fuel oil to the nozzle is shut off so that the very low pressures do not spray oil into the combustion chamber. (At low pressures, the oil is not atomized and if sprayed into the combustion chamber causes smoke. Also, the burner may shut down with a "boom.") The oil shutdown feature of the fuel pump can be checked by placing a gauge in the fuel line to the oil nozzle. When the burner is shut off, the pressure should drop to approximately 85 psig and hold.

A fast oil shutdown can also be accomplished by using a small solenoid valve in the line between the pump and the nozzle, as shown in Figure 12-53. This solenoid can be wired in parallel with the oil pump motor so that it opens and closes upon a call for heat, as seen in Figure 12-54.

An oil pump runs at a speed of 1750 or 3450 RPM. The pump also has either clockwise or counterclockwise rotation. Therefore, the correct replacement pump must be selected when a pump is changed. When changing an oil pump, be sure to take the old pump to the supply house and ask for one just like the old one. While there, place the pumps side by side and compare the oil fittings, bypass plug combinations, and rotation speed and direction.

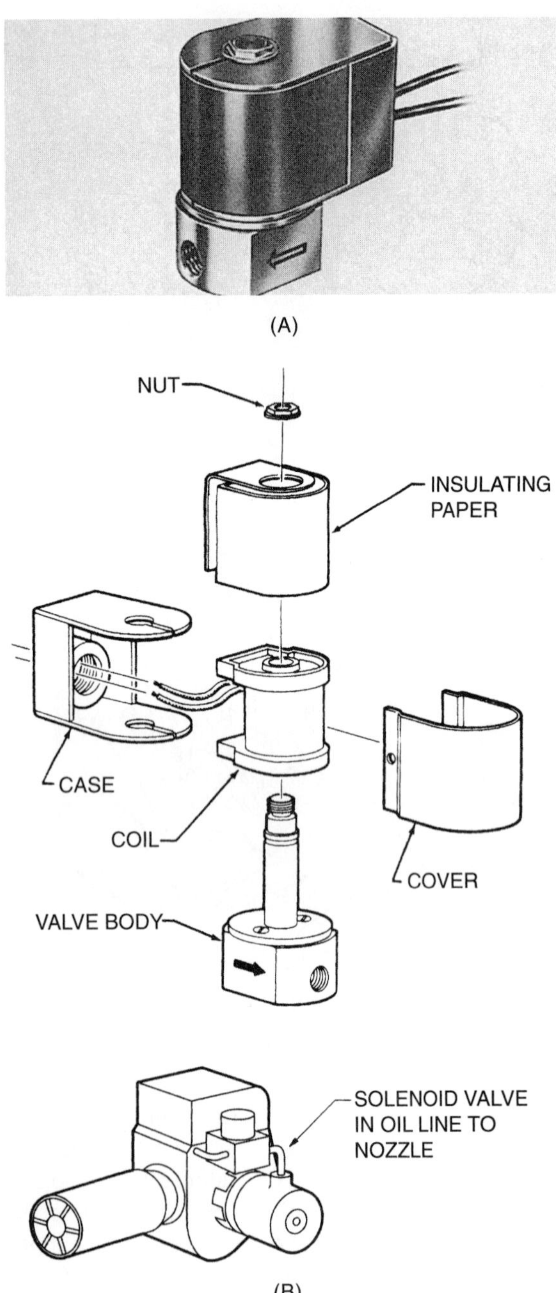

Figure 12-53. (A) This solenoid valve is used to shut off the oil when the internal oil shutoff is not functioning. *Courtesy Honeywell, Inc., Residential Division* (B) Solenoid valve location on pump.

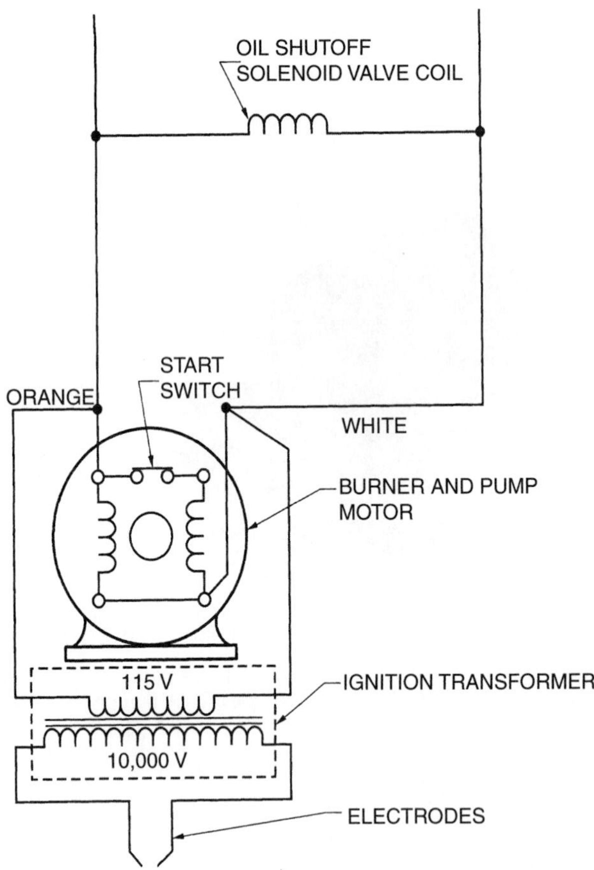

Figure 12-54. The solenoid valve is wired in parallel to the pump motor.

Nozzle

The nozzle prepares the fuel oil for combustion by atomizing, metering, and patterning the fuel oil. It does this by breaking the fuel into tiny droplets. The smallest of these droplets are ignited. The larger droplets (there are more of these) provide more heat transfer to the heat exchanger when they are ignited. Atomization of the fuel oil is a complex process. The lateral movement of the fuel oil must be changed to a circular motion opposite that of the airflow from the combustion head. The fuel enters the orifice of the nozzle through the swirl chamber. Figure 12-55 shows a typical nozzle and its swirl chamber.

The bore size of the orifice in the nozzle is designed to allow a certain amount of fuel through at a given pressure to produce the fuel flow desired. The fuel flow controls

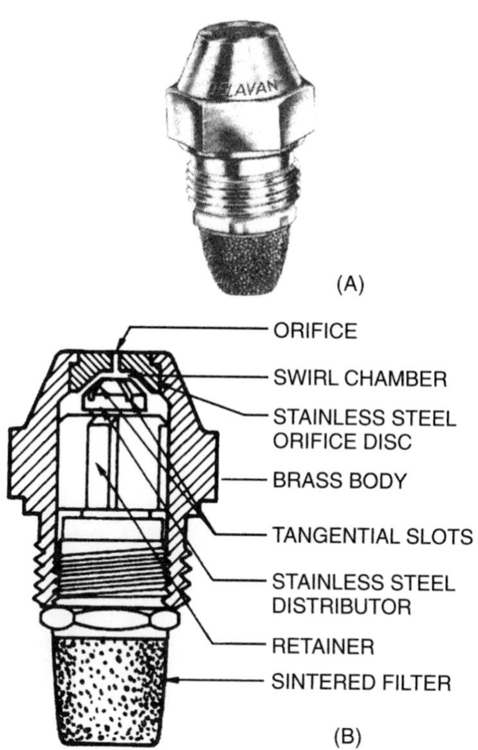

Figure 12-55. (A) Typical oil nozzle. *Courtesy Delavan Corporation* (B) Cross section of the oil nozzle.

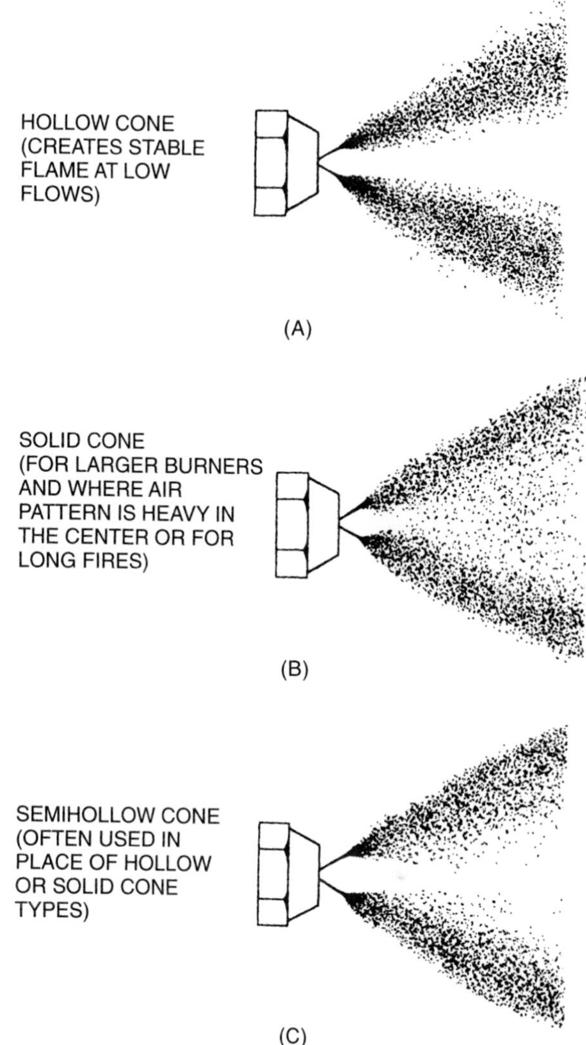

Figure 12-56. Spray patterns. (A) Hollow pattern. (B) Solid pattern. (C) Semihollow pattern. *Courtesy Delavan Corporation*

the Btu input to the appliance. Each nozzle is marked as to the amount of fuel it will deliver. A nozzle marked 1.00 delivers 1 gallon of No. 2 fuel oil per hour with an input pressure of 100 psig and an oil temperature of 60°F. Approximately 140,000 Btu/h of heat are input to the furnace. A nozzle marked 0.8 gph delivers 112,000 Btu/h (140,000 × .8 = 112,000). Other nozzle capacities can be calculated in the same manner.

As mentioned previously, the flame pattern must fit the combustion chamber. The nozzle must be designed so that the spray ignites smoothly and provides a steady, quiet fire that burns cleanly and efficiently. The nozzle must have a uniform spray pattern and an angle that is best suited to the requirement of the specific burner. There are three basic spray patterns: hollow, semihollow, and solid, as shown in Figure 12-56. Each pattern can have a spray angle of 30 degrees to 90 degrees.

Hollow cone nozzles usually produce a more stable spray angle and pattern than solid cone nozzles of the same flow rate do. These nozzles are often used when the flow rate is less than 1 gph. Solid cone nozzles distribute droplets fairly evenly throughout the pattern. These nozzles are often used in larger burners. Semihollow cone nozzles are often used instead of hollow or solid cone nozzles. The higher flow rates produce a more solid spray pattern, and the lower flow rates produce a hollower spray pattern.

The oil-burner nozzle is a precision device that must be handled carefully. These nozzles are not intended for cleaning or rebuilding in the field. They are too inexpensive. Cleaning with wire brushes or metal instruments will distort the passages in the nozzle. Handling the inlet filter will leave deposits from your hands that can be pushed through to the orifice, particularly in smaller-bore nozzles.

Caution: New nozzles must be kept in their shipping containers and not allowed to roll around in the tool-box. Otherwise they will be damaged.

A special tool, a nozzle wrench, should be used to remove the nozzle to prevent damage. This tool is used instead of the two wrenches that would otherwise be required to hold the nozzle and its adapter for removal. Figure 12-57 shows a nozzle wrench.

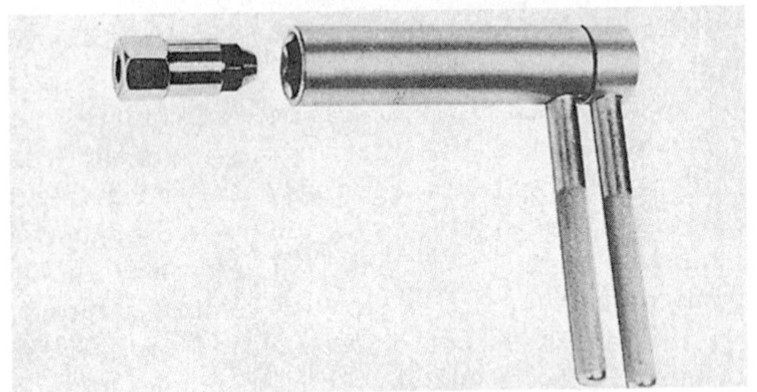

Figure 12-57. Nozzle wrench. *Courtesy Delavan Corporation*

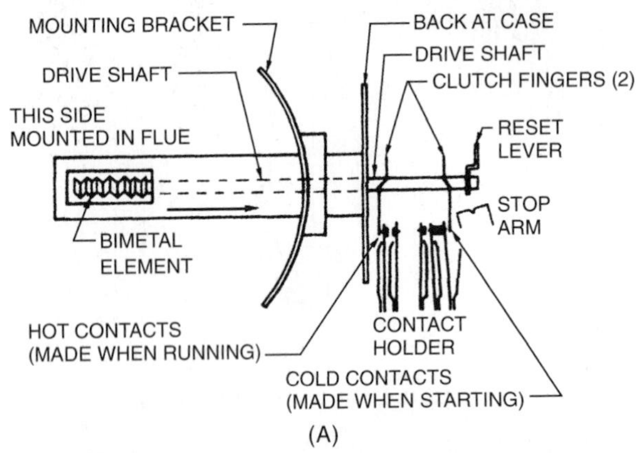

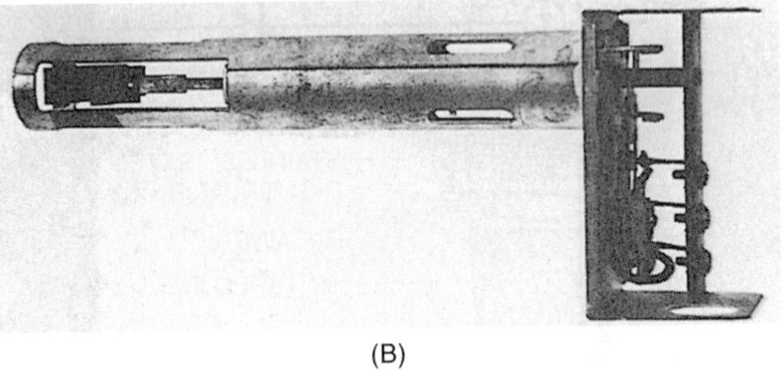

Figure 12-58. (A) Illustration of a stack switch. *Courtesy Honeywell, Inc., Residential Division* (B) Stack switch safety device. *Photo by Bill Johnson*

Caution: The nozzle wrench must be kept clean, or it will do more harm than good. Debris often packs into this wrench and can be transferred to the end of the nozzle. Check the wrench often.

Primary Control Unit

The burner primary control unit is used to regulate the burner sequence. It contains the burner flame safety feature for the unit. The flame safety feature ensures that the flame is established within a reasonable length of time, usually 90 seconds. If the burner does not ignite within 90 seconds after the thermostat calls for heat, or if the burner goes out for 90 seconds, the burner motor is stopped and must be manually reset. This safety feature prevents an excess of oil from building up in the combustion chamber.

There are two basic types of flame safety controls: the stack switch, or stack relay, and the cad cell, or light-sensing control. The *stack* switch has been used for many years and consists of a bimetal relay located in the flue just as it exits the appliance, as shown in Figure 12-58. The bimetal stack relay senses heat in the flue and expands, pushing the drive shaft in the direction of the arrow, as shown in Figure 12-58A. The drive shaft then closes the hot contacts.

Figure 12-59 is a wiring diagram showing the current flow during the initial start-up of the oil burner. The 24-volt room thermostat calls for heat. This call energizes the 1K coil, which closes the 1K1 and 1K2 contacts. Current flows through the safety switch heater and cold contacts. The hot contacts remain open. The closing of the lK1 contacts provides current to the oil burner motor, the oil valve (optional oil shutoff valve), and the ignition transformer.

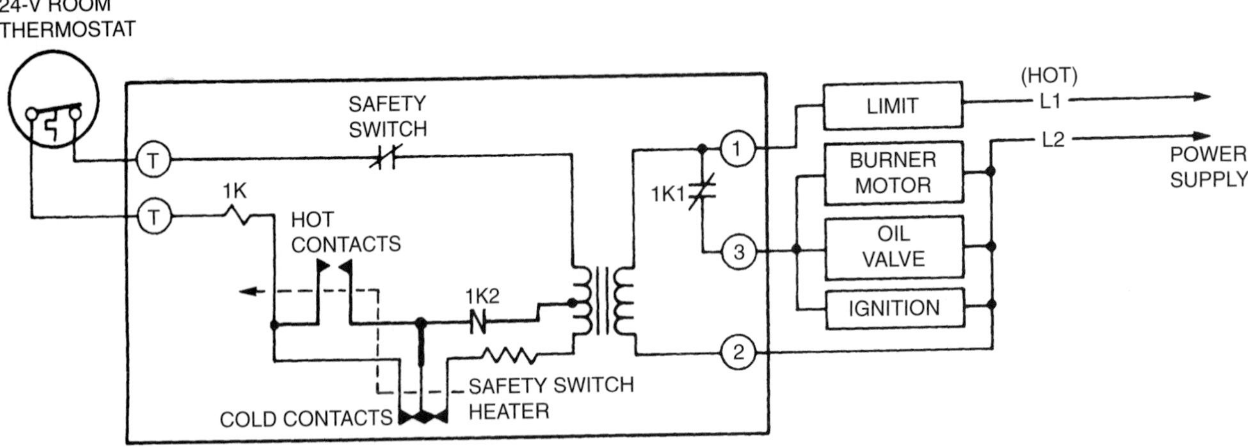

Figure 12-59. Wiring diagram showing the initial start-up of an oil burner.

The circuit is completed through the 1K2 contact, the hot contacts to the 1K coil. The 1K1 coil remains closed, and the furnace continues to run as in a normal safe start-up. The safety-switch heater is no longer in the circuit.

If there is no ignition and consequently no heat in the stack, the safety-switch heater remains in the circuit. In approximately 90 seconds, the safety-switch heater opens the safety switch, which causes the burner to shut down. To start the cycle again, you must manually shut off the system, then depress the reset button.

The *cad (cadmium sulfide) cell* safety device senses the light of the flame (i.e., there is ignition) and offers a low resistance. It has a very high resistance when there is no light (no flame). Figure 12-60 shows the resistance being measured in both cases. The cad cell must be positioned properly to sense sufficient light for the burner to operate efficiently. In most installations the cad cell is preferred. It acts faster, has no moving mechanical parts, and consequently is considered more dependable.

In a standard modern primary circuit the cad cell can be coupled with a triac, which is a solid-state device designed to conduct current when the cad cell circuit resistance is high (no flame).

In Figure 12-61 the circuit shows the current flow in a normal start-up. There is low resistance in the cad cell. The triac does not conduct readily; therefore, the current bypasses the safety-switch heater, and the burner continues to run.

In Figure 12-62 there is no flame and the resistance is high in the cad cell. By design, the triac conducts current through the safety-switch heater. If this current continues through the heater, it opens the safety switch and causes the burner to shut down. To recycle the system, depress the reset button.

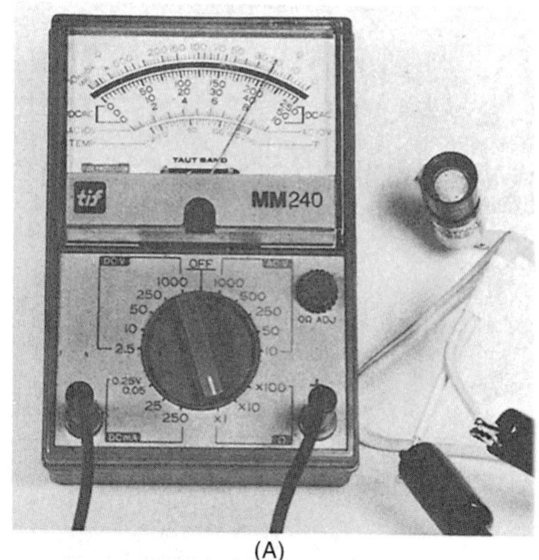

(A)

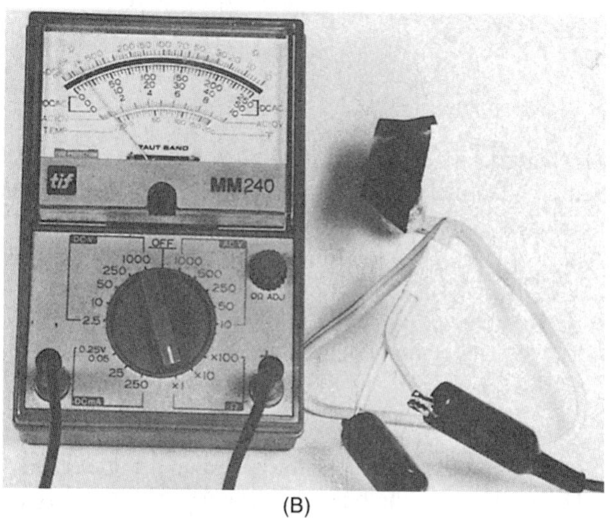

(B)

Figure 12-60. The cad cell has a high resistance when there is no light from the flame. (A) Measuring resistance when light is present. (B) Measuring resistance when there is no light.

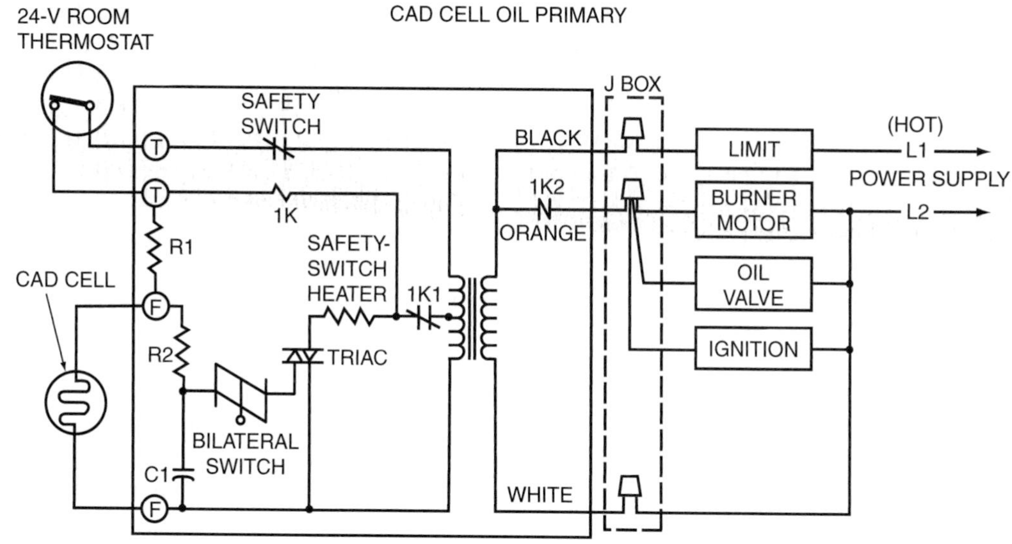

Figure 12-61. Wiring diagram for a normal start-up.

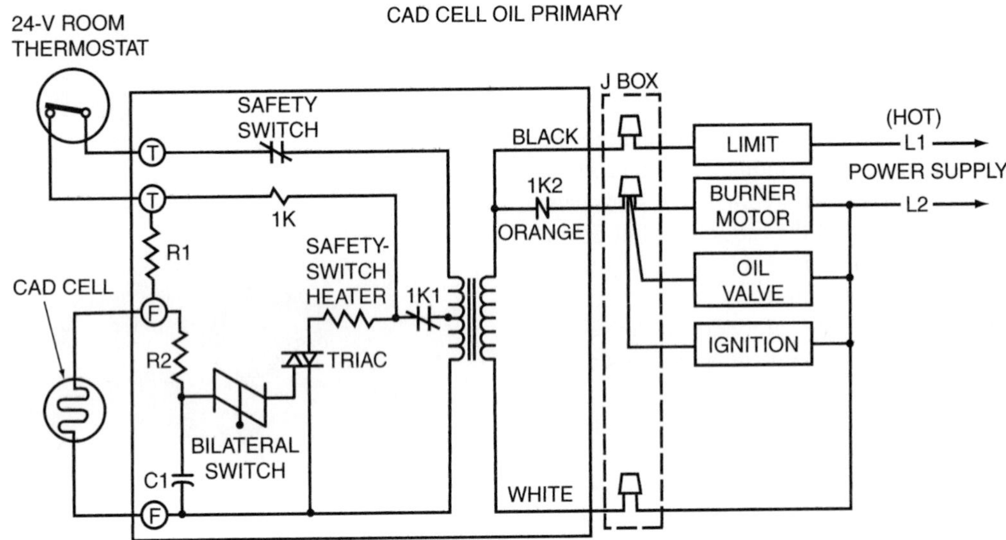

Figure 12-62. Wiring diagram when there is no flame.

Warning: Do not reset the primary control too many times because unburned oil can accumulate after each reset. If this happens and ignition occurs, the surplus oil will generate a lot of heat. Stop the burner motor, but allow the furnace fan to run. Close the air shutter to reduce the air to the fire, and notify the fire department. Do not try to open the inspection door to put out the fire. Let it burn itself out from reduced air.

Mounting Flange

The mounting flange fastens the burner to the appliance. It is designed so that the burner can be easily removed from the appliance. With some flanges, you must loosen the nuts, rotate the burner slightly, and remove the flange over the nuts. Some burner flanges require you to remove the nuts completely and slide the burner off the bolts. Either type of flange, with a gasket, should fasten the burner tightly to the appliance to prevent air leaks.

Summary

- Oil heat is popular where gas mains are not available and in the colder climates, where heat pumps are not practical.
- Oil heat is reliable and simple heat to maintain.
- One gallon of No. 2 fuel oil contains approximately 140,000 Btu of heat energy.
- Combustion is rapid oxidation, or burning, and is used to convert the oil to heat.
- Combustion takes place in the combustion chamber in an oil-burning appliance.
- The combustion chamber confines and controls combustion by heating some of the droplets of oil that are not ignited during the atomization process.
- Oil droplets must not touch the surface of the combustion chamber, or they will cool and create smoke.
- The shape of the combustion chamber must fit the shape of the flame for the best results.
- Combustion chambers were made of fire brick for many years, but they were heavy and likely to break. More modern, lightweight materials are available today.
- The lighter materials used make combustion chambers heat faster and create the hot reflective surface needed to help combustion.
- The heat exchanger surrounds the combustion chamber and transfers heat from the by-products of combustion into the medium to be heated: air or water.
- Heat exchangers for forced-air furnaces are made of stamped steel that is welded at the seams.
- The heat exchanger must be airtight to prevent the by-products of combustion from mixing with the room air. The by-products of poor combustion contain carbon monoxide, which is poisonous.
- The heat exchanger extracts as much heat as practical from the by-products of combustion.
- The heat exchanger cannot extract all the heat because some of it pushes the flue gases up the flue.
- The inside design of the heat exchanger causes the by-products of combustion to stay in contact and exchange as much heat as practical.
- In larger boilers, the heat exchangers are made of cast iron or steel. The by-products of combustion are on one side, and water is on the other side.
- Burners for oil-fired equipment are called gun burners and are mounted either on the front of the appliance or on pedestals on the floor for larger applications.
- The burner motor is usually a fractional-horsepower motor mounted to the burner housing with bolts.

It turns the oil pump and a small fan for forcing air into the burning oil.

- The burner motor runs at a speed of 1750 or 3450 RPM, and it turns clockwise or counterclockwise for different applications.
- A flexible coupler at the end of the motor shaft transfers the motor rotation to the oil pump, and the fan is fastened to the motor shaft with a set screw.
- The motor has a manual reset feature that protects the motor in case of overload.
- The oil is ignited with electrodes that are powered by a high-voltage transformer. Approximately 10,000 volts are used for ignition.
- The 10,000-volt ignition is much like the ignition system on an automobile. The voltage will not injure you, but it can give you an unpleasant electrical shock. This voltage creates an arc that should span a gap of approximately ½ inch.
- Electrodes are routed through ceramic insulators carrying the high-voltage arc to the vicinity of the oil for ignition. The electrode gap must be correct for proper ignition.
- The electrodes are not in the flame, but the arc at the end of the electrodes is blown into the flame by the air for combustion.
- The burner air-tube assembly carries air alongside the electrodes to the vicinity of the flame.
- The end of the burner air tube contains a swirl mechanism, called the air cone, which mixes the air with the oil. The air spins in one direction and the atomized oil in the opposite direction to improve the mixing of oil droplets and air.
- There are two types of air cones: the standard type that creates a lazy flame and the newer, more efficient flame-retention cone that holds the flame much tighter to the burner head.
- Approximately 1,540 cubic feet of air are required to completely burn 1 gallon of No. 2 fuel oil. However, normal combustion uses nearly 2,000 cubic feet of air. The excess air passes through the process but is necessary for safe combustion.
- Too little air causes smoke in the by-products of combustion, and too much air causes heat to escape up the flue, which is inefficient.
- Oil pumps are positive-displacement pumps that can pump high pressures. The oil pump has a capacity of approximately 2.5 gallons per hour, which is oversized for most systems. The excess capacity is bypassed to the pump inlet or the tank, depending on the system.
- The oil pump can be either a single-stage or a two-stage pump, depending on the application. When the fuel tank is above the pump, the pump can be single stage. When the tank is below the pump, a two-stage pump must be used. The second stage of the pump delivers the oil to the pump inlet (gravity does this with the tank above the pump).
- The pump system can be either a one-pipe or a two-pipe system, again depending on the application. The second pipe delivers the unused oil to the tank instead of to the pump inlet. This aids in priming a system that has run out of oil.
- The pump has a bypass plug that must be in place for a two-pipe application and that must be removed for a single-pipe installation. If the bypass plug is left in place in a one-pipe application and the pump is started, excess pressure is imposed on the pump seal and can create a leak.
- Oil pumps are either rotary or cam type pumps. They are not usually serviced by the technician, but they are replaced when problems occur.
- The typical two-stage oil pump with a two-pipe system can pull oil from as far as 15 feet below the pump. When the tank is nearly empty, the oil is at its lowest point.
- Approximately 1 inch of vacuum is created for each foot that the pump must lift the oil.
- If the oil is more than 15 feet below the tank, a secondary pump should be used to lift the oil to the pump or to a secondary storage tank near the appliance pump.
- Each pump has a built-in, adjustable regulating valve for setting the oil pressure. The typical oil pressure is 100 psig.
- The oil pump also has an oil-pressure shutoff to prevent low-pressure oil from being sprayed into the combustion chamber at the beginning and the end of the cycle. Low-pressure oil creates smoke. The shut-off valve keeps the pressure above 85 psig.
- An oil pump runs at either 1,750 or 3,450 RPM, and its rotation is clockwise or counterclockwise, so the correct replacement pump for an application must be chosen. You should take the old pump to the supply house and compare it with the replacement for best results.
- The oil pump prepares the oil at 100 psig for the oil nozzle.
- The oil nozzle prepares the oil for burning by metering the correct amount and creating the correct pattern for burning. It changes the the oil to a swirling pattern of tiny droplets to meet the air from the air cone.
- Each nozzle is marked to show the flow rate in gph (gallons per hour), the angle of spray, and the spray pattern.
- The spray pattern is either hollow, semihollow, or solid, depending on the application.
- The nozzle is a precision device that must be treated carefully. Do not try to clean a nozzle and its parts with a wire brush.
- A special tool called a nozzle wrench must be used to install nozzles.
- The burner primary control ensures that the flame ignites within a reasonable length of time, usually 90 seconds, and stays lit. If the flame does not ignite within 90 seconds or goes out for 90 seconds, the primary control and its circuits shut off the burner, which must then be manually reset.

- There are two types of primary controls: a cad (cadmium sulfide) cell that senses the yellow-orange burner flame and a stack switch, which senses the temperature in the flue pipe leaving the appliance. The cad cell is the most popular in modern equipment.
- Resetting an oil burner repeatedly without ignition causes oil to build up in the combustion chamber. When it is then ignited, more heat is produced than the appliance can handle.
- The mounting flange holds the burner assembly to the front of the appliance. It has a gasket to prevent uncontrolled air from leaking into the system during combustion.

Review Questions

1. Why is oil heat popular? Give two reasons.
2. What is the most-used fuel oil?
3. Is oil delivered in the solid, liquid, or vapor state?
4. Is oil burned in the solid, liquid, or vapor state?
5. Combustion chambers can be made of which two materials?
6. What is the purpose of the combustion chamber?
7. What is the purpose of the heat exchanger?
8. Heat exchangers for forced-air furnaces are made of what material?
9. Heat exchangers for boilers can be made of which two materials?
10. How many revolutions per minute can a typical oil burner motor turn?
11. What fastens the oil pump to the burner motor on a gun burner assembly?
12. What is the approximate voltage of the ignition transformer on a gun burner?
13. Where is the electrical arc for ignition on a gun burner?
14. What is the name of the newer type of air cone for a gun burner?
15. How many cubic feet of air are used to burn 1 gallon of fuel oil?
16. What are the symptoms of too little combustion air for a gun burner? too much air?
17. The bypass plug must be in place with a two-pipe system. True or false?
18. What is the purpose of the second pipe in a two-pipe system?
19. When should a two-stage pump be used?
20. What is the typical oil pressure for a gun burner?
21. What prevents the oil from burning at lower pressures when the unit shuts down?
22. What are three ways in which the oil nozzle prepares the oil to burn?
23. Why is it best to change an oil-burner nozzle rather than try to clean it?
24. How long does the oil burner run without ignition before the safety device shuts it down?
25. What are two types of safety devices used to shut off the burner if ignition does not occur?

13 Oil-Burning Equipment

Objectives

Upon completion of this unit, you should be able to

- **describe the upflow furnace versus the downflow or horizontal furnace.**
- **describe underground oil storage tanks and their problems.**
- **explain how water affects fuel oil.**
- **discuss above ground oil storage.**
- **explain the difference in a 1 and 2 pipe oil system.**

13.1 The Oil Supply System

Fuel oil must be supplied to the burner pump as a clean liquid, at the correct pressure. The fuel oil supply system can do this if it is properly designed and installed according to the governing codes and appropriate installation practices. This supply system consists of the oil-storage tank and the piping system.

At least two basic types of oil-storage systems are used: the central storage tank with supply tanks and the supply tank as the only tank. The central storage system is used when one system includes a number of furnaces (e.g., an apartment house). This type of system may have an auxiliary oil pump to lift the oil to the supply tank from the storage tank. Figure 13-1 shows an auxiliary oil pump. On the other hand, when the oil is pulled directly from the storage tank, the tank is called the *supply tank.*

Figure 13-1. Auxiliary oil pump. *Courtesy Suntec Industries Incorporated, Rockford, Illinois*

Oil-Storage Tanks

The storage system consists of one or more tanks installed correctly for the application according to the governing codes. Typical steel oil-tank capacities are 275, 550, and 1,050 gallons. When more capacity is needed, the tanks can be manifolded together if the local codes are followed. Tanks can be underground or above ground, as shown in Figure 13-2. Where local code permits, some

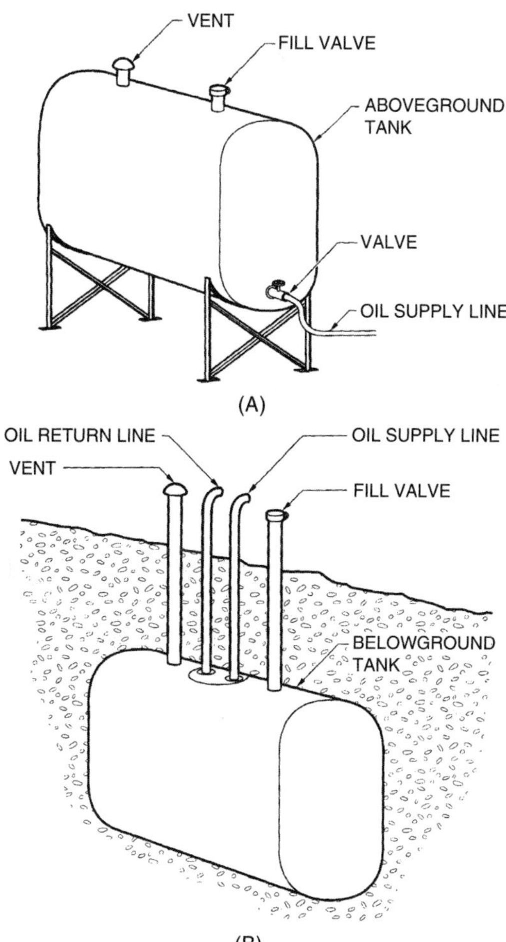

Figure 13-2. Storage tanks can be (A) above ground or (B) below ground.

aboveground tanks can be installed inside buildings as shown in Figure 13-3. Materials other than steel can be used for oil storage; check the code for your area to learn which materials are acceptable.

For many years, underground storage tanks have been buried by preparing the proper hole, coating the tank with asphalt (tar) to prevent corrosion, and covering the tank. It is customary to bury the tank at least 2 feet below the surface to prevent damage from surface digging. The tank should be positioned in the hole on a slant, away from the oil-line pickup connections. This slant allows any water or sludge to drain to the far end of the tank, as shown in Figure 13-4.

The fill pipe must be at least 2" pipe and be run as directly as possible to the tank. If possible, the fill pipe should be run straight into the tank so that the opening can be used to measure the oil level, as shown in Figure 13-5.

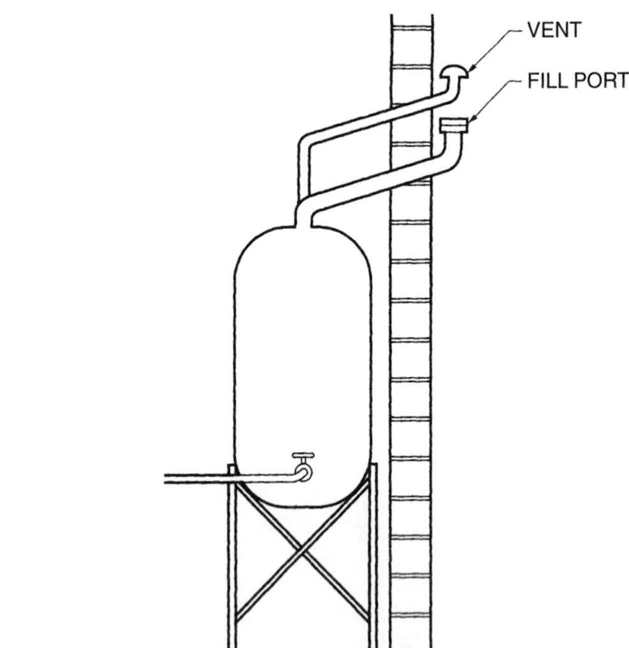

Figure 13-3. Aboveground tank installed inside a building.

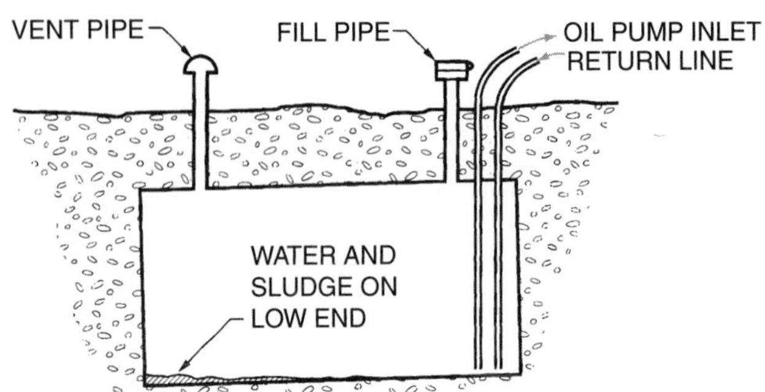

Figure 13-4 The tank should slant away from the oil-line pickup.

In some cases, the fill pipe may have to be offset with the tank. It should slant toward the tank if there is any horizontal distance to the pipe, as shown in Figure 13-6.

The tank must have a vent to allow air to escape as the fuel oil is added. The vent pipe must be 1¼" pipe and often has a whistle that sounds when oil is added to the tank,

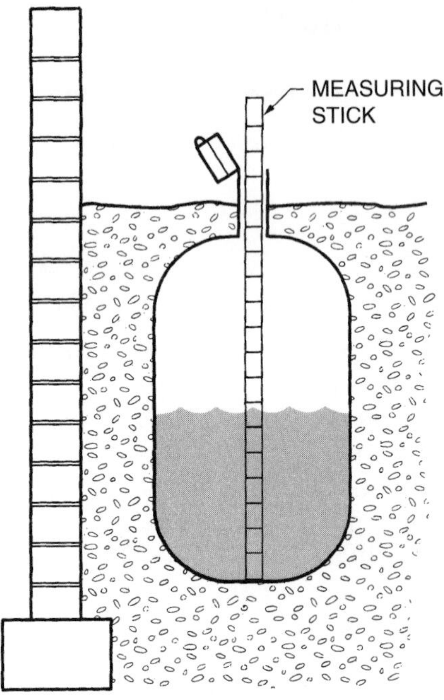

Figure 13-5. When the fill pipe is straight, the oil can be measured with a measuring stick.

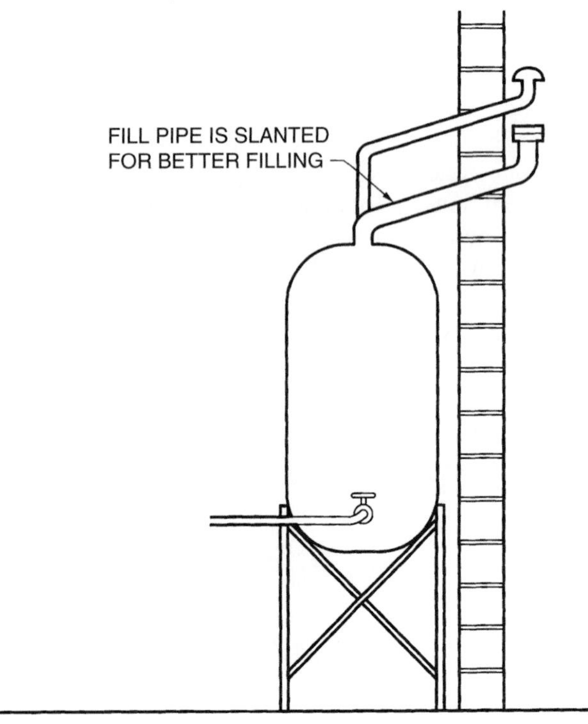

Figure 13-6. If the fill pipe must have bends, it must slant downward toward the tank.

as shown in Figure 13-7. The whistle is activated from air escaping the tank as the fuel is added. When the whistle stops, the tank truck driver knows to shut off the fuel when the whistle stops sounding, as shown in Figure 13-7.

A special fitting called *a duplex bushing* is used to fasten the field lines to the tank. This bushing eliminates the need to put fittings inside the tank, where a leak would be hard to detect. This bushing also allows the supply and return piping length to be adjusted. The supply line (pump suction pickup) should be approximately 3 inches off the bottom of the tank to allow for sludge buildup. The return line should be approximately 6 inches off the bottom of the tank. Figure 13-8 shows supply and return line positioning.

When an underground tank has a leak and the level of the oil is above the leak, oil will seep out. If the oil level is below the leak, water may seep in as shown in Figure 13-9. Either instance is a problem. Ground pollution is illegal, and an oil burner will not burn with water in the fuel. The technician can discover water in the fuel by using a special paste on the end of the oil measuring stick, as shown in Figure 13-10. When water is discovered, the underground tank should no longer be used, and the local authorities should be consulted about what to do. In some localities, the authorities may require you to dig up the tank and, if oil has leaked, clean the ground around the tank. Some authorities may even require you to fill the tank with a wet mixture of cement to prevent the tank from being reused.

Aboveground tanks are mounted on legs or stands. The legs are available in different lengths for different tank heights. Aboveground tanks may be installed outside or

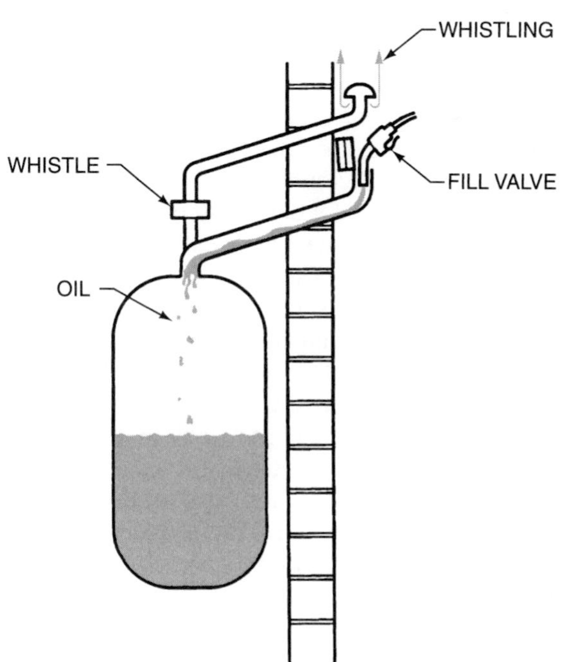

Figure 13-7. The whistle helps the tank truck driver know when the tank is full.

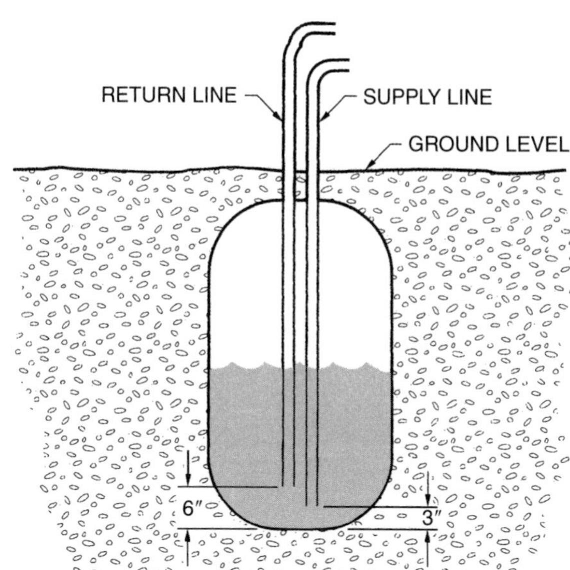

Figure 13-8. The oil return pipe on a two-pipe system should be approximately 6 inches off the bottom of the tank. The supply pipe should be 3 inches off the bottom.

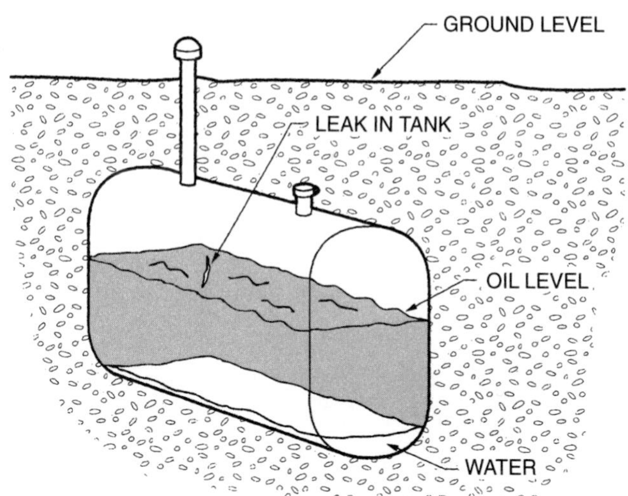

Figure 13-9. A leak in an underground tank. The oil level is below the leak, so water seeps in.

inside, depending on the code. When installed outside, they should be located where they are protected from physical damage, such as away from a drive-way as shown in Figure 13-11. The legs or stand must be located on a firm foundation so that the tank does not settle and topple over. The tank should be slightly high on the oil pickup end so that water and sludge can settle in the low end of the tank. The low end of the tank should have a drain valve for draining the water and sludge.

Like dew on a car window, water collects in aboveground tanks from condensation. During the night, the tank and its contents cool, and air is drawn into the tank. This air contains moisture that collects on the cold surfaces of the

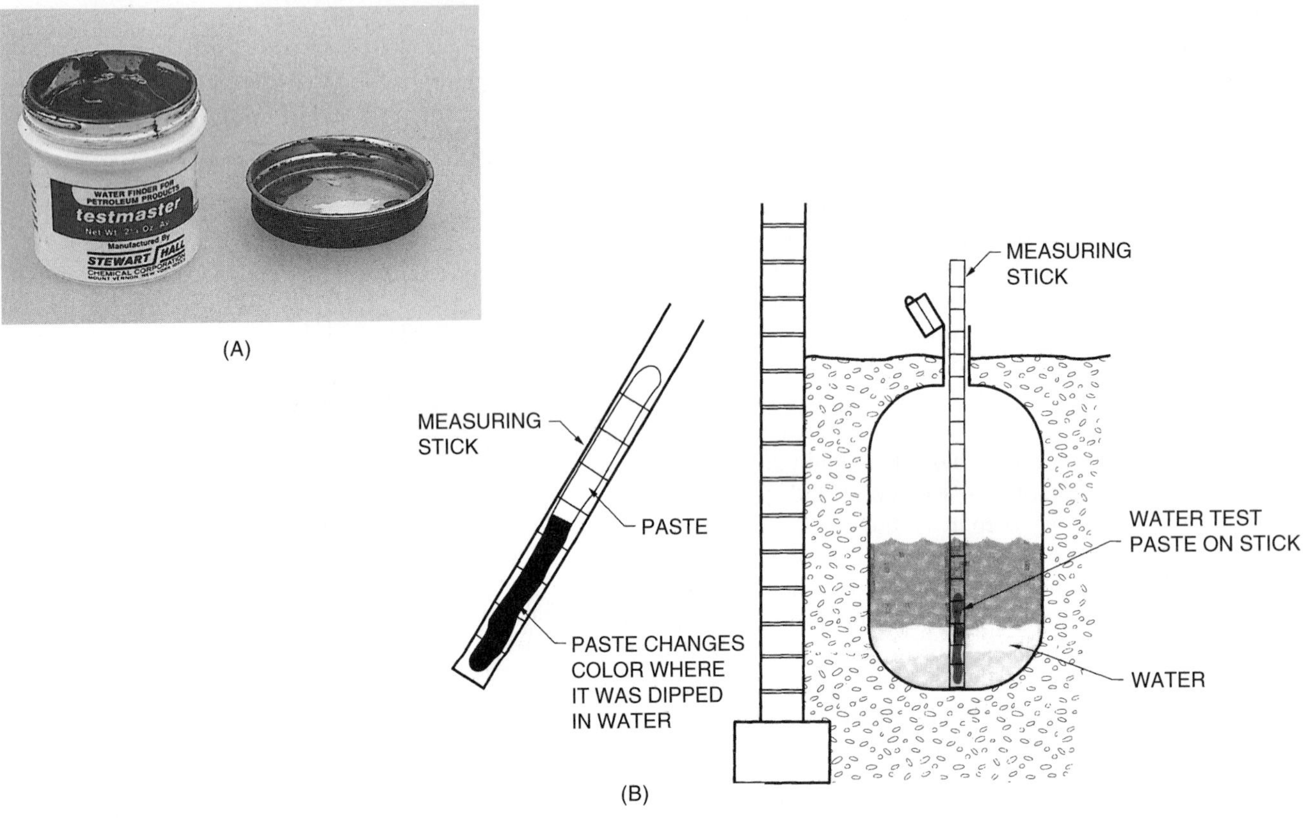

Figure 13-10. (A) Test paste. *Photo by Bill Johnson.* (B) Testing for water in an oil tank.

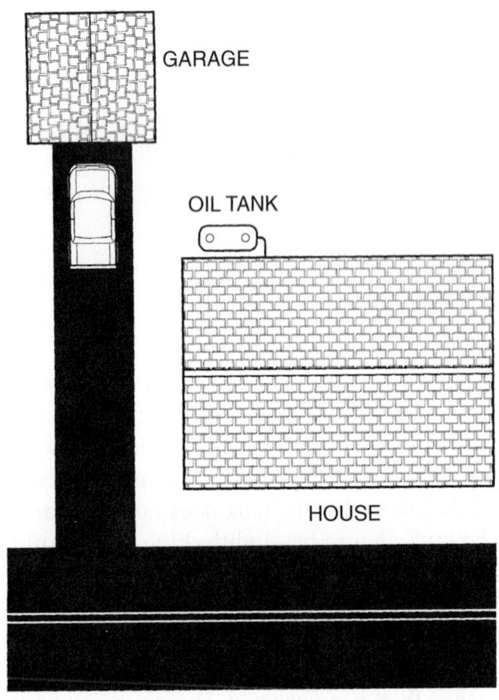

Figure 13-11. Aboveground tank location.

oil and the tank sides, as shown in Figure 13-12. During the day, the warmer temperatures heat the tank, and some of the air moves out to be replaced the next night with new moisture-laden air. Moisture accumulates in the tank in this way with time.

When tanks are permitted to be installed inside buildings, the code usually requires that storage of as much as 275 or 550 gallons be allowed, for safety reasons. This inside storage may be necessary in densely populated areas, where there is no place for outside storage. The local code may also insist that the tank be installed in a concrete or brick vault for protection from fire. Figure 13-13 shows an example of a vault. Tanks located in basements are not as likely to sweat and accumulate moisture because the temperature is more constant.

When the tank is located above the burner, it is often required that the line leaving the tank exit from the top with an antisiphon valve located at the highest point in the tank. This valve prevents oil from siphoning from the tank if a line breaks, as shown in Figure 13-14. The antisiphon valve requires a lot of suction to open, and it should be used only when required.

Note: Whenever you must install a storage tank, save yourself time by asking the local authorities for the proper location and procedures.

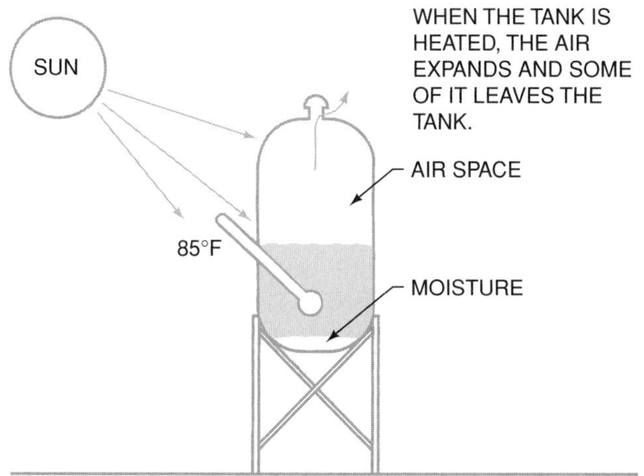

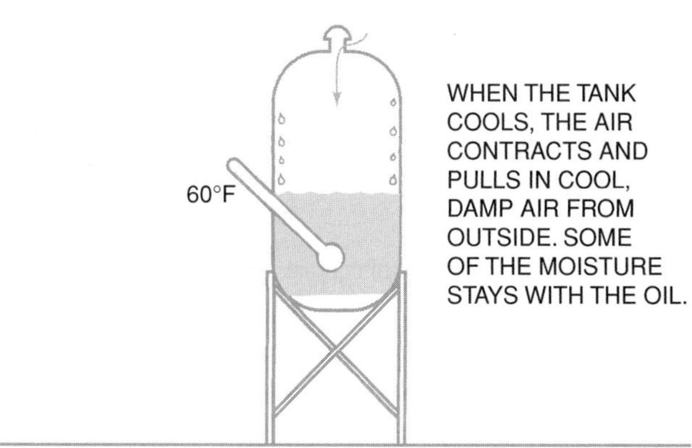

Figure 13-12. Aboveground tanks breathe.

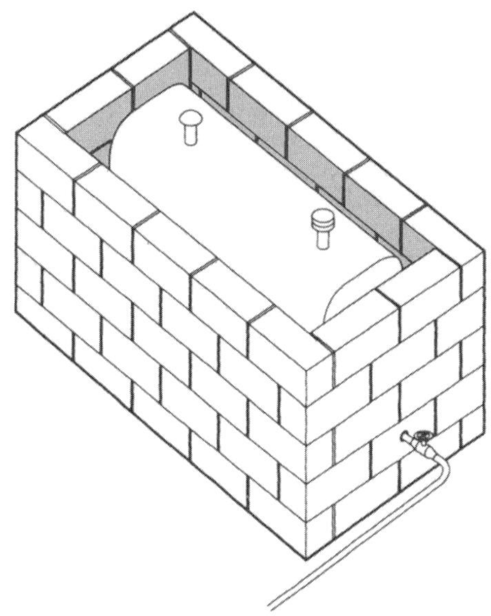

Figure 13-13. Tank inside a concrete vault.

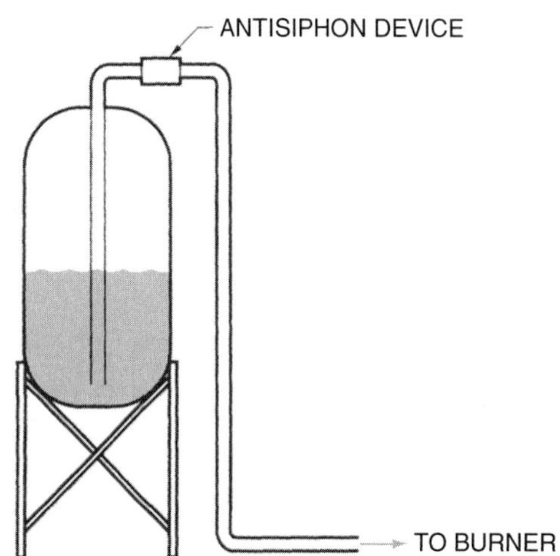

Figure 13-14. Antisiphon device.

The Piping System

Most piping systems for oil burners are run with copper tubing because it is easy to install and is economical. Hard pipe is used in some installations, but the system must be tight on the pump suction side because it is in a vacuum when the tank is below the burner. Because vacuum leaks are difficult to find, installing a system with pipe that is leak free is also difficult.

Most systems are run with ⅜-inch OD soft copper for the pump supply (suction) and the return. As the suction lift becomes greater and the run becomes longer, the supply line may need to be sized up to ½-inch OD copper. Figure 13-15 shows one manufacturer's recommendations for line sizing for one series of pumps. If you have questions about a line size that is to be installed or that is already installed and not performing, request specific line-size recommendations from the pump manufacturer.

The copper tubing connections should be made with flare fittings. Proper flaring practices must be followed, or leaks will occur. These practices include reaming the burr from the tubing before flaring or using a deburring flaring tool, as shown in Figure 13-16. The tubing should be routed in a direct line where possible with as few turns as possible. Often the piping must be run across a concrete floor. The tubing must be protected so that damage does not occur, so the floor may need to be grooved and the tubing recessed in the grooves, or the piping may be protected with runners, as shown in Figure 13-17.

If the tank is below the burner, the tubing should not be routed above the burner to avoid crossing the floor. Routing the tubing above the burner creates a loop in which air can become trapped, see Figure 13-18. One way to clear the air trap is to use positive pressure to pump oil through the top of the loop, with a check valve

MODEL A SINGLE-STAGE TWO-STEP • TWO-PIPE MAXIMUM LINE LENGTH (H + R)

Lift "H" Figure 4	3450 RPM					
	3/8" OD Tubing		1/2" OD Tubing			5/8" OD Tubing
	10 GPH	16 GPH	10 GPH	16 GPH	23 GPH	23 GPH
0′	33′	29′	100′	100′	72′	100′
1′	31′	27′	100′	100′	66′	100′
2′	28′	25′	100′	98′	59′	100′
3′	25′	23′	100′	89′	53′	100′
4′	23′	20′	92′	80′	46′	100′
5′	21′	18′	82′	72′	40′	100′
6′	18′	16′	72′	63′	34′	100′
7′	16′	14′	62′	55′	27′	88′
8′	13′	12′	52′	46′	20′	72′
9′	11′	9′	43′	37′	14′	56′
10′	—	—	33′	29′	8′	39′

MODEL B TWO-STAGE TWO-STEP AND TWO-STAGE HIGH-PRESSURE • TWO-PIPE MAXIMUM LINE LENGTH (H + R)

Lift "H" Figure 4	3450 RPM					
	3/8" OD Tubing		1/2" OD Tubing			5/8" OD Tubing
	10 GPH	16 GPH	10 GPH	16 GPH	23 GPH	23 GPH
0′	70′	60′	100′	100′	100′	100′
2′	64′	55′	100′	100′	100′	100′
4′	58′	50′	100′	100′	100′	100′
6′	52′	44′	100′	100′	100′	100′
8′	45′	39′	100′	100′	100′	100′
10′	39′	34′	100′	100′	100′	100′
12′	33′	28′	100′	100′	94′	100′
14′	27′	23′	100′	91′	76′	100′
16′	21′	18′	81′	70′	59′	100′
18′	—	—	57′	49′	41′	100′

*Maximum firing rate not to exceed maximum nozzle capacity or strainer rating, whichever is less. A greater firing rate requires a suitable external strainer.

Figure 13-15. Oil-line sizing. *Courtesy Suntec Industries Incorporated, Rockford, Illinois*

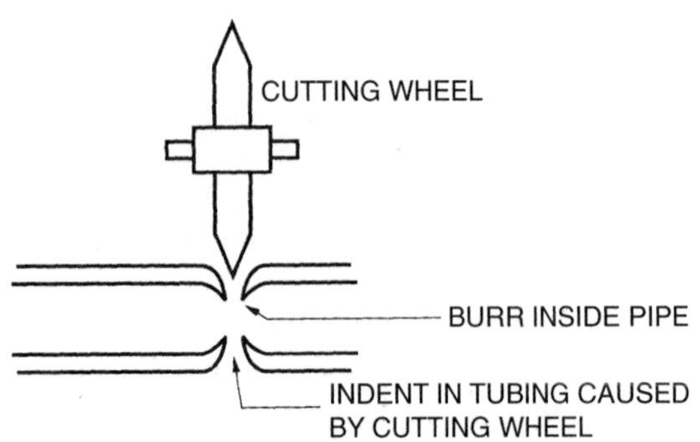

Figure 13-16. This burr must be removed.

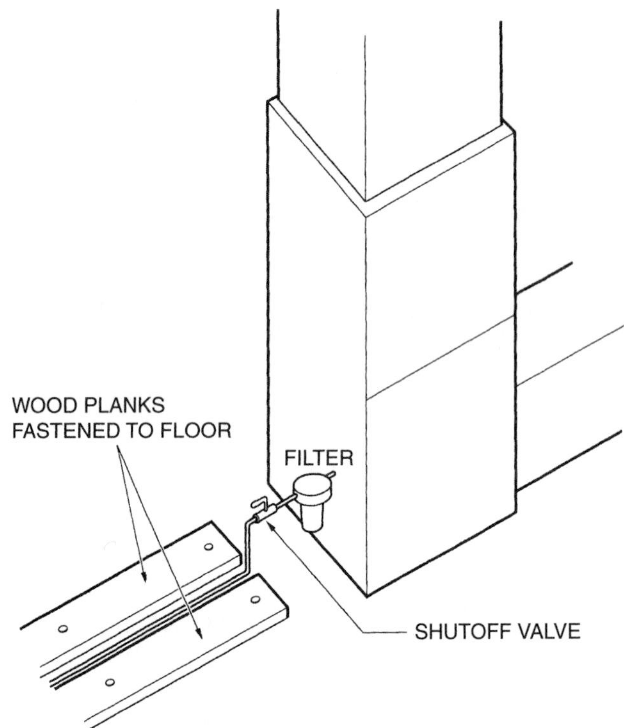

Figure 13-17. Protecting tubing that is routed across the floor.

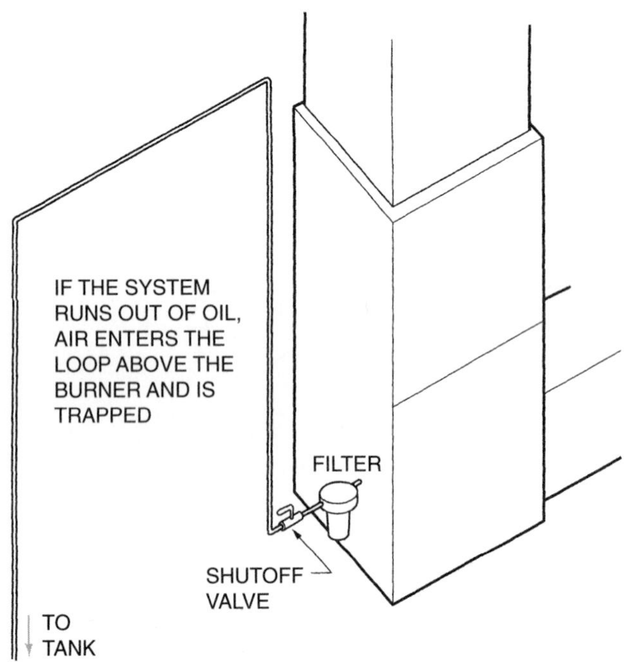

Figure 13-18. If the tank is below the burner, tubing should not be routed above the burner, or air traps will occur.

preventing the oil from siphoning back to the tank. Figure 13-19 shows this setup. This procedure must be repeated each time the tank runs dry, so proper tank monitoring is necessary.

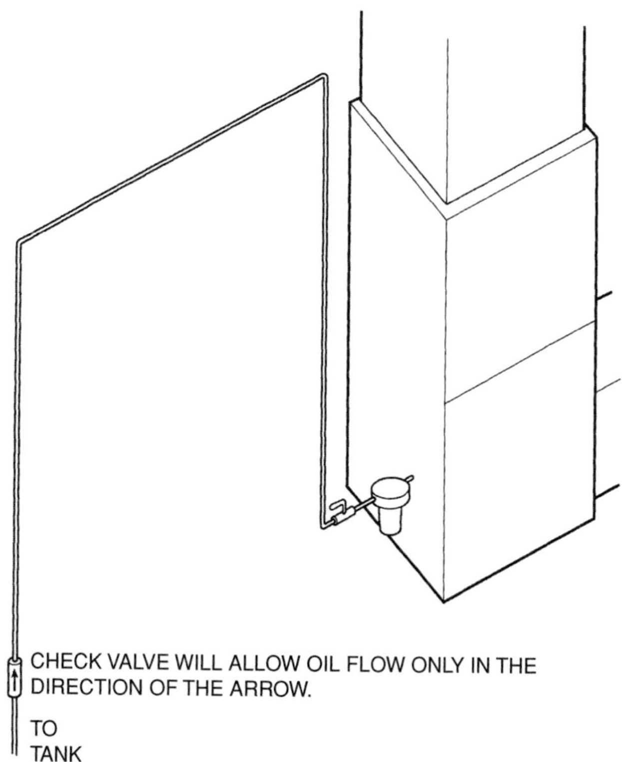

Figure 13-19. A check valve prevents oil from siphoning back to the tank when the line is routed above the burner.

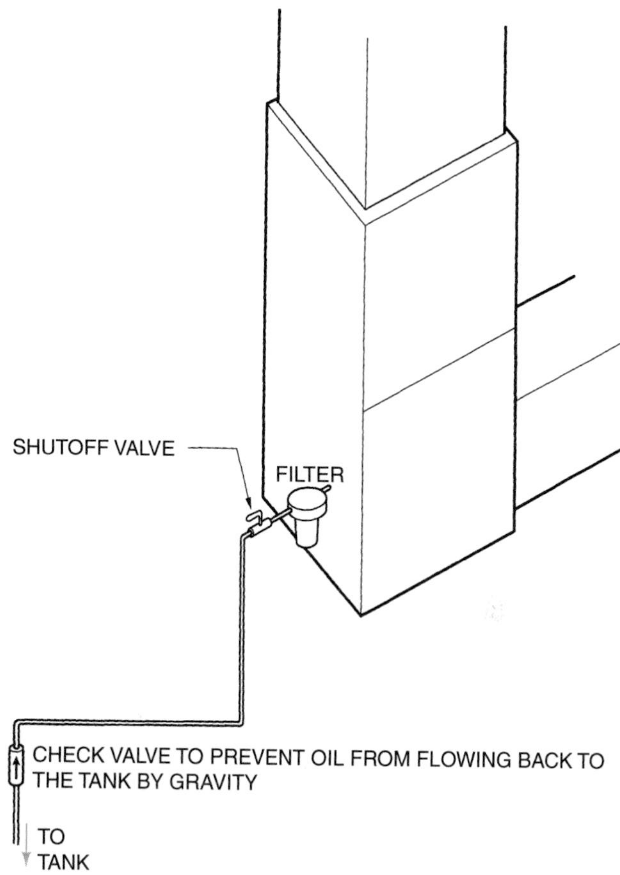

Figure 13-20. A check valve prevents the pump from losing prime.

Pumps

When the pump is above the tank, the pump may lose prime when the burner is shut off. A loss of prime occurs when the oil in the line runs back into the tank as a result of gravity and is replaced with air. The pump may not pull oil from the tank on start-up in time to prevent the system from shutting off because of the safety feature. (Remember, the pump has only 90 seconds before the system will shut down.) This problem can be remedied with a check valve in the pump supply (suction) line, as shown in Figure 13-20. There are several types of check valves: the swing check, the spring check, and the magnetic check. Figure 13-21 shows spring and magnetic check valves. The swing check valve probably has the least amount of pressure drop. The pressure drop in the spring and magnetic check valves can cause the pump to operate into an excess vacuum. The pressure drop of the check valve is added to the vacuum required to lift the oil.

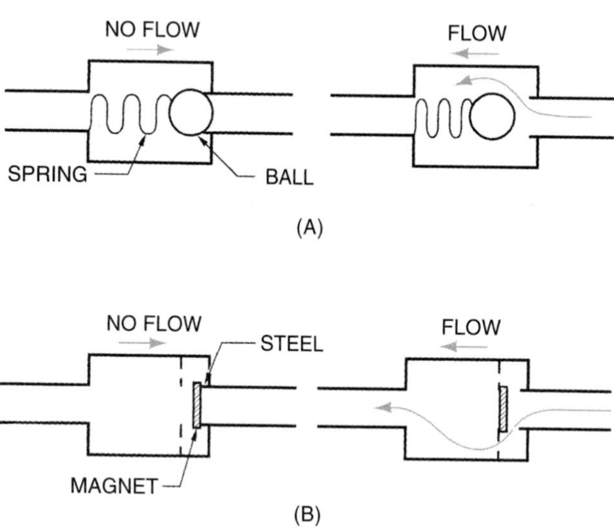

Figure 13-21. (A) Spring check valve and (B) magnetic check valve.

Fittings

All fittings in a piping system create turbulence and a pressure drop in the oil flow and must be kept to a minimum. However, some are necessary. An aboveground tank should have an oil shutoff valve if the oil line exits the bottom of the tank or if it exits the top and there is no antisiphon device, see Figure 13-22. The type of valve used at the tank determines the pressure drop through the valve. Figure 13-23 shows a gate valve and a ball valve. These valves have the least amount of pressure drop. Globe valves are often used, as shown in Figure 13-24,

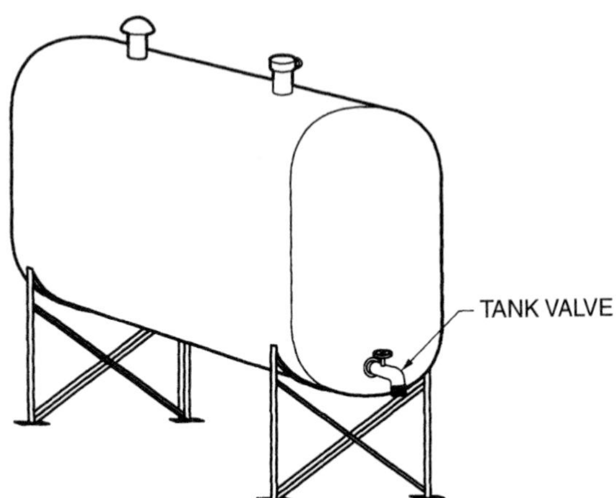

Figure 13-22. Oil shutoff valve at the tank.

but they have more of a pressure drop. Pressure drop is expressed in equivalent feet of pipe for valves and fittings. Figure 13-25 shows the pressure drops for several types of valves and fittings.

Figure 13-24. The globe valve has more of a pressure drop than other valves. *Photo by Bill Johnson*

Filters

Oil filters are necessary in the oil supply lines. Oil tanks create dirt particles from the moisture in the tank. These particles can restrict the small passages in the oil nozzle and must be removed. In the main oil line, in-line oil filters with changeable cartridges are common, as shown in Figure 13-26. These filter cartridges should be changed as part of the annual service. When the filter is serviced, it is a good practice to fill the filter casing with oil before putting the cartridge into the casing so that the pump does not have to pump air until it is primed when you start it, see Figure 13-27.

(A)

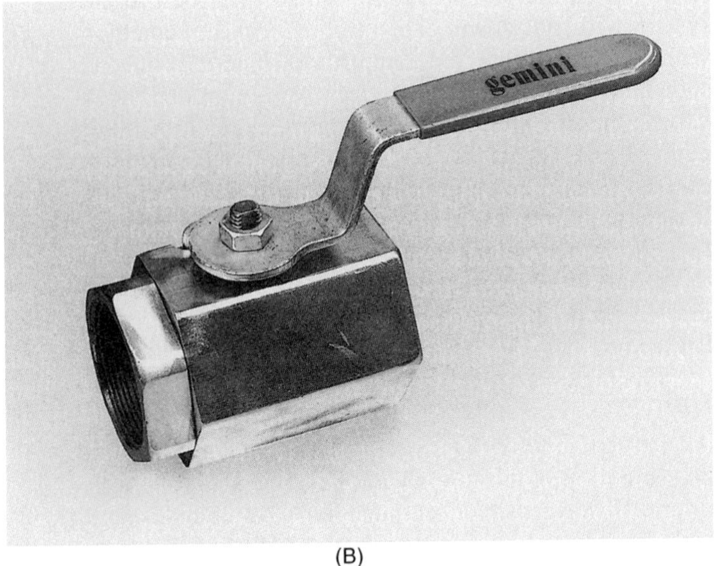

(B)

Figure 13-23. Gate valves and ball valves have the least pressure drop. (A) Gate valve. (B) Ball valve. *Photos by Bill Johnson*

Table C-2

		Screwed fittings[2]				90° welding elbows and smooth bends[3]					
		45° ell	90° ell	180° close return bends	Tee	R/d = 1	R/d = 1 1/3	R/d = 2	R/d = 4	R/d = 6	R/d = 8
	k factor =	0.42	0.90	2.00	1.80	0.48	0.36	0.27	0.21	0.27	0.36
	L/d' ratio[5] *n* =	14	30	67	60	16	12	9	7	9	12
Norminal pipe size, in.	Inside diam. *d*, in., Sched. 40[7]										
		L = equivalent length in feet of Schedule 40 (standard weight) straight pipe[7]									
1/2	0.622	0.73	1.55	3.47	3.10	0.83	0.62	0.47	0.36	0.47	0.62
3/4	0.824	0.96	2.06	4.60	4.12	1.10	0.82	0.62	0.48	0.62	0.82
1	1.049	1.22	2.62	5.82	5.24	1.40	1.05	0.79	0.61	1.79	1.05
1 1/4	1.380	1.61	3.45	7.66	6.90	1.84	1.38	1.03	0.81	1.03	1.38
1 1/2	1.610	1.88	4.02	8.95	8.04	2.14	1.61	1.21	0.94	1.21	1.61
2	2.067	2.41	5.17	11.5	10.3	2.76	2.07	1.55	1.21	1.55	2.07
2 1/2	2.469	2.88	6.16	13.7	12.3	3.29	2.47	1.85	1.44	1.85	2.47
3	3.068	3.58	7.67	17.1	15.3	4.09	3.07	2.30	1.79	2.30	3.07
4	4.026	4.70	10.1	22.4	20.2	5.37	4.03	3.02	2.35	3.02	4.03
5	5.047	5.88	12.6	28.0	25.2	6.72	5.05	3.78	2.94	3.78	5.05
6	6.065	7.07	15.2	33.8	30.4	8.09	6.07	4.55	3.54	4.55	6.07
8	7.981	9.31	20.0	44.6	40.0	10.6	7.98	5.98	4.65	5.98	7.98
10	10.02	11.7	25.0	55.7	50.0	13.3	10.0	7.51	5.85	7.51	10.0
12	11.94	13.9	29.8	66.3	59.6	15.9	11.9	8.95	6.96	8.95	11.9
14	13.13	15.3	32.8	73.0	65.6	17.5	13.1	9.85	7.65	9.85	13.1
16	15.00	17.5	37.5	83.5	75.0	20.0	15.0	11.2	8.75	11.2	15.0
18	16.88	19.7	42.1	93.8	84.2	22.5	16.9	12.7	9.85	12.7	16.9
20	18.81	22.0	47.0	105	94.0	25.1	18.8	14.1	11.0	14.1	18.8
24	22.63	26.4	56.6	126	113	30.2	22.6	17.0	13.2	17.0	22.6

1. Values for welded fittings are for conditions where bore is not obstructed by weld spatter or backing rings. If appreciably obstructed, use values for "Screwed fittings."
2. Flanged fittings have three-fourths the resistance of screwed elbows and tees.
3. Tabular figures give the extra resitance due to curvature alone to which should be added the full length of travel.
4. Small size socket-welding fittings are equivalent to miter elbows and miter tees.

(A)

Table C-2 (Continued)

Miter elbows[4] (No. of miters)					Welding tees		Valves (screwed, flanged, or welded)			
1–45°	1–60°	1–90°	2–90°	3–90°	Forged	Miter[4]	Gate	Globe	Angle	Swing check
0.45	0.90	1.80	0.60	0.45	1.35	1.80	0.21	10	5.0	2.5
15	30	60	20	15	45	60	7	333	167	83
			6	6						
L = equivalent length in feet of Schedule 40 (standard weight) straight pipe[7]										
0.78	1.55	3.10	1.04	0.78	2.33	3.10	0.36	17.3	8.65	4.32
1.03	2.06	4.12	1.37	1.03	3.09	4.12	0.48	22.9	11.4	5.72
1.31	2.62	5.24	1.75	1.31	3.93	5.24	0.61	29.1	14.6	7.27
1.72	3.45	6.90	2.30	1.72	5.17	6.90	0.81	38.3	19.1	9.58
2.01	4.02	8.04	2.68	2.01	6.04	8.04	0.94	44.7	22.4	11.2
2.58	5.17	10.3	3.45	2.58	7.75	10.3	1.21	57.4	28.7	14.4
3.08	6.16	12.3	4.11	3.08	9.25	12.3	1.44	68.5	34.3	17.1
3.84	7.67	15.3	5.11	3.84	11.5	15.3	1.79	85.2	42.6	21.3
5.04	10.1	20.2	6.71	5.04	15.1	20.2	2.35	112	56.0	28.0
6.30	12.6	25.2	8.40	6.30	18.9	25.2	2.94	140	70.0	35.0
7.58	15.2	30.4	10.1	7.58	22.8	30.4	3.54	168	84.1	42.1
9.97	20.0	40.0	13.3	9.97	29.9	40.0	4.65	222	111	55.5
12.5	25.0	50.0	16.7	12.5	37.6	50.0	5.85	278	139	69.5
14.9	29.8	59.6	19.9	14.9	44.8	59.6	6.96	332	166	83.0
16.4	32.8	65.6	21.9	16.4	49.2	65.6	7.65	364	182	91.0
18.8	37.5	75.0	25.0	18.8	56.2	75.0	8.75	417	208	104
21.1	42.1	84.2	28.1	21.1	63.2	84.2	9.85	469	234	117
23.5	47.0	94.0	31.4	23.5	70.6	94.0	11.0	522	261	131
28.3	56.6	113	37.8	28.3	85.0	113	13.2	629	314	157

For SI units: 1 foot = 0.305 m

5. Equivalent resistance in number of diameters of straight pipe computed for a value of — 0.0075 from the relation — /4.
6. For condition of minimum resistance where the centerline length of each miter is between and 2 1/2.
7. For pipe having other inside diameters, the equivalent resistence may be computed from the above values.

(B)

Figure 13-25. Pressure drop expressed in equivalent feet for valves and fittings. *Courtesy American Society of Heating, Refrigerating and Air-Conditioning Engineers*

(A)

(B)

Figure 13-26. (A) In-line oil filter and cartridge. (B) Filter housing. *Photos by Bill Johnson*

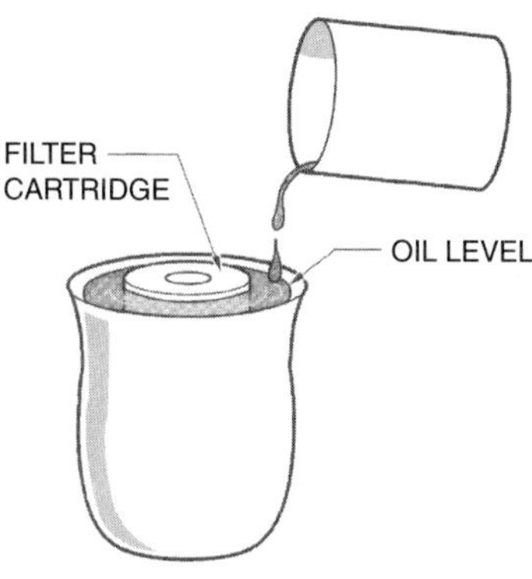

Figure 13-27. Fill the oil filter casing with oil after servicing it to aid in priming the pump.

A small oil filter can also be placed in the line between the pump and the nozzle, as shown in Figure 13-28. This filter removes pump shavings before they reach the strainer at the nozzle inlet.

Caution: Keeping the oil clean and free of particles is essential.

13.2 Automatic Control for Oil Heat

Automatic control for oil heat involves controlling forced-air furnaces or boilers. Forced-air furnace control differs from boiler control in that air-conditioning can be added to the furnace and must be integrated into the low-voltage control circuit.

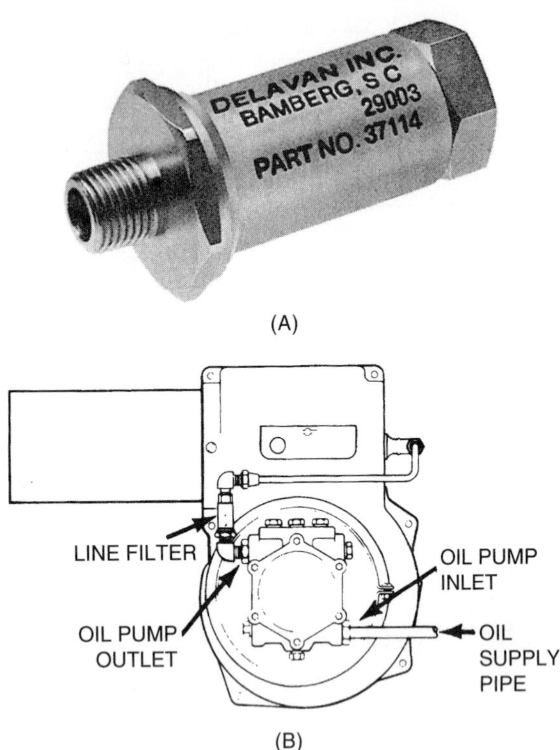

Figure 13-28 (A) Photo of small oil filter. *Courtesy Delavan Corporation.* (B) Small oil filter in the oil line to the nozzle.

Low-Voltage Control

The main power supply to all residential oil furnaces is 115 volts and is 230 volts to many commercial furnaces. Typically, the line voltage is reduced to 24 volts of control voltage. The low control voltage is used in all residential and many commercial installations because of safety: Customers are not likely to be injured from 24 volts if they tamper with the controls. Many local codes allow the heating technician to run low-voltage control wiring without a license because the hazard is minimal.

A typical oil forced-air furnace that is used exclusively for heating has a wiring diagram like the one shown in Figure 13-29. This system has 115-volt and 24-volt controls, but only the 24-volt circuit goes to the room thermostat. Notice that only two wires are required for the low-voltage control circuit. They fasten to the two terminals marked *T* on the primary. The thermostat has only two wires, so wiring is simple for heating-only systems and gives the service technician the opportunity to control the oil burner from the primary control location. Often this location is under a house and the technician need not make a trip to the thermostat. The other two terminals marked *F* on the primary are the connections for the cad cell. If the technician wants the burner to start, only a jumper from T to T is required, provided that the burner does not need to be reset. Figure 13-30 shows this jumper.

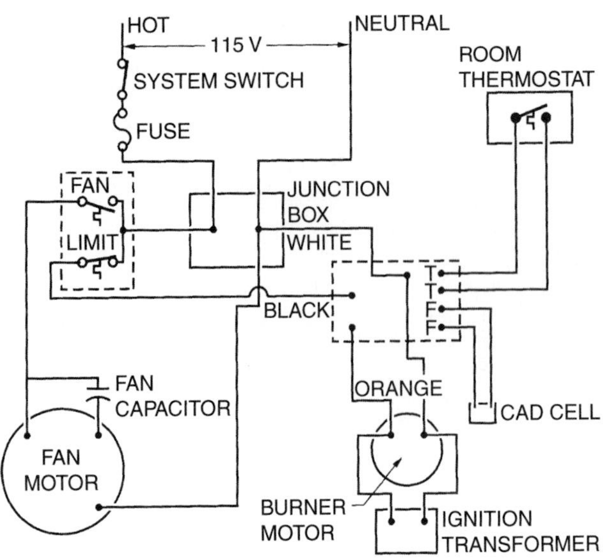

Figure 13-29 Wiring diagram for a furnace used exclusively for heating.

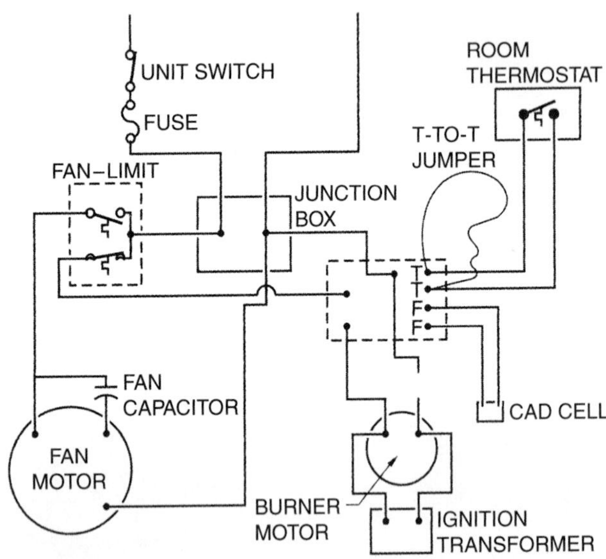

Figure 13-30 The technician needs only a T-to-T jumper to start the burner.

When this control system is used where air-conditioning is required, an *isolated subbase thermostat* may be used. This thermostat has two control circuits wired into the subbase—one for heating and one for cooling—as shown in Figure 13-31. The low-voltage power supply for the oil furnace is a separate, isolated circuit. The cooling system must have its own control-circuit power supply.

One type of oil-burner primary control has a control transformer designed for heating and air-conditioning that is built in for convenience. The compressor contactor for air-conditioning requires more capacity from the control transformer, so a larger transformer is required. A typical heating-only control transformer may have a capacity

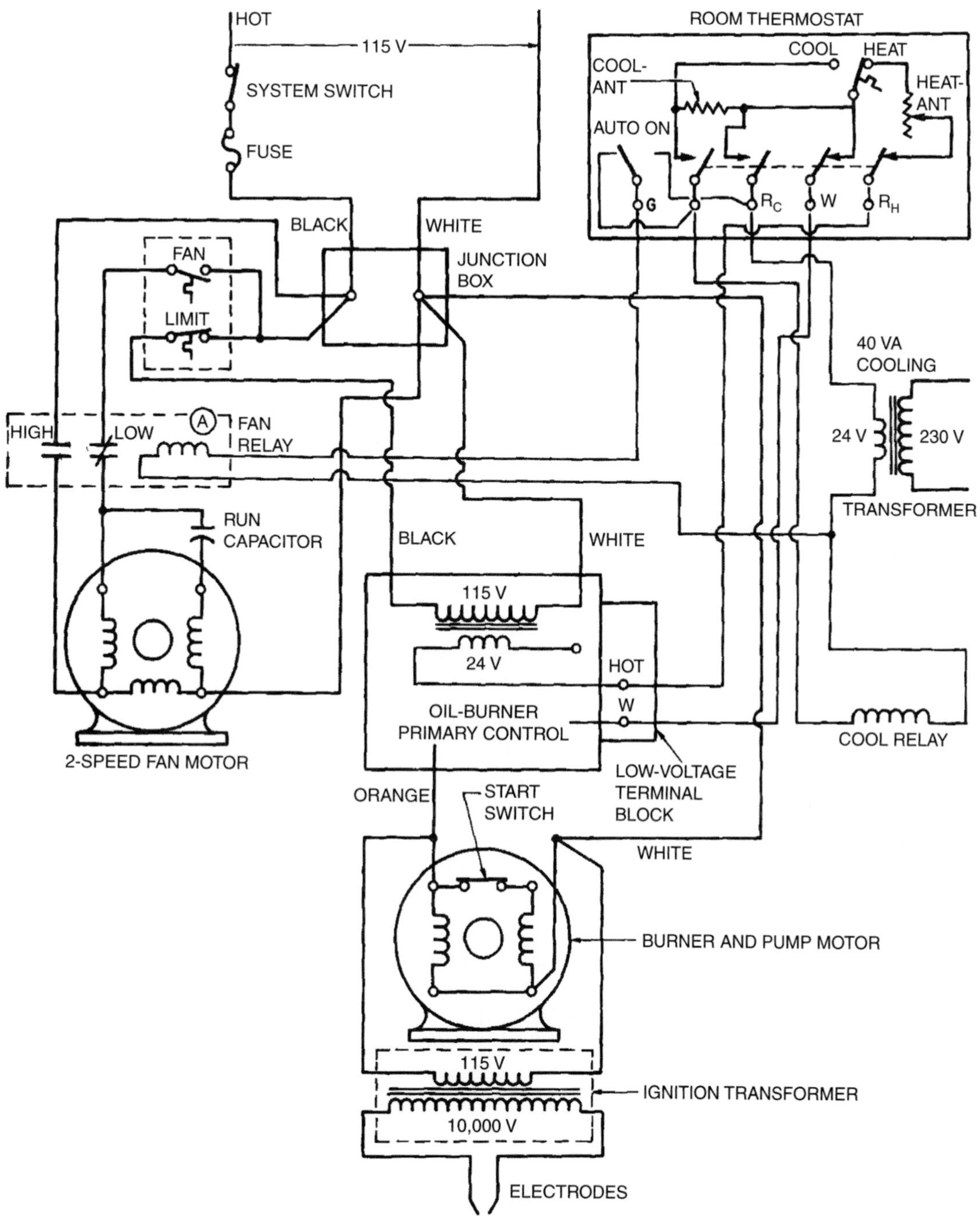

Figure 13-31 Isolated subbase thermostat diagram.

of 25 VA (volt–amperes), whereas a 40-VA transformer may be required for air-conditioning. Note the wiring diagram shown in Figure 13-32 for the combination primary control.

The primary control must turn the burner on and off in response to the low-voltage operating controls (the thermostat). When this thermostat closes its contacts, low voltage is supplied to the coil that energizes and closes a switch by a magnetic force. This switch transfers line voltage to the burner circuit to start the burner motor and energize the ignition transformer. (Remember that there are two types of primary control circuits: the cad cell and the stack switch.)

Fan Circuits

The fan circuit in an oil furnace starts the fan only after the heat exchanger is hot. This circuit also keeps the fan running after the burner is turned off until most of the heat is out of the heat exchanger. If the fan were to be started when the burner is started, the first air out of the duct would be cool compared with body temperature,

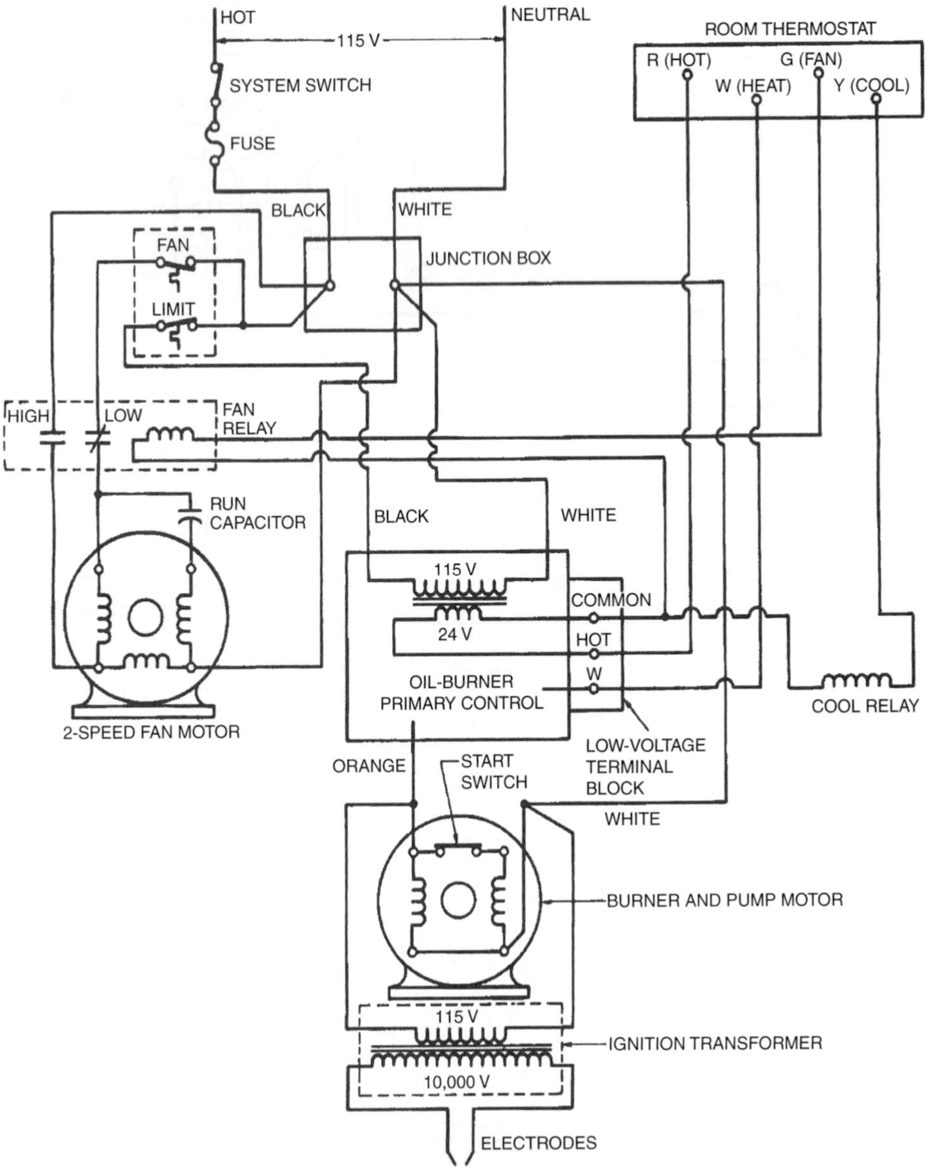

Figure 13-32 The combination primary control allows air-conditioning to be added to the system.

and customers would complaint about drafts. If the fan were turned off with the burner, the heat left in the heat exchanger would rise up the flue and be wasted, as shown in Figure 13-33.

The fan in an oil forced-air furnace is controlled by temperature, much as the fan in a gas furnace is. A temperature-controlled fan has a fan–limit control with a bimetal sensing element located where it can sense the temperature of the heat exchanger, Figure 13-34. The power from the power supply goes to the fan–limit control before going to the primary control. If the unit is overheated, the limit portion of the fan–limit control does not allow power to reach the primary control, and the burner does not fire, see Figure 13-35. The fan portion of the fan–limit control starts the fan long before the limit portion shuts off the burner. The horizontal or downflow furnace may have an auxiliary limit

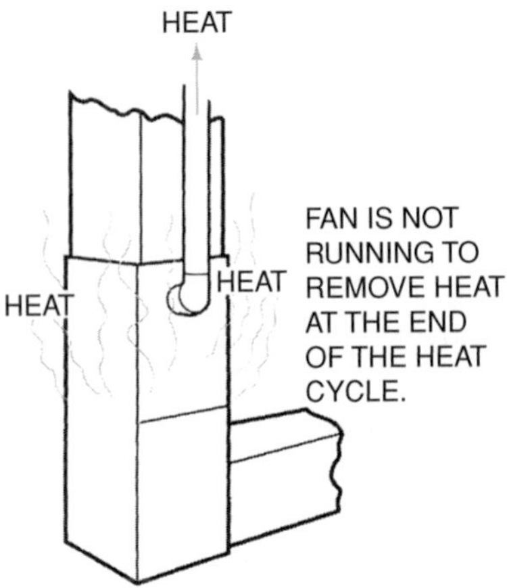

Figure 13-33. The fan should run after the burner shuts off, or the residual heat will rise up the flue.

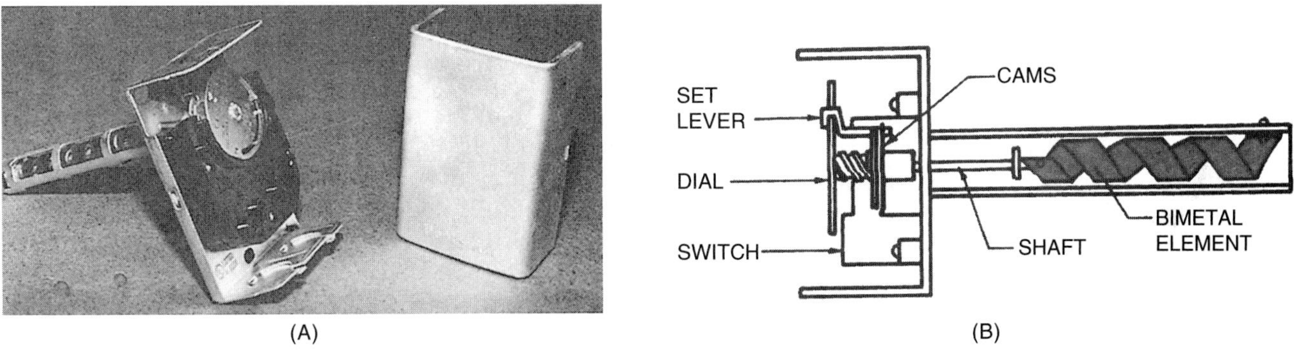

Figure 13-34. (A) Thermostatic fan control. *Photo by Bill Johnson.* (B) Location of bimetal element.

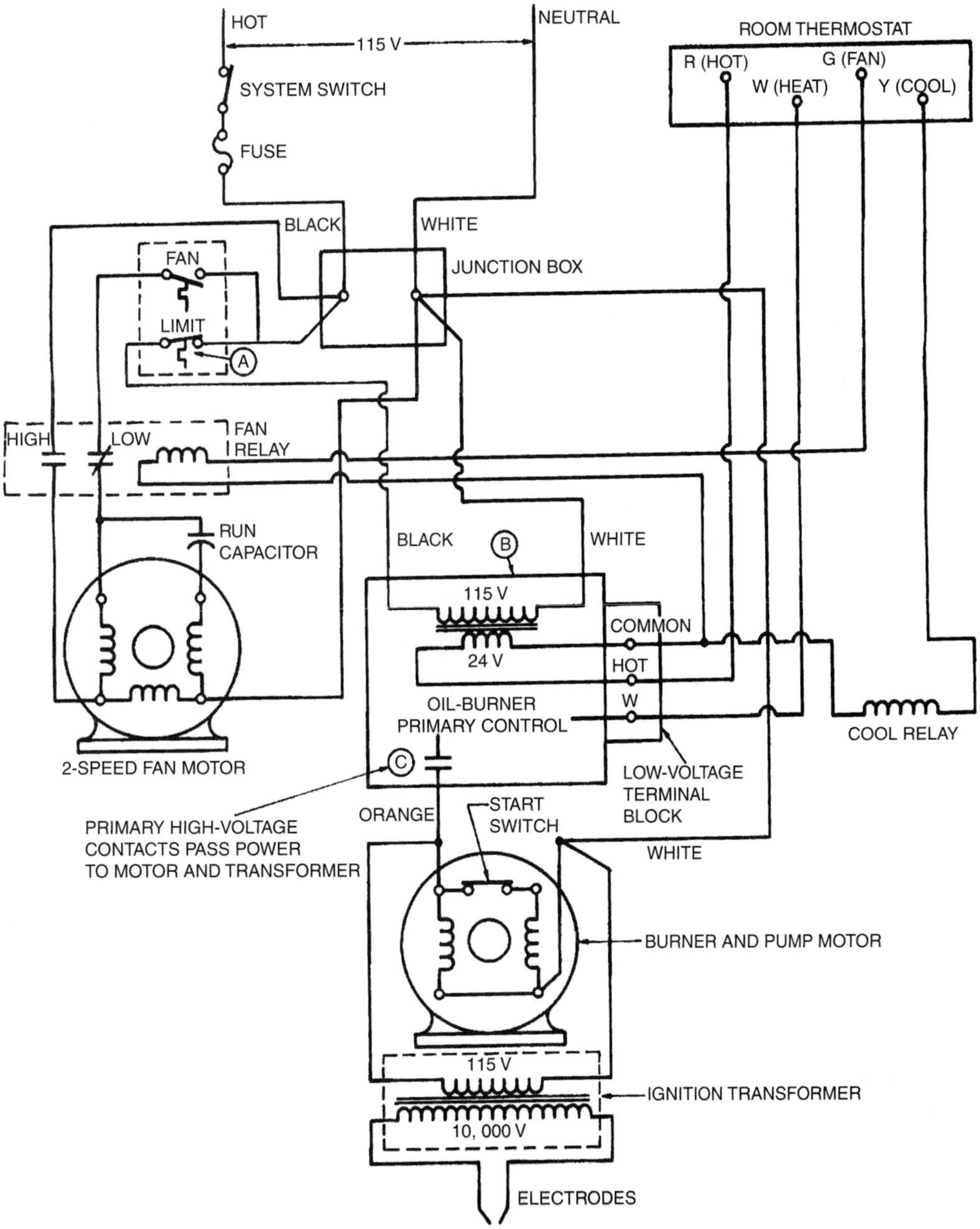

Figure 13-35. The limit portion of the fan–limit control shuts off the burner in case of overheating.

control because the temperature at the heat exchanger is more difficult to detect.

Oil Boiler Control

The burners in oil boilers are the same as those used in forced-air furnaces. They have either cad cell or stack switch safety devices. The room thermostat controls the water pump and allows the boiler to constantly remain hot when the system has more than one zone and pump or has a built-in hot-water heater, as shown in Figure 13-36. If the hot-water system has one pump, the room thermostat stops and starts the boiler and water pump or allows the water pump to run and only cycles the boiler off, Figure 13-37. Cycling the boiler on and off can cause noise in the system as a result of pipe expansion and contraction.

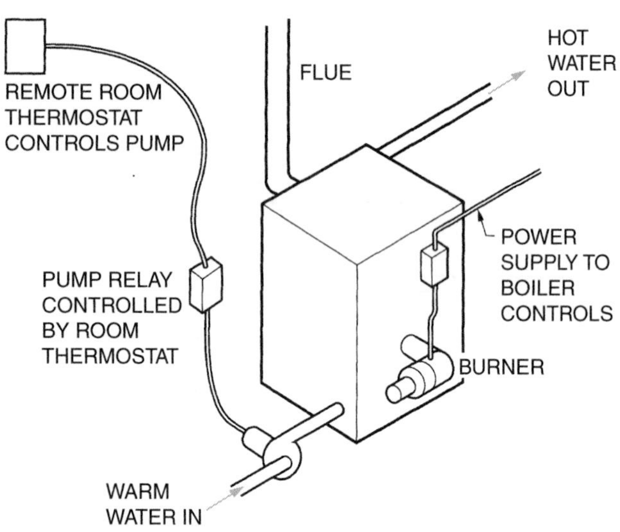

Figure 13-36. Hot-water system controls.

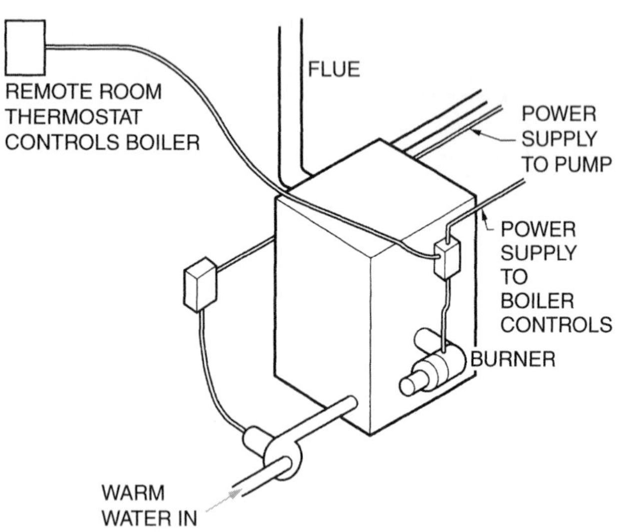

Figure 13-37. The room thermostat stops and starts the boiler in this diagram.

Unit 8 discusses gas boilers, which are much like oil boilers. Often the manufacturer makes one boiler shell and either a gas burner or an oil burner is fitted to the shell.

Steam boilers are controlled with pressure switches that stop and start the boilers according to the steam pressure. Figure 13-38 shows a steam boiler pressure control. The steam rises from the boiler to the room heating units, where it is turned to condensate. The condensate is returned to the boiler by a float-actuated condensate pump in some installations. Figure 13-39 shows a steam condensate pump.

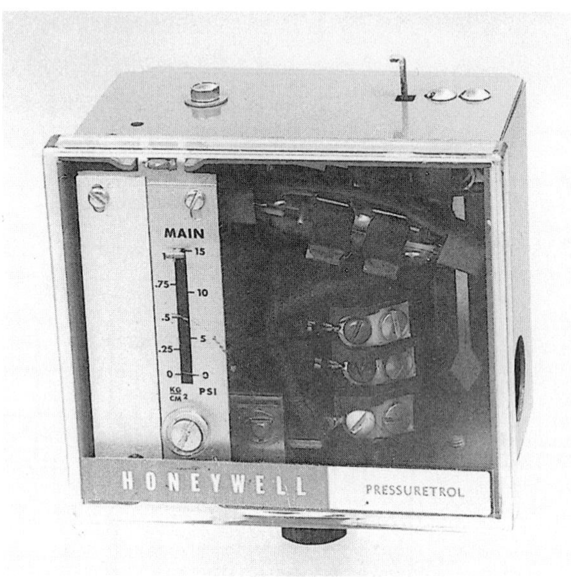

Figure 13-38. The steam boiler is controlled with a pressure switch. *Photo by Bill Johnson*

Figure 13-39. A float-actuated sump pump pumps the condensate back to the boiler. *Photo by Bill Johnson*

The point to remember is that the oil burners are the same for forced-air furnaces as for boilers and have the same control designations. The T terminals stop and start the burner. The control used to stop and start the boiler is found in the circuit between the T terminals. The cad cell or stack switch functions in the same manner as it does in the forced-air furnace.

13.3 Forced-Air Furnaces

Oil forced-air furnaces are much like gas furnaces except the heat exchanger is more heavy duty. The heat exchangers are typically much thicker for oil-burning equipment because of the increased combustion temperatures for oil. The thinner-gauge metals are more subject to stress cracks. Because oil heat exchangers are so thick, they are often welded at the seams and filler welding rods are used on the more heavy-duty furnaces, as shown in Figure 13-40.

Airflow

Oil furnace airflow can be upflow, downflow, or horizontal. See Figure 13-41.

UPFLOW FURNACE

High-Profile Furnace The high-profile upflow furnace is designed to be installed in a basement or a hall closet with the supply duct exiting the top of the furnace. Many homes have basements with ceilings high enough for the duct to be installed under the floor of the upper story, see Figure 13-42. Notice that the ceiling must be more than 8 feet high for full clearance below the duct.

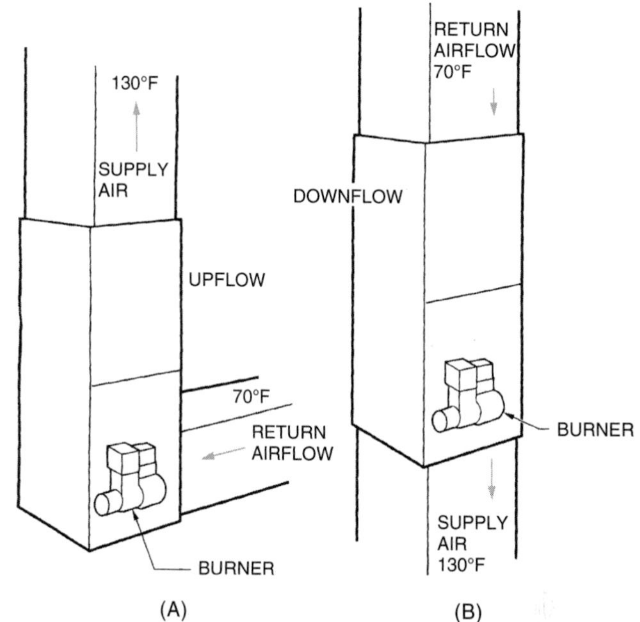

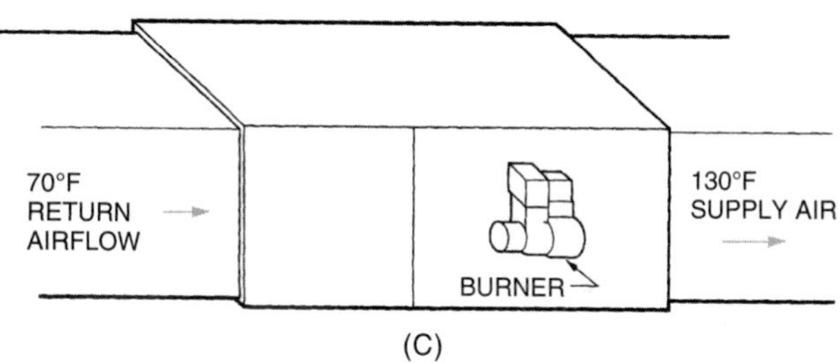

Figure 13-41. Oil-furnace airflow can be (A) upflow, (B) downflow, or (C) horizontal.

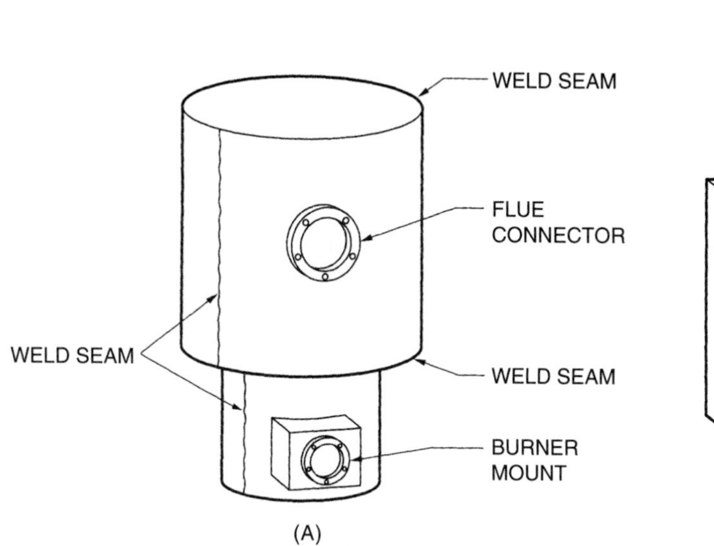

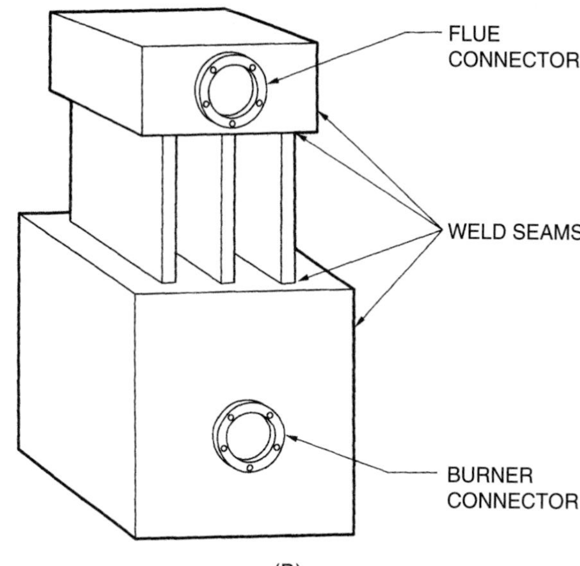

Figure 13-40. Many oil-furnace heat exchangers are welded at the seams. (A) Round heat exchanger. (B) Square heat exchanger.

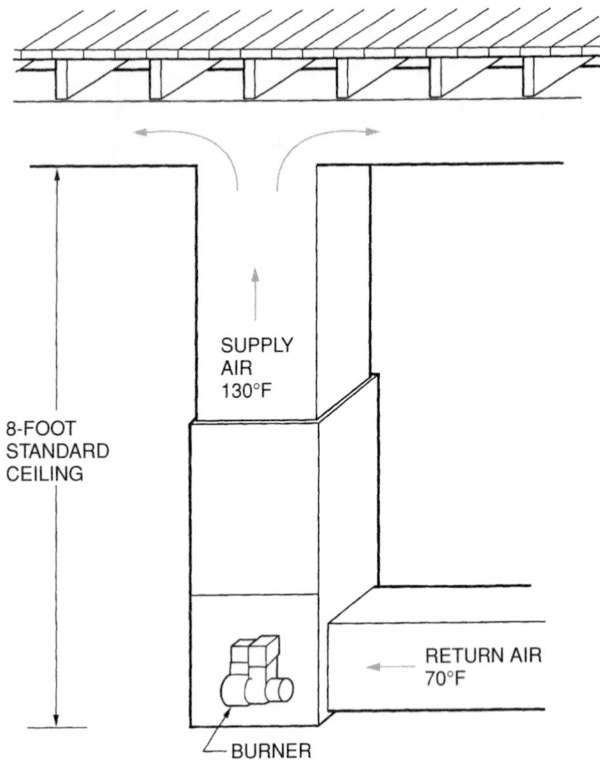

Figure 13-42. This furnace is located in the basement with the duct run overhead. There is still headroom for people to walk around in the basement.

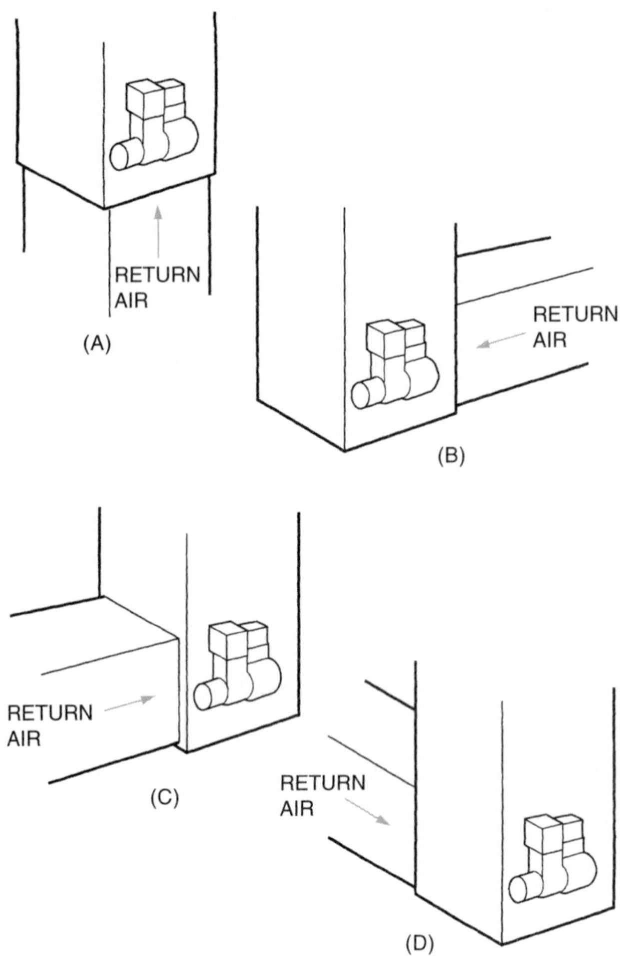

Figure 13-44. Different return air inlets for an upflow furnace. (A) On the bottom, (B) on the right side, (C) on the left side, and (D) on the back of the furnace.

The filter for the furnace is located at the furnace common return in a filter rack. Alternatively, it may be located in a return air-filter grille when a few common returns are used.

The furnace has a flange on the discharge air outlet so that the installer can fasten the duct system to the furnace, as shown in Figure 13-43. This flange must be used to ensure a tight connection for the duct.

The return air inlet may be installed on the bottom, on either side, or on the back of the typical upflow furnace, as shown in Figure 13-44. The manufacturer often provides knock-out panels at these locations for the installer. As with gas furnaces, the furnace manufacturer does not know where the return air will be fastened, so an auxiliary filter rack may be furnished with the furnace.

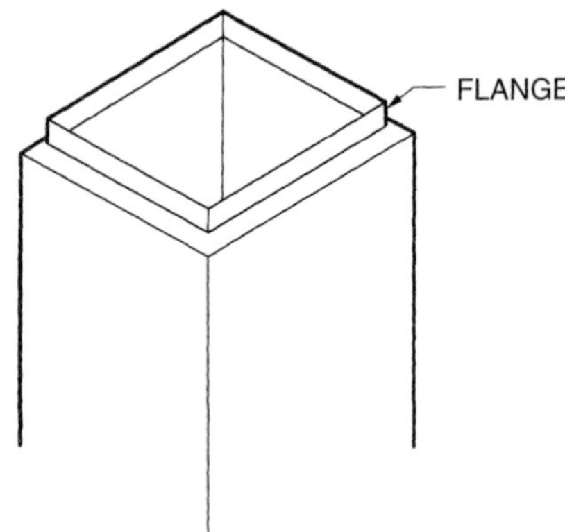

Figure 13-43. This flange on the furnace outlet allows the installer to fasten the duct to the furnace with an airtight connection.

Low Profile The low-profile upflow furnace can be installed where ceiling height is too low for a high-profile furnace. The fan and return air are in the side or back of this furnace, with the return air entering and exiting the top, as shown in Figure 13-45. This furnace requires more floor space than the high profile but less headroom. The filter is usually located in the return air portion of the furnace. Referring to Figure 13-46, note the location of the flue pipe in the return airstream in the furnace. The manufacturer wanted to extract the most heat possible from the heat exchange.

DOWNFLOW FURNACE. The downflow furnace looks just like an upflow furnace except the furnace fan is on top of the heat exchanger and blows the air downward.

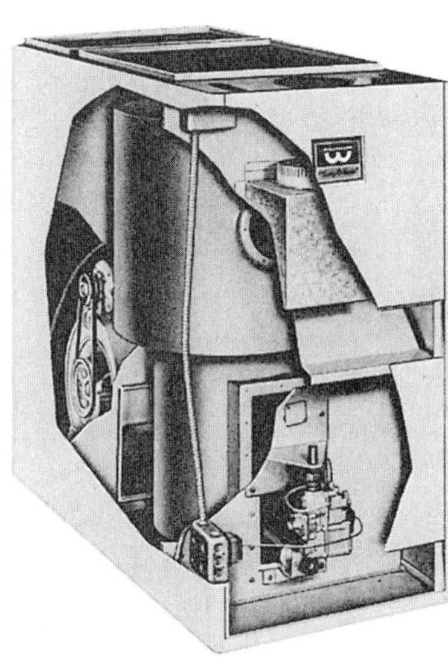

Figure 13-45. The fan and return air are in the side of the low-boy furnace. *Courtesy The Williamson Company*

Figure 13-46. The flue is in the return air to extract more heat from the by-products of combustion. *Courtesy ThermoPride*

The air inlet is at the top, so the return air is taken from above. These installations would have the return air duct in the attic and the supply duct below the floor in a single story structure.

HORIZONTAL FURNACE. The horizontal furnace is designed to be installed in an attic or in the crawl space under a house, as shown in Figure 13-47. Sometimes the furnace is installed above the ceiling in a shop area for a commercial application, as shown in Figure 13-48. The burner is on the side of the furnace, with the flue attached to either side, as shown in Figure 13-49. The flue can be on the same side as the burner for ease in venting the appliance.

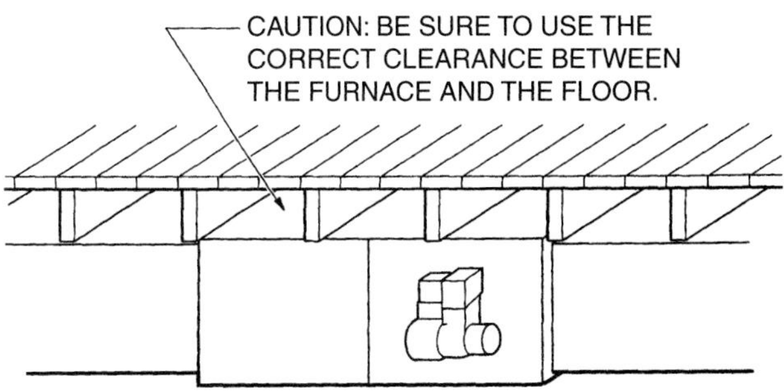

Figure 13-47. This horizontal furnace is installed in a crawl space below the floor.

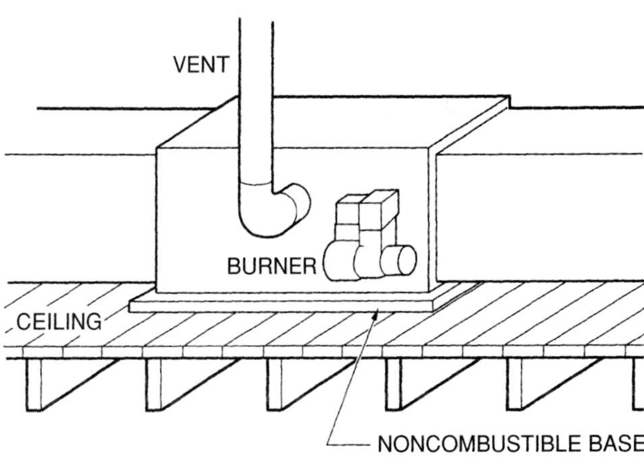

Figure 13-48. This horizontal furnace is installed in an attic.

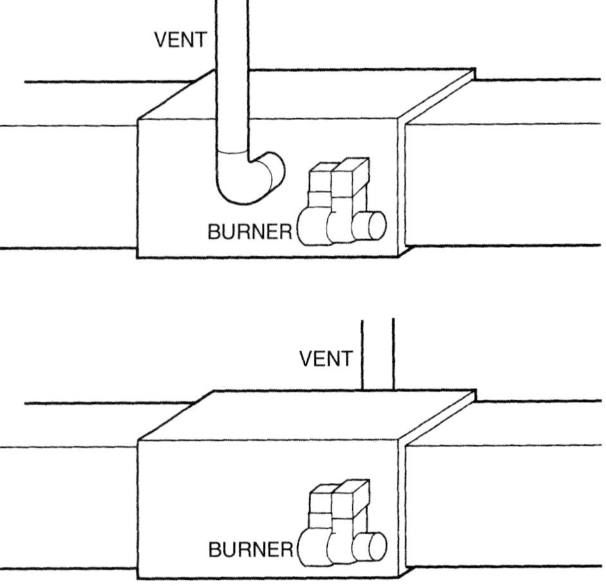

Figure 13-49. The flue can be attached to either side of the horizontal furnace.

Unit Heaters

Oil unit heaters are manufactured to free-blow air into such applications as warehouses and large-volume buildings. Figure 13-50 shows a typical unit heater. These heaters have prop fans or centrifugal fans, see Figure 13-51. The heaters do not have ducts and must be used as free-blowing.

Regulation of Equipment

All furnaces are approved for safety and capacity by regulatory agencies. Underwriters Laboratories (UL) lists all oil furnaces approved for safety with a stamp on the nameplate. The Department of Energy (DOE) determines the capacity and efficiency of oil furnaces.

Unlike gas furnaces, which are rated by the input, oil furnaces are rated by the *bonnet capacity*, which is the actual unit output. When operating correctly, an oil furnace with a flame-retention burner and a 1-gph (gallon per hour) nozzle should have an input of 140,000 Btu/h and a bonnet capacity of approximately 119,000 Btu/h (85% efficiency). On the other hand, an oil furnace with a conventional burner and a 1-gph nozzle should have an input of 140,000 Btu/h and a bonnet capacity of approximately 98,000 Btu/h (70% efficiency). Consequently, many service technicians recommend changing conventional burners to retention-head burners to improve system efficiency.

The Air Side of a Furnace

The air side of an oil furnace must move the return air into the furnace and distribute the warm air to the conditioned space. The furnace does this with a centrifugal fan, or blower, as shown in Figure 13-52. The centrifugal fan furnishes the pressure difference to create air movement across the heat exchanger. The inlet to the blower is in a negative pressure in

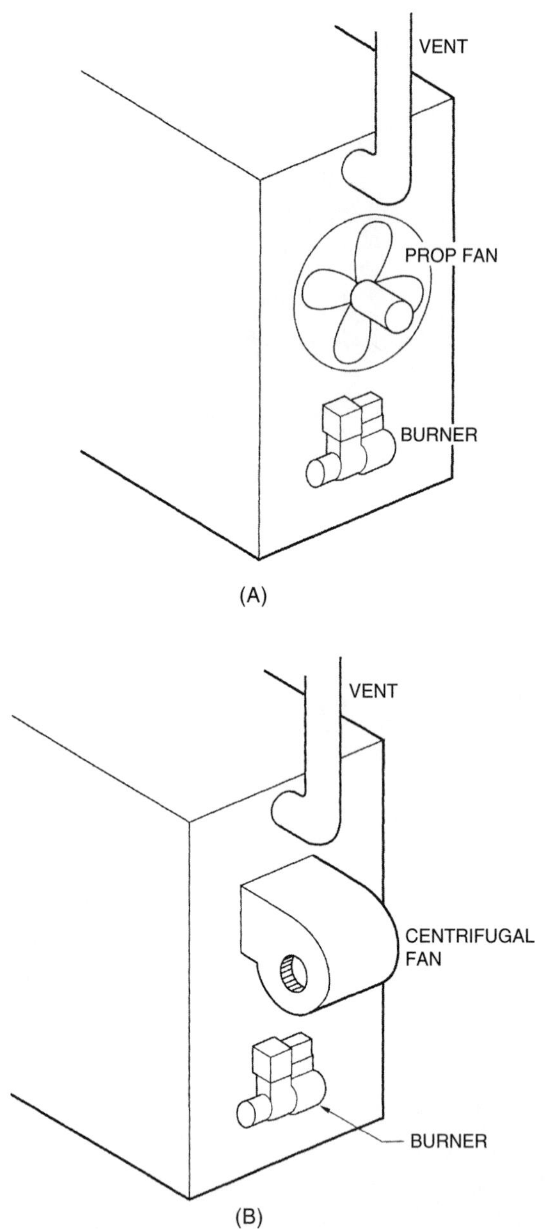

Figure 13-51. The unit heater has either (A) a prop fan or (B) a centrifugal fan.

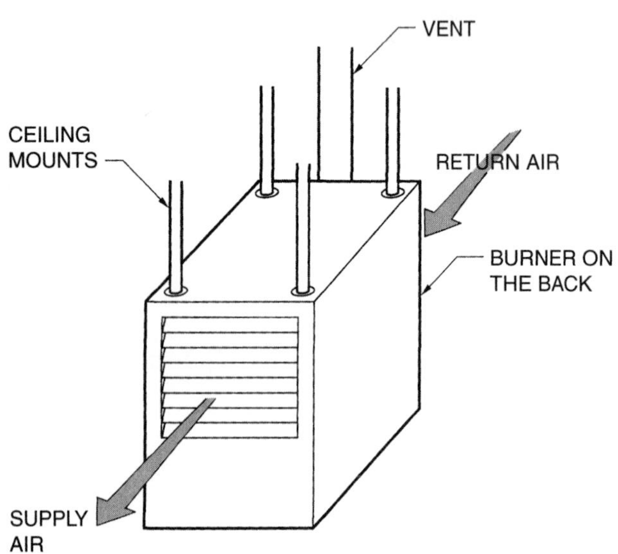

Figure 13-50. The unit heater is suspended from the ceiling.

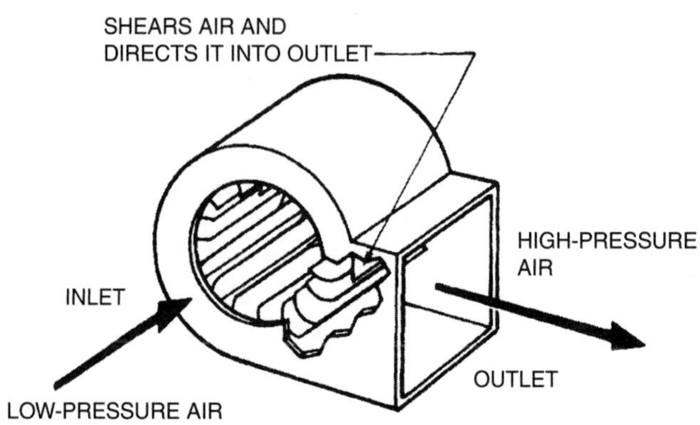

Figure 13-52. Oil furnaces use centrifugal fans.

relation to the room pressure, and the outlet to the blower is in a positive pressure, as shown in Figure 13-53.

There are two types of drives for the fan wheel: belt drive and direct drive as shown in Figure 13-54. The belt-drive blower was used for many years, but most small equipment (up to approximately 1 horsepower) now use the direct-drive blower. The reason for this is that the belt-drive system has many moving parts—two motor bearings, two fan shaft bearings, two pulleys, and a belt, Figure 13-55—that can rub together and wear out. On the other hand, the direct-drive fan motor has only two bearings and no belt, and the fan wheel is mounted directly on the fan shaft as shown in Figure 13-56. Consequently, this motor requires less maintenance and is more efficient than the belt-drive motor.

The fan must create enough pressure difference to move the air through the filter to the blower inlet, then

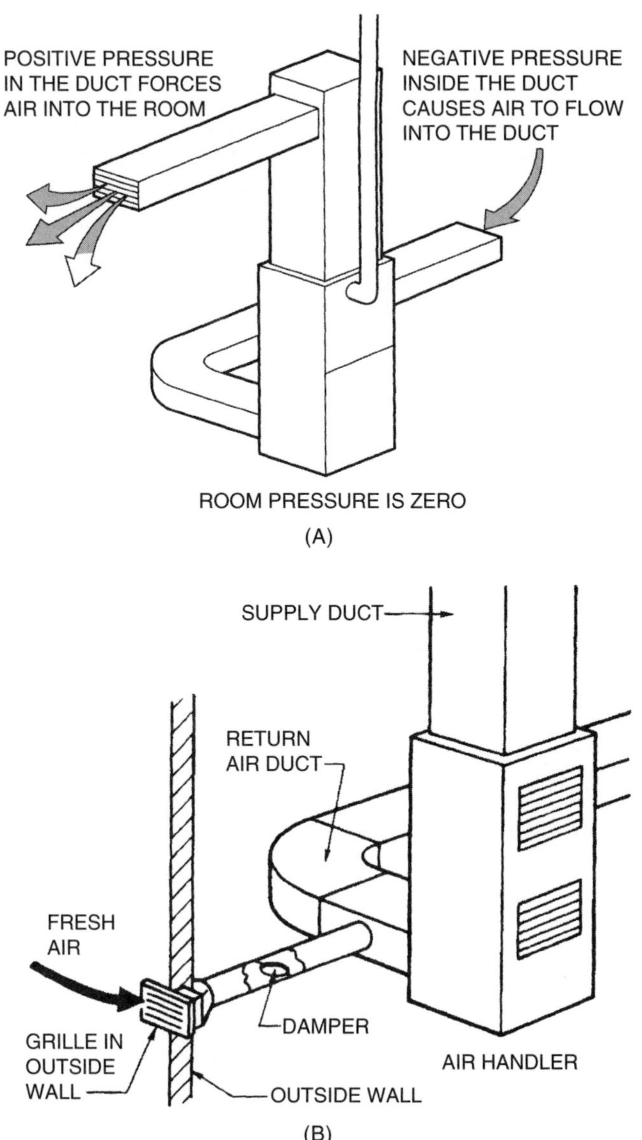

Figure 13-53. The blower creates a negative pressure at the return air grille and a positive pressure at the supply registers.

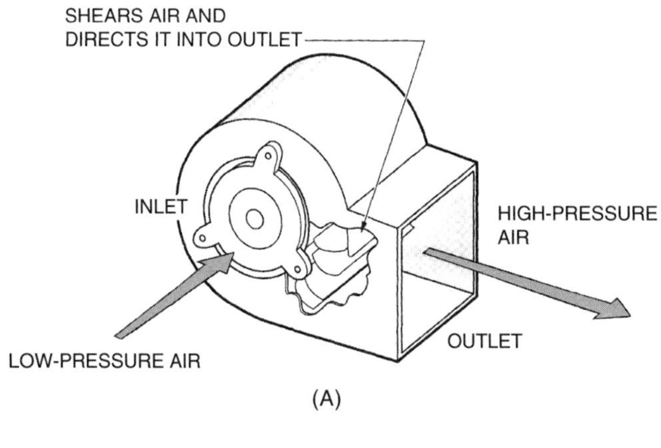

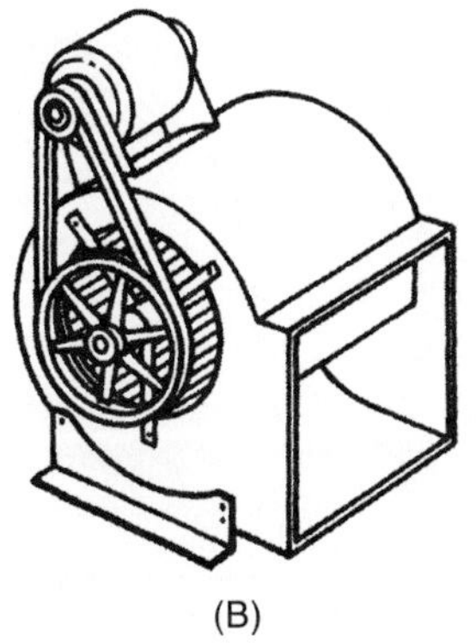

Figure 13-54. (A) Direct-drive fan, or blower. (B) Belt-drive fan.

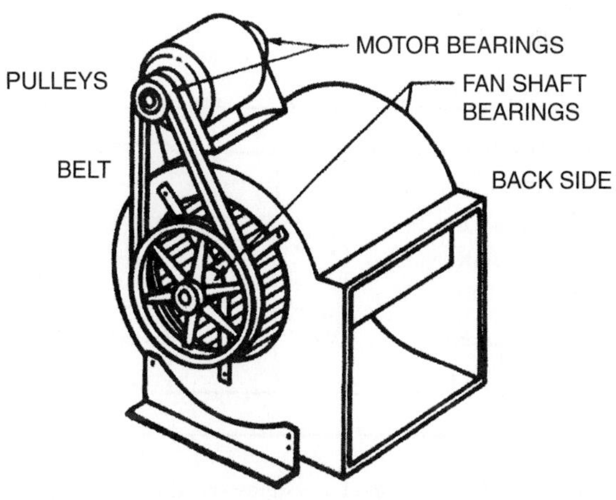

Figure 13-55. The belt-drive blower has many components.

to force the air to the furnace outlet. In addition to this force, the fan must overcome the friction loss of the air moving through the return and supply duct system, which includes the ducts, grilles, and registers, see Figure 13-57. Duct systems are discussed in more detail in Unit 22.

The manufacturer designs the furnace to have the correct airflow across the heat exchanger. The return air should sweep the entire heat exchanger to prevent hot spots from occurring on the heat exchanger surface, as

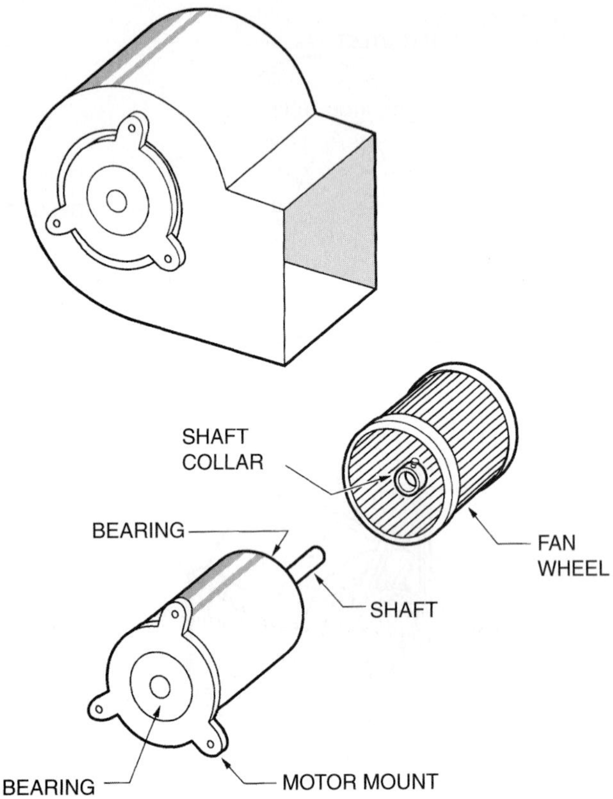

Figure 13-56. The direct-drive blower is simpler than the belt-drive blower.

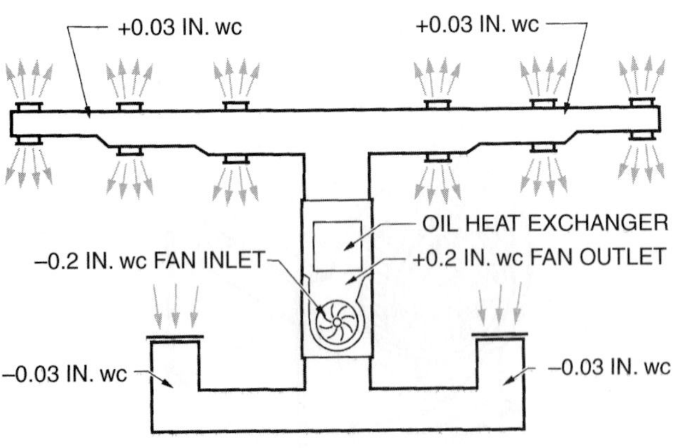

Figure 13-57. Simple duct systems showing air pressure.

shown in Figure 13-58. The technician should check the airflow at start-up and anytime reduced airflow problems are suspected, such as when the furnace burner operation is interrupted because of the high limit. This interruption is not particularly noticeable because the system automatically resets. However, the owner may notice that the burner shuts off and that the structure is not heated to a comfortable level in cold weather. When a heat exchanger is cracked, low air flow may be suspected. Low air flow causes an overheated heat exchanger and possible stress cracks.

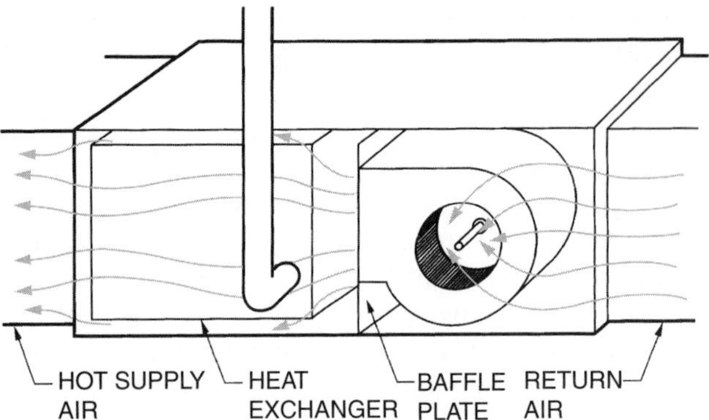

Figure 13-58. The return air should sweep the entire heat exchanger to prevent hot spots.

The maximum air temperature at the furnace supply duct should be printed on the furnace nameplate. Typically this temperature is 200°F. Actual air temperature should be considerably less than this because restricted air filters can reduce airflow or the customer may shut off air registers. A better air temperature to maintain is 150°F.

The technician may need to know the airflow in cfm (cubic feet per minute) and can check it using the same methods as used for gas furnaces. The energy input into the airstream and the temperature difference must be known. The temperature difference can be measured by placing a temperature probe in the return and supply air ducts, as shown in Figure 13-59.

Note: Do not place the probe directly over the heat exchanger, or it will be influenced by the radiant heat from the heat exchanger.

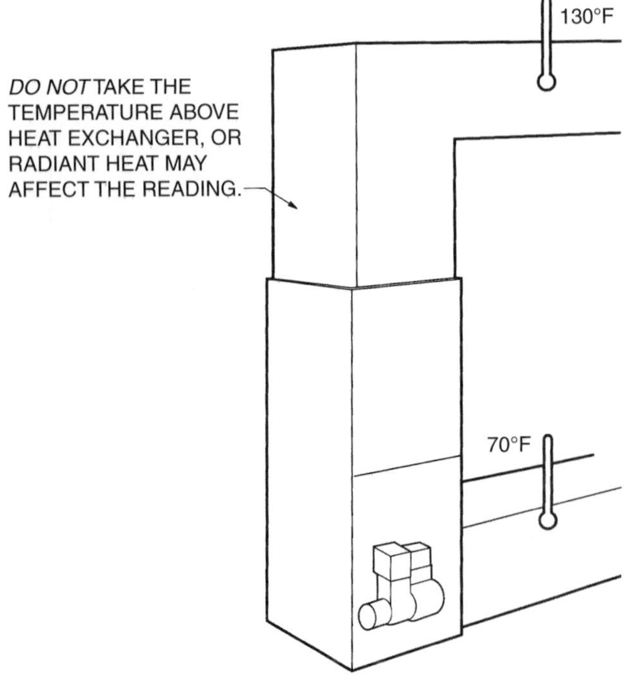

Figure 13-59. Taking the air temperature rise across a furnace.

The following formula can be used to calculate the airflow in cfm:

$$\text{cfm} = Q/(1.08 \times \text{TD})$$

where cfm = Airflow in cubic feet per minute.

Q = Heat input to the airstream in Btu/h.

1.08 = A constant that changes pounds of air to cfm.

TD = Temperature difference in degrees Fahrenheit.

Suppose that a technician needs to know if the air-flow is sufficient to add a duct to an addition to a house. The technician just performed the annual start-up checkup of the furnace. The efficiency was 82% (a 1.25-gph nozzle was used). The output of heat to the airstream was 143,500 Btu/h (140,000 Btu × 1.25 gph = 175,000 Btu/h input × .82 efficiency = 143,500 Btu/h bonnet capacity). The temperature entering the furnace was 70°F, and the temperature leaving the furnace was 160°F. The high leaving temperature indicates that more airflow will help the furnace if the main duct is large enough to add a branch duct for the addition.

$$\begin{aligned} \text{cfm} &= Q/(1.08 \times \text{TD}) \\ &= 143{,}500/(1.08 \times 90) \\ &= 1{,}476 \end{aligned}$$

It is common for the designer to size the ducts for heating-only systems as small as possible because of economics. The technician must pay attention to the furnace nameplate for the maximum allowable air temperature rise. If there is not enough airflow, the technician can increase the speed of the fan, and more ducts can be added. When the return air duct is too small, adding to the supply duct does not increase the airflow. Unit 22 discusses how to evaluate the system and shows how the system can be changed to correct the airflow.

The Vent System

The vent system for an oil-burning appliance must work together with the combustion chamber and the heat exchanger. The system starts with the combustion air. As mentioned in Unit 11, when a gallon of fuel oil is burned, it must have air for oxygen to support combustion. Every time a gallon of fuel oil is burned, approximately 2,000 cubic feet of air are required. Two kinds of air are discussed: ventilation air to keep the room cool and combustion air, as shown in Figure 13-60.

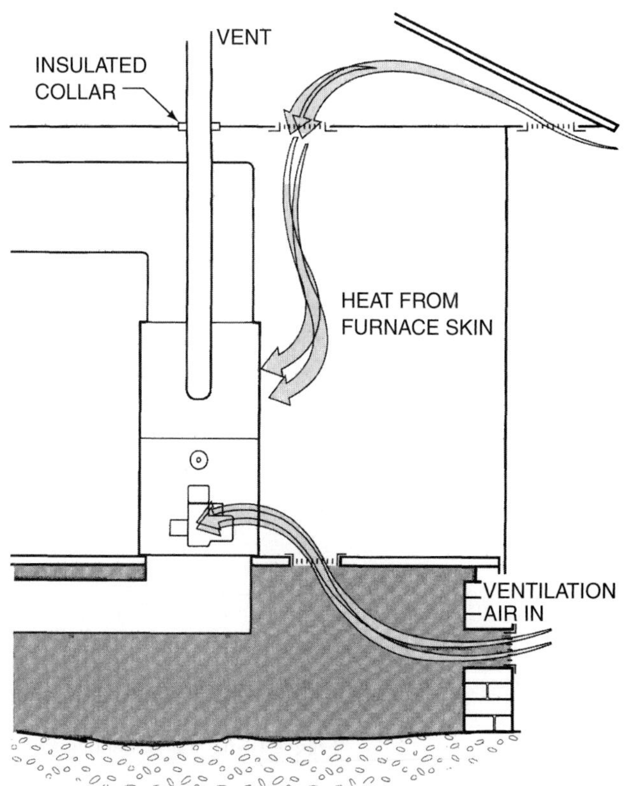

Figure 13-60. The furnace must have two kinds of makeup air: ventilation and combustion.

VENTILATION AND COMBUSTION AIR. The installation situations described next are from pamphlet 31 by the National Fire Protection Association (NFPA). This pamphlet discusses fuel oil-burning appliances. Not all situations are discussed here, only enough to help you realize that there are national codes and that much thought has been put into making installations safe. You should research the local and state codes that apply to where you install systems.

Some buildings, such as warehouses, are large and have enough infiltration through cracks around the doors and windows that no extra air is required, Figure 13-61. Other cases require that air from the outside be introduced to the building by various means and various amounts. Volume 2 of the National Fire Protection Association (NFPA) discusses these options. The paragraph numbers from this Volume 2 will be referred to in the following examples. The furnace room may be a confined space and need ventilation and combustion air from within the main building, as shown in Figure 13-62A. According to paragraph 5.4.1 of the NFPA when this is the case, there shall be two openings: one high and one low. Each opening shall have a free area of not less than 1 square inch per 1,000 Btu/h (which is 140 square inches per gallon of fuel oil burned per hour) of the total input rating of all appliances in the enclosure. When a building has adequate makeup air for the burner, no special preparations must be made; the ventilation air just leaks into the building. However, when the building is large but tight, provision must be made for outdoor air for combustion. In this case, according to paragraph 5.4.3 NFPA, a permanent opening or openings for combustion

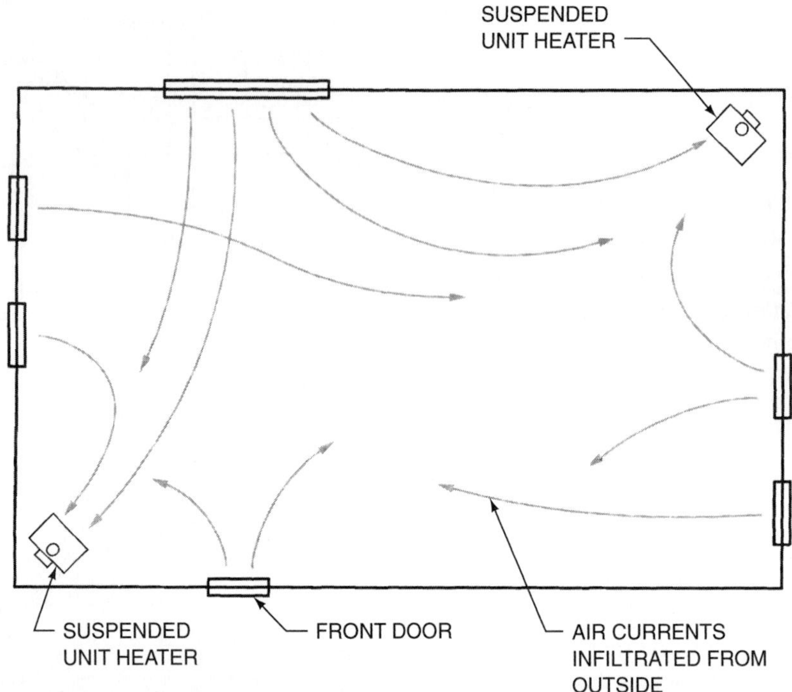

Figure 13-61. Large buildings take in air for combustion through infiltration.

air having a total free area of not less than 1 square inch per 5,000 Btu/h (which is 28 square inches per gallon of oil burned per hour) must be provided.

Any residence built since the mid-1970s may require planned ventilation or makeup air for the oil burner. The structure is tight, and the furnace may be located in the middle of the house in a closet. The furnace may also be located on the middle floor of a multifamily home or a large house, where there is no easy access for makeup air.

According to paragraph 5.4.2, NAPA when an appliance is located in a confined space and all the combustion and ventilation air must come from the outside, there shall be two openings that communicate directly, or by means of duct, with the outdoors or to such spaces (crawl or attic) that freely communicate with the outdoors. When directly communicating with the outdoors or by means of vertical ducts, as shown in Figure 13-62B, each opening shall have a free area of not less than 1 square inch per 4,000 Btu/h (which is 35 square inches per gallon of oil burned per hour) of the total input rating of all appliances in the enclosure. If horizontal ducts are used as in Figure 13-62C, each opening shall have a free area of not less than 1 square inch per 2,000 Btu/h (which is 70 square inches per gallon of oil burned per hour) of the total input rating of all appliances in the enclosure.

Some installations use ventilation air from inside the building and combustion air from the outside. The enclosure shall be provided with two openings for ventilation, located and sized as described in paragraph 1-5.4.1. In addition, there shall be one opening directly communicating with the outdoors or to such spaces (crawl or attic) that freely communicate with the outdoors. This opening shall have a free area of not less than 1 square inch per 5,000 Btu/h (which is 28 square inches per gallon of oil burned per hour) of the total input rating of all appliances in the enclosure.

LOUVERS AND GRILLES. The technician should pay close attention to the louvers and grilles used for ventilation and combustion air. Note that all the NFPA specifications called for *free area* of the grille or register. The area of the opening is not the free area; part of the opening is taken up with the louvers, as shown in Figure 13-63. When the louvers and grille are made of wood, 20% to 25% of the area is free, and 75% to 80% of the area is taken up with the louvers. When the louvers and grille are made of metal, 60% to 75% of the area is free. For example, suppose that a grille measures 20 inches by 20 inches. The total area is 400 square inches, but the free area is only 240 to 300 square inches (400 × .60 = 240 sq in., and 400 × .75 = 300 sq in.), Figure 13-64. If you have no data on the grille, be conservative and decide that the free area is 60%, or 240 square inches. When you need to know the size of the louver or grille to purchase for a certain free area, divide the free area needed by the percentage of free area of the louver or grille type chosen. For example, suppose that a free area of 200 square inches is needed. What size metal grille should be purchased if it has a free area of 75%? To solve the problem, divide 200 square inches by .75. The grille should be 267 square inches.

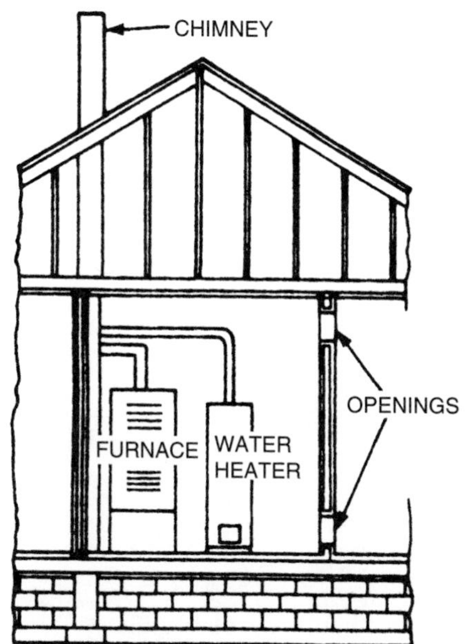

Note: Each opening shall have a free area of not less than 1 sq in. per 1000 Btu per hr (140 sq in. per gal per hr) of the total input rating of all appliances in the enclosure.

(A)

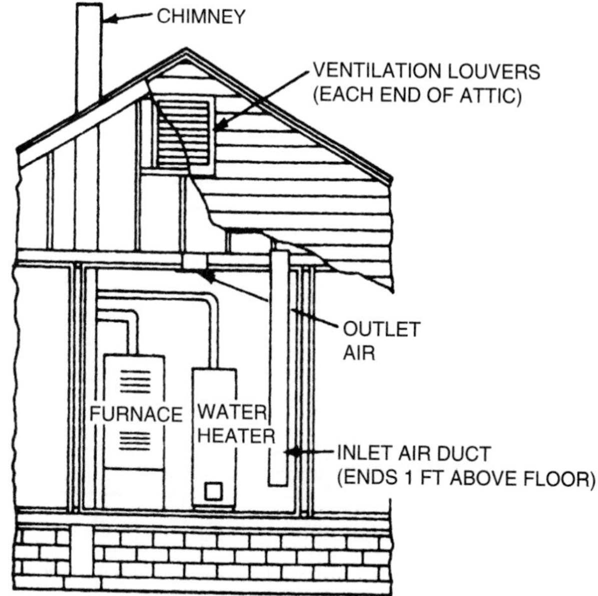

Note: The inlet and outlet air openings shall each have a free area of not less than 1 sq in. per 4000 Btu per hr (35 sq in. per gal per hr) of the total input rating of all appliances in the enclosure.

(B)

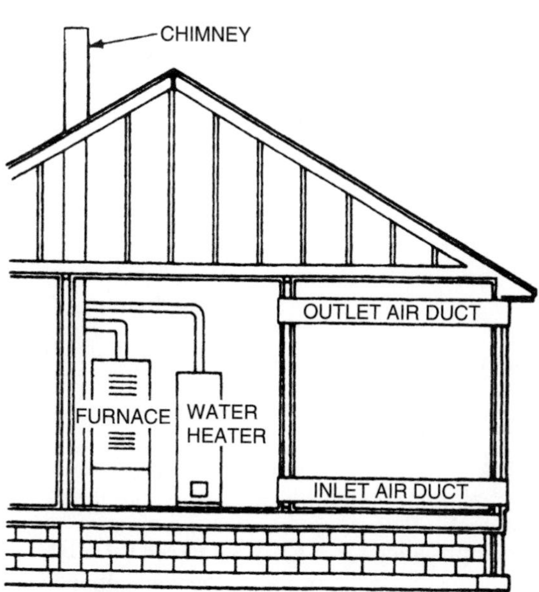

Note: Each air duct opening shall have a free area of not less than 1 sq in. per 2000 Btu per hr (70 sq in. per gal per hr) of the total input rating of all appliances in the enclosure.

(C)

Figure 13-62. Ventilation air from inside the building and combustion air from outside. Appliance is in a confined space and (A) ventilation and combustion air come from inside the building; (B) ventilation and combustion air come from outdoors through a ventilated attic; (C) ventilation and combustion air come from outdoors, horizontal ducts. *Courtesy National Fire Protection Association*

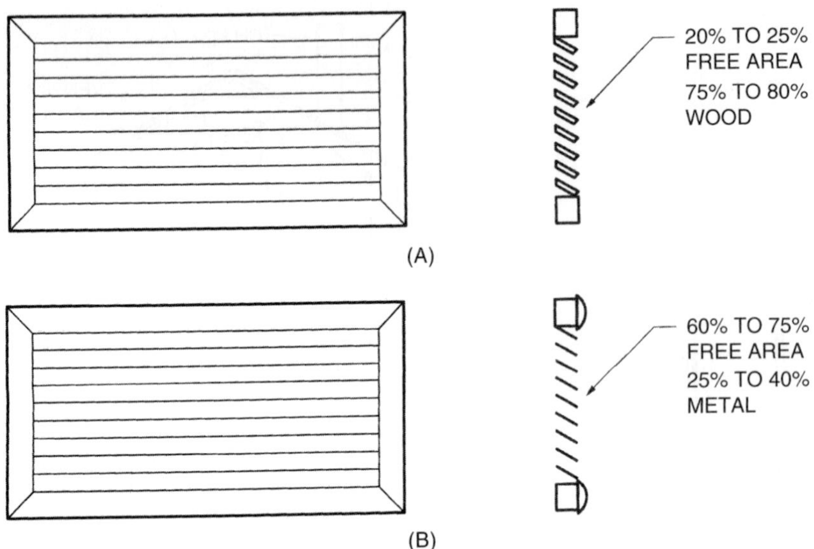

Figure 13-63. Part of the grille opening is taken up with the louvers. (A) Wood grille. (B) Metal grille.

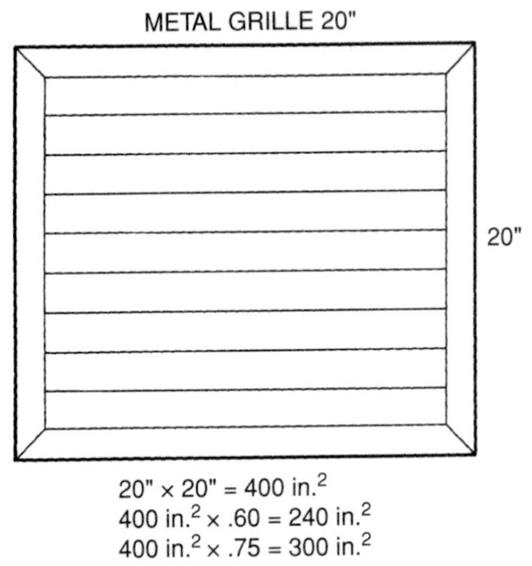

Figure 13-64. Calculating the free area of a grille.

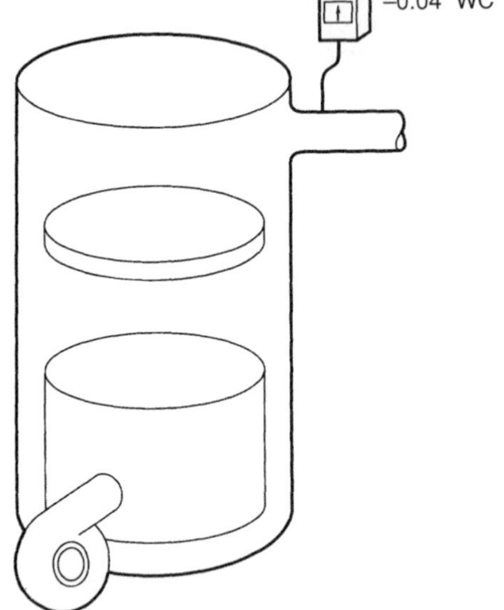

Figure 13-65. The draft at the breach should be approximately –0.04 inch of wc.

Caution: The technician should be aware that owners of equipment may notice the cold air coming in the combustion air vents and try to shut them off, thinking they are saving money. If this is discovered, the technician should talk to the owner and explain why the vents are there. Without makeup air, the structure is liable to start operating in a vacuum, which may down-draft the appliance causing CO (carbon monoxide) to enter the structure.

DRAFT PRESSURE. As mentioned previously, the air for the oil burner is blown into the combustion chamber, but the flue system is not in a positive pressure. The blower supplies only air, not pressure. A typical oil burner will have a negative draft pressure over the fire of −0.01 to −0.02 inch of water column (wc). The draft at the breach of the system, leaving the heat exchanger, should be approximately −0.04 inch of wc, as shown in Figure 13-65. The vent system is responsible for maintaining these negative pressures over the fire and at the breach.

VENT CONNECTOR. The next portion of the vent system is the vent connector. This horizontal length of pipe shall be as short as practical and shall not exceed 10 feet in length. Thus, the installer must position the appliance as close to the vertical section of the system as possible. The connector length shall not be longer than 75% of the vertical length of the flue system.

There are many more regulations that determine how to route the vent connector, such as a connector may not pass through a floor or a ceiling. The vent connector material must be approved. The connector must be at least as large as the connection on the appliance unless the appliance manufacturer approves of a smaller size. A vent connector may not pass through a combustible wall without proper connections. The vent connector must be properly fastened to a masonry chimney. The connector shall be pitched upward in the direction of the flow of the by-products of combustion. Figure 13-66 illustrates some of these requirements.

Note: Check your local code for all venting practices.

CHIMNEY. After the flue gases meet the vertical rise of the system, they may be routed through a metal or masonry chimney. Some chimneys are factory made and tested. These chimneys must be approved for use. The size of the vertical rise should be at least the size of the connector and the connection on the furnace. For example, if a furnace has a 6-inch outlet, the connector and the chimney should be at least 6 inches unless otherwise approved by the furnace manufacturer. Whatever the size of the chimney or vertical rise of the system, it should provide the negative draft mentioned previously, −0.04 inch of wc at the outlet of the appliance.

Often the draft is too great because a chimney is too tall or is oversized. A draft regulator, also called a *barometric damper*, shall be used to prevent overdraft and heat wasted up the flue. Figure 13-67 shows a typical barometric damper. The regulator has adjustments that may be made to set the draft at the breach of the appliance. To do so, as shown in Figure 13-68, place a draft gauge in the flue as close to the furnace as possible and adjust the regulator until this gauge reads −0.04 inch of wc.

The vent system should terminate correctly at the top of the structure. The NFPA code calls for the vent to extend through the roof 3 feet and to extend at least 2 feet above any portion of the roof that is within 10 feet, as shown in Figure 13-69. Proper vent termination helps ensure proper venting.

Figure 13-67. Draft regulator, or barometric damper.

Figure 13-68. Adjusting the draft regulator.

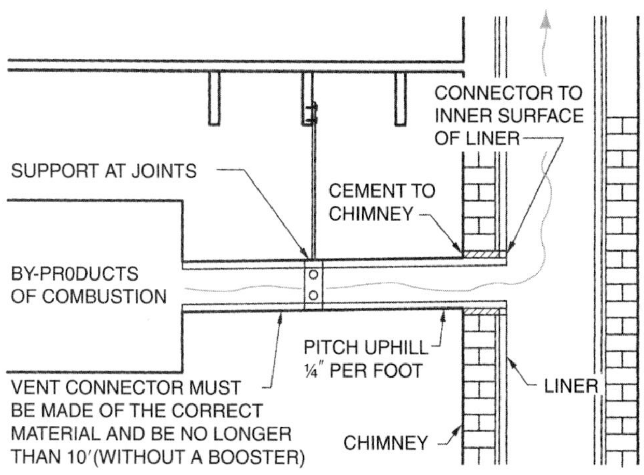

Figure 13-66. Vent connectors.

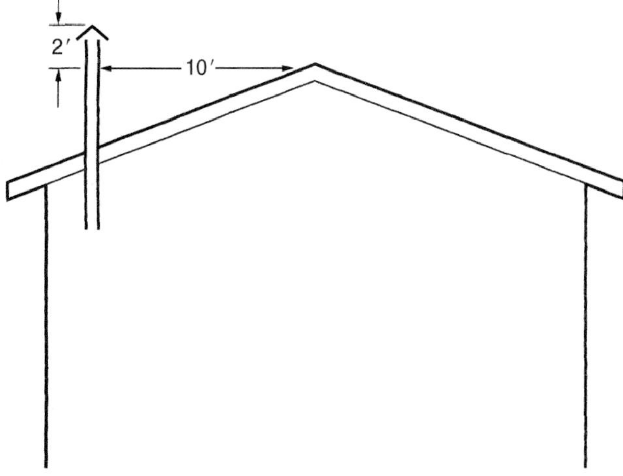

Figure 13-69. The vent must extend the proper distance through the roof.

CORRECT DRAFT. The vent system should have a negative pressure at the breach and over the combustion chamber even before the fire is ignited. When there is not a negative draft, smoke and soot will puff out the burner inspection door and around the transformer mount. When the oil burner is a forced-air furnace, some of this smoke may be pulled into the fan compartment if the door is not tight, as shown in Figure 13-70. Many oil furnaces have been mistakenly replaced because soot was drawn past the fan door and blown into the house. The technician should verify the draft with a cold furnace before replacing it. Otherwise, the new furnace may do the same thing.

When a furnace does not have the correct draft, the technician should look at the complete system, starting with the flue connector, which should be checked for soot buildup. If soot is in the connector, soot is probably in the heat exchanger. The furnace may need a good cleaning. All inspection doors must be removed and wires and brushes used to break the soot loose. A vacuum cleaner should be used to remove the soot from the furnace and the surrounding area. Be sure that the vacuum has a good filter, or the soot will just blow through the vacuum, as shown in Figure 13-71. It is a good practice to pipe the vacuum discharge out of the house in case the vacuum cleaner filter bursts, as shown in Figure 13-72.

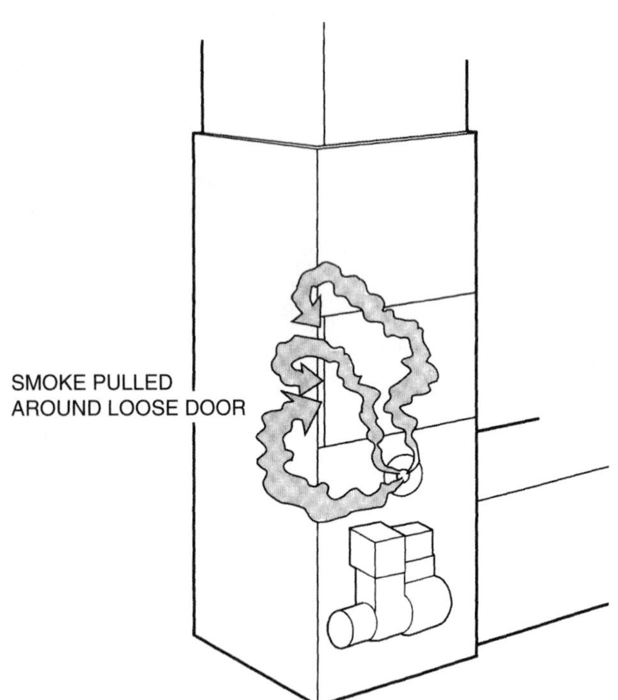

Figure 13-70. Smoke may be pulled into the fan compartment if the door is not tight.

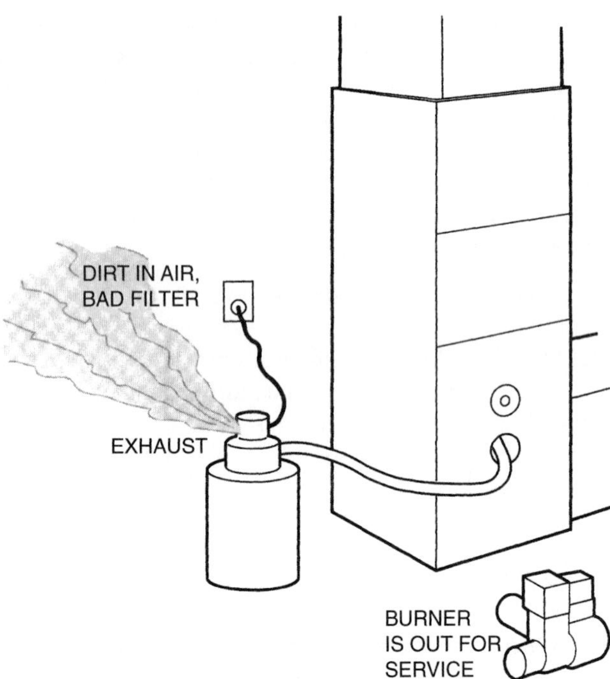

Figure 13-71. Make sure that the vacuum cleaner has a good filter.

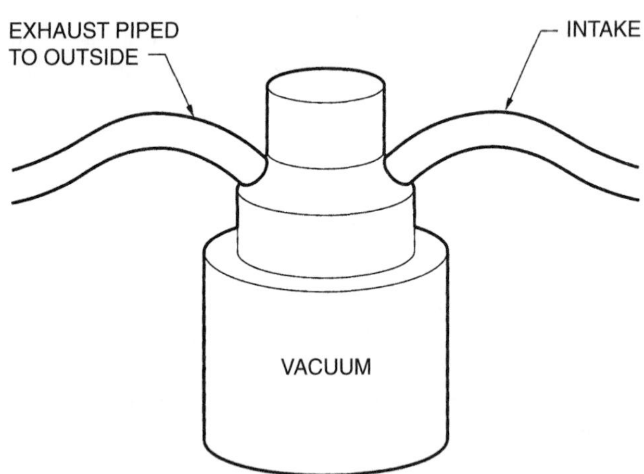

Figure 13-72. The vacuum exhaust may be vented to the outside if a lot of soot exists.

Summary

- Clean fuel oil must be supplied to the oil pump by the best piping and storage system.
- There are two basic types of storage systems: a central storage tank with supply tanks and the supply tank as the only tank.
- The storage tanks must be approved for the application.
- Tanks may be above or below ground. When underground, the tanks must be properly protected from corrosion.
- The proper fill and vent system must be applied to each tank. The vent system allows air to escape when the tank is being filled.

- Underground tanks are subject to water leaking in and oil leaking out of holes in the tank. Either case is undesirable.
- Aboveground tanks can be mounted on legs for elevation.
- Aboveground tanks sweat on the inside in some climates, and water builds up in the tanks.
- All tanks should be installed with a slight pitch away from the oil pickup pipe so that water and sludge drain away from the oil pickup line.
- In densely populated areas, some tanks are installed inside buildings.
- A tank must be installed according to the codes for the locality.
- Most piping systems are piped with ⅜-inch copper pipe. Longer runs may be of ½-inch copper.
- Flare fittings are usually used. It is important to use correct flaring and tubing practices for good results.
- An oil filter is piped into the system to remove sludge and scale from the tank.
- Automatic control for oil appliances involves controlling forced-air furnaces and boilers.
- The power supply for forced-air furnaces in the home is 115 volts; 230 volts are used for some commercial applications. The control voltage is usually 24 volts.
- Some primary controls used on forced-air systems allow air-conditioning to be added to the systems.
- The fan must be controlled for forced-air systems, and the pump circuits must be controlled for boilers.
- The fan circuit is usually temperature controlled to start after the furnace is heated and to continue to run after the burner is shut off until the furnace cools. This timing prevents drafts at the beginning of the cycle and removes the excess heat at the end of the cycle.
- Oil furnaces are upflow, downflow, and horizontal. Low-profile furnaces are used for upflow in low-headroom applications. Another forced-air oil appliance is the unit heater, which is suspended from the ceiling in some applications.
- Underwriters Laboratories (UL) maintains a list of all furnaces in regard to safety. The Department of Energy (DOE) determines the capacity and efficiency of furnaces.
- Forced-air furnaces have two types of drives for fans: belt and direct.
- The technician should verify the correct airflow for all installations. Too much airflow can cause condensation of moisture in the flue gases. Too little air can cause the furnace to cycle because of excess temperature. Either instance shortens the life of the furnace.
- The formula for airflow is $\text{cfm} = Q/(1.08 \times \text{TD})$.
- The vent system for forced-air furnaces involves the combustion chamber, the heat exchanger, and the flue, or chimney. The local code must be followed to ensure correct venting of flue gases.
- Close attention must be paid to the air for combustion; 2,000 cubic feet of air must be supplied for each gallon of fuel oil burned.
- Even though the air is blown into the combustion process, the burner is in a negative pressure over the fire and in the flue. This negative pressure is created by the heat from the by-products of combustion and is called *draft.* The draft over the fire should be approximately −0.02 inch of wc. The draft at the flue connection should be approximately −0.04 inch of wc.
- A draft regulator should be installed after each appliance—furnace or boiler—and adjusted to maintain the correct draft, −0.04 inch of wc.
- Vent connectors and chimneys must be the correct size (at least as large as the connection on the appliance) and made of the correct material, with a minimum of elbows. The vent must pitch upward in the direction of flow.
- The chimney must terminate through the roof according to local code.
- The technician should always be aware of the correct draft in an oil-burning appliance.

Review Questions

1. What is the purpose of the vent on a fuel oil tank?
2. How does moisture enter an aboveground tank?
3. How can the technician detect the water level in a fuel oil tank?
4. What is the typical copper line size for a small fuel oil system?
5. What happens if too little air flows through a forced-air furnace?
6. Suppose that a system has a bonnet capacity of 120,000 Btu/h and an air temperature rise of 60°F. What is the airflow in cfm?
7. What should a typical draft be above the fire in an oil-burning appliance? at the vent?
8. What device is used to prevent too much draft in an oil-burning appliance?

14 Burning Oil as a Fuel

Objectives

Upon completion of this unit, you should be able to

- **describe combustion fuel oil boiler.**
- **perform combustion analysis on an oil boiler system.**

14.1 Fuel Oil Boilers

The fuel oil boiler burns fuel oil and transfers the heat into water instead of into air. There are two types of boilers: hot water and steam. Hot-water boilers are liquid full of water, and the water is circulated through the system to terminal units that transfer the heat to room air. Hot-water heat is discussed in detail in Unit 17. The steam boiler is similar to the hot-water boiler, except it has a vapor space at the top of the boiler where the steam collects as shown in Figure 14-1.

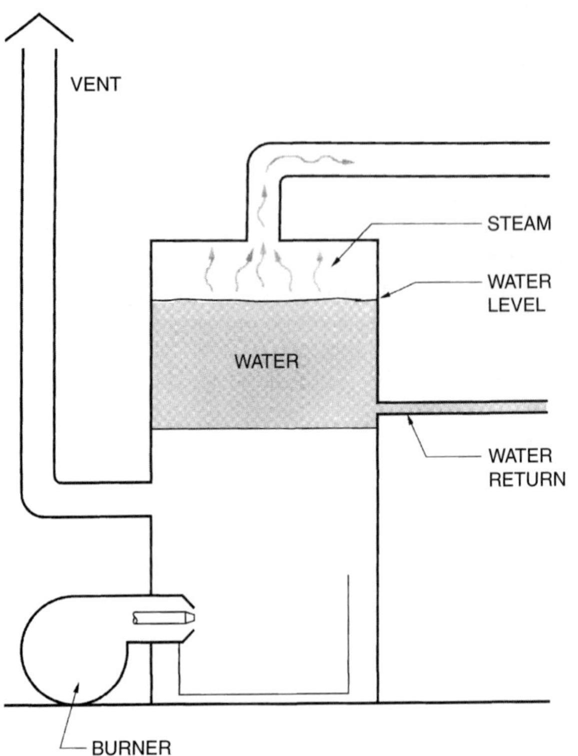

Figure 14-1. Steam collects at the top of the boiler, above the water.

Steam is circulated through the system to the terminal units where heat is removed from the steam, which then changes back to water.

The fuel oil boiler uses the same burner as the furnace, except the burner may be pedestal mounted in front of the boiler. The larger the boiler, the more likely the burner is mounted on a pedestal. Figure 14-2 shows a pedestal.

Oil boilers are made of cast iron or steel. Cast-iron boilers are sectional. The boiler has two end sections and at least one middle section. As more capacity is needed, sections are added to the middle of the boiler. Figure 14-3 shows a typical sectional boiler. The manufacturer needs to make only end sections and middle sections for a range of boiler sizes. When a larger range of sizes is needed, the manufacturer makes larger end and middle sections for the new range. The outside of the boiler sections is covered with insulation and a sheet-metal skin to hold the insulation in place and make the boiler more decorative as shown in Figure 14-4. When a boiler requires more capacity than the sectional boiler can offer, the fire-tube boiler is used. It has a main fire tube where combustion takes place. The by-products of combustion turn at the end of the fire tube and make one or two more passes through the other tubes in the boiler to the vent system

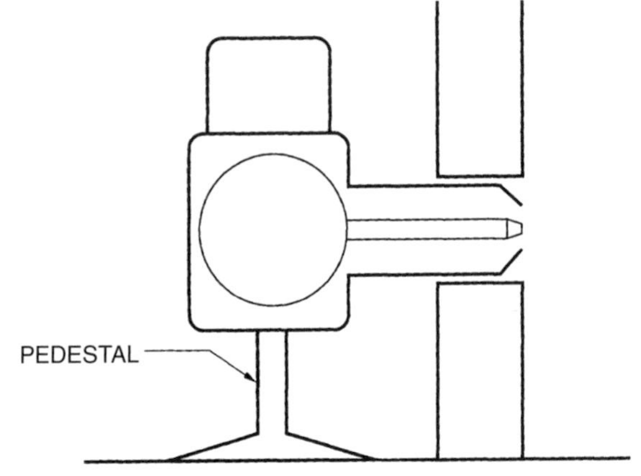

Figure 14-2. The burner of the larger boiler is mounted on a pedestal.

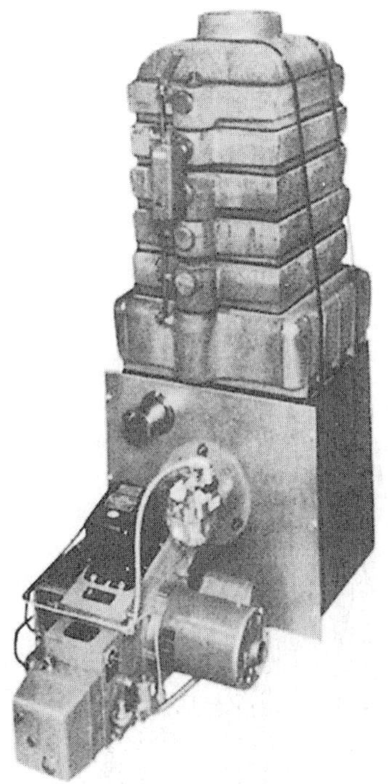

Figure 14-3. A sectional boiler can be expanded with more middle sections. *Courtesy Hydro Therm*

as shown in Figure 14-5. By this time, as much heat as practical has been removed from the by-products of combustion and the remaining heat is used to push the flue gases up the flue and out of the structure.

Water boilers have safety controls other than the flame safety control. Like the forced-air furnace, the boiler must have a limit control to prevent it from becoming too hot. The limit control is inserted into the water in the boiler by means of a dry well, as shown in Figure 14-6. This control may have one set of contacts to control the boiler water temperature and another used as the high limit. The control sequence is as follows: As the water temperature begins to rise, the low limit (acting as the thermostat) shuts off the boiler at the set point (e.g., 190°F). If the low limit fails, the upper limit contacts open and stop the boiler at 210°F. The pressure relief valve on the boiler protects it when both the operating control (low limit) and safety control (high limit) fail. This valve can be set to relieve the pressure on the boiler if the limit allows the burner to continue to run. The usual setting is 30 psig for a water-heating boiler. Figure 14-7 shows a relief valve mechanism.

Steam boilers operate from pressure, so pressure switches are used for control and safety. These are usually Bourdon tube controls with mercury switches as shown in Figure 14-8. One control is used as an operating control

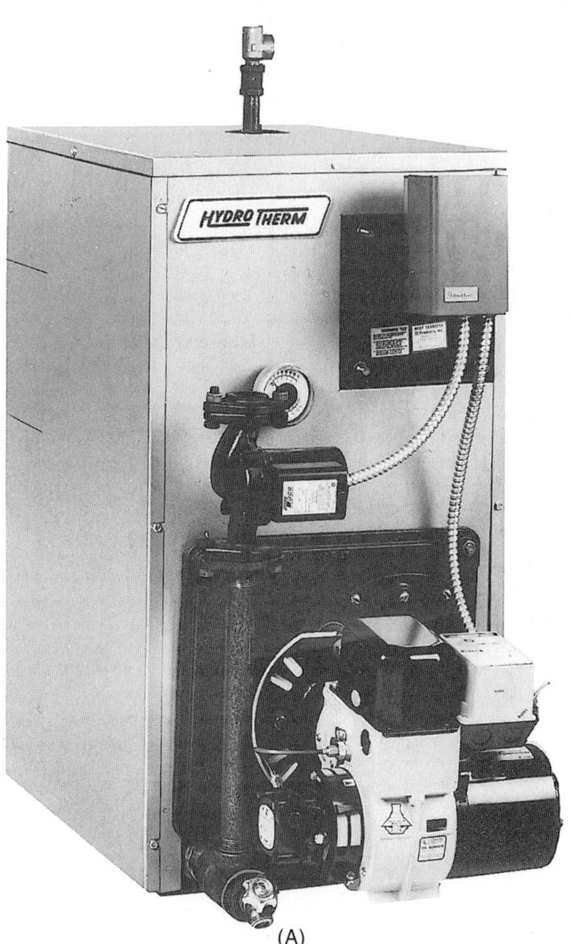

(A)

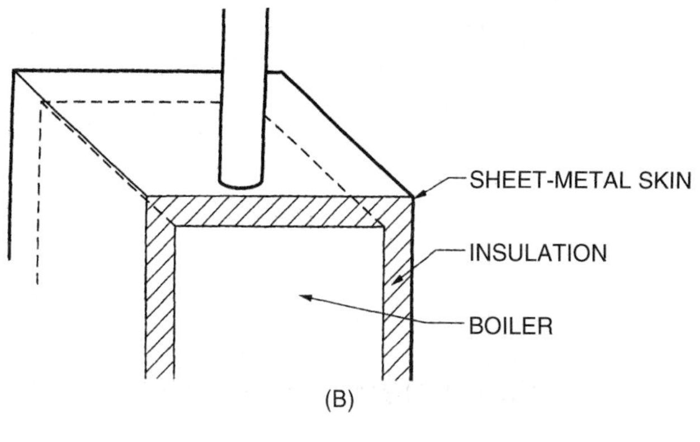

(B)

Figure 14-4. (A) The boiler skin makes the boiler more decorative. *Courtesy Hydro Therm* (B) The boiler skin holds the insulation.

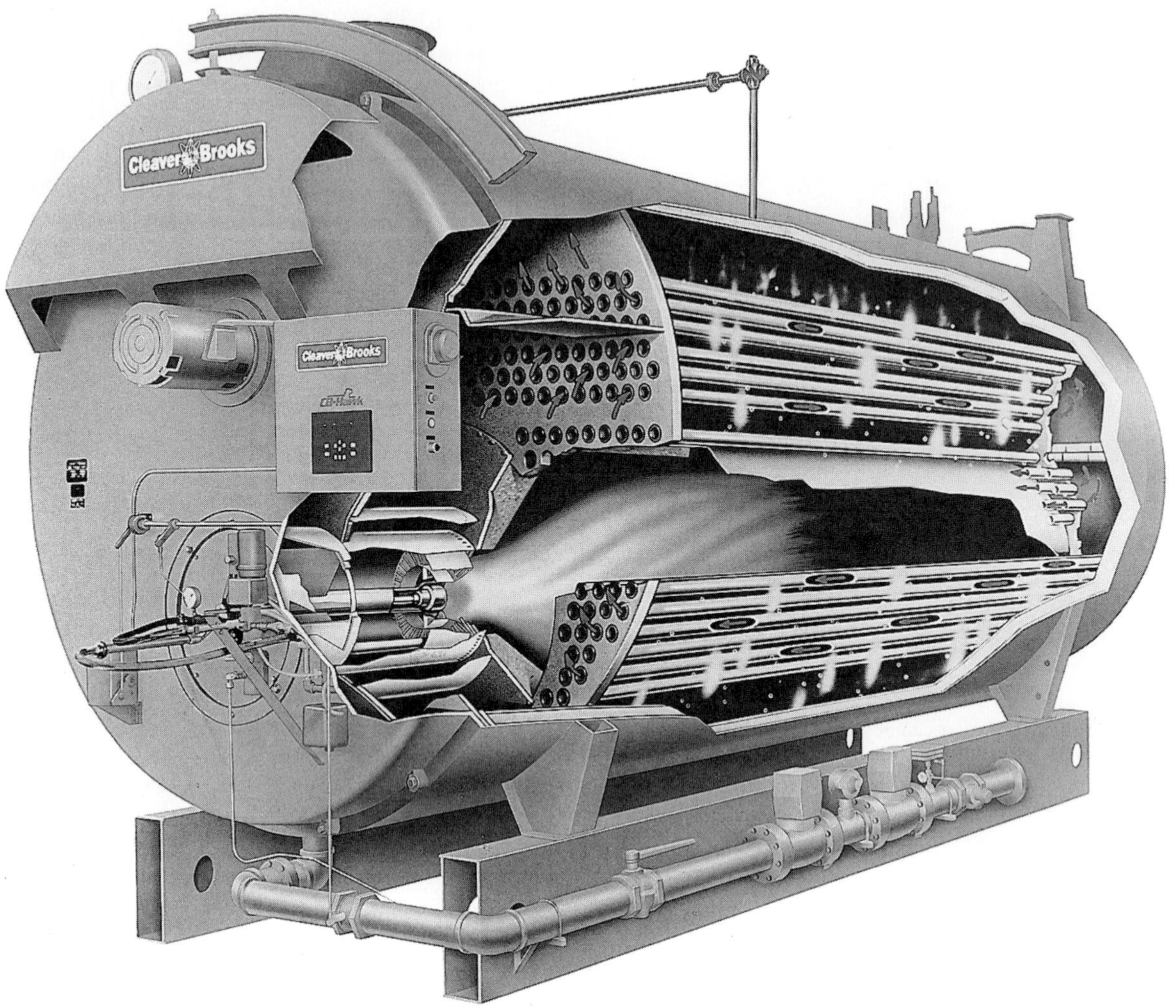

Figure 14-5. Fire-tube boiler. *Courtesy Cleaver Brooks*

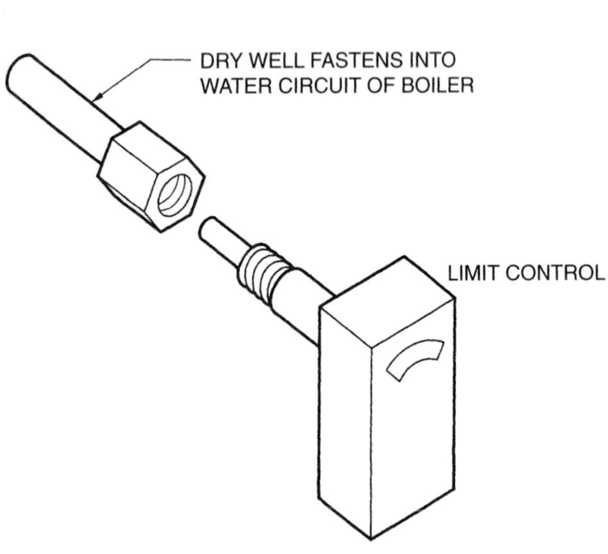

Figure 14-6. The limit control is inserted into a dry well.

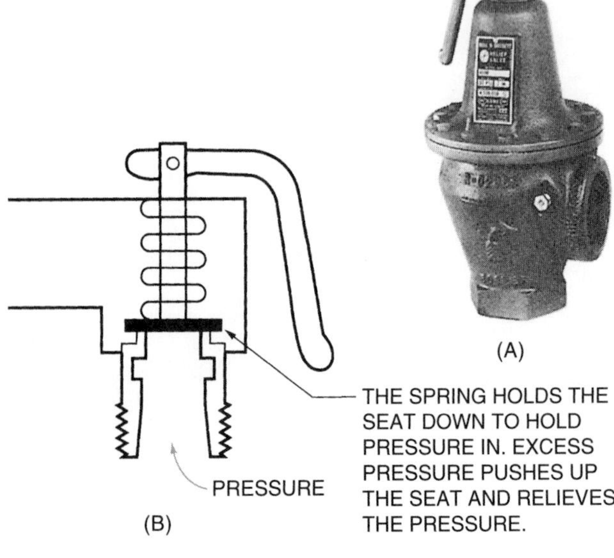

Figure 14-7. (A) The pressure relief valve for a water boiler. *Photo by Bill Johnson* (B) Cross section of a safety valve.

and is set to shut off the burner at 10 psig steam; the other control is set at 12 psig to act as the limit control. The safety relief valve relieves the pressure from the boiler if the pressure rises to 15 psig as shown in Figure 14-9.

14.2 Combustion Analysis

Oil-burning equipment can benefit more from adjustment than gas-burning equipment can because the combustion characteristics can be changed more in oil burners. The technician can make some rather dramatic differences in the performance of the customer's equipment. The technician can tune up an oil burner to burn clean and not waste fuel. If the burner does not burn clean, the heat exchanger becomes dirty and less efficient. Chimneys and flue pipes become clogged.

Many technicians begin the annual furnace checkup by turning on the furnace and running a combustion analysis, then making adjustments and running another combustion analysis for the customer. If the technician can show that a furnace is operating at 75% efficiency before the tune-up and at 82% efficiency after, the customer will save 7% that winter. If the fuel bill is usually $600 per year, the customer saves $42 ($600×.07 = $42). This savings alone pays for some of the service call.

It is a good practice to change the in-line filter and nozzle every year whether they appear to be functional or not. This practice is particularly true for a residential system because the nozzle clearance is so small and only a small amount of debris restricts it. Oil should be delivered clean and dry (without moisture), but this is not always the case. Old fuel tanks and piping systems also contain rust and sludge. A filter cartridge and new nozzle are inexpensive insurance.

An example of a typical furnace combustion analysis for a modern furnace is given next to explain how combustion analysis should occur. Then, we analyze an oil boiler where changes must be made. The combustion analysis is included in a complete inspection of the system. Remember: The components of the system must work together. Figure 14-10 shows a chart that plots CO_2 readings and stack temperatures for determining the combustion efficiency.

Suppose that a furnace is 2 years old and is known to have a flame-retention head burner. The owner requests an annual service call, including combustion analysis before and after adjustments are made. This service is usually provided in the fall before winter start-up. Following are the steps that the service technician would take:

1. Move the proper tools and parts to the furnace area. These include hand tools, pressure gauge, oil filter and combustion analysis kit, as shown in Figure 14-11.
2. Inspect the complete system, starting with the tank. Check for tank corrosion in aboveground tanks. Check the piping system for kinks or damage. Use a

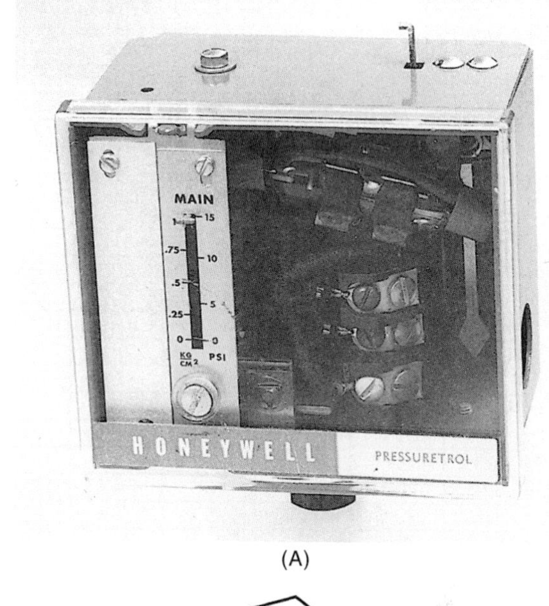

(A)

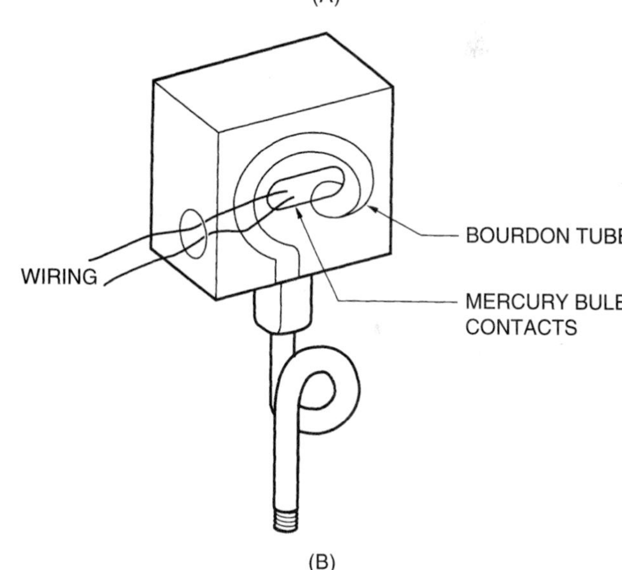

(B)

Figure 14-8. (A) The pressure controls for a steam boiler are usually mercury-tube controls. *Photo by Bill Johnson* (B) Control components.

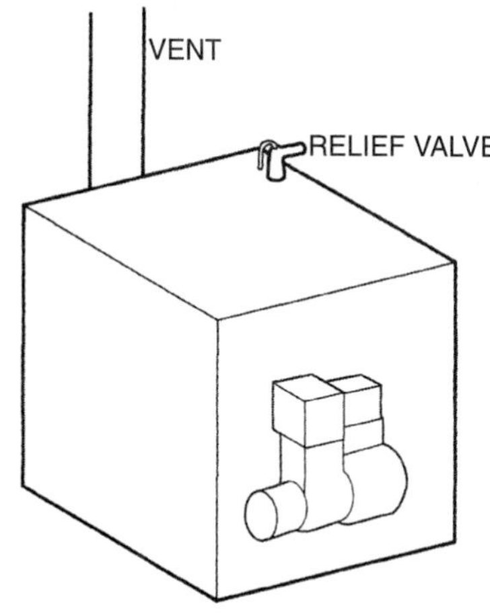

Figure 14-9. Location of a relief valve on a boiler.

Net Stack Temp. (degrees F.)

Percent CO_2	300°	350°	400°	450°	500°	550°	600°	650°	700°	750°	800°	850°	900°
15	87½	86½	85¼	84¼	84¼	82	81	79¾	78¾	77½	76½	75½	74¼
	87½	86¼	85	84	83	81¾	80¾	79¼	78½	77¼	76	75	73¾
14	87¼	86	84¾	83¾	82¾	81½	80¼	79	78	76¾	75½	74½	73
	87	85¾	84½	83½	82½	81¼	80	78¾	77½	76¼	75¼	74	72¼
13	86¾	85½	84¼	83¼	82	80¾	79½	78¼	77	75¾	74½	73½	71¾
	86½	85¼	84	83	81½	80¼	79	77¾	76½	75¼	73¾	72¾	71
12	86¼	85	83¾	82½	81¼	79¾	78½	77¼	75¾	74½	73	71½	70¼
	86	84¾	83½	82	80¾	79¼	78	76½	75¼	73¾	72¼	70¾	69½
11	85¾	84½	83	81½	80¼	78¾	77¼	75¾	74½	73	71½	70	68½
	85½	84	82½	81	79½	78	76½	75	73¾	72	70½	69	67½
10	85	83½	82	80½	78¾	77¼	75¾	74¼	72¾	71	69½	68	66¼
	84½	83	81½	79¾	78	76½	75	73¼	71¾	70	68¼	66¾	65
9	84	82¼	80¾	79	77¼	75¾	74	72¼	70¾	68¾	67	65¼	63½
	83½	81¾	80	78¼	76½	74¾	73	71¼	69½	67½	65½	63¾	62
8	83	81	79¼	77½	75½	73¾	71¾	70	68	66	64	62	60
	82¼	80¼	78½	76½	74½	72½	70½	68½	66½	64¼	62¼	60	58
7	81½	79½	77¼	75¼	73¼	71	69	67	64¾	62½	60¼	57¾	55½
	80¾	78½	76¼	74	71¾	69½	67¼	65	62¾	60¼	57¾	55½	53
6	79¾	77¼	75	72½	70	67¾	65¼	62¾	60¼	57½	55	52½	50
	78½	76	73½	71	68	65½	63	60¼	57½	54½	51¾	49	46½
5	77¼	74½	71¾	69	65¾	63	60	57	54	51	48	45½	42½
	75½	72½	69½	66¼	63	60	56¾	53½	50¼	47	43½	40¼	36¾
4	73¼	69¾	66¼	62¾	59¼	55¾	52	48½	45	41¾	37½	33¾	30

CO_2 Measurements

Figure 14-10. A chart for plotting furnace efficiency by using the CO_2 content of the flue gases. *Courtesy R.W. Beckett Corporation*

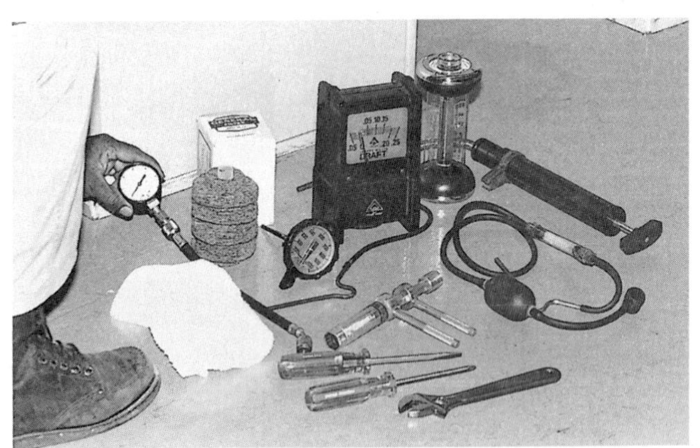

Figure 14-11. Hand tools and supplies for servicing an oil burner. *Photo by Bill Johnson*

water detector to check for water in the tank as shown in Figure 14-12. (The water detector is a paste that is smeared on the oil-level measuring stick.)

3. Make sure that all furnace vents are open. Examine the flue pipe for corrosion or damage. Make sure that the flue outlet is clear. Figure 14-13 shows soot in a flue outlet.
4. Examine the furnace on the outside for corrosion. Look for smoke streaks around the inspection door as shown in Figure 14-14. Make sure that all components (such as the transformer, the inspection door, and the burner-to-appliance gasket) are tight so that air leaks do not affect the test.

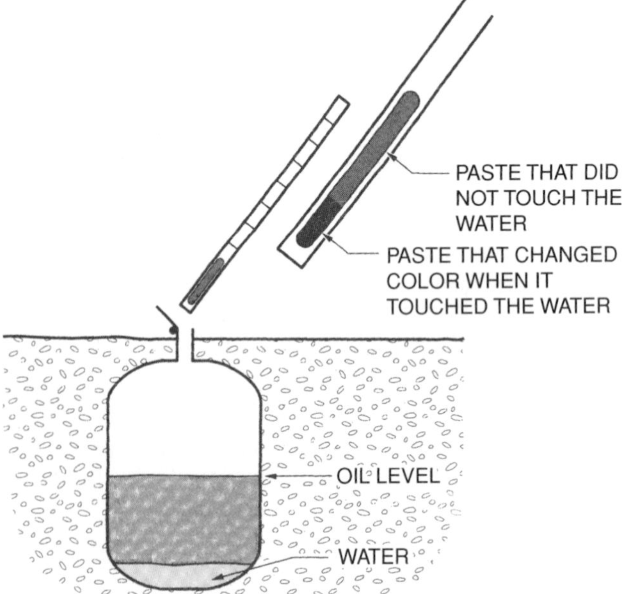

Figure 14-12. Checking for water in an oil tank.

5. Insert the stack thermometer in the flue, before the draft regulator. Start the furnace and allow it to warm for 10 minutes or until the stack temperature stops rising as shown in Figure 14-15.
6. Check the draft over the fire and set the draft regulator to maintain the draft over the fire to -0.02 inches of water column for the proper over fire draft.

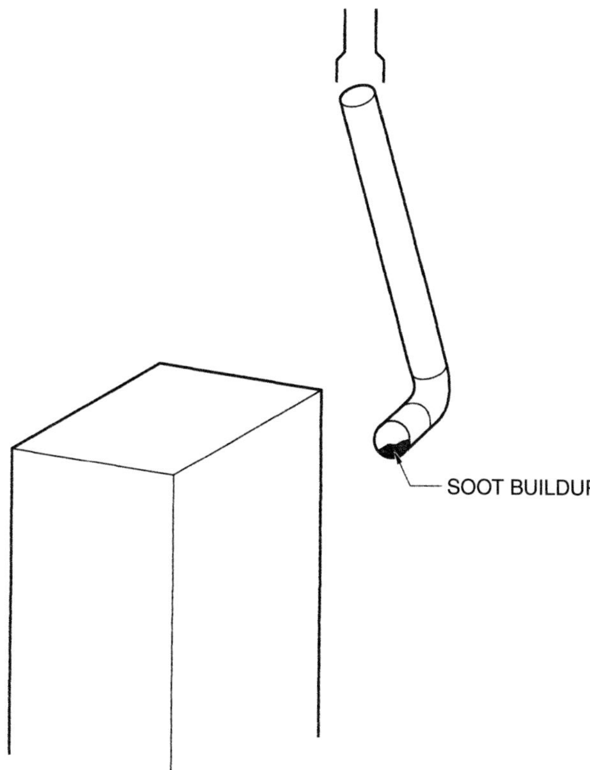

Figure 14-13. The flue must be clear of soot.

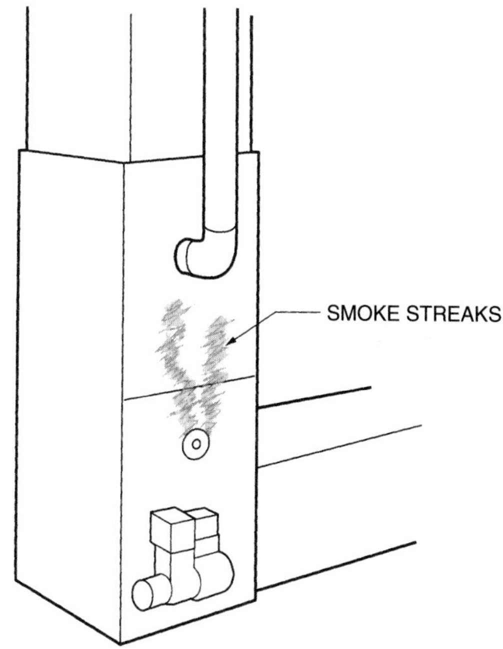

Figure 14-14. Examine the outside of the furnace for smoke streaks and corrosion.

7. Perform a smoke test, as shown in Figure 14-16. Suppose that the smoke test shows a smoke level of No. 2.
8. Perform a CO_2 level test, as shown in Figure 14-17. Suppose that the CO_2 level is 8.5%. An experienced

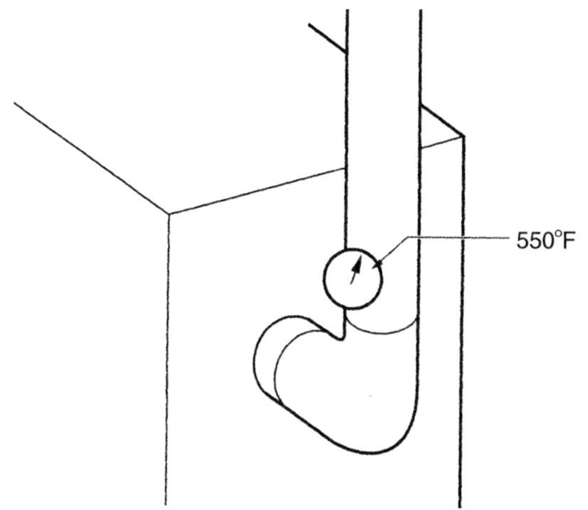

Figure 14-15. Allow the furnace to run until the stack temperature stops rising.

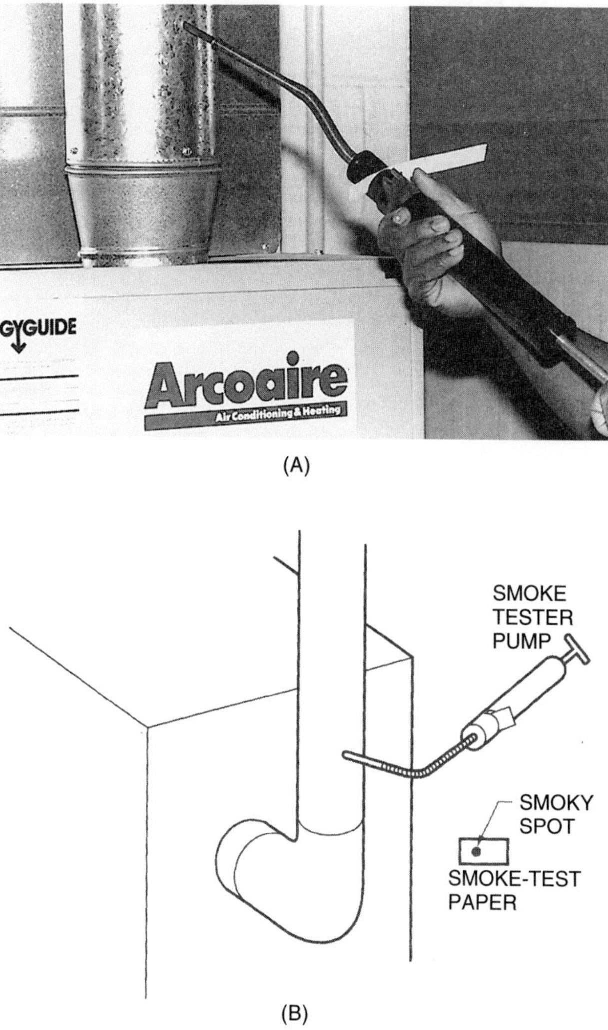

Figure 14-16. (A) Smoke tester. *Photo by Bill Johnson* (B) Smoke test.

technician would look at these results—a smoke level of No. 2 and a CO_2 level of 8.5%—and deduce that the fire is smoking, but not from a shortage of air. Adding air reduces CO_2.

9. Change the in-line oil filter, remove the nozzle assembly, check the electrode settings, change the nozzle, and replace the assembly.
10. Install an oil-pressure gauge, then start the system. Figure 14-18 shows an oil-pressure gauge.
11. Check the oil pressure and set it at 100 psig if needed. Recheck the stack temperature, then perform a smoke test and a CO_2 level test. Suppose that the oil pressure is 100 psig, the new stack temperature is still 450°F (net), the new smoke level is No. 0, and the new CO_2 level is 10.5%. The furnace efficiency is now 81%.

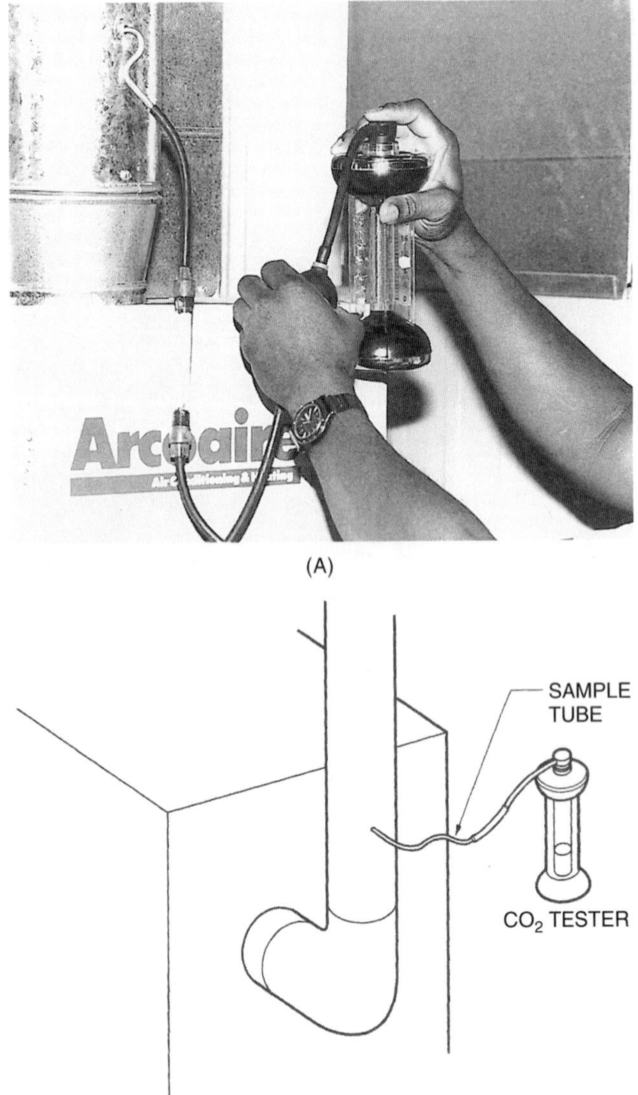

Figure 14-17. (A) CO_2 tester. *Photo by Bill Johnson* (B) CO_2 level test.

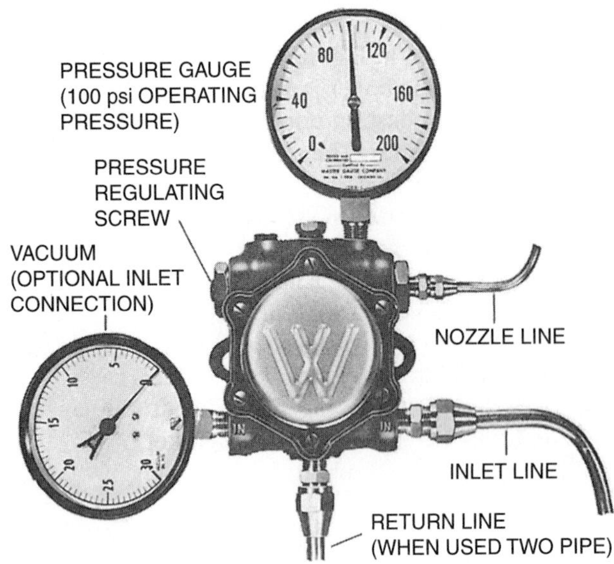

Figure 14-18. Oil-pressure gauge. *Courtesy Webster Electric Company*

Something caused the efficiency to change. Probably the oil nozzle was beginning to show signs of a poor oil pattern. The service call is a success because the efficiency is improved and the oil nozzle would only get worse if unchanged. The customer is satisfied with the improved efficiency.

Suppose next that a technician is called on to check out an old oil boiler for efficiency and to tune it up. This boiler has a standard gun burner without a flame-retention head. The technician would take the following steps:

1. Move the correct tools to the boiler location.
2. Start the boiler and bring it up to temperature. When it is almost to the cutout point, the stack temperature should be steady. Take a smoke test. Suppose that the smoke level is No. 4 and that the stack temperature is 600°F.
3. Take a CO_2 level reading. Suppose that the CO_2 level is 4%. These readings do not correlate, so something must be wrong. If air is added to clear the smoke, the CO_2 reading will become even lower. The stack temperature is too high probably because of too much primary air or a dirty heat exchanger. Look over the system for problems at this time.
4. Find that several air leaks around the burner components are causing excess air, the transformer is not sitting down tight, and the gasket around the burner opening is leaking air. There are smoke streaks around the burner and the inspection door, where smoke has been puffing out on start-up. A draft problem is suspected.
5. Remove the flue pipe. It is partially blocked with soot, as shown in Figure 14-19. Clean this pipe and

reinstall it. The boiler is a small fire-tube boiler, so open the combustion doors (see Figure 14-20), and inspect the tubes. They are coated with soot, as shown in Figure 14-21. Clean the tubes with steel brushes. When the inside of the boiler is clean, close the doors and check them for a tight fit.

6. Remove the burner. Change the nozzle, set the electrodes, and clean the insulators. Clean the cad cell of soot. Remove and clean the fan wheel. It has dirt buildup in the blades. This buildup will reduce the airflow because the fan will put out less air. Reinstall the burner with a new gasket.
7. Change the in-line oil filter.
8. Inspect the piping and the tank. The system is ready to restart.
9. Install an oil-pressure gauge and start the burner. Adjust the oil pressure from 85 psig to 100 psig. When the boiler is up to temperature and the stack temperature is steady, it reads 500°F. The reduction in stack temperature means less heat up the flue. Perform a new smoke test. The smoke level is now No. 1. The new CO_2 reading is 8%. From the chart in Figure 14-10, the new efficiency is 75.5%. This is a noticeable improvement. If this boiler is maintained, the efficiency will remain at this level.
10. Now check the draft to see if it is marginal. The draft over the fire is –0.02 inch of wc, and the draft at the flue outlet to the boiler is –0.04 inch of wc. The draft regulator is set correctly, and the boiler is operating correctly.

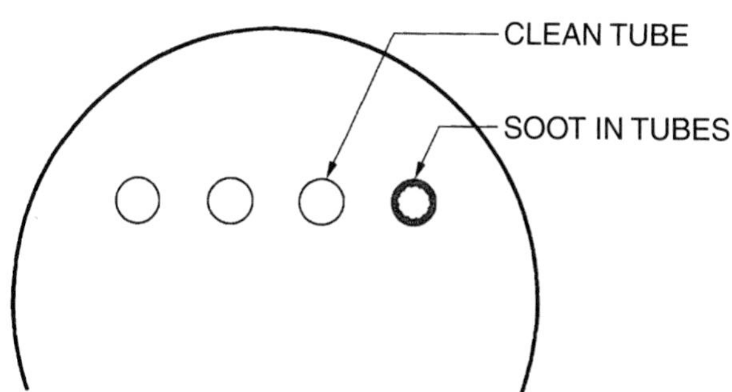

Figure 14-21. One of these tubes has a coating of soot.

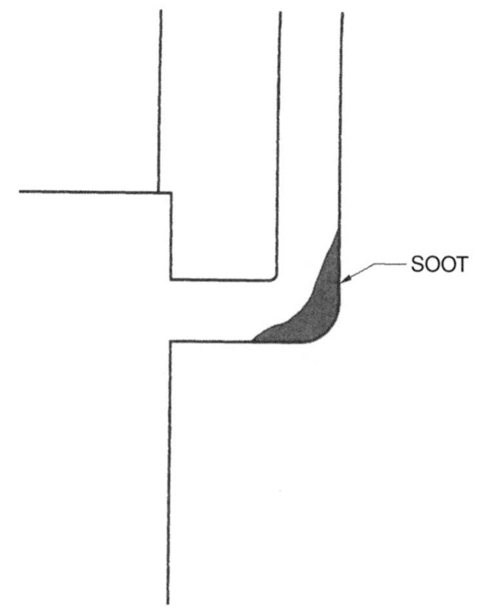

Figure 14-19. This flue pipe is partially blocked with soot.

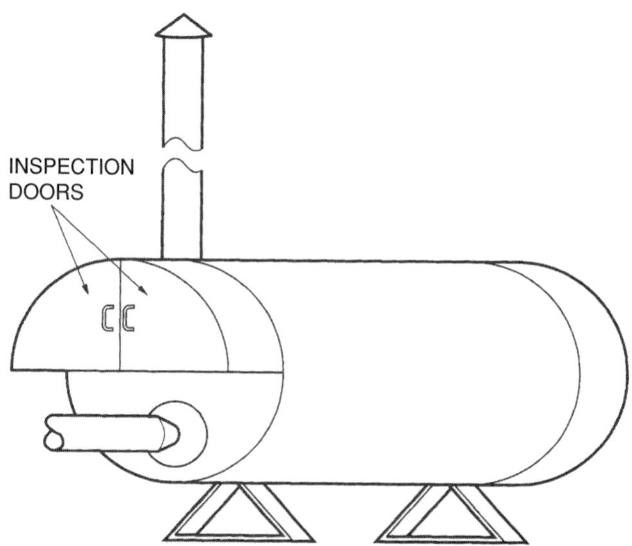

Figure 14-20. This small fire-tube boiler has doors that allow access to the tubes.

If the efficiency of the old burner could not be adjusted to more than 70%, the technician may recommend that it be replaced with a flame-retention head burner. The new burner would improve the efficiency, and the boiler would provide many more years of good service, provided that it is in good shape.

The CO_2 method of checking efficiency is not the only method. Technicians often use the oxygen content of the flue gases to determine the combustion efficiency. The oxygen content can be determined by using an oxygen tester, as shown in Figure 14-22. When the percentage of oxygen is determined, it can be plotted on a chart similar to the CO_2 chart. Figure 14-23 shows an oxygen combustion analysis chart. The technician should keep in mind that when air is added, the oxygen content increases and the CO_2 content decreases.

The tests just described were performed with fluid combustion analysis. However, many technicians use the electronic analyzer because of its ease of use and convenience. Figure 14-24 shows a typical electronic analyzer. These testers are more expensive but have many features

Net Stack Temp. (degrees F.)

Percent Oxygen	300°	350°	400°	450°	500°	550°	600°	650°	700°	750°	800°	850°	900°
15	75½	72½	69½	66¼	63	60	56¾	53½	50¼	47	43½	40¼	36¾
14	77¼	74½	72¾	70	68	64¼	61½	58¾	55¾	52¾	49¼	47¼	44½
13	79¾	77¼	75	72½	70	67¾	65¼	62¾	60¼	57½	55	52½	50
12	80¾	78½	76¾	74¾	72½	70¼	68¼	66	63¾	61½	59	56¾	54¼
11	82¼	80¼	78½	76½	74½	72½	70½	68½	65¾	64¼	62¼	60	58
10	83	81	79¾	77¾	76	74¼	72½	70¾	68¾	67	64¾	63	61
9	84	82¼	80¾	79	77¼	75¾	74	72¼	70¾	68¾	67	65¼	63½
8	84¾	83	81¾	80¼	78½	77	75½	73¾	72¼	70½	69	67½	65¾
7	85½	83¾	82½	80¾	79¼	77¾	76¼	74¾	73¼	71½	70	68½	67
6	85¾	84½	83	81½	80¼	78¾	77¼	75¾	74½	73	71½	70	68½
5	86	85	83¾	82¼	81	79½	78¼	77	75½	74	72½	71¼	70
4	86½	85¼	84	83	81½	80¼	79	77¾	76½	75¼	73½	72¾	71
3	87	85¾	84½	83½	82¼	81	79¾	78½	77¼	76	74¾	73¾	72
2	87¼	86	84¾	83¾	82¾	81½	80½	79	78	76¾	75½	74½	73
1	87½	86½	85	84¼	83¼	82	81	79½	78¾	77½	76¼	75¼	74

Oxygen Measurements

Figure 14-23. Oxygen combustion analysis chart. *Courtesy R.W. Beckett Corporation*

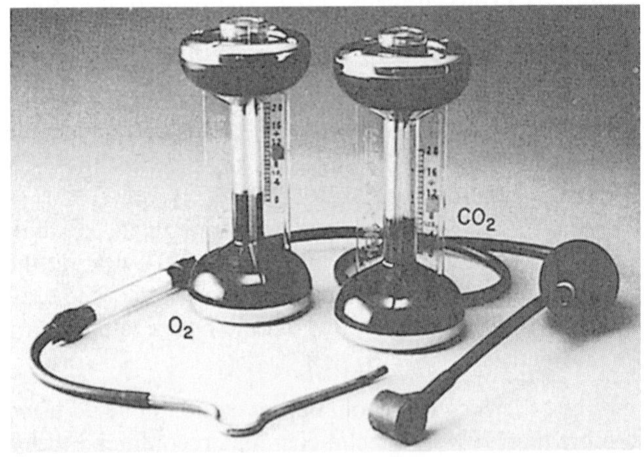

Figure 14-22. Combustion can be analyzed with an O_2 tester. *Courtesy United Technologies Bacharach*

that make them worthwhile if you perform many tests each year.

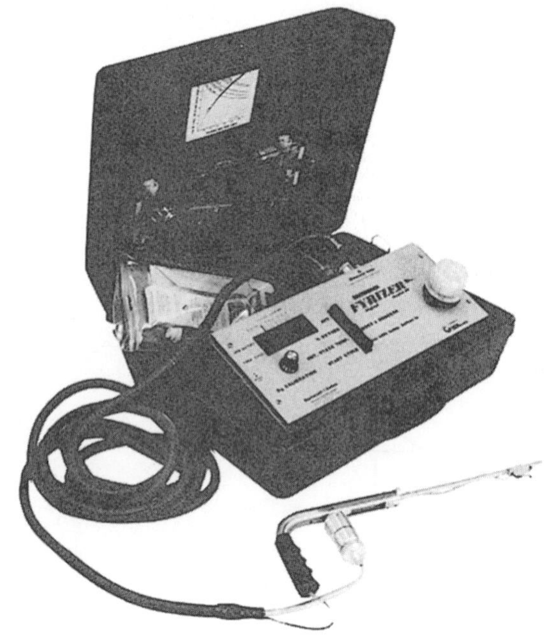

Figure 14-24. Electronic combustion analyzer. *Courtesy United Technologies Bacharach*

Summary

- Boilers are either hot water or steam.
- The burner for an oil boiler is the same as for a furnace.
- Boilers are made of cast iron or steel with sheet-metal covers. Insulation is installed between the cover and the boiler.
- Water boilers are liquid full of water, and steam boilers are partially filled with water, with steam at the top.
- Boilers have overtemperature controls to prevent them from overheating.
- Water boilers are temperature controlled with a thermostat and a high limit control. Steam boilers are controlled by steam pressure. Both boilers have relief valves to vent pressure in case of overtemperature.
- Combustion analysis is important for oil-burning equipment because it can vary in efficiency more than gas burners do.

- It is a good practice to perform a before-and-after combustion analysis each year on oil-burning equipment. The before analysis is performed, then the system is tuned up, and the after analysis shows the improvement after service is performed.
- The oil filter, the air filter, and the nozzle should be routinely replaced.
- When the efficiency of an older burner cannot be improved beyond 70%, the technician should consider recommending that the burner be replaced with a flame-retention burner, if the appliance is otherwise in good shape.

Review Questions

1. What type of control shuts off a hot-water boiler and a steam boiler?
2. Why is it a good practice to perform before-and-after combustion analysis on oil-burning appliances during the annual service call?
3. What prevents pressure buildup in a boiler?
4. What would be an advantage in a flame retention burner?
5. What could be the cause of a high stack temperature?

15 Troubleshooting and Maintenance for Oil-Fired Equipment

Objectives

Upon completion of this unit, you should be able to

- **perform basic electrical analysis on the control system.**
- **troubleshoot basic problems in the mechanical portions of the system.**

15.1 Troubleshooting Oil-Fired Equipment

Oil-fired equipment can have either electrical or mechanical problems. Electrical problems occur in the electrical circuits, such as the controls or the fan circuit. With a boiler, the electrical problems may be in the pump circuit.

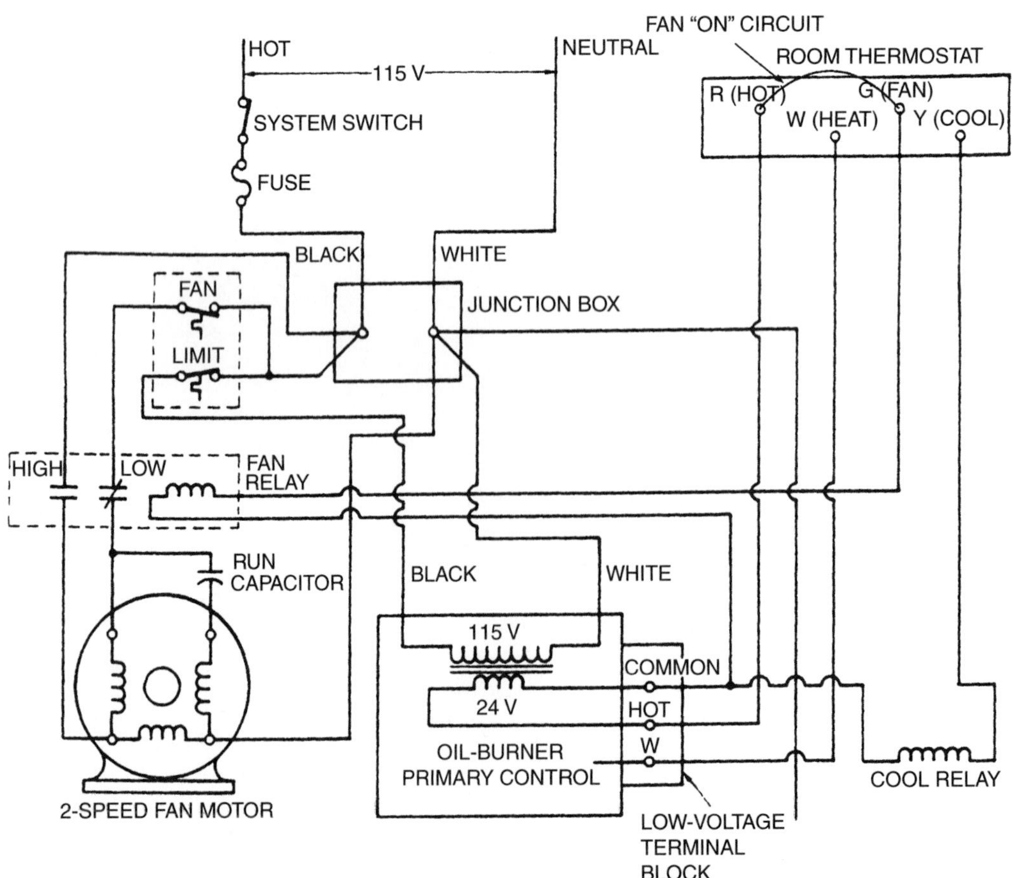

Figure 15-1. If the fan starts when the thermostat fan switch is turned on, there is control voltage.

Mechanical problems can occur in the oil supply system, the combustion process, or the flue system. Usually you will encounter only one problem, or, if several problems exist, they often have one source. For example, a combustion problem can cause soot, which clouds the cad cell and causes the electrical circuit to shut off the burner.

Furnace Electrical Problems

The oil-furnace electrical system is not difficult to troubleshoot. As mentioned, the electrical problem may be in the fan or in the burner circuit. The first thing a technician must do when an electrical problem is suspected, is determine whether the unit has power. If the unit also has air-conditioning, this is simple; the technician must only turn the fan switch on the room thermostat to the on position. If the fan starts from the fan relay circuit, the furnace has power as shown in Figure 15-1.

If the system is used exclusively for heating, the technician should instead check for power at the furnace disconnect (if one exists), as shown in Figure 15-2. If there is no disconnect at the furnace, the unit should have a switch with a removable cover. You can use the switch connections to check for power, as shown in Figure 15-3.

Caution: Use care when placing a meter probe on the switch connections, or you will short the meter lead to the housing as shown in Figure 15-4.

Another alternative is to remove the junction box in the furnace and check for power under the connectors in the box, as shown in Figure 15-5.

Figure 15-2. Checking for power at the unit disconnect.

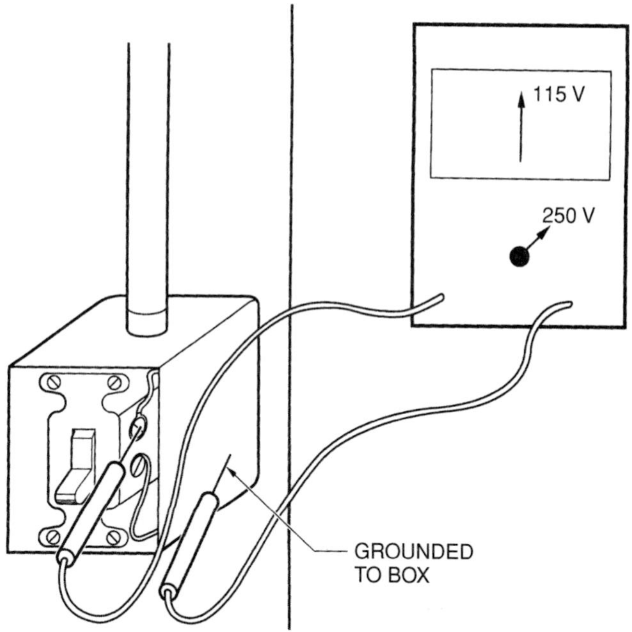

Figure 15-3. Checking for power at the unit switch.

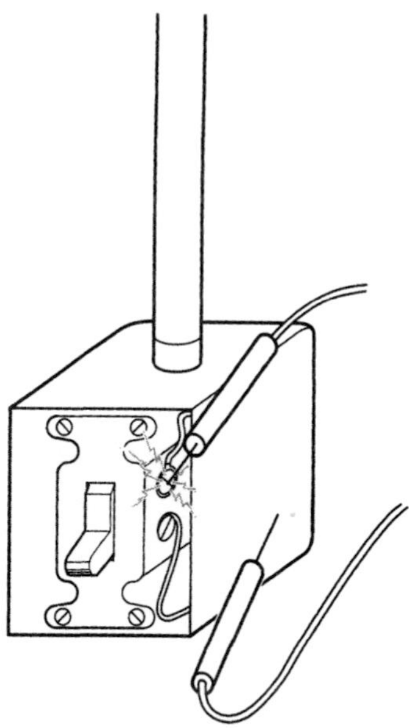

Figure 15-4. Watch the meter leads when checking for power at the system switch.

Caution: Turn off the switch while removing the cover to the junction box and finding the correct terminals. When you locate the correct terminals, place your meter leads on the terminals and turn the power on to read the meter.

After you establish that the furnace has power, any problem must be in the furnace.

The technician can make some deductions about the controls without using a meter. For example, suppose that for some reason the furnace fan does not start. Upon a call for heat, the burner ignites and heats the furnace to the point when the fan should start. When it does not start,

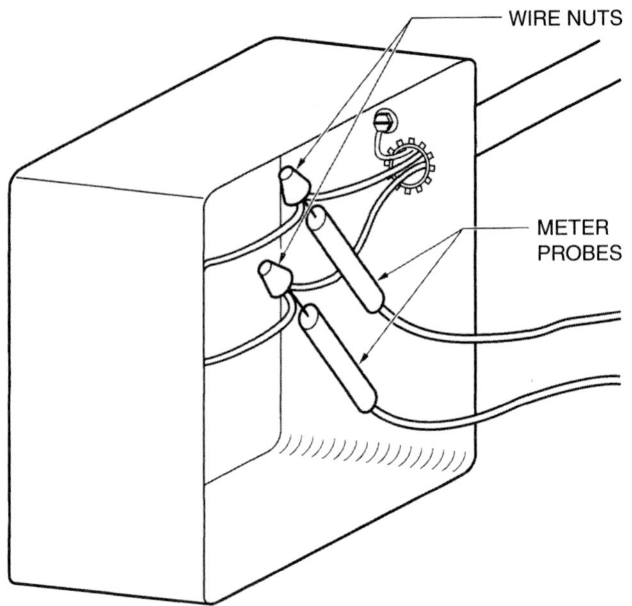

Figure 15-5. Checking for power at the junction box.

the limit control shuts off the furnace. The technician who approaches the furnace will notice the heat of the furnace and know why the burner is stopping and starting. It would be pointless to troubleshoot the limit circuit when it is obvious that a fan problem exists.

Let us use the diagram in Figure 15-6 to locate a fan problem. The technician will use a meter in this case. The fan is not running because an open circuit is in the fan winding. The technician notices that the furnace is hot; the furnace is heating, so it must have power. The technician deduces that there is a problem with the fan circuit and takes the following actions:

1. Removes the front of the furnace for access to the fan–limit control.
2. Removes the cover to the fan–limit control.
3. Turns the meter to the 250-volt scale and places one meter lead on the furnace frame and the other on the fan portion to the limit control feeding the fan motor. The meter registers 115 volts. The circuit is calling for the fan to run, but it is not running.

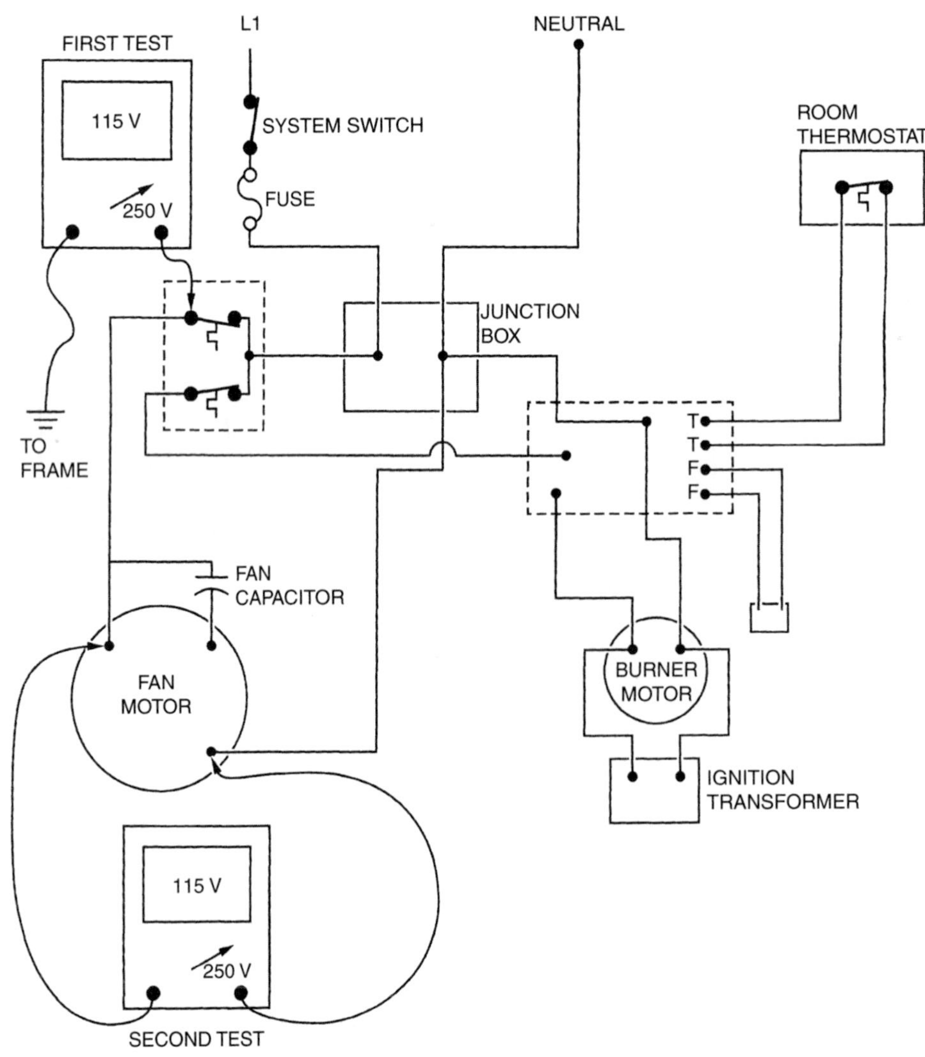

Figure 15-6. Wiring diagram showing the fan circuit.

4. Shuts off the power to the furnace and removes the fan compartment door. Carefully touches the fan motor and discovers that it is not even warm. If power is reaching the motor terminals, there is no winding in the motor, or heat would be generated.
5. Removes the cover to the motor terminals and connects the meter leads to common and hot, the wires feeding the fan motor. Turns on the power, and the leads have power. The fan is getting power but will not start.
6. Turns off the power and removes the meter leads, then disconnects the wires leading to the motor. Reconnects the meter leads and switches the meter to the R×1 scale; there is no reading. The motor has an open circuit as shown in Figure 15-7. This motor must be repaired or replaced.

Finding a problem with no ignition involves a similar process. Suppose that the furnace does not fire. The technician should take the following steps:

1. Check for furnace power at the furnace disconnect; the meter registers 115 volts as shown in Figure 15-8.

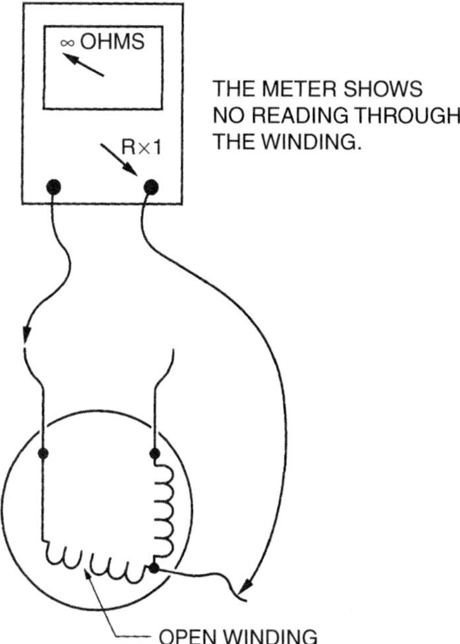

Figure 15-7. This motor has an open circuit.

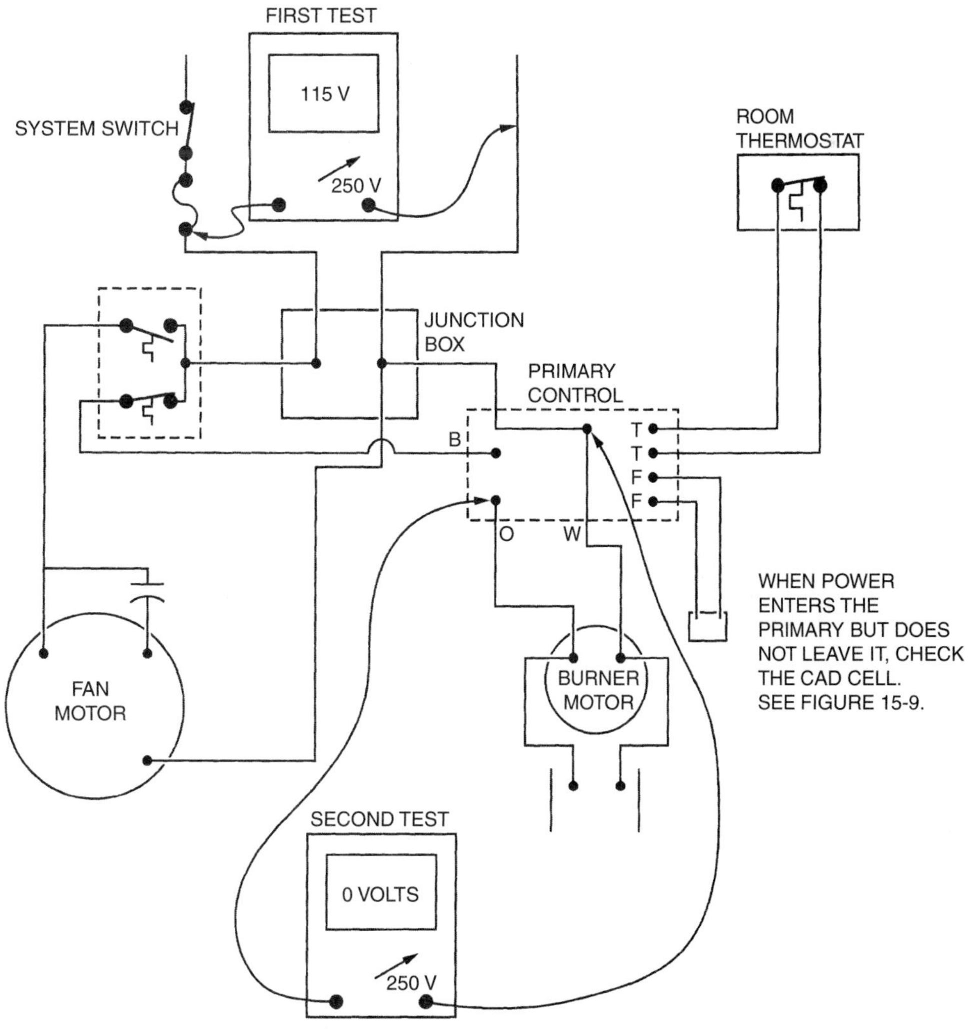

Figure 15-8. At the furnace disconnect the meter registers 115 volts.

2. Look at the diagram and realize that the primary control acts as a switch to start the burner motor. The only external controls that would affect it are the limit switch, the cad cell, and the room thermostat.
3. Press the reset button to see if the furnace will start.

Caution: It is always best to examine the combustion chamber for excess oil before pushing the reset button. The owner may have reset the furnace several times and accumulated oil in the combustion chamber. This oil must be removed before start-up.

4. Check that the thermostat is calling for heat by jumping the terminals between T and T at the primary control. The furnace still does not start.
5. Turn off the power, remove the cad cell, and check it with a VOM, using the ohm scale, as shown in Figure 15-9. The cad cell is operable.
6. Assume that the primary control has an open circuit. Looking at the diagram in Figure 15-10, determine that power comes into the primary control on the black wire and that the white wire is neutral. When the primary control is working correctly, power should pass through the primary control wire to the orange wire, with the white wire remaining neutral. The primary control acts as a switch to the circuit. The switch is open circuit and will not close when the reset button is pushed. The primary control is defective and must be replaced.

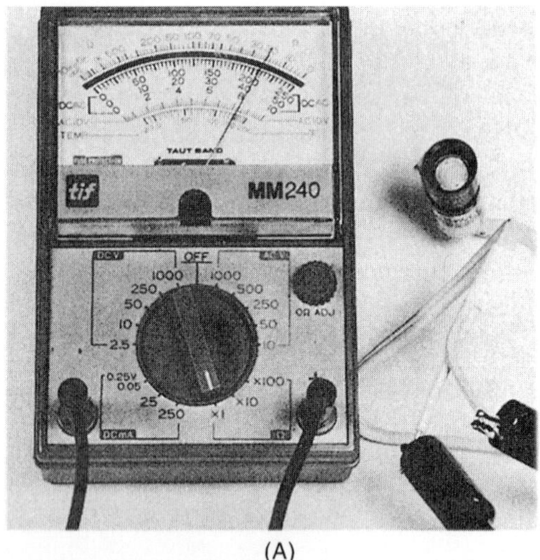

(A)

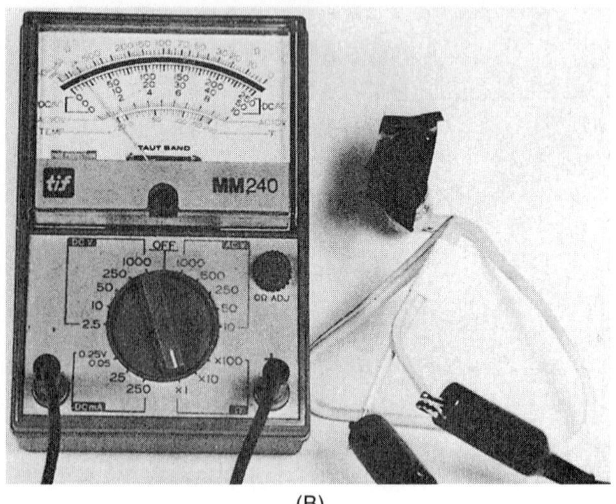

(B)

Figure 15-9. Checking a cad cell. The cad cell changes resistance with light. (A) The cad cell with light shining in and a resistance of 20 ohms. (Note: The meter is on the R×1 scale.) (B) The cell eye covered with electrical tape. The reading changes to 620 ohms. This cad cell is used to sense the flame in an oil furnace. *Photos by Bill Johnson*

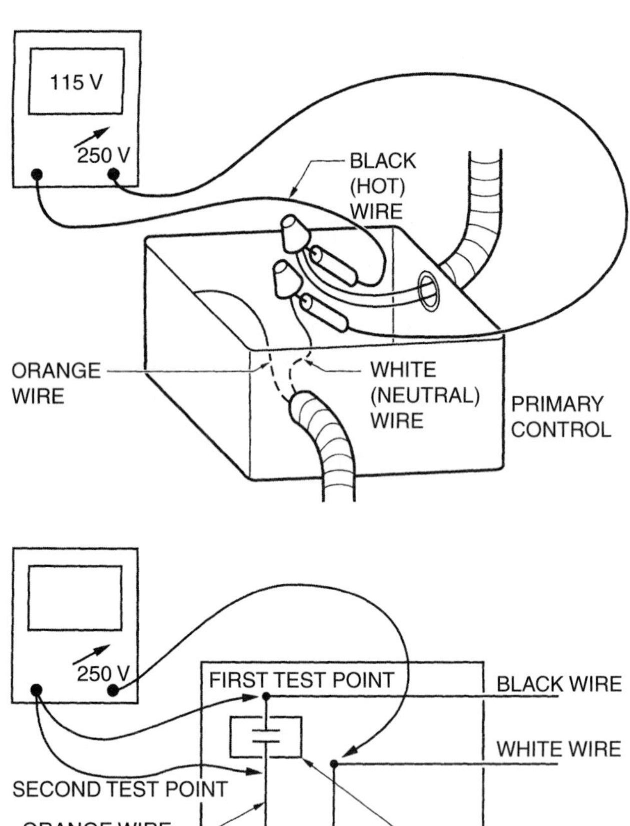

Figure 15-10. Troubleshooting the primary control as a switch.

Boiler Electrical Problems

Because boilers use the same burners as furnaces, the electrical problems are much the same, except for the pump. The pump is usually operated from a relay in the primary control. If the primary control has power and the pump will not run, the pump circuit must be checked. Suppose that the boiler is up to temperature and the pump will not run. The technician should do the following:

1. Remove the cover to the pump electrical terminals and check the voltage. If there is none, look at the

wiring diagram or follow the wiring back and find the relay that should operate the pump.
2. Check for power into and out of the relay. If power is going in but no power is going out, the relay is defective and must be replaced.

Furnace Mechanical Problems

Oil-burning forced-air furnaces are much like gas furnaces, except for the combustion chamber. Oil-burning furnaces are much more likely to gather soot than gas-burning furnaces are.

The combustion chamber must be in good condition and be able to contain the fire of the burner. The chamber must be complete, without holes in the wall or crumbled brick as shown in Figure 15-11. Combustion chambers can often be repaired with wet-blanket refractory material. Older furnaces and boilers may have baffles to force the combustion gases closer to the walls of the heat exchanger. These baffles may be deteriorated or may have been removed by a previous service technician for unknown reasons and may have to be replaced to improve the efficiency of the appliance as shown in Figure 15-12.

The heat exchanger often gathers soot when the burner is out of adjustment. If so, it must be cleaned. With a boiler, the technician can easily open the doors and clean the tubes. Cleaning a furnace is more difficult. Many manufacturers do not provide ready access to the fire side of the heat exchanger, and cleaning is a dirty job. If there are access ports, these can be removed and brushes or wire used to loosen the soot. The flue pipe connection on both sides of the furnace can be removed for access and cleaning. Brushes can be used to clean between the surfaces, and the soot can then be removed with a vacuum cleaner. If the furnace is in a house, the vacuum cleaner exhaust should be piped to the outside to prevent soot from being blown into the house because of a defective filter as shown in Figure 15-13.

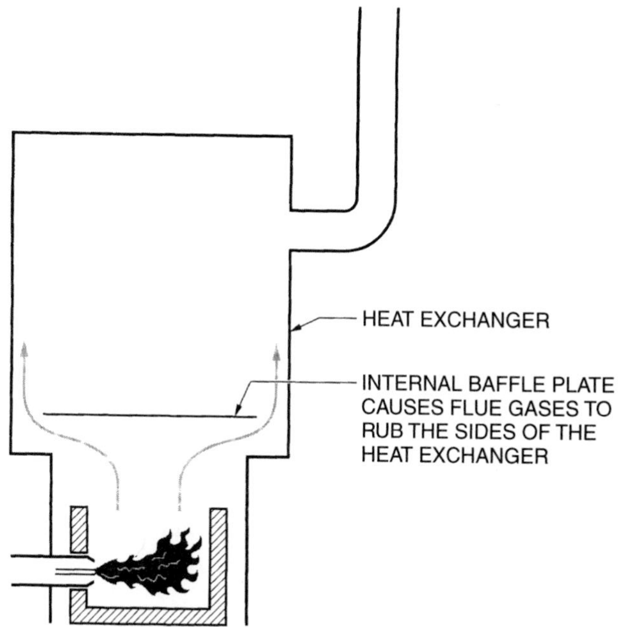

Figure 15-12. The proper baffles must be in place in the furnace for it to have the best efficiency.

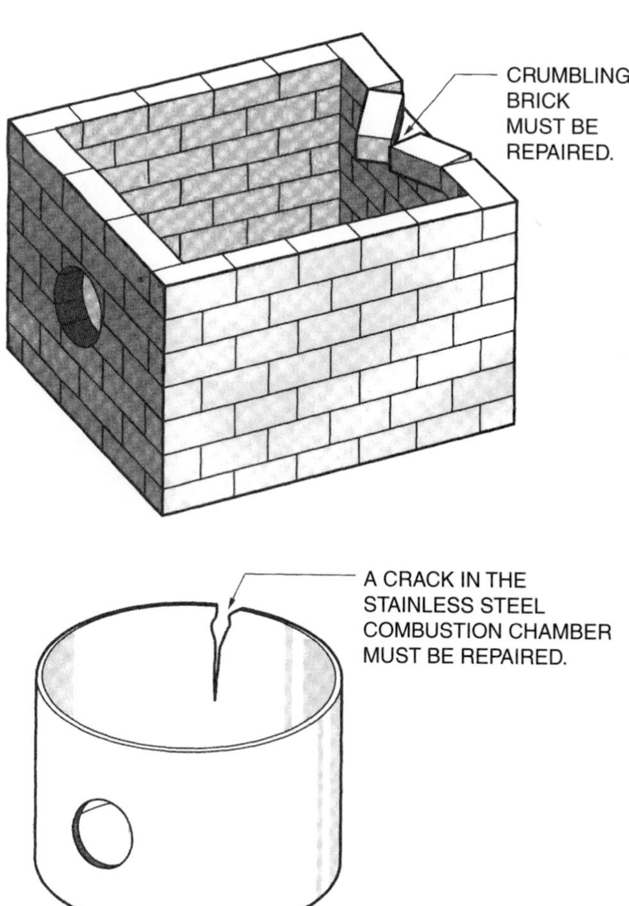

Figure 15-11. The combustion chamber must be in good condition.

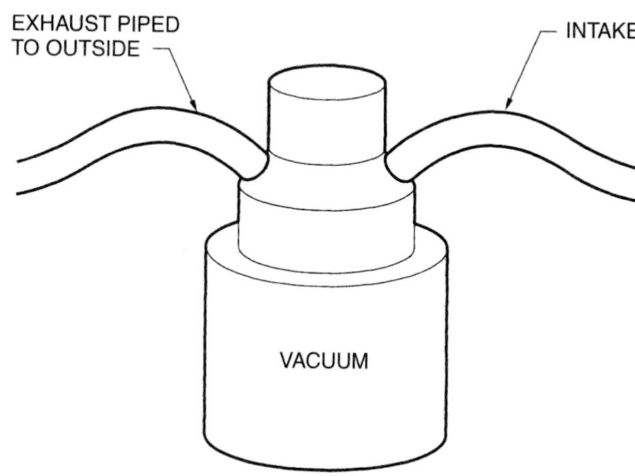

Figure 15-13. Piping the vacuum exhaust to the outside prevents the vacuum discharge from creating dust in the house.

Burner assemblies should be removed and the electrodes and insulators cleaned of soot. Soot is carbon and therefore a good conductor of electricity. It must be removed, or the high voltage from the electrodes will be shorted to the frame ground, as shown in Figure 15-14.

Cad cells are eyes and become soot coated in furnaces that are out of adjustment as shown in Figure 15-15. They must be cleaned.

Caution: When working with cad cells, never handle the connection pins on the back of the cell as shown in Figure 15-16. The cad cell circuit is a very low-voltage circuit, and oil from your skin can cause enough corrosion to create a problem. If you handle these pins, wipe them clean before reinstalling the cad cell.

A large amount of energy is released in the vicinity of the burner, so problems most often occur in the burner assembly. Air leaking around the burner components is one problem that occurs if the service technician is careless while servicing the system. All components should be fastened tightly, especially the transformer and the gasket between the burner and the furnace. The inspection door should also be shut tightly as shown in Figure 15-17. This tightness should be checked before you run a final flue gas analysis, or the results of the analysis may be poor and you will have to tighten the components and repeat the procedure.

Problems on the fuel side of the system occur because of plugged filters, water in the oil, nozzle problems, and leaks in the oil inlet line that is operating in a vacuum when the tank is below the burner. Changing filters and nozzles solves some of these problems. Finding vacuum leaks in the supply line is more difficult. First, look over the system for obvious problems. For example, if the leak just started, there may be physical damage to the oil line. The owner may have crimped the line or otherwise damaged it. Examine it. If examination does not reveal the problem, the line may be pressurized with compressed air, as shown in Figure 15-18.

Caution: Never pressurize the filter to more than 12 psig; this is the working pressure of the filter.

The filter can also be removed from the line for a high-pressure test, as shown in Figure 15-19. Some technicians also install a set of gauges on the supply line and operate

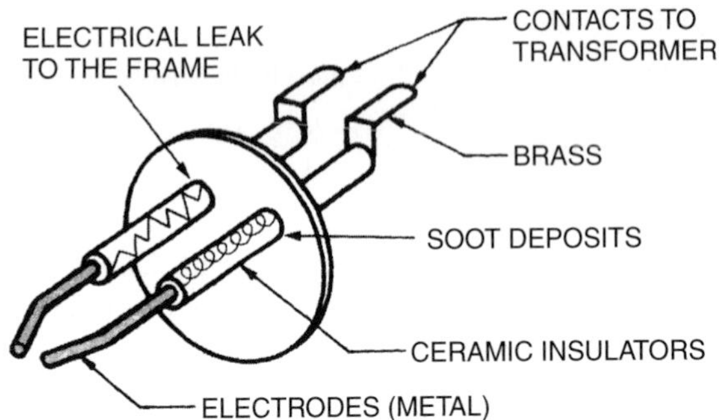

Figure 15-14. Soot creates a path along which the high voltage can leak to ground.

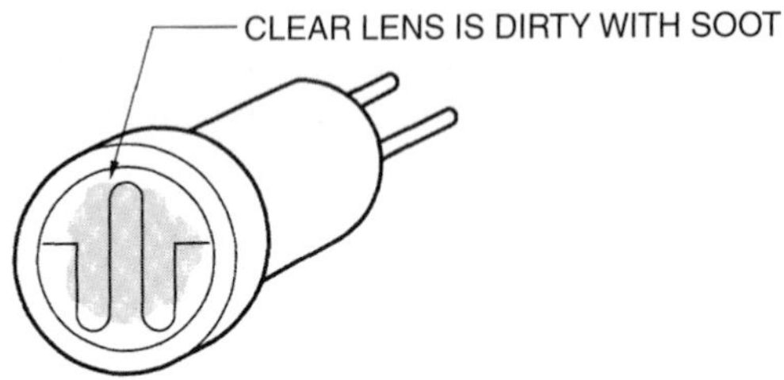

Figure 15-15. Soot-coated cad cell.

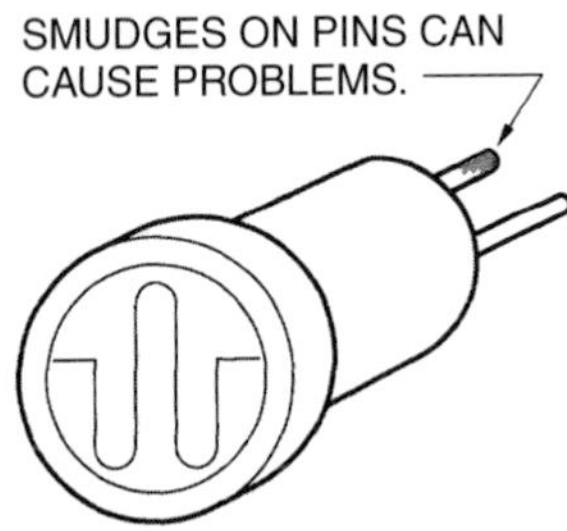

Figure 15-16. Do not handle the cad cell connectors.

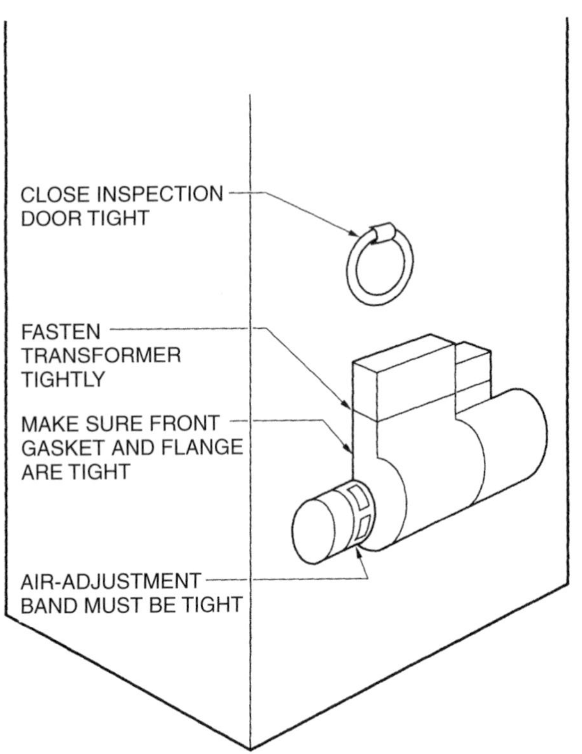

Figure 15-17. The components must be fastened tightly, or air will leak into the combustion process.

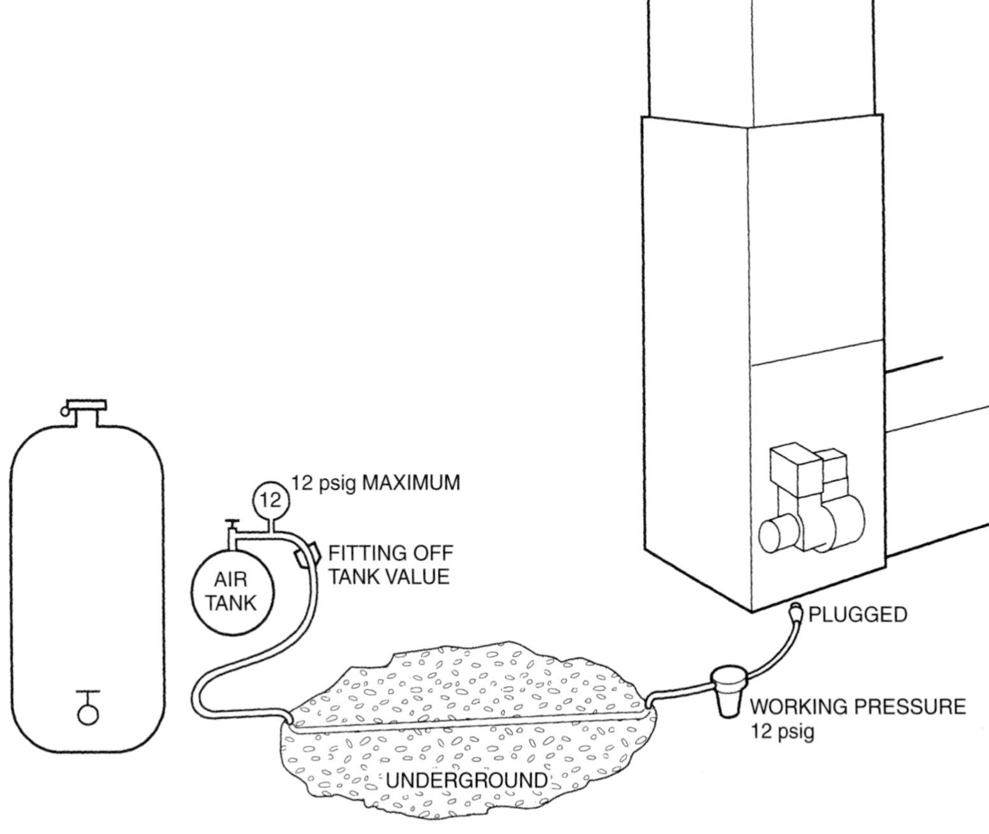

Figure 15-18. Pressurizing the oil line.

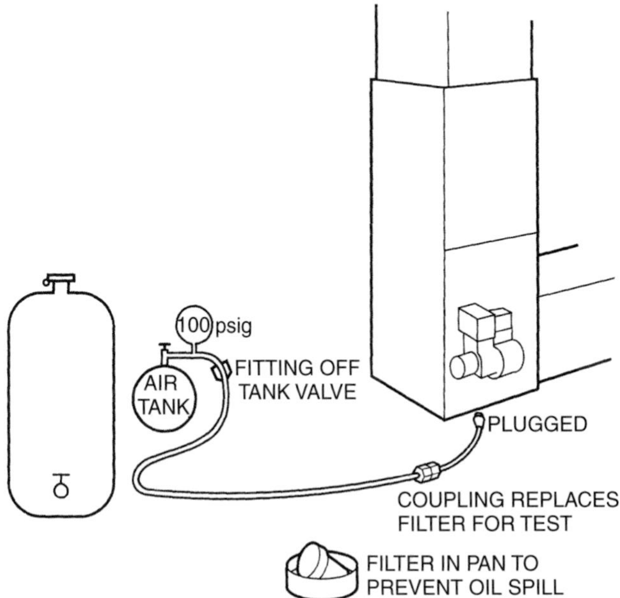

Figure 15-19. Remove the in-line oil filter for a high-pressure test.

the oil pump with the tank valve off so that the line will operate in a vacuum, then shut the pump off and see if the vacuum holds as shown in Figure 15-20. If it does not hold, the supply line has a leak. The line then must be pressurized to find the leak. Perhaps starting with a pressurized leak check is a shortcut because you cannot find the leak while the line is in a vacuum.

In extremely cold climates, cold oil may cause ignition problems and burning problems. When oil is cold, it does not atomize as easily as warmer oil. Many technicians offset the cold oil by increasing the nozzle pressure. The greater pressure atomizes the oil into finer droplets, but also may overfire the furnace or boiler, so a smaller nozzle must be substituted for the standard nozzle. The technician must perform combustion analysis tests and experiment with the correct nozzle and pressure for the best performance. Figure 15-21 shows a chart that can be used to calculate the overfire rate for a nozzle substitution.

Boiler Mechanical Problems

Boiler mechanical problems are much like furnace mechanical problems as far as combustion is concerned. However, the water side of the boiler presents a different set of problems. Water contains minerals and air, and both can cause problems in a boiler. The minerals collect at the hottest point in the system, which is on the heat-exchange surface between the water and the combustion by-products. This collection of minerals can become thick and deter heat exchange if not treated. Each time

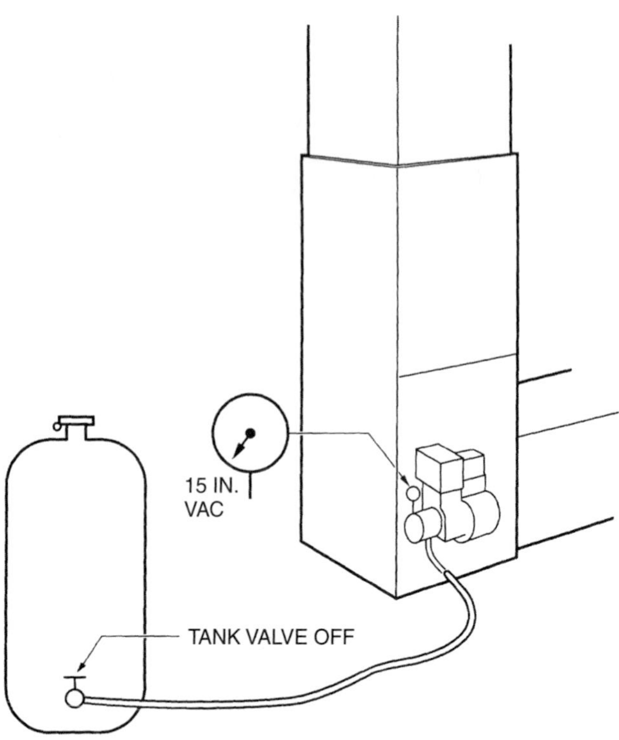

Figure 15-20. A vacuum test of the oil supply line.

Effects of Pressure on Nozzle Flow Rate

NOZZLE RATING AT 100 psi	NOZZLE FLOW RATES IN GALLONS PER HOUR (Approx.)					
	120 psi	145 psi	160 psi	175 psi	200 psi	300 psi
.40	0.44	0.48	0.51	0.53	0.57	0.69
.50	0.55	0.60	0.63	0.66	0.70	0.86
.60	0.66	0.72	0.76	0.79	0.85	1.04
.65	0.71	0.78	0.82	0.86	0.92	1.12
.75	0.82	0.90	0.95	0.99	1.05	1.30
.85	0.93	1.02	1.08	1.12	1.20	1.47
.90	0.99	1.08	1.14	1.19	1.27	1.56
1.00	1.10	1.20	1.27	1.32	1.41	1.73
1.10	1.21	1.32	1.39	1.46	1.55	1.90
1.20	1.31	1.44	1.51	1.59	1.70	2.08
1.25	1.37	1.50	1.58	1.65	1.76	2.16
1.35	1.48	1.63	1.71	1.79	1.91	2.34
1.50	1.64	1.81	1.90	1.98	2.12	2.60
1.65	1.81	1.99	2.09	2.18	2.33	2.86
1.75	1.92	2.11	2.22	2.32	2.48	3.03
2.00	2.19	2.41	2.53	2.65	2.82	3.48
2.25	2.47	2.71	2.85	2.98	3.18	3.90
2.50	2.74	3.01	3.16	3.31	3.54	4.33
2.75	3.00	3.31	3.48	3.64	3.90	4.75
3.00	3.29	3.61	3.80	3.97	4.25	5.20
3.25	3.56	3.91	4.10	4.30	4.60	5.63
3.50	3.82	4.21	4.42	4.63	4.95	6.06
4.00	4.37	4.82	5.05	5.29	5.65	6.92
4.50	4.92	5.42	5.70	5.95	6.35	7.80
5.00	5.46	6.02	6.30	6.61	7.05	8.65

Figure 15-21. This chart shows the overfire rate for nozzle substitution. *Courtesy Delavan Fuel Metering Products*

Figure 15-22. The bubbles in this glass of water result from oxygen released from within the water.

a pound of water is introduced into the system, it brings with it some minerals. Some collect on the boiler surface. If a system is allowed to leak water for a long time, the mineral deposits are added to the heat exchange on a continuous basis if the water is not treated with the correct chemicals. Water-treatment companies can test the water for minerals and suggest the correct combination of chemicals for the system. The larger the system, the more important the treatment because of the higher cost of replacing the boiler.

When water is allowed to leak from a system for a long time, rust will also occur. Each pound of fresh water contains a small amount of oxygen. The bubbles that collect in a glass of water after it stands for a while originate from inside the water as shown in Figure 15-22. Thus, air will also collect inside a hot-water or steam system. With new water comes new air, containing new oxygen that will cause more rust in the system. Water-treatment companies can provide chemicals to help prevent rust and corrosion. However, preventive measures are best: Do not allow a system to leak and take in new water on a continuous basis. Find and stop the leak.

15.2 Preventive Maintenance

The Combustion Side of the System

The oil side of the system is the same for forced-air and boiler systems, and it must be maintained for the best continued performance. Oil systems are much more likely than gas systems to collect soot on the fuel side of the system. It is easier to prevent soot buildup than to remove it every year, so changing the oil filters, changing the oil nozzle, and checking for water in the oil annually is a good practice for all oil-burning equipment. When the nozzle is changed, a combustion analysis should be performed and left with the customer for the system records. As explained, when the customer is shown a before-and-after analysis, credibility is given to the service call.

Oil systems should also be kept clean of oil to prevent fire hazards and odors. The technician should clean up after every call or risk being called back to finish the job.

The Air Side of the System

The air side of the system requires that the air filters be changed on a regular basis, depending on the amount of dust and debris. The filters should typically be changed every 3 months until the technician determines that longer periods between changes are allowed. The technician should also be aware of air registers turned off in unused rooms and make sure that enough registers are open. If the duct work is on the borderline of being too small, closing even one or two registers can cause the system to trip off because of the high limit control. When the system does this, it overheats the furnace and causes premature heat-exchanger failure. If the duct system is adequate, as many as 25% of the air registers can be closed before a problem occurs. The only way that the technician can know whether too many registers are closed is to run an air temperature rise check on the furnace to be sure that the temperature rise across the furnace is not too high as shown in Figure 15-23. The temperature rise limits will be printed on the furnace nameplate.

The Water Side of Boilers

The water side of a boiler requires regular maintenance. If the boiler is a hot-water boiler, the water treatment must be kept up to standard as recommended by the water-treatment company. The larger and more expensive the system, the more attention must be given to the treatment of the water.

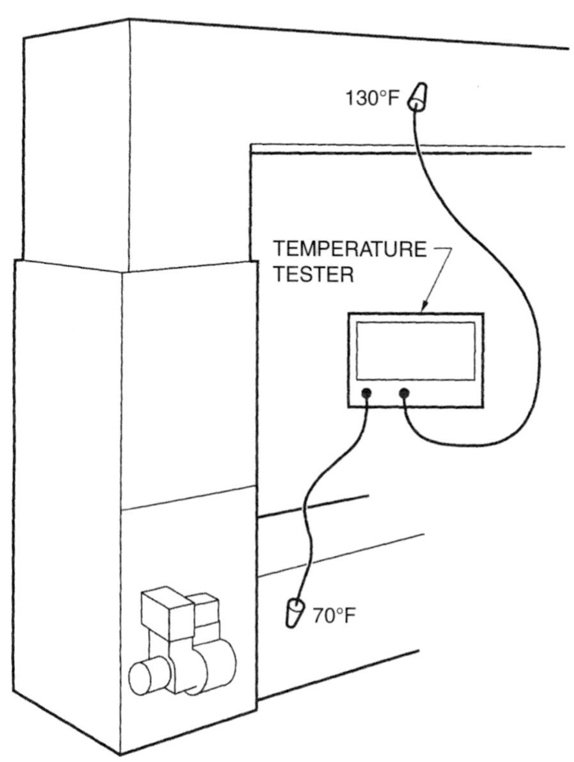

Figure 15-23. Determining the air temperature rise across an oil furnace.

Water-treatment companies often recommend an automatic-feed chemical system that must be checked at the proper intervals for large systems.

The system will have a relief valve to relieve pressure in the boiler if the internal pressure becomes excessive. Many service technicians pull the lever on the relief valve from time to time to clear the seat from possible rust that could prevent the valve from opening under high pressure. Often when the lever is pulled on an old system that has not been recently serviced, the seat will not seal tightly, and the lever must be pulled again. On rare occasions, the seat will not seal back at all and will need to be changed. By raising the valve off the seat, you can cause it to leak, but if it does, you have also found a malfunction.

Hot-water boilers are controlled with a thermostat and a high limit or only a thermostat in some cases. If the boiler is controlled with only a thermostat and it is cutting off and on, it is operating correctly. If the boiler has a high limit set above the thermostat, you do not know whether the high limit will function if needed. You may want to check it by not allowing the thermostat to shut off the boiler. When the boiler temperature reaches the set point of the high limit, it should shut off; if not, the relief valve will relieve the boiler pressure. In this event, the high limit must be changed.

The water pump must be lubricated and the pump couplings checked for vibration. If air is in the system and it reaches the water pump, the air passing through the pump will be noisy. You will need to bleed the air from the system and look for the source of the air.

The electrical portion of the system should be checked for frayed wires and pitted contacts on the pump starter.

Steam boilers have a water line in the boiler and a low-water cutoff at the water line, as shown in Figure 15-24. All contamination collects at the water line and must be bled from the system regularly. In commercial applications, a technician usually does a boiler *blow down* once a day or every 2 days to remove the contamination from the

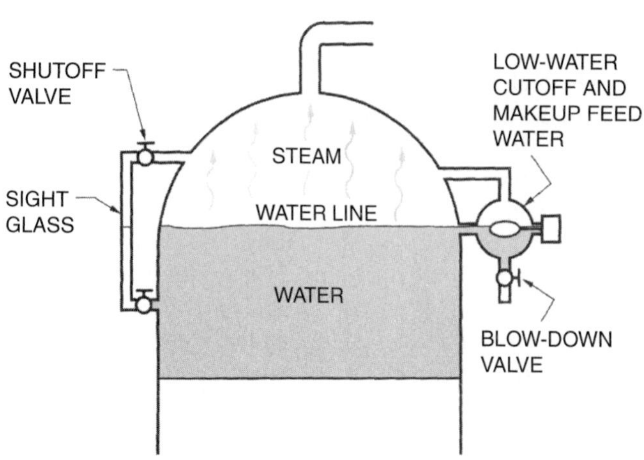

Figure 15-24. The water line in a steam boiler.

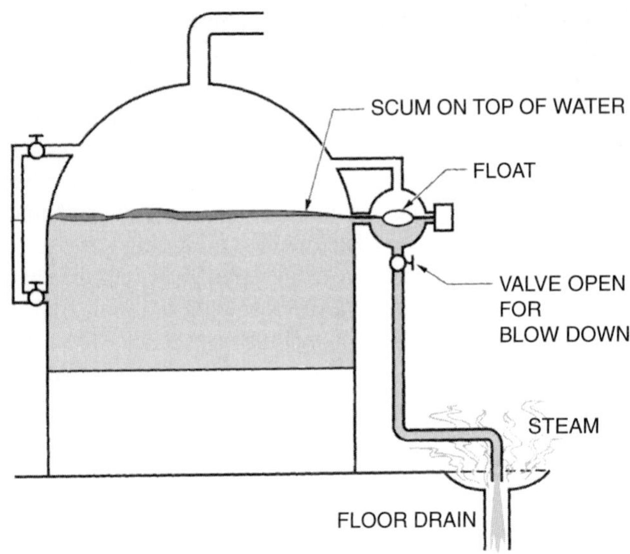

Figure 15-25. Blowing down a steam boiler.

system. Each residential application should also require a blow down but not as often. To blow down the system, the technician opens the valve on the bottom of the low-water cutoff and allows a small amount of water and steam to escape down the drain as shown in Figure 15-25. The water and steam carry some of the contamination out of the system. Steam boilers should also have a water-treatment program to prevent corrosion.

15.3 Technician Service Calls

Service Call 1

A new customer calls. The customer wants a complete checkup of the oil furnace, including an efficiency test. This customer will watch the complete procedure.

The technician arrives and explains to the customer that the first thing to do is to run the furnace and perform an efficiency test. The thermostat is set to approximately 10°F above the room temperature to allow the furnace to warm up while the technician sets up and to ensure that the furnace will not shut off during the test.

The technician takes the proper tools to the basement, where the furnace is located. The customer is already there and asks if the technician would object to an observer. The technician explains that a good technician should not object to being watched.

The technician inserts the stack thermometer in the flue and observes the temperature. It is no longer rising, and the indoor fan is running. This is a sign that the system is up to operating temperature and efficiency. A draft test over the fire shows 0.02 in. of water column. A sample of the combustion gas is taken and checked for efficiency. A smoke test is also performed. The unit is operating with a slight amount of smoke, and the test shows 65% efficiency, which is 10% lower than it should be for this burner.

The technician then removes a low-voltage wire from the primary, which shuts off the oil burner and allows the fan to continue to run and cool the furnace. The technician then removes the burner nozzle assembly and replaces the nozzle with a new one. Before reinstalling the nozzle assembly, the technician also sets the electrode spacing, then changes the oil filter and the air filter and oils the furnace blower and fan motors. The furnace is ready to start again.

The technician starts the furnace and allows it to heat up to temperature. When the stack temperature stops rising, the technician takes another sample of flue gas. The furnace is now operating at 67% efficiency and a little smoke is still present in the system. The technician checks the furnace nameplate and notices that this furnace needs a 0.75 gallon-per-hour (gph) nozzle. It has a 1-gph nozzle. Evidently the last service technician installed the wrong nozzle. It is common (but a poor practice) for some technicians to use what they have on the truck, even if they must use wrong nozzles.

The technician removes the nozzle, then installs the correct nozzle, starts the unit, and checks the oil pressure. The last technician reduced the pressure to 75 psig to correct for the oversized nozzle. The technician changes the pressure to 100 psig (the correct pressure) and restarts the furnace.

The air to the burner is adjusted. The stack thermometer reads 660°F. This is a net stack reading of 600°F (660°F – 60°F room temperature = 600°F net stack temperature). The technician runs another smoke test, which now shows minimal smoke. The combustion analysis now reveals that the system is running at 75% efficiency.

The customer was informed throughout the process and is surprised to learn that oil-burner service is so exact. The technician sets the room thermostat to normal and leaves a satisfied customer.

Service Call 2

A customer calls. The oil furnace in the basement is making a noise when it shuts off. The oil pump is not shutting off the oil fast enough when the room thermostat is satisfied.

The technician arrives, goes to the room thermostat, and turns it above room temperature to keep the furnace running. This furnace was serviced within the past 60 days, so the nozzle and oil filter do not need to be changed. The technician goes to the furnace and removes the low-voltage wire from the primary control to stop the burner. The fire does not extinguish immediately but shuts down slowly with a rumble.

The technician installs a gauge on the pump pressure side of the pump and starts the burner. The oil pressure is normal, 100 psig. When the furnace is shut off, the pressure slowly bleeds down to approximately 15 psig before the pump shutoff stops the oil flow. The pump shutoff has a malfunction. Either the pump should be changed or a solenoid should be installed to shut off the oil. The technician chooses the solenoid.

The technician installs the solenoid in the small oil line that runs from the pump to the nozzle and wires the solenoid coil in parallel with the burner motor so that it will be energized only when the burner is operating. The technician disconnects the line where it enters the burner housing and inserts it into a bottle to catch any oil that may escape. The burner is then turned on for a few seconds. This clears any air that is trapped in the solenoid from the line leading to the nozzle.

The technician reconnects the line to the housing and starts the burner. After it runs for a few minutes, the technician shuts down the burner. It has a normal shutdown. The start-up and shutdown procedure is repeated several times. The furnace operates correctly each time. The pressure gauge is removed, and the technician sets the room thermostat to the correct setting before leaving.

Service Call 3

A customer calls. The customer smells smoke when the furnace starts up. This customer does not have the furnace serviced each year. The oil filter and nozzle are in such poor condition that the furnace has soot buildup in the heat exchanger and the flue. This buildup restricts the flue gases from leaving the furnace. When the burner starts, it puffs smoke out around the burner until the draft is established.

The technician goes to the furnace in the basement, and the customer follows. The technician turns off the system switch, goes back upstairs, and sets the thermostat 5°F above the room temperature so that the furnace can be started from the basement. The technician returns to the basement, starts the furnace, and observes the puff of smoke. The technician inserts a draft gauge in the burner door; the reading is +0.01 inch wc (positive pressure). The burner is not venting. The technician shuts down the furnace before it becomes too hot to work with.

Next, the technician explains to the customer that the furnace must have a complete service and cleaning and that waiting to have the furnace serviced until it is in such poor condition makes the furnace work harder, increases the fuel bill, and can result in costly repair bills. The technician wants the customer to understand that it is a good investment to have an oil furnace serviced each year: the nozzle and the filter changed, and the electrodes adjusted. The technician proves this by letting the customer see the inside of the furnace.

Warning: Be sure to wear breathing protection for the following procedure.

The technician removes the burner, opens the inspection covers to the heat exchanger and the flue pipe, then brings a canister vacuum and a set of small brushes in from the truck. The technician starts by vacuuming the combustion chamber, then the flue area. Next, small brushes are used and passed between the heat exchanger walls to loosen the soot. The soot is then vacuumed. Coat-hanger wire is pushed into every conceivable crevice of the flue passages of the furnace. The customer is amazed at the amount of soot that comes out of the furnace.

The technician then loosens the flue pipe at the chimney and cleans the horizontal run for the flue pipe and the bottom of the chimney. Soot buildup closes nearly half the connection. The burner nozzle is replaced, the motor is oiled, and the electrodes are aligned. Then the oil filter and the air filter are changed. The furnace is then reassembled. The entire procedure has taken approximately 1½ hours. It would take even longer if the furnace were a horizontal furnace in a crawl space under a house.

The technician installs a pressure gauge to monitor the oil pressure, starts the furnace, and points out that the draft over the fire is now −0.02 inch of wc. The furnace is now venting correctly. There is no smoke. When the furnace has warmed up, a flue gas analysis shows that the furnace is operating at 72% efficiency.

The customer now knows that it is not worthwhile to wait until a furnace is in poor condition before having it serviced. The technician sets the room thermostat to the original setting and leaves the job.

Service Call 4

A customer from a duplex apartment calls. There is no heat. The customer is out of fuel. The customer had fuel delivered last week, but, unknown to the customer, the driver filled the wrong tank. This system has an underground tank, and sometimes it is difficult to get the oil pump to prime (pull fuel oil to the pump with an empty line). The technician will be fooled for some time, thinking that there is fuel and that the pump is defective.

The technician arrives and goes to the room thermostat. It is set at 10°F above the room temperature. The technician goes to the furnace, which is in the garage at the end of the apartment. The furnace is off because the primary control tripped. The technician takes a flashlight and examines the heat exchanger for any oil buildup that may have accumulated because the customer had reset the primary. If a customer repeatedly resets the primary trying to get the furnace to fire and the technician then starts the furnace, this excess oil is dangerous when it catches fire. There is no excess oil in the combustion chamber, so the technician pushes the reset button. The burner does not fire.

The technician suspects that the electrodes are not firing correctly or that the pump is not pumping. The technician fastens a hose to the bleed port on the side of the oil pump. The hose is inserted in a bottle to catch any oil that may escape, then the bleed port is opened and the primary control is reset. This is a two-pipe system and should prime by itself. (Connecting the bleed hose only verifies that the oil is actually at the burner.) The burner and pump motor are then started. No oil comes out of the hose, so the pump must be defective, the technician thinks. Before changing the pump, the technician decides to make sure that oil is in the line. The line entering the pump is removed. There is no oil in the line. The technician opens the filter housing and finds very little oil in the filter. The technician now decides to check the tank and borrows the customer's measuring stick. The stick is pushed through the fill hole to the bottom of the tank. When it is removed, the stick is dry; there is no oil.

The technician tells the customer, who calls the oil company. The driver who delivered the oil is nearby and is dispatched to the job. The driver points to the tank that was filled yesterday, the wrong one. Then the correct one is filled.

The technician starts the furnace and bleeds the pump until a solid column of oil flows from the bleed port line, then closes the line valve. The burner ignites and goes through a normal cycle. This furnace is not under contract for maintenance, so the technician suggests to the occupant that a complete service call with a nozzle change, electrode adjustment, and filter change be done as soon as it can be scheduled. The occupant agrees. The technician completes the service call, turns the thermostat to the normal setting, and leaves.

Service Call 5

A customer calls. There is no heat in a small retail store. The cad cell is defective, and the furnace will not fire.

The technician arrives, goes to the room thermostat, which is set 10°F higher than the room temperature. The thermostat is calling for heat. The technician goes to the furnace in the basement and discovers that it needs to be reset in order to run. The technician examines the heat exchanger with a flashlight and finds no oil accumulation. The primary control is then reset. The burner motor starts, and the fuel ignites. It runs for 90 seconds and shuts down. Something is wrong in the flame-proving circuit, the primary, or the cad cell. The technician turns off the power, removes the burner assembly, and examines the cad cell. It is not dirty, so the resistance through it should be checked. The cad cell is checked with room light shining on it. The resistance through a properly functioning cad cell under room light is approximately 200 ohms. This cad cell shows 10,000 ohms resistance and is defective. The technician replaces the cad cell. While the burner assembly is out, the technician replaces the nozzle, then sets the electrodes. Before refiring the furnace, the technician replaces the oil filter and the air filters and oils the pump and furnace fan motor.

When the furnace is started, the burner lights and stays on. The technician runs an efficiency check as a routine part of the call. The furnace is running at 72% efficiency, which is acceptable for this older burner. The technician resets the room thermostat and leaves the job.

Service Call 6

A customer calls. There is no heat. The primary control is defective. The furnace does not start when the reset button is pushed.

The technician arrives and checks to see that the thermostat is calling for heat. The set point is much higher than the room temperature. The technician goes to the garage, where the furnace is located, and presses the reset button. Nothing happens. The primary control is carefully checked to see whether it has power. It has 115 volts. Next, the circuit leaving the primary, the orange wire, is checked. (The technician realizes that power enters the primary on the white and black wires and that the white wire is neutral, which leaves the orange wire.) There is power on the white to black, but none on the white to orange, as shown in Figure 15-26. The

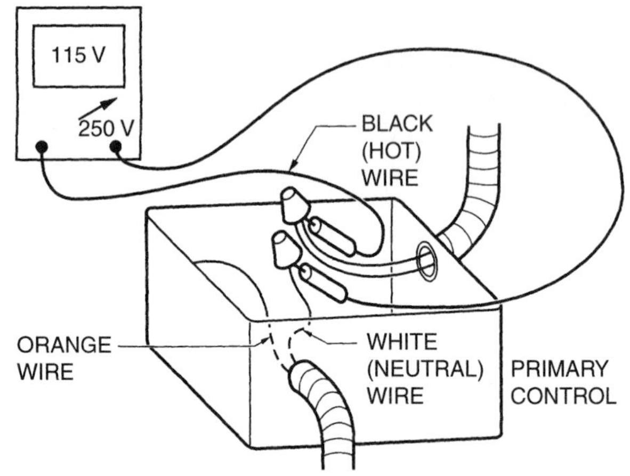

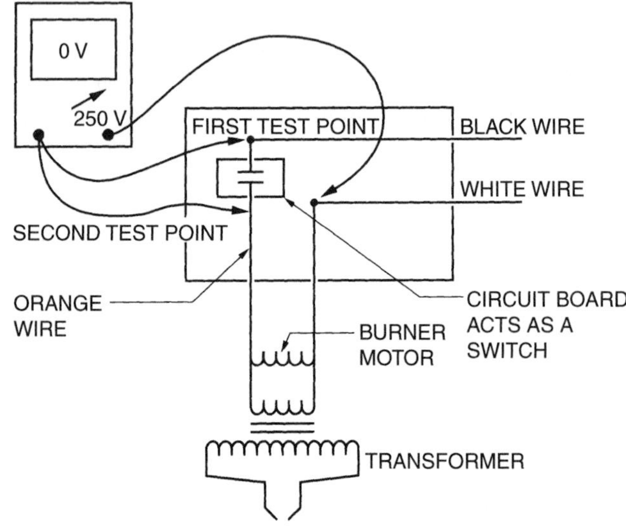

Figure 15-26. There is power on the white to black wires but none on the orange to white.

primary is not allowing power to pass. The technician shuts off the power and replaces the primary control. Power is resumed, and the furnace starts. This furnace had a complete checkup within the past 60 days, so the technician has no reason to do anything else to the system so he resets the thermostat before leaving.

Summary

- Most problems with oil-burning appliances are either electrical or mechanical. Because the appliance is controlled electrically, the problem almost always seems to be electrical, even though a mechanical problem may have started the sequence. For example, a dirty cad cell eye can be caused by smoke from poor combustion, but it will shut off the unit at the primary control, and the unit will have to be reset with an electrical control.
- The electrical diagram of the furnace can be used to troubleshoot electrical problems.
- Power into the appliance is checked at the disconnect or at the switch entering the unit.
- Think of the primary control as a switch. Power enters it on the black wire, whereas the white wire is neutral. Power must leave the primary control on the orange wire; the white wire remains neutral.
- The furnace fan is on a separate circuit; so is the pump on a boiler.
- Furnace mechanical problems involve combustion and fuel supply problems.
- Boiler mechanical problems involve combustion problems and water-circulation problems.
- Air in the water side of the system is a common problem.
- Minerals also cause problems in the water side of the system.
- The burner requires more attention than any other part of an oil-burning appliance.
- The air side of a furnace system requires maintenance.
- The water side of a boiler requires maintenance.

Review Questions

1. State the two categories that problems occur in oil burning equipment.
2. What is the purpose of the cad cell in an oil burner?
3. What may cause the cad cell to not function correctly?
4. What would the technician suspect if the burner on a forced air furnace would start and run for awhile then shut off and the fan would not start?
5. What control operates the fan on a forced air furnace?
6. Why must the combustion chamber be in good condition on an oil burning appliance?
7. How do minerals enter the water side of an oil boiler?
8. How are these minerals dealt with in a boiler?
9. What is the result of soot in an oil burning appliance?
10. What routine procedure must be performed on a steam boiler low water cut-off?

16 Properties of Water

Objectives

Upon completion of this unit, you should be able to

- **explain why water is an excellent material for the transfer of heat.**
- **calculate the amount of heat stored in a substance.**
- **described how air is dissolved into and/or released from water.**
- **determine when water will boil based on pressure and temperature.**
- **explain how a fluid's dynamic viscosity affects its flow rate in a water-based heating system.**

Table 1 Specific Heat of Various Materials

Material	Specific Heat (Btu/lb °F.)	Density (lb/cubic ft.)
Water	1	62.4
Concrete	0.21	140
Steel	0.12	489
Wood	0.65	27
Ice	0.49	57.5
Air	0.24	0.074
Gypsum	0.26	78
Sand	0.1	94.6
Ethyl Alcohol	0.68	49.3

16.1 Specific Heat

Technicians must have a good understanding of several properties and principles of water before they can effectively design, install, and/or maintain water-based heating systems. Among these properties is the heating capacity and specific heat of water and other substances. Specific heat is the measurement of the amount of heat energy required to raise the temperature of one pound of a substance by one degree Fahrenheit. For water this is typically one Btu/pound degree fahrenheit. Consider raising the temperature of 223 pounds of water 15 degrees—the specific heat required to accomplish this is found by multiplying 223 pounds by 15 degrees, thus resulting in 3345 Btu.

When the substance encounters a phase change, the specific heat also changes. A phase change is when the substance changes—for example, from a solid (ice) to a liquid or from a liquid to a gas (steam). The specific heat for ice is 0.48 Btu/lb °F; and the specific heat for water vapor is 0.489 Btu lb/°F. The following table shows various materials and their specific heats. It should be apparent that water has one of the highest specific heats and is therefore an excellent median in the transfer of heat.

16.2 Sensible Heat and Latent Heat

The amount of heat absorbed by a substance while it is in a particular phase is referred to as sensible heat. However, when the substance changes states while absorbing heat (from a solid to a liquid) and its temperature remains constant, then the heat absorbed by the substance is referred to as latent heat. If you understand the concept of latent heat, you can understand why it will snow sometimes when the temperature is 32°F and at other times it will rain at the same temperature, see Figure 16-1.

16.3 Sensible Heat Calculations

To calculate the quantity of heat stored within a given material that is undergoing a specific temperature change, the following equation can be used. The equation is a general equation and can be used for water as well as other substances and materials.

Equation 16.1 $$Q = mc(\Delta T)$$

where Q = quantity of heat absorbed or released from the material (Btu)

m = weight of the material (pounds)

c = Specific heat of the material (Btu/lb/°F)

ΔT = temperature change of the material (°F)

This equation can be simplified by inserting the specific heat value of water (which is 1) into the equation to yield:

Equation 16.2 $$Q = m \times 1 \times \Delta T \text{ or } Q = w(\Delta T)$$

Keeping in mind that water weighs 8.33 pounds per gallon (62.4 pounds per cubic foot), the equation can once

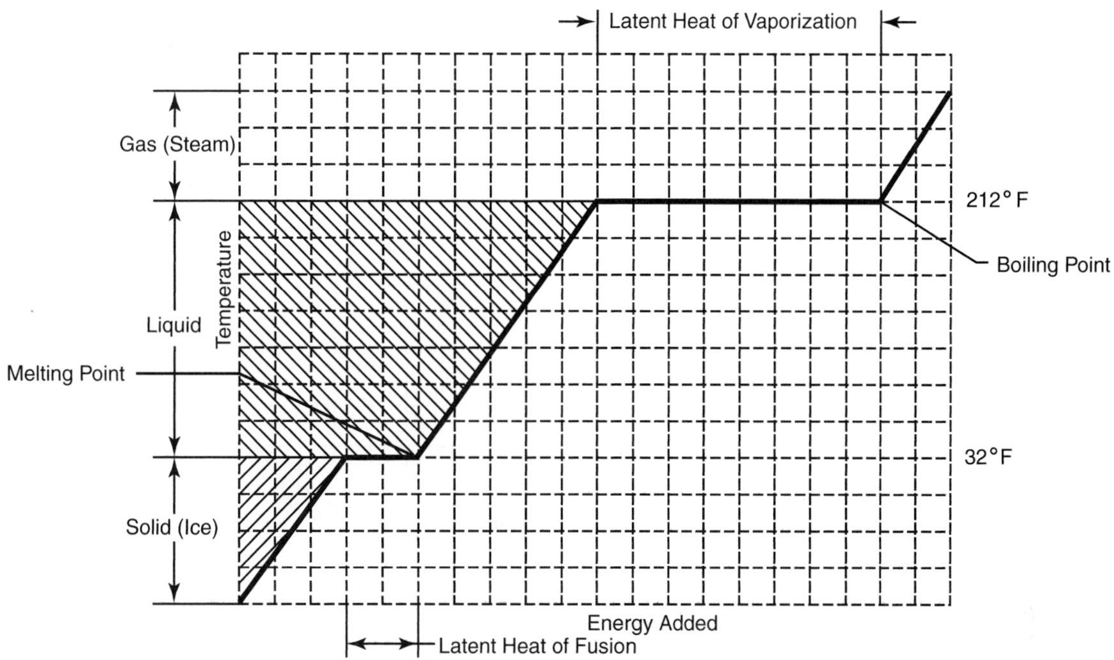

Figure 16-1. Latent heat and sensible heat.

again be changed for calculating the sensible heat by substituting the weight of water per gallon into the equation and adding a new variable *v* for the volume of water in gallons. This equation is shown in the following equation.

Equation 16.3 $$Q = 8.33v(\Delta T)$$

where Q = quantity of heat absorbed or released from the material (Btu)

v = volume of water (gallons)

ΔT = temperature change of the material (°F)

8.33 = weight of water lbs/gal

For example, calculate the amount of heat required to raise the temperature of a storage tank from 70°F to 150°F if the tank has a 375 gallon capacity.

1. Determine the change in temperature.

$$150°F - 70°F = \Delta T$$

$$\Delta T = 80°F$$

2. Calculate the amount of heat required.

Using Equation 16.3 yields the following:

$$Q = 8.33v(\Delta T)$$

$$Q = 8.33\,(375)(80°F)$$

$$Q = 249900 \text{ Btus}$$

16.4 Density

The density of a substance is mass per unit volume, that is, how tightly packed an object is. It takes about 62.5 pounds of water at 50°F to fill a one cubic foot container (put another way, 64.2 lb/cubic foot). However, as the temperature of the water increases, its density decreases. This is because as energy is added to water (in the form of heat), the molecules start to expand and require more space. Though this might sound like an unimportant fact, it must be considered when designing water-based heating systems because this expansion is extremely powerful and could cause pipes and tanks to rupture. This relationship between water and temperature and density is outlined in Figure 16-2.

16.5 Calculating Density of Water at Various Temperatures

The density of water can be determined using either a graph, as shown in Figure 16-2, or it can be calculated using the formula shown in Equation 16.4.

Equation 16.4 $$D = 62.56 + 3.413 \times 10^{-4} \times T - 6.255 \times 10^{-5} \times T^2$$

where D = density

T = temperature

For example, calculate the density of water at 75°F and 215°F.

To calculate the density of water at 75°F:

$$D = 62.56 + 3.413 \times 10^{-4} \times T - 6.255 \times 10^{-5} \times T^2$$

$$D = 62.56 + 3.413 \times 10^{-4} \times (75) - 6.255 \times 10^{-5} \times (75)^2$$

$$D = 62.234 \text{ lb/ft}^3$$

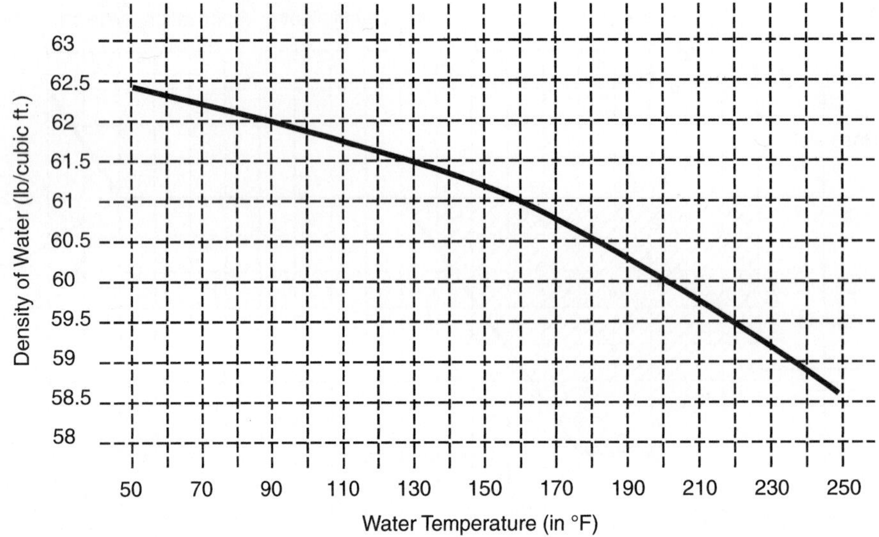

Figure 16-2. Water/temperature and density relationship.

To calculate the density of water at 215°F:

$$D = 62.56 + 3.413 \times 10^{-4} \times T - 6.255 \times 10^{-5} \times T^2$$

$$D = 62.56 + 3.413 \times 10^{-4} \times (215) - 6.255 \times 10^{-5} \times (215)^2$$

$$D = 59.74198425 \text{ lb/ft}^3$$

As per the previous discussion, it is apparent that as the temperature of water increases, the density decreases and therefore it exerts more pressure on a container's walls.

Density also comes into play when we are calculating the rate at which heat is transferred. This will be discussed in more detail in the next section.

16.6 Sensible Heat Transfer Rate

In the previous discussion, the topic of determining the amount of sensible heat required to raise or lower the temperature of a storage tank was introduced. However, the rate at which heat is transferred from a stream of water to a heat exchanger is critical information for a designer when designing a water-based heating system. This is typically done by using Equation 16.1 and introducing time into the equation. In addition to introducing time into the equation, density and specific heat must also be added. This yields Equation 16.5.

Equation 16.5 $$Q = (8.01Dc)f(\Delta T)$$

where Q = rate of heat transfer into or out of the water stream (Btu/hr)

8.01 = a constant based on the units used

D = density of the fluid (lb/cubic ft)

c = specific heat of the material (Btu/lb °F)

f = flow rate of water through the device (gallons per minute)

ΔT = temperature change of the water through the device (°F)

Because the density of the water varies at different temperatures, the average temperature of the water entering and leaving the process should be used to determine the density.

For example, calculate the rate of heat transfer for a system in which water enters the system at 200°F and leaves at 150°F if the flow rate is 5.5 gpm and 6.4 gpm.

1. Calculate the average temperature.

   200°F
   150°F
   350°F

   350/2 = average temperature

   average temperature = 175°F

2. Calculate the density of the average temperature.

Using Equation 16.4, the density of the water can be calculated.

$$D = 62.56 + 3.413 \times 10^{-4} \times T - 6.255 \times 10^{-5} \times T^2$$

$$D = 62.56 + 3.413 \times 10^{-4} \times 175 - 6.255 \times 10^{-5} \times 175^2$$

$$D = 60.704 \text{ lb/ft}^3$$

3. Calculate heat transfer rate for 5.5 gpm.

$$Q = (8.01Dc)f(\Delta T)$$

$$Q = (8.01 \times 60.704 \times 1)5.5(200 - 150)$$

$$Q = 133{,}716 \text{ Btu/hr}$$

4. Calculate the heat transfer rate for 6.4 gpm.

$$Q = (8.01Dc)f(\Delta T)$$

$$Q = (8.01 \times 60.704 \times 1)6.4(200 - 150)$$

$$Q = 155{,}596.50 \text{ Btu/hr}$$

16.7 Vapor Pressure

Because water can exist as either a liquid or a gas, an important property must be introduced, that is, vapor pressure. Vapor pressure is the minimum amount of pressure that must be applied to a liquid's surface to prevent it from evaporating. In other words, it is an indication of a liquid's evaporation rate. Therefore, the higher the vapor pressure of a substance, the lower the boiling point. This is because it requires more pressure to keep the substance from boiling. Also, the vapor pressure of a substance is strongly associated with the temperature of a substance. The higher the temperature of the substance, the higher the vapor pressure and thus the easier it is for the substance to boil. This explains why water boils at different temperatures at different elevations. At sea level, water boils at 212°F because the atmospheric pressure at sea level is 14.7 psi. However, when water is heated at a higher elevation—where the atmospheric pressure is much lower—the temperature at which water will boil is much less. Consider a pot of water heated at sea level that will boil at 212°F. However, if the same pot of water is heated at 5000 feet above sea level (e.g., in Denver), then it will start to boil at 202°F. The relationship between the vapor pressure of water and temperature can be seen in Figure 16-3. When a substance has a high vapor pressure at normal temperatures, then that substance is called volatile.

Because water is the preferred medium for the production of electricity (water and steam turbines) and because it is used extensively in commercial and residential heating applications, the relationship between vapor pressure and temperature and pressure have been well documented in the form of charts and graphs. However, this relationship can be calculated using Equation 16.6.

Equation 16.6 $$Pv = 0.771 - 0.0326(T) + 5.75 \times 10^{-4}(T)^2 - 3.9 \times 10^{-6}(T)^3 + 1.59 \times 10^{-8}(T)^4$$

where Pv = vapor pressure of water (psia)

T = temperature of the water

For example, calculate the vapor pressure of water at 70°F and again at 130°F.

1. Calculate the vapor pressure of water at 70°F. Using equation 16.6

$$Pv = 0.771 - 0.0326(T) + 5.75 \times 10^{-4}(T)^2 - 3.9 \times 10^{-6}(T)^3 + 1.59 \text{ x } 10^{-8}(T)^4$$

$$Pv = 0.771 - 0.0326(50) + 5.75 \times 10^{-4}(50)^2 - 3.9 \times 10^{-6}(50)^3 + 1.59 \times 10^{-8}(50)^4$$

$$Pv = 0.19 \text{ psia}$$

2. Calculate the vapor pressure of water at 130°F. Using equation 16.6

$$Pv = 0.771 - 0.0326(T) + 5.75 \times 10^{-4}(T)^2 - 3.9 \times 10^{-6}(T)^3 + 1.59 \times 10^{-8}(T)^4$$

$$Pv = 0.771 - 0.0326(130) + 5.75 \times 10^{-4}(130)^2 - 3.9 \times 10^{-6}(130)^3 + 1.59 \times 10^{-8}(130)^4$$

$$Pv = 2.223 \text{ psia}$$

The subject of water vapor pressure is important when working with hydronic and other water based heating systems in which pump and valves are used. When the absolute pressure of the water falls below the water vapor pressure, then the problem of cavitation in the pumps and valves can occur.

16.8 Viscosity and Water

All liquids have a natural tendency to resist flow; some liquids have a higher resistance, whereas others have a

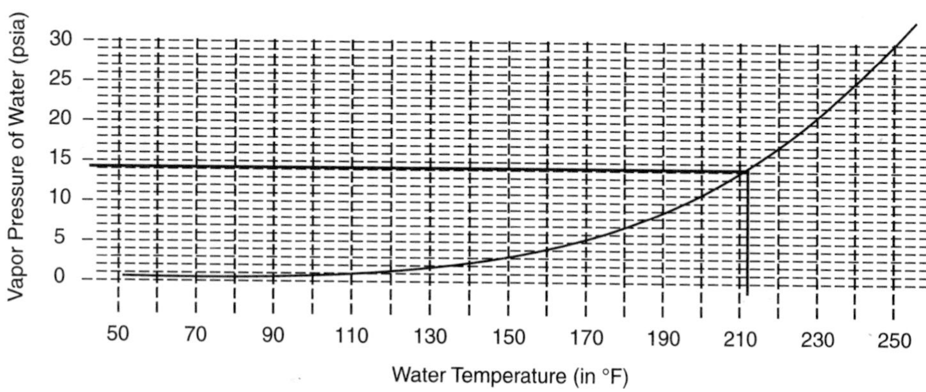

Figure 16-3. Water vapor pressure vs. temperature graph.

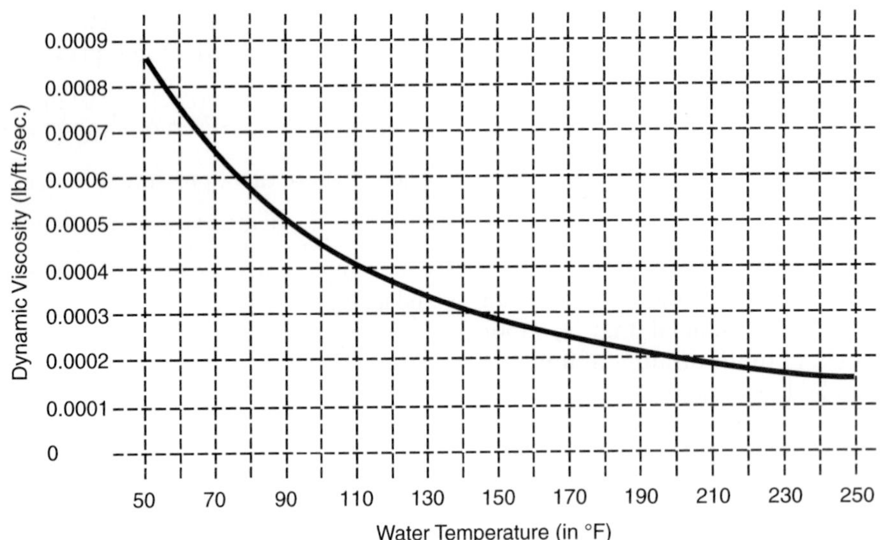

Figure 16-4.

lesser resistance. This tendency of a liquid to resist flow is known as viscosity. The greater the viscosity, the more resistance the fluid faces as it flows through a piping system and its components. It also stands to reason that the higher the viscosity, the more power it will take to move the fluid through the piping system.

Like the other properties of water discussed up to this point, viscosity is also affected by temperature. As the temperature of water increases, its viscosity decreases and thus it becomes easier for the water to flow. The viscosity of water can be determined using a graph similar to the one shown in Figure 16-4, or it can be calculated using the formula shown in Equation 16.7.

Equation 16.7 $$\mu = 0.001834 - 2.73 \times 10^{-5}(T) + 1.92 \times 10^{-7}(T)^2 - 6.53 \times 10^{-10}(T)^3 + 8.58 \times 10^{-13}(T)^4$$

where μ = dynamic viscosity of water

T = temperature of the water

For example, compute the viscosity of water at 65°F and 200°F.

1. Calculate the vapor pressure of water at 65°F. Using Equation 16.7:

$$\mu = 0.001834 - 2.73 \times 10^{-5}(T) + 1.92 \times 10^{-7}(T)^2 - 6.53 \times 10^{-10}(T)^3 + 8.58 \times 10^{-13}(T)^4$$

$$\mu = 0.001834 - 2.73 \times 10^{-5}(65) + 1.92 \times 10^{-7}(65)^2 - 6.53 \times 10^{-10}(65)^3 + 8.58 \times 10^{-13}(65)^4$$

$$\mu = 0.00071$$

2. Calculating the vapor pressure of water at 200°F. Using Equation 16.7:

$$\mu = 0.001834 - 2.73 \times 10^{-5}(T) + 1.92 \times 10^{-7}(T)^2 - 6.53 \times 10^{-10}(T)^3 + 8.58 \times 10^{-13}(T)^4$$

$$\mu = 0.001834 - 2.73 \times 10^{-5}(200) + 1.92 \times 10^{-7}(200)^2 - 6.53 \times 10^{-10}(200)^3 + 8.58 \times 10^{-13}(200)^4$$

$$\mu = 0.0002 \text{ lb}_{mass}/\text{ft/sec}$$

In the preceding calculations, the viscosity of the water decreases as the temperature increases, thus proving the previous statement that viscosity is affected by temperature.

Summary

- Specific heat is the measurement of the amount of heat energy required to raise the temperature of one pound of a substance by one degree Fahrenheit.
- When the substance encounters a phase change, the specific heat also changes.
- A phase change occurs when the substance changes, for example, from a solid (ice) to a liquid or from a liquid to a gas (steam).
- The amount of heat absorbed by a substance while it is in a particular phase is referred to as sensible heat.
- When the substance changes states while absorbing heat (from a solid to a liquid) and its temperature remains constant, then the heat absorbed by the substance is referred to as latent heat.
- The density of a substance is mass per unit volume.

- Vapor pressure is the minimum amount of pressure that must be applied to a liquid's surface to prevent it from evaporating.
- All liquids have a natural tendency to resist flow; some liquids have a higher resistance, whereas others have lesser resistance.
- This tendency of a liquid to resist flow is known as viscosity.

Review Questions

1. How many gallons of water would be required in a storage tank to absorb 250,000 Btu while undergoing a temperature change from 120°F to 180°F?
2. Water enters a baseboard radiator at 180°F and at a flow rate of 2.0 gpm. Assuming the radiator releases heat into the room at a rate of 20,000 Btu/hr, what is the temperature of the water leaving the radiator?
3. How much pressure must be maintained on water to maintain it as a liquid at 250°F? State your answer in psia.
4. If you wanted to make water boil at 100°F, how many psi of vacuum (below atmospheric pressure) would be required? *Hint*: psi vacuum = 14.7 − absolute pressure in psia.
5. Determine the air flow rate (in cubic feet per minute) necessary to transport 40,000 Btu/hr based on a temperature increase from 65°F to 135°F. Base the calculation on a specific heat for air of 0.245 Btu/lb °F and an assumed density of 0.07 lb/ft^3.
6. Assume you need to design a hydronic system that can deliver 80,000 Btu/hr. What flow rate of water is required if the temperature drop of the distribution system is to be 10°F? What flow rate is required if the temperature drop is to be 20°F?
7. Water flows through a series-connected baseboard system as shown in Figure 4-8. The inlet temperature to the first baseboard is 165°F. Determine the outlet temperature of the first baseboard, and both the inlet and outlet temperatures for the second and third baseboards. Assume the heat output from the interconnecting piping is insignificant. *Hint*: Use the approximation that the inlet temperature for a baseboard equals the outlet temperature from the previous baseboard.
8. At 50° F, 1000 lbs. of water occupies a volume of 16.026 ft^3. If this same 1000 lbs. of water is heated to 200°F, what volume will it occupy?
9. How much heat is required to raise the temperature of 10 lbs. of ice from 20°F to 32°F? How much heat is then required to change the 10 lbs. of ice into 10 lbs. of liquid water at 32°F? Finally, how much heat is needed to raise the temperature of this water from 32°F to 150°F? State all answers in Btus.
10. A swimming pool contains 18,000 gallons of water. A boiler with an output of 50,000 Btu/hr is operated 8 hours per day to heat the pool. Assuming all the heat is absorbed by the pool water with no losses to the ground or air, how much higher is the temperature of the pool after the 8 hours of heating?

17 Hot-Water Heat

Objectives

Upon completion of this unit, you should be able to

- **describe the different types of hot-water heating systems.**
- **recognize the different piping accessories in a hot-water system.**
- **discuss the principles of a centrifugal water pump.**
- **distinguish the types of valves used in hot-water systems.**
- **identify three types of control systems for hot-water heat.**
- **calculate the pressure drop in a hot-water piping system.**
- **fill a hot-water system to the correct pressure.**
- **check a relief valve for a hot-water system.**

17.1 Introduction to Hot-Water Heat

One of the first forms of home heating was the fireplace. However, over time it was discovered that a fireplace is the most inefficient way of heating. In addition, it proved to be one of the most dangerous forms of heating because it exposes a dwelling and its occupants to open flames, dangerous fumes, and gases. Fireplaces also provide uneven heating, meaning that different areas of the home and/or room are cooler than others. Fireplaces require a regular, high level of maintenance—wood must constantly be carried in or stored near the fireplace. Finally, one of the byproducts of burning wood—ash—must be removed and safely disposed of.

By the late 1800s, heat produced in a central area and moved throughout the residence began to replace heat produced by a fireplace in each room. Naturally, central heating systems for large homes and businesses followed. In the early systems, the source of the heat was often hot water. In such a system, heat produced by burning fossil fuels, wood, or electricity is transferred to the water. That water is then circulated through the different areas of the home to the terminal units.

The central heating unit, usually a boiler, could be installed away from the family activity in the home or away from the public activity in a building. The wood or coal used to fire the boiler could be carried to a basement or an outbuilding, and the mess resulting from combustion could be contained within this area. Only the heat-laden water needed to be piped to the occupied space.

Early hot-water systems used large pipes and large room radiators to carry water. The boiler was below the room radiators and piping system. When the water in the boiler was heated, it expanded, rose to the radiators above, and circulation was established. The expanding water required an expansion tank at the highest point in the system, usually in the attic. This tank typically had a float valve attached to the city water main. This valve allowed water to flow into the tank to maintain the water level in the system. Because the tank was in a remote location, little maintenance was performed on it. Float valves often allowed too much water into the tank, so an overflow mechanism was often used.

These *gravity hot-water systems* were simple and reliable, and many are still in service in older homes. Because they are so simple, servicing them is easy. However, these systems are not installed today because the cost of the large pipes and terminal units is not economical.

Modern systems have smaller boilers, pipes, and radiators. Other types of heat-exchange devices are now used, and circulator pumps, instead of gravity, circulate the water. These newer systems, which are discussed in this text, are much easier to install, and they provide better temperature regulation with zone control.

Modern hot-water heating systems cost more initially than some other types of heat, but they provide good, reliable heat. Furthermore, occupants of a structure with a well-designed and installed system will never know that it is operating.

Hot-water heat in homes occurs most often in the northern states, where summer cooling is not needed. Hot-water heat in buildings is usually installed along with the summer cooling system, which may use chilled water.

17.2 Hot-Water Boilers

The hot-water heating system starts with the boiler. Boilers are discussed in detail in the units on gas, oil, and electric heat. Each fuel requires a different type of boiler; however, oil and gas boilers are similar. In some cases, for large installations, boilers are dual fuel (use either oil or gas, depending on which is the most economical at the time). The purpose of the boiler in the system is to allow the energy in the fuel to be transferred to water, to make circulating hot water.

Hot-water boilers are operated liquid full of water because the system is liquid full of water. The boiler may be located in the basement of the installation in a boiler room, where the full weight of the water in the system is exerted downward on the boiler, as shown in Figure 17-1. On the other hand, the boiler room may be on the top floor of the building to prevent the weight of the water from pressurizing the boiler, Figure 17-2. Another way to alleviate the building pressure on the boiler is to use a heat exchanger to exchange heat from the boiler circuit to the building piping circuit, Figure 17-3. Regardless of the system, it is liquid full of water to the topmost component, except the expansion tank, which is discussed in Section 17.6.

When a water boiler has an operating pressure of less than 30 psig, it is a *low-pressure boiler.* Low-pressure boilers are classified as having a working pressure of 30 psig at 250°F. However, these boilers must be located where the pressure on the boiler is less than the 30 psig working pressure, or the relief valve will relieve the pressure. When these boilers are in high-rise buildings, they are often on the upper floors so that high-pressure boilers are not needed. Note that a column of water exerts a pressure of 0.43 psig per foot of height. This limits the building height to 69.8 feet above the low-pressure boiler (30 psig ÷ 0.43 psig/ft = 69.8 ft), or approximately six stories. However, this height does not allow for the extra pressure required when the system is filled. Consequently,

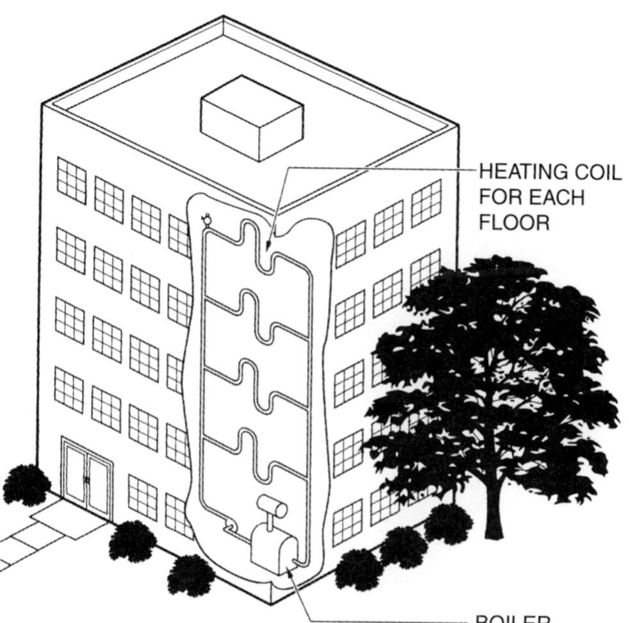

Figure 17-1. The weight of the water in the system is on the boiler, creating pressure.

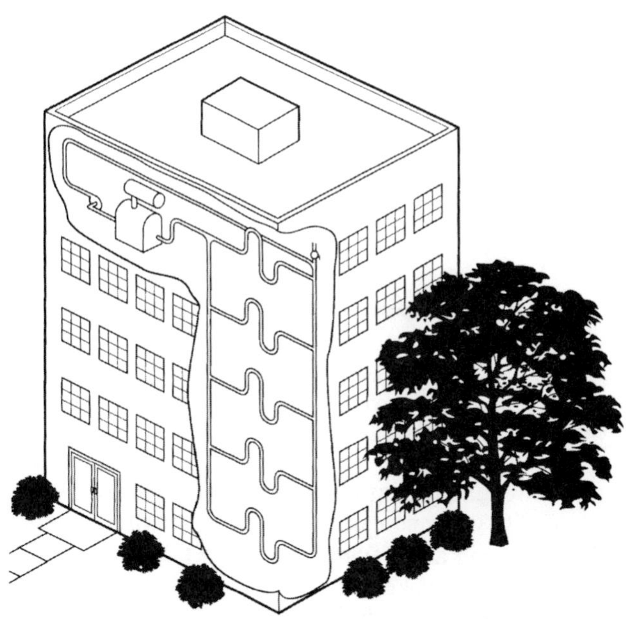

Figure 17-2. This boiler is on the top floor of the building to prevent the pressure from weighing on the boiler.

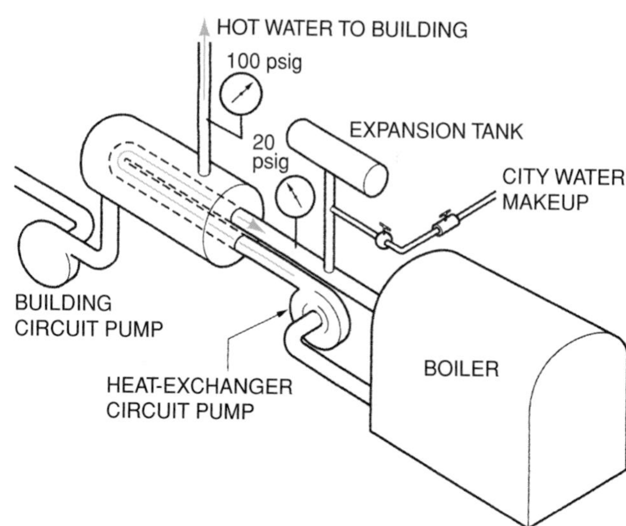

Figure 17-3. The heat exchanger keeps the building pressure off the boiler.

other methods are often used to transfer heat from the boiler water to a secondary water circuit, such as a heat exchanger. The heat exchanger works well, except some temperature loss occurs in it, thus the circulating water can never be as hot as the boiler water, Figure 17-4.

Boilers are natural air separators. When water with air is heated in a boiler, the air separates from the water. This air is either allowed to escape through an automatic bleed system, or it is allowed to enter the system water and is later separated and vented from the system. Air can enter the water side of a hot-water system during the initial installation, or it can be introduced into the

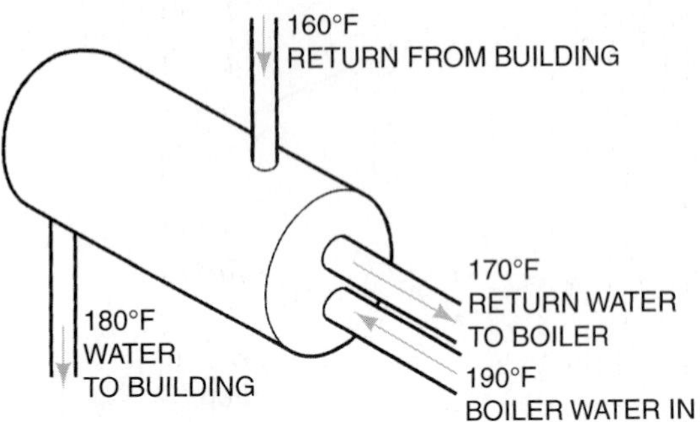

Figure 17-4. A heat exchanger loses some heat exchange capacity in the heat exchange due to temperature difference during each heat exchange step.

system through makeup water when water leaks out of the system. City water contains air, as can be seen by filling a glass with water and leaving it in the open for a few hours, Figure 17-5. All makeup water contains this air, which is 21% oxygen. The air is harmful to the system in two ways: It interferes with water flow and causes corrosion in the system. When water is circulated through the system, the water rises to the top and water flow stops in this portion of the system until the air is relieved. An air bleed method must be provided at the high point in each hot-water system, Figure 17-6. Water treatment should be provided to prevent corrosion.

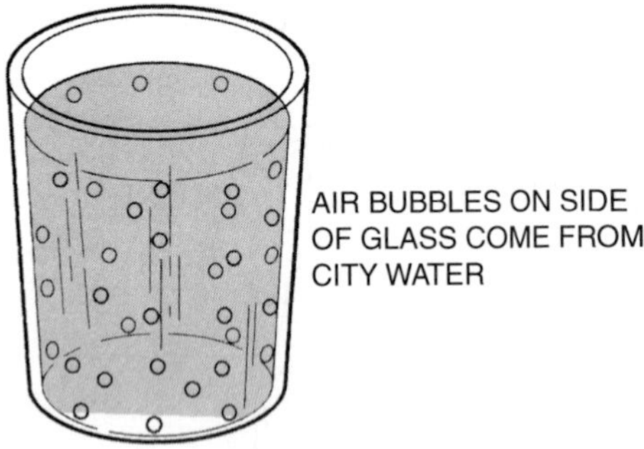

Figure 17-5. City water contains air.

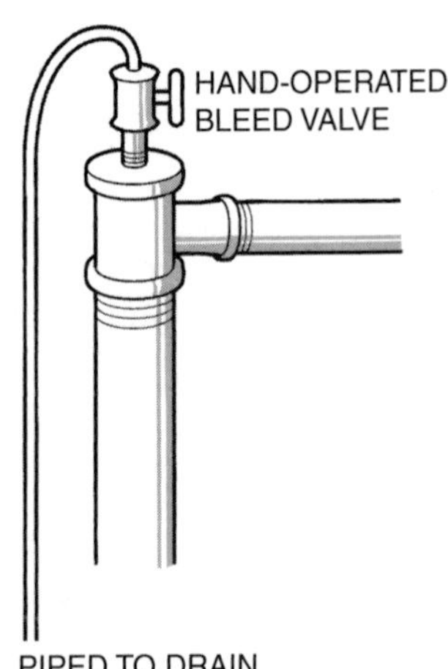

Figure 17-6. An air bleed. It must be installed at the high point in the system.

17.3 Water Pumps

The centrifugal water pump is the most common type of water pump used in hot-water systems; it is also called a *circulator.* This pump forces the hot water from the boiler through the piping to the heat transfer units and back to the boiler. Centrifugal force is used in the pump to create a pressure difference to circulate the water through the system. Centrifugal force is generated whenever an object is rotated around a central axis. The object or matter being rotated moves away from the center because of its weight, much like a rock on the end of a string that is swung, as shown in Figure 17-7. This force increases proportionally with the speed of the rotation. The concept illustrated in Figure 17-7 can also be applied to pumping water.

To understand how a centrifugal water pump works, imagine a vegetable can with holes drilled around the bottom. When water pours in from the top, it runs out the holes, Figure 17-8. Now, suppose that the can is rotated. The water slings out from the holes, as shown in Figure 17-9. The faster the can is turned, the farther the water slings. If the water is captured in a housing, the centrifugal force of the water can be guided to a pipe outlet, and the water will have pressure. The centrifugal action of the turning water is converted to pressure that can be used to circulate water in a piping system, Figure 17-10. Figure 17-11 shows a typical centrifugal pump.

Figure 17-7. A rock on a string is swung to illustrate centrifugal force.

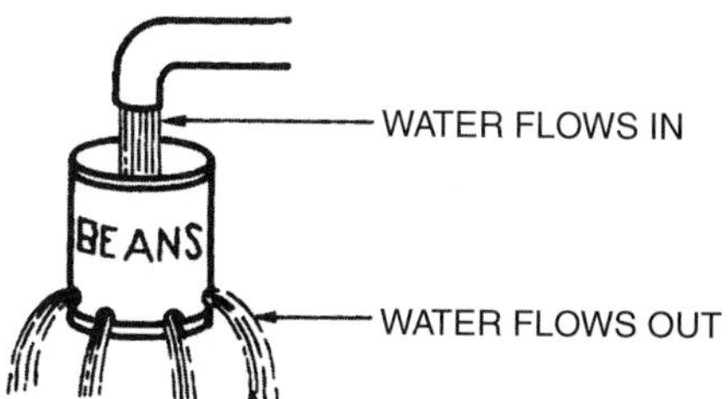

Figure 17-8. A vegetable can is used to illustrate the action of a centrifugal pump.

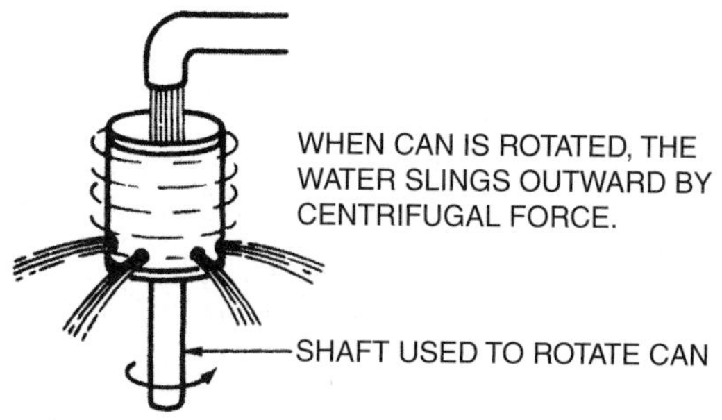

Figure 17-9. The can is rotated to impart centrifugal energy to the water.

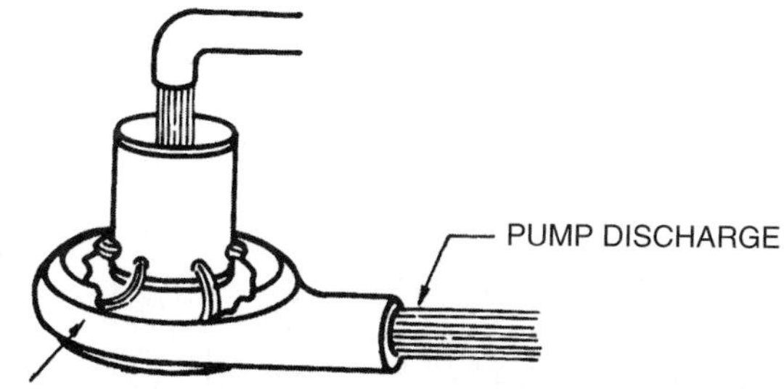

Figure 17-10. Pressure created by centrifugal action.

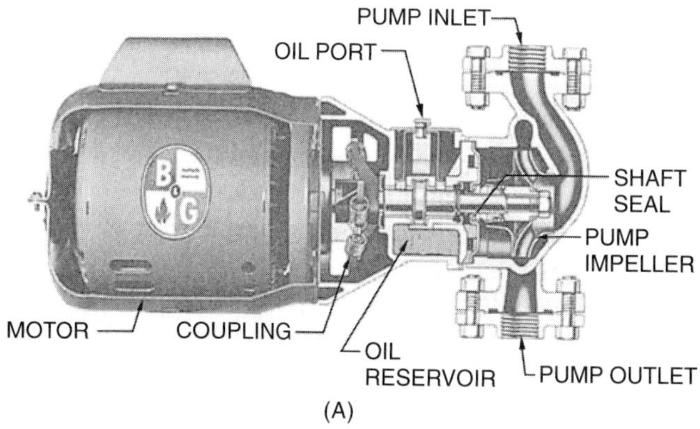

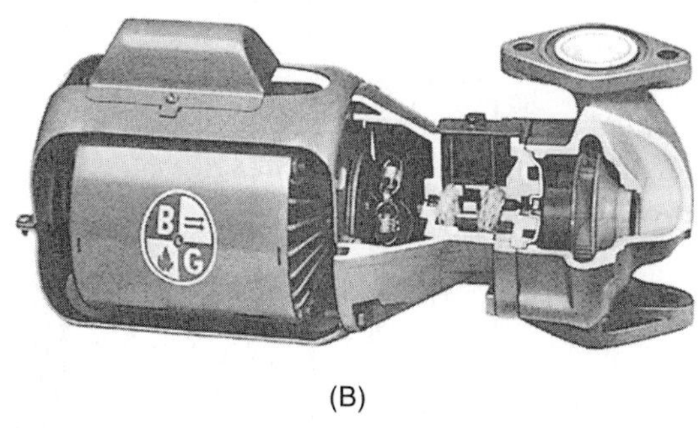

Figure 17-11. (A) Cross section of a centrifugal pump. (B) Centrifugal pump. *Courtesy ITT Fluid Handling Division*

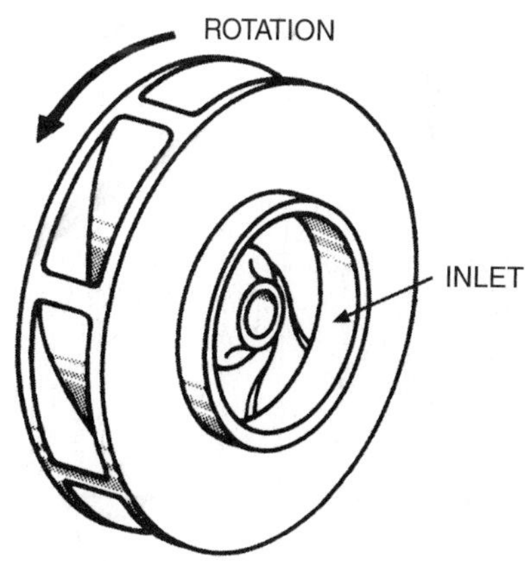

Figure 17-12. The impeller in a centrifugal pump.

The impeller, shown in Figure 17-12, is the part of the pump that spins and forces the water through the system. Proper impeller rotation direction is essential. The vanes, or blades, in the impeller must slap and throw the water into the volute (spiral casing), which directs the water to the pump outlet.

Pump capacities range from small to large. Small pumps have small impellers. A larger diameter impeller imparts more velocity to the water and therefore creates more of a pressure difference from the inlet to the outlet. As the impeller opening becomes wider, more gallons of water flow can be moved through the pump and more of a pressure difference can be expected, see Figure 17-13. If the same impeller is turned at a faster speed, more water is pumped, and a greater pressure is imparted to the water. Manufacturers make pumps with a complete range of capacities, which are rated according to gallons per minute (gpm) and pressure difference.

The capacity of a centrifugal pump is plotted on a graph. The following example demonstrates how a centrifugal pump performance curve is developed. This is a progressive example that can be followed in Figure 17-14.

1. A pump is connected to a reservoir that maintains the water level just above the pump inlet. The pump meets no resistance to the flow, and maximum water flow, 80 gpm, exists.

6"
4"
NOTICE THAT THE INLET
IS THE SAME SIZE.

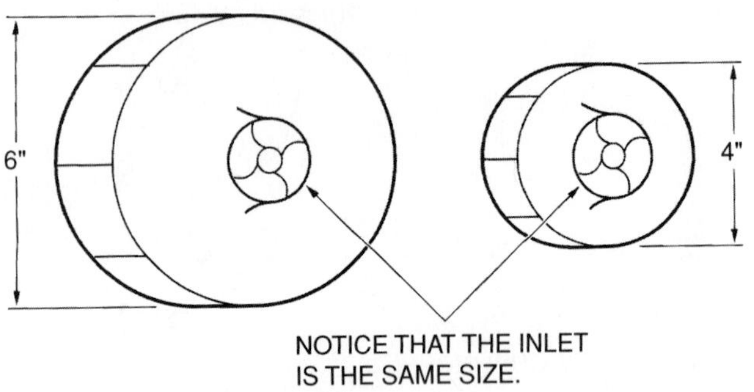

Figure 17-13. The larger the diameter of the impeller, the more pressure imparted to the water.

2. A pipe is extended to 10 feet high (known as feet of head, or pressure), and the pump begins to meet some pumping resistance. The water flow reduces to 75 gpm.
3. The pipe is extended to 20 feet, and the flow reduces to 68 gpm.
4. When the pipe is extended to 30 feet, the water flow reduces to 63 gpm.
5. The pipe is extended to 40 feet. The pump is meeting more and more resistance to flow. The water flow is now 58 gpm.
6. At 50 feet, the flow slows to 48 gpm.
7. At 60 feet, the flow slows to 38 gpm.

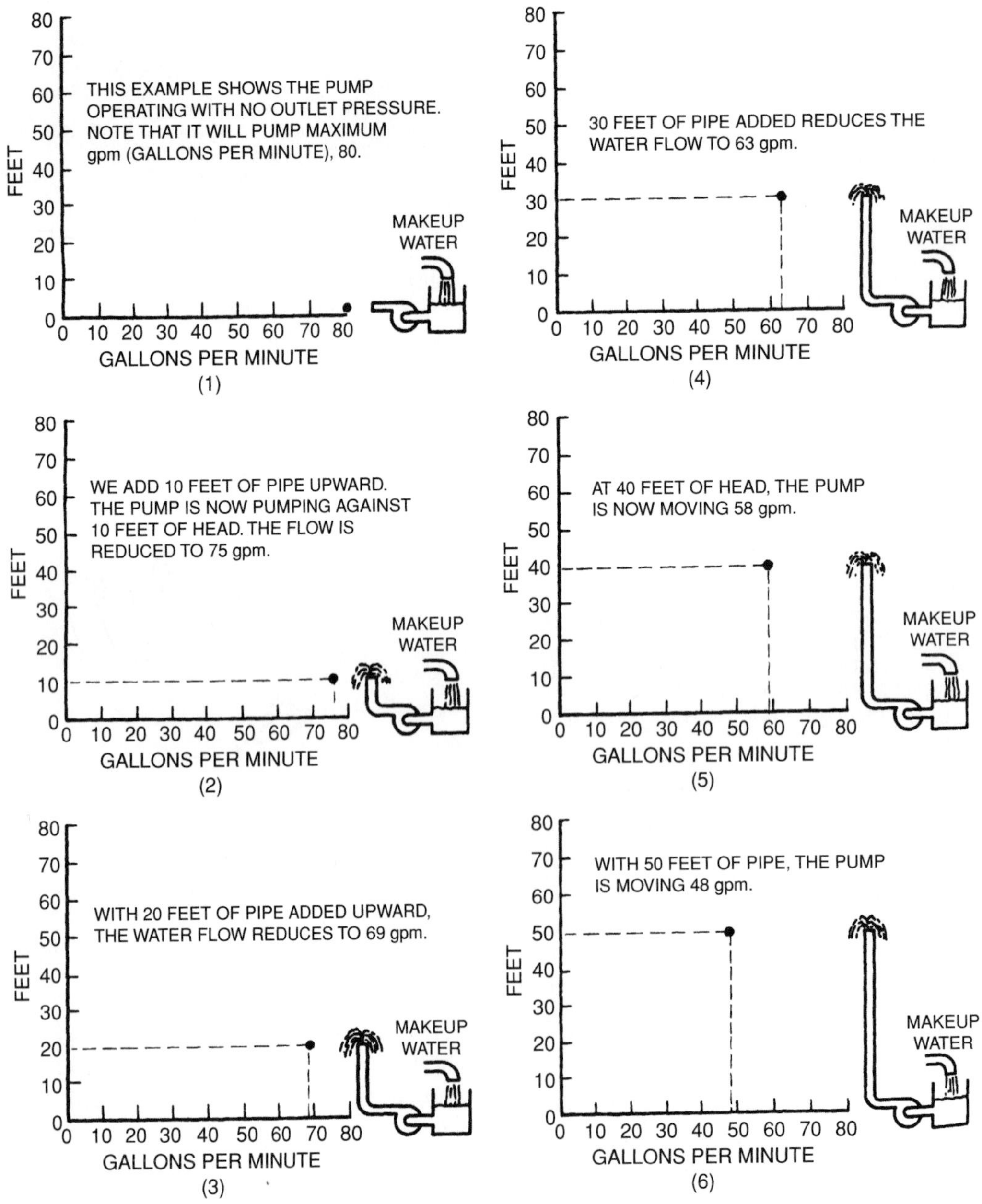

THIS EXAMPLE SHOWS THE PUMP OPERATING WITH NO OUTLET PRESSURE. NOTE THAT IT WILL PUMP MAXIMUM gpm (GALLONS PER MINUTE), 80.
FEET
GALLONS PER MINUTE
MAKEUP WATER
(1)
WE ADD 10 FEET OF PIPE UPWARD. THE PUMP IS NOW PUMPING AGAINST 10 FEET OF HEAD. THE FLOW IS REDUCED TO 75 gpm.
(2)
WITH 20 FEET OF PIPE ADDED UPWARD, THE WATER FLOW REDUCES TO 69 gpm.
(3)
30 FEET OF PIPE ADDED REDUCES THE WATER FLOW TO 63 gpm.
(4)
AT 40 FEET OF HEAD, THE PUMP IS NOW MOVING 58 gpm.
(5)
WITH 50 FEET OF PIPE, THE PUMP IS MOVING 48 gpm.
(6)

Figure 17-14. Centrifugal pump curve developed.

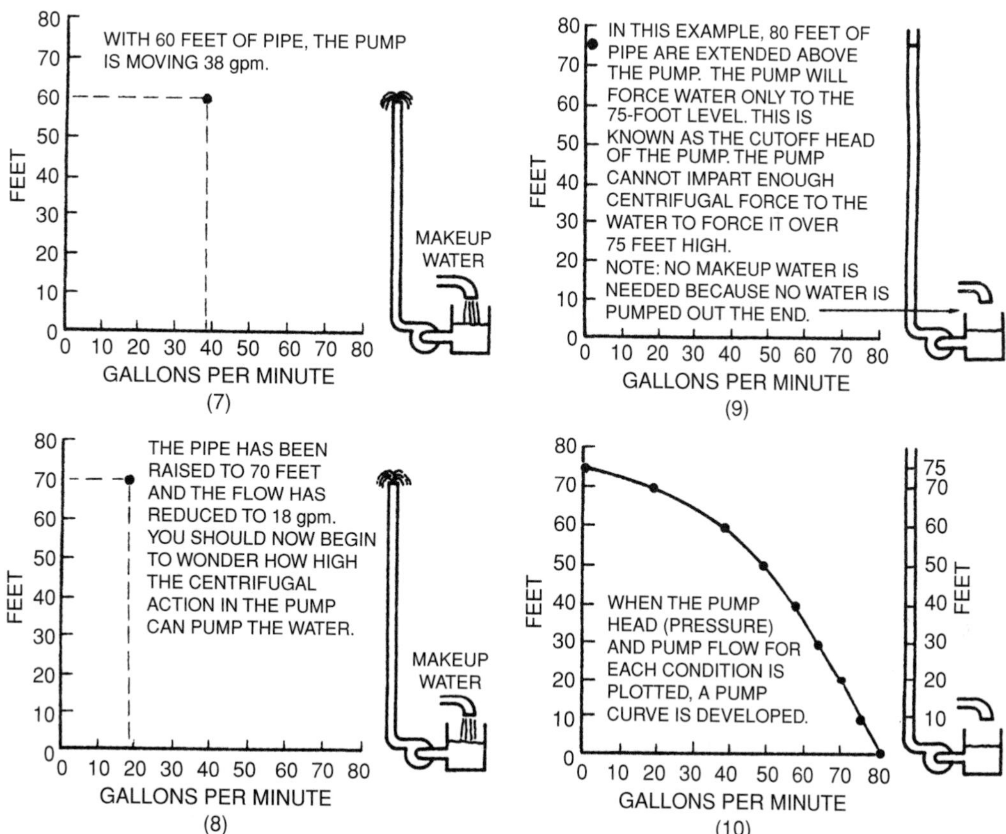

Figure 17-14. *(Continued)*

8. At 70 feet, the water flow reduces to 18 gpm. The pump has almost reached its pumping limit and can impart only a fixed amount of centrifugal force to the water. Therefore, it can lift the water only to a certain height. A pump with a larger impeller diameter would be able to pump the water higher.
9. When 80 feet of pipe are added to the pump outlet, the water rises to the 75-foot level. If you could look into the pipe, you would see the water gently moving up and down in the pipe but not rising any farther. The pump has reached its pumping capacity. This is the shutoff point, or cutoff head, of the pump.
10. By following steps 1 through 9 and plotting the results on a graph, the manufacturer develops the pump curve.

The curves for a family of actual pumps look more like those in Figure 17-15. Notice that these curves are a little different from the curve in the example. The example curve is a theoretical curve. Each curve in the family starts with some pump head at the maximum gpm because there must be a pressure rise across the pump for flow to occur.

Some pump curves have different profiles. Some are described as *more flat* and some are *steep*, as shown in Figure 17-16. Some designers use the differences in curves to choose different pumps. Some designers prefer a pump with a more flat curve and some prefer the steep curve. When the curve is more flat than steep, the gpm changes more with small pressure changes across the pump, as shown in Figure 17-17.

Designers should choose a pump from the center one-third of any pump curve to avoid problems. For example, if the designer chooses a pump on the curve close to the shutoff head, and the system has more of a pressure drop than expected, the pump will not move enough water. The pump also will not be as efficient. If the pump is chosen on the low head end of the curve and the system has less head than expected, the pump will overload by pumping too much water. Valves could correct the overload, but the pump would be inefficient.

Note in the example in Figure 17-14 that enough pipe was added to the piping circuit so that the pump could not overcome the height of the water column, and no damage occurred. A pump can pump against a dead head for a short time without damage when there is no place for the water to move to. When the vertical head was added to the pump discharge, the pumping action slowed and finally stopped, even though the pump was still turning. No water left the pipe. The pump was turning, but nothing seemed to be happening. Actually, the water was spinning around in the pump. Some of the mechanical

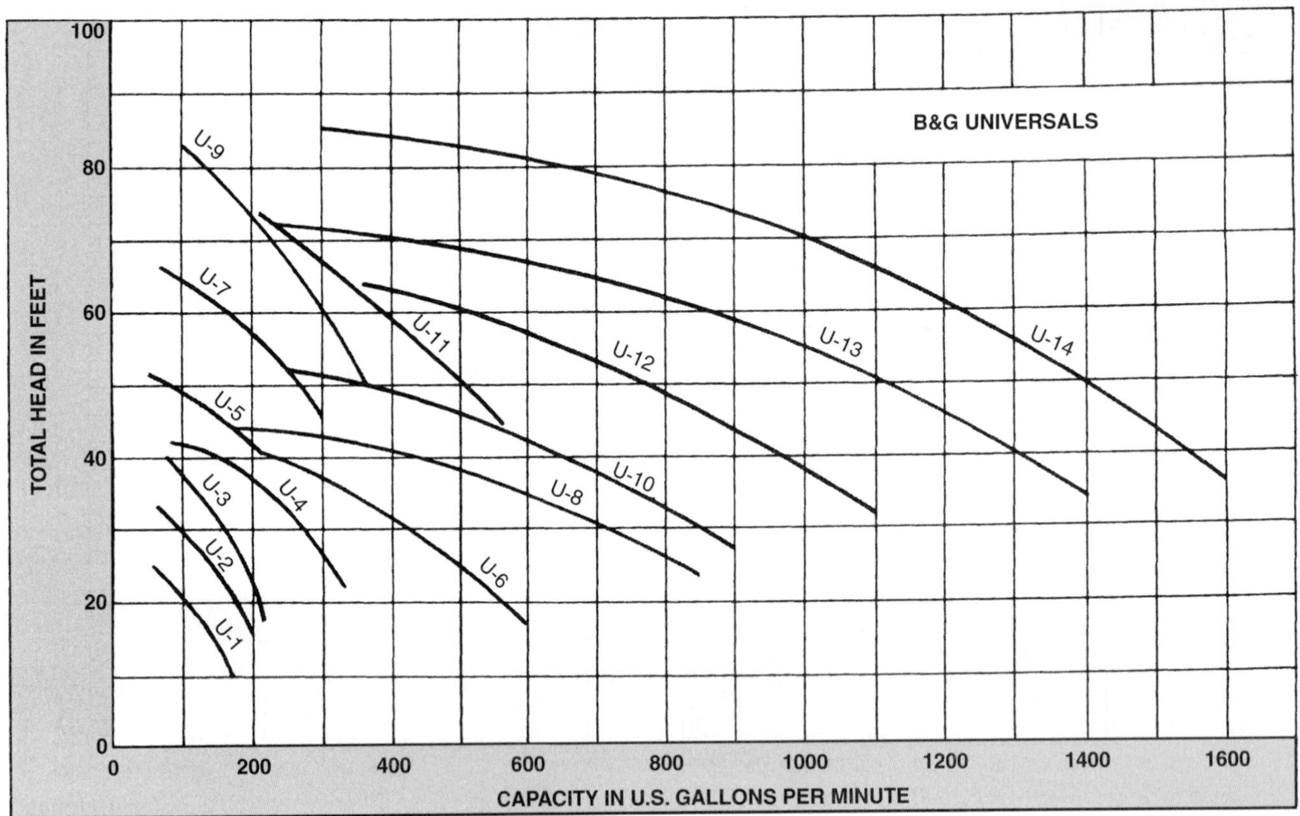

Figure 17-15. Family of centrifugal pump curves for base-mounted pumps. *Courtesy ITT Fluid Handling Division*

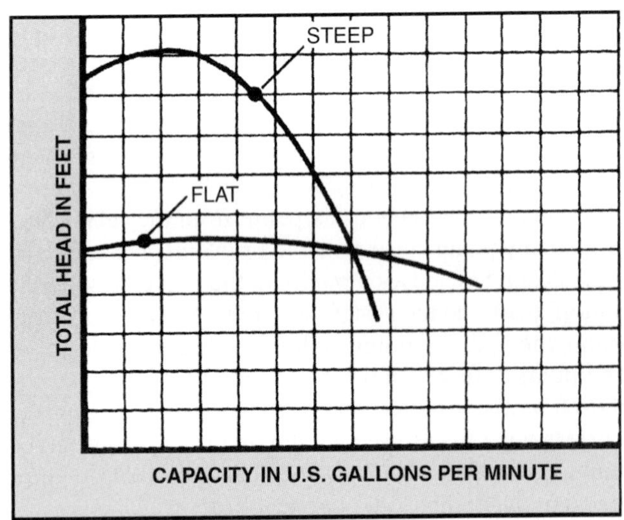

Figure 17-16. Flat and steep pump curves. *Courtesy ITT Fluid Handling Division*

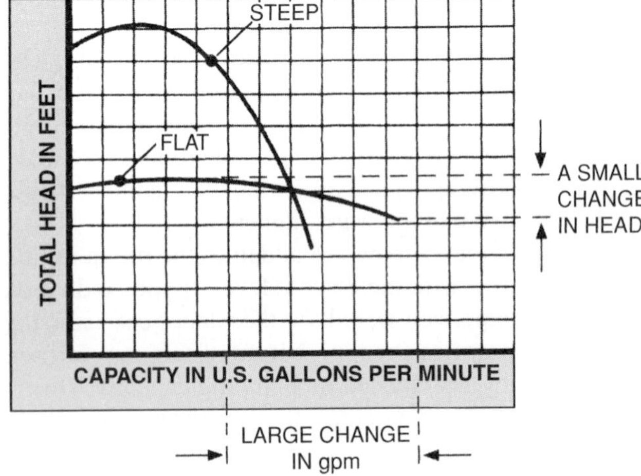

Figure 17-17. Characteristics of a flat curve. *Courtesy ITT Fluid Handling Division*

energy turned to heat energy, and the water in the pump would have become extremely hot if the pump had been allowed to run in this condition for long.

It would be awkward for the manufacturer to add vertical pipe to all size pumps so that the pumping head can be checked. Instead, the pumping head can be simulated by using valves at the inlet and outlet of the pump.

Pressure is force per unit of area. It is commonly expressed in psig (pounds per square inch gauge). Pressure can also be expressed in feet of head of water. A water manometer can be used to measure air pressures at low values. It is a U tube of glass that is open on both ends and is half full of water. When a water manometer is connected to a pipe with 1 psig of air pressure in it, the pressure of the air in the pipe will push the water column in the manometer toward the atmosphere. The column on the left will drop and the column on the right will rise. A pressure of 1 psig will create a 27.7-inch difference in the height of the water column, as shown in Figure 17-18. In other words, 1 psig of pressure will support a water column that is 27.7 inches (or 2.31 feet) high. Likewise, a water column that is 2.31 feet high will exert a pressure of 1 psig downward. The feet of head shown in Figure 17-14 can be converted to psig.

When a pump is checked in the field, the pressure can be taken on each side of the pump and converted to feet of head for use on the manufacturer's pump curves. It is a good practice to use the same gauge for checking inlet and outlet pressure so that a gauge error does not enter into the reading. For example, if one gauge reads 1 psig high and the other reads 1 psig low, a 2 psig error is introduced into the pressure drop. Using a single gauge ensures that the actual pressure difference is recorded. Figure 17-19 shows how a single gauge can be used to check the pressure difference.

The characteristics of centrifugal pump power usage seem the reverse of common sense because the maximum power used in the example in Figure 17-14 occurred when the pressure difference across the pump was least. It would seem as if more of a pressure difference across the pump would require more power, but this is not so. The more water pumped by volume, the more power required. When vertical head is added to the pump outlet, the pump consumes less power. This is true until the shutoff head is reached, when the pump uses the least amount of power. The way to remember the pumping and power requirements is to remember this: The pump power requirement is proportional to the quantity of water that moves through the pump. When the pressure drop across the pump increases, the power requirement reduces because less water is pumped.

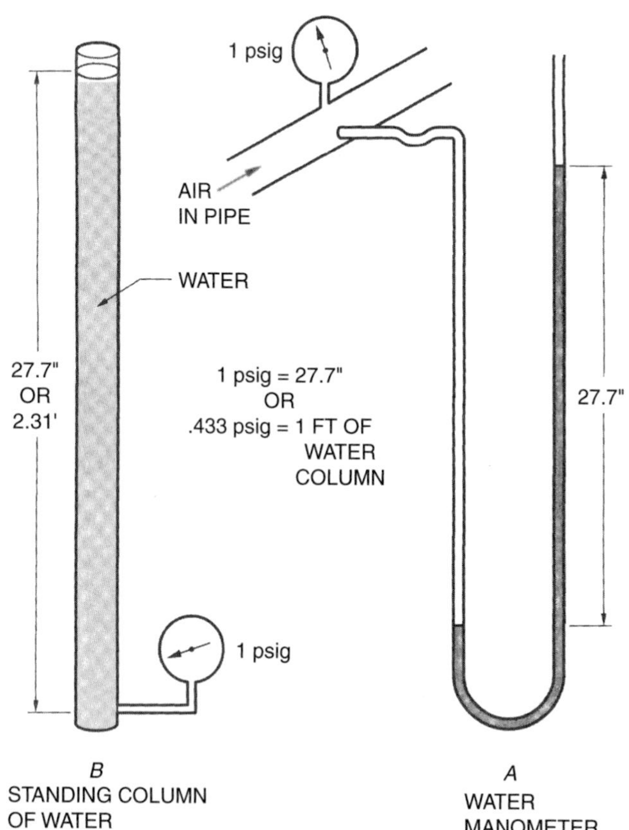

Figure 17-18. The pressure created by a column of water.

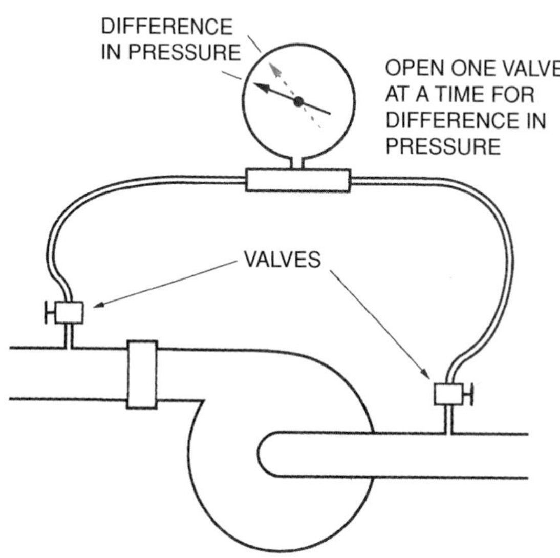

Figure 17-19. Using a single gauge to check the drop in water pressure.

Pump curves furnished by the manufacturer usually include the power requirements for the pump on the pump curve graph. The pump horsepower curve is somewhat in the reverse of the pumping curve because as the pump head is increased, the horsepower requirement becomes less. Less water is moving through the pump.

Pump efficiency also enters into choosing a pump. The pump has a different efficiency at each pumping rate. The pump efficiency is the horsepower imparted into the water divided by the actual horsepower at the pump shaft. Only the horsepower transferred into the water creates pumping action. Some of the horsepower is converted into heat and friction loss in the pump. This loss does not contribute to the water velocity, which is the pressure difference across the pump. A pump efficiency curve looks like those in Figure 17-20. Notice that the efficiency curve crosses the pump gpm curve. This intersection point indicates the most efficient impeller size to choose for the pump. A motor chosen to pump its rated horsepower at close to this point is the best motor to use with the pump. Figure 17-21 is a graph of a manufacturer's pump, horsepower, and efficiency curves. It shows that a 9½-inch impeller and a 20-horsepower motor are a good combination.

The small pump used in a small system has only a small pressure rise from the inlet to the outlet. These small pumps are known as *circulators* because of the small pressure difference. They seem only to stir the water, not

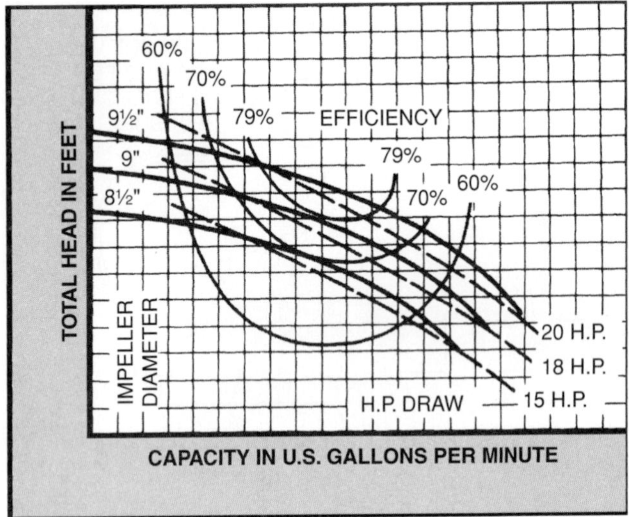

Figure 17-20. Pump efficiency curves. Notice that there are three impeller diameters. *Courtesy ITT Fluid Handling Division*

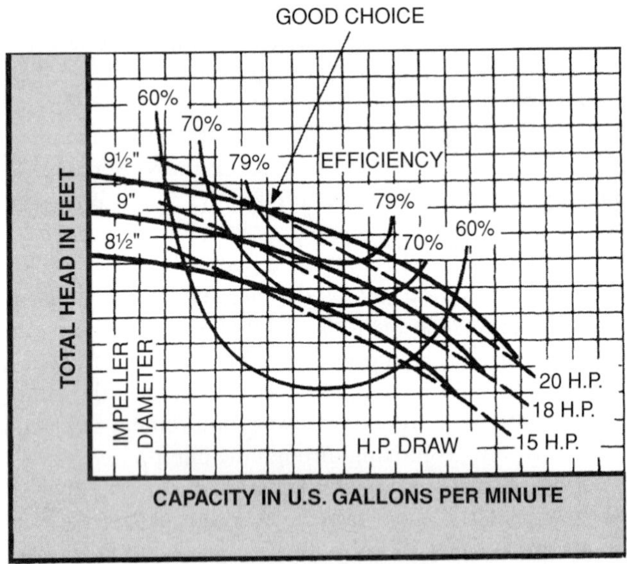

Figure 17-21. Choosing the best pump and motor from the performance curves. *Courtesy ITT Fluid Handling Division*

to create a lot of flow, and therefore operate quietly. The technician must sometimes listen closely in a noisy equipment room to detect the sound of one of these pumps when it is running.

Considering the amount of pressure in a high-rise building, we might think that the system pump has a lot of work to do. However, in this type of situation, the system is a closed-loop system. A column of water exerts pressure down on the pump inlet equal to the pressure that the pump must overcome on the vertical discharge. So, the pump need only overcome the friction of the water moving against the pipe walls and any turbulence in the pipe.

A larger system with a long piping run requires a different type of pump, the rotary vane pump with positive displacement. Unlike the centrifugal pump, this pump can create tremendous pressure differences. When the cylinder or chamber of one of these pumps is full of water and the motor is turning, the pump will empty the cylinder, or stall.

Different pumps are made of different materials. Pumps are subject to the corrosion in the system as well as hot circulating water. Some of them are made of steel or cast iron, and some are made of nonferrous metals, such as brass. Brass pumps often have stainless steel shafts for turning the impeller, with a brass or bronze pump housing to reduce corrosion.

A pump can be mounted in line with the motor, or it can be belt driven, see Figure 17-22. More pumps are mounted in line (direct drive) than are belt driven. The direct-drive pump can be bought as an assembly, with the pump motor and frame assembled together at the factory, or the pump can be purchased separately and mounted by the installation technician, Figure 17-23. In both cases, the pumps have similar characteristics.

The direct-drive pump must have a coupling between the pump and the motor. The coupling helps compensate for slight differences in shaft alignment, Figure 17-24. Two shafts can never be in perfect alignment and even if they were, they would not stay aligned when the system is started because of the temperature difference between the pump and the motor.

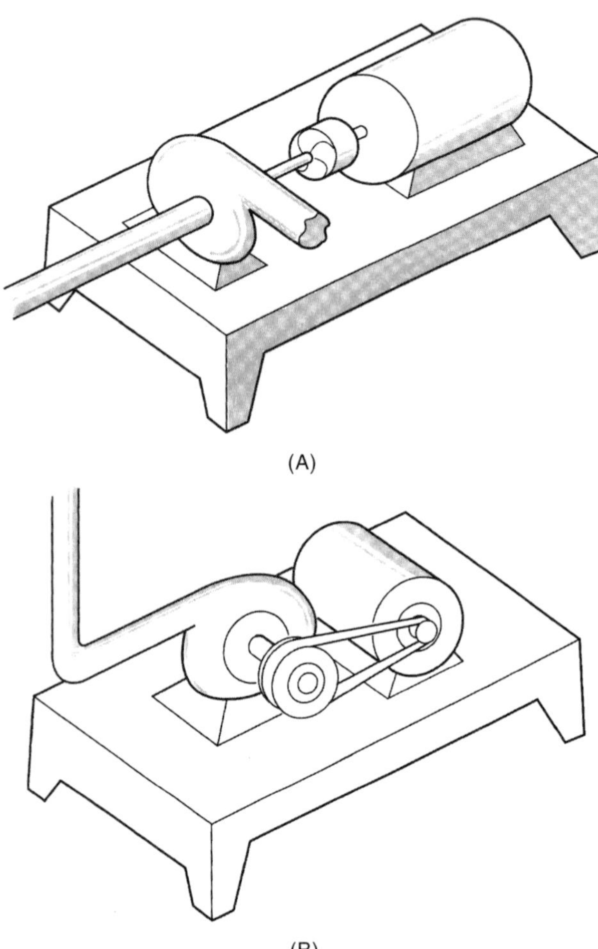

Figure 17-22. (A) Direct-drive pump. (B) Belt-drive pump.

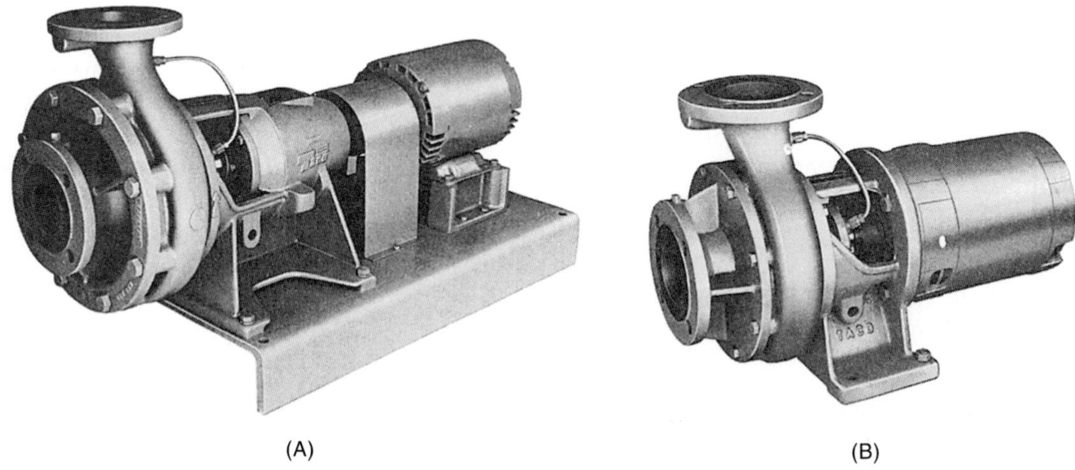

Figure 17-23. Two types of direct-drive pumps. (A) Base mounted (B) Frame mounted. *Courtesy Taco Inc.*

A hot-water pump is much warmer than the motor. If the shafts were aligned while cold, they would not be aligned after running until they were up to operating temperature.

Several kinds of flexible couplings can couple the motor to the pump. Figure 17-25 shows a spring-type coupling often used with a smaller pump. A larger pump requires a more heavy-duty coupling because it is under more load. Figure 17-26 shows one of these couplers.

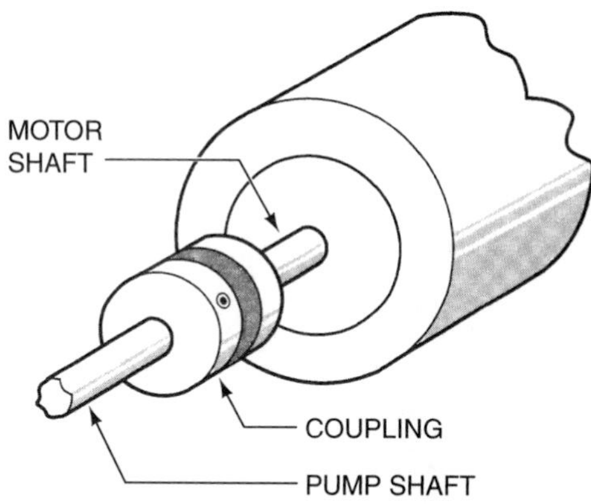

Figure 17-24. The coupling between the pump and the motor corrects a slight misalignment.

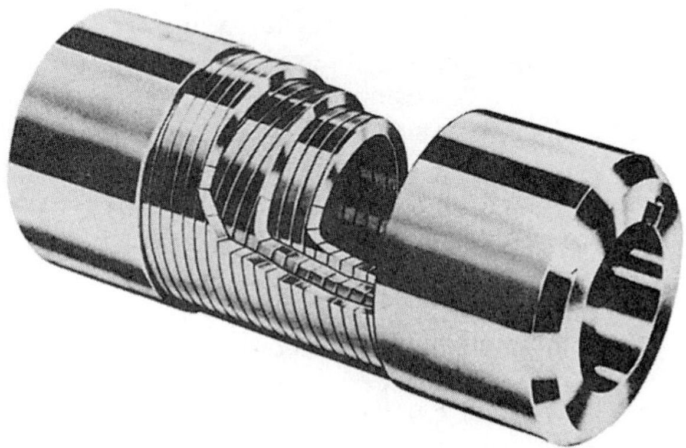

Figure 17-25. Spring-type pump-to-motor coupling. *Courtesy Lovejoy, Inc.*

Each pump must be able to keep the water in the pump while the motor drives the impeller. The shaft seal is used to do this. There are two types of shaft seals: a packing-gland type, in which a packing is tightened down on the rotating shaft, and a spring-loaded shaft seal, or *mechanical seal*, that pushes against a shoulder. The packing-gland shaft seal requires more maintenance than the spring-loaded seal does. If the packing nut is too loose, excess water will leak out; if it is too tight, it will overheat. Figure 17-27 shows how a

Figure 17-26. Heavy-duty pump coupling. *Courtesy Lovejoy, Inc.*

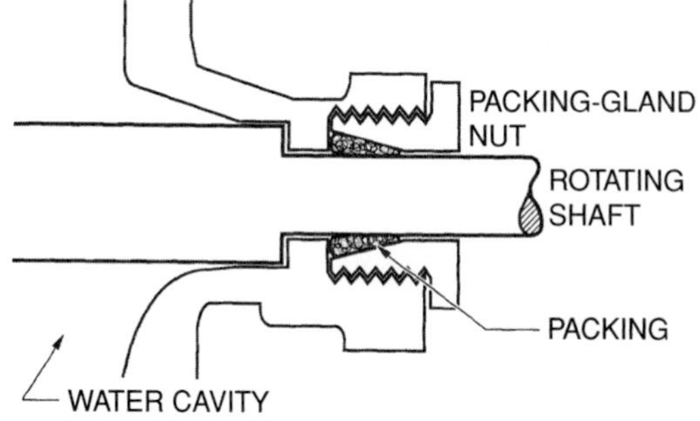

Figure 17-27. The packing-gland nut must be tightened correctly.

packing-gland seal attaches to a shaft. The technician must tighten the seal correctly and check the packing gland regularly. When the nut is correctly tightened, only a small amount of water leaks from the packing gland to lubricate the seal.

The mechanical seal with the spring requires no maintenance. It must only be installed correctly. It is lubricated with water in the vicinity of the seal.

Both shaft seals require correct pump-shaft to motor-shaft alignment. The correct alignment is described in each manufacturer's literature because each type of pump has a different requirement. All alignments are close, within 1/1,000 of an inch.

Caution: Sighting a shaft alignment by eye is not practical and does not work. A misaligned shaft can cause seal or bearing failure before you can shut off the pump.

Shaft alignment is accomplished with precision instruments called dial indicators applied to the pump and motor shafts. The dial indicator will detect any misalignment. Either the pump or the motor is raised to accomplish the alignment. When the pump or motor is raised, shim stock (very thin metal) is placed under the mounts and the pump or motor is tightened down on the shim stock to hold it in place. Many times it is desirable to re-check the shaft alignment after the components have run long enough to become normal operating temperature. This allows for any expansion during running.

Some small pumps may not have shaft seals because the shaft does not protrude into the water cavity. Instead, the pump has a magnetic drive that turns the impeller with a magnet from the outside, Figure 17-28. Consequently, no seal maintenance is required.

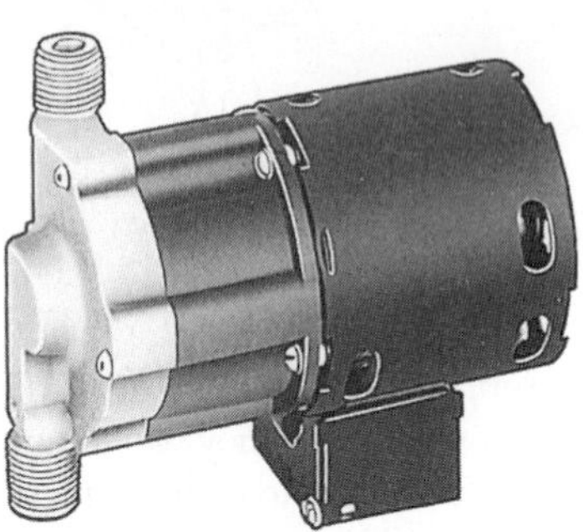

Figure 17-28. Magnetic-drive pump. *Courtesy Grainger Division, W. W. Grainger, Inc.*

Figure 17-29. Double-inlet impeller. *Courtesy Taco, Inc.*

Figure 17-30. Pump with a double-inlet impeller. *Courtesy Taco, Inc.*

The water entering a centrifugal pump tends to push the impeller down the shaft and away from the motor. This is called *thrust* and is countered with a thrust bearing in the single-inlet pump. Larger pumps have double-inlet impellers, see Figure 17-29 for more capacity, and the thrust is canceled by each thrust pushing against the other. A double-inlet impeller requires a different type of pump housing that splits the water stream to an impeller inlet on each side of the impeller, Figure 17-30. This type of pump is used in larger commercial installations.

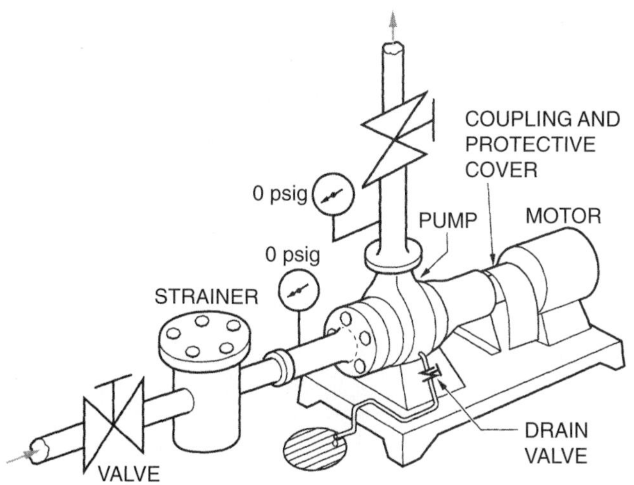

Figure 17-31. Valves used to service the pump and strainer.

All pump housings must be made so that they are serviceable. The smaller pump is typically built with the pump and motor on the same frame. Often a pump is suspended by the piping in the system. The technician can completely disassemble the pump without removing it from the piping. When a valve is furnished on each side of the pump, the piping system need not be drained so that the technician can service the pump. The valves must be closed, then the pump can be drained and disassembled. Figure 17-31 shows these valves.

The larger pump can be close coupled (the pump is mounted on the motor housing), or the pump can be mounted on a base at the end of the motor shaft. See Figure 17-32. The close-coupled pump with the motor mounted on the pump housing does not need to be aligned. The base-mounted pump that is mounted away from the motor must have shaft alignment if the pump is removed from the base. When this type of pump is installed, the pump and base should be mounted on a solid base and then aligned. When the alignment is complete, tapered dowel pins are then driven into the motor base and the pump base to hold them steady, as shown in Figure 17-33. Both pump-and-motor can be rebuilt in place without removal. The motor as well as the pump can be disassembled while in place. The technician should not remove them from their base unless absolutely necessary.

All pumps and motors have bearings for the turning shafts. There are two types of bearings: sleeve, and ball or roller bearings. Sleeve bearings for pumps are made of porous bronze or babbitt metal (an alloy of tin, copper, lead, and antimony) and support the weight of the motor shaft while it turns. Sleeve bearings operate quietly and are preferred where noise and light loads are found.

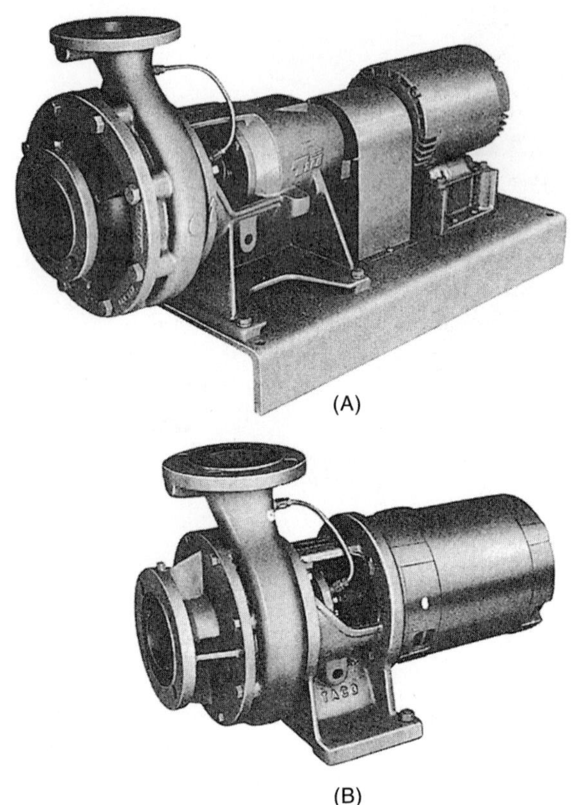

Figure 17-32. (A) Base-mounted pump, (B) Frame-mounted pump. *Courtesy Taco, Inc.*

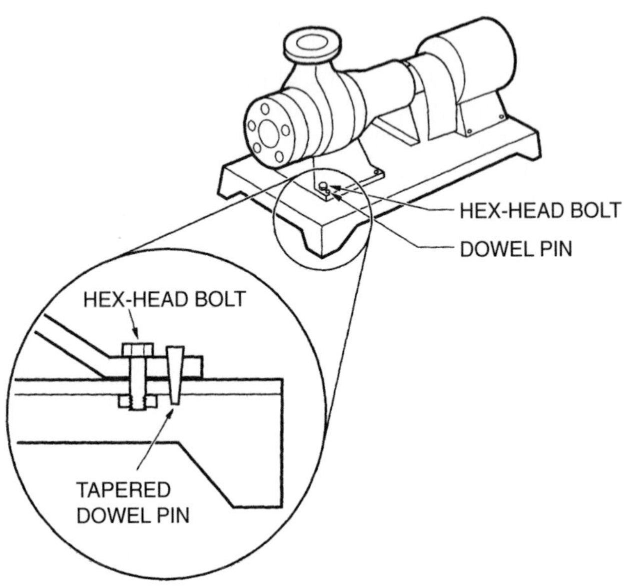

Figure 17-33. Pump and motor are often fastened to the base with dowel pins.

There must be a film of oil between the motor shaft and the sleeve bearing, as shown in Figure 17-34.

Caution: The oil must be the correct weight (viscosity), or the shaft will make contact with the sleeve and begin to score the bearing surface. This damage may cause the bearing to deteriorate quickly or slowly, depending on the amount of damage and the lubrication.

Often the manufacturer provides an oil reservoir that allows oil to slowly feed the bearings. This reservoir must be maintained with the weight and type of oil recommended by each manufacturer.

The ball bearing and the roller bearing can carry more of a load than the sleeve bearing can, but they make more noise. Figure 17-35 shows some ball bearings. Both types of bearings must be lubricated with the proper type and weight of grease. Proper care must be taken when you lubricate these bearings. A grease fitting should be located near the bearing with a relief plug on the other side of the bearing. The relief plug must be removed when you fill the bearing with grease so that the surplus grease can move out of the bearing. Figure 17-36 shows grease relief plugs for ball bearings and roller bearings.

Caution: If the relief plug is not removed, the grease will rupture the seals at the end of the bearing, and grease will escape to the atmosphere or into the water in some cases.

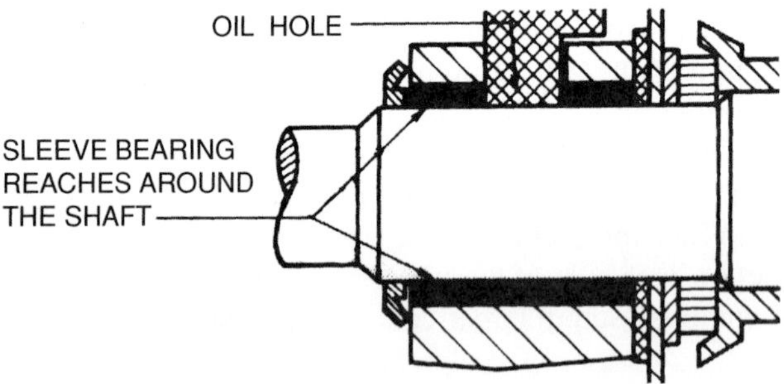

Figure 17-34. There must be an oil film between the sleeve bearing and the shaft.

Figure 17-35. Ball bearings. *Courtesy Century Electric, Inc.*

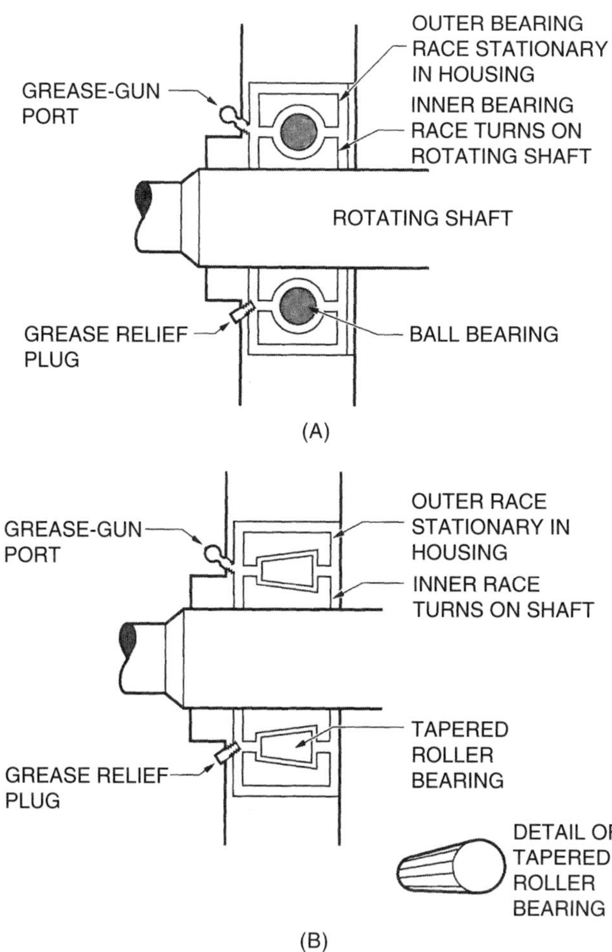

Figure 17-36. (A) Lubricating ball bearings. (B) Lubricating roller bearings.

The technician will often encounter poor pumping practices with centrifugal pumps and must be able to recognize them. For instance, when too much water is circulating in a system, the technician should *never* throttle the inlet water to the centrifugal pump—particularly in a hot-water system. This throttling causes a pressure drop between the valve and the impeller. If enough of a pressure drop occurs, the hot water will flash to steam. This is called *cavitation.* Vapor passing through the pump will cause noise and will eventually damage the impeller. The water pump is designed to pump water, not steam. Any pressure drop at the pump inlet can lead to cavitation. A restricted strainer just before the pump or throttled pump suction will often cause hot water to cavitate, Figure 17-37.

Circulating air in a water system will also make noise at the centrifugal pump. Often air is introduced at the centrifugal pump because the inlet at the seal is in a vacuum and the seal has a leak. The leak allows atmospheric air to enter in small amounts. For this reason, it is a good idea to keep the pump suction in a positive pressure. Air can also enter the closed system when water leaks out and city water is added to the system to maintain the water level.

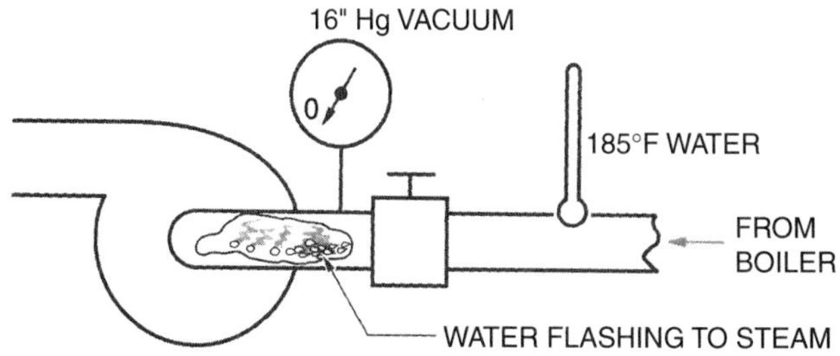

Figure 17-37. Starved-pump suction causes cavitation.

As mentioned previously, city water contains air, which is released into the system.

Because a centrifugal pump will not pump air, any system with a centrifugal pump must be properly primed before the pump is started. To prime a system is to fill it completely with water. This is accomplished in closed hot-water systems with city water. City water pressure is added until the system is liquid full with approximately 5 psig at the highest point in the system, where the bleed valves are located. The air can then be allowed to escape the system at this point, Figure 17-38. The pump is then primed and full of water. Circulation will occur when the pump is started.

Often an installation will have a bleed port on the pump for allowing air to escape at this point. Air is hard to remove from a system. The piping designer should provide bleed ports at every high point in the system where air may accumulate. The designer must also furnish a good method for filling the system.

The piping entering and leaving the pump should be routed correctly, particularly the piping entering the pump. For example, if there are sharp turns at the pump inlet, water will tend to load the impeller on one side, so the pipe should not

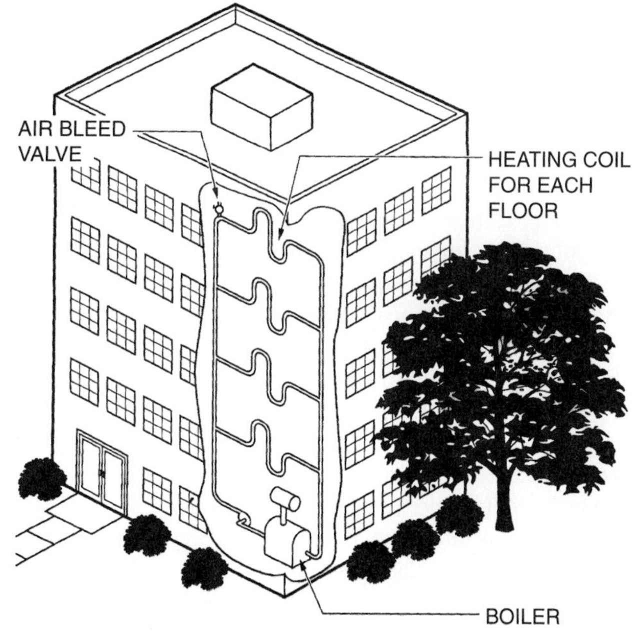

Figure 17-38. Air bleed at the highest point.

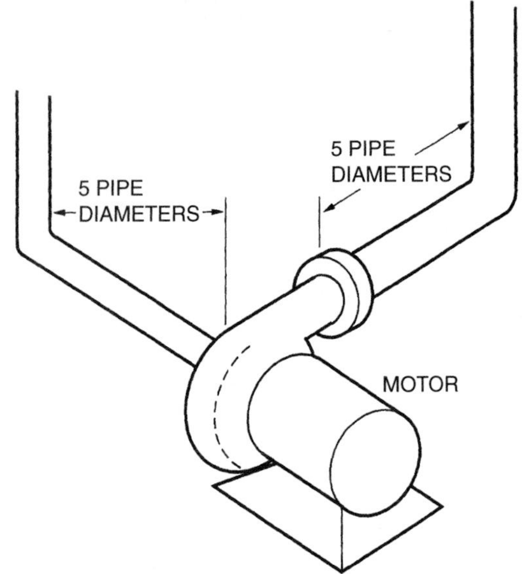

Figure 17-39. The piping entering and leaving the pump should be installed correctly.

have an elbow at the pump inlet. There should be a distance of approximately five pipe diameters between the pump inlet and the first elbow to ensure that the water enters the pump with minimum turbulence. For best pump and piping performance, the piping leaving the pump should also leave in a straight line and not make any fast turns, Figure 17-39.

17.4 Terminal Heating Units

When the hot water is pumped to the conditioned space, a heat-exchange unit must transfer the heat to the conditioned space. These units are either natural-convection or forced-convection units.

Water Flow

Before terminal units are discussed further, water flow must be discussed because it is used in sizing the units. Much of the following information is available in the *IBR* (Institute of Boiler and Radiator Manufacturers) *Guide 250.* IBR is now the Hydronics Institute in Berkeley Heights, New Jersey. The Hydronics Institute researches heating and cooling equipment used in hydronic (water- and steam-circulating) systems. This institute also tests and rates equipment for the industry on a nonbiased basis. IBR ratings are used throughout this unit.

Water flow through a typical small system is expressed in pounds per hour (lb/h). The two typical water flow rates are 500 and 2,000 pounds per hour. This is the same as saying 1 gpm or 4 gpm (gallons per minute). When 1 gallon of water is circulated per minute, 8.33 pounds of water per minute are circulated because 1 gallon of water weighs 8.33 pounds. When 8.33 pounds per minute are circulated for 1 hour, 499.8 pounds are circulated (8.33 lb × 60 min/h = 499.8 lb/h), rounded to 500 pounds per hour; likewise, 4 gpm times 500 pounds per hour equals 2,000 pounds per hour.

Because adding 1 Btu of heat to 1 pound of water raises the temperature of the water 1°F, when the temperature of 500 pounds of water is raised 1°F, 500 Btu of heat have been added. In a circulating hot-water unit, the water temperature drops as the room air takes heat from the water. This temperature drop is used to rate hot-water systems because a 1°F drop in water temperature with a flow rate of 500 pounds per hour will give the unit a 500 Btu per hour capacity. This can be restated: A unit will have a capacity of 500 Btu per hour for each 1°F temperature drop, with a flow rate of 500 pounds per hour (or 1 gpm). The following table illustrates how equipment is rated:

500 lb/h (1 gpm) Flow Rate

10°F drop = 10 × 500 = 5,000 Btu/h capacity
20°F drop = 20 × 500 = 10,000 Btu/h capacity
30°F drop = 30 × 500 = 15,000 Btu/h capacity
40°F drop = 40 × 500 = 20,000 Btu/h capacity
50°F drop = 50 × 500 = 25,000 Btu/h capacity

2,000 lb/h (4 gpm) Flow Rate

10°F drop = 10 × 2,000 = 20,000 Btu/h capacity
20°F drop = 20 × 2,000 = 40,000 Btu/h capacity
30°F drop = 30 × 2,000 = 60,000 Btu/h capacity
40°F drop = 40 × 2,000 = 80,000 Btu/h capacity
50°F drop = 50 × 2,000 = 100,000 Btu/h capacity

Different types of terminal units have built-in features that limit the temperature drop through the units. Following are the recommendations of the *IBR Guide 250*:

1. *Cast-iron radiators:* Not more than a 30°F drop.
2. *Forced-air convectors:* Temperature drop depends on the manufacturer's ratings; it may be a 10°F, 20°F, or 30°F drop.
3. *Unit heaters:* Temperature drop can be 50°F or more, depending on the manufacturer.
4. *Baseboard or commercial finned-tube units:* As much as a 50°F drop, provided that minimum flow rates are maintained. The minimum flow rates are established to ensure a flow rate of more than the laminar flow rate, which is so slow that proper heat exchange does not occur. See the following flow rate table from the *IBR Guide 250*:

Pipe or Tube Size in Element (inches)	Minimum Design (gpm)
½	0.3
¾	0.5
1	0.9
1¼	1.6

Natural-Convection Units

Natural-convection units are available as baseboard heaters, panel heaters, and radiators. Figure 17-40 shows a baseboard heater and a radiator.

BASEBOARD HEAT. The baseboard heater is much like the electric baseboard heater discussed in Unit 3. It is mounted at the baseboard of the conditioned space, usually along the wall. As air is heated in the unit, the air expands and rises because it becomes less dense as shown in Figure 17-41. An air current is started that allows the warm air to rise up the cool wall, and a gentle current moves across the ceiling and back across the floor, as shown in Figure 17-42. Baseboard units are rated in Btu per foot of length for two flow rates: 500 and 2,000 pounds per hour. All that the designer must know are the room heat loss in Btu per hour, the design water temperature (and drop), and the flow rate of the system being designed. For example, suppose that a room has a heat loss of 12,500 Btu per hour, the design water temperature is 180°F, and the system flow rate is 500 pounds per hour. The table in Figure 17-43B indicates that the heat requirement can be fulfilled with 22.3 feet of a 500 pound per hour heater rated at 560 Btu per hour per linear foot (12,500/560 = 22.3 ft).

The different trim features of baseboard units allow them to be installed around corners, both on the outside and inside of the corner. Units can also be connected for longer lengths.

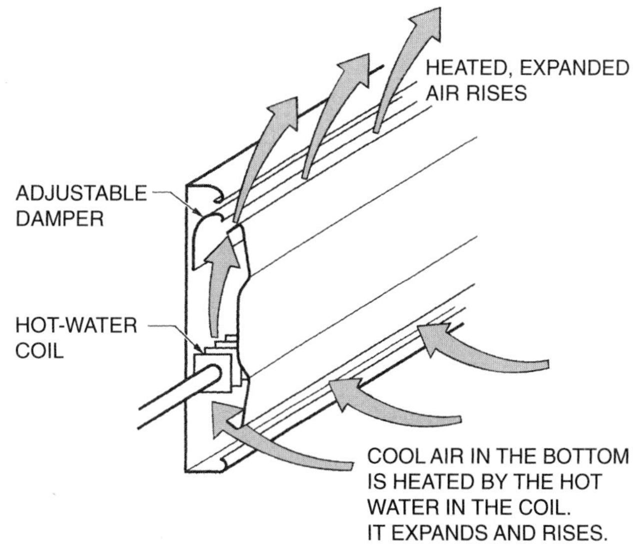

Figure 17-41. Principles of a baseboard heater.

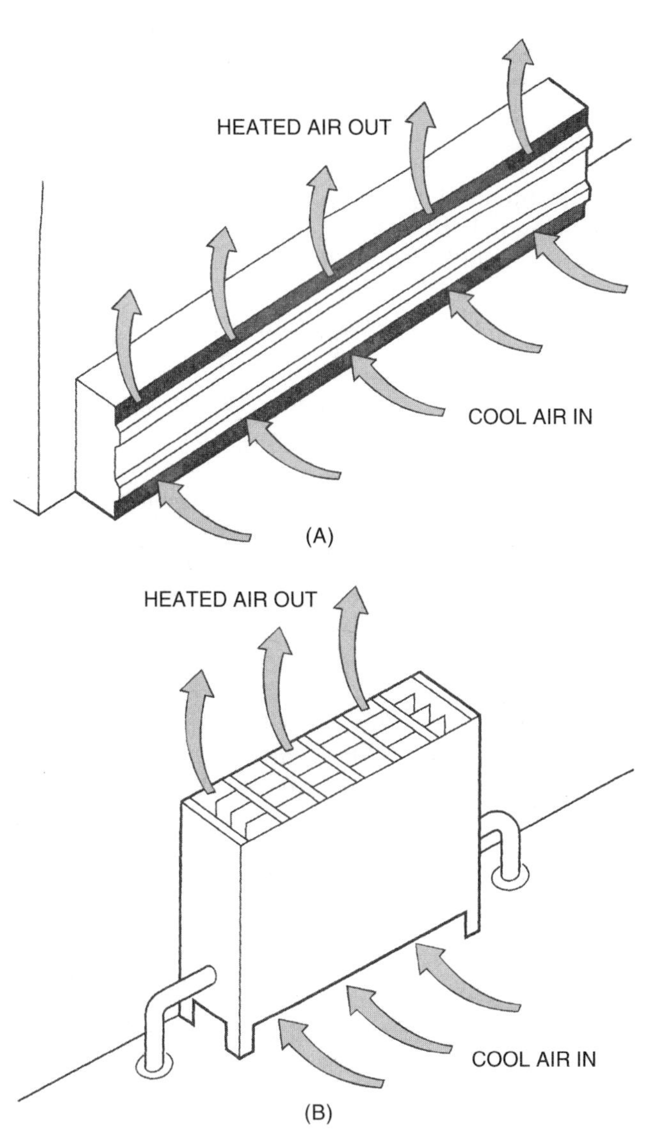

Figure 17-40. Natural-convection units. (A) Baseboard heater. (B) Radiator.

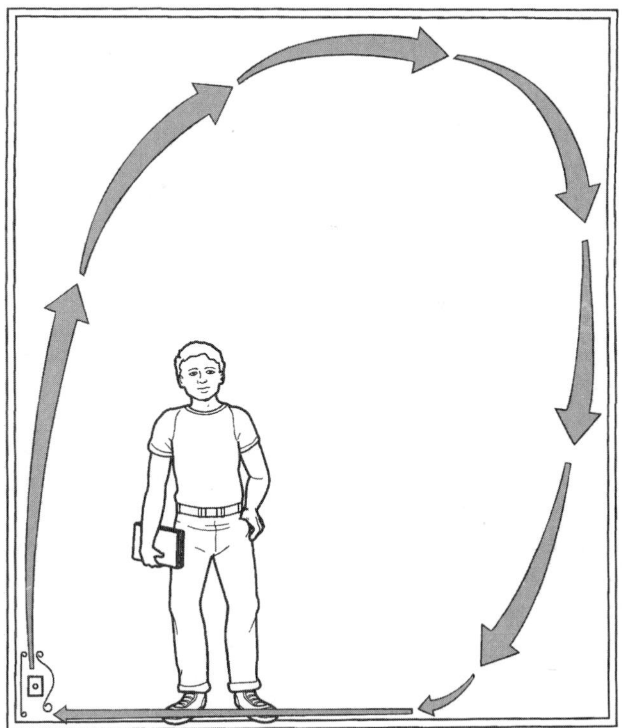

Figure 17-42. Air current started by a baseboard heater.

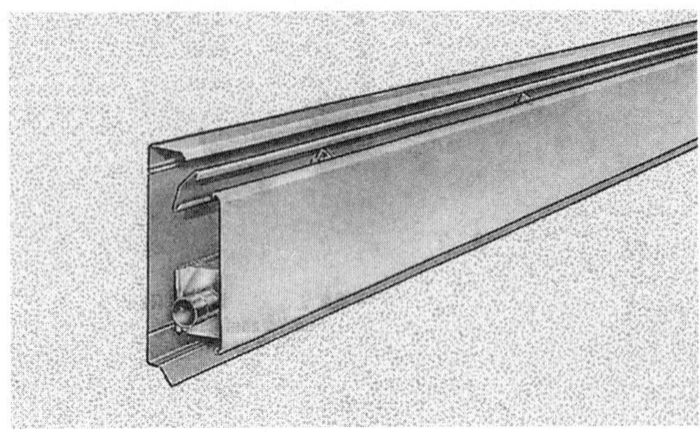

(A)

All Models	Water Flow Rate Lb/Hr	Hot Water—BtuH/Hr/Lin Ft								
		140°	150°	160°	170°	180°	190°	200°	210°	220°
	2000 Lbs/Hr	335	400	460	530	590	670	730	800	870
	500 Lbs/Hr	310	370	430	500	560	630	690	760	820

(B)

Figure 17-43. (A) Cross section of a baseboard heater. *Courtesy Heatrim America* (B) Rating table for baseboard heat. *Courtesy Grainger Division, W W. Grainger, Inc.*

Baseboard convectors are known for their even and quiet operation. Because they have no fans, they make no noise. The only disadvantage of baseboard heat as a system is that because there is no common airstream, such as in a forced-air system, there is no central method to filter the air or add humidity to the space.

PANEL HEAT. Panel heat can be in the ceiling, walls, or floor of the structure. This method of warming a room does not cause the air movement that is associated with forced-air systems, so drafts are not a problem, and this type of heat is very comfortable. When panel heat is in the ceiling, coils of pipe are routed under the ceiling material, usually plaster on metal lath. The ceiling above the heat is well insulated to prevent heat from rising into the attic. The room load can be calculated for a particular room, and the correct amount of water flow at the design temperature is circulated to add the heat to the room. The water temperature is usually less than 140°F to establish a panel surface temperature of approximately 120°F. The 120°F surface temperature will warm the entire room, making the floor, walls, and furnishings more pleasant to be in contact with as shown in Figure 17-44.

Panel heat in walls is similar to panel heat in ceilings.

Panel heat in the floor operates a little differently than ceiling or wall heat. Copper, steel, or plastic pipes are buried in concrete flooring, and water is circulated in the pipes. The mass of the floor is much greater than that of a plaster wall or ceiling and does not respond to temperature changes as fast as a wall or ceiling panel does. For this reason, space temperature thermostats are not used to control room temperature. Outdoor temperature is typically used to control the floor temperature. As the outdoor temperature falls, the slab temperature is allowed to warm to a maximum of approximately 80°F. The floor cannot be allowed to become as hot as a ceiling or a wall panel, or the floor will be too hot to walk on. One problem with floor panel heat is outdoor temperature swings. If a cold night is followed by a warm summer day, the floor may stay warm long enough to overheat the room.

RADIATORS. Radiators are room heaters that have hot water circulating in them. They are typically located along walls. These radiators can be part of an old system that was updated from a natural-convection system to a forced-convection system with the addition of a water pump. They may even be part of an old steam system converted to forced convection. However, they are still room heaters. These old radiators are large for the typical rooms that they may be heating, so they often have decorative covers that allow air to circulate over them. Modern radiators are freestanding, Figure 17-45 or wall mounted. The earlier radiators were made of cast iron and were large and bulky. Modern radiators are made of stamped steel.

Forced-Convection Units

A forced-convection heater is a hot-water coil in a forced-air furnace or in a duct system furnished with a central fan, a room forced-air convector, or a unit heater.

HOT-WATER COIL. When a hot-water coil is in a furnace, it is a finned-tube coil in the airstream. The fins on the coil give the tubes more surface area for more efficient heat transfer. During manufacturing, the fins are typically placed on the coil piping first, then the coil piping is expanded into the fins for a tight fit so that heat exchanges to the fins more efficiently as shown in Figure 17-46.

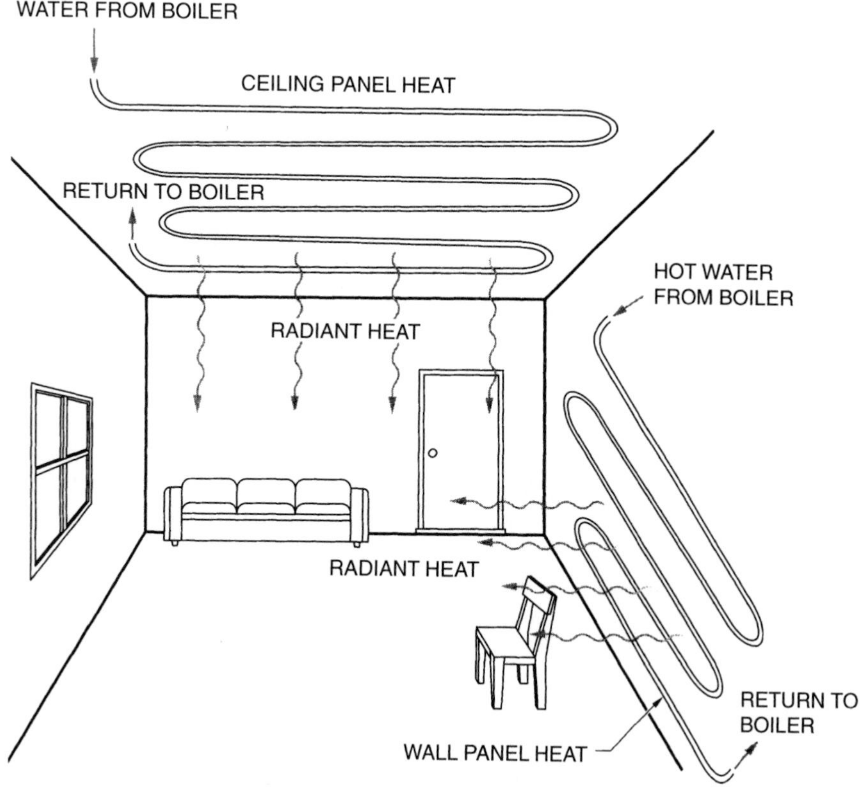

Figure 17-44. Panel heat.

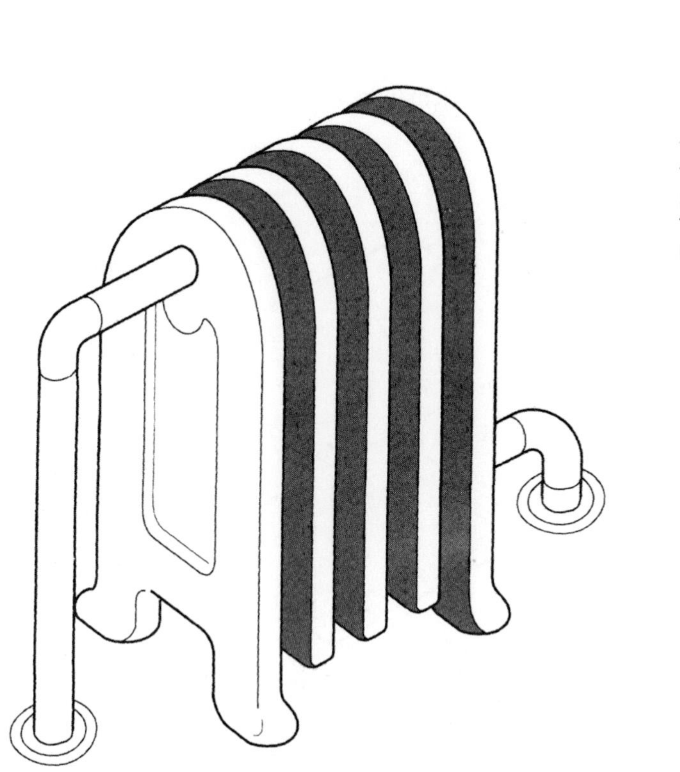

Figure 17-45. Radiator heater.

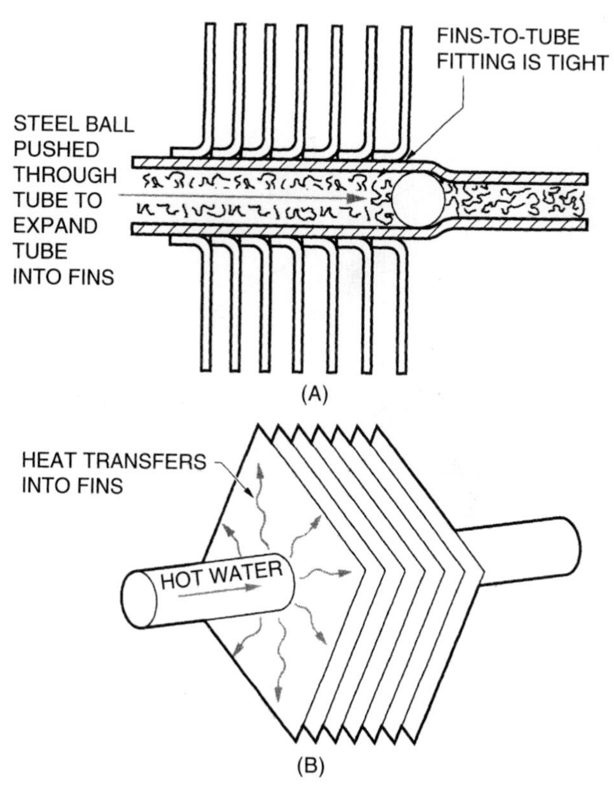

Figure 17-46. (A) The fins must be fastened tightly to the tube. (B) Heat transfer to the fins.

These coils may have several rows of tubing through which the water can flow, and the coil must be located correctly in the airstream. There are two possibilities for positioning the coil in the airstream: parallel flow and counterflow. With parallel flow, the air enters the side with the entering water. With counterflow, the air enters the side with the leaving water. The counterflow coil extracts more heat from the coil, and the leaving air temperature is closer to the entering water temperature. The reason for this is the difference in temperature between the water and the air. Remember that temperature difference is the driving force for heat exchange. With the hottest water exchanging heat with the hottest air, more heat will exchange as seen in Figure 17-47. Coils are often piped backward by accident and when this happens, the coil is short of capacity.

The designer plans a specified gpm of water flow at a specified design temperature with a specified airflow to ensure the correct coil output. For example, a certain coil may have an entering water temperature of 180°F with a design temperature drop of 20°F. The design temperature for the coil selection is 170°F. This was determined by finding the average temperature in the coil: (180°F entering water + 160°F leaving water)/2 = 170°F. The flow rate is 2,000 pounds per hour (4 gpm). The capacity of this coil is 40,000 Btu per hour.

Larger coils are calculated differently because of increased flow rate. Suppose that a large coil has a flow rate of 10 gpm with a 20°F temperature drop. The capacity of this coil is 100,000 Btu per hour. You can follow this calculation in Figure 17-48. The capacity was calculated from the following formula:

$$q = 500 \times \text{gpm} \times \text{TD}$$

where q = Heat content in Btu/h.

500 = Btu/1 gal of flow for 1 hour.

TD = Temperature difference.

$$\begin{aligned} q &= 500 \times \text{gpm} \times \text{TD} \\ &= 500 \times 10 \times 20 \\ &= 100{,}000 \text{ Btu/h} \end{aligned}$$

When all this heat is transferred into an airstream, the amount of heat available to the conditioned space is 100,000 Btu per hour minus any heat that may be lost through the duct work to unconditioned space.

These coils mounted in duct work can be used for large heating capacities. There are two advantages to forced-air systems: The air can be filtered because it is a moving airstream, and humidity can be added to the air easily, as shown in Figure 17-49.

ROOM FORCED-AIR CONVECTORS. Room forced-air convectors are often used to heat individual rooms. They can be seen in motels, school classrooms, and offices. The unit often sits on the floor, next to the outside wall, and is designed to match the room decor. A hot-water coil in the unit adds heat to the air, Figure 17-50. A fan distributes the heated air to the room. When one unit does not have enough capacity for a room, multiple units can be used.

In addition to heating, these units often have cooling coils for air-conditioning. The cooling coil may be the same coil circulating cold water or a separate chilled-water cooling coil. Some units have refrigeration units installed inside. More features give the unit more versatility for different seasonal demands. For example, an office building

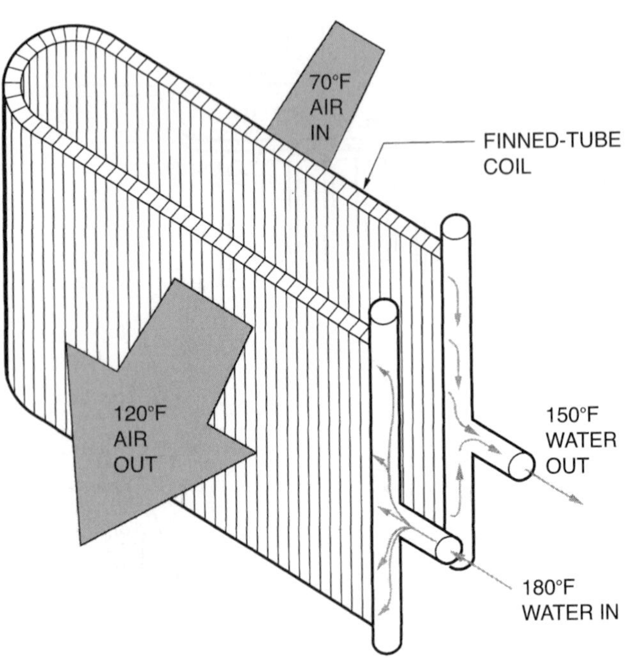

Figure 17-47. Heat exchange between the air and the water in a forced-air coil.

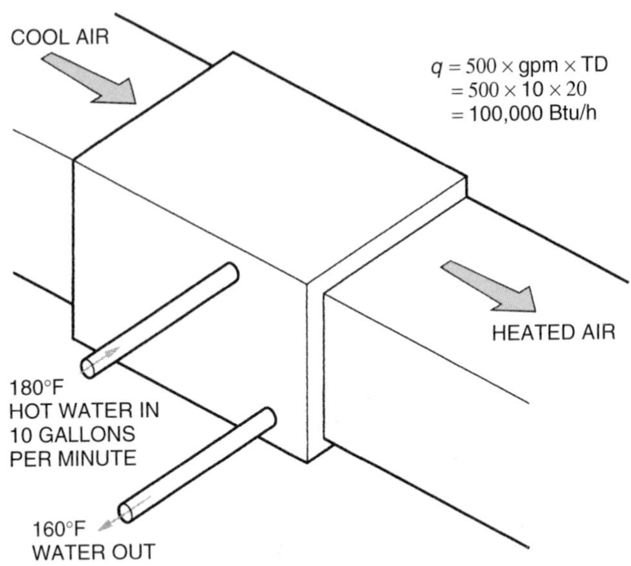

Figure 17-48. Large forced-air coil capacity.

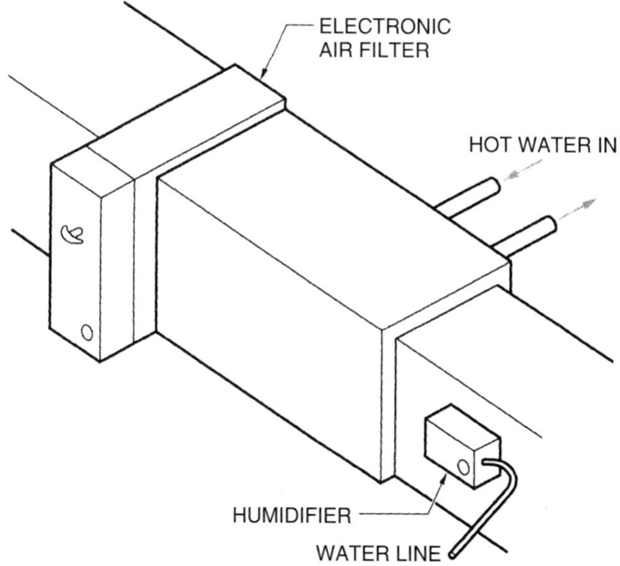

Figure 17-49. Air can be filtered and humidified in a forced-air system.

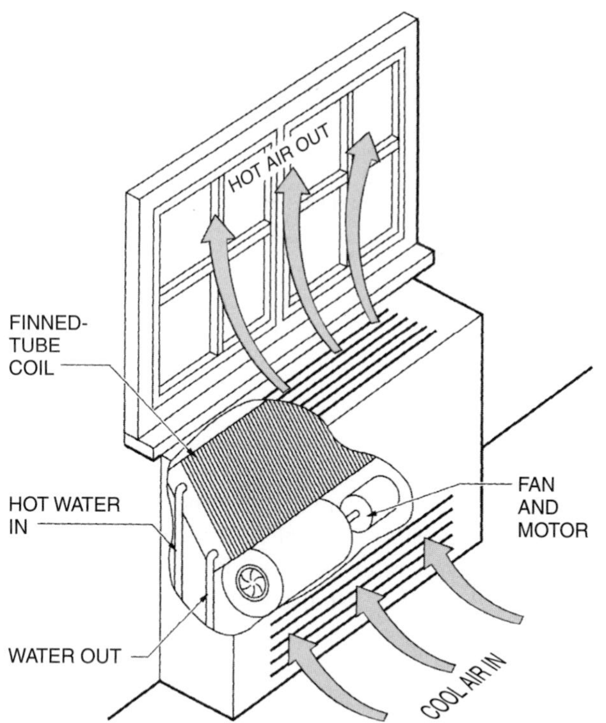

Figure 17-50. Room forced-air convector.

may need heat on one side because it is in the shade and cooling on the other because of the sun.

UNIT HEATERS. A unit heater hangs from the ceiling and has a hot-water coil and a fan that moves the heat into the conditioned space. This type of heater is often seen in large rooms, such as gymnasiums, warehouse spaces, and large lobbies of buildings. Figure 17-51 shows a forced-air unit heater.

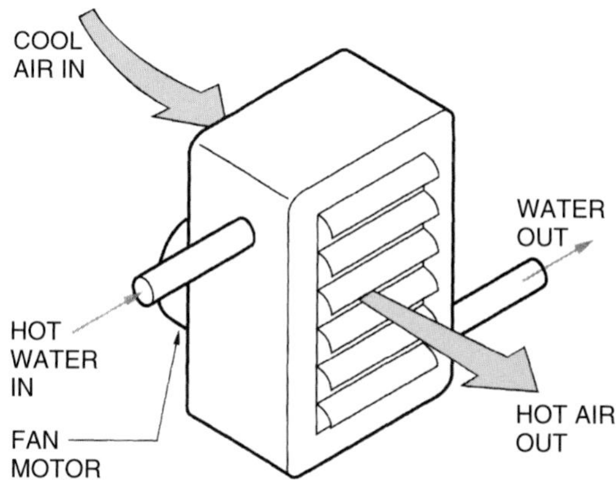

Figure 17-51. Forced-air unit heater.

17.5 Piping Systems

Hot water can be routed to the various types of heat exchangers or terminal units by one of several methods. The designer determines these methods, and the service technician must work with them. The common pipe configurations supply hot water to the units and return the water to the boiler for more heat to be added. The four common types of piping systems are these:

1. one-pipe series loop
2. one-pipe system with flow tees
3. two-pipe direct-return system
4. two-pipe reverse-return system

These four systems are pictured in Figure 17-52.

The two-pipe system can be used in larger applications to incorporate chilled water for air-conditioning into the system with the hot-water piping. Either hot or cold water is circulated in the two-pipe system, depending on the season. Another type of system has four pipes with one complete two-pipe system for hot water and the other for chilled water, as shown in Figure 17-53.

Note: The designer tries to fit the job and comfort-level requirements into a budget for the application. The larger the application and the more sophisticated the job, the more costly the job. The designer must determine from the type of system and the data for the units what the system water temperature drop will be. The system temperature drop is the difference between the water temperature leaving the boiler and the water temperature returning to the boiler. This temperature drop may be as much as 50°F. From the system temperature drop (this is the same as the boiler temperature rise), the designer determines the system design temperature.

We start our discussion of piping systems with a simple one-pipe series system. This system, which can be used in a small residence, has one boiler and one pump, as shown in Figure 17-54. This system works well, but

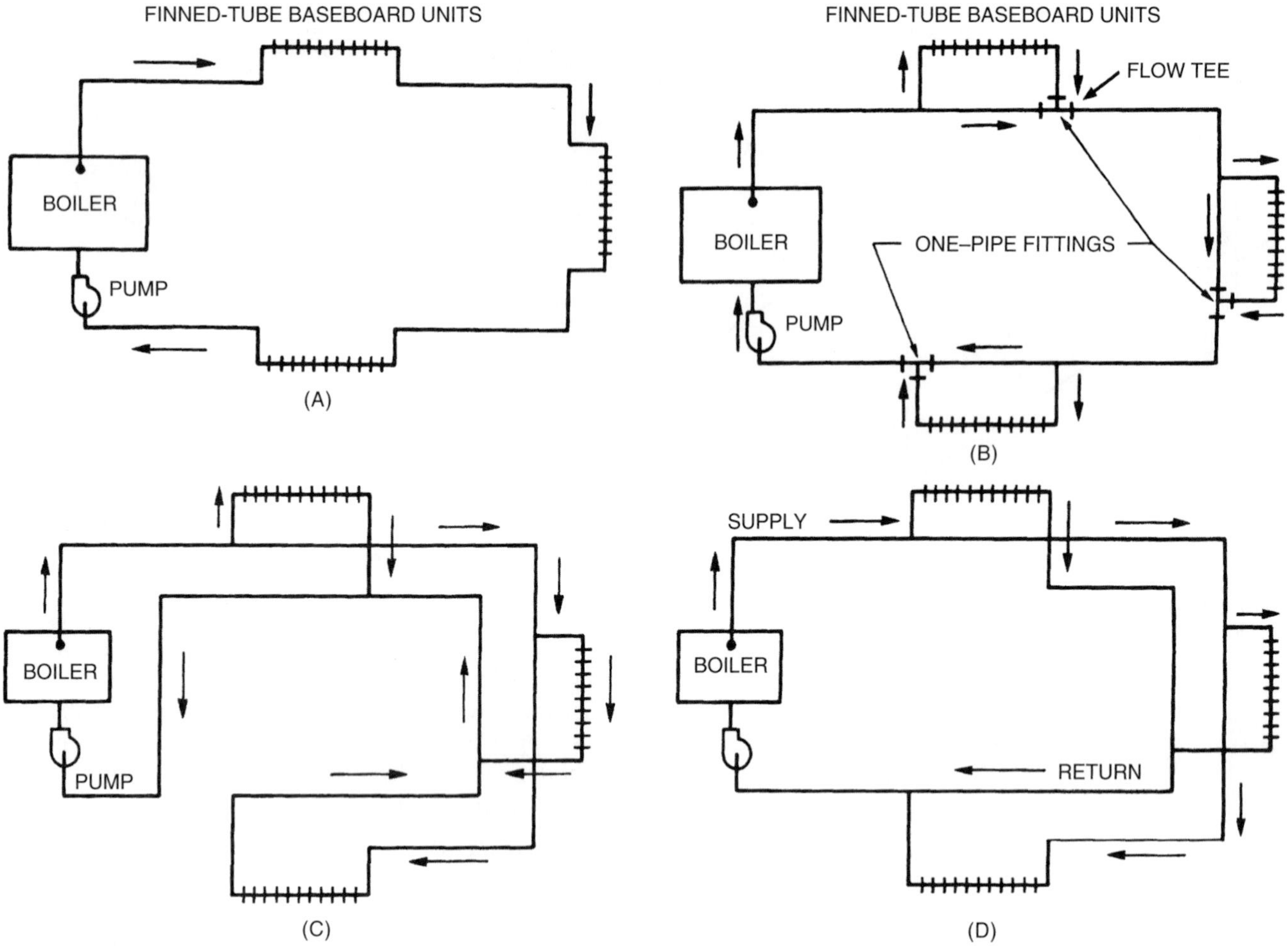

Figure 17-52. Four piping systems for hot-water heat. (A) One-pipe series loop. (B) One-pipe system with flow tees. (C) Two-pipe system with direct return. (D) Two-pipe system with reverse return.

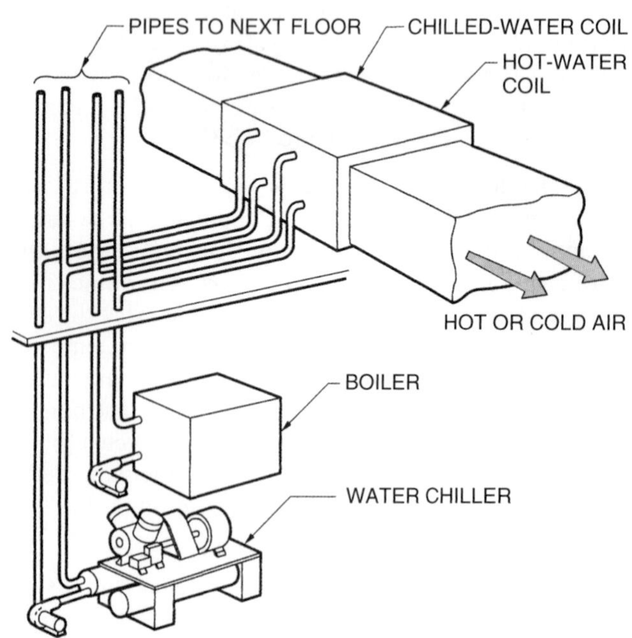

Figure 17-53. Four-pipe system with heating and cooling.

it has only one zone. The same water circulates through each unit. As the water proceeds around the circuit, its temperature reduces. The total system may have a temperature drop of 40°F from the time the water leaves the boiler until it returns to the boiler. It should be noted that as the water temperature reduces, it has less capacity to heat. The convectors must also be longer the farther they are from the boiler.

A larger home may have several circuits that are in parallel with one another, but each circuit may have several convectors in series with one another, Figure 17-55. This system has some zone control in that the individual circuits can serve different zones of the house according to occupancy. For example, one zone may be the kitchen, den, and living room. Another zone may be the sleeping quarters and still another zone the recreation area. In this type of system, the pumps can be controlled with room thermostats in the various zones, so the temperature can be adjusted, as shown in Figure 17-56.

Another one-pipe system uses a special tee fitting at the coil inlet or outlet that causes water to flow through the coil. The tee connects the coil to the one-pipe main to pass

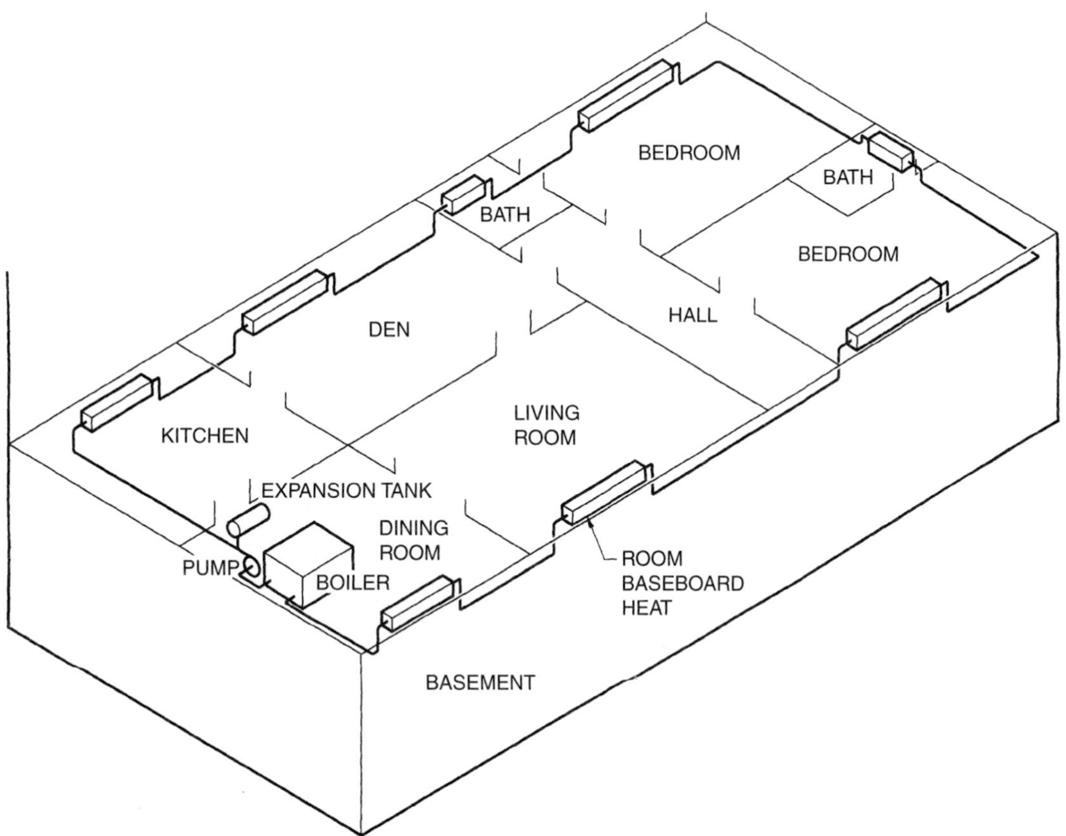

Figure 17-54. One-pipe system in a home.

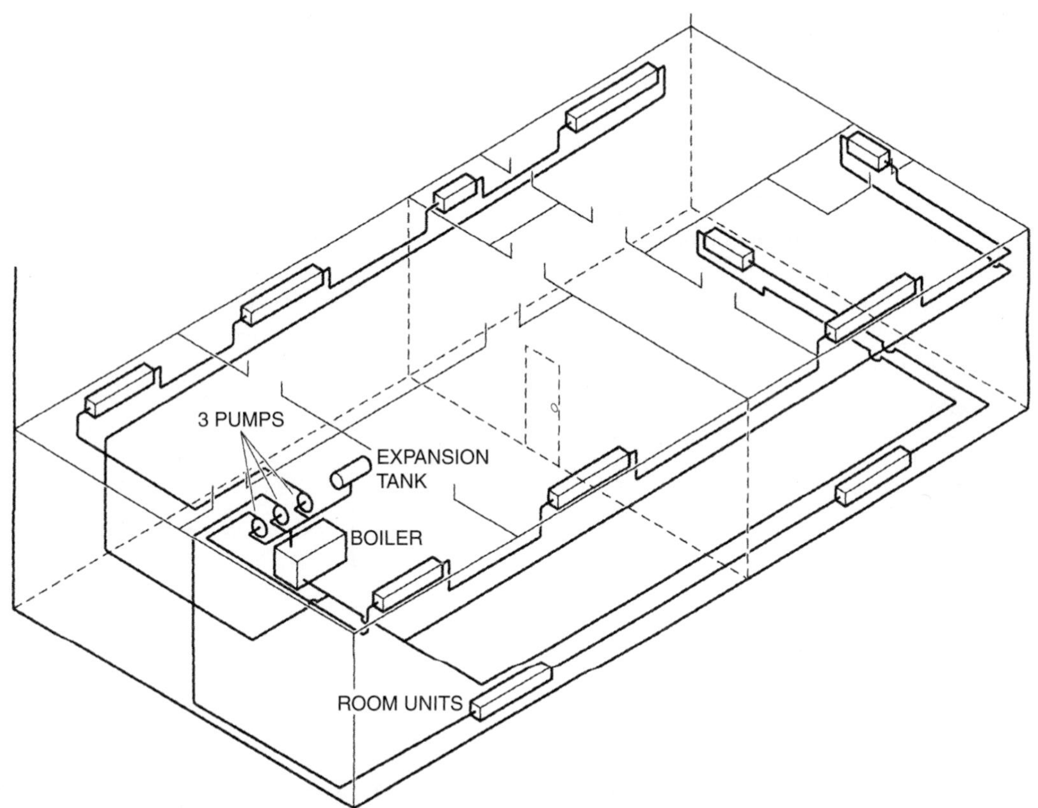

Figure 17-55. Larger home with several circuits.

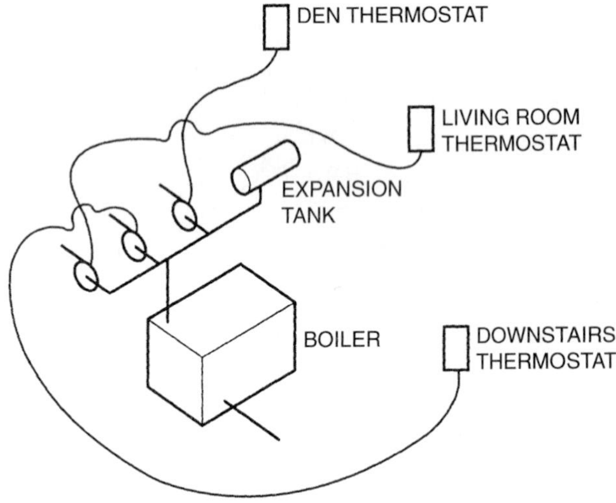

Figure 17-56. Room thermostats control the pumps in the various zones.

water on to the next unit, as shown in Figure 17-57. The mix water at the end of the coil is slightly lower than the entering water. The economical advantage of this system is that only one pipe circulates the water, so less piping is purchased.

Another type of system, the two-pipe system, has a main pipe for a supply and a main returning the water to the boiler. This type of system has a direct- or reverse-return main. The direct-return main can be identified by the placement of the terminal units. If the closest unit to the boiler in the supply piping is the closest unit to the boiler in the return piping, the system is a direct-return system. This system typically requires more balancing of the water in the circuit to achieve the correct water flow to each unit because the shortest circuit will tend to have the most water flow. (There is less resistance in the short circuit.) The circuit water flow can be balanced with valves, as shown in Figure 17-58.

The two-pipe system may instead have a reverse-return piping system, in which the unit that is closest to the boiler in the supply piping is the farthest from the boiler in the return piping. This system balances the flow with the resistance to flow in the water piping. However, it is not used exclusively because it is expensive. It requires more pipe, as shown in Figure 17-59.

Piping Materials

The systems are piped with steel, wrought iron, or copper pipe. The designer determines the material when planning the structure. Steel pipe has welded connections and is used for large installations. Wrought-iron pipe is black iron or galvanized, depending on the designer's choice. Galvanized pipe has a zinc coating that protects the pipe from corrosion and rust. Wrought-iron pipe uses threaded connections to fasten the pipe, as shown in Figure 17-60 and is

HOW TO USE B & G MONOFLO FITTINGS

- Be sure the RING is between the risers OR
- Be sure the SUPPLY arrow on the supply riser and the RETURN arrow on the return riser point in the direction of flow.

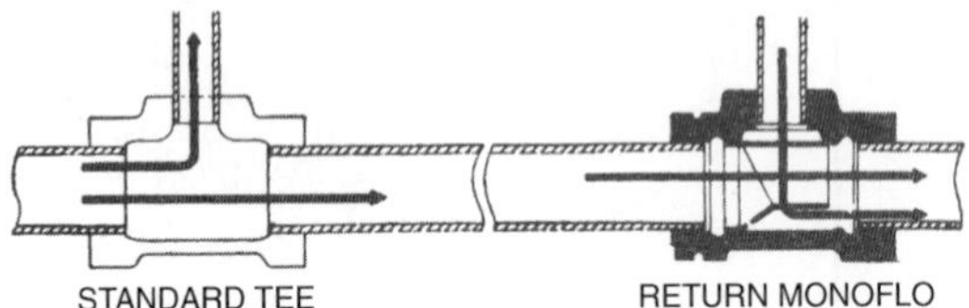

STANDARD TEE RETURN MONOFLO

For radiators above the main–normal resistance

For most installations where radiators are above the main, only one Monoflo Fitting need be used for each radiator.

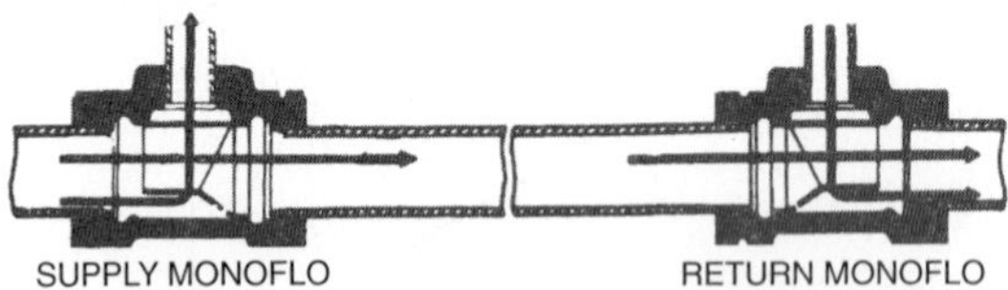

SUPPLY MONOFLO RETURN MONOFLO

For radiators above the main–high resistance

Where characteristics of the installation are such that, resistance to circulation is high, two Fittings will supply the diversion capacity necessary.

SUPPLY MONOFLO RETURN MONOFLO

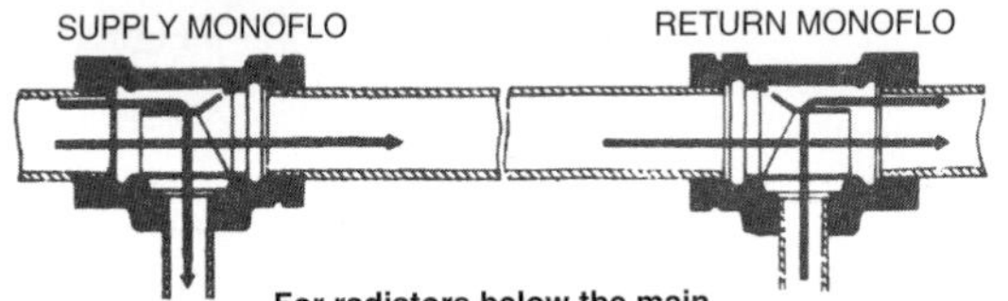

For radiators below the main

Radiators below the main require the use of both a Supply and Return Monoflo Fitting, except on a 3/4" main use a single return fitting.

(A)

(B)

Figure 17-57. (A) A tee for a one-pipe system. *Courtesy ITT Fluid Handling Division* (B) Diagram of a one-pipe system.

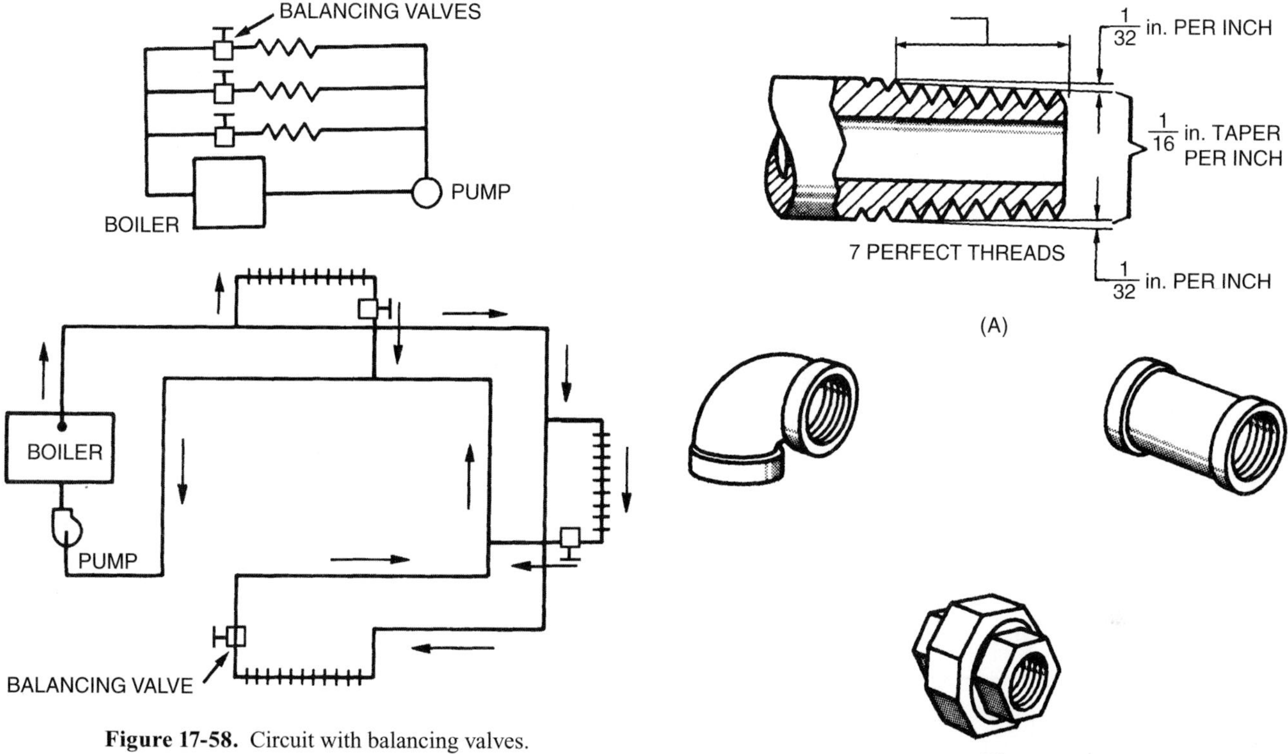

Figure 17-58. Circuit with balancing valves.

Figure 17-60. Wrought-iron pipe requires threaded connections. (A) Threads on a connection. (B) Connectors.

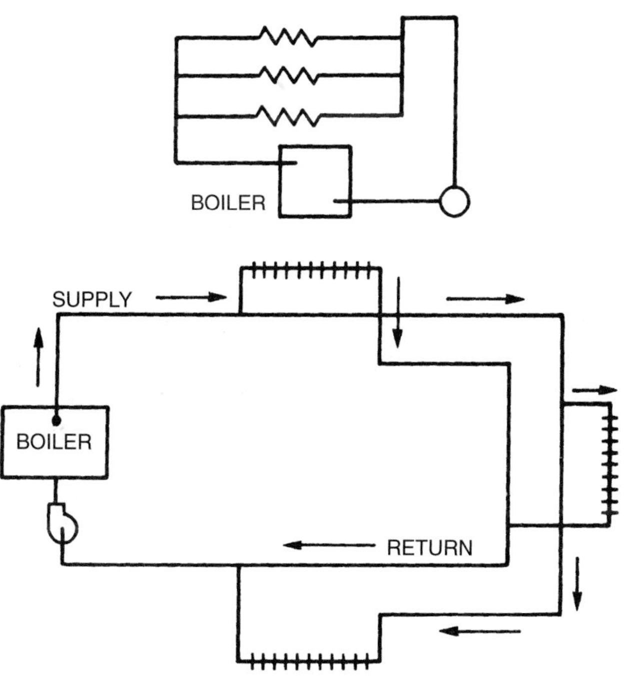

Figure 17-59. A reverse-return system requires more pipe.

either schedule 40 or 80. Schedule 40 pipe is standard pipe with a typical working pressure of 125 psig, and schedule 80 is heavy-duty pipe for higher working pressures.

Copper pipe is often used for routing hot water and either low- or high-temperature solder is used to join the pipe with "sweat" connections. Copper pipe is rated in letters: K, L, and M. The letter value describes the wall thickness and therefore the working pressure. Figure 17-61 shows a table of properties for copper pipe.

Pipe Sizing

As water flows farther in a piping system toward the pump inlet, the pressure becomes less and less, as shown in Figure 17-62. This *pressure drop* results from the friction of the water flowing in the pipe. Different pipe materials have different friction losses because of the interior finishes of the pipes. For example, the inside of copper pipe is much smoother than that of wrought-iron pipe, so water flows through it more easily with less pressure drop.

Pressure drop in a system depends on the amount of water that is flowing in the system. For example, it is easy to reason that when 20 gpm are pumped through a 2-inch pipe, the water will not move as fast as if 40 gpm were pumped through the same pipe. As more water is pumped through the same pipe, the velocity—and the pressure drop—increase. This pressure drop is expressed in several terms in the industry: feet of head and pounds per square inch drop, which is the difference between the pressure at the pump inlet and the pressure at the pump outlet, as shown in Figure 17-63. The industry typically

TABLE 3—PHYSICAL PROPERTIES OF COPPER TUBING

CLASSIFICATION	NOM. TUBE SIZE (in.)	OUTSIDE DIAM (in.)	STUBBS GAGE	WALL THICK-NESS (in.)	INSIDE DIAM (in.)	TRANS-VERSE AREA (sq in.)	MINIMUM TEST PRESSURE (psi)	WEIGHT OF TUBE (lb/ft)	WT OF WATER IN TUBE* (lb/ft)	OUTSIDE SURFACE (sq ft/ft)
HARD	¼	⅜	23	.025	.325	.083	1000	.106	.036	.098
	⅜	½	23	.025	.450	.159	1000	.144	.069	.131
	½	⅝	22	.028	.569	.254	890	.203	.110	.164
	¾	⅞	21	.032	.811	.516	710	.328	.224	.229
	1	1⅛	20	.035	1.055	.874	600	.464	.379	.295
	1¼	1⅜	19	.042	1.291	1.309	590	.681	.566	.360
Govt. Type "M" 250 Lb Working Pressure	1½	1⅝	18	.049	1.527	1.831	580	.94	.793	.425
	2	2⅛	17	.058	2.009	3.17	520	1.46	1.372	.556
	2½	2⅝	16	.065	2.495	4.89	470	2.03	2.120	.687
	3	3⅛	15	.072	2.981	6.98	440	2.68	3.020	.818
	3½	3⅝	14	.083	3.459	9.40	430	3.58	4.060	.949
	4	4⅛	13	.095	3.935	12.16	430	4.66	5.262	1.08
	5	5⅛	12	.109	4.907	18.91	400	6.66	8.180	1.34
	6	6⅛		.122	5.881	27.16	375	8.91	11.750	1.60
	8	8⅛		.170	7.785	47.6	375	16.46	20.60	2.13
HARD	⅜	½	19	.035	.430	.146	1000	.198	.063	.131
	½	⅝		.040	.545	.233	1000	.284	.101	.164
	¾	⅞		.045	.785	.484	1000	.454	.209	.229
	1	1⅛		.050	1.025	.825	880	.653	.358	.295
	1¼	1⅜		.055	1.265	1.256	780	.882	.554	.360
Govt. Type "L" 250 Lb Working Pressure	1½	1⅝		.060	1.505	1.78	720	1.14	.770	.425
	2	2⅛		.070	1.985	3.094	640	1.75	1.338	.556
	2½	2⅝		.080	2.465	4.77	580	2.48	2.070	.687
	3	3⅛		.090	2.945	6.812	550	3.33	2.975	.818
	3½	3⅝		.100	3.425	9.213	530	4.29	4.000	.949
	4	4⅛		.110	3.905	11.97	510	5.38	5.180	1.08
	5	5⅛		.125	4.875	18.67	460	7.61	8.090	1.34
	6	6⅛		.140	5.845	26.83	430	10.20	11.610	1.60
HARD	¼	⅜	21	.032	.311	.076	1000	.133	.033	.098
	⅜	½	18	.049	.402	.127	1000	.269	.055	.131
	½	⅝	18	.049	.527	.218	1000	.344	.094	.164
	¾	⅞	16	.065	.745	.436	1000	.641	.189	.229
	1	1⅛	16	.065	.995	.778	780	.839	.336	.295
	1¼	1⅜	16	.065	1.245	1.217	630	1.04	.526	.360
Govt. Type "K" 400 Lb Working Pressure	1½	1⅝	15	.072	1.481	1.722	580	1.36	.745	.425
	2	2⅛	14	.083	1.959	3.014	510	2.06	1.300	.556
	2½	2⅝	13	.095	2.435	4.656	470	2.92	2.015	.687
	3	3⅛	12	.109	2.907	6.637	450	4.00	2.870	8.18
	3½	3⅝	11	.120	3.385	8.999	430	5.12	3.890	.949
	4	4⅛	10	.134	3.857	11.68	420	6.51	5.05	1.08
	5	5⅛		.160	4.805	18.13	400	9.67	7.80	1.34
	6	6⅛		.192	5.741	25.88	400	13.87	11.20	1.60
SOFT	¼	⅜	21	.032	.311	.076	1000	.133	.033	.098
	⅜	½	18	.049	.402	.127	1000	.269	.055	.131
	½	⅝	18	.049	.527	.218	1000	.344	.094	.164
	¾	⅞	16	.065	.745	.436	1000	.641	.189	.229
	1	1⅛	16	.065	.995	.778	780	.839	.336	.295
	1¼	1⅜	16	.065	1.245	1.217	630	1.04	.526	.360
Govt. Type "K" 250 Lb Working Pressure	1½	1⅝	15	.072	1.481	1.722	580	1.36	.745	.425
	2	2⅛	14	.083	1.959	3.014	510	2.06	1.300	.556
	2½	2⅝	13	.095	2.435	4.656	470	2.92	2.015	.687
	3	3⅛	12	.109	2.907	6.637	450	4.00	2.870	.818
	3½	3⅝	11	.120	3.385	8.999	430	5.12	3.890	.949
	4	4⅛	10	.134	3.857	11.68	420	6.51	5.05	1.08
	5	5⅛		.160	4.805	18.13	400	9.67	7.80	1.34
	6	6⅛		.192	5.741	25.88	400	13.87	11.2	1.60

*To change "Wt of Water in Tube (lb/ft)" to "Gallons of Water in Tube (gal/ft)," divide values in table by 8.34.

Figure 17-61. Properties of copper pipe for types K, L, and M. *Courtesy Carrier Corporation*

describes pressure drop in feet of water. (Remember that 1 psig is equal to 2.31 feet, or 27.7 inches, of water.) Feet of water is commonly called feet of head, or friction loss, when the pressure drop in a pipe is referred to.

Four important terms are used when water flow in pipes is discussed:

1. flow rate in gallons per minute (gpm)
2. pipe size in inches

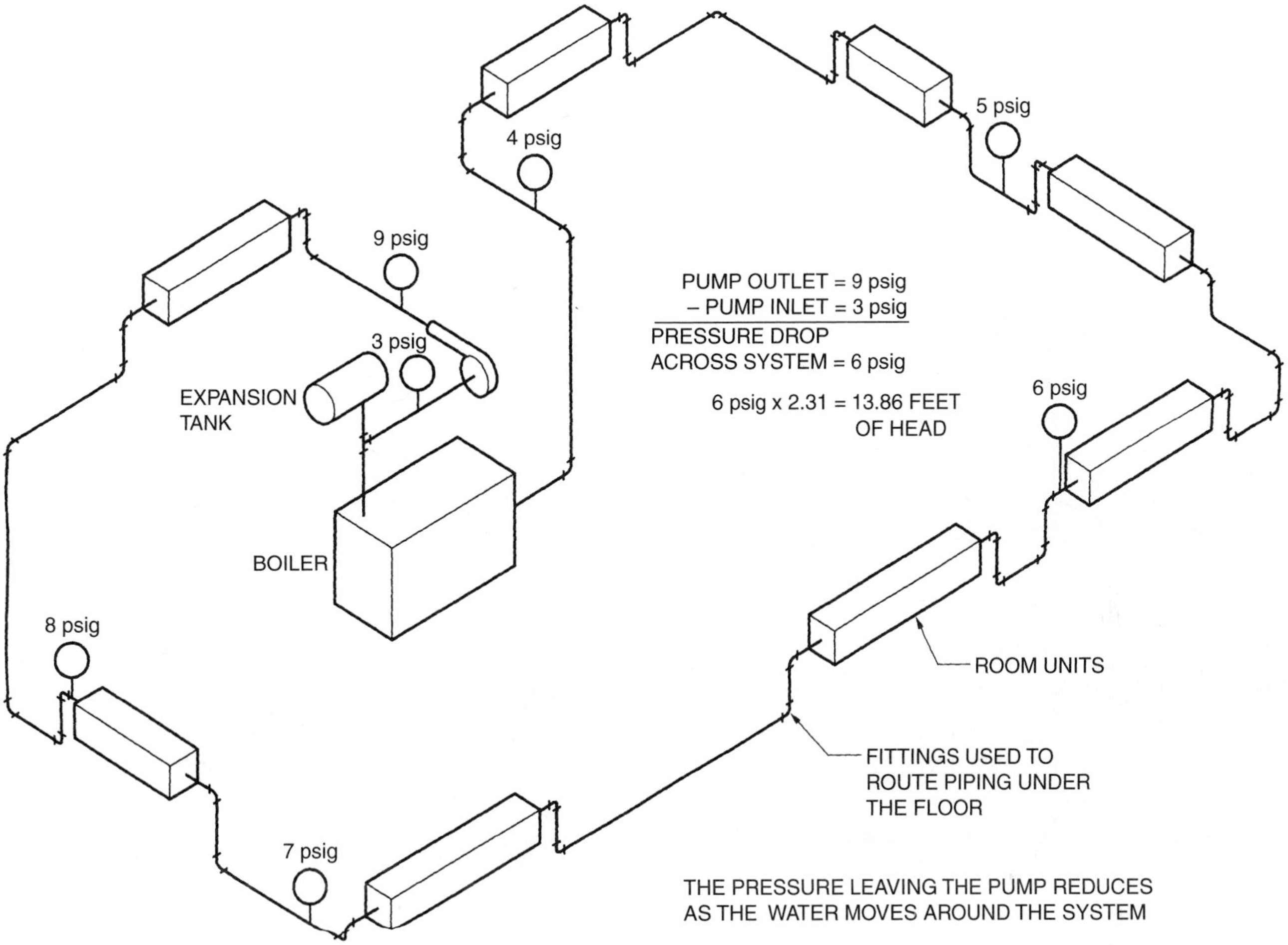

Figure 17-62. Pressure drop in system piping.

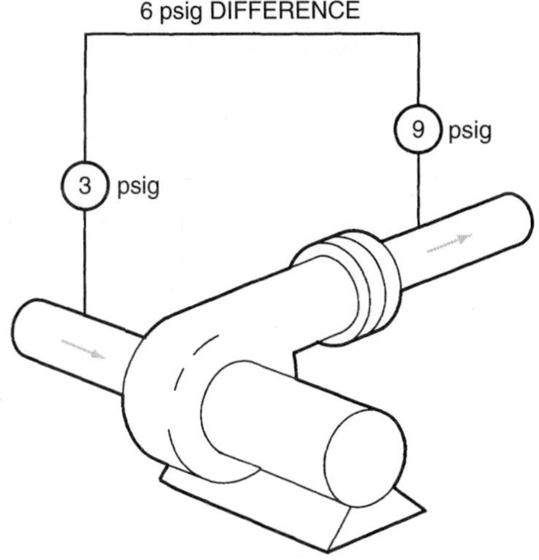

Figure 17-63. Pressure difference across a pump.

3. friction loss expressed in feet of loss per 100 feet of pipe run
4. water velocity in feet per second (fps)

If you know any two values, the other two can be found from a pipe sizing chart, like the one shown in Figure 17-64. For example, if a 2-inch wrought-iron pipe has 20 gallons of water flowing through it, the velocity of the water is 1.9 fps and the friction loss is 0.87 foot of water per 100 feet of pipe. As mentioned previously, when 40 gpm are moved through the same 2-inch pipe, the numbers change to a velocity of 3.9 fps with a friction loss of 3.1 feet of water per 100 feet of pipe, Figure 17-65.

When a designer sizes pipe, the gpm of water are known, and the maximum velocity is decided by the designer from experience and engineering manuals such as the ASHRAE (American Society of Heating, Refrigerating and Air-Conditioning Engineers) Fundamentals guide. This guide recommends that hot-water systems be designed with velocities between 1 and 4 fps, with 2.5 being typical. The faster the water moves through the pipe, the more noise it makes. Also, as the pressure drop increases, the pumping costs rise. The designer must determine the best velocity for the application. For example, suppose that the designer wants to circulate 60 gpm, and the pipe is schedule 40 wrought iron for general service, such as

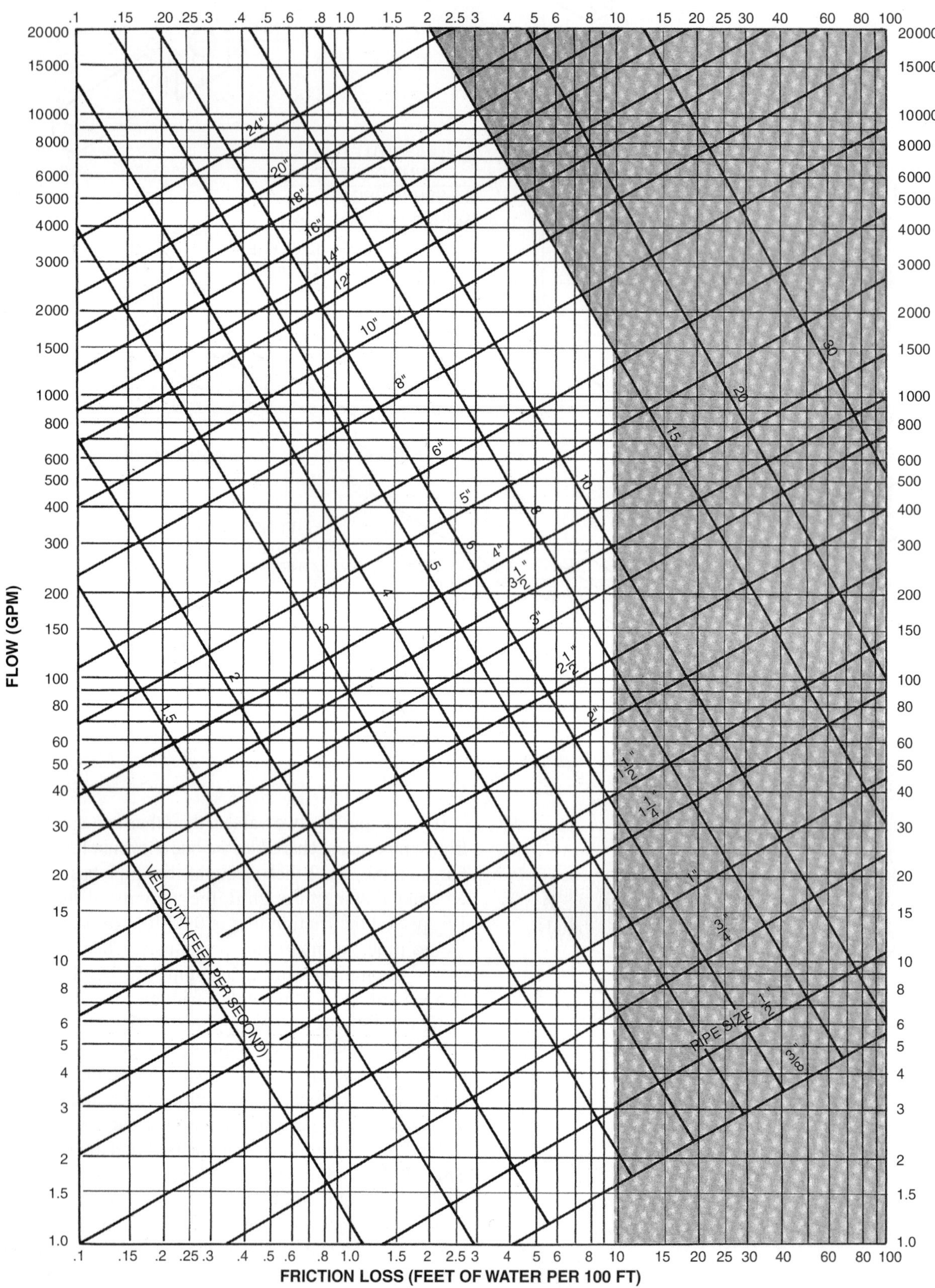

Figure 17-64. Pipe sizing chart. *Courtesy Carrier Corporation*

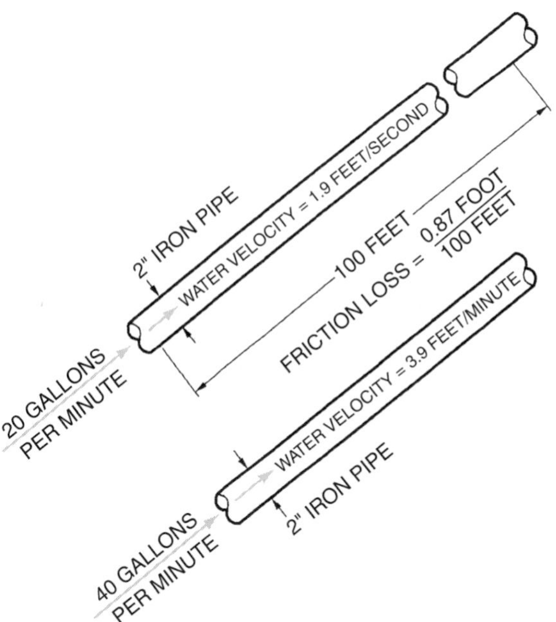

Figure 17-65. Sample piping problem.

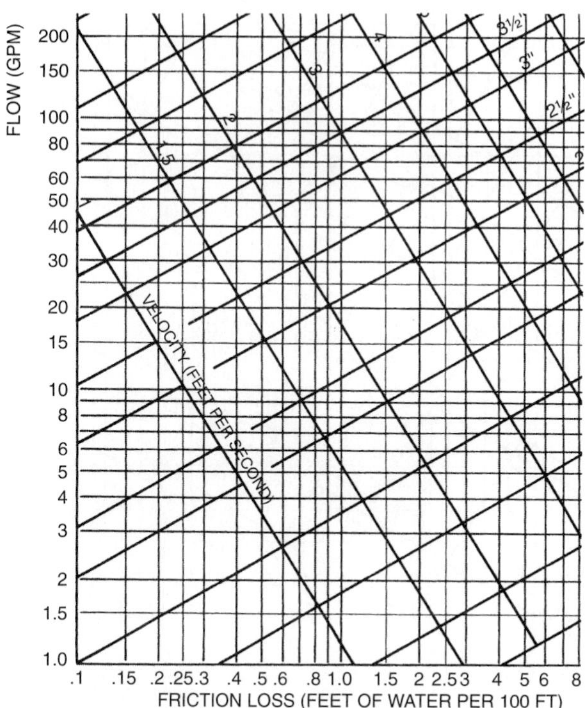

Figure 17-66. Three-inch pipe sizing example. *Courtesy Carrier Corporation*

the main serving a building. Noise is to be kept low, so a velocity of 3 fps is chosen. Referring to the pipe sizing chart, we find that the pipe size is a little less than 3 inches. The designer would round up to 3 inches and use this size; the other numbers would change to those for a 3-inch pipe, Figure 17-66. The friction loss would be approximately 0.95 foot of water for each 100 feet of pipe.

If the pipe run were only 50 feet, the loss would be only 0.475 foot. If the run were 150 feet, the loss would be 1.425 feet, as shown in Figure 17-67. The formula shown in this figure can be used to find the friction loss per 100 feet for any length of pipe.

For example, suppose that 10 gpm are flowing through a 1¼-inch wrought-iron pipe, and we want to know the friction loss for 135 feet of pipe. As shown in Figure 17-68, the friction loss is calculated as follows:

$$\begin{aligned}\text{Friction loss} &= (\text{Loss per 100 feet} \times \text{Length})/100 \\ &= (1.75 \times 135)/100 \\ &= 2.36 \text{ feet}\end{aligned}$$

Note: The 1.75 was found on the wrought-iron pipe chart, as shown in Figure 17-64 by first finding the function of 10 gpm and 1¼-inch pipe, then looking at the bottom of the chart. Also notice that the velocity of the water in this pipe is approximately 2.25 feet per second.

This example is a simplified version of how a designer may size the pipe in a wrought-iron pipe system. Figure 17-69 shows the pipe sizing chart for copper.

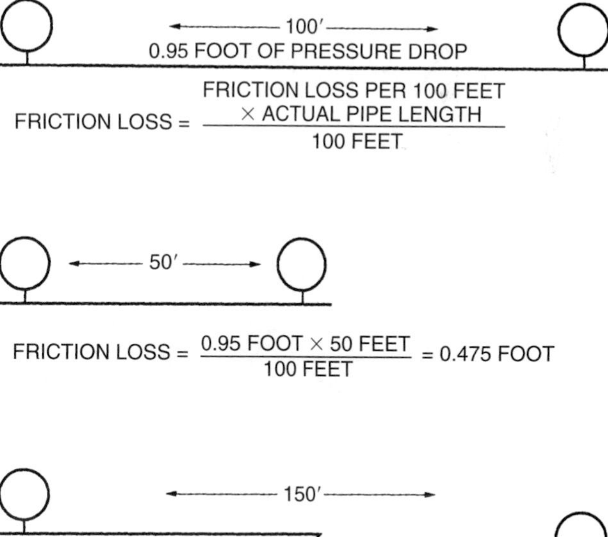

Figure 17-67. Different pipe lengths and pressure drops.

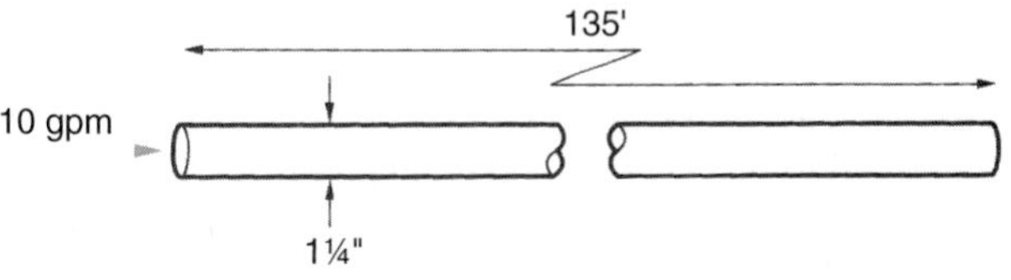

Figure 17-68. Piping example.

CHART 5—FRICTION LOSS FOR CLOSED AND OPEN PIPING SYSTEMS

Copper Tubing

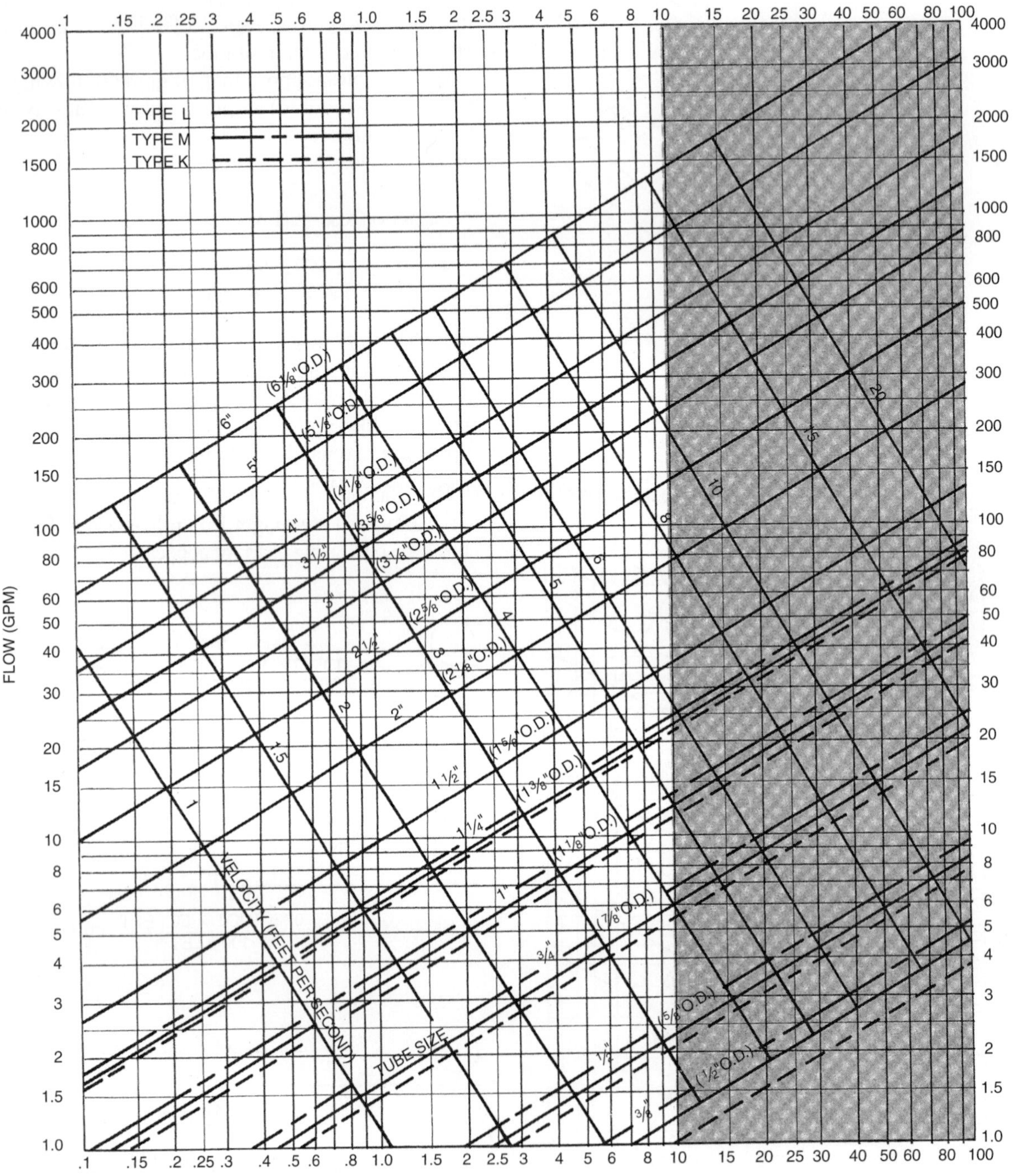

Figure 17-69. Pipe sizing chart for copper. *Courtesy Carrier Corporation*

The pipe fittings, boiler, and convectors also create a pressure drop in the water flow. Fittings cause more of a pressure drop than straight pipe because of the turbulence in the fittings. As water turns in an elbow, a large amount of turbulence is created, as shown in Figure 17-70. This turbulence causes a pressure drop, which must be accounted for in the pipe sizing procedures. A T fitting has one of three pressure-drop ratings, depending on the flow pattern through the T. The water may flow straight through (called a *running T*), through the branch outlet, or into the branch fitting (called a *bull-head T*). Figure 17-71 shows these three tees.

Pipe-fitting pressure drop is described in terms of equivalent feet of pipe. For example, a fitting that has an equivalent feet of pipe pressure drop of 3.3 feet has the same pressure drop as 3.3 feet of pipe, see Figure 17-72.

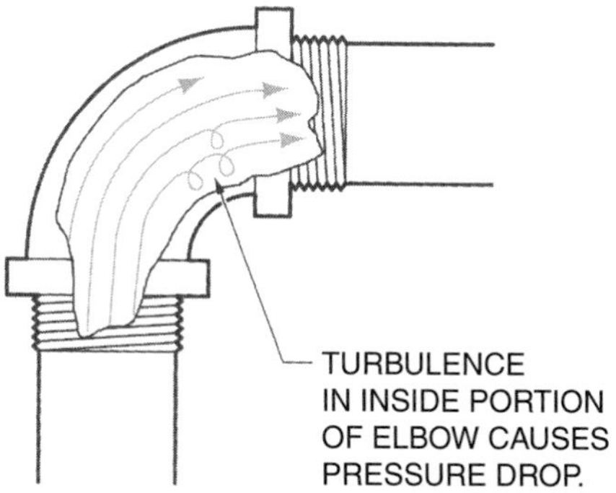

TURBULENCE
IN INSIDE PORTION
OF ELBOW CAUSES
PRESSURE DROP.

Figure 17-70. Pressure drop in an elbow.

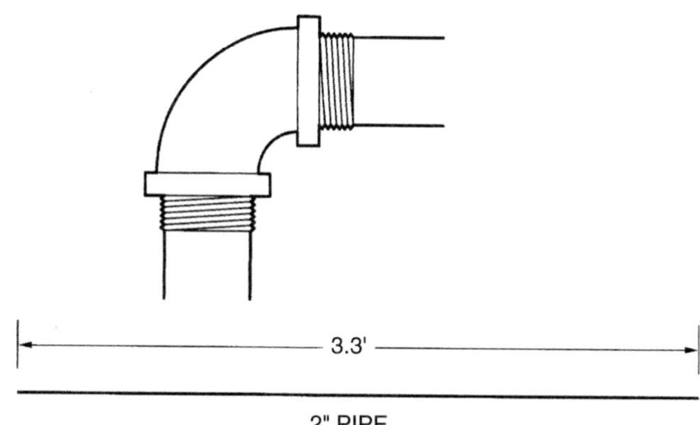

Figure 17-72. A 2-inch elbow has the same pressure drop as 3.3 feet of pipe.

As the water flow through fittings increases, the pressure drop increases. Most designers use a pressure-drop table, such as those shown in Figure 17-73, for fittings. (Note that these tables cover valves *and* fittings. Valve construction is discussed in detail later in this unit, in Section 17.6.)

When a designer designs a system, several pipe sizes may be considered before the correct size is chosen. By considering several sizes, the designer chooses the most economical pipe size. If three pipe sizes will carry the water, the smaller of the three will probably have too much of a pressure drop. The larger will likely be oversized and not economical. The size in the middle will probably be the best choice.

Designing usually starts with a trial pipe size. The trial size is found by using the system gpm and the desired velocity and locating these coordinates on the pipe sizing chart. The system is designed around the trial pipe size, then it is checked to ensure that the pressure drops work out correctly. Many designers have found that if they add 50% to the pipe length—to compensate for the fittings, valves, and accessories—they arrive at the approximate pressure drop for the system. For example, suppose that the total length of a system is 175 feet, including the pressure drop in the heat exchanger. The designer can use a 263-foot pressure drop for the trial pipe size (175 × 1.5 = 263 feet). NOTE: 88 feet has been added for estimated fittings. By referring to the pipe sizing chart with the gallons of water per minute, the designer can choose the pipe size and then calculate the total system pressure drop before choosing the correct pump.

Suppose that the design water flow is 25 gpm and the desired velocity is 3 fps for a wrought-iron system. According to the pipe sizing chart, the size for the system is between 1½ and 2 inches, Figure 17-74A. If the designer rounds down to 1½-inch pipe, the velocity will be approximately 3.9 fps. This number is a little high if the

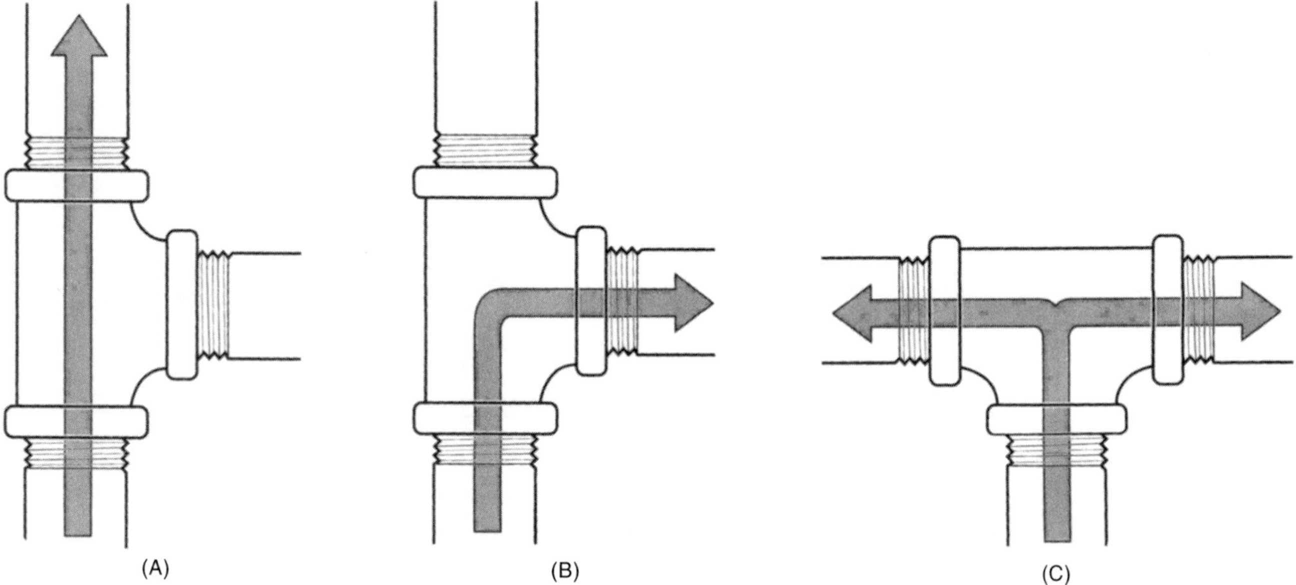

Figure 17-71. Three directions in which water can flow through a tee connection. (A) Straight, or running T, flow. (B) Side outlet, or branch, flow. (C) Into branch fitting, or bullhead T, flow.

FITTING LOSSES IN EQUIVALENT FEET OF PIPE

Screwed, Welded, Flanged, Flared, and Brazed Connections

NOMINAL PIPE OR TUBE SIZE (in.)	SMOOTH BEND ELBOWS						SMOOTH BEND TEES			
	90° Std*	90° Long Rad.†	90° Street*	45° Std*	45° Street*	180° Std*	Flow-Thru Branch	Straight-Thru Flow		
								No Reduction	Reduced ¼	Reduced ½
⅜	1.4	0.9	2.3	0.7	1.1	2.3	2.7	0.9	1.2	1.4
½	1.6	1.0	2.5	0.8	1.3	2.5	3.0	1.0	1.4	1.6
¾	2.0	1.4	3.2	0.9	1.6	3.2	4.0	1.4	1.9	2.0
1	2.6	1.7	4.1	1.3	2.1	4.1	5.0	1.7	2.3	2.6
1¼	3.3	2.3	5.6	1.7	3.0	5.6	7.0	2.3	3.1	3.3
1½	4.0	2.6	6.3	2.1	3.4	6.3	8.0	2.6	3.7	4.0
2	5.0	3.3	8.2	2.6	4.5	8.2	10	3.3	4.7	5.0
2½	6.0	4.1	10	3.2	5.2	10	12	4.1	5.6	6.0
3	7.5	5.0	12	4.0	6.4	12	15	5.0	7.0	7.5
3½	9.0	5.9	15	4.7	7.3	15	18	5.9	8.0	9.0
4	10	6.7	17	5.2	8.5	17	21	6.7	9.0	10
5	13	8.2	21	6.5	11	21	25	8.2	12	13
6	16	10	25	7.9	13	25	30	10	14	16
8	20	13	—	10	—	33	40	13	18	20
10	25	16	—	13	—	42	50	16	23	25
12	30	19	—	16	—	50	60	19	26	30
14	34	23	—	18	—	55	68	23	30	34
16	38	26	—	20	—	62	78	26	35	38
18	42	29	—	23	—	70	85	29	40	42
20	50	33	—	26	—	81	100	33	44	50
24	60	40	—	30	—	94	115	40	50	60

(A)

VALVE LOSSES IN EQUIVALENT FEET OF PIPE*

Screwed, Welded, Flanged, and Flared Connections

NOMINAL PIPE OR TUBE SIZE (in.)	GLOBE †	60° - Y	45° - Y	ANGLE †	GATE ††	SWING CHECK ‡	LIFT CHECK
⅜	17	8	6	6	0.6	5	Globe & Vertical Lift Same as Globe Valve**
½	18	9	7	7	0.7	6	
¾	22	11	9	9	0.9	8	
1	29	15	12	12	1.0	10	
1¼	38	20	15	15	1.5	14	
1½	43	24	18	18	1.8	16	
2	55	30	24	24	2.3	20	
2½	69	35	29	29	2.8	25	
3	84	43	35	35	3.2	30	
3½	100	50	41	41	4.0	35	
4	120	58	47	47	4.5	40	
5	140	71	58	58	6	50	
6	170	88	70	70	7	60	
8	220	115	85	85	9	80	
10	280	145	105	105	12	100	
12	320	165	130	130	13	120	Angle Lift Same as Angle Valve
14	360	185	155	155	15	135	
16	410	210	180	180	17	150	
18	460	240	200	200	19	165	
20	520	275	235	235	22	200	
24	610	320	265	265	25	240	

45° OR 60°

*Losses are for all valves in fully open position.

†These losses do not apply to valves with needle point type seats.

‡Losses also apply to the in-line, ball type check valve.

**For "Y" pattern globe lift check valve with seat approximately equal to the nominal pipe diameter, use values of 60° "Y" valve for loss.

††Regular and short pattern plug cock valves, when fully open, have same loss as gate valve. For valve losses of short pattern plug cocks above 6 ins. check manufacturer.

(B)

Figure 17-73. (A) Pressure-drop table for fittings. (B) Pressure-drop table for valves. *Courtesy Carrier Corporation*

design goal is 3 fps. If the size is rounded up to 2 inches, the velocity will be 2.5 fps. Either is a good choice, but a velocity of 2.5 fps is closer to the design goal of 3 fps. If 2½-inch pipe would be chosen, the velocity would be only 0.5 fps. This pipe size would probably be too large to be economical.

The total pressure drop in the system can be found by using the 25 gpm and 2-inch pipe and looking at the bottom of the table for the pressure drop for 100 feet of pipe. This drop is 1.3 feet of water per 100 feet of pipe. Consequently, the pressure drop for 263 feet is 3.4 feet: (263 × 1.3)/100 = 3.4 feet.

Next, the designer can determine the actual feet of head by using the fitting and valve tables from Figure 17-73. For example, refer to Figure 17-74B and suppose that the system has twenty 2-inch long radius elbows, two 45-degree angle globe valves, and a Y strainer. Adding the values from the tables, we can see that the valves and accessories have an actual pressure drop of 128 feet:

20 × 3.3 = 66 for the 2-inch radius elbows
2 × 24 = 48 for the 2-inch globe valves
Total = 114 feet

The designer originally used 175 feet plus 88 feet estimated for fittings, for a total of 263 feet. The 50% figure for estimating the fittings (88) is close to the actual figure (114) ft. The designer can now recalculate the pressure drop to the actual feet of head: 175 + 114 = 289 feet. The system pressure drop is 3.8 feet: (303 × 1.3)/100 = 3.8 feet. The designer can next choose a pump that can pump 25 gpm against 3.8 feet of head (friction loss).

A residential system with convectors and a series loop is sized like this: Suppose that a system circulates 4 gpm of water and uses 125 feet of pipe, including the convectors, which are the same size as the pipe. The system has five convectors that each require four copper radius elbows, for a total of 20 elbows. The system also has two globe valves, as shown in Figures 17-75 and 17-76.

For the trial pipe size, we use 188 equivalent feet (125 × 1.5 = 187.5, rounded to 188). According to Figure 17-77, the pipe size for Type L tubing at 4 gpm and a velocity of 2.8 fps is ¾-inch ID (⅞-inch OD). If we go down in pipe size to: ½-inch ID (⅝-inch OD), the velocity is out of range at 5.75 fps. If we go up in pipe size to 1-inch ID (1⅛-inch OD), the velocity drops to approximately 1.5 fps, which is too low to be economical.

Now, let us calculate the actual feet of head for the system:

125 feet of pipe = 125
28 radius elbows × 1.4 = 39
2 globe valves × 22 = 22
Total = 155 feet

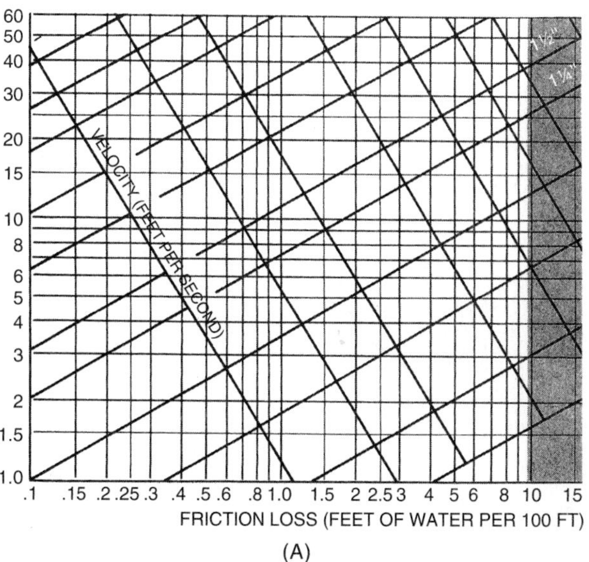

(A)

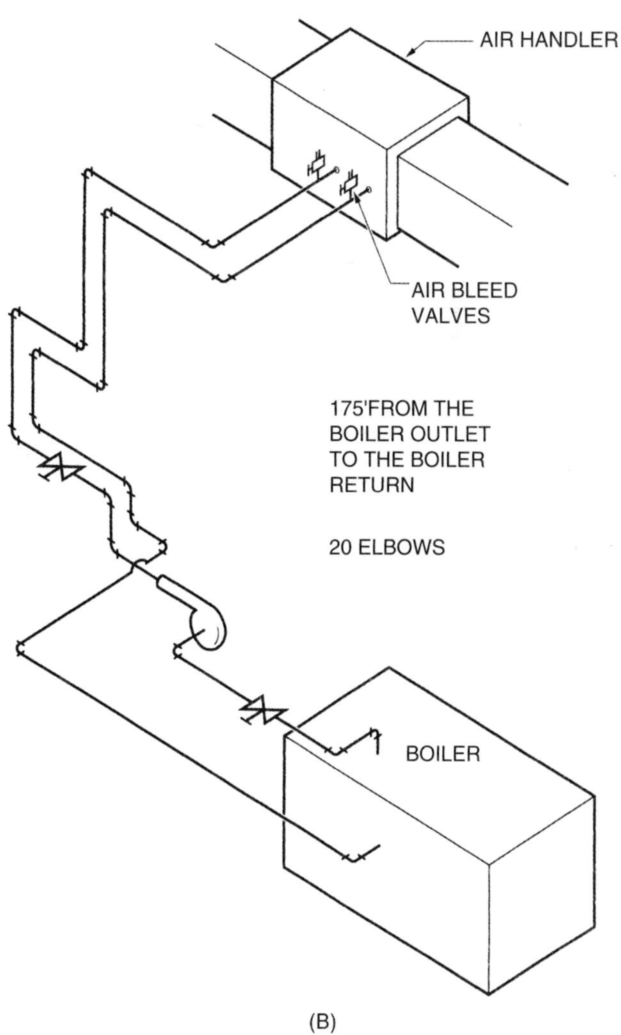

(B)

Figure 17-74. (A) Graph for pipe sizing example. (B) Diagram for pipe sizing example. *Courtesty Carrier Corporation*

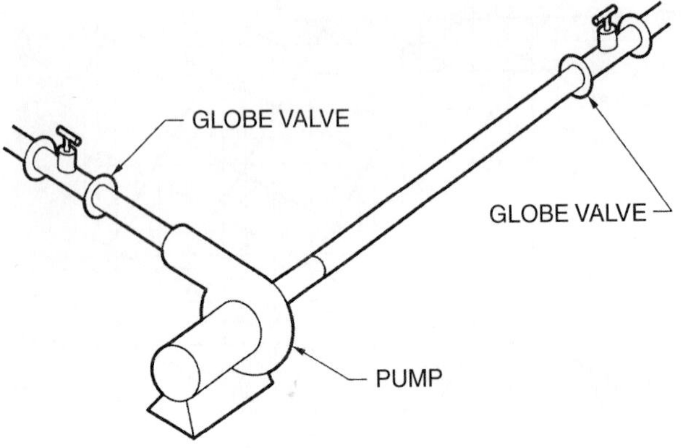

Figure 17-75. Residential system with two global valves.

This amount is close to the trial pipe size length of 188, so the pressure drop for the system can be found by referring to the piping chart at ¾-inch copper pipe and 4 gpm and looking down to the bottom of Figure 17-77 to find 5.2 feet of water per 100 feet. The total pressure drop is 8.1 feet: $(5.2 \times 155)/100 = 8.1$ feet. The designer would choose a pump that could pump 4 gpm with a head of 8.1 feet.

Many concepts are involved in designing hot-water system piping. The method just discussed can be used to check the pipe size for a typical system in the field or to design a system. The *IBR 250 Guide* discusses another method for designing residential systems that includes a data sheet on which you can keep track of all the data for a complex residence. The IBR method is highly recommended for anyone who designs systems on a regular basis.

17.6 Valves and Accessories

Each system has an assortment of valves and accessories to control the system flow and to isolate the system for service. The first two parts of this discussion involve the types of hand valves and control valves used to isolate or vary the flow of water for balancing purposes. The last part discusses accessories.

Hand Valves

Five types of hand valves are used in typical systems: plug cock, globe, gate, ball, and butterfly. Most of these valves are made of cast brass to prevent corrosion. If valves and accessories are made of cast iron or steel, they rust and become hard to function properly. Therefore, most valves and accessories are brass.

Each valve is discussed in detail because the technician must be familiar with all types of valves. Valves cause pressure drops in systems, just as pipe lengths do. The pressure drop in a valve results from the turbulence of the water moving through the valve. The more streamlined the valve, the less the pressure drop. Because piping

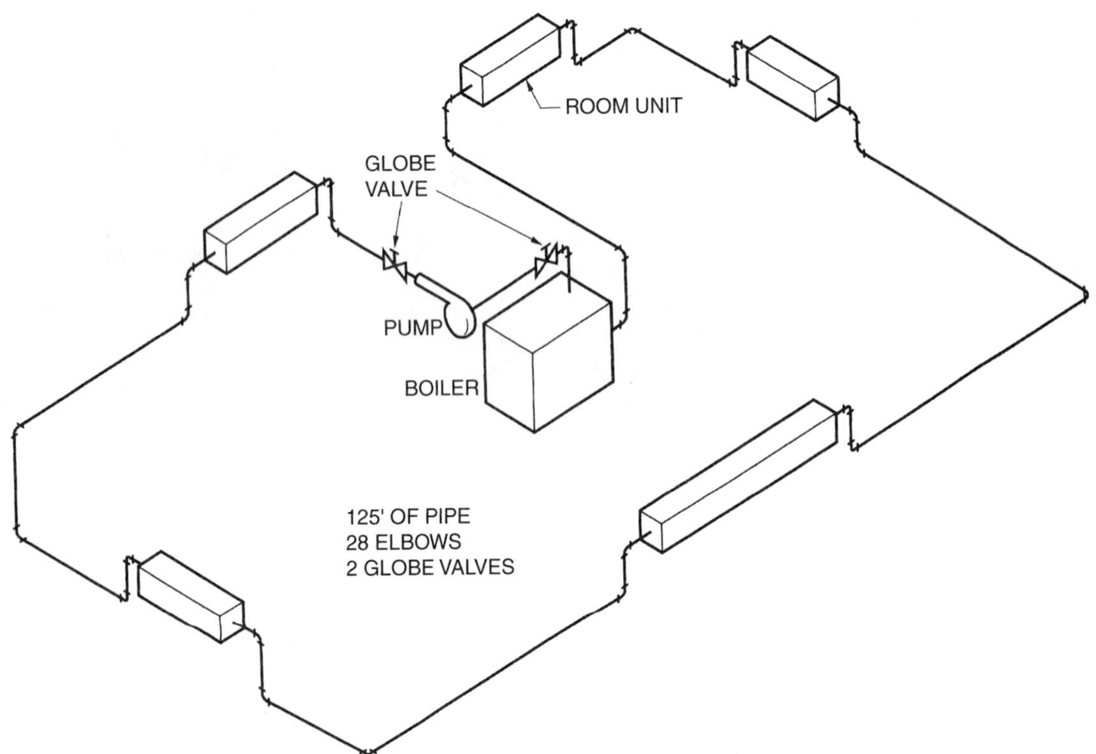

Figure 17-76. Determining the trial pipe size.

CHART 5—FRICTION LOSS FOR CLOSED AND OPEN PIPING SYSTEMS

Copper Tubing

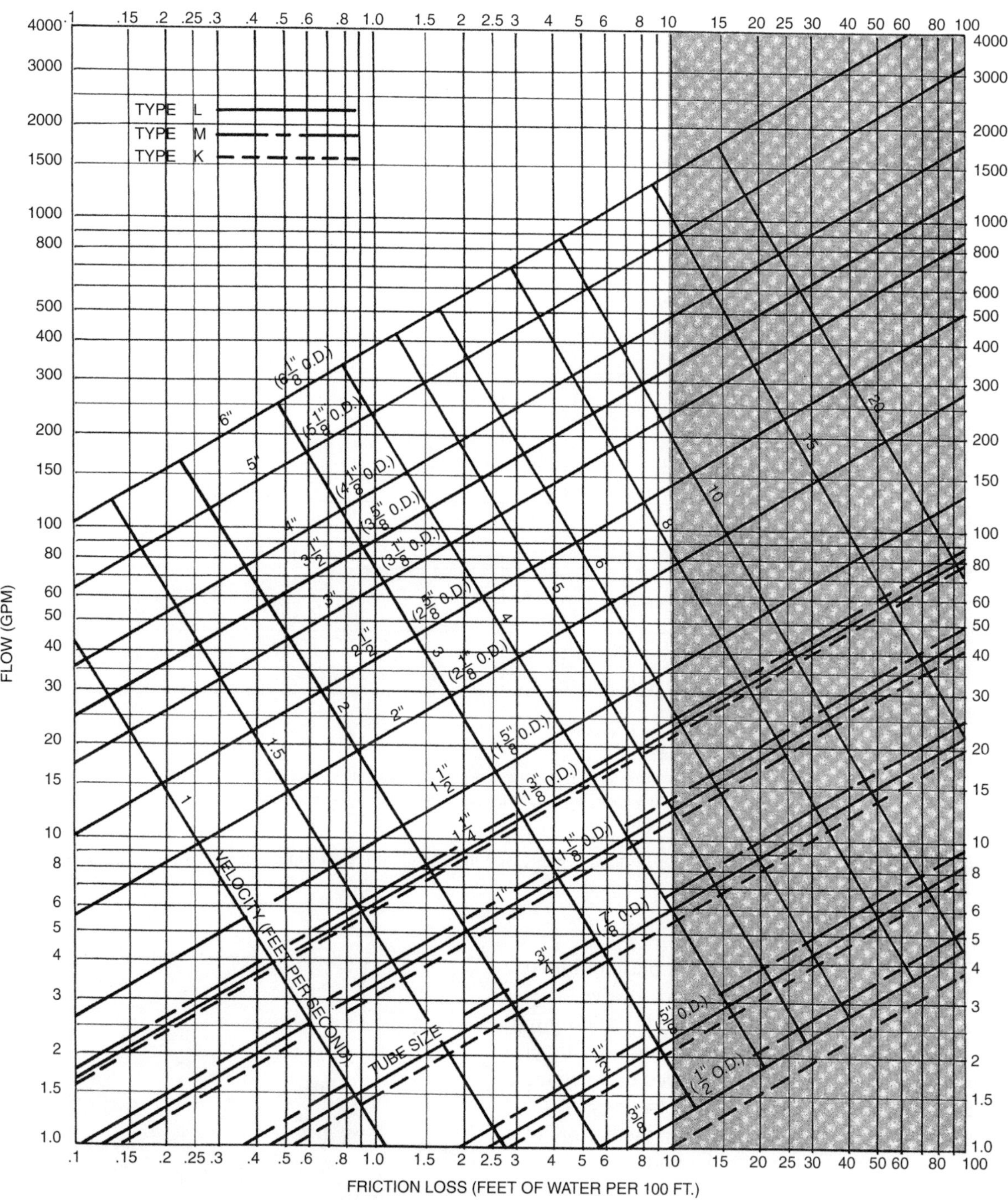

Figure 17-77. Three-quarter-inch ID copper pipe with a flow of 4 gpm has a pressure drop of 5.2 feet per 100 feet. *Courtesy Carrier Corporation*

pressure drop is given in feet of head, it is convenient to show the pressure drops for valves in a comparable term, equivalent length of pipe. Figure 17-78 shows equivalent feet of pipe for various valves. For example, a 1-inch globe valve is the same as 29 feet of pipe. Equivalent length of pipe conveniently allows you to calculate the pressure drop for the complete system. (Remember that feet of head can also be converted to psig.)

VALVE LOSSES IN EQUIVALENT FEET OF PIPE*

Screwed, Welded, Flanged, and Flared Connections

NOMINAL PIPE OR TUBE SIZE (in.)	GLOBE†	60° - Y	45° - Y	ANGLE†	GATE††	SWING CHECK ‡	LIFT CHECK
		45° OR 60°					
⅜	17	8	6	6	0.6	5	Globe & Vertical Lift Same as Globe Valve**
½	18	9	7	7	0.7	6	
¾	22	11	9	9	0.9	8	
1	29	15	12	12	1.0	10	
1¼	38	20	15	15	1.5	14	
1½	43	24	18	18	1.8	16	
2	55	30	24	24	2.3	20	
2½	69	35	29	29	2.8	25	
3	84	43	35	35	3.2	30	
3½	100	50	41	41	4.0	35	
4	120	58	47	47	4.5	40	
5	140	71	58	58	6	50	
6	170	88	70	70	7	60	
8	220	115	85	85	9	80	
10	280	145	105	105	12	100	Angle Lift Same as Angle Valve
12	320	165	130	130	13	120	
14	360	185	155	155	15	135	
16	410	210	180	180	17	150	
18	460	240	200	200	19	165	
20	520	275	235	235	22	200	
24	610	320	265	265	25	240	

*Losses are for all valves in fully open position.

†These losses do not apply to valves with needle point type seats.

‡Losses also apply to the in-line, ball type check valve.

**For "Y" pattern globe lift check valve with seat approximately equal to the nominal pipe diameter, use values of 60° "Y" valve for loss.

††Regular and short pattern plug cock valves, when fully open, have same loss as gate valve. For valve losses of short pattern plug cocks above 6 ins. check manufacturer.

Figure 17-78. The pressure drop for valves (as for fittings) is expressed in equivalent feet of pipe. *Courtesy Carrier Corporation*

PLUG COCK. The plug cock is the simplest and least expensive valve. This valve is constructed of a valve body, the plug, and the plug fastener, which holds the plug in the valve body. The plug is tapered to match the taper in the valve body. A special lubricant is used between the plug and the body that helps seal liquids (or vapors) inside the valve. As it is tightened, the nut on the bottom of the valve pulls the plug tighter into the body. This valve is often used to balance the flow in a circuit by partially shutting off the valve and securing it by tightening the nut on the bottom. Only one-quarter of a turn is required to open or close the valve because of the way the plug is made. In valve sizes of ⅜ inch and larger, the plug cock does not usually have a handle and must be operated with a wrench, which is placed on the square shoulder on top of the valve, as shown in Figure 17-79. Typically, the plug can be rotated 360 degrees, and the valve will open twice and close twice during the rotation. For this reason, a mark

Figure 17-79. Square-shoulder plug cock is operated by turning it with a wrench. *Photo by Bill Johnson*

on the top of the plug indicates whether the valve is open or closed. If the mark is crossways of the valve body, the valve is closed. If the mark is running with the valve, the valve is open, as shown in Figure 17-80.

Smaller versions of the plug cock can be found in the air bleed portions of many systems. These valves are typically ¼ inch in pipe size and often have a wing-type top that can be hand operated.

Plug cock valves are not designed to be opened and closed every day; they should be used in applications that require only intermittent use. The plug cock valve has very little pressure drop and is not listed in the pressure-drop table. It is much like the gate valve in pressure drop.

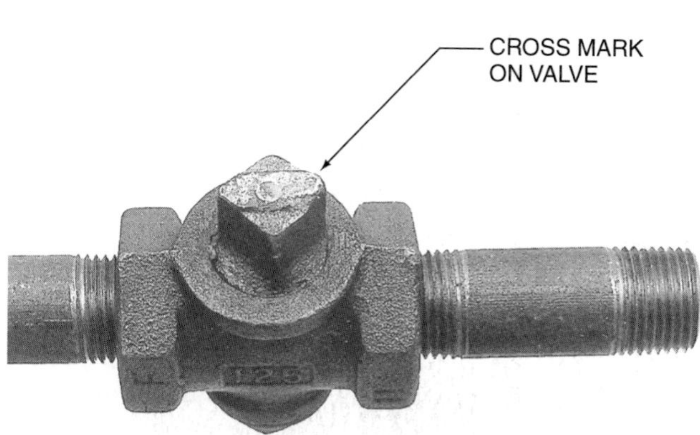

Figure 17-80. The mark on the top of the plug cock designates the position of the valve. *Photo by Bill Johnson*

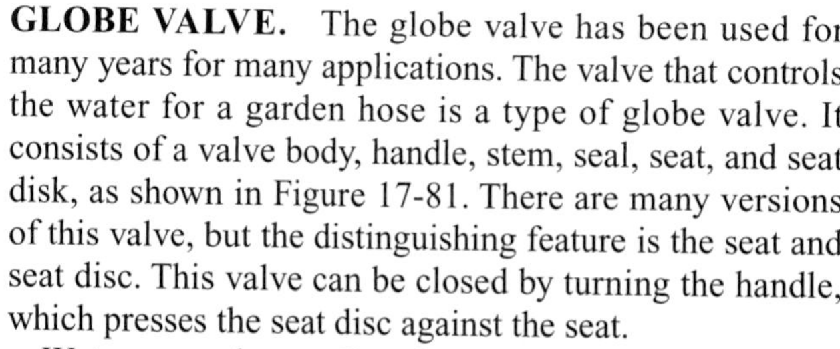

GLOBE VALVE. The globe valve has been used for many years for many applications. The valve that controls the water for a garden hose is a type of globe valve. It consists of a valve body, handle, stem, seal, seat, and seat disk, as shown in Figure 17-81. There are many versions of this valve, but the distinguishing feature is the seat and seat disc. This valve can be closed by turning the handle, which presses the seat disc against the seat.

Water must change direction in order to move through the globe valve, Figure 17-82. This change in direction causes a pressure drop.

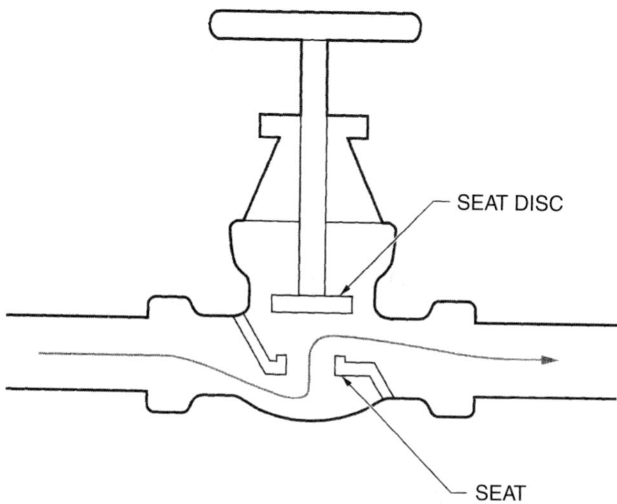

Figure 17-82. Water flow through a globe valve.

(A)

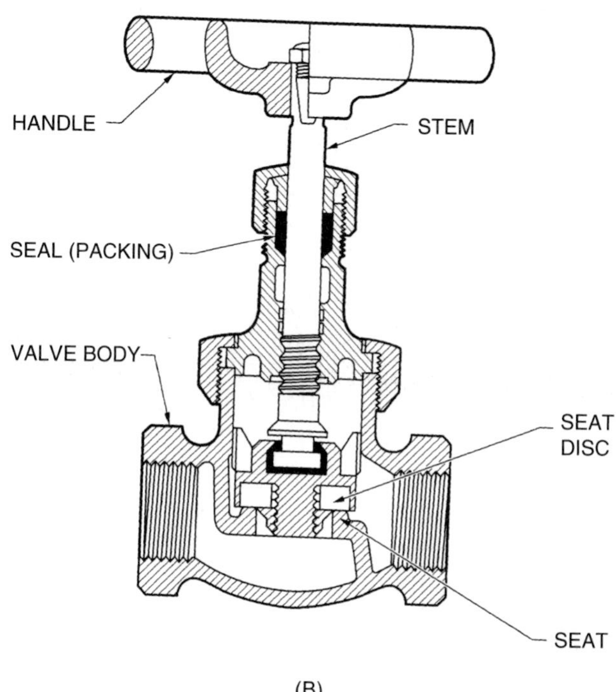

(B)

Figure 17-81. (A) Globe valve. *Photo by Bill Johnson* (B) Cross section. *Courtesy Nibco*

The globe valve is manufactured either straight through or on an angle. The typical home water faucet (called a *hose bib*) that controls the water to a hose is an angle or a straight-through globe valve. The angle globe valve has approximately one-half the pressure drop of the straight-through globe valve, so it is popular for some applications.

Globe valves are directional in their piping arrangement. They must be located with the pressure to be contained on the bottom of the seat, as shown in Figure 17-83. If installed backward, they "chatter" when they are closed and there is flow. The chatter is a result of valve disc vibration while water passes through the valve.

As with all valves, containing the water in the valve while the valve stem is turned is important. Most globe valves have packing-gland seals. The packing-gland seal tightens a packing around the valve stem shaft when the stem packing nut is turned, as shown in Figure 17-84. This type of seal allows the technician to rebuild the packing as long as the valve is seated and the pressure is relieved on the outlet side of the valve.

The packing used to seal the valve is a form of rope or Teflon. Graphite rope is used for high-temperature applications, waxed rope is used for low-temperature applications, and Teflon packing is also used. As the packing nut is tightened, it forces the packing to tighten around the stem. This type of leak prevention is adequate for valves that are not opened and closed regularly. Each time the valve stem moves through the packing, the packing wears and eventually it needs to be replaced.

Seals for the valve stem can be made with O rings around the stems on some small valves. The O-ring seal can be opened and closed many times before a leak will occur.

Another type of seal that is sometimes used for the globe valve is the diaphragm. This seal can also be opened and closed many times before it fails.

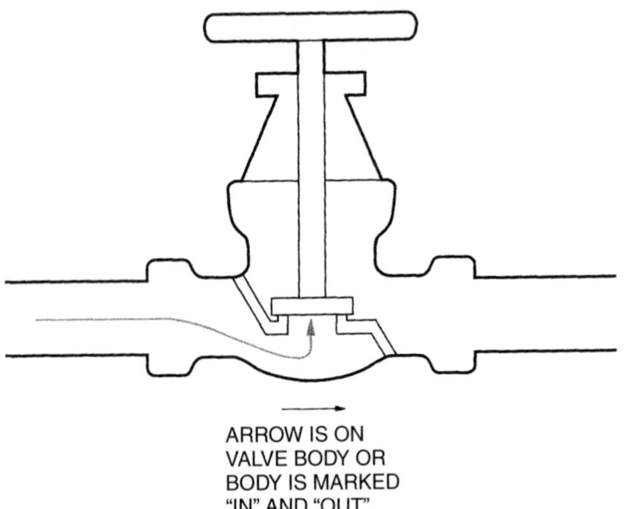

Figure 17-83. A globe valve must be installed in the proper direction.

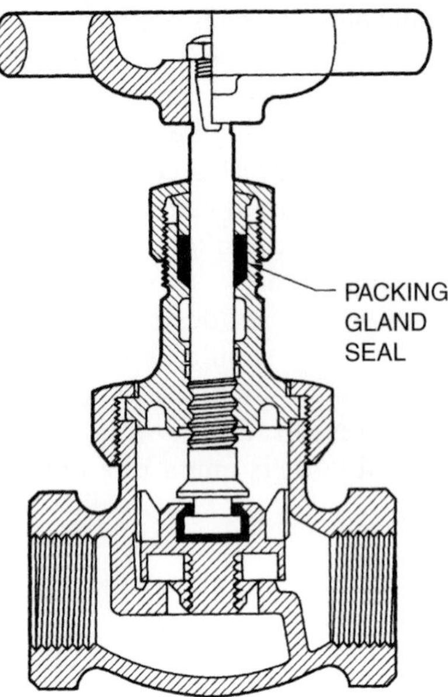

Figure 17-84. Packing-gland seal for a valve stem. *Courtesy Nibco*

Because globe valves have some pressure drop, they are good as balancing valves, valves that are installed in the circuit to create a pressure drop to control water flow.

GATE VALVE. Externally, the gate valve looks similar to the globe valve. However, inside the gate valve is a rising gate that opens to allow flow. The valve body must have room for this gate to rise, so the gate valve can be identified from the outside by its larger size. This type of valve has very little pressure drop and is often used as a shutoff valve.

The valve mechanism inside the gate valve is a wedge-shaped device that is forced into the valve seat when the valve is closed. Figure 17-85 shows a typical gate valve. When the seat is clean, this valve can be closed tightly for a good seal. The gate must be protected when the valve is disassembled so that it is not scratched. It will not seal correctly if it is scratched. Deposits of dirt or rust can build up under the gate and prevent the gate from closing tightly, and therefore allow leakage. The valve can be disassembled so that the valve body can be cleaned, Figure 17-86.

Because the gate must open, there must be room in the valve body for the gate to rise. Gate valves are *rising stem* and *nonrising stem valves.* The rising stem valve must have clearance above the valve for the stem to rise when the valve is opened. The designer must plan this clearance into the installation. The nonrising stem gate valve raises the gate within the valve. This valve is often more expensive but can be used when not much clearance is available.

The gate valve must have a seal around the stem that raises the gate. The packing gland is often used as well as "O" rings.

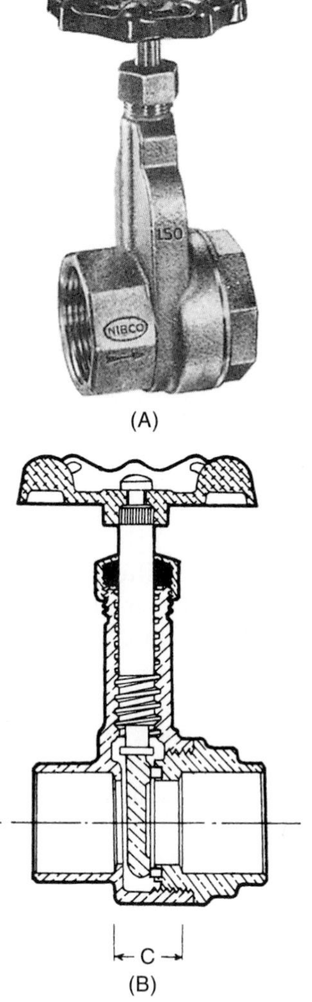

Figure 17-85. (A) Gate valve. (B) Cross section. *Courtesy Nibco*

Figure 17-86. The gate valve can be disassembled for cleaning. *Photo by Bill Johnson*

Note: Gate and globe valves both have handles that open the valves. When the valve is opened, the handle should never be allowed to stay in the completely open position because the next person to approach the valve may assume that the valve is closed and try to force it open when it is already open. It is a good practice to open the valve until it touches the back stop and then turn the handle one turn toward closed so that the next operator will not mistake it for closed. The next operator may be you.

BALL VALVE. The ball valve is a straight-through valve because there is little resistance inside the valve. Actually, the only resistance is the piping connection. It is a quick-close valve because it turns only 90 degrees from on to off. This valve can be more expensive than the gate valve. It takes up little room in the piping. Figure 17-87 shows a typical ball valve.

BUTTERFLY VALVE. This valve is built much like a damper in a round duct. When the handle of the valve is turned, a round disc inside the valve turns 90 degrees to close the valve. The disc seals against the sides of the valve when it is closed. This valve requires little space in the piping. The butterfly valve also is a quick-closing valve. Figure 17-88 shows a typical butterfly valve.

Control Valves

Control valves stop, start, divert, mix, or modulate the flow of water. To modulate a water circuit, you must be able to partially close a valve according to the needs of the control circuit. The control system controls these valves, and operators perform the action of the valves. This action can be controlled by electrical circuits, electronic circuits, or pneumatic air. For example, a room thermostat can be used to control the water flow for the various zones in a house. The room thermostat could vary a signal to the valve to vary the amount of water flowing to the hot-water coil. Upon a call for full heat, the valve would be wide open. A call for partial heat would partially close the valve. Figure 17-89 shows a modulating valve. These valves can be controlled with electronic motors, or variable air pressure can control air-operated valves.

Different valves are used for different reasons. If the flow through the system can vary, a straight-through valve,

Figure 17-87. Ball valve. *Photo by Bill Johnson*

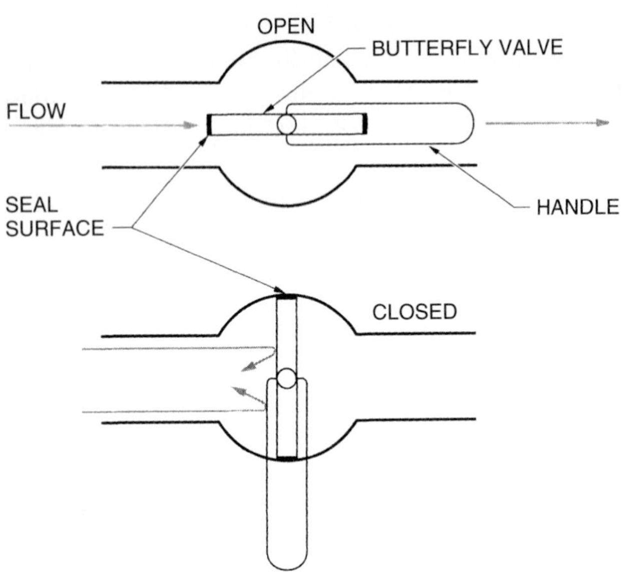

Figure 17-88. Butterfly valve.

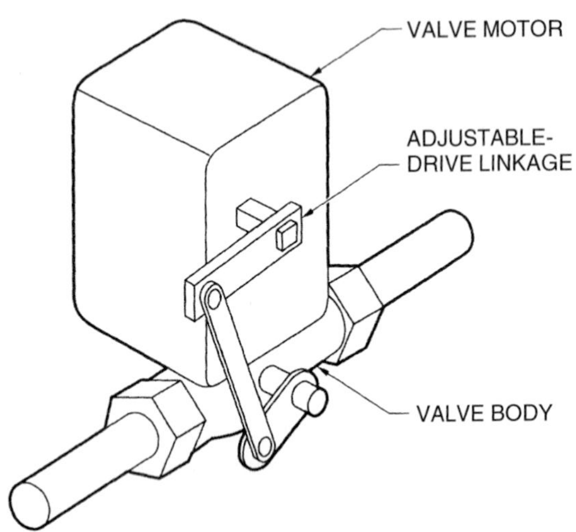

Figure 17-89. Modulating valve.

as shown in Figure 17-90, can be used. However, if all the valves are straight through and close because all the room thermostats are satisfied, all flow will stop. The water will just move around and around in the pump, and the pump will overheat. Other valve arrangements are used to bypass water around the coil. These are discussed later in this section.

Valves that are controlled with electric circuits open or close, or divert water; they cannot modulate. The electric valve may be a solenoid valve. This valve has a magnetic coil that pulls a plunger up into the coil when it is energized. These valves often make noise when they are energized, and this noise travels throughout the piping system. They are not often used in water systems. Often this valve has a spring in the top of the plunger chamber to soften the noise.

Figure 17-90. Typical hot-water valve that is either open or closed. *Photo by Bill Johnson*

Water valves often have different types of operators to make them more quiet. The motorized valve with a small electric motor, shown in Figure 17-91, is often used. This valve drives the valve seat open against a spring tension and holds the valve open. The motor is in locked rotor amperage while the valve is open. Locked rotor amperage for this small motor is only approximately 10 watts, so the motor will not damage itself. When the thermostat satisfies, the motor is turned off, and the spring closes the valve.

Another type of silent-operation valve uses a rodand-tube operator, as shown in Figure 17-92. The rod is made of a metal with a high coefficient of expansion, whereas the tube has a low coefficient of expansion. A small electric heater is wrapped around the tube and when energized opens the valve with the expansion of the rod. This valve is slow acting because it relies on the expansion

Figure 17-91. Motorized valve. *Photo by Bill Johnson*

Figure 17-92. Rod-and-tube valve. *Photo by Bill Johnson*

and contraction of the tube. Sometimes the valve has a thermostat that shuts off the heater for intervals when the heater is energized for a long period. This feature conserves power and prevents the heater from overheating.

A valve that stops or starts the flow of water does exactly what it says, it can only stop or start the water flow. This is a basic valve and is the action of all ordinary electric valves.

Installations often require some of the water to bypass the coil when the thermostat is satisfied. Bypassing is accomplished with a diverting valve or a mixing valve. The diverting valve has three ports; two are outlet ports. As shown in Figure 17-93, in one position the valve allows full flow through the coil. In the other position, water flows around the coil. Notice that this valve is positioned at the coil inlet. The mixing valve is positioned at the coil outlet, and it also has three ports; 2 inlets and 1 outlet. Figure 17-94 shows the water flow when a mixing valve is used. A mixing valve is not inter-changeable with a diverting valve. If they are inter-changed, valve chatter will occur.

Pneumatic valves are operated by air pressure. These valves are not discussed in depth in this text because the pneumatic subject is broad. These valves operate from 0 to 15 psig in pressure. As the pressure is increased on the diaphragm, the valve either closes or opens, depending on whether it is a direct-acting valve or a reverse-acting valve. Because the volume of air is large for the larger valve, it often has a pilot positioner to govern the air to the valve diaphragm from a main air supply. Figure 17-95 shows the pilot positioner operation.

Each control valve has a rating called the Cv factor of the valve. This factor is related to the amount of flow through the valve with a 1 psig pressure drop (2.31 feet of head) through the valve. For example, a valve with a Cv factor of 5 passes 5 gallons of water per minute when the valve has a 1 psig pressure drop. However, the valve may have more or less of a pressure drop than 1 psig. The following formulas help explain the Cv factor:

$$\text{Cv factor} = \text{gpm flow at 1 psig pressure drop}$$

$$\text{Flow rate in gpm} = \text{Cv} \times \sqrt{\text{Pressure drop}}$$

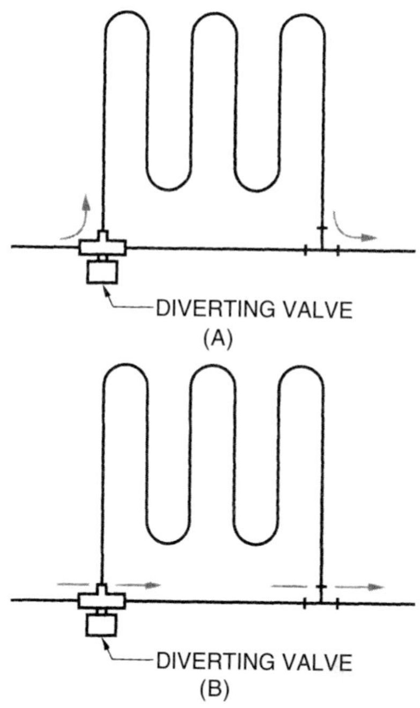

Figure 17-93. Diverting valve with three ports. (A) Water flow through the coil. (B) Water flow diverted around the coil.

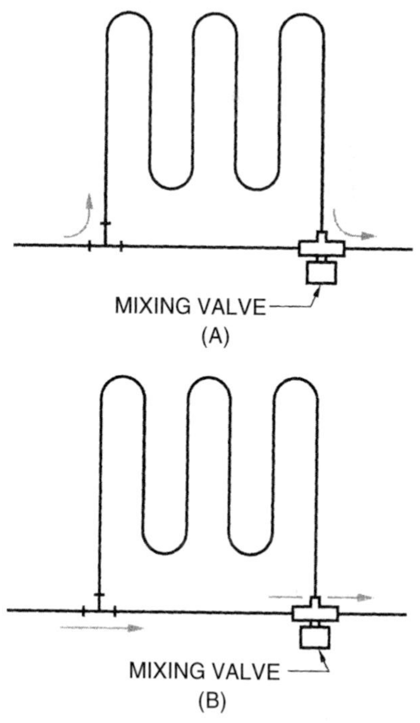

Figure 17-94. Mixing valve with three ports. (A) Water flow through the coil. (B) Water flow around the coil, through the valve.

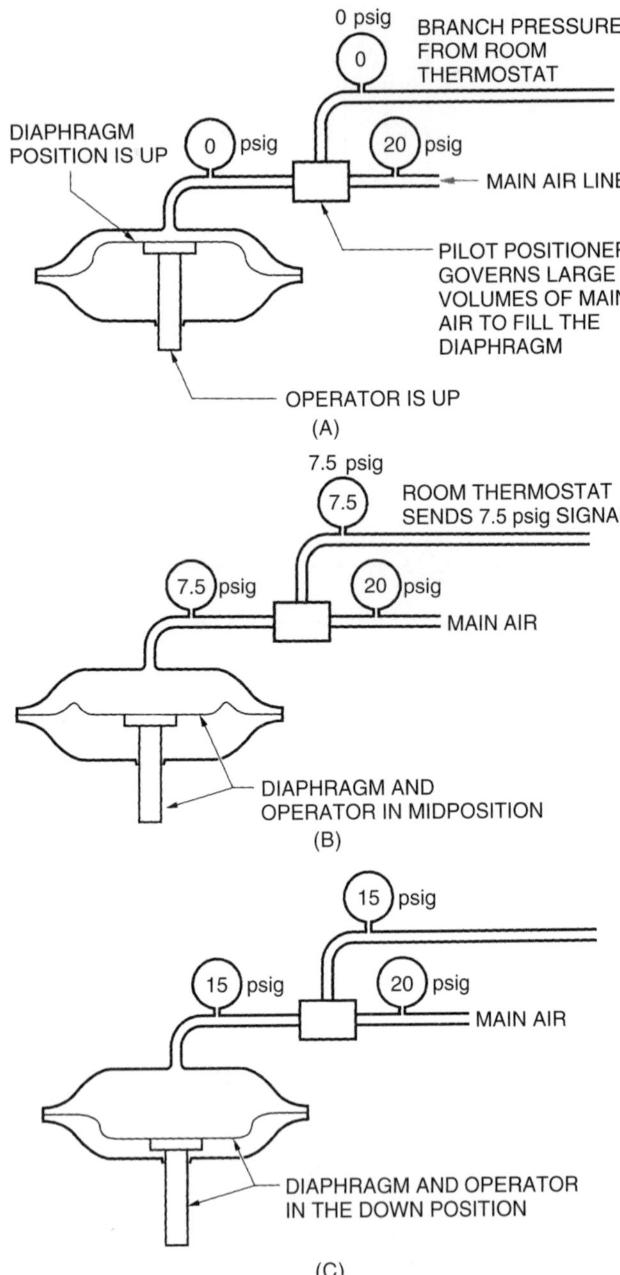

Figure 17-95. Large pneumatic valve with a pilot positioner. Diaphragm and operator (A) in up position, (B) in midposition, (C) in down position.

For example, suppose that a valve has a pressure drop of 1.9 and a Cv factor of 3.5. The flow rate of water at this pressure drop is 4.8 gpm as illustrated in Figure 17-96:

$$\text{gpm} = \text{Cv} \times \sqrt{\text{Pressure drop}}$$
$$= 3.5 \times 1.38 \text{ (the square root of 1.9)}$$
$$= 4.8 \text{ gpm}$$

The Cv factor gives the designer a method of finding the flow rate through a valve at different pressure drops than may be shown in the manufacturer's catalog. The designer may also need to know the Cv factor of a valve for a particular design flow rate. For example, suppose that a pressure drop of 2.5 psig is needed and a flow rate of 6 gpm is needed. The designer will choose a valve with a Cv factor of 3.8:

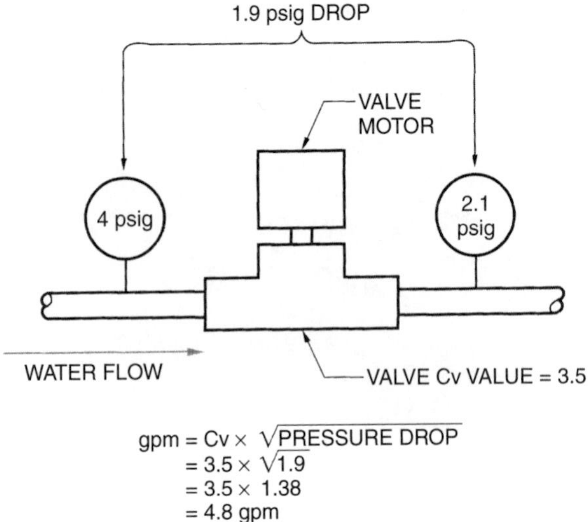

Figure 17-96. Calculating the Cv factor of a valve.

$$\text{gpm} = \text{Cv} \times \sqrt{\text{Pressure drop}}$$
$$\text{Cv} = \text{gpm}/\sqrt{\text{Pressure drop}}$$
$$= 6/1.58 \text{ (the square root of 2.5)}$$
$$= 3.79, \text{ or } 3.8$$

The technician will not often use the Cv factor for control valves, but you should be familiar with it. When you install a system, the prints may mention the Cv factor. Being able to converse with the engineer who designed the system is advantageous.

Accessories

Accessories in the system are used to improve system performance or aid the technician. The accessories discussed in this section are check valves, flow control valves, flow adjustment valves, flow-checking devices, air-control fittings, expansion tanks, fill valves, low-water cutoffs, relief valves, strainers, and instrumentation.

CHECK VALVES. Check valves are valves in the circuit that allow flow in only one direction. There are several types: the swing check, the ball check, and the magnetic check.

The swing check valve has a hinged gate mounted on an angle inside. When the flow tries to reverse, the gate swings shut, as shown in Figure 17-97. This is the simplest check valve and has only a few moving parts. It can be mounted horizontally or vertically if it is mounted with the flow upward. If mounted in the vertical position with the flow downward, the gate will not close, as

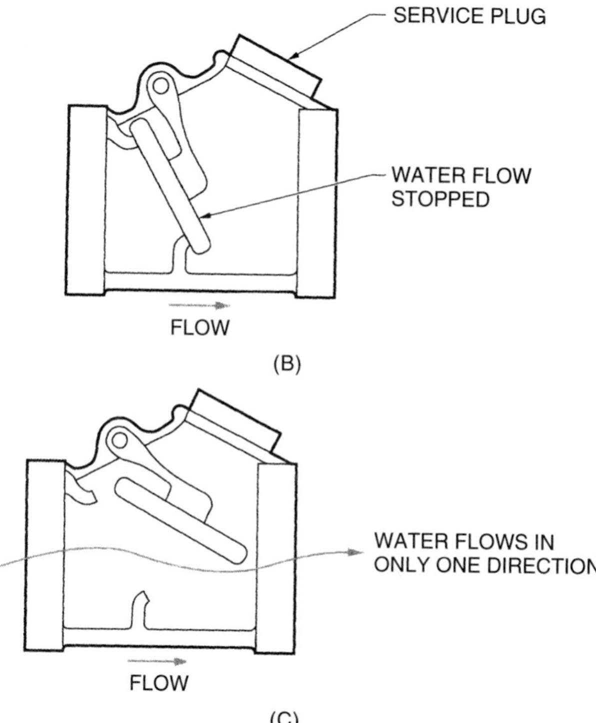

Figure 17-97. (A) Swing check valve. *Photo by Bill Johnson* (B) Water flow stopped. (C) Water flowing.

shown in Figure 17-98. This valve can be disassembled for servicing or cleaning by removing the top and lifting out the gate.

The ball check valve has a ball that moves back and forth. When it moves in the direction of the arrow on the valve body, flow can occur, as shown in Figure 17-99. A spring behind the ball ensures that the ball seats when the flow reverses. This spring can rust, so this type of valve is not often used in water systems. This valve can be mounted in any position because position does not affect the spring. The spring causes this valve to have more of a pressure drop than that of the swing check valve.

The disc in a magnetic check valve is attracted to the seat by a magnet. When flow starts, the pressure of the

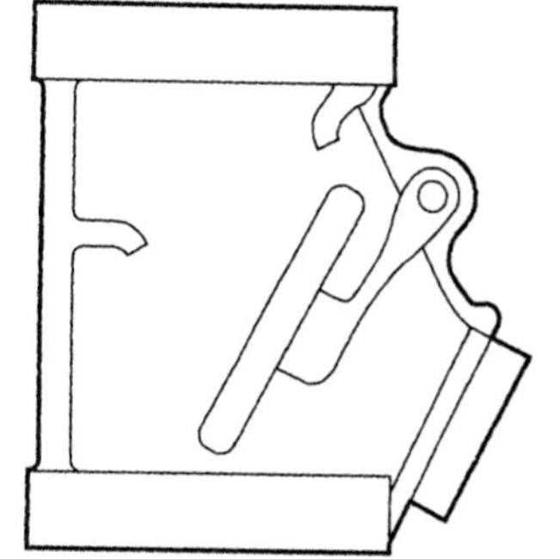

Figure 17-98. The swing check valve must be mounted correctly.

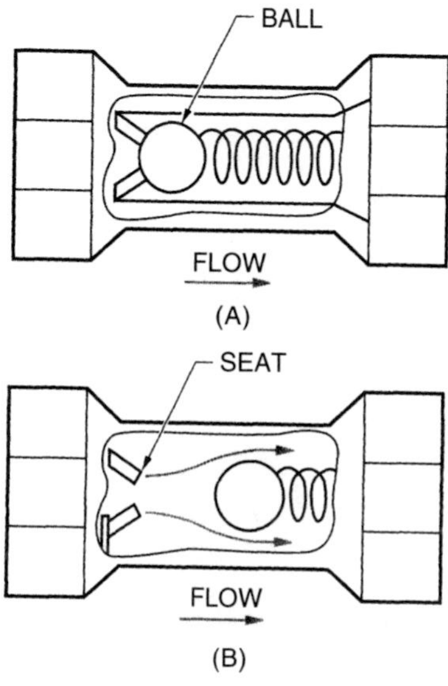

Figure 17-99. Ball check valve operation.

moving liquid moves the magnetic disc off the seat, as shown in Figure 17-100. This valve may be installed in any position and has a slight pressure drop.

FLOW CONTROL VALVES. Flow control valves prevent water from moving through the system during the off cycle when the pump is off but the boiler is still hot, as shown in Figure 17-101. For example, a system may have several circuits, with a pump for each. The boiler remains hot and operates on its own thermostat. Some of the circuits may not be calling for heat, but the water will migrate in the system as a result of gravity. Remember that as the boiler heats the water, the water becomes less dense and rises through the system to the convectors. The centrifugal water pump offers no resistance to this flow, so these zones will still have some heat by gravity.

The flow control valve has a fitting on top that allows the technician to open the valve wide so that if the system loses power or pump pressure, some heat can be extracted from the system by gravity (provided that the boiler is on and hot). Some gas boilers may have thermopile control and need no external power.

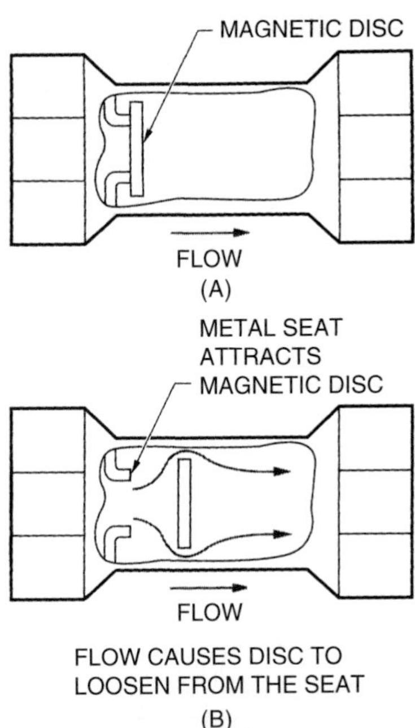

Figure 17-100. Magnetic check valve operation.

Figure 17-101. Two flow control valves. Flow control valves stop hot-water migration during the off cycle. *Courtesy ITT Fluid Handling Division*

FLOW ADJUSTMENT VALVES. Flow adjustment valves are used to balance the flow through the different circuits. This type of valve has one pressure tap on the inlet and one on the outlet of the valve and a corresponding pressure-drop chart. Figure 17-102 shows a typical flow

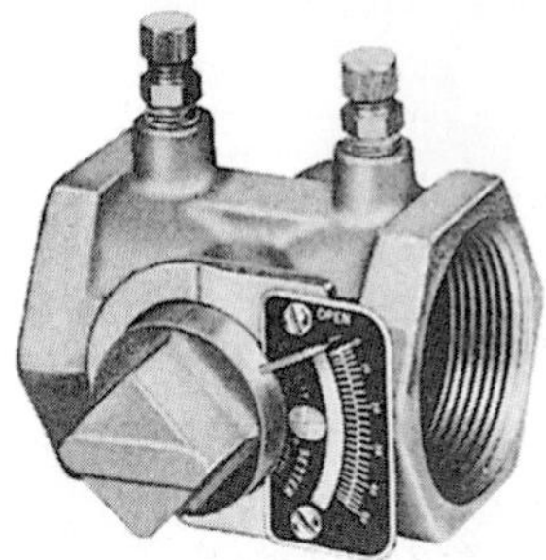

Figure 17-102. Flow adjustment valve. *Courtesy ITT Fluid Handling Division*

adjustment valve. The valve is a precision device. It is placed in the circuit, and the technician checks the pressure drop across the valve and refers to a graph to determine the gpm. The graph lists inches of water on one side and gpm on the other (remember that 27.7 in. of water = 1 psig, or 2.31 ft of water). The valve is gradually closed until the correct pressure drop corresponding to the desired gpm is reached. A special gauge is often used to measure the pressure drop. This gauge has a needle that can travel all around the dial. When pressure is applied to both sides of the gauge, the gauge reads the difference in pressure, illustrated in Figure 17-103. Notice that it reads in inches of water. The technician then knows how many gallons of water are flowing per minute in the circuit. This valve is a circuit gpm tester and a balancing valve in one fitting; it saves time and money for the installation.

FLOW-CHECKING DEVICES. The flow-checking device is a precision orifice with a pressure tap at the inlet and another at the outlet. It is much like the flow adjustment valve, except it does not have the built-in

Figure 17-103. Differential pressure gauge. *Photo by Bill Johnson*

valve. Figure 17-104 shows a flow-checking device. A graph is furnished with the device; this graph lists pressure-drop readings on one side (expressed in inches of water) and gpm on the lines of the graph. These flow-checking devices can be placed in individual circuits or in the main circuits to determine the total water flow in the system. These devices are quite accurate when installed and used properly. Good installation practices call for any circuit pressure-drop measuring device to be installed in a straight portion of the piping, where the flow is calm and not turbulent. Installing the device at least five pipe diameters from an upstream fitting and three pipe diameters from a downstream fitting ensures that the valve will give accurate readings, Figure 17-105. Notice that these fittings are rated in inches of water column instead of feet; they create little pressure drop in the circuit.

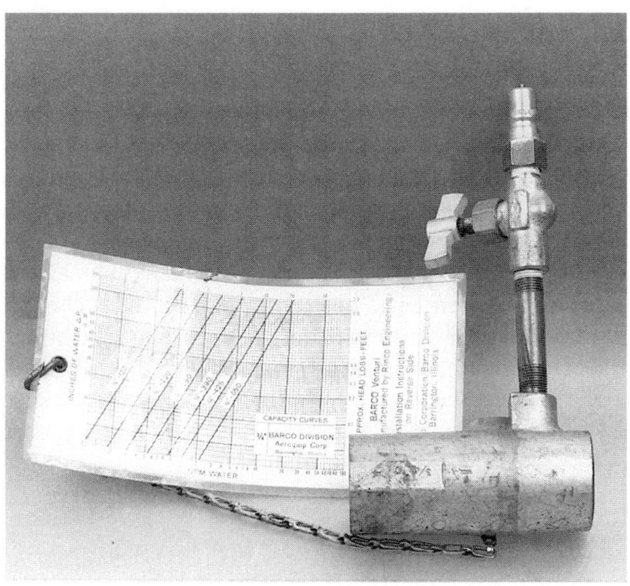

Figure 17-104. Flow-checking device. *Photo by Bill Johnson*

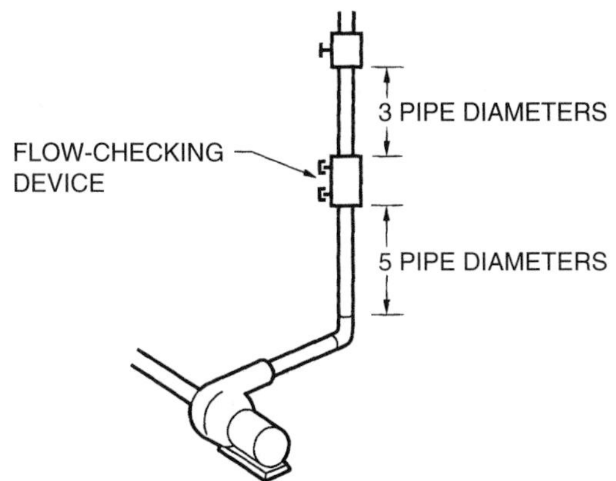

Figure 17-105. A properly installed flow-checking device.

AIR-CONTROL FITTINGS. Air circulating in a water system causes noise and corrosion and will impede water flow to the point of stopping it. As mentioned previously, air enters the system when the system is filled initially, when pump seals operate in a vacuum, and when city water is introduced into the system because of leaks. Each hot-water system must have a method for dealing with air that accumulates in the system. Air can be released from the system at any high point in the system. All high points should have air bleed fittings. Most convectors or heat-exchange devices need air bleed fittings. There are two types of air bleeds: manual and automatic.

Manual devices are small valves that open the circuit to the atmosphere. This valve often has a square head for a wrench or a winged head for hand operation, Figure 17-106. Some bleed valves, particularly those on convectors, require a screw driver to operate. Figure 17-107 shows one of these valves. This design prevents the public from easily opening the valves.

The automatic bleed valve has a float mechanism that closes the valve when it is full of water. If air passes through the valve, the float drops and bleeds out the air. When the air is bled, the float drops and closes the valve. Figure 17-108 shows an automatic bleed valve. It is a good idea to place these valves in all high points in the system. When the valve is located above a ceiling or in a place where water might damage something, the exhaust for the valve should be piped to a drain, as shown in Figure 17-109. If the valve begins to leak, the water will then go down the drain.

Figure 17-106. Manual bleed valve. *Photo by Bill Johnson*

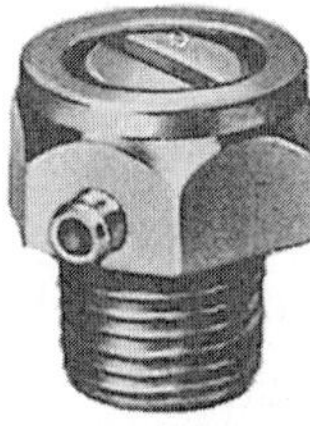

Figure 17-107. The technician must use a screwdriver to open this manual bleed valve. *Courtesy ITT Fluid Handling Division*

Figure 17-108. Automatic bleed valve. *Courtesy ITT Fluid Handling Division*

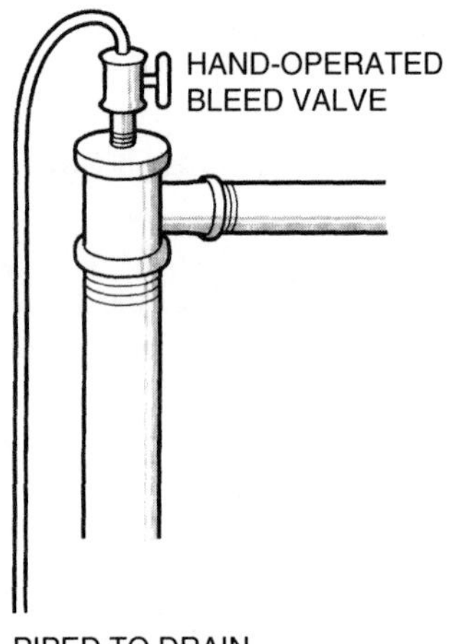

Figure 17-109. A line to a drain prevents possible leak damage.

When automatic bleed valves are used in a system, the operating technician may not know when a water leak occurs. For example, if a leak develops in a coil that has a drain pan, water leaks down the drain pan, air-laden city water makes up the water level on a regular basis, and no one is aware of the leak. A manual bleed system may be more advantageous because the technician should be aware of leaks.

Manual shutoff valves should always be positioned before automatic bleed valves in the line for servicing purposes. Some service technicians prefer to only open the manual valves located before the automatic bleed valves while filling the system, instead of leaving them open all the time because automatic bleed valves also serve as vacuum breakers. If the system pressure is reduced, air will enter the system through the automatic bleed valves. With the manual valves closed, air will not enter the system.

EXPANSION TANKS. When water is heated, it expands; when it is cooled, it contracts. A system must have a method for dealing with this expansion and contraction. The expansion tank is the device that compensates for expansion and contraction. It is often called a *compression tank.* See Figure 17-110.

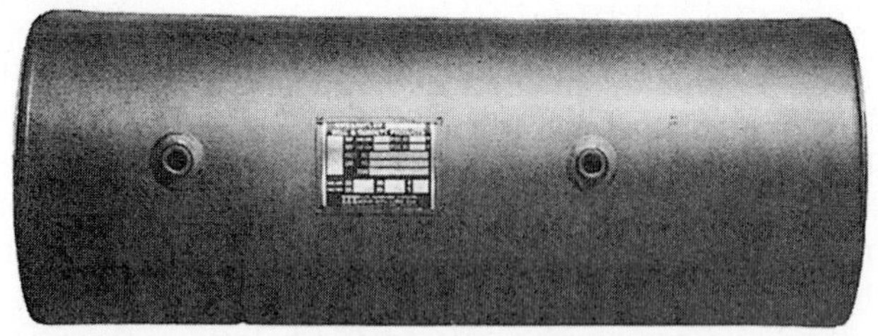

Figure 17-110. Expansion tank, often called a compression tank. *Courtesy ITT Fluid Handling Division*

The expansion tank must be the correct size. If the tank is too large, it will be expensive. If the tank is undersized, the expansion of the water will pressurize the system, and the relief valve will relieve the pressure.

When water is heated from 40°F to 200°F, the volume increases by approximately 4%. Consequently, if the system has 500 gallons of water at 40°F, it will have an additional 20 gallons of water at 200°F ($500 \times .04 = 20$). A portion of the expansion tank is filled with air. The 20 gallons of expanded water move into the expansion tank and compress this air. The tank must be much larger than 20 gallons because of the air.

When pressure is increased on the air in the tank, the air compresses, as shown in Figure 17-111. Because air is added to the system anytime city water is added, the added air can be moved to the expansion tank with an air-separation system. The boiler is a perfect air separator. The added water slows when it passes through the boiler, and it is heated at the same time. This causes a natural

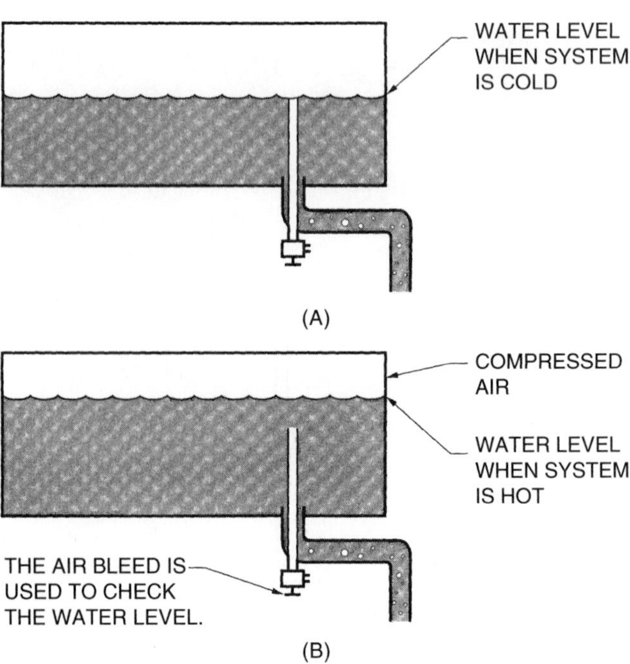

Figure 17-111. The air in an expansion tank compresses. (A) System is cold. (B) System is hot.

separation of air and water. Special fittings on top of the boiler that trap air and route it to the expansion tank are used on most systems, as illustrated in Figure 17-112.

The water for the system is removed from down in the boiler with a dip tube to pick up the system water down in the boiler. As the boiler is stopped and started, the water in the expansion tank moves up and down, from a pressure difference, making room for the new air, Figure 17-113. Air in the top of the boiler migrates to the expansion tank with this constant movement of cooling and heating.

The volume of a system can be determined by calculating the volume of the piping, heat exchangers, and boiler.

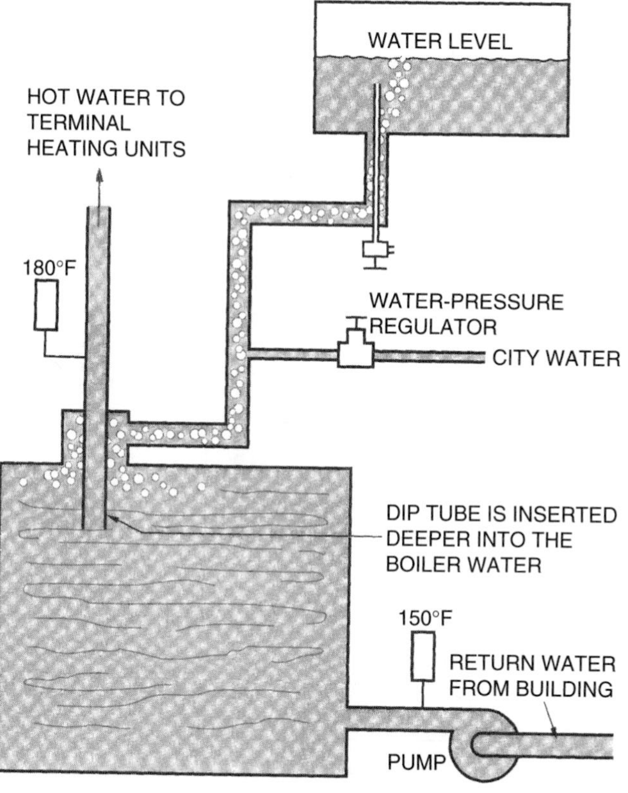

Figure 17-112. Air-separating device on the boiler.

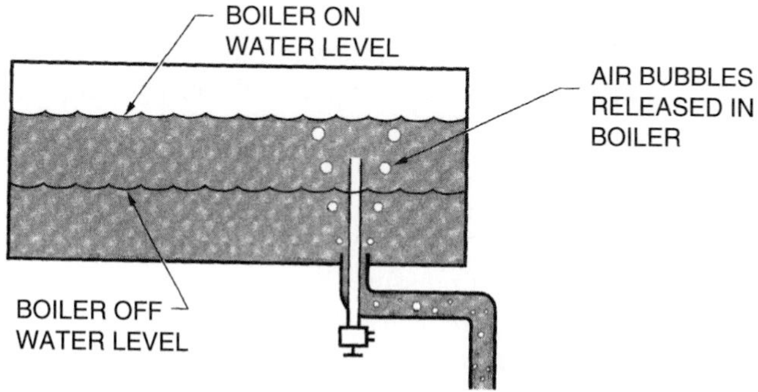

Figure 17-113. The water in the boiler moves up and down in the expansion tank.

The table in Figure 17-114 shows the volume of various pipe sizes. To find the most accurate volume of the boiler and the heat-exchange units, you must get in touch with the manufacturers. The table in Figure 17-115 is often used to calculate the system volume for typical systems.

The technician should know the water level in the expansion tank when the system is full and up to temperature. Sometimes a sight glass on the end of the expansion tank allows you to observe the water level, as shown in Figure 17-116. Sometimes a tank has a special fitting on the bottom that has an air bleed port for checking the water level, as shown in Figure 17-117. When the system is filled and up to temperature and there is an excess of air in the expansion tank, the technician should open the bleed port on the bottom of the tank and allow air to escape until water begins to come out of the port. The water level is then at the correct level in the tank. See Figure 17-118. Expansion tanks can become air-bound from air entering the system. The correct level can be maintained with the bleed port.

Another type of expansion tank has a flexible container inside called a *bladder.* The bladder is made of a rubber-like material and is filled with air. When water expands into the bladder area of the tank, the air in the bladder compresses, as illustrated in Figure 17-119.

FILL VALVES. Each hot-water system must have an automatic method of filling the system because water can be lost while the system is operating in the automatic mode. Automatic fill valves operate according to system pressure. If the system pressure reduces, water is allowed to enter the system from the city water supply. These valves maintain the water pressure in the system at a preset value. The system operating pressure is determined by the height of the system in the building. The fill pressure is enough to fill the system to 4 or 5 psig at the highest point in the system but cannot exceed the boiler relief pressure of 30 psig, Figure 17-120. Water fill valves are all that stands between the system water and city drinking water pressure. They must have reverse-flow check valves in them to prevent system water from flowing back into

PIPE DIA. SCH. #40	GAL. PER LIN. FT.	PIPE DIA. SCH. #40	GAL. PER LIN. FT.
½"	.016	3"	.38
¾"	.028	4"	.66
1"	.045	5"	1.04
1¼"	.078	6"	1.5
1½"	.106	8"	2.66
2"	.17	10"	4.2
2½"	.25	12"	5.96

TABLE C – WATER VOLUME PER FOOT OF PIPE

Figure 17-114. Table for estimating the water volume per foot of pipe. *Courtesy ITT Fluid Handling Division*

Basis For Component Volumes of Table B

Boilers: Average water content expressed in Gals per Net BTU/HR rating. Based on data received from various boiler mfrs.

Radiation: (Cast Iron) Large Tube = .114 Gals/sq. ft.
Thin Tube = .056 Gals/sq. ft.
Convectors = 1.5 Gal/10,000 BTU/HR. @200°F.
Baseboard = 4.7 Gal/10,000 BTU/HR. @200°F.

Radiation: (Non-Ferrous) Convectors = .64 Gal/10,000 BTU/HR. @200°F.
Baseboard = .37 Gal/10,000 BTU/HR. One Tube ¾" @200°F.
Fan Coil & = 0.2 Gal/10,000 BTU/HR. Unit Htr. @180°F.

Note: Values in this table are only averages. For closer estimate of system volume, use manufacturer's data. Also see pipe volume table

	BOILERS Consult Mfr. When Pos.		RADIATION See Basis For Tables Above							Average Piping Systems Based on Average Pipe Sizes for 20°F Temp. Drop Design		
			CAST IRON				NON-FERROUS					
MBH 1000's BTU/HR.	Conven-tional (Not Water Tube)	Small Flash Type	Large Tube	Thin Tube	Convector	Baseboard	Convector	Baseboard	Fan Coil & Unit Htr.	One Pipe	Two Pipe	Radiant Panel
50	12	5.4	28.5	15.7	7	23.5	3.2	1.9	1.0	7.8	10.5	11
60	14.5	6.4	34.2	16.8	7.5	28.3	4.0	2.2	1.2	9.0	12.5	13.5
70	17	7.2	40	19.6	8.7	33	4.5	2.6	1.4	10.0	14.5	16
80	19.5	8.0	45.6	22.4	10	37.8	5.2	3.0	1.6	11.5	17	18.5
90	22	8.6	51.4	25.2	11.25	42.5	5.8	3.3	1.8	13	18	21
100	25	9.4	57	28	12.5	47.1	6.5	3.7	2.0	14	21	24
125	30	11	71	39	17.7	58.8	8.1	4.6	2.5	18	27	30
150	36	13	85.5	47.8	21	70.8	9.7	5.5	3.0	22	34	37
175	42	14.5	99	55	24.7	82.8	11.3	6.5	3.5	26	40	44
200	48	16	114	63.8	28.5	94.3	13	7.4	4	30	47	50
250	60		142	79.8	35.6	117.8	16.2	9	5	39	62	65
300	70		171	95.7	42.7	141.5	19.5	11	6	49	78	70
350	84		199	111	49.7	165	22.7	13	7	58	94	
400	95		228	127	57	188.5	26	15	8	70	110	
450	107		256	143	64	212	29.2	17	9	80	130	
500	120		285	157	71	235.8	32.5	19	10	90	150	
600	140		342	191	85.5	282.8	39	22	12	115	190	
700	160		399	223	98.7	330	45.5	26	14	140	225	
800	190		456	251	114	377	52	30	16	165	260	
900	210		513	287	128	424.3	58.5	33	18	195	310	
1000	235		570	319	142	476.5	65	37	20	225	360	
2000	480		1140	638	285	942	130	74	40	540	900	
3000	720		1710	957	427	1414	195	110	60	960	1500	
4000	960		2280	1276	560	1885	260	150	80	1350	2100	
8000	1200		4560	2553	1140	3770	520	300	160	3400	5000	
12000	3000		6840	3830	1710	5655	780	445	240	5000	9000	
16000	3700		9120	5107	2280	7542	1040	580	320	8300	13000	

TABLE B – AVERAGE WATER VOLUME OF SYSTEM COMPONENTS (IN GAL OF WATER)

Figure 17-115. Table for estimating the average water volume of system components (in gallons of water). *Courtesy ITT Fluid Handling Division*

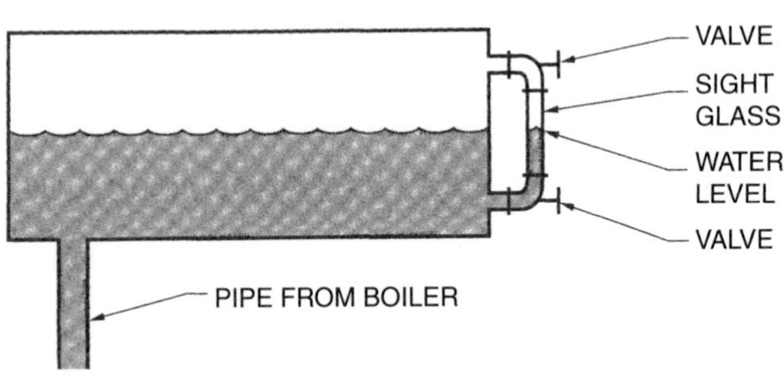

Figure 17-116. Sight glass in an expansion tank.

the city water main. Otherwise, water treatment from the system could contaminate city water if the city water pressure dropped below the system pressure. Figure 17-121 shows a reverse-flow check valve.

The fill valve (often called a *pressure regulator*) has an adjustment for setting the system pressure. The technician can adjust the system pressure, but the correct operating pressure must be known. Most systems operate at no more than 12 psig in the area of the boiler for a one- or two-story building. Remember that the boiler relief valve for low-temperature boilers is set at 30 psig. Often the boiler is on

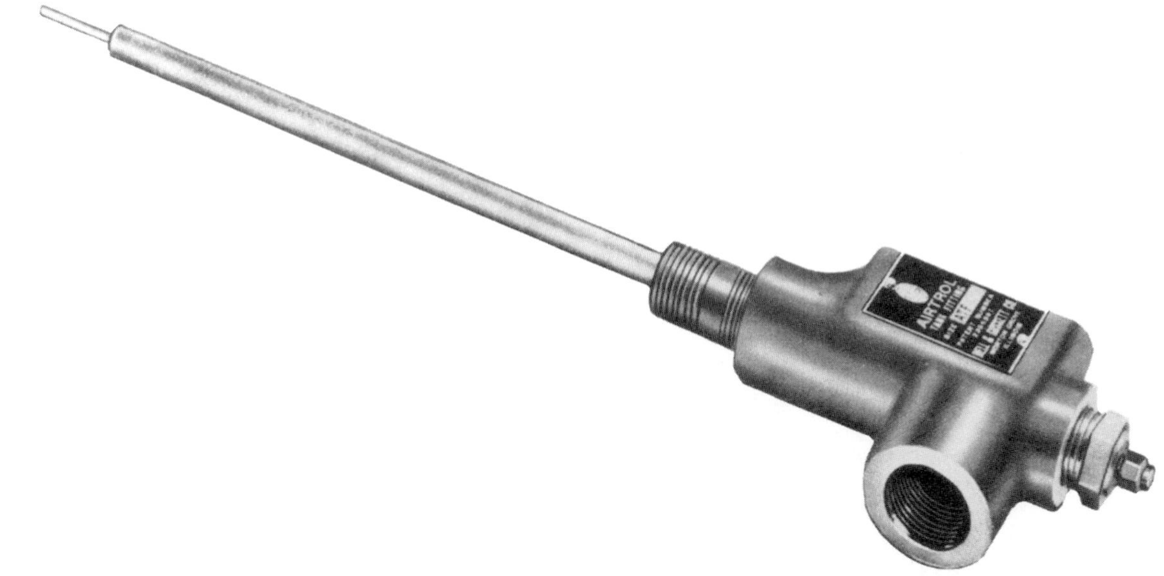

Figure 17-117. Special fitting for determining the water level in an expansion tank. *Courtesy ITT Fluid Handling Division*

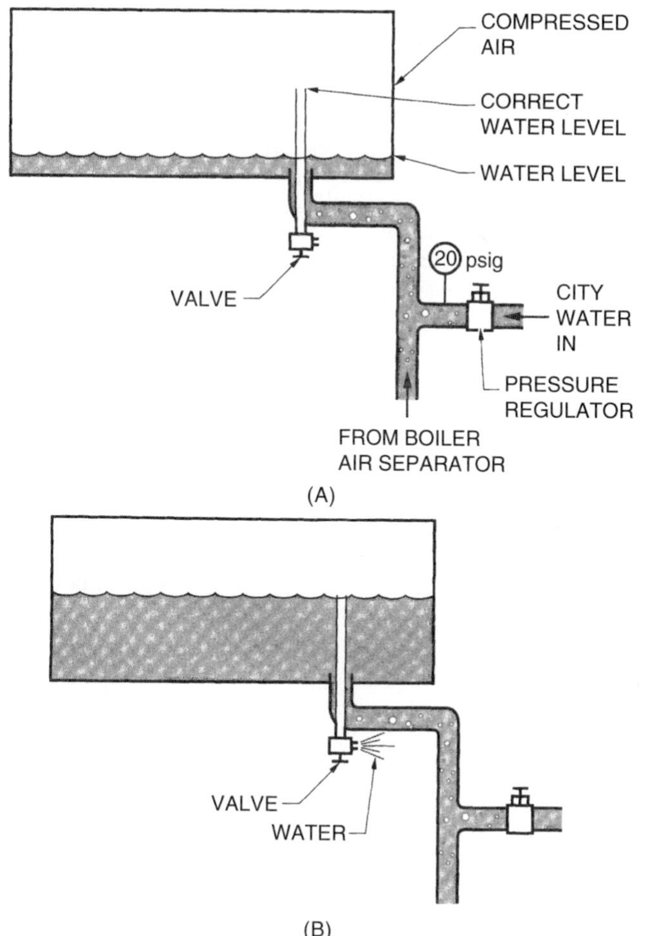

Figure 17-118. Air can be vented from this valve until water appears, then the water level is known. (A) The system is filled to system pressure. The air is compressed in the tank, but the tank is not filled to the correct level. (B) Air is allowed to escape until water escapes the test valve. When water appears, the water level is correct.

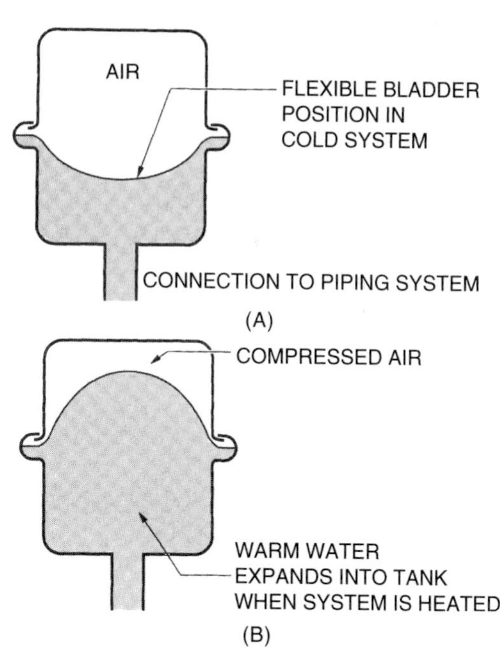

Figure 17-119. Bladder expansion tank. (A) Cold system. (B) Heated system.

the top floor of a building, so the standing head of water for the building will not pressurize the boiler. The fill valve often has a relief valve at the outlet in case of valve failure. This is a *system* relief valve, as shown in Figure 17-122, not to be confused with the *boiler* relief valve.

Filling an empty system by using the pressure-regulator fill valve would be slow because the valve creates a pressure drop from the city water to the system water. Instead, a bypass hand valve is often used to fill a system that is completely empty. This valve allows the technician to turn city

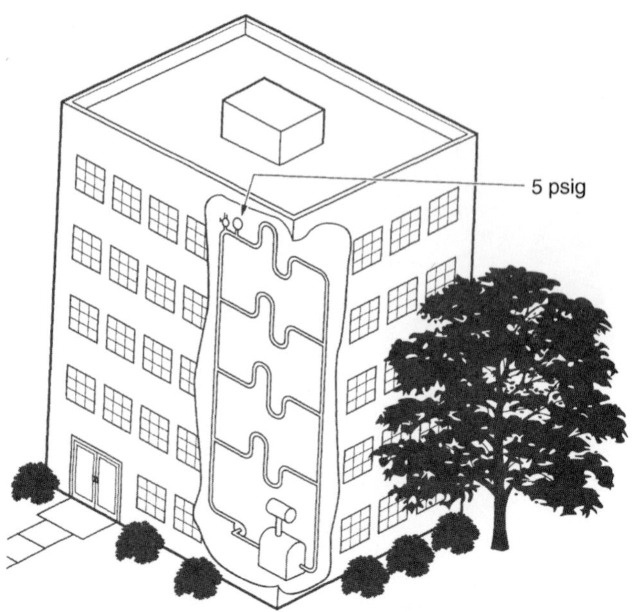

Figure 17-120. The fill pressure should be enough to fill the system with approximately 5 psig at the top of the system.

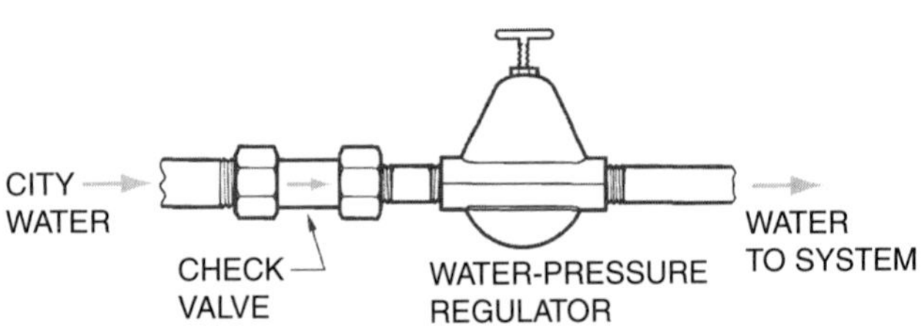

Figure 17-121. Reverse-flow check valve.

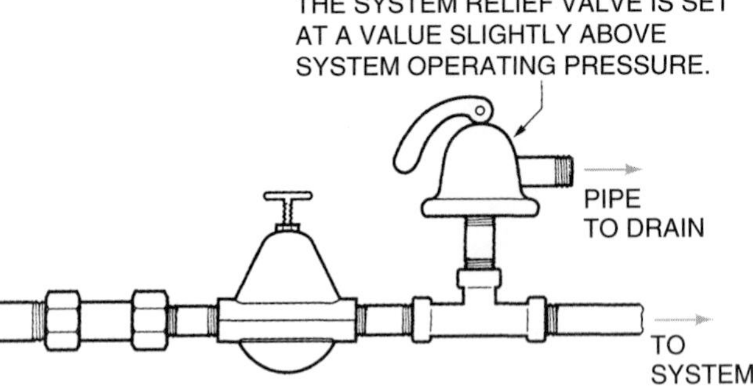

Figure 17-122. System relief valve.

Figure 17-124. Quick-fill handle. *Courtesy ITT Fluid Handling Division*

water pressure into the system for fast filling, Figure 17-123. If the technician were filling a system and looked away from the system when it was nearly full, either the system relief valve or the boiler relief valve would release the excess pressure. Often a system fill valve has a quick-fill handle to allow an empty system to be filled quickly. Figure 17-124 shows a quick-fill handle on a pressure regulator.

The system relief valve must have a termination point for any water that is relieved from the system. This water can be hot and must be directed down a drain and away from any possible place where personal injury may occur, Figure 17-125.

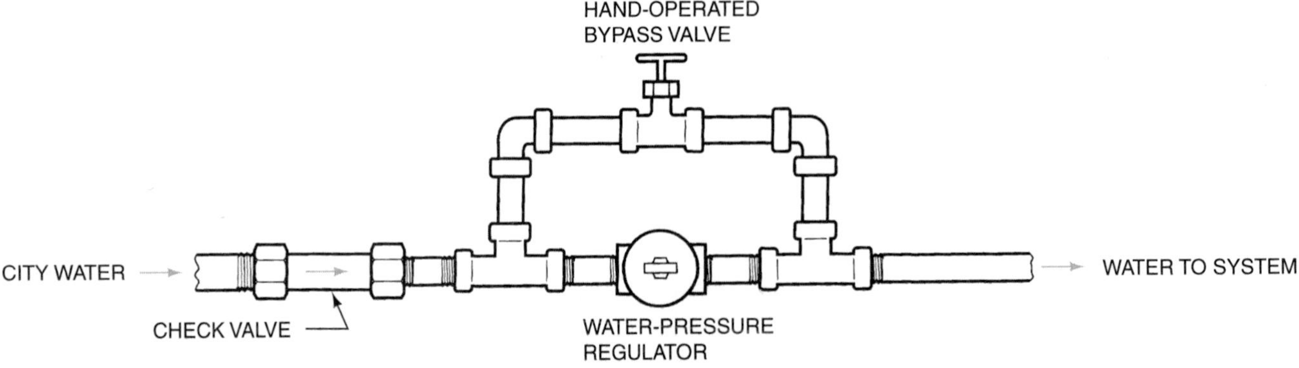

Figure 17-123. Bypass valve arrangement for filling a system.

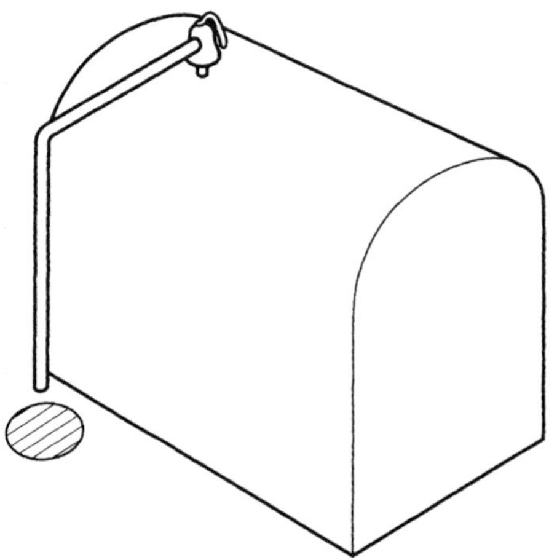

Figure 17-125. Terminal discharge point for system relief valve.

LOW-WATER CUTOFFS. Each hot-water system should have a method for shutting off the fuel supply if the system water becomes too low. The low-water cutoff is necessary because water may, for some reason, be escaping faster than it can be replaced by the makeup water system. One reason may be that the relief valve sticks open, or a large system leak may occur. The low-water shutoff can be placed in the system at any point above the minimum water level for the system, which is at some point above the boiler. If the boiler is electric, this point is above the highest heating element, as shown in Figure 17-126. If the boiler is gas or oil, the minimum point is above any point where heat is added to the water. Often the low-water cutoff is placed in the piping adjacent to the boiler.

RELIEF VALVES. There are two relief valves: one in the system and one on the boiler. When in the system, the valve is set above the system working pressure. The system working pressure can vary but is typically less than 30 psig when the boiler is in the circuit and the system relief valve is near the boiler, see Figure 17-127. When the system relief valve is lower than the boiler, such as in a high-rise system with the boiler on the top floor, the relief valve may be set at a higher setting, as shown in Figure 17-128.

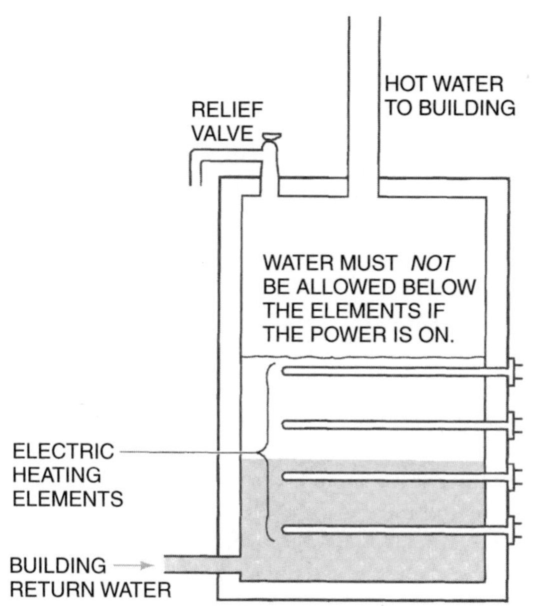

Figure 17-126. Low point of water in an electric boiler. The low-water shutoff should be above this point.

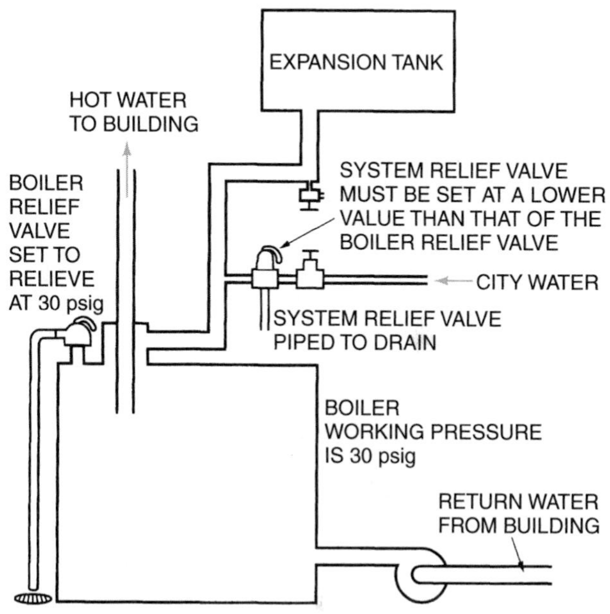

Figure 17-127. The system working pressure must not exceed 30 psig.

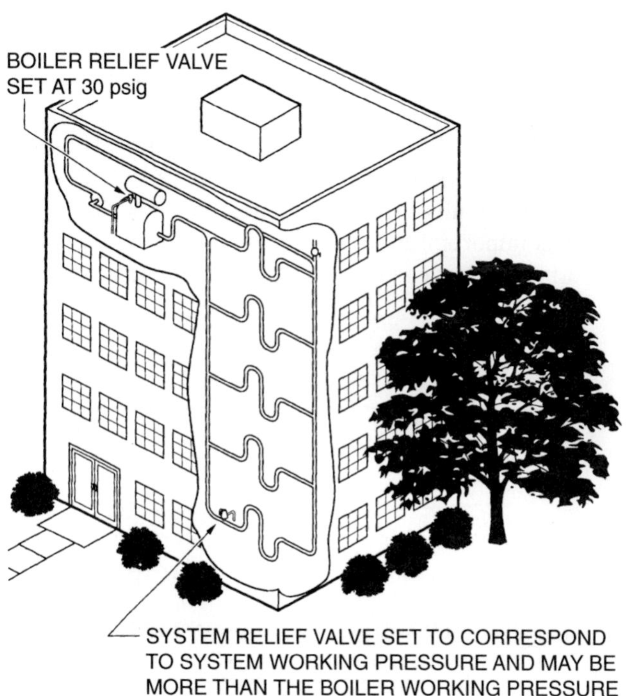

Figure 17-128. The system relief valve is set at a higher value if the boiler is on the top floor of the structure.

The boiler relief valve must be set at the working pressure of the boiler. As mentioned before, most boilers in low water temperature systems are tested for a working pressure of 30 psig, and the boiler relief valve is chosen for this pressure.

The relief valve on the boiler must have enough capacity to maintain the boiler pressure at no more than 30 psig with the fuel supply heating at the maximum. For example, a boiler that has a capacity of 400,000 Btu per hour must have a relief valve that will not allow the pressure in the boiler to rise above 30 psig if the fuel supply continues to heat the boiler (e.g., if the boiler thermostat and high limit control fail and allow the boiler to continue to heat).

Caution: When the technician replaces a relief valve, the proper replacement must be used. It must have at least the capacity of the manufacturer's defective valve.

STRAINERS. A strainer is important in a hot-water system because it removes any rust and scale that may be circulating in the system. When a system is first started, you can expect to catch almost anything in the strainers, particularly in a large system. Workmen have been known to leave lunch boxes in large pipes, and the lunch boxes were caught in the strainers. The strainer should be positioned just prior to the pump to stop the particles from moving into the impeller. A strainer is typically shaped in a Y with a removable end that allows you to clean the strainer, see Figure 17-129. A properly designed system will have a valve before and one after the strainer so that the complete system does not have to be drained to service the strainer. Often a valve before the strainer and one after the pump serve as service points for the strainer and the pump, as shown in Figure 17-130.

INSTRUMENTATION. All systems must have some instrumentation so that a technician can know at a glance what is happening in the system. The most basic home system may have only an altitude gauge on the boiler, Figure 17-131. All water boilers have this gauge. The more complicated and larger the system, the more important instrumentation is. A pressure gauge on each side of the pump is important in larger systems for checking pump performance. Both gauges should be at the same height so that the standing pressure head does not cause an error in the gauges. If one gauge is 3 feet off the floor and another is 5.31 feet off the floor, a 1 psig error is automatically built into the gauge reading. The error occurs because a column of water 2.31 feet high has a standing pressure of 1 psig, as shown in Figure 17-132. Another method of taking gauge readings is to install one gauge with a valve on each side so that pressure from the pump inlet or the pump outlet can be checked with the same gauge, Figure 17-133. This is the most accurate method for checking the pressure drop. If the gauge has a slight error, the error will not affect the pressure-drop reading because a pressure difference is being recorded on 1 gauge. Correctly installed gauges can be used with the pump curve to determine whether the correct amount of water is moving in the circuit.

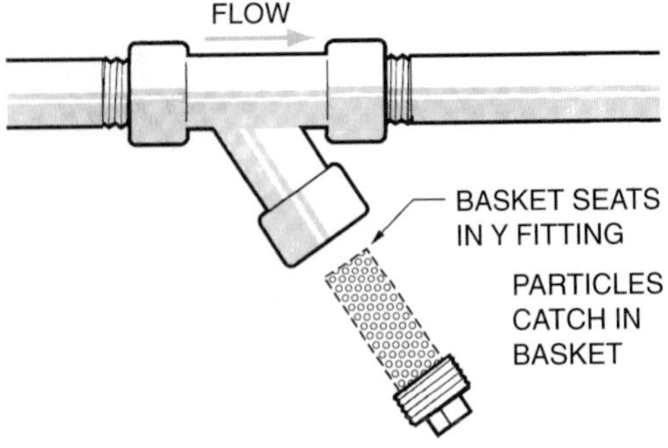

Figure 17-129. Y strainer with removable basket.

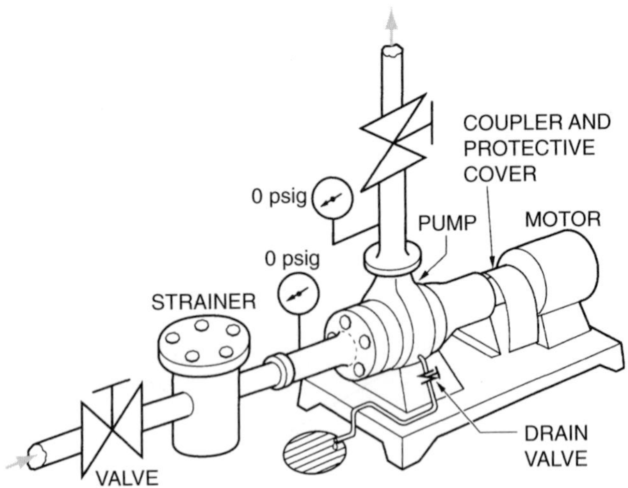

Figure 17-130. A valve before the strainer and one after the pump aid in servicing the strainer.

Figure 17-131. Boiler altitude gauge. *Photo by Bill Johnson*

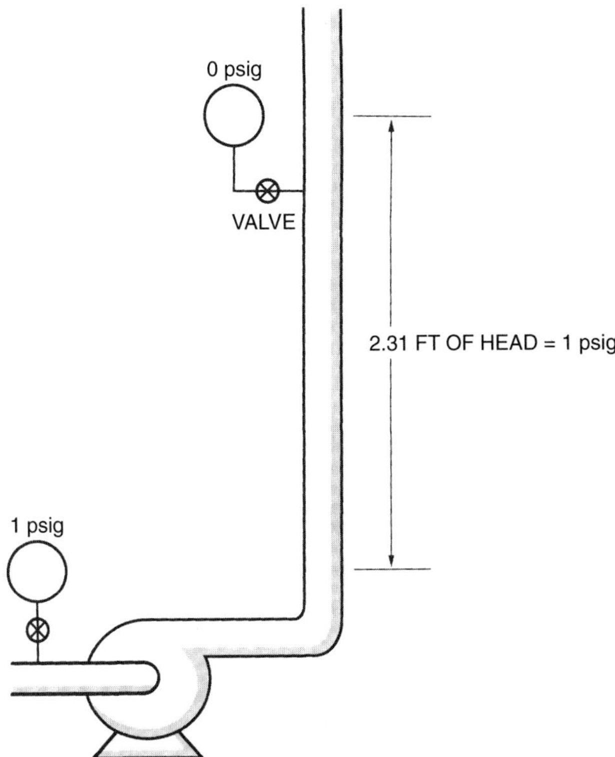

Figure 17-132. A column of water that is 2.31 feet high exerts a pressure of 1 psig.

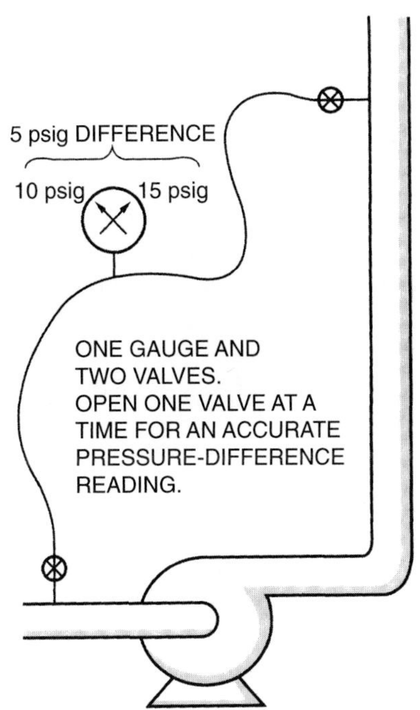

Figure 17-133. One gauge for both readings.

Pressure gauges can also be used to detect restricted strainers. If one gauge is fastened before and one after the strainer and the strainer becomes restricted, a pressure drop will occur, as shown in Figure 17-134.

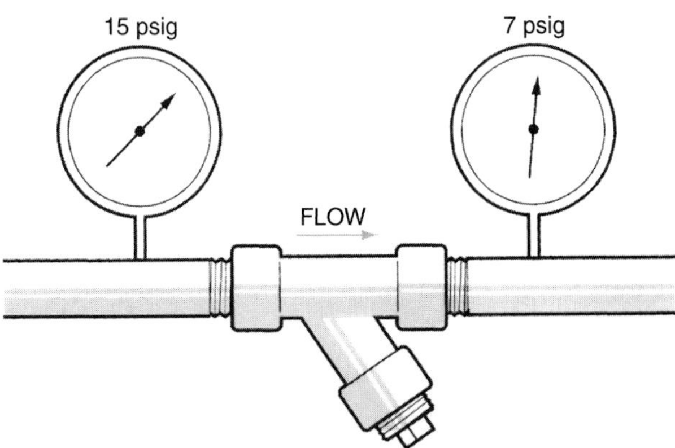

Figure 17-134. A gauge reading on each side of a restricted strainer will reveal the problem.

Thermometers are also installed in the larger, more complex systems, Figure 17-135. These gauges tell the technician the temperature of the water leaving the boiler and the return water temperature. As shown in Figure 17-136, thermometers are installed in wells in the piping so that they can be removed for calibration or servicing. The thermometer

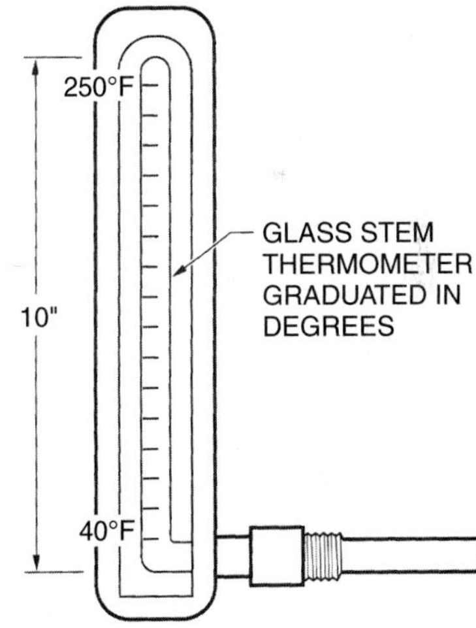

Figure 17-135. A thermometer is used to detect system temperature.

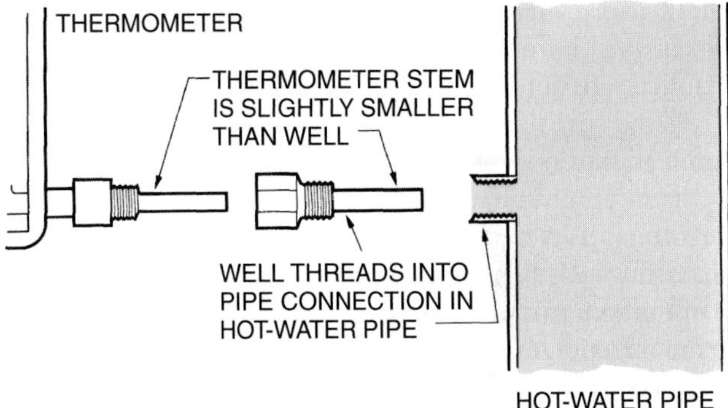

Figure 17-136. Thermometer well.

should be installed at the correct height for the technician to read the temperatures, with the thermometer face turned so that it can be easily read. Only high-quality thermometers should be used so that the technician can rely on the readings.

Glass stem thermometers maintain accuracy for longer periods than dial thermometers, but they are harder to read. Thermometers that read in the range of the system water temperatures provide the most accurate readings. For a hot-water system, a thermometer that has a 10-inch scale and reads from 40°F to 250°F is adequate. A thermometer that has a 4-inch scale and reads from 0°F to 300°F is not a good choice because the scale is too short for accurate readings.

17.7 Piping Practices

The piping system must be installed correctly so that the system has the best chance of operating correctly. Factors such as expansion and contraction of the pipe when the system is heated and cooled must be taken into consideration before the installation is started. All piping expands when the system is heated from room temperature to the temperature of the circulating water. Copper pipe has a much higher coefficient of expansion than iron or steel pipe does. According to the *IBR 200 Guide*, a 100-foot length of steel pipe expands 1 inch when heated from room temperature to 200°F, whereas a 100-foot length of copper pipe expands 1.5 inch under the same conditions. Long piping runs expand so much that leaks can occur if expansion joints are not provided. Expansion and contraction also cause noise as the system warms up to temperature and cools down. Room for expansion can be provided in piping runs with expansion loops or expansion joints, such as those shown in Figure 17-137.

Baseboard convectors are limited to length because of expansion. Manufacturers' recommendations must be followed when you select baseboard radiation because these systems may have long runs of units. Expansion joints may be recommended at intervals.

The pipe must be reamed at all connections where it was cut during installation. The burr made by the pipe cutter restricts flow by causing turbulence and must be eliminated, as shown in Figure 17-138. A complete piping system will have many of these restrictions if good piping practices are not followed.

Thread seal should be applied only to external threaded connections, as depicted in Figure 17-139. If it is applied to the internal threaded connections, the thread seal will turn loose and be carried through the piping and cause problems such as plugged heat exchangers and strainers.

Water pipes in the system should be exactly level or rise in the direction of the flow so that air that enters the system can move with the water, see Figure 17-140. If the pipes are pitched downward in the direction of the water flow, air will trap in the high spots and restrict the water flow.

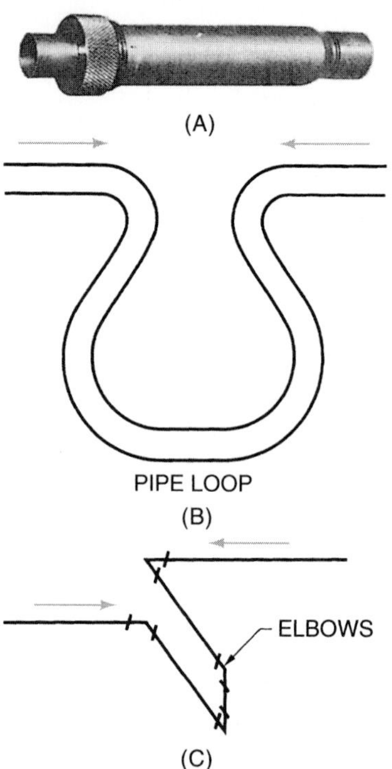

Figure 17-137. Room for pipe expansion is necessary. (A) Factory expansion connector. *Courtesy Edwards Engineering Corporation* (B) Copper pipe is bent for expansion loop. (C) Pipe leg allows for expansion.

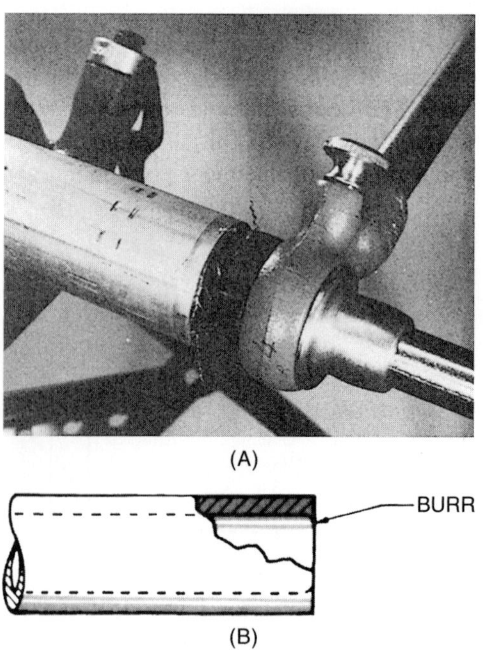

Figure 17-138. (A) Using a reamer to remove a pipe cutting burr. *Photo by Bill Johnson* (B) Burr.

All pipes that are routed through unconditioned spaces must be insulated, or heat will escape the system. This lost heat is expensive. When insulation is removed so that pipes can be serviced, it must be replaced as soon as possible.

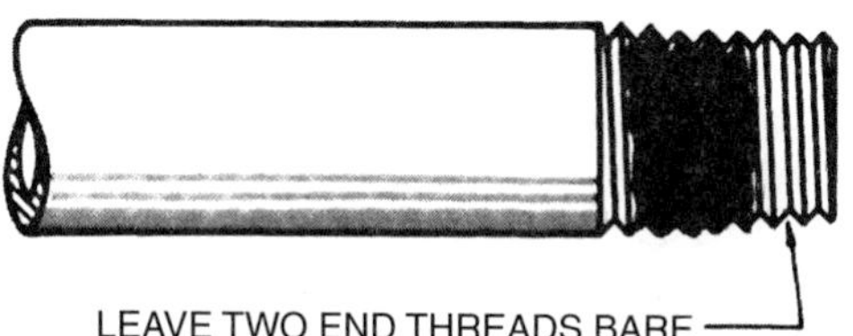

Figure 17-139. Thread seal must be applied only to the external portion of the threads.

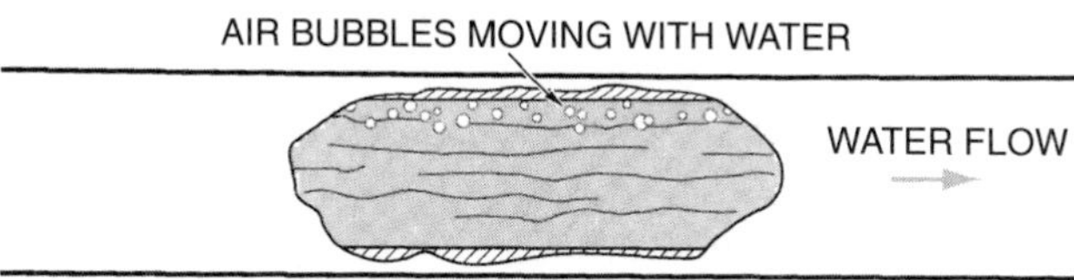

Figure 17-140. Water pipes must be level or pitched upward in the direction of the water flow so that the air can move with the water.

Each water system must have a method for draining the complete system in case of freezing weather when the heating system may not be functioning. For example, if a building will be vacant for a long time, the complete system must be drained. Drain valves at all low points with a method of breaking the vacuum while draining will ensure that the system is empty. The vacuum break must be at the high point in the system. When a valve on the bottom of a system is opened to drain the system, a vacuum forms at the high points in the system as the water tries to drain. It is possible that all the water will not drain; some may be suspended in the system. These places where water remains will freeze in cold weather. Automatic air bleed valves can provide the necessary vacuum break, Figure 17-141. If there are no automatic bleed valves, the technician must open all manual bleed valves in order to drain the system completely.

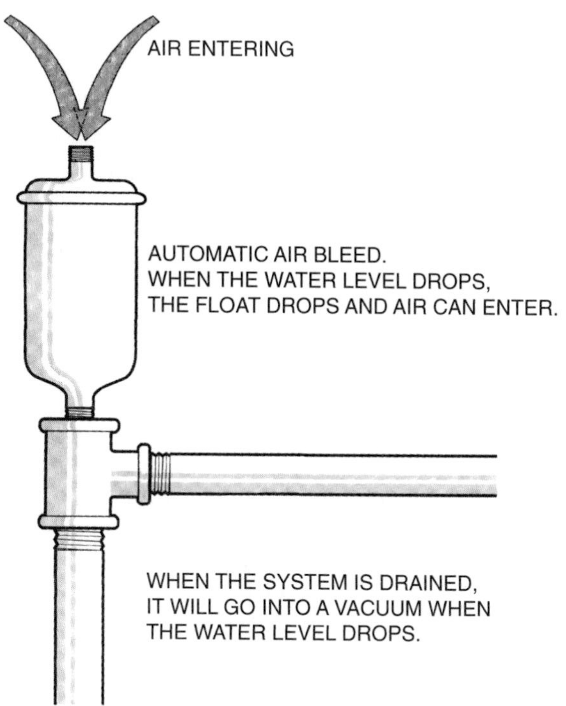

Figure 17-141. Automatic air bleed valve also serves as an automatic air vacuum break.

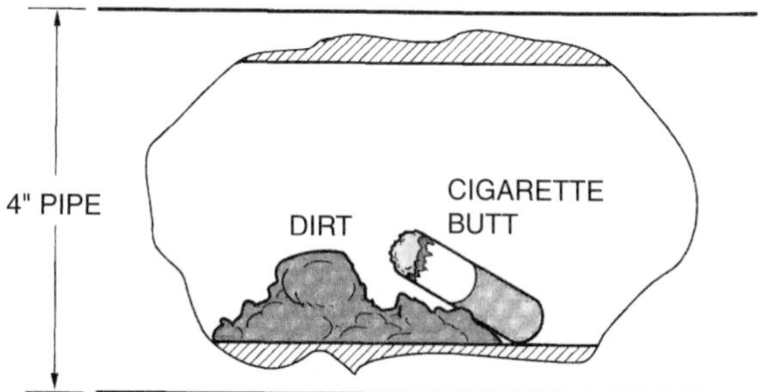

Figure 17-142. Construction dirt will be in the pipe of a new job.

After a system is installed, the inside of the piping must be cleaned. Construction dirt, oil from threading the pipe, and pipe dope will be inside the piping, Figure 17-142. The boiler will have an oil coating on the inside that must be removed. It is recommended that the system be operated for several days after it is filled and leak tested, then a special detergent should be added and circulated in the system for several more days. The system should then be drained and refilled several times to clean out the detergent and suspended oil. Finally, the system should be filled with water and water treatment provided by a company that specializes in water treatment. The water treatment contains additives for preventing rust and corrosion.

17.8 System Balancing

After the system is installed and started, it must be balanced. Balancing is the part of the process that ensures that the correct amount of heat is released in each zone of the structure. Balancing starts with a room-by-room load calculation of the structure. A detailed load calculation determines the heat loss for each room. Heat is continuously escaping to the outside of the structure, and the heating system replaces this heat. The designer must know how much heat is escaping in order to replace the correct amount. The technician should know the heat loss for each room and should obtain a report describing the water flow for each room or circuit. The larger the installation, the more data will be furnished.

Each zone should have a method for determining the water flow and a method for adjusting it to the correct flow. As mentioned previously, orifice pressure-drop devices are often used, Figure 17-143. A circuit adjusting, or flow adjustment, valve that has a known flow of water is often used in each circuit or zone. Figure 17-144 shows a typical circuit adjusting valve.

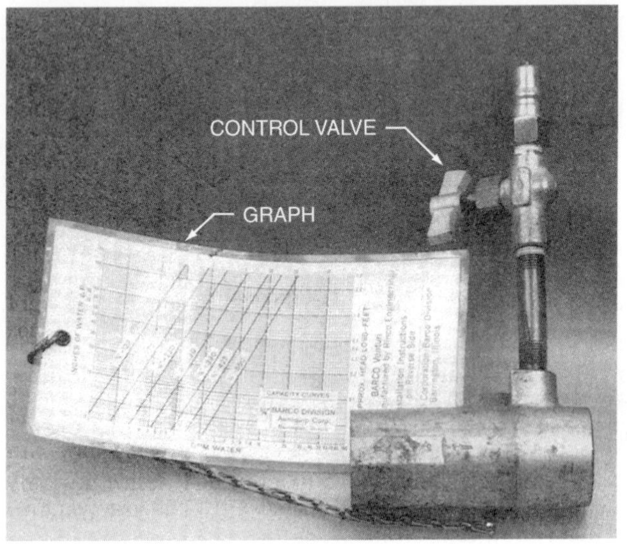

Figure 17-143. Orifice pressure-drop device. *Photo by Bill Johnson*

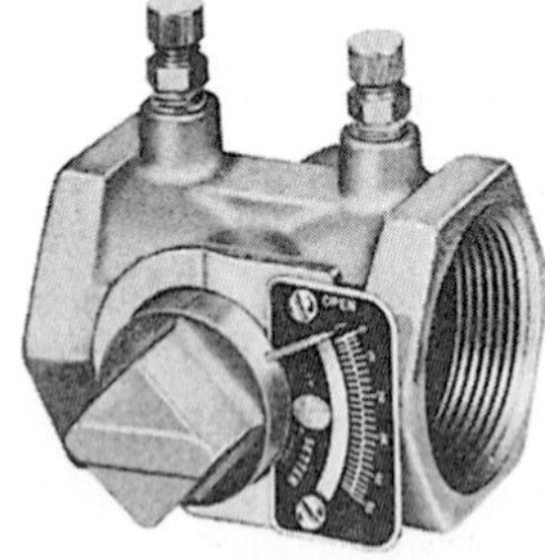

Figure 17-144. Circuit adjusting, or flow adjustment, valve. *Courtesy ITT Fluid Handling Division*

The technician should start the balancing process by opening all valves in the system. When full flow is established, the pump amperage must be checked, Figure 17-145. If there is too much flow, caused by oversized piping, there will be too much amperage, as shown in Figure 17-146. The pump curves and the pressure drop across the pump can be used to establish the correct total system flow with a system that has a pressure checkpoint at the pump. The technician must check the pressure drop across the pump and use the pump curves to determine the actual flow, Figure 17-147. If there is too much flow, the flow should be throttled at the pump outlet. If there is not enough flow, the system should be checked to determine whether all possible restrictions are open (such as the system strainer and the balancing valves) and to make sure that no air is in the system. If the system is correct, and full flow cannot be established, the pump may be too small or may have an internal problem. A rag could be restricting the flow in the pump impeller, as shown in Figure 17-148.

When full system flow is established, the technician should then establish the flow in each zone. The flow should be balanced at each unit. After the system is balanced, the technician should check the total system flow again to make sure that it is still correct.

Hot-water systems are somewhat forgiving of poor design. If the boiler is large enough and the heat level can be raised a few degrees, more heat can be added to the various zones and some rebalancing performed to help deficient zones.

With forced-air coils, when the airflow is known across a water coil, the capacity of the coil can be checked by using the sensible heat formula: $q = 1.08 \times \text{cfm} \times \text{TD}$. For example, suppose that the capacity of a room unit is in question. The technician believes that there is not enough water flow, and there are no pressure checkpoints for determining the flow. The technician can use

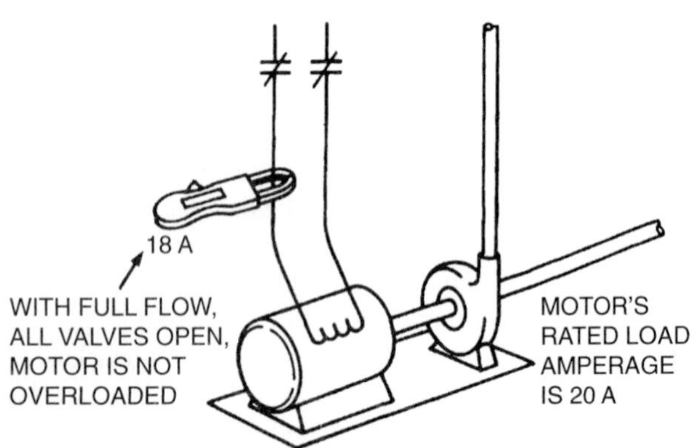

Figure 17-145. Establish full flow and check the pump amperage.

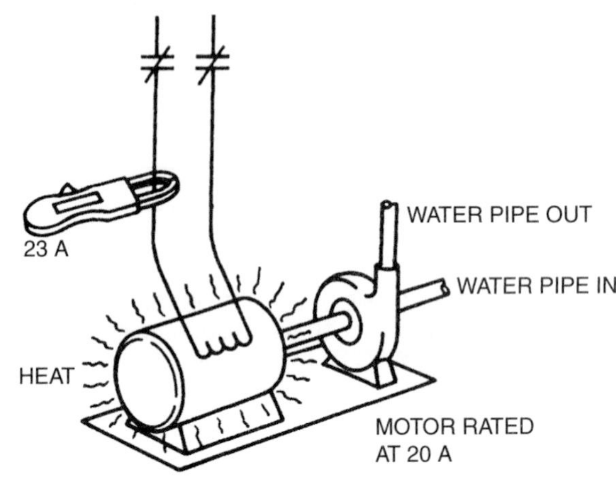

Figure 17-146. Too much flow shows as overamperage at the motor.

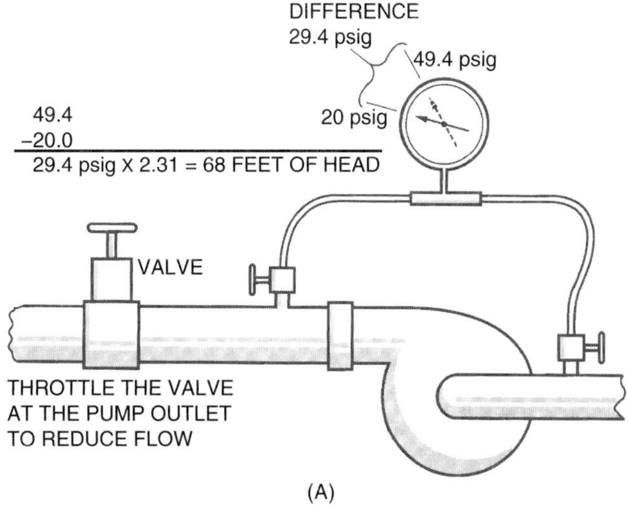

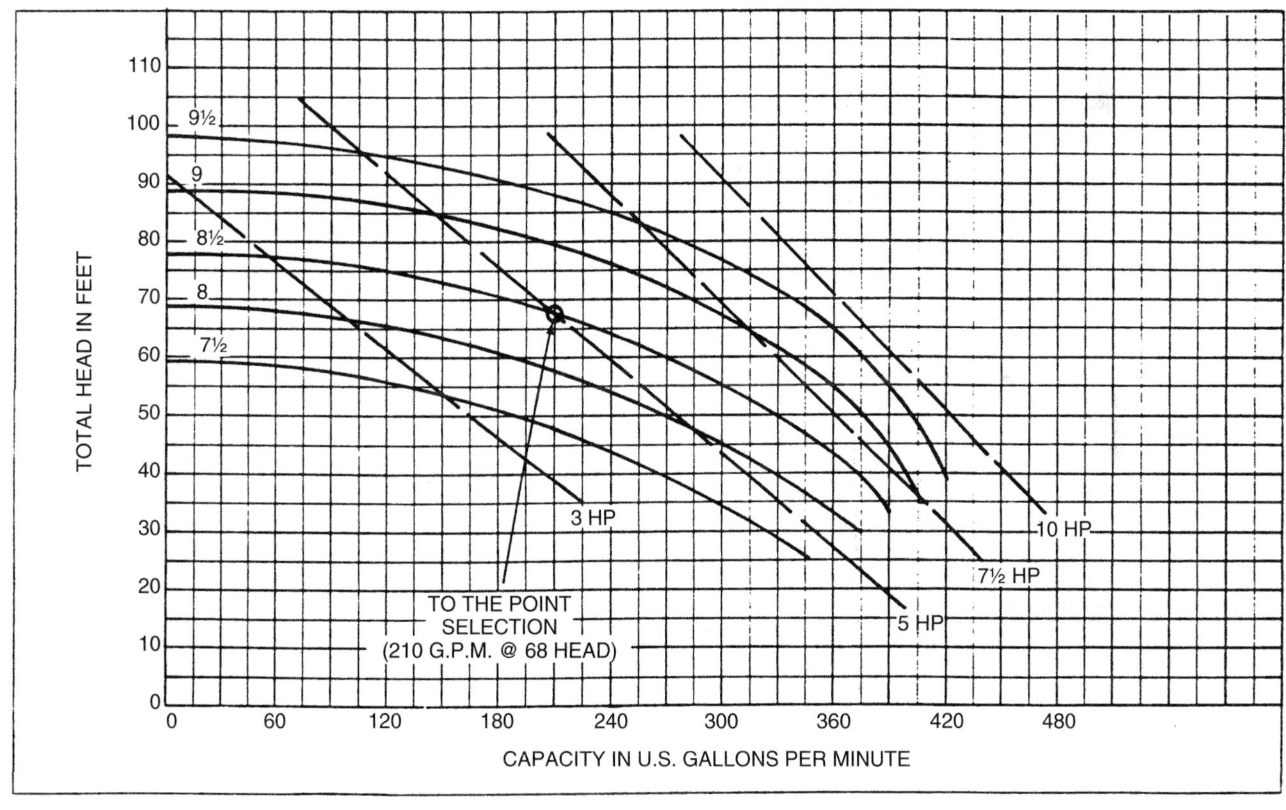

Figure 17-147. Check (A) the pressure drop across the pump against (B) the pump curves. *Courtesy ITT Fluid Handling Division*

the manufacturer's data to determine the airflow across the unit, then take the temperature rise of the air and calculate the water flow in gpm.

Suppose that the unit has an airflow of 500 cfm and an air temperature rise of 40°F. The capacity of the coil can be calculated as follows:

$$q = 1.08 \times \text{cfm} \times \text{TD}$$
$$= 1.08 \times 500 \times 40$$
$$= 21{,}600 \text{ Btu/h}$$

We know the amount of heat that the system is putting out. Now we can use the following formula to determine the amount of water that is circulating:

$$q = 500 \times \text{gpm} \times \text{TD}$$

This formula can be rearranged to solve for gpm:

$$\text{gpm} = q/(500 \times \text{TD})$$
$$= 21{,}600/(500 \times 40)$$
$$= 1.08 \text{ gpm}$$

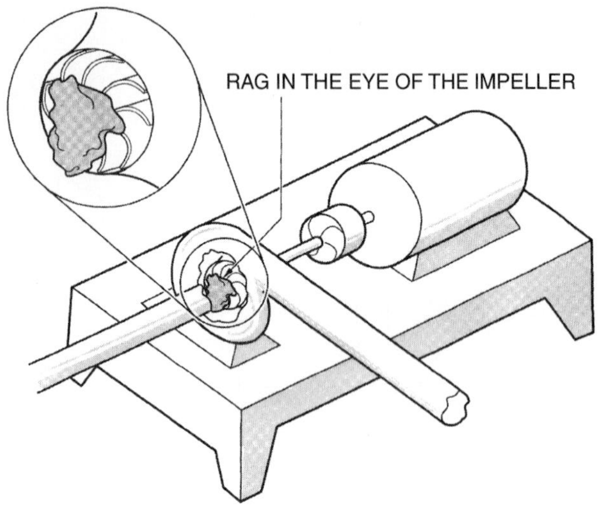

Figure 17-148. A rag in the pump impeller will restrict the flow.

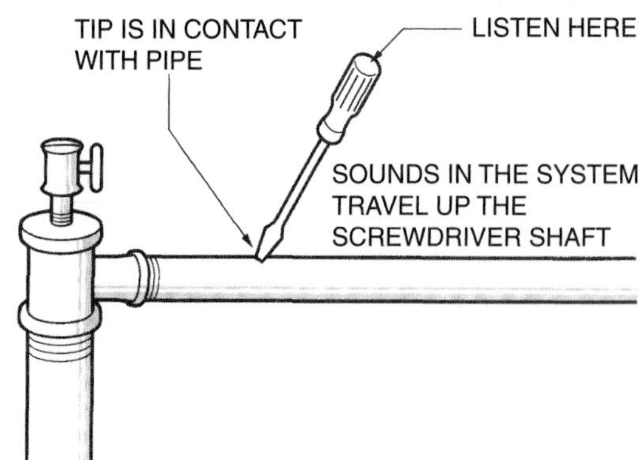

Figure 17-149. How to listen to a system with a homemade stethoscope.

This value can be compared with the specifications for the job to determine whether it is the correct amount.

A hot-water system that is properly designed and installed will work correctly. When the system is properly maintained and the correct water treatment is used, the system will last the life of the structure.

17.9 Troubleshooting Hot-Water Systems

Troubleshooting the hot-water system involves solving problems with the system or the heat source, the boiler. System problems arise in the piping and in the terminal units.

Air probably causes more trouble in hot-water systems than anything else. If the system has been in operation for many years and then quits heating, the technician should first make sure that the pump is operating. As explained previously, often a small system does not have pressure gauges and thermometers in the system. The technician must rely on experience to determine whether air in the system is preventing the water from moving.

Water moving through a system makes little noise. The technician can place the blade tip of a screwdriver on the pipe and can listen at the screwdriver handle to hear whether the water is moving in the pipe, Figure 17-149. The screwdriver acts as a stethoscope by transmitting sound up the shaft.

If the pump is turning and part of the system has heat but the rest of the system does not, the system almost certainly has some air in it, and part of the system is air-bound. Figure 17-150 illustrates an air-bound system.

The air can be removed by opening the vent valves at the highest point of the system. The technician should then investigate the system to determine where the air is entering the system. Air entering a system can have three sources:

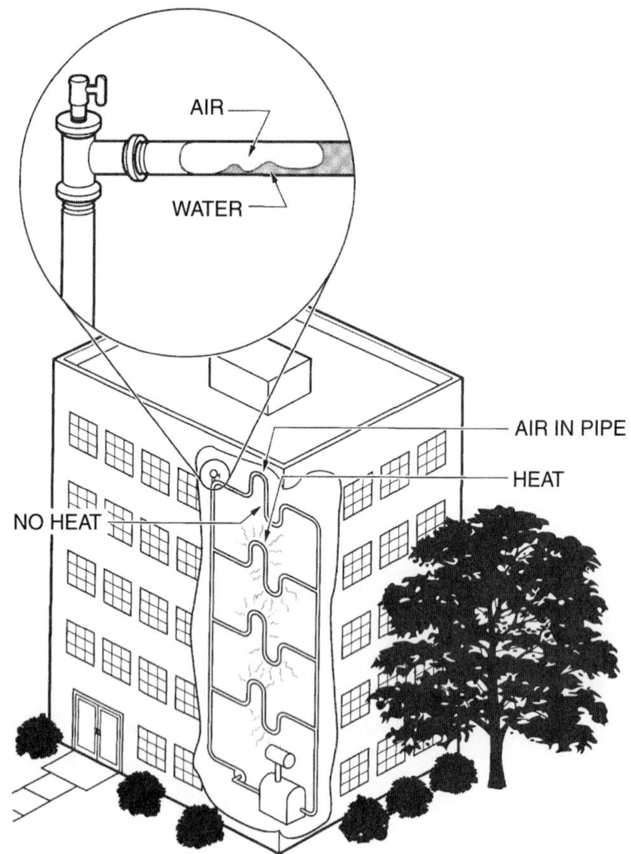

Figure 17-150. Air-bound system.

1. makeup water from the city main
2. draining and refilling the system
3. any part of the system that is operating below atmospheric pressure, such as the pump seal

If the air enters the system in the makeup water, the system may have a leak. Losing water on a regular basis is detrimental to the system because the fresh water contains oxygen, which causes rust and corrosion. The leak should be found and repaired.

If the pump inlet is operating in a vacuum and the seal leaks, air will enter the system, as shown in Figure 17-151. The seal can be repaired. The low-pressure area can sometimes be brought up to atmospheric pressure by raising the system pressure. The pump water flow can be throttled with a valve in the pump inlet, as shown in Figure 17-152, but doing so is not a good practice because cavitation can occur. The pump outlet should instead be used to throttle the water volume.

Problems at the terminal unit involve either the water side of the unit or the air side. Drapes or other obstructions may block the air on the air side of base-board convection system. Baseboard convectors may also have dirt and lint buildup on the coils that reduces the airflow.

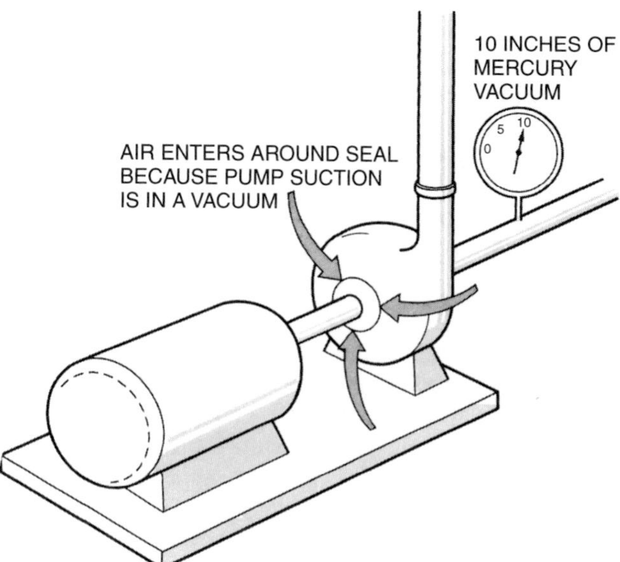

Figure 17-151. Pump inlet operating in a vacuum.

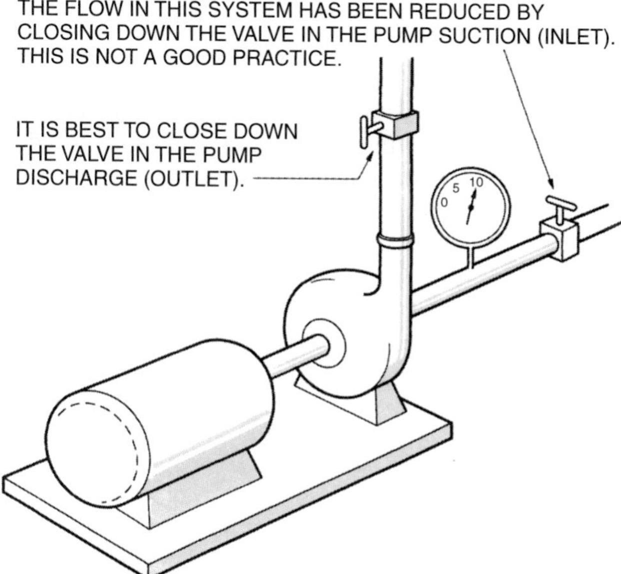

Figure 17-152. The water flow is throttled at the inlet of this pump. This is an incorrect pumping practice for a centrifugal pump.

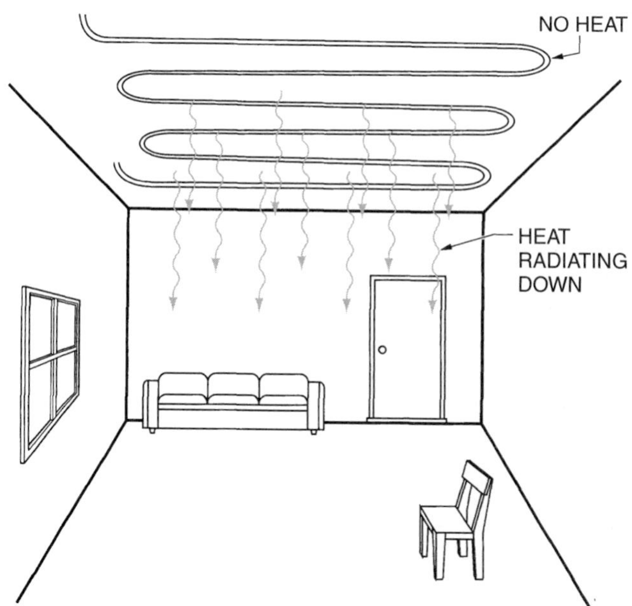

Figure 17-153. This panel heater has heat in only part of the coil. Suspect air in the water system.

Radiant heat systems are much the same. The technician can trace the water circuit by touching the panel. If the circuit is hot only part of the way around the system, suspect air in the circuit, Figure 17-153.

When water is flowing through the system and the system is not heating to capacity, the technician should determine whether the water is hot enough. Typically, the water temperature leaving the boiler may be as low as 180°F. This can be increased to 200°F in most systems. It is always best to know what the system designer had in mind and what the initial design temperature was. This information can be obtained from the prints and specifications for the building.

When bitterly cold weather occurs, some systems may not have enough capacity to heat the structure. Often the heating system will heat a structure according to the design outdoor temperature (e.g., 0°F), and the temperature may drop below 0°F for a long time. If the boiler is cycling off and on, more heat is available, but the system is not moving the heat to the structure. Often the boiler thermostat can simply be turned to a higher temperature.

Caution: Look at the tag on the relief valve for a maximum temperature. Do not turn the thermostat past this temperature, or the boiler relief valve may relieve. Do not exceed 200°F.

Often a system has too much heat in one area and not enough in another. Office floor plans are frequently rearranged. The rearrangement may cause the heating load to shift from one area to another. If so, the technician must rebalance the load by changing the water circuit (when convectors or panel heat is used). If a forced-air distribution system is used, the air may need to be redirected to move the correct amount of heat to each area. Either way, the technician must balance the system. This is often accomplished by trial and error. Any major changes should be noted on the building plans.

The technician should always look for anything unusual: closed air registers, blocked return grilles, dampers that have been changed. False loads on a building may be discovered, such as fresh-air dampers that are wide open rather than at the correct settings. These open dampers will cause too much cold air to pass over the coil. Exhaust fans that run all the time and are not needed will bring in too much fresh air; so will doors that are left open while someone unloads items from a truck and brings them into the building. Windows that are left open will also increase the load. On the other hand, an open trap door to an attic will act as a chimney and move air from the structure. This air must be made up with fresh air.

A building usually has an energy-saving night set-back thermostat that maintains only a minimum temperature in the building at night. Most building operators use 55°F as the minimum night setback temperature. If the furnishings, floors, and walls in a building are allowed to become colder than 55°F, they will be cold for several hours after the heating system restarts. As a result, the occupants will be uncomfortable. Typically, the building operator changes the temperature to night setback at approximately 4:30 P.M. on a working day and allows the system to resume the comfort temperature at 6:30 A.M. so that the building is up to temperature by 8:30 A.M. In cold weather, the heating system may need to be started earlier than usual. The same is true after a weekend, when the system is off from 4:30 P.M. on Friday until 6:30 A.M. on Monday.

The technician can use thermometers to verify that an air handler is functioning to capacity. With the air-flow and the water flow obtained from the building prints, the technician can check the capacity.

17.10 Preventive Maintenance for Hot-Water Systems

A convector or a radiant panel system requires little maintenance on the air side of the system. A convector can have an airflow blockage, such as drapes in the convector or objects dropped into it. Some convectors have dampers that regulate the amount of airflow across the convectors. The technician should determine whether the damper is functional. Convection unit coils can become clogged with dirt and lint. If they are dirty, the covers should be removed and the elements cleaned to get the most efficiency from them. In addition, air-vent valves can be opened so that the technician can be sure that only water is in the pipes.

The forced-air system has a fan, filters, a coil, duct work, return and supply grilles, and registers that must be maintained and checked. When reduced heat is detected, the technician should make sure that the fan is operating at the correct speed. A fan belt may be loose if a belt-drive system is used.

Caution: Do not overtighten the fan belt; doing so will cause motor overload and possibly bearing wear.

When the fan is direct drive, the fan motor may be a permanent split-capacitor motor. A defective capacitor will cause the motor to turn slower than usual and possibly trip off because of overload.

Good filter maintenance is necessary, or dirt will pass through the filters and stop at the next component, the coil. Reduced airflow that results from dirty filters or a dirty coil reduces the heating capacity of the unit. This problem may not be noticed until the first cold day after the airflow is reduced.

Coils cannot be cleaned with only a vacuum cleaner. They must be cleaned with water and a detergent approved for coil cleaning. Consult your supply house for a cleaner that is compatible with the copper coil and the aluminum fins.

The fan wheel may become clogged with dirt in extreme cases. If this occurs, the blades must be cleaned or the fan capacity will be reduced. The fan wheel should be removed for cleaning. Otherwise, if the dirt is scraped out of the fan wheel while it is in place, much of the dirt will fall into the fan housing and when the fan is started, the dirt will blow into the duct.

Boiler maintenance involves the controls, the burner (in a fossil fuel system), contactors (for electric heat), the flue, and the water side of the system. The various burner and element maintenance procedures are discussed in detail in Units 5, 10, and 14, the units that cover gas, oil, and electric heat.

The controls must be checked for loose connections and frayed wires. All safety controls should be checked to verify that they will shut off the system when needed. The boiler thermostat can be checked by firing the boiler and waiting until it reaches the set-point temperature. At this point, the thermostat should shut off the boiler. If the boiler has an upper (or high) limit control, make sure that it is wired into the circuit and will shut off the boiler.

Check the flue to make sure that it is venting correctly. To do so, hold a match at the draft diverter. Air should be moving up the flue.

On the water side of the system, the low-water cutoff requires a blow down on a regular basis. With a large system, the technician should perform a blow down daily. Also, if the system is to be drained for a particular reason, the low-water cutoff should be checked to ensure that it will shut off the boiler fuel supply in case of low water. It is impractical to drain a system just to check the low-water cutoff.

The boiler relief valve should be opened to clear any rust from around the seat. To do so, raise the lever all the way open, then allow it to spring closed.

Warning: Make sure that you stand away from the exhaust of the valve, or you may be burned.

If the valve continues to drip because of a leak, open it again. Something may be under the valve seat that will blow out when the valve is opened again.

The water in all hot-water systems should be treated. Smaller systems are more likely not to be treated, even though they should be. Water treatment, in which the water is checked for the mineral content and the acid level, should be performed by a qualified member of a professional water-treatment company. The water-treatment professional will add the correct amount of chemicals to the water to make it almost neutral in acid content. This prevents corrosion. Minerals in the water may be kept in suspension with other chemicals. Without these chemicals, the minerals will deposit on the hottest part of the system, the boiler tubes. Rust inhibitors are also added to the water to slow rust and corrosion.

All pumps must be checked to ensure that they are lubricated and operating correctly. Any pump that is noisy must be checked. Pump or motor bearings may be nearly worn out. The time to replace them is before they fail, if you can. It is much easier to change a pump or a motor when the weather is mild, rather than on a Saturday night in the middle of winter.

Strainers should be checked each year before the system is started. With proper valve placement, this is easy. Shut off the valves and drain the pump and strainer. Remove the strainer basket, clean it, and replace it. Open the valves and allow water to enter the pump and strainer. Bleed all the air from the pump and strainer at the pump if there is a bleed port.

Check all valve stems for leaks. Repair them as needed.

In general, a visual check of the complete system often uncovers many small system problems in time for them to be corrected before they malfunction.

17.11 Installation and Startup of a Hot-Water System

Once the hot water system has been designed the next step in the process is installing, testing, and finally starting up the system. If the system was designed by an engineer or a design company, before starting the installation process always check the drawings and/or specifications for the type and quantity of the materials to be used. Never start an installation project without having the materials readily available. Never under any circumstances should you deviate from the design submitted without first consulting the individual and/or company responsible for designing the system. The installation process can be accomplished in several steps:

- Installing the heat source
- Installing the piping and testing for leaks
- Wiring the unit
- Filling the system
- Starting the system

Installing the heat source

When installing the heat source the following guidelines should be followed:

- Make sure that the heat source is level, since it may not operate as intended if it is not level.
- If the heat source uses a combustible fuel like gas or oil, try to locate the heat source as close to the chimney as possible.
- Always check all local and state building codes for fuel piping requirements. In addition always follow all manufacturer specifications and recommendations—except when they are in conflict with building codes (this should never happen but if there is a discrepancy, contact your local building code enforcement office and/or the manufacturer for clarification).
- Verify that the amount of air introduced into the area around the heat source is adequate to support combustion.
- Verify that all packing materials and/or shipping materials and constraints have been properly removed.
- Install the pressure relief in the tap specified by the manufacturer.
- Verify that there is adequate clearance around the heat source for service, especially around service panels and connections.
- Install a disconnect either on or within 2–3 feet of the heat source.

Installing the piping and testing for leaks

When installing the piping the following guidelines should be followed:

- All vertical pipes in the circuit should be kept vertical and all horizontal pipes should be kept horizontal. The exception to this is when a pipe must be pitched for drainage. In this case, the pipe should have a minimum slope of ¼ inch per horizontal foot. For example, if a 10-foot pipe is required to be sloped for drainage, the amount of drop in the pipe can be calculated as follows:

$$10 \text{ feet} \times 0.25 \text{ inches/foot} = 2.5 \text{ inches of drop}$$

- Try to plan piping layouts that minimize the number of joints and fittings. If at all possible use 20-foot straight sections of pipe on long runs.
- Try to minimize the number of changes in direction.
- All components that may require servicing should have adequate space for servicing.

- When laying out a piping system always use the centerline of the pipe rather than its edges to determine dimensions and measurements.
- Piping layouts should be done with respect to major pieces of equipment such as boilers, circulators, mixing valves, etc.
- Provide adequate bracing and support for the piping system in accordance with all local, state, and/or manufacturer requirements and recommendations. It is extremely important that there NOT be any sag in the piping.
- When a pipe has to go through a structural support, always consult local and state building codes relating to drilling recommendations and guidelines. If guidelines are not provided, then the optimum location for a hole in a structural framing member is centered between the bottom and top edges of the structural member.
- Never notch into the bottom of floor joists, ceiling joists, and/or trusses to accommodate piping. This will compromise the strength of the structural member and can result in structural failure.
- Dry-fit pipe fittings prior to soldering. This can help avoid excessive measuring, recutting, and soldering.
- Leak check all tubing and piping, especially if the piping and/or tubing is to be placed in concrete.

Wiring the unit

When wiring the unit the following guidelines should be followed:

- All wires and connections should be labeled to avoid confusion. An incorrectly connected electrical system can cause unpredictable results, equipment damage, and injury or death (depending upon the amount of current provided to the circuit).
- Never jumper overload protection devices. If it is necessary to jumper a device, it should be done only on a temporary basis and only according to manufacturer recommendations and/or specifications.
- Carefully read and follow all wiring diagrams supplied by the manufacturer.
- Ensure that the controls being used are designed to operate and/or perform the desired task.
- Follow all local and state building codes as well as fire codes regarding wire sizes, connections, device locations, etc.
- Locate all electrical wiring, connections, devices, etc. above the piping. This will help prevent the possibility of electrical shock if the system develops a leak.
- Ensure that all equipment is properly grounded and GFI circuitry is installed in accordance to manufacturer specifications and/or local and state regulations.

Filling the system

Before the system can be filled the desired system pressure must first be determined. This is done by first calculating the amount of water pressure necessary to lift the water to its highest point in the system. Measure the distance from the circulator to the highest point in the piping system. In general one pound of water pressure will produce 2.3 feet of lift. Once the amount of pressure required to reach the highest point in the system has been determined, to have adequate pressure at the highest point in the system an additional 3 psi must be added. If the system contains a diaphragm-type expansion tank, check the air pressure inside the compression tank. The air pressure inside should equal the system pressure calculated in the previous step. If necessary the pressure inside the expansion tank can be adjusted using a small battery-operated pump or a bicycle pump.

To help purge unwanted air in the system close the main shutoff valve and open the boiler drain on the first supply-pipe tee. Keep all zones closed except one. Use the fill valve to blow the air from that zone. The water flows first through the system and then into the bottom of the boiler. The air will be pushed through the system's piping, into the bottom of the boiler, and then out the boiler's drain in the top of the boiler.

Starting the system

Most boiler manufacturers will supply a recommended startup procedure with the equipment. When one is available from the manufacturer it should be followed. The installer should call the manufacturer for clarification before deviating from any of the manufacturer's startup procedures. If a startup procedure is not supplied with the equipment, however, the technician should keep the following recommendations in mind:

- Check the operation of all safeties on initial startup.
- Test the operation of all circulators, zone valves, thermostats, etc.
- Make certain that the thermostats control the operation of the correct zone valves and circulators.
- Make certain that the boiler cycles on and off on the aquastat.
- Test the operation of all safety controls.
- Test the combustion efficiency of the boiler (if a combustible fuel is used).
- Determine the carbon dioxide and carbon monoxide levels and compare them with acceptable levels.

Summary

- Hydronic heat is hot-water or steam heat.
- Hot water is moved to the conditioned space from a central boiler.
- Early hot-water systems used large pipes, and gravity moved the water to the conditioned space. The boilers were large compared with modern boilers, which are smaller and more efficient. The older boilers used open expansion tanks with floats to make up for water lost to leaks.
- Servicing the earlier systems is simple.

- The heating system starts with the boiler.
- Boilers use the heat from gas, oil, or electricity and add it to water.
- Water boilers and systems are liquid full of water.
- When a boiler is located on the bottom floor of a building, the full weight of the system water exerts pressure on the boiler. A boiler is often located on the top floor of a building so that the standing pressure of the building is not on the boiler.
- Water exerts a pressure of 0.43 pound per vertical foot. A column of water 2.31 feet high exerts a pressure of 1 psig.
- Low-pressure boilers have working pressures of as much as 30 psig and their temperatures should not exceed 250°F.
- In a large building, a heat exchanger is often used to transfer the heat from the boiler to the building circuit, so the boiler may be a low-pressure boiler.
- Air suspended in water tends to separate from the water in the boiler.
- Air in a hot-water system is detrimental. It causes corrosion and can actually block the flow of water and stop the heating capacity of the system. Air moves to the high point in the system.
- A centrifugal water pump circulates water in a typical hot-water system and is often called a circulator.
- Centrifugal force moves the water through the pump. The centrifugal force of the water moving to the outside of the spinning impeller is converted to pressure difference from the pump inlet to the pump outlet. The water is trapped in the volute (casing) of the impeller, where the impeller is connected to the system piping.
- The pumping capacity of a pump can be increased by using a different impeller. The larger the impeller diameter, the more pressure difference the pump can handle. If the width of the impeller is increased, the pump will pump more volume.
- Centrifugal pump capacities are plotted on a pump curve. As the pressure across the pump (called head and expressed in feet of water) is increased, the capacity of the pump is reduced. As the capacity is reduced, the power to operate the pump is reduced. It does not matter how the pump capacity is varied, throttling the inlet or the outlet will cause the same results.
- The power consumption of a centrifugal pump is in direct proportion to the quantity of water moving through the pump.
- Some pump curves are steep and some are more flat.
- The designer who selects a pump typically selects a pump in the middle one-third of the pump curve to allow for possible mistakes in the pipe sizing and in case the system has more or less pumping resistance than expected.
- The shutoff head of a pump is obtained when the pump discharge is restricted to the point that no water flows through the pump.
- Feet of head can be converted to psig: 2.31 ft of head = 1 psig. Consequently, a technician can check the pressure on the inlet and the outlet of a pump, then convert it to feet of head and determine the capacity of a pump. Selective placement of the gauges is desired. Many technicians use one gauge and two valves to determine the pressure drop across a pump.
- The pump efficiency is the horsepower of energy imparted to the water divided by the horsepower applied to the pump shaft. Each pump has an efficiency curve. The designer should choose the best efficiency possible for a pump and system.
- The small pump (called a circulator) in a residential system has a small pressure rise across the pump from inlet to outlet. This pump seems to only stir the water; therefore, it is very quiet in operation.
- Pumps are made of several types of material. Cast iron and steel pumps are subject to rust and corrosion more than brass or bronze pumps are. These pumps often have stainless steel shafts for the impellers.
- A pump can be mounted in line with the motor (direct drive) or can be belt driven. An in-line pump may be factory mounted on a base or field mounted.
- A direct-drive pump must have a flexible coupling between the pump and the motor to compensate for slight misalignments. If the shafts are in perfect alignment when cold, they will not be in alignment after the system warms up.
- Each pump must have a method of containing the water in the system while turning the impeller. A shaft seal can keep the water inside the pump housing. The shaft seal must seal the rotating shaft of the pump. The seal is a packing-gland seal or a spring-loaded seal that pushes against the shaft. Packing-gland seals are adjustable and require a slight amount of leakage for proper seal lubrication.
- A small pump can use a magnetic drive that turns the impeller with a magnet on the outside of the pump housing.
- Larger pump impellers are used for increased pumping capacities. As impellers become larger, more force is applied to the pump shaft toward the impeller inlet. This force imposes a sideways thrust that must be counteracted with a thrust bearing. Often a pump has a double-inlet impeller to increase capacity and counter this thrust.
- Each pump housing is made so that the pump is serviceable. A small pump can be piped into the system and suspended by the system piping.
- A pump can be mounted on the motor housing, or it can be base mounted. A pump that is mounted on the motor housing is easier to align after the pump is serviced because the housing keeps the alignment straight.
- A large pump is typically fastened to a concrete base or other substantial base, then aligned. When alignment is complete, the pump base and the motor base are drilled and dowel pins are driven in to secure them.

- There are two types of bearings for pumps and motors: sleeve, and ball or roller. Sleeve bearings are typically made of porous bronze or babbitt metal. The bearings must support the weight of the motor shaft while turning.
- Sleeve bearings are quiet, and the motor shaft floats on a thin layer of oil. The correct weight of oil must be used to prevent the shaft from coming in contact with the sleeve, or damage will occur.
- Ball or roller bearings can carry more of a load than sleeve bearings can, but they make more noise. These bearings are lubricated with grease through a fitting on the bearing housing. A grease relief plug must be removed from the far side of the bearing to allow grease to enter the bearing and to prevent excess pressure on the grease seals.
- Proper practices should be followed for the centrifugal pump to perform correctly. The water volume through a centrifugal pump should be throttled at the pump discharge (outlet), not at the suction (inlet). Throttling at the pump suction may reduce the pump inlet pressure below the boiling point of hot water and cause pump cavitation. Cavitation is water flashing to steam.
- Small amounts of air may be introduced into the system through the pump seal if the pump is throttled on the suction side because this portion of the pipe may be in a vacuum.
- The centrifugal pump will not pump air, so all air must be bled from the centrifugal pump before start-up.
- City water pressure is used to fill a new system to a pressure of approximately 5 psig at the highest point. This is necessary to push air out of the system at the highest point.
- The piping entering and leaving the pump should have as few bends as possible to improve pump performance.
- The terminal unit transfers the heat to the conditioned space. These units are either natural or forced draft.
- The design engineer determines the water flow through terminal units. The water flow through small systems is rated in pounds per hour: either 500 or 2,000 pounds per hour. One gallon per minute (gpm) is equal to 500 pounds per hour, and 4 gpm equal 2,000 pounds per hour. Adding 1 Btu to 1 pound of water raises the temperature 1°F. When 500 pounds of water are raised 1°F, 500 Btu of heat are added to the water (8.33 lb/gal $\times$ 60 min/h $\times$ 1 Btu/lb = 499.8, rounded to 500).
- Different types of terminal units require different rates of circulation. The correct flow must be accomplished. Flow that is too fast causes noise; flow that is too slow, called laminar flow, affects proper heat exchange.
- Natural-convection units are baseboard, panel, or radiator units.
- The baseboard convector mounts next to the wall at the baseboard. As air is heated, it becomes less dense and rises, and new, cool air takes its place. A gentle air current is started. These units have decorative trim.
- Baseboard units are rated as a certain amount of Btu per foot at one of the two water flow rates: 500 or 2,000 pounds per hour.
- With panel heat, pipes are routed under the room surface material: in the walls, ceiling, or concrete floor. Ceiling and wall systems operate at higher temperatures than floor systems do.
- Radiators are room heaters with hot water circulating through them. They are typically located along outside walls. Some radiators are cast iron and some are steel.
- A forced-convection system has a hot-water coil and a fan to force air over a finned-tube coil. The system may or may not have duct work to carry the heated air to remote places.
- The location of the coil in the airstream is important in order to have the best coil efficiency. The coil can be placed so that the room return air is in contact with the water leaving the coil (counterflow) or the water entering the coil (parallel flow). Counterflow installations are the most efficient.
- The room forced-air convector has a fan coil unit that distributes air in the room. No duct work is used. This unit often has a chilled-water coil or a refrigeration unit to operate the air-conditioning for the room.
- The unit heater is another forced-convection type of heat. It is typically located in the ceiling and has a fan to distribute air. It is usually used in large rooms, such as warehouses and gymnasiums.
- The design engineer controls the type of piping system that is used to carry the water to the terminal units. Four types of pipe routing systems are discussed in this text: the one-pipe series loop system, the one-pipe system with flow tees, the two-pipe direct-return system, and the two-pipe reverse-return system.
- A larger building usually has air-conditioning and often has chilled water incorporated into the piping system.
- The designer tries to make the job economical along with providing comfort to all zones in the structure.
- System temperature difference is the difference between the water temperature entering the boiler and the water temperature leaving the boiler.
- A small home typically has a one-pipe system with a series loop. As the water progresses around the system, it begins to cool. The total system water flows through each unit, so there is no zone control; there is only one zone.
- A larger home may have several circuits or zones with a pump for each zone. The house may be heated in the areas that are occupied with a correctly chosen zone system.
- In a two-pipe system with a direct return, the unit that is closest to the boiler outlet is also closest to the boiler inlet. This system may have too much flow through the closest unit, whereas the farthest unit may be starved for water. This system requires more balancing to achieve correct water flow.
- A two-pipe reverse-return system requires more piping but is self-balancing for the correct water flow.
- Systems are typically piped with steel pipe (welded connections for larger systems), wrought-iron pipe (black or galvanized with threaded connections), or copper tubing (soldered connections).

- As the water progresses around the system, the pressure begins to drop as a result of friction loss in the piping. Friction loss is expressed in feet of head.
- Pipe is sized from a pipe sizing chart that plots friction loss, gpm, pipe size, and water velocity. If you know any two of these, the other two can be found on the chart. The gpm is usually a known, and the friction loss is obtained from a table.
- Each system has an assortment of valves and accessories to control flow and to allow convenient servicing.
- Five types of hand valves are commonly used: plug cock, globe, gate, ball, and butterfly.
- The plug cock is the simplest and has very little pressure drop. It usually has a square head and is operated with an adjustable wrench. To open the valve, the operator need only turn the valve one-quarter turn. A mark on the top indicates whether the valve is open or closed. If the mark is crossways, the valve is closed; if it is longways, the valve is open.
- The globe valve is common and is constructed with a valve disc and seat. The seal is either a packing gland, an O ring, or a diaphragm. This valve has some pressure drop because the water must make turns moving through the valve. It is used as the hose bib on a home water system.
- The gate valve has a wedge-shaped gate inside that seats closed when it is down and opens when raised. These valves have either rising stems or nonrising stems. Room must be allowed at the top of a rising stem valve so that the stem can rise to open the valve.
- The technician should never open the globe or gate valve completely and leave it that way because the next operator may think that the valve is closed and do damage to it by trying to open it. The proper method is to open the valve all the way, then turn the valve handle one turn toward closed.
- The ball valve is a quick-closing valve that has little pressure drop.
- The butterfly valve has a round disc inside that closes when it is turned crossways of the flow. This valve is a quick-close valve with little pressure drop.
- Control valves in the system stop, start, divert, mix, or modulate the flow of water.
- To modulate is to partially close a valve according to the needs of the system. Most automatic valves either open or close; they cannot modulate.
- Control valves are controlled by electrical circuits, electronic circuits, or pneumatic air.
- All valves respond to temperature-control thermostats.
- Most valves are on or off valves. These valves use a solenoid coil, a motorized actuator, or a rod-and-tube actuator. The motorized and rod-and-tube actuators are more silent in operation than the solenoid is.
- Some system valves either divert or mix. Both have three ports: two inlet ports and one outlet port.
- Pneumatic valves are air operated, using a pressure of 15 psig.
- The Cv factor of a valve is related to the amount of flow through the valve at 1 psig pressure drop.
- Accessories in the system include check valves, flow control valves, flow adjustment valves, flow-checking devices, expansion tanks, strainers, fill valves, air-control fittings, relief valves, low-water cutoffs, and instrumentation.
- Check valves allow flow in only one direction. There are several types: swing check, ball check, and magnetic check.
- The swing check valve must be mounted horizontally or vertically with the flow upward. If mounted vertically with the flow downward, it will not check flow.
- Spring and magnetic check valves have more of a pressure drop than the swing check has.
- Flow control valves prevent hot water from migrating into the circuit during the off cycle when the boiler may still be hot. A fitting on top of this valve allows the technician to clean the valve and manually over-ride it for heat when a pump fails or when the power is off.
- The flow adjustment valve has a known flow rate at a particular pressure drop with a built-in valve for adjustment. The characteristics of the valve pressure drop are printed on a graph furnished with the valve. These characteristics are used to check and balance the flow in a circuit.
- The flow-checking device has a fixed orifice and a pressure port on each side of the orifice that is used to determine the flow of water in the circuit. A graph is furnished with the device for determining the flow rate.
- Air-control fittings are devices used to manually or automatically remove air from systems. The manual valve is turned by hand or with a wrench; the automatic valve has a float mechanism to bleed air. The float drops and opens the valve when air is present. The exhaust from an automatic bleed valve should be piped to a drain if the exhaust water will damage the surroundings.
- Water expands when it is heated from room temperature to the typical system operating temperature. The expansion is taken up in a closed system by a device called an expansion tank, or a compression tank. This tank is positioned above the boiler, and small amounts of air that enter the system can migrate to this tank.
- There are two types of expansion tanks: one has an air pocket common to the system and the other is known as a bladder tank and has an air-filled bladder. A bleed valve is used to set and determine the water level in the air-pocket tank.
- The volume of a system must be calculated in order to properly size the expansion tank. An oversized tank is too expensive, and an undersized tank will cause the boiler to relieve pressure when the system is heated from room temperature to operating temperature.
- An automatic system must have a method to automatically add water to the system when water leaks out. Automatic fill valves, often called system pressure regulators, operate according to the system pressure. If the system pressure drops below the set point, the valve

allows water to enter and the system pressure to build up to the set point. This valve must have a method of preventing system water, which may contain chemicals, from moving into the city water system if the city water pressure drops below the system pressure.

- Often a system has a bypass valve around the automatic fill valve that quickly allows the system to fill. An automatic fill valve may have a quick-fill feature that is hand operated.
- Each system should have a low-water cutoff to prevent the boiler from operating if the system runs out of water. This cutoff may be positioned anywhere above the boiler heat-exchanger line.
- Relief valves are installed in two typical places in the hot-water system: on the boiler and in the system. The typical small system working pressure is less than 30 psig. The boiler relief valve must not exceed the working pressure of the boiler, 30 psig for low pressure. The relief valve on the boiler must have enough capacity so that the boiler will not build pressure with the fuel supply adding maximum heat, such as when the high limit is stuck closed;
- It is important that the technician replace any relief valve with the correct replacement.
- All systems must have some instrumentation. The larger the system, the more instrumentation. Minimum instrumentation is the boiler gauge, which shows temperature and pressure. More instrumentation involves pressure and temperature instruments at important parts of the system, such as before and after the pumps, on the boiler, and in important circuits.
- Pressure gauges are used to detect excess pressures or pressure drops. Using one gauge and two valves is the preferred method of checking the pressure drop.
- Temperature indicators are used to test system performance. Glass stem thermometers with long (10-inch) scales are the best choice.
- Good piping practices must be followed for good system performance. Expansion of the pipes must be taken into account with expansion loops and room for movement where pipes are routed through the floor.
- All threaded connections must have the correct number of threads, according to the pipe size. The proper thread seal must be used on the external threads only.
- All connections where the pipes are cut must be reamed to remove the thread cutting burrs.
- The pipes should be exactly level or rise slightly in the direction of the water flow so that air can move through the system.
- All pipes routed through unconditioned spaces must be insulated.
- Each water system must have a method for draining the system to prevent freezing if the building is not occupied for long time periods.
- After the system is installed, it must be cleaned and flushed to remove any oil or dirt from construction. Then the proper water treatment should be added to protect the system from mineral deposits, rust, and corrosion.
- After a new system is started, it must be balanced to ensure the correct water flow to the correct circuit. Most technicians start by opening all valves and verifying the total water flow. Then each circuit may be balanced to the correct water flow. The last procedure is to again verify the total water flow at the pump.
- Troubleshooting involves solving either hot-water system or heat-source (boiler) problems.
- Air in the system is probably the most common problem in a hot-water system. Water alone moving in the system makes little noise; air makes noise.
- If excess air is detected in the system, it should be investigated. Air may enter from fresh makeup water, draining and refilling the system, or any part of the system that is operating in a vacuum.
- When water only is flowing in the system and it is not keeping the space up to temperature, the technician should investigate the system for proper water temperature and balance.
- Problems at the terminal unit are either on the water or the air side of the system. Water-side problems are insufficient water flow or water that is not up to temperature. Air-side problems are insufficient airflow or a dirty heat exchanger.
- Often when a system does not keep the structure warm enough in abnormally cold weather, the boiler temperature can be increased. If the boiler is cycling off and on, there is more capacity in the boiler. Do not turn the temperature up higher than the rated temperature of the relief valve temperature setting.
- Too much heat in one area and not enough in another is a balancing problem. The floor plans in office buildings are often rearranged.
- The technician should look for anything unusual, such as closed air registers, dirty air filters, blocked return grilles, and dampers that have been changed.
- False loads on the building, such as doors or windows left open, often occur.
- An incorrect night setback procedure will cause a building to be cold in the morning.
- Water flow and air temperature rise tests may be performed on a forced-air unit if the technician knows the design data.
- Coils must be cleaned when needed with the correct detergent and water.
- Fan wheels may also need cleaning.
- Boiler controls and water treatment must be maintained.
- The technician can check the boiler temperature and high limit controls by making the controls function. If they do not function correctly, they must be replaced.
- The low-water cutoff should be blown down regularly.
- All pumps and fan motors must be checked and lubricated regularly.

Review Questions

1. What two problems are caused by air circulating in a hot-water system?
2. When the pressure rise across a centrifugal pump is increased, does the volume of water moving through the pump increase or decrease?
3. When the water discharge line of a centrifugal pump is throttled down, does the current draw of the pump motor increase or decrease?
4. What three methods are used to prevent water from leaking from water pumps in the drive system?
5. A hot-water terminal unit needs 2,000 pounds of water per hour. This converts to how many gpm?
6. What are two disadvantages of a one-pipe series loop system?
7. What is the main advantage of a two-pipe system with a reverse return rather than a direct return?
8. What procedure should be used when you open a gate valve or a globe valve?
9. How does a flow-checking fixed-orifice device function?
10. What two types of expansion tanks are commonly used?
11. What are the maximum working pressure and the maximum temperature of a low-pressure hot-water boiler?
12. What is the proper location of the low-water cutoff in a hot-water system?
13. What is the relief valve setting for a low-pressure hot-water boiler?
14. What five types of hand valves are commonly used in hot-water systems?
15. Which of the five valves has (or have) the least amount of pressure drop?
16. What is the purpose of flow control valves in a hot-water heating system?
17. How is an expansion tank in a hot-water system sized?
18. What are three sources of air in a hot-water system?

18 Steam Heat

Objectives

Upon completion of this unit, you should be able to

- **use the gauges on a basic steam system to determine the pressure.**
- **describe three basic condensate traps.**
- **inspect a condensate trap for problems.**

18.1 Introduction to Steam Heat

When central heating systems originated, not only was hot water used to provide heat, but also steam was often used. Together, steam heat and hot-water heat are called *hydronic*, or wet, *heat*.

Steam heat is similar to hot-water heat, except the water is boiled into a vapor that collects in the top of a boiler. The steam is then circulated through the conditioned space in pipes. At terminal units, air passes over the steam heat exchanger, removing the latent heat, and condenses it. Heat is added to the air, as shown in Figure 18-1, and the condensate is returned to the boiler. A steam condensate trap is usually placed at the terminal unit outlet to create a liquid seal between the terminal unit and the boiler return piping. Figure 18-2 shows a condensate trap. Large pipe

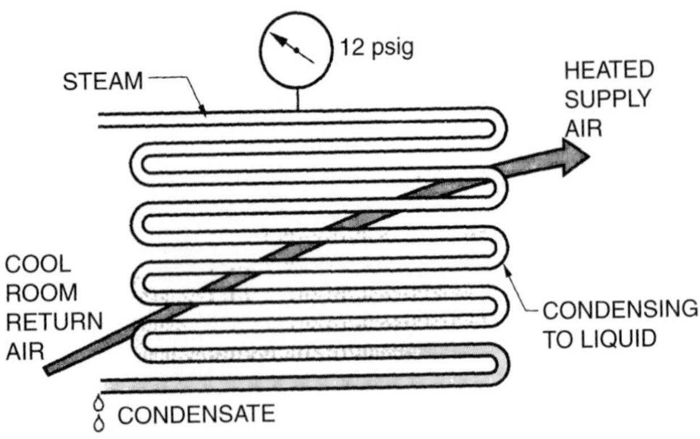

Figure 18-1. When steam is condensed, heat is added to the room air.

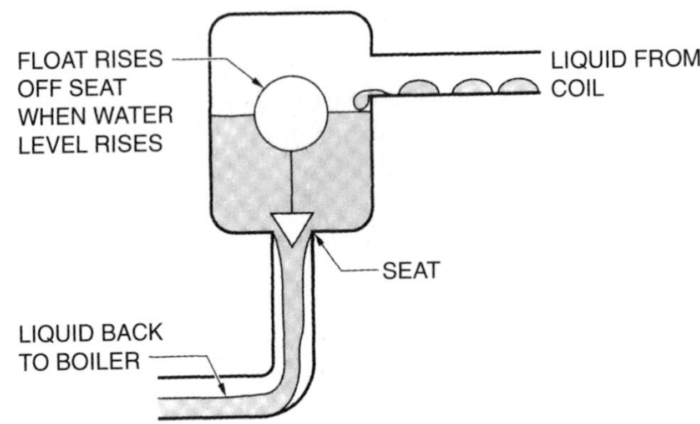

Figure 18-2. Simplified diagram of a condensate trap.

is used in steam systems for the steam vapor, and smaller lines are used for the condensate return to the boiler.

The latent heat contained in the water vapor is 970 Btu per pound at atmospheric pressure. Low-pressure steam systems operate between atmospheric pressure and 15 psig. There is little difference in the latent heat at the different pressure levels, so 970 Btu per pound is the figure that we use in this text. When the temperature of 1 pound of water is cooled from 200°F to 160°F, only 40 Btu of heat are removed. Condensing 1 pound of 212°F steam removes 970 Btu of heat, Figure 18-3. Consequently, much less steam than water must be circulated to heat a space.

The vapor line on steam systems is larger than the liquid condensate return line because water occupies much more space in the vapor state than in the liquid state. One pound of water yields approximately 16 cubic feet of vapor at 10 psig. Steam piping requires special considerations that are discussed in section 18.4.

A steam system with the boiler below the heat-exchange (terminal) units uses a gravity return for the condensate and needs no booster pump. Many older systems were gravity systems, and some are still in use today. When the boiler is above the units, a pump must be used to pump the condensate back to the boiler.

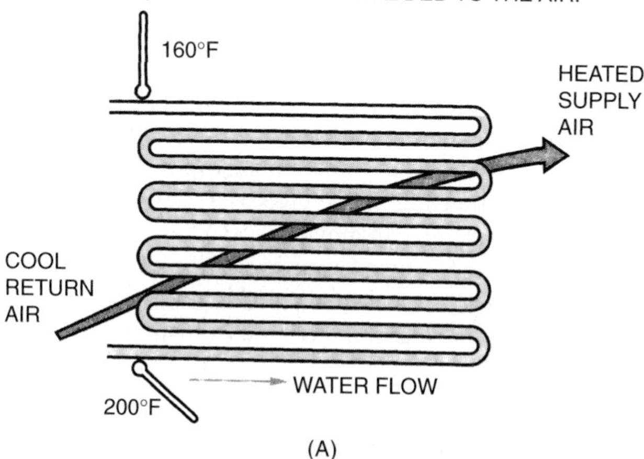

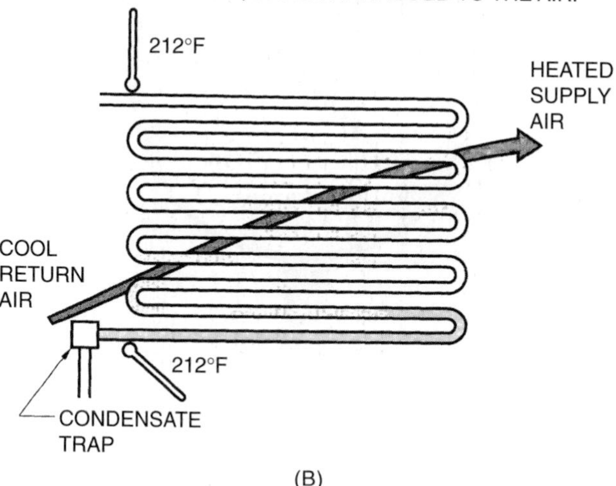

Figure 18-3. (A) Cooling a pound of water versus (B) condensing a pound of water.

18.2 Steam Boilers

Steam boilers are discussed in detail in the various units for fuels (gas, oil, and electric heat, respectively).

Boiler Appearance

The steam boiler looks much the same as the hot-water boiler used for gas and oil heat, but it is different when electricity is used. Electric steam boilers are larger than electric hot-water boilers because of the volume of steam at the top of the boiler. However, electric steam boilers are typically small in comparison with gas and oil steam boilers. Gas and oil boilers also have flues that must be vented to the outside.

Externally, steam boilers look much the same as hot-water boilers do. On the inside, however, there must be room for steam to collect. This volume must be above the point in the boiler where heat is added to the water. This room is called the *steam dome.* The steam leaves the boiler at the top of the dome and moves into the system, as shown in Figure 18-4.

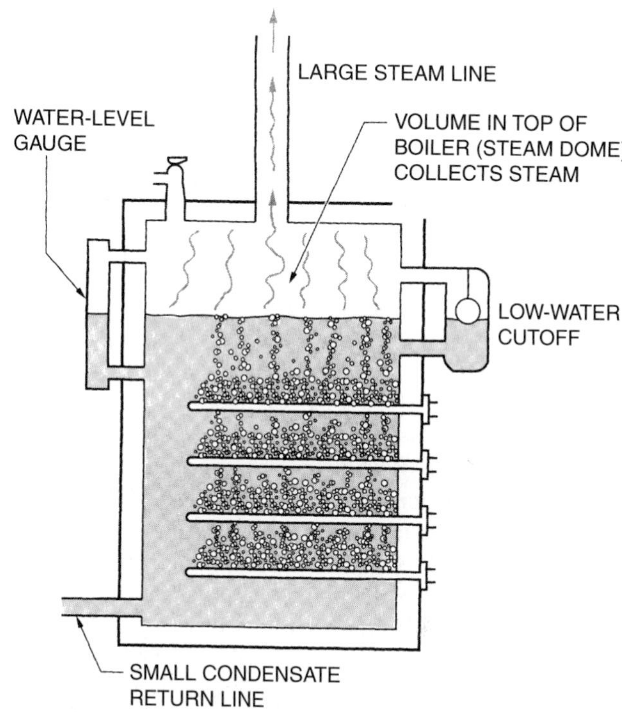

Figure 18-4. The volume in the top of a boiler, called the steam dome, collects the steam.

Boilers are typically placed in categories according to the following characteristics:

- The operating pressure
- The operating temperature
- Type of fuel consumption
- Installation

The Operating Pressure

A low pressure hot water boiler has a maximum pressure of 30 psig, whereas a low pressure steam boiler has a maximum allowable working pressure of about 15 psig. When a steam boiler operates at a pressure greater than 15 psi and over 6 boiler horsepower, it is considered a low pressure boiler. When a water boiler operates at a pressure greater than 30 psi, it is considered to be high pressure.

The Operating Temperature

When a boiler operates at a temperature below 250°F, it is considered to be a low temperature boiler. However, when a boiler operates at a temperature between 250°F–300°F, it is considered to be a medium temperature boiler. Finally when a boiler operates at a temperature range greater than 350°F it is considered to be a high temperature boiler.

Installation

Boiler installation is generally divided into two categories: package boilers and field erected boilers. A package boiler is shipped almost entirely assembled. The only things necessary to start operating the boiler are a foundation, water supply, electrical supply, fuel supply (for fossil fuel burning equipment), and steam connections. A field erected boiler must be constructed on-site because of its size and complexity. Field erected boilers are typically used in large-scale commercial applications.

Boiler Fittings

Low-pressure steam boilers and systems are discussed in this text. High-pressure systems are used in large commercial applications, which are outside the scope of this text.

A typical low-pressure steam boiler has a maximum operating pressure of 15 psig. The operating pressure is limited by the materials and construction of the boiler, which are regulated by the American Society of Mechanical Engineers (ASME). Each steam boiler must have a safety valve to ensure that the boiler pressure does not exceed the working pressure. The safety valve is a spring-loaded relief valve that pops open when the pressure under the seat is exceeded. It does not ease open as a water relief valve does. The boiler safety valve is set to relieve the pressure at 15 psig. Figure 18-5 shows a typical boiler safety relief valve for a steam system. Each steam boiler must have a group of controls that make the system safe; the safety valve is the most important of the group.

Steam boilers range in size from very small home applications to large industrial boilers. All the pressure in a steam system is created by the steam in the boiler. A steam boiler does not have the liquid head (pressure) that a water boiler in a tall building has because the steam rises and moves to the heat-exchange terminal unit where it cools and condenses. Because water takes up less space than steam, the condensing steam creates a low-pressure area to which more steam can move.

Each steam boiler must have a pressure gauge so that the technician can determine the pressure in the boiler. These gauges usually have a range of 0 to 30 psig. Figure 18-6 shows one of these gauges.

Figure 18-6. Steam pressure gauge for a low-pressure boiler range of 0 to 30 psig. *Photo by Bill Johnson*

(A)

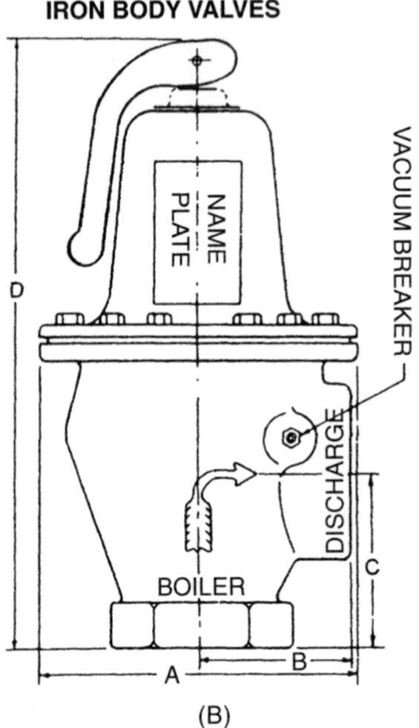

(B)

Figure 18-5. (A) Boiler safety relief valve. (B) Cross section. *Courtesy ITT Fluid Handling Division*

The low-pressure boiler often has a compound gauge to indicate vacuum when the boiler pressure is less than atmospheric pressure. A compound gauge has two scales; one below atmospheric pressure and one above,with 0 being atmospheric, as shown in Figure 18-7. All gauges must be high-quality so that the technician can rely on them to indicate the status of the boiler.

The gauge mechanism must be protected from the steam, so a fitting is installed in front of the gauge. Two types of fittings are used. One is called a *pigtail.* Water condenses in the pigtail and does not allow hot steam to reach the Bourdon tube in the gauge. Another type of protection is the U tube fitting, which has the same function. Both are shown in Figure 18-8.

A gauge valve should also be placed between the pigtail or U tube and the gauge, as shown in Figure 18-9. This valve allows the gauge to be removed from the system and serviced or checked without reducing the boiler pressure.

Figure 18-7. A compound gauge has a scale above and below atmospheric pressure. *Photo by Bill Johnson*

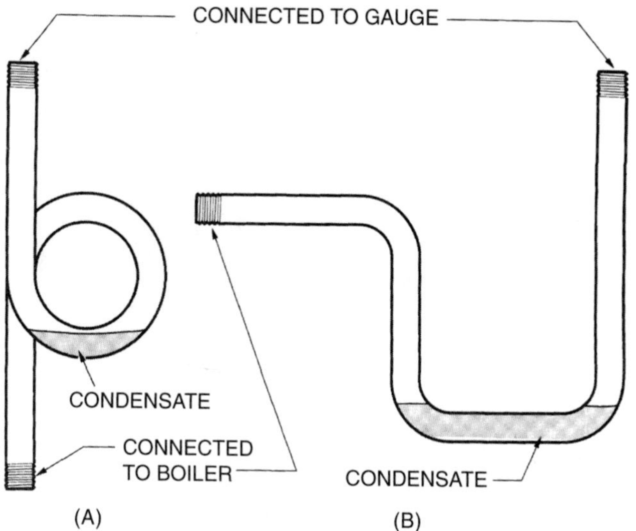

Figure 18-8. (A) Pigtail and (B) U tube protection for a gauge.

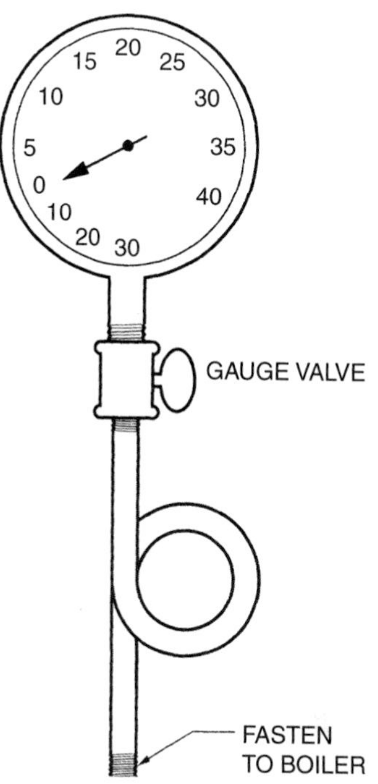

Figure 18-9. The gauge valve is positioned before the gauge so that the gauge can be serviced.

The technician needs to know the water level in the boiler, so a gauge glass is mounted on the side of the boiler to show this level. A gauge glass is used instead of a sight glass in the side of the boiler because the water in the boiler moves up and down while boiling and the exact level would be hard to detect with a sight glass. The external gauge glass is much more stable because no heat is being added to the water in the glass. The glass has a blow-down valve on the bottom to allow the technician to flush water and sediments out of the inside of the glass. The gauge glass must have a valve at the inlet and one at the outlet to allow the technician to isolate the gauge glass from the boiler for servicing, and in case the glass breaks, the glass port can be isolated and the boiler will not have to be shut off.

Water level is critical in a steam boiler because the heat exchanger is only slightly below this level in the boiler. If the water level drops below the heat exchanger, the boiler can be damaged. If the boiler runs dry and the burner fires, the boiler components will quickly overheat and be damaged.

Warning: Never add water to a hot, dry boiler. An explosion will likely occur if you do.

Because the water level is so important, a low-water fuel cutoff control is installed at the water level of each steam boiler. This control has a float that operates a switch to shut off the control circuit to the fuel supply if the water level becomes too low. Figure 18-10 shows a typical low-water cutoff control.

Sludge, foam, or lightweight trash in a.steam system floats on top of the water, so the float mechanism is exposed to the trash of the system. Figure 18-11 shows a

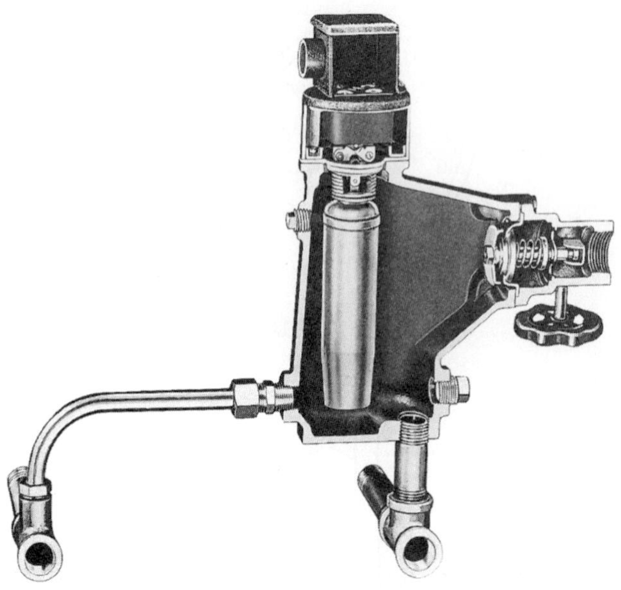

Figure 18-10. Low-water cutoff control. *Courtesy McDonnell and Miller ITT*

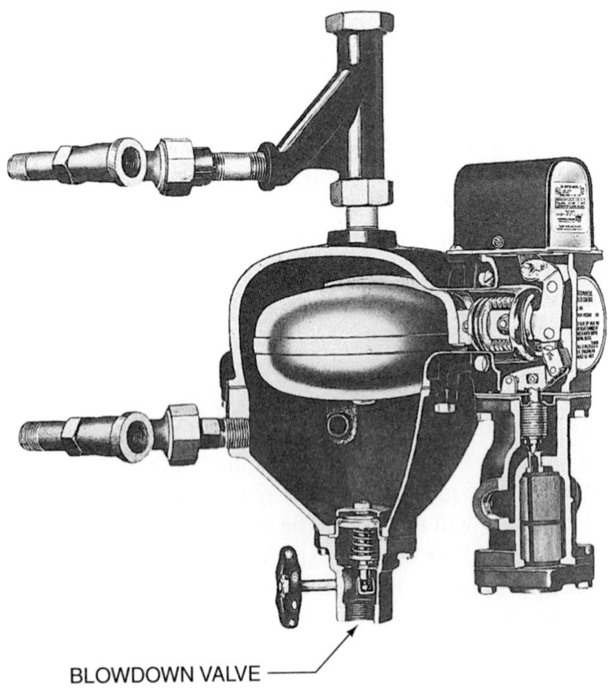

Figure 18-11. Blow-down valve. *Courtesy McDonnell and Miller ITT*

blow-down valve, which is located at the bottom of the low-water cutoff. The technician should periodically open this valve and blow water and steam out the valve. A blow down should be done daily on a large system, but a blow down is not usually performed each day on a residential system. The blow-down valve is a special valve that opens and closes quickly. Only one-quarter of a turn will open it, and when the handle is released, it closes automatically. It closes in this manner because it is opened against spring pressure. The blow-down exhaust pipe must be directed into a drain so that the hot water and steam do not blow onto the technician.

Warning: Do not leave a boiler that does not have adequate low-water protection, or a fatal accident may occur.

The technician is responsible for verifying this control.

When water leaks from a steam system, it must be made up, or the system water level will drop. A combination low-water cutoff and feedwater control, as shown in Figure 18-12, can make up lost water. This control has a float with two switches: One energizes a solenoid to allow city water to enter the boiler. The other is set at a lower level and shuts off the boiler fuel supply if this level is reached.

There may be times when the pressure in the boiler circuit is more than that in the city water main. For example, suppose that the fresh-water makeup valve opens when the city water pressure is at 0 psig. If the boiler has a pressure of 10 psig, water and steam will flow into the city water supply. The water may contain water-treatment chemicals that are harmful to people. Consequently, a check valve must be installed in the line with the solenoid to prevent boiler water from backing up into the city water system. This check valve is often called a "back flow prevention valve."

Every boiler must have a pressure control to stop and start the boiler according to the amount of steam pressure. This control is typically a mercury bulb contact control. The electrical circuit consists of electrical contact made by mercury rolling to one end or the other of the control bulb, much as a room thermostat operates. When the mercury rolls to the end that puts it in contact with the control wiring, the boiler fires. Figure 18-13 shows the operation of this control.

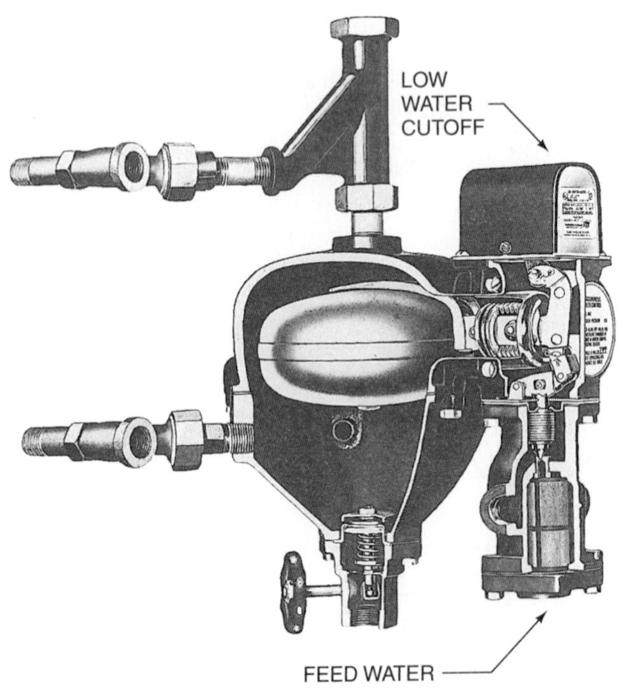

Figure 18-12. Combination low-water cutoff and feedwater control. *Courtesy McDonnell and Miller ITT*

Often a boiler has a low-fire and a high-fire set of contacts in the control. They are two-stage controls that maintain different steam pressures, according to demand. The boiler initially fires at high fire when it is cold. As the pressure increases with the heat from the energy source, the boiler reduces to low fire. For example, when a cold boiler fires, the pressure may be 0 psig. When the boiler reaches 7 psig, the high-fire contacts may open and reduce the burner to low fire. If the boiler pressure were to rise to 10 psig due to light load, the contacts would stop the boiler fire completely, and after the steam pressure in the boiler dropped to approximately 5 psig, the boiler would fire at low fire.

The boiler control has a Bourdon tube that senses pressure and tilts the mercury tube. The Bourdon tube and the mercury tube are mounted on the same plane. Like the pressure gauge mentioned previously, the Bourdon tube must be protected from direct steam. A pigtail connection is often used, as with the pressure gauge. The relationship between the Bourdon tube and the pigtail is important. The mercury tube must be mounted exactly level, as you look into the front of the control. If the pigtail is mounted curled on the same plane as the Bourdon tube, the control may become out of calibration when the pigtail is heated because it will expand slightly and cause the control to be off level. The Bourdon tube must be at a right angle to the pigtail for the control to remain accurate, Figure 18-14.

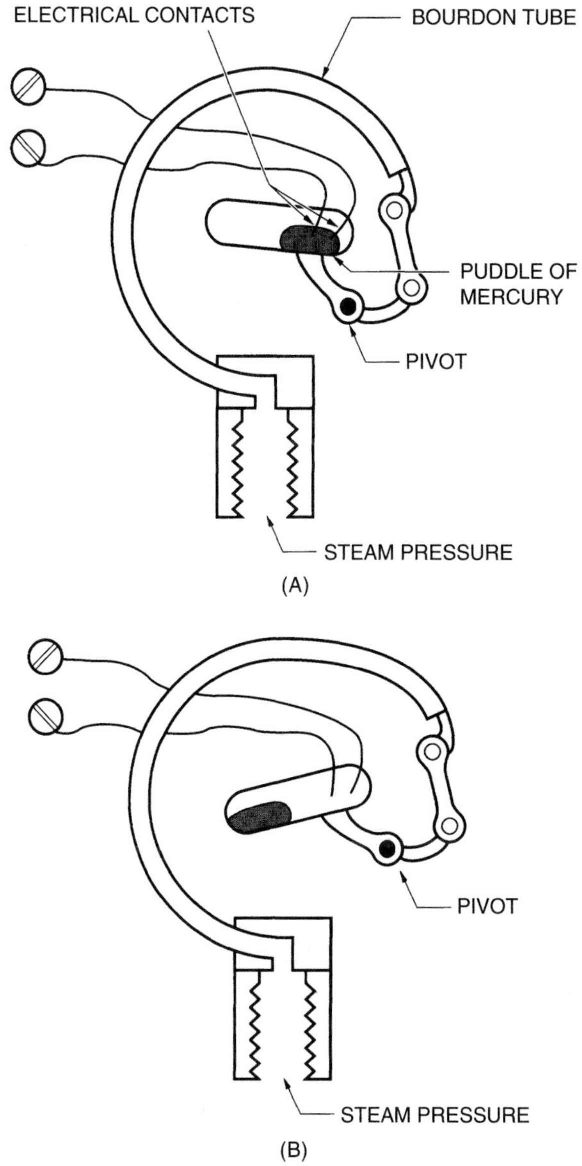

Figure 18-13. Boiler pressure control. (A) The pressure is below the set point, so the mercury puddle makes the circuit between the contacts and starts the boiler. (B) When the steam pressure rises, the Bourdon tube tries to straighten, which turns the mercury bulb on a pivot and causes the mercury to roll away from the contacts. This stops the boiler.

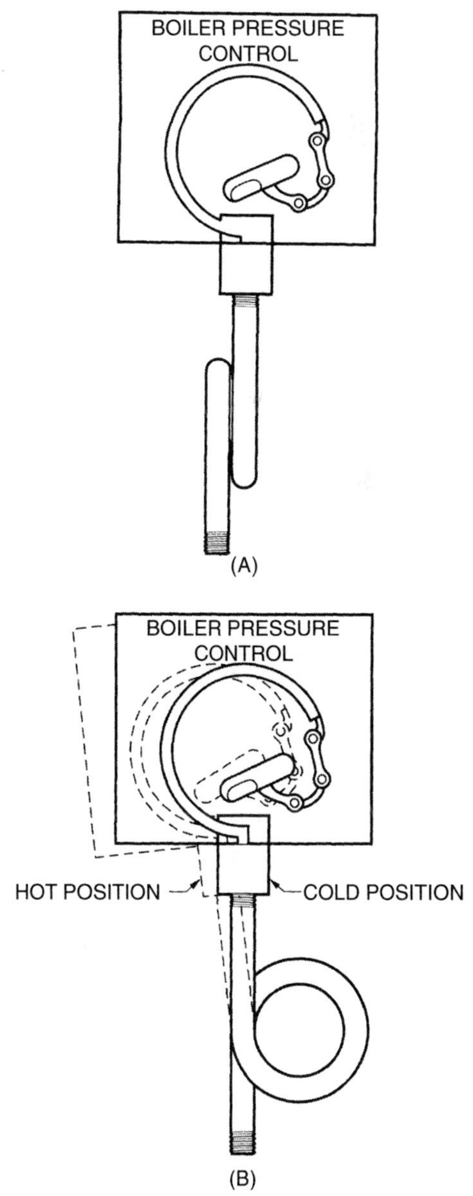

Figure 18-14. Pressure control and pigtail (A) mounted correctly and (B) mounted incorrectly. When the pigtail is heated, it expands. If the control is mounted incorrectly, it will not be accurate. The control can be bent slightly backward and not be affected.

Part of the steam system is liquid and part is vapor. Therefore, pressure differences can occur in the system because of the water seal between the liquid and vapor portions of the system. These pressure differences can cause the water level in the boiler to fluctuate so much that part of the boiler tubes may be exposed to the heat source. If this happens, damage can occur. So that this does not happen, a piping configuration called a *Hartford loop* is used for steam boilers to prevent the pressure difference. With the Hartford loop, which is piped next to the boiler, the water level in the boiler remains level and constant, as shown in Figure 18-15.

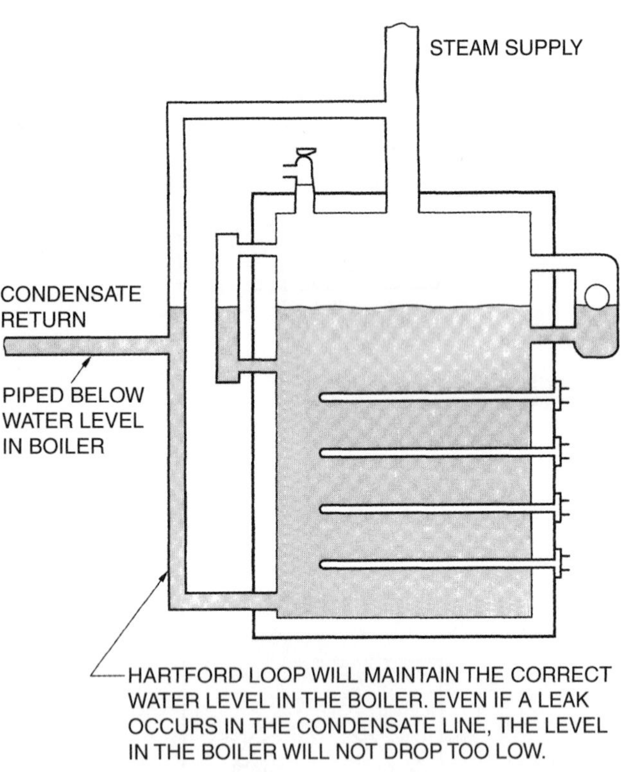

Figure 18-15. Hartford loop.

18.3 Steam-System Terminal Heating Units

As with hot-water heat, several types of heating units add heat to the conditioned space. Some units are room radiators, similar to those used for hot-water heat, see Figure 18-16. All terminal units receive steam and change it to condensate. Some terminal units are forced-air systems, with coils in an airstream, as depicted in Figure 18-17. A terminal unit may also be a heat exchanger for a hot-water system, as shown in Figure 18-18. The heat-exchange unit can be recognized as the point where the large steam line enters and a small condensate line leaves the unit,

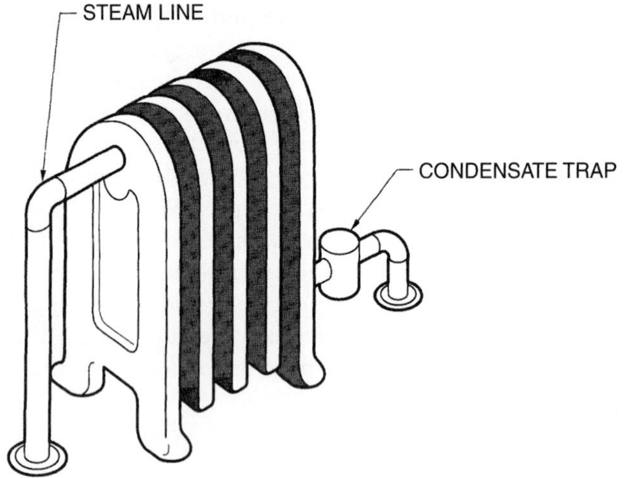

Figure 18-16. Steam radiator terminal unit.

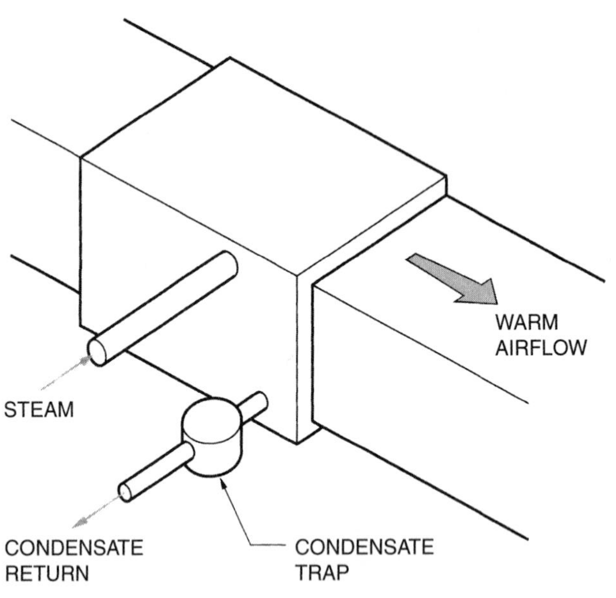

Figure 18-17. Forced-air terminal unit.

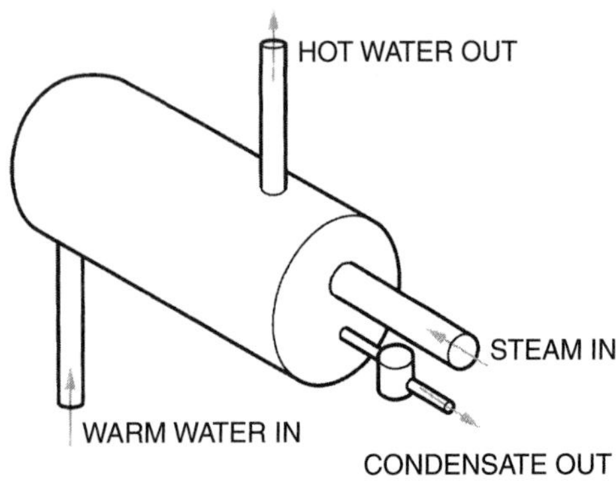

Figure 18-18. Hot-water heat exchanger.

Figure 18-19. All steam heat exchangers change steam to condensate. Because vapor enters the unit and liquid leaves it, there must be a division point. The division point occurs at the steam condensate trap, as shown in Figure 18-20. For best efficiency, the liquid at the bottom of the steam coil must be minimal. Any liquid buildup in the coil can cause water hammer and possibly damage the coil. To understand water hammer, imagine a pound of water moving at a high velocity when it hits an elbow. When it hits, it sounds like a hammer striking a surface, Figure 18-21. Because of its weight, water moving at a high velocity can do damage when it reaches a turn in the pipe.

Terminal units using steam are rated by *square feet of steam*. There are 240 Btu per square foot of steam. This has nothing to do with the area of the appliance, but applies to the rating of the appliance. A unit with 24,000 Btu per hour of capacity is the same as 100 square feet of steam.

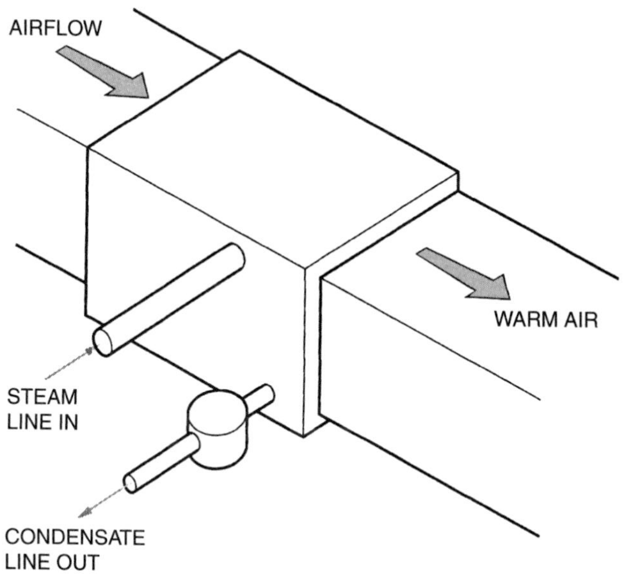

Figure 18-19. The large line is the steam vapor line, the small line is the condensate return.

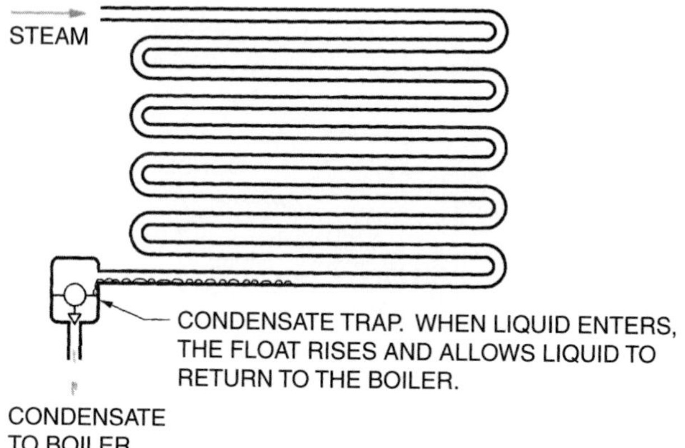

Figure 18-20. The steam condensate trap creates a liquid seal between the vapor and the liquid in the system.

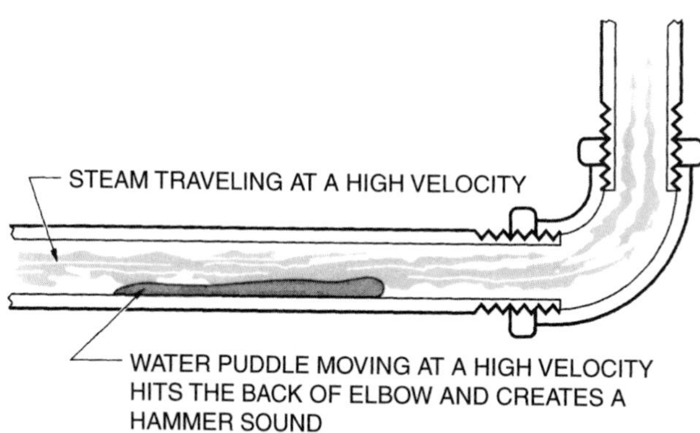

Figure 18-21. Water hammer.

There are several types of steam-system condensate returns. All have the same purpose: to return the liquid to the boiler. For now, understand that each steam terminal unit must have a liquid seal method to prevent steam from returning to the boiler. Sometimes a system uses a liquid seal farther from the terminal unit to reduce pipe costs. This type of system, a one-pipe system, is discussed in Section 18.4.

Most often, a system uses a condensate trap. This trap is extremely important to the system because it maintains a liquid seal to prevent raw steam from moving to the condensate return mains. Because a pressure difference moves the steam in the system, if steam moves into the condensate return system, the pressure difference is reduced, thus the flow of steam in the system is also reduced. With the reduction in steam flow, the system capacity reduces. A system with a condensate pump in the return system will lose raw steam to the atmosphere if steam is returned to the pump reservoir. The trap also traps any noncondensable vapors in the terminal unit, and the air vent releases them to the atmosphere, as shown in Figure 18-22. If the trap holds back too much water, water will back up in the coil and reduce the efficiency of the coil, as shown in Figure 18-23.

Several types of condensate traps are currently used: mechanical, thermostatic, and kinetic. Let us first discuss the mechanical trap. The condensate trap must be able to handle the large volume of condensate while the system is heating up to temperature as well as the reduced flow when the system is up to temperature. The simplest mechanical trap is a float trap. When the system starts, the trap contains a lot of condensate because the cool terminal unit will condense more steam to condensate. The float opens and allows this liquid to flow through the condensate return piping. When the system begins to need less heat, steam tries to return. When the steam reaches the trap, the float drops and does not allow it to pass, as shown in Figure 18-24.

Two problems can occur in a float trap system. Air can be forced through the system with the first steam, and this air does not leave the trap by itself. The air does not allow the float to open, so the trap remains closed. The air could be bled manually, but this would not be an automatic operation. One solution is the mechanical combination trap, which has a thermostatic element added for air elimination. (This is different from the thermostatic condensate trap described later in this section.) The other problem that can occur inside the float trap is wear on the float. If the float rubs the walls of the trap, the constant movement of the float can rub a hole in it. The float then fills with water and sinks. When it sinks, it closes the trap and does not allow the condensate to return, as shown in Figure 18-25.

The inverted bucket trap is much like the float trap. It has an inverted chamber. When liquid enters the body of the trap, the bucket falls and opens a port. When steam enters, the bucket rises and closes the port, as shown in Figure 18-26. A small air bleed port allows air to escape the inverted chamber. Without this air bleed, the air would displace the steam. When air or condensate moves through the trap, the inverted bucket drops and allows the air or condensate to move into the condensate line. The small bleed hole in the top of the bucket prevents air and other noncondensables from remaining in the trap and blocking

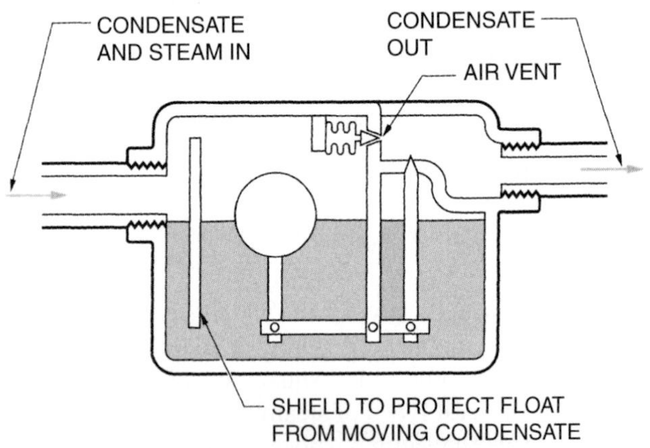

Figure 18-22. The air vent in the condensate trap allows air to escape the system.

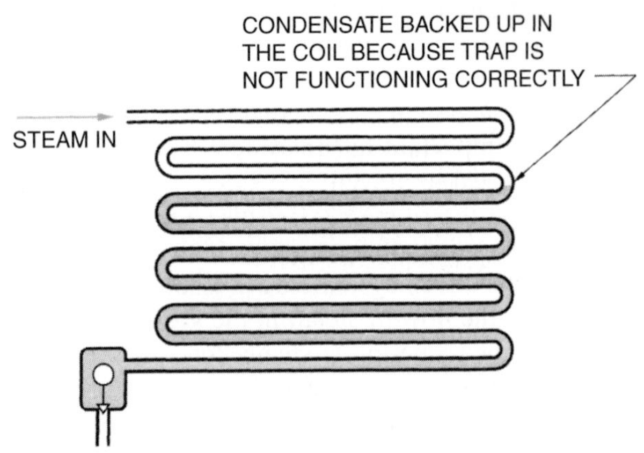

Figure 18-23. This condensate trap is not functioning properly.

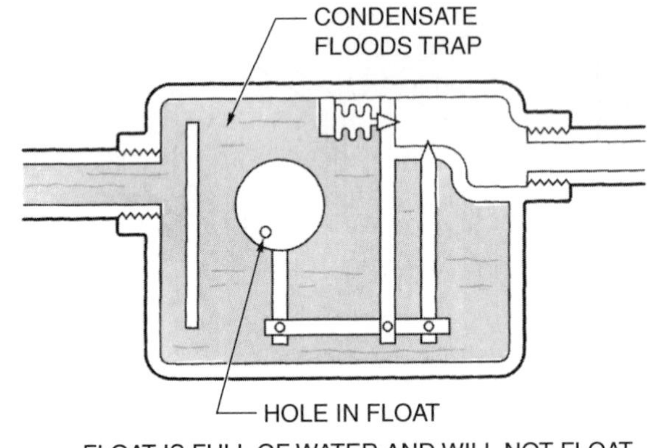

Figure 18-25. This float has a hole.

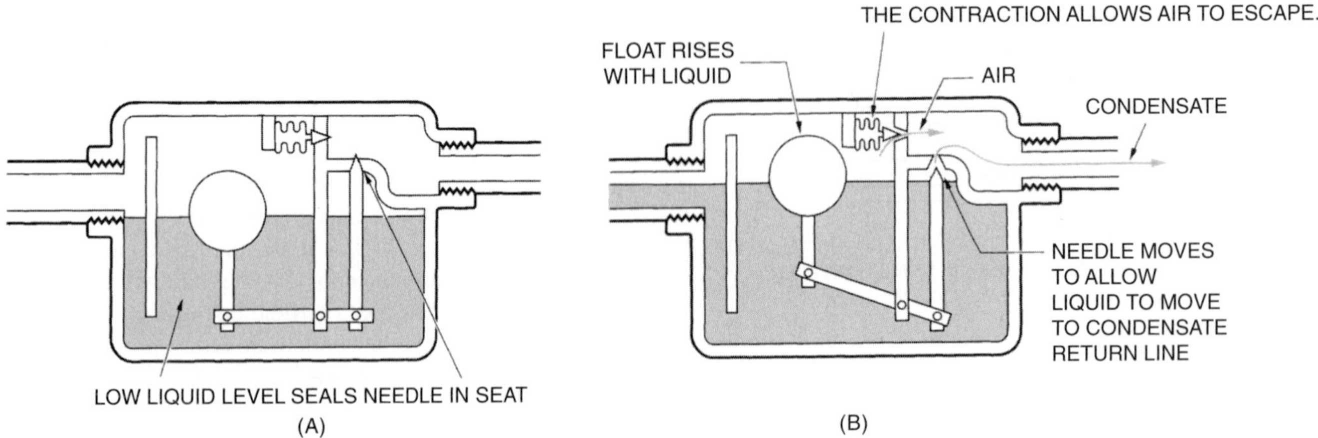

Figure 18-24. How a float trap with thermostatic air elimination works. (A) Steam cannot return to boiler. (B) Condensate and air return to boiler.

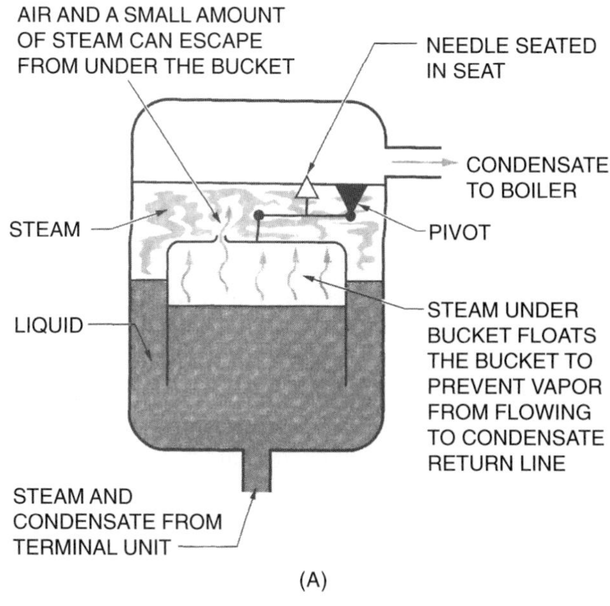

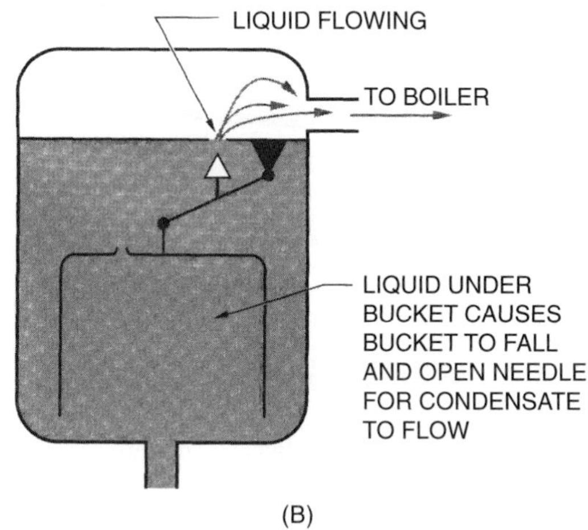

Figure 18-26. Inverted bucket trap operation. (A) Bucket rises and prevents steam from returning to boiler. (B) Bucket falls and port opens so that condensate can return to boiler.

condensate flow. When air or noncondensables gather in the top of the float, they escape slowly to the orifice of the trap.

The thermostatic trap is located away from the unit, with several feet of pipe between the trap and the unit. This trap functions by the difference in the steam temperature and the condensate temperature. It has a chamber filled with a fluid to sense the temperature of the condensate or steam that is entering the trap. When the element is hot because steam is present, it expands and shuts off the flow out of the trap. When condensate that is cooled below saturation temperature (*subcooled*) is present, the element contracts and allows the contents to flow out of the trap. See Figure 18-27.

The third type of trap is called a kinetic, or impulse, trap. This trap has a movable disc inside that works on the principle that moving steam has less energy than moving air or liquid does. When air or condensate is moving, the trap opens and allows flow.

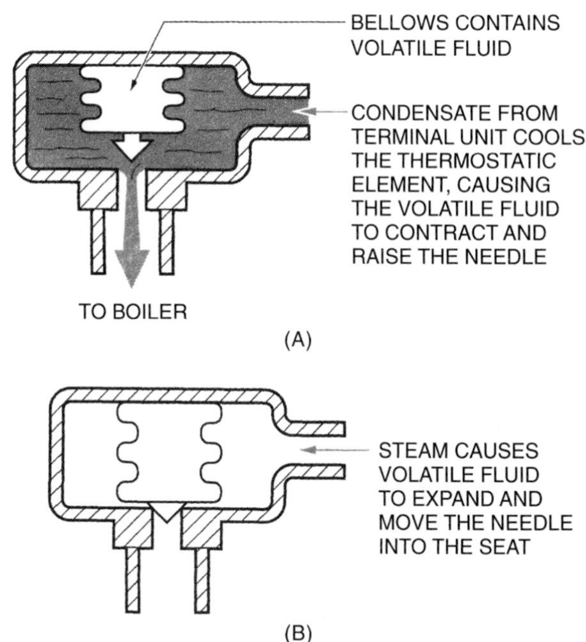

Figure 18-27. Thermostatic trap operation. (A) Condensate can flow. (B) Steam cannot flow.

Regardless of its type, the condensate trap has a very small bore in the regulating device that can be blocked by any particle that may be moving in the system. A fine mesh strainer must be located before each trap to stop migrating particles. The typical strainer is called a Y strainer. Proper valve placement allows the technician to isolate the strainer and service it. The strainer has a removable end.

Warning: Remove the strainer cover only after the proper valves are closed and the pressure is bled out of the trap.

The terminal heating unit and the steam trap work together as a system, served by the steam piping entering and the condensate pipe leaving. Figure 18-28 shows this relationship.

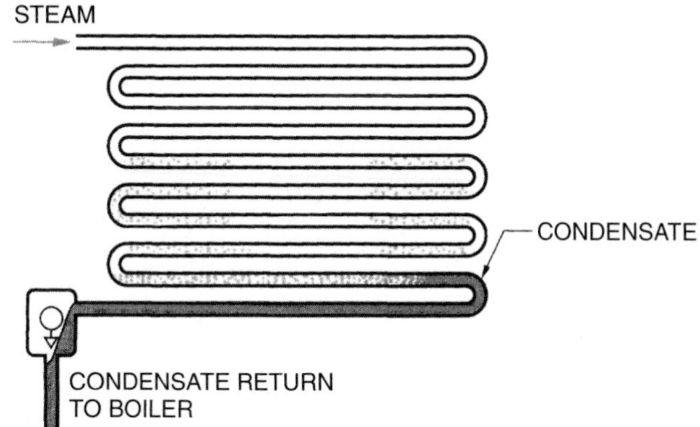

THE CONDENSATE TRAP MUST MAINTAIN A LIQUID SEAL WITHOUT ALLOWING TOO MUCH CONDENSATE TO BACK UP IN THE COIL.

Figure 18-28. The condensate trap and the steam coil work together.

18.4 Steam-Heat Piping

Steam-heat piping is either black iron pipe with threaded connections or steel pipe with welded connections. All threaded connections should have the correct number of threads per pipe size with the correct pipe compound, or pipe dope, applied to the external threads of the connection, as shown in Figure 18-29. Welded systems are erected by professional pipe fitters who must follow the correct piping practices.

Special Considerations

Because part of the steam heating system is vapor and part is liquid, special considerations must be made for each part of the system. Steam naturally rises, so most steam systems are designed with the steam lines pitched upward, as shown in Figure 18-30. The pressure difference between the boiler and the terminal unit creates the flow of steam. This pressure difference is created by condensing the steam in the terminal heating unit. When the steam condenses, the volume is reduced. The reduced volume creates a low-pressure area into which more steam can move, as depicted in Figure 18-31.

When a system is first started in the morning or at the beginning of the season, the pipes are cold and much of the initial steam entering the system condenses in the cold pipes. If the pipes are pitched upward in the direction of the steam flow, any condensation taking place near the boiler can run back to the boiler, as shown in Figure 18-32. Larger systems are more complex. When the steam mains are farther from the boiler, it is not practical to drain the line condensate back to the boiler. Instead, the condensate is allowed to move with the steam to the next riser. (A riser is a vertical main or branch line.) The condensate is then allowed to fall to the bottom of the riser, where it is collected and trapped. It is

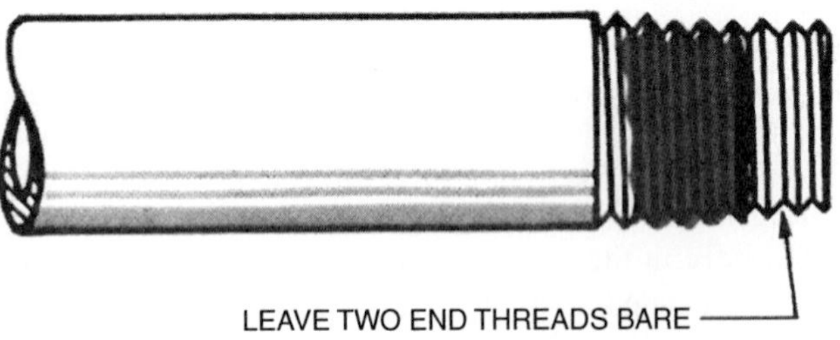

Figure 18-29. Pipe compound, or pipe dope, must be applied to the external threads.

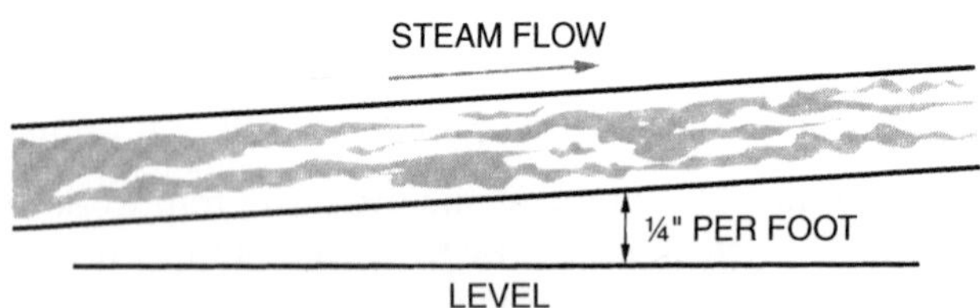

Figure 18-30. Steam line pitched upward.

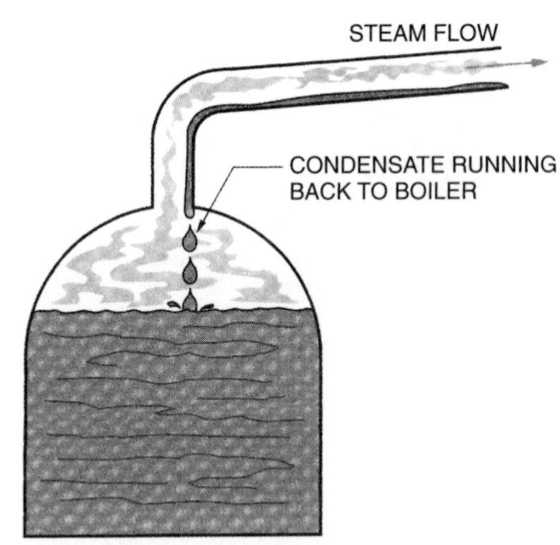

Figure 18-32. Condensate runs back to the boiler in the steam line in this example.

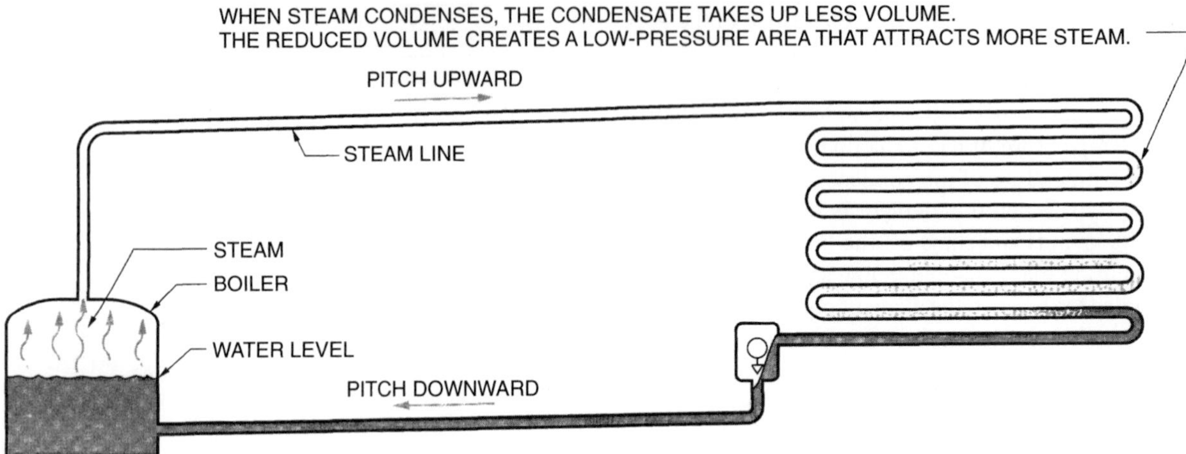

Figure 18-31. The condensing steam creates a low-pressure area that attracts more steam.

then drained back to the condensate main piping to be returned to the boiler, Figure 18-33. These traps operate just as the traps in the terminal unit do. They maintain a liquid seal and allow liquid to return to the boiler.

Each portion of the steam piping system must be carefully planned. Steam piping expands more than piping in a typical hot-water system does because there is more of a temperature difference in a steam system. When pipes are heated, they expand. When they cool, they contract. Space must be provided for the expansion. Several methods are used to provide this space. Figure 18-34 shows some of these methods. If room for expansion is not provided, leaks will occur where the pipe is stressed. Noise from dimension changes in piping is also a factor in steam piping.

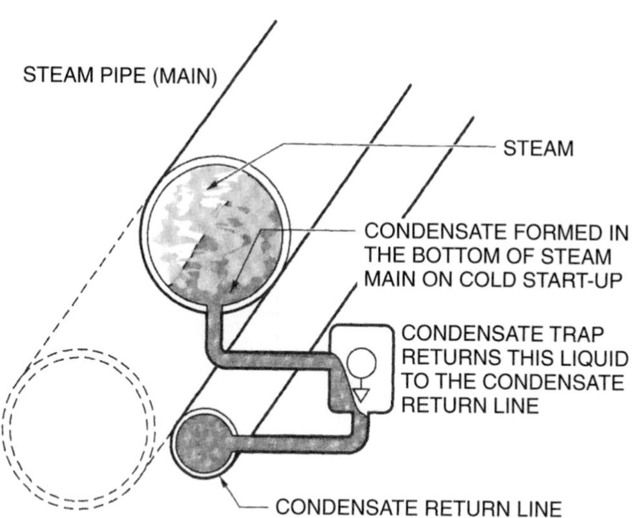

Figure 18-33. Condensate traps in the steam mains function the same as those at the coil.

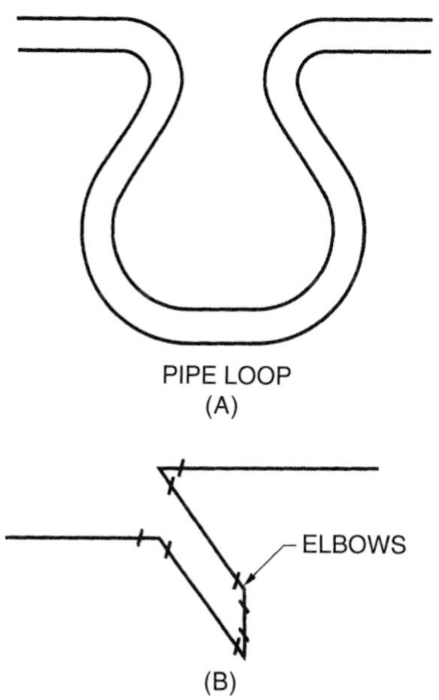

Figure 18-34. Provision for expansion. (A) Pipe loop. (B) Pipe leg allows for expansion.

Special fittings are used in the steam mains that encourage condensate movement. These fittings are called *eccentric fittings.* Referring to Figure 18-35, you can easily see that without these fittings condensate puddles would form along the main piping when condensation is taking place. Steam velocity could move one of these puddles of water at a high rate of speed and cause water hammer.

When steam mains pitch upward, they are called *non-dripped lines,* which means that any condensate that forms will run opposite the direction of the steam. Steam mains that are pitched downward in the direction of the steam are called *dripped lines.* Figure 18-36 illustrates both types of lines. Condensate lines must always pitch downward in the direction of the flow.

When a steam main meets an obstacle, such as steel beams, the piping must be routed specially. A vapor line must be routed over the obstacle, and a small condensate line underneath. Figure 18-37 illustrates this routing.

The condensate return line may meet a similar obstacle, and special routing must be used for it. A small line must be routed over the top of the obstacle for air or steam vapor that may move down the line, and a larger line must be routed underneath to carry the condensate, Figure 18-38.

Valves must be placed at points where service is needed, such as before branch circuits in large complex systems, before and after terminal units and their traps and strainers, and at the boiler, Figure 18-39. Note that some of these valves are in the condensate lines. All these valves allow the technician to isolate the system for routine or emergency service. The larger the system, the more valves needed. Valves are discussed in detail in Unit 17.

Pipe Sizing

Pipe is sized for a complex steam system by the engineering firm that works up the mechanical system for the building. Two pipe systems must be sized for a steam system: a vapor line and a condensate return line. The technician should be able to check these pipe sizes for field troubleshooting purposes.

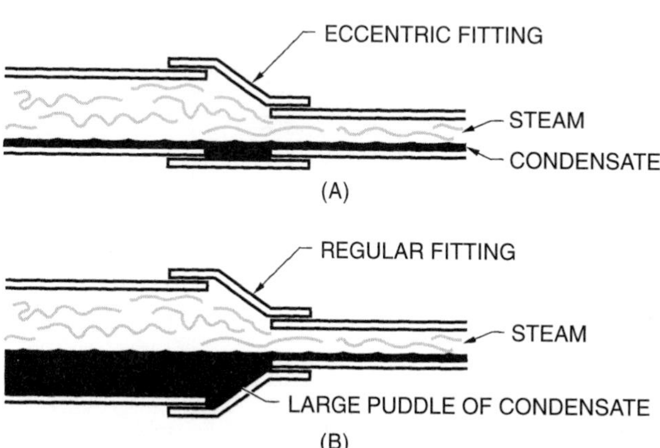

Figure 18-35. (A) Eccentric fitting prevents water from forming in the line. (B) Regular fitting allows condensate puddles to form.

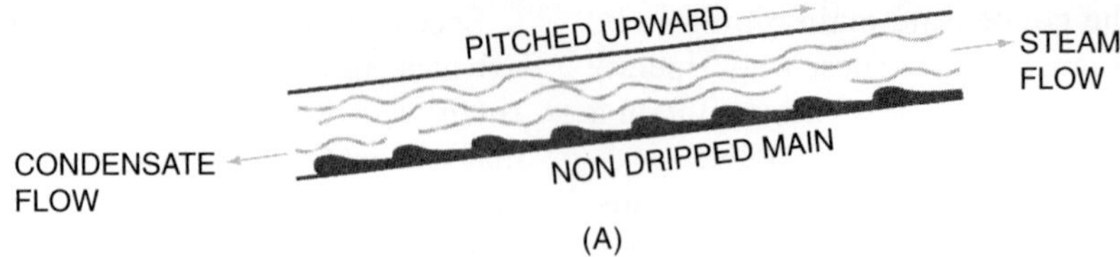

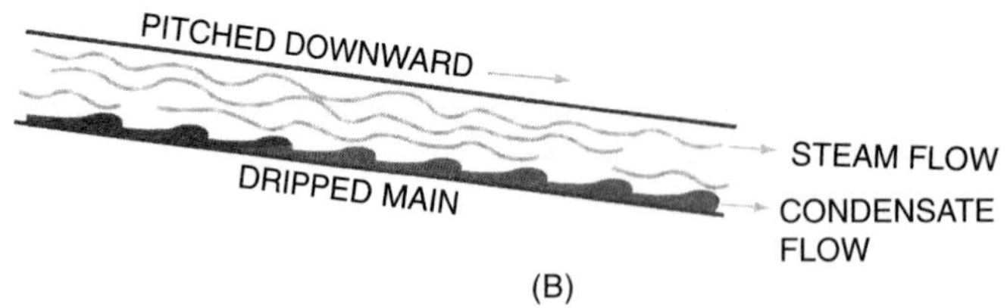

Figure 18-36. (A) Nondripped steam main. (B) Dripped steam main.

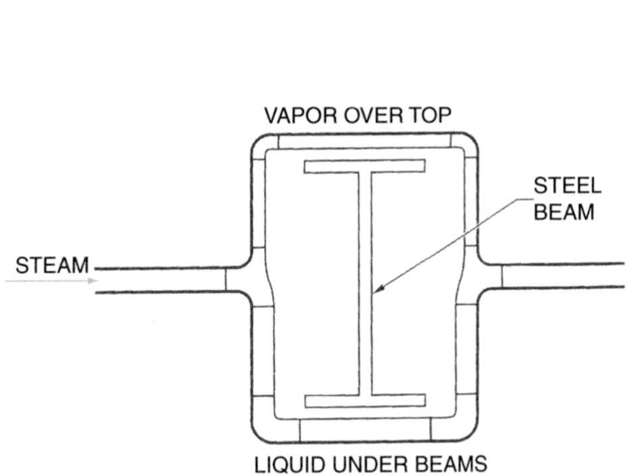

Figure 18-37. Routing a steam line over a beam.

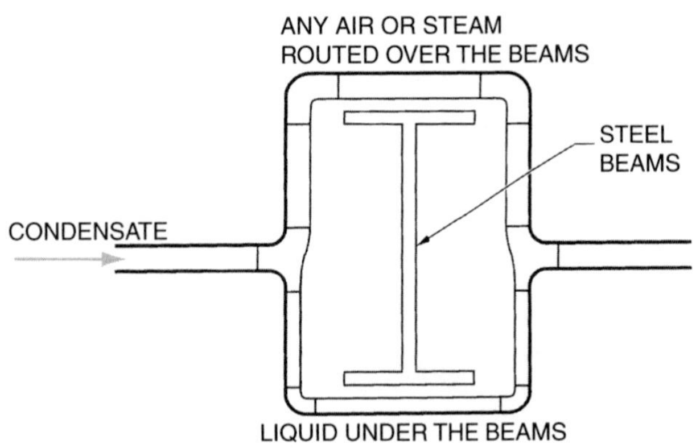

Figure 18-38. Routing a condensate line under a beam.

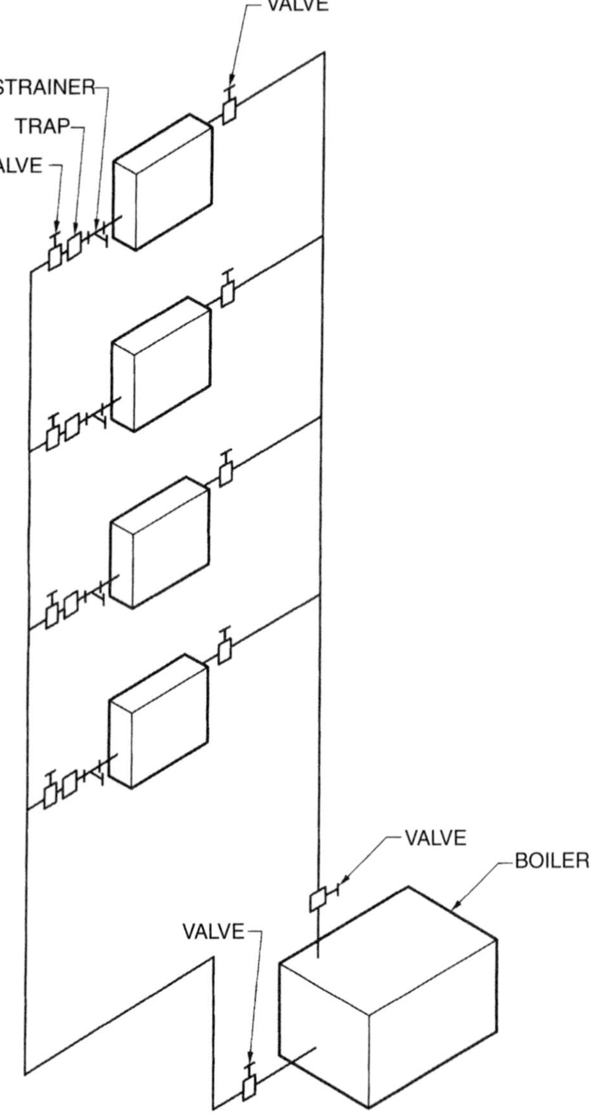

Figure 18-39. Value locations for service and convenience.

The designer must balance the cost of the pipe for the system against the possible noise that the velocity of the steam will make if the pipe is too small. Two situations may occur for sizing the vapor line. One is when the condensate flows in the direction of the steam—a dripped system. The other is a system in which the condensate flows counter to the steam flow—a nondripped system—shown in Figure 18-40. In a nondripped system, the steam velocity must allow the condensate to flow counter to the steam. Research sponsored by the American Society of Heating, Refrigerating and Air-Conditioning Engineers (ASHRAE) found that the velocity in pipes in which the condensate flows counter to the steam can be 12,000 to 15,000 feet per minute (fpm). Higher velocities than 15,000 fpm will slow the condensate in counterflow systems and create large puddles of water that can then cause water hammer, as shown in Figure 18-41.

The flow rates for low-pressure dripped systems for different pipe sizes are found in Figure 18-42. Note that the pipe sizes are in the left-hand column and flow is expressed in pounds per hour of steam. The table lists several pressure drops for either 3.5 or 12 psig steam pressure. Remember that the heat content of steam is 970 Btu per pound.

The flow rates for low-pressure nondripped systems are found in Figure 18-43. A nondripped system is a one-pipe system. Actually, a complete system is installed as a one-pipe system, with all the condensate draining back to the boiler through the steam main. These systems must be small, with short runs of pipe, such as those in residences. Portions of dripped systems can also be one pipe in nature, with only one pipe running to the terminal unit from the dripped main, as shown in Figure 18-44.

The condensate return line is much smaller than the steam vapor line. This line must pitch downward all the

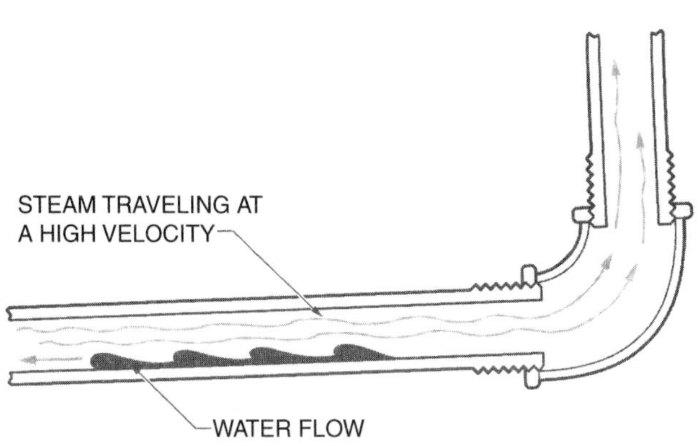

Figure 18-40. Nondripped steam main.

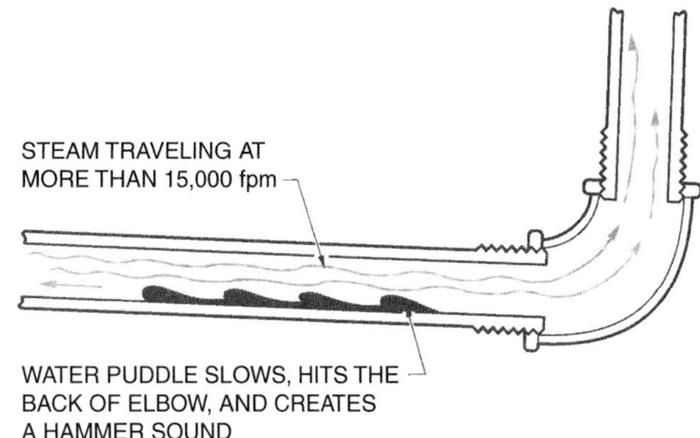

Figure 18-41. Puddles of water can create water hammer.

Table 13 Flow Rate of Steam in Schedule 40 Pipe[a] at Initial Saturation Pressure of 3.5 and 12 Psig[b,c]
(Flow Rate Expressed in Pounds per Hour)

Nom. Pipe Size inches	Pressure Drop—Psi Per 100 Ft in Length: 1/16 Psi (1 oz) Sat. press. psig 3.5	1/16 Psi (1 oz) 12	1/8 Psi (2 oz) 3.5	1/8 Psi (2 oz) 12	¼ Psi (4 oz) 3.5	¼ Psi (4 oz) 12	½ Psi (8 oz) 3.5	½ Psi (8 oz) 12	¾ Psi (12 oz) 3.5	¾ Psi (12 oz) 12	1 Psi 3.5	1 Psi 12	2 Psi 3.5	2 Psi 12
¾	9	11	14	16	20	24	29	35	36	43	42	50	60	73
1	17	21	26	31	37	46	54	66	68	82	81	95	114	137
1¼	36	45	53	66	78	96	111	138	140	170	162	200	232	280
1½	56	70	84	100	120	147	174	210	218	260	246	304	360	430
2	108	134	162	194	234	285	336	410	420	510	480	590	710	850
2½	174	215	258	310	378	460	540	660	680	820	780	950	1150	1370
3	318	380	465	550	660	810	960	1160	1190	1430	1380	1670	1950	2400
3½	462	550	670	800	990	1218	1410	1700	1740	2100	2000	2420	2950	3450
4	640	800	950	1160	1410	1690	1980	2400	2450	3000	2880	3460	4200	4900
5	1200	1430	1680	2100	2440	3000	3570	4250	4380	5250	5100	6100	7500	8600
6	1920	2300	2820	3350	3960	4850	5700	5700	7000	8600	8400	10000	11900	14200
8	3900	4800	5570	7000	8100	10000	11400	14300	14500	17700	16500	20500	24000	29500
10	7200	8800	10200	12600	15000	18200	21000	26000	26200	32000	30000	37000	42700	52000
12	11400	13700	16500	19500	23400	28400	33000	40000	41000	49500	48000	57500	67800	81000

[a]Based on Moody Friction Factor, where flow of condensate does not inhibit the flow of steam.

[b]The flow rates at 3.5 psig can be used to cover sat. press. from 1 to 6 psig, and the rates at 12 psig can be used to cover sat. press. from 8 to 16 psig with an error not exceeding 8%.

Figure 18-42. Flow rates for dripped steam lines. *Courtesy American Society of Heating, Refrigerating and Air-Conditioning Engineers*

Table 11 Comparative Capacity of Steam Lines at Various Pitches for Steam and Condensate Flowing in Opposite Directions[a]

(Pitch of Pipe in Inches per 10 Ft. Velocity in Ft per Sec)

Pitch of Pipe	¼ in.		½ in.		1 in.		1½ in.		2 in.		3 in.		4 in.		5 in.	
Pipe Size Inches	Capacity	Max. Vel.	Capacity	Max. Vel.	Capacity	Max. Vel.	Capacity	Max. Vel.	Capacity	Max. Vel.	Capacity	Max. Vel.	Capacity	Max. Vel.	Capacity	Max. Vel.
Capacity Expressed in Pounds per Hour																
¾	3.2	8	4.1	11	5.7	13	6.4	14	7.1	16	8.3	17	9.9	22	10.5	22
1	6.8	9	9.0	12	11.7	15	12.8	17	14.8	19	17.3	22	19.2	24	20.5	25
1¼	11.8	11	15.9	14	19.9	17	24.6	20	27.0	22	31.3	25	33.4	26	38.1	31
1½	19.8	12	25.9	16	33.0	19	37.4	22	42.0	24	46.8	26	50.8	28	59.2	33
2	42.9	15	54.0	18	68.8	24	83.3	27	92.9	30	99.6	32	102.4	32	115.0	33

Figure 18-43. Flow rates for nondripped steam mains. *Courtesy American Society of Heating, Refrigerating and Air-Conditioning Engineers*

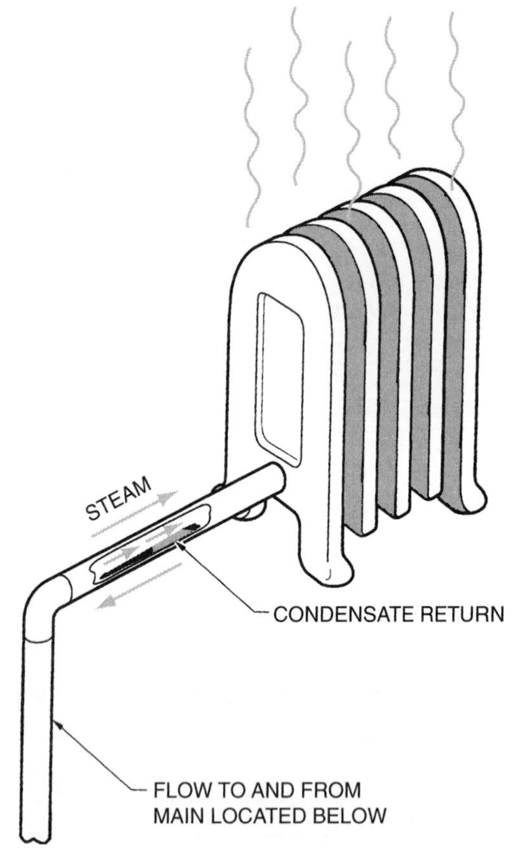

Figure 18-44. One-Pipe nondripped line to the terminal unit.

way to the boiler unless a condensate pump is used to lift the hot condensate to the boiler.

If the condensate trap is suspected as a problem, the manufacturer must be consulted for the capacity of the trap. Do not overlook something as simple as a restricted strainer.

Good piping practice calls for the installation crew to ream all piping connections, to keep the pipe clean of thread cutting oil and debris, and to use the correct thread seal on each connection. Proper location of valves and unions for servicing must be practiced.

All steam piping mains should be insulated unless they are routed through the conditioned space. Insulation is desirable even in the conditioned space because uncontrolled condensation will occur (particularly on start-up) and possibly cause water hammer. All condensate return piping that is routed through the unconditioned space should be insulated to prevent loss of heat from the system.

18.5 Troubleshooting Steam Systems

Steam-system troubleshooting is much like troubleshooting hot-water systems. The problem is either on the air side or on the steam side of the system.

The air side of the system may be either natural draft or forced draft. The same problems occur as with the hot-water system, except instead of a fall in water temperature, steam condenses. With a steam coil, the air-flow can be measured, the temperature rise can be measured, and the sensible heat formula can be used to calculate the amount of steam that has condensed. For example, suppose that a steam coil has 10,000 cubic feet per minute (cfm) of air moving across it and a temperature rise of 40°F. The heat content entering the airstream can be calculated from the following equation:

$$q = 1.08 \times \text{cfm} \times \text{TD}$$

where q = sensible heat transferred to the air.
1.08 = Constant.
cfm = Airflow in cubic feet per minute.
TD = Temperature difference.

To calculate the heat content:

$$\begin{aligned} q &= 1.08 \times \text{cfm} \times \text{TD} \\ &= 1.08 \times 10{,}000 \times 40 \\ &= 432{,}000 \text{ Btu/h} \end{aligned}$$

This value can be converted to pounds of steam per hour by dividing it by 970 Btu per pound:

$$432{,}000 \text{ Btu/h} \div 970 \text{ Btu/lb} = 445.36 \text{ lb/h}$$

If this value is close to the specifications, the coil is performing up to capacity. If not, the first thing to suspect is the possibility of air in the coil. Manually bleed the air from the coil.

If this does not improve the performance of the coil, you may want to use a temperature tester with a surface probe to determine whether the condensate trap is backing condensate up into the coil. If condensate is backing up in the coil, the last portion of the coil will be lower in temperature than the portion of the coil where the saturated steam is circulating. The coil should be at saturation temperature until very close to the end of the coil. Float traps can be checked to see if they are functioning by using the temperature probe on each side of the trap. There should be a temperature difference of a few degrees across the trap. If saturation temperature is found on both sides of the trap, steam is moving through the trap, as shown in Figure 18-45.

Thermostatic traps have more of a temperature drop because the condensate must be cooler than saturation temperature for the thermostat to open the trap. Look for approximately 5°F or a 6°F difference in thermostatic traps.

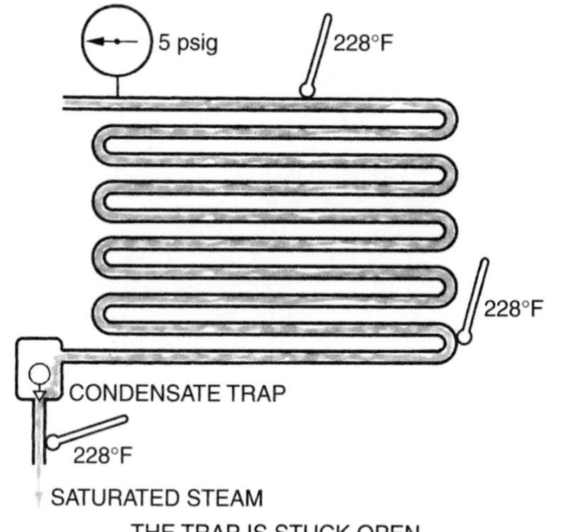

Figure 18-45. Checking a float trap by using temperature difference.

The distribution pipes inside larger steam coils can develop mineral buildup. These pipes can be checked by isolating the coil with valves and removing the inspection plugs.

Warning: Make sure that all the pressure is out of the coil before removing the plugs. Room air passing over the coil will quickly reduce the pressure by condensing the steam.

The pressure in low-pressure systems will go below atmospheric pressure when the boiler is cold if the system vent is not functioning. The purpose of the vent is to prevent the system and boiler from going into a vacuum when the system is cold. This vent is also called a *vacuum break*. A vacuum occurs if the system is airtight because as the steam in the system condenses when the system is cooling, the condensate takes up less space; therefore the system pressure drops below atmospheric pressure. The system vent also serves a second purpose: to allow air to escape from the system, Figure 18-46.

During start-up of a cold system, most of the steam that enters the system for the first few minutes condenses in the mains. Traps should be in the mains to aid in returning the condensate during start-up. These traps must function, or the system will become waterlogged and no heat will reach the terminal units. It may take hours to push the water out of the mains through the traps at the terminal units. When a waterlogged system is encountered, first check the traps in the mains. Make sure that the isolation valves for the traps are open.

Condensate pumps that return the condensate to the boiler must work on demand. There are usually two pumps in a pump housing. One is a backup pump. If the first pump fails, the other one removes the condensate. Each pump is operated by a float. When the water level (condensate) rises, the pump starts and pushes the water

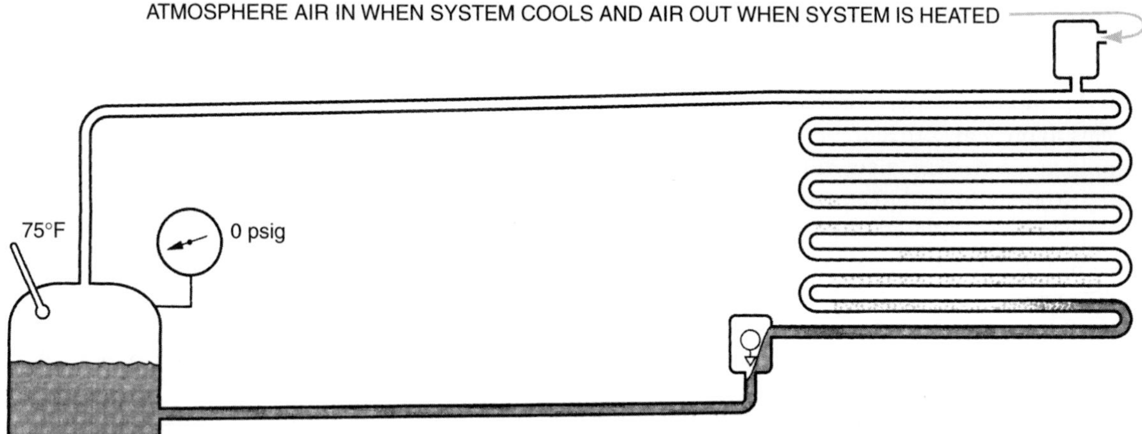

Figure 18-46. System vent can be a vacuum breaker or can allow air to escape the system when it is started up from cold and is full of air.

out of the reservoir back to the boiler. If both pumps fail, water will pour out onto the floor in the equipment room. There must be a floor drain close to the pumps, or water will accumulate on the floor if both pumps fail. If water is flowing over the reservoir into the equipment room, the technician will notice steam in the equipment room. This steam is not hot, like steam coming from the system; it is more like a cloud of warm vapor.

18.6 Preventive Maintenance for Steam Systems

The steam system, like the hot-water system, requires air-side maintenance and water-side maintenance.

Hot-water air-side maintenance problems are discussed in Unit 17, Section 17.10. The steam air side requires the same type of maintenance.

Water-side maintenance involves the condensate return method or the boiler.

System maintenance usually involves the traps. The traps should be checked monthly to ensure that they are functioning correctly. A temperature probe can be used until the technician gains enough experience to listen to the traps and carefully touch test the pipes to make sure that condensate is flowing correctly. In some systems, the strainers must be cleaned and the traps checked regularly. Steam lost to the condensate system through traps that are stuck open is expensive.

Some technicians listen to a trap for the passage of steam through the trap. Steam is enclosed in the pipe and the trap and is not easy to hear. A screwdriver or wooden stick can be used like a homemade stethoscope to listen to the trap and the pipe, as shown in Figure 18-47. When steam is passing through the trap, it will make a hissing sound, much like air moving through a small nozzle. If the trap is passing steam, it must be repaired.

Infrared scanners can be used to detect faulty steam traps in a large system. These scanners show different colors for different temperatures. The technician can stand on the floor and check a steam trap next to the ceiling in a matter of minutes, Figure 18-48. These scanners are expensive but are worthwhile for some systems.

The water in the boiler must be treated with the correct chemicals to prevent corrosion of the boiler parts and the system piping. Water treatment must be taken care of by specialists who are trained to analyze the water and to suggest the chemicals required to keep the water in the correct chemical balance.

The boiler controls should be checked to make sure that they are functioning properly. The control checkout starts with the pressure control that stops the boiler when the correct pressure level is reached. This is the operating control of the boiler. For example, if the control is set to stop the boiler fuel supply at 12 psig, make sure that it does by observing it shutting off the boiler.

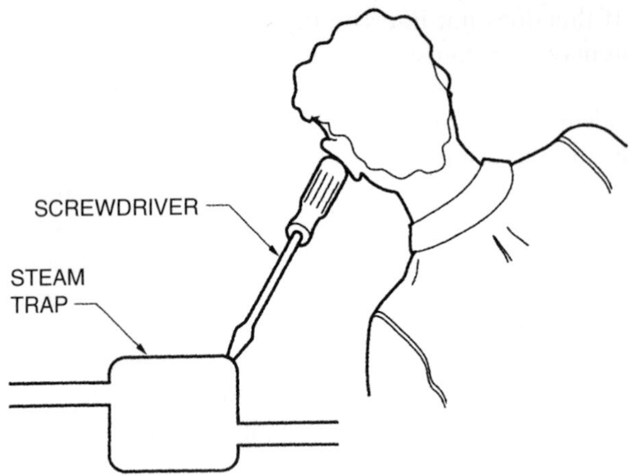

Figure 18-47. A homemade stethoscope is used to listen to the steam flow in the system.

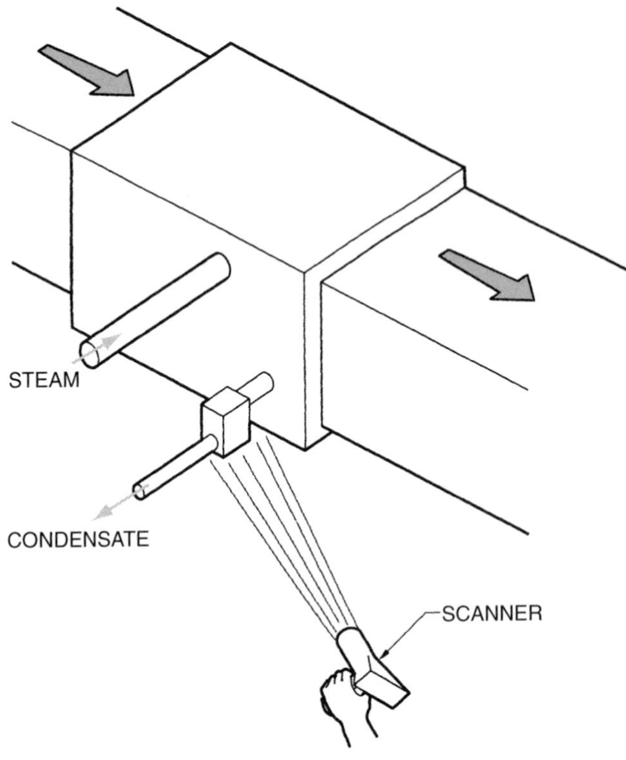

Figure 18-48. Checking a steam trap with a scanner.

The low-water cutoff chamber should be blown down daily. The condensate pump float chamber should be inspected periodically to ensure that no trash or obstruction is in the reservoir or float chamber. The pump or motor may require oil or grease at this time. The condensate pump is often stopped and started with a contactor. The contacts should be inspected and replaced when they become pitted.

At the beginning of the heating season, all piping insulation should be inspected for deterioration. If the piping system was worked on, the insulation may have been removed and should be replaced. It is important that the insulation be kept in place, or heat energy will be lost.

Summary

- Steam heat is similar to hot-water heat, except the latent heat of the steam vapor contains a larger amount of heat.
- Steam is circulated through the system as a vapor to the system terminal unit, where it is condensed to a liquid and returned to the boiler.
- The steam vapor line is larger than the liquid condensate return line.
- In a typical older system, the boiler was below the terminal heat-exchange units, and the steam supply line was pitched upward toward the unit. The condensate returned down the steam line to the boiler by gravity. The more modern system has a condensate trap to provide a liquid seal between the outlet of the terminal unit and the boiler.
- Gas- and oil-fired steam boilers are much like hot-water boilers. Electric steam boilers are typically larger than electric water boilers because of the need for a steam dome.
- The low-pressure steam boiler has a working pressure of 15 psig with a safety relief valve that is quick opening to relieve the pressure.
- The steam boiler does not have the liquid head that the hot-water boiler has in a tall building.
- Each boiler must be equipped with a pressure gauge and a pressure control for shutting off the boiler when the proper pressure is reached.
- The pressure control and pressure gauge are installed with a connector between them that prevents raw steam from reaching the indicator mechanism, which is a Bourdon tube. The connector is called a pigtail or a U tube. A gauge valve should be positioned in the line to allow service to the gauge.
- Each steam boiler has a boiler vent that allows air to escape the boiler on start-up and that breaks the vacuum when the boiler cools to room temperature. When the boiler cools, the steam inside condenses. The condensation reduces the volume. The boiler pressure would reduce to a vacuum without the vent.
- A gauge glass should be provided with each steam boiler so that the technician can see the boiler water level at a glance. This glass has valves so that it can be isolated for servicing and in case of breakage.
- Water level is critical in a steam boiler because the heat-exchange surface is just below the operating water level. Damage to this surface occurs quickly without enough water.
- A low-water fuel cutoff control is located at the water level to prevent the boiler from overheating. The low-water cutoff may contain a water makeup switch that operates a solenoid to add water when the float drops.
- The low-water cutoff is at the water level of the boiler, where light floating debris collects. This control should be blown down daily to remove any foreign particles.
- There must be a check valve or a back-feed device to prevent water in the boiler from moving into the fresh-water piping in the event that boiler pressure is greater than city water pressure (e.g., when city water is shut off for some time).
- System pressure differences in a steam system will cause the water level in a boiler to fluctuate up and down without a special piping arrangement called a Hartford loop.
- Steam terminal units receive steam vapor and condense it while adding heat to something, usually the conditioned space. A terminal unit can exchange heat to a hot-water circuit, a steam radiator, or a forced-air coil.
- Terminal units are rated in square feet of steam; 1 square foot of steam equals 240 Btu of heat per hour. When a pound of steam is condensed, 970 Btu of heat are removed and added to the heated substance, air or water.
- There are several types of steam-system condensate return methods. All return condensate to the boiler.
- A system usually has a steam trap at the heat-exchange outlet that creates a liquid seal between the vapor and the liquid parts of the system. If raw steam is allowed to enter the liquid portion of the system, there will be no pressure difference to encourage steam to flow.
- The steam trap also traps noncondensable gases, such as air, that have moved through the heat exchanger and vents them to the atmosphere.
- The steam trap must create the liquid seal without reducing the efficiency of the heat exchanger by holding back too much condensate.
- Three types of common steam traps are mechanical, thermostatic, and kinetic.
- The mechanical trap is either a float or an inverted bucket. The mechanical trap can have an additional thermostatic feature that vents trapped air to the atmosphere.
- The thermostatic trap is farther downstream from the terminal unit to allow the condensate to cool and thus attain a lower temperature difference than the saturated steam. When the thermostatic element senses condensate, it opens and allows the subcooled liquid to move into the condensate main until the element senses hot steam, then the trap closes.
- The kinetic trap has a movable disc that allows air and condensate to pass, but not steam. It operates according to the difference between the energy of the moving steam and the energy of the moving air and liquid.
- Each trap must be able to move more condensate during a cold start-up than during normal operation because a large volume of condensate develops in the cold pipes and coils when steam is first turned into the system.
- Each trap has a very small bore through which the condensate passes. This bore must be protected with a strainer to prevent pipe scale and other foreign objects from blocking it.

- The terminal unit and the steam trap must work together.
- Steam-heat piping is either black iron pipe with threaded connections or steel with welded connections.
- Threaded connections must be correctly made with the correct pipe thread compound applied only to the external threads.
- The steam pipes in many systems pitch upward in the direction of the steam flow.
- The pressure difference between the boiler and the terminal unit creates the steam flow.
- The condensate line must pitch downward in the direction of the flow to encourage condensate movement.
- Condensate flows back to the boiler by gravity in a system in which the boiler is below the terminal units. When the boiler is above the terminal units, a condensate pump is used to move the condensate back to the boiler.
- During a cold start-up, condensate may run in the opposite direction of the steam and drip back to the vertical riser because the main is pitched upward. This is known as a nondripped main. When the steam line pitches downward in the direction of the steam flow, the condensate moves with the steam. This line is known as a dripped main.
- Condensate movement with the steam can cause water hammer because the condensate travels at a high velocity and will make a noise when the line changes direction.
- The steam main must have a special route when it meets an obstacle. The main is routed differently from the condensate return line.
- Valves that allow servicing should be located in places that may need to be isolated, such as terminal units, strainers, traps, or individual circuits.
- The engineering firm that designs the building sizes the steam pipes. The service technician should know enough about pipe sizing to recognize a problem.
- There are two pipes to size in a two-pipe system: the steam line and the condensate line. A one-pipe system, or a nondripped main, requires special sizing because the steam flows counter to the condensate.
- The designer must balance the cost of an oversized system against the possibility of a noisy system if it is undersized. Undersizing causes excess steam velocities.
- Acceptable steam flow rates are from 12,000 to 15,000 feet per minute.
- Steam-system forced-air coils are checked for capacity by taking the air temperature rise and the airflow across the coil to determine the quantity of steam that is being condensed.
- Preventive maintenance for steam systems involves the air side or the steam side of the system. Much of the steam side of the maintenance involves checking and servicing the condensate traps. They can be checked by using thermometers or scanners. Some technicians listen to the system through a screwdriver.
- Water in the boiler must be treated with the correct chemicals to prevent rust, corrosion, and mineral deposits.
- The boiler controls should be checked to make sure that they function correctly.
- The system condensate pump should be checked and lubricated as needed. The float chamber should be inspected for deterioration.

Review Questions

1. Which is larger: the steam vapor line or the condensate return line?
2. What is the working pressure of a low-pressure steam boiler?
3. What is the purpose of a sight glass on a steam boiler, and why is one not needed on a hot-water boiler?
4. What is the purpose of the Hartford loop in a steam boiler?
5. What is the purpose of the vent on a steam boiler?
6. Why is the water level critical in a steam boiler?
7. What is the purpose of the check valve in the makeup water line in a boiler?
8. What are three kinds of condensate traps?
9. What is the purpose of the condensate trap?
10. A steam terminal unit needs to put out 50,000 Btu of heat per hour. How many square feet of steam is this?
11. How does a nondripped steam main operate?
12. Why must the condensate main pitch downward in the direction of the flow?

19 Technician Service Calls for Hot-Water and Steam Systems

Objectives

Upon completion of this unit, you should be able to

- **perform basic service procedures in a logical order.**
- **determine when a hot-water system has a problem with air in the system.**
- **perform basic troubleshooting on a steam condensate trap.**

19.1 Troubleshooting Process

Some consider troubleshooting to be an art form; whereas, others consider it to be science. Actually it is a combination of art and science—trying to solve a problem using a pure scientific approach may not always work. In any case, some steps can be followed (especially by new technicians) to assist in the development of troubleshooting skills. One of the most important of these—establishing a good rapport with the customer—was briefly discussed in Unit 10. Often what the customer perceives as the ultimate problem is nothing more than a symptom. Often when the technician establishes a good rapport with the customer, it is easier to get the actual information needed to determine the actual problem. However, as mentioned earlier, rapport is only a small portion of the troubleshooting process. There are actually two major phases that can be associated with the troubleshooting process: the identification process and the repair process.

The identification process is more than just shining a flashlight into a boiler or heat exchanger and trying to spot a defect or a malfunctioning part. The identification process can be divided into several key phases or steps. They are

1. Gather information
2. Verify the issue
3. Look for quick fixes
4. Perform the appropriate diagnostics
5. Use additional resources to research the issues (if necessary)
6. Escalate the issue (if necessary)

Gathering Information

When starting to troubleshoot a system, the first step is to gather the information necessary to correctly identify the problem. This is often done by simply asking the homeowner a few simple questions. However, when questioning the homeowner, it is necessary to keep two general rules in mind. They are

1. Start with open questions such as "What is the issue?" Open questions cannot be answered with a "yes" or "no".
2. Let the customer explain in their own words what they have experienced/are experiencing. *Never interrupt* a customer or add comments to what they are telling you.

Verifying the Issue

As stated earlier, the situation or problem that the customer describes is often not the actual problem. Therefore, always verify that the problem described by the customer is the actual problem of the system and not just a symptom.

Looking for Quick Fixes

Although in many cases the actual fix is more involved than simply resetting or changing the battery in a thermostat, there are still cases in which the simplest and/or most obvious fix corrects the problem. For example, suppose that you were called to look at a customer's gas central heating unit because it would not ignite. If the customer is using an electronically controlled thermostat, it might be useful to check its battery before starting to break the furnace down.

Performing the Appropriate Diagnostics

If the quick fix does not resolve the issue, then it will be necessary to perform more thorough diagnostics. Often the equipment manufacturers will supply troubleshooting charts and information to help diagnose the equipment.

Using Additional Resources to Research the Issues

If you have never encountered a problem like the one that is currently before you, and you are having trouble locating the issue, don't be afraid to go to the Internet, a distributor, or even a fellow colleague to help resolve the issue.

Escalating the Issue

If you are working for a large company and continue to have problems locating and correcting the problem, the issue can often be escalated to a service manager for assistance. If you are self-employed or working for a small company, when you encounter a problem that you cannot resolve, the equipment manufacturer can often be of assistance. In any case, though, you should never escalate an issue unless you are truly stumped.

Once the issue has been correctly identified, the repair process can proceed. Like the identification process, the repair process also involves several steps. They are

1. Repair or replace the faulty item and/or equipment.
2. Test the system thoroughly to verify that the repair actually corrected the issue.
3. Educate the homeowner about the nature of the problem and the action(s) taken to correct the problem.
4. Complete all administrative paperwork.

Verifying that the Repair Actually Corrected the Problem

This is one of the most critical steps in the process. Never leave a customer site without first testing the repair and/or installation to confirm that you actually corrected the problem.

Educating the Homeowner about the Nature of the Problem

Always show the customer the worn and replaced parts and explain to them why the old parts are defective. If a customer understands the issue and the corrective action taken, they are less likely to become dissatisfied with the repair job.

Completing all Administrative Paperwork

This is especially important when dealing with warranty work. If the necessary paperwork is not completed correctly and on time, then there will be a delay in the service company receiving their payment. In some cases, the claim may even be denied.

19.2 Service Call 1

A customer calls. There is no heat at a motel. The problem is on the top two floors of a four-story building. The system has just been started for the first time this season, and air is in the circulating hot water. The automatic vent valves have rust and scale in them, which resulted from lack of water treatment and lack of proper maintenance.

The technician arrives and goes to the boiler room. The boiler is hot, so the heat source is working. The water pump is running because the technician can hear water circulating. The technician decides to go to the top floor to one of the room units and removes the cover; the coil is at room temperature. The technician listens to the coil closely and can hear nothing circulating. There may be air in the system, so the bleed ports should be located. The technician knows that these ports must be at the high point in the system, usually on top of the water risers from the basement.

The technician goes to the basement where the water lines start up through the building and finds a reference point. The pipes rise through the building next to the elevator shaft. The technician goes to the top floor to the approximate spot and finds a service panel in the ceiling in the hall. Using a ladder, the technician locates the automatic bleed port for the water supply. This is the pump discharge line.

The technician removes the rubber line from the top of the automatic vent. The rubber line carries any water that bleeds to a drain. There is a valve stem in the automatic vent (as in an automobile tire). The technician presses the stem, but nothing escapes. There is a hand valve under the automatic vent, so the valve is closed and the automatic vent is removed. The technician then carefully opens the hand valve and air begins to flowout. Air is allowed to bleed until water starts to run. The water is very dirty, much like dirty water in a drainage ditch.

Warning: Care should be used around hot water.

The technician then bleeds the return pipe in the same manner until all the air is out, and only water is at the top of both pipes. The technician then goes to the heating coil where the cover was previously removed. Hot water is now circulating through the coil.

The technician takes the motel manager to the basement, drains some water from the system, and shows it to the manager. The water should be clear or slightly colored with water treatment. The technician suggests that the manager call a water-treatment company that specializes in boiler water treatment, or the system will develop major problems in the future.

The technician then replaces the two automatic vents with new ones and opens the hand valves so that they can operate correctly. The technician reminds the manager to call a water-treatment expert, then leaves.

19.3 Service Call 2

A customer in a small building calls. The system is heating well in most of the building, but there is no heat in one portion. The building has four hot-water circulating pumps, and one of them is locked up—the bearings are seized. These pumps are 230-volt, three-phase, 2-horsepower pumps.

The technician arrives at the building and consults the customer. They go to the part of the building where there is no heat. The thermostat is calling for heat. They go to the basement, where the boiler and pumps are located. They can tell from examination that the number 3 pump, which serves the part of the building that is cool, is not turning. The technician carefully touches the pump motor; it is cool and has not been trying to run. Either the motor or the pump has problems.

The technician chooses to check the voltage to the motor first, then goes to the disconnect switch on the wall and measures 230 volts across phases 1 to 2, 2 to 3, and 1 to 3. This circuit is a 230-volt, three-phase motor with a motor starter. The motor's overload protection is in the starter, and it is tripped.

The technician still does not know if the motor or the pump has problems. The motor is checked using an ohmmeter. All three windings have equal resistance, and there is no ground circuit in the motor. The power is turned on, an ammeter is clamped on one of the motor leads, and the overload reset is pressed. The technician can hear the motor trying to start, and it is pulling locked-rotor current. The disconnect is pulled to stop the power from continuing to the motor. It is evident that either the pump or the motor is stuck.

The technician returns to the pump. It is on the floor next to the others. The guard over the pump-shaft coupling is removed. The technician tries to turn over the pump by hand (remember, the power is off). The pump is difficult to turn. The question now is whether the pump or the motor is tight.

The technician disassembles the pump coupling and tries to turn the motor by hand. It is free. The technician then tries the pump; it is too tight. The pump bearings must be defective. The technician gets permission from the owner to disassemble the pump. The technician shuts off the valves at the pump inlet and outlet, removes the bolts around the pump housing, then removes the impeller and housing from the main pump body. The technician takes the pump impeller housing and bearings to the shop and replaces them, installs a new shaft seal, and returns to the job site.

While the technician is assembling the pump, the customer asks why the whole pump was not taken instead of just the impeller and the housing assembly. The technician shows the customer where the pump housing is fastened to the pump with bolts and dowel pins. The relationship of the pump shaft and motor shaft have been maintained by not removing the complete pump. They will not need to be aligned after they are reassembled. The pump is manufactured to be rebuilt in place to avoid realignment.

After completing the pump assembly, the technician turns it over by hand, assembles the pump coupling, and affirms that all fasteners are tight. Then the pump is started. It runs and the amperage is normal. The technician turns off the power, assembles the pump coupling guard, and leaves.

19.4 Service Call 3

A customer calls. One of the pumps beside the boiler is making a noise. The structure is a large home with three pumps, one for each zone. The pump coupling is defective. It is a spring-type coupling and is quickly wearing out.

The technician arrives and goes to the basement, hears the coupling, and notices metal filings. The power is shut off with the disconnect, and a resistance test is performed on the motor. It now sounds good. The technician writes on the report that this pump is several years old and that if another coupling fails, the pump cradle mount (rubber) or neoprene may need to be replaced. If the rubber or neoprene is soft, misalignment will occur, and coupling failure will result. The technician lubricates the motors and pump bearings before leaving the job.

19.5 Service Call 4

A customer calls from an apartment house. There is no heat in one of the apartments. This apartment house has 25 fan coil units with a zone control valve on each and a central boiler. One of the zone control valves is defective.

The technician arrives and goes to the apartment with no heat. The technician sets the room thermostat higher than the usual room setting. The technician goes to the fan coil unit in the hallway ceiling and opens the fan coil compartment door. It drops down on hinges. The technician stands on a short ladder to reach the controls. There is no heat in the coil, and no water is flowing.

This unit has a zone control valve with a small heat motor. The technician checks for power to the valve coil; it has 24 volts, as it should. This valve has a manual feature, so the valve is opened by hand. The technician can hear the water starting to flow and can feel the coil getting hot. The valve heat motor or the valve assembly must be replaced. The technician chooses to replace the assembly because the valve and its valve seat are old.

The technician obtains a valve assembly from the stock of parts at the apartment and replaces the old one. The technician shuts off the power and the water to the valve and water coil. A plastic drop cloth is then spread to catch any water that may fall. Then a bucket is placed under the valve before the top assembly is removed. The old top is removed, and the new top is installed. The electrical connections are made, and the valves are opened.

The technician turns on the power and sees the valve begin to move after a few seconds (remember, this is a heat motor valve, and it responds slowly). The technician can now feel heat in the coil. The compartment door is closed. The plastic drop cloth is removed, the room thermostat is set back to the usual setting, and the technician leaves.

19.6 Service Call 5

A home owner calls. Water is on the floor around the boiler in the basement. The relief valve is relieving because the gas valve is not shutting off tightly and the boiler is overheating.

The technician goes to the basement and sees the boiler relief valve seeping. The boiler gauge reads 30 psig, which is the rating of the relief valve. The system usually operates at 12 psig. The technician looks at the flame in the burner section and sees that the burner is burning at a low fire. A voltmeter check shows that there is no voltage to the operating coil of the gas valve. This means that the valve is partially stuck open.

The technician shuts off the gas and the power and replaces the valve. When this is accomplished, power and gas are restored. The boiler control is set higher, the gas valve opens, and the burner lights. When the technician resets the boiler control to normal, the burner goes out. The relief valve has stopped leaking, so the technician leaves the job.

Warning: Care should be taken when working with hot-water systems to ensure that all controls function normally. An overheated boiler can easily explode.

19.7 Service Call 6

A customer who owns a multistory office building has no heat on the third floor. The weather is extremely cold for the season. The heating system is a hot-water system with a fan coil on each floor. The problem is a dirty coil that resulted from poor filter maintenance.

The technician arrives and goes first to the boiler. It has cycled off because the thermostat is satisfied. The water temperature leaving the boiler is 200°F, which is normal for this job.

The technician then goes to the second floor and determines that the air temperature is 65°F. This is too low for an office. The technician then goes to the equipment room where the air handler is located. The thermometer entering the air handler reads 198°F (some heat is lost in the piping from the boiler), but the water temperature leaving is only 188°F. The airstream is not taking enough heat out of the water. The water temperature leaving the coil should be approximately 180°F, which would be a 20°F temperature drop across the coil. The technician then takes the air temperature difference across the coil. The air entering is 65°F, and the air leaving is 75°F. The air should have a 30°F rise across this coil according to previous records. It is obvious to the technician that a heat-exchange problem exists. If there was a reduced airflow problem, such as slipping belts, a dirty air filter, or closed air registers, the temperature difference across the coil would be higher than normal rather than lower than normal. With restricted airflow and a coil that is functioning correctly, less airflow would receive the same amount of heat and cause the outlet air to be abnormally hot.

The technician turns off the fan and removes one of the filters so that the hot-water coil can be observed with a strong flashlight. The coil appears to be dirty, so further investigation is prompted. Panels are removed from the unit, and a good look at the coil reveals it to be extremely dirty. This coil must be cleaned with detergent and water. Doing so will require several hours, so the technician vacuums the coil surface to remove as much dirt as possible. Then the technician changes the filters, replaces the unit panels, puts the unit back in service, and leaves after telling the building manager that the coil must be cleaned later, in the evening.

The technician returns that evening, after the office has closed, with a helper and a high-pressure washer with the proper detergent to clean the coil. The proper panels are removed so that the coil can be saturated with the detergent. After the coil is saturated, the detergent needs time to soak into the dirt, so the technician connects the washer to a power supply and prepares the room for moving wash water to the floor drain. Many items on the floor must be moved to higher places.

After the detergent has soaked in for approximately 30 minutes, the technician and helper start the high-pressure washer and wash the coil from the front side, in the opposite direction of the airflow. This is called *back-washing.* Both are amazed at the amount of dirt that flushes out of the core of the coil. When no more dirty water flushes out, the panels are replaced.

The technician decides that looking at the fan wheel may be a good idea because so much dirt was in the coil. The fan compartment panels are removed so that the fan wheel can be examined. It is very dirty.

The fan wheel is cleaned by scraping the dirt from the blades. The dirt is then vacuumed from the blades and from inside the fan scroll. The removed fan dirt must be swept up, or it will blow into the building.

All panels are replaced and the system is started. The fan motor and shaft are lubricated.

The technician returns to the building the next morning to check the system again. The second floor is warm, as it should be. The technician meets with the building manager and discusses a service contract on the complete system. The manager is interested. The technician explains that the fan rooms on the other 10 floors of the building are probably in much the same shape as the one on the second floor was. The system must be in good working order before a service agreement can be reached. They both agree, and time to inspect, clean, and lubricate all fans, motors, and pumps is allotted and agreed on. It is good business for a service technician to help acquire service contracts, rather than only repair equipment. The technician should keep an eye on future business.

19.8 Service Call 7

A customer calls and says that the shop behind the office is not heating. This is the first cold morning of the season. The problem is that the boiler in the steam-heat system is in the office portion of the system, and the shop is across the road. The steam has a long way to travel through cold pipes. Usually a lot of condensation in the steam mains must be returned through condensate traps in the mains. Someone during the summer shut off the condensate trap valve that allows the condensate to return to the boiler without traveling through the coils. This valve must be open for quick condensate return, as shown in Figure 19-1.

The technician arrives shortly after lunch, after handling several other emergency calls. The shop is just beginning to heat, so the system seems to be working. The technician talks to the shop foreman, and they agree that there seems to be no problem. The technician tells the foreman that a return trip in the morning when the system first starts may tell them more, so the technician leaves.

The next day, the technician arrives before the workers and meets the foreman. The boiler is up to pressure, but there is no heat in the shop. The technician checks the system by carefully touching the pipes. The steam line entering the first unit heater is at room temperature; no steam is reaching the building. The technician follows the steam line to the point where it goes underground, then the technician goes outside. There is a manhole cover next to the building.

The technician removes the manhole cover and finds a condensate trap in the manhole for draining excess condensate back to the condensate return main. The hand valve used for isolating the trap is closed, so the technician opens the valve. Condensate can be heard flowing through the trap.

The technician is sure that this closed valve is the problem but decides to wait to make sure. While waiting, the technician questions the foreman as to how the valve became closed. The foreman says that the steam lines were dug up and replaced during the summer because of a leak, and someone must have closed the valve and forgotten to open it.

After approximately 15 minutes, the steam line entering the shop becomes very hot, and the system begins to heat. The technician leaves, telling the foreman to call if the system does not function correctly in the future.

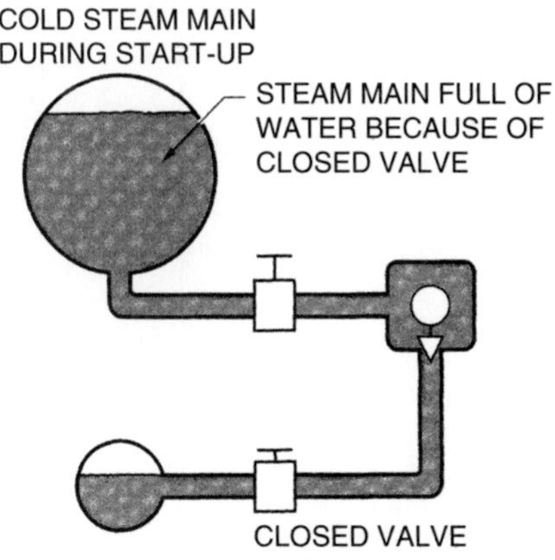

Figure 19-1. The condensate trap in the steam main must be open for rapid condensate return on start-up.

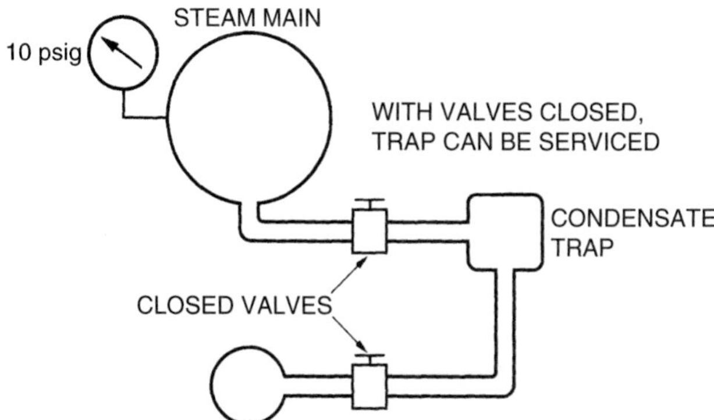

Figure 19-2. The condensate trap is isolated with valves so that it can be serviced.

19.9 Service Call 8

A customer calls and says that one portion of a small office building is not heating. The problem is that the building has five air handlers with a steam coil in each, and one of the air handlers has a condensate trap that will not open.

The technician arrives and goes to the air handler serving the area that is not heating. The steam line entering the coil is not hot. The technician touches the steam trap; it is cool. A screwdriver placed to the ear is used to listen to the trap, and it is determined that the trap is not opening. This trap is a float trap, so the technician raps on the side of the trap with a block of wood and listens again. The trap is now passing water. The float was stuck in the down position and shut.

The system begins to heat. The technician must decide whether to shut off the system and repair the trap now, or to allow the space to become warm and repair the trap later. The office manager asks the technician to come back later and change the trap. The old trap may work for hours or days but will surely fail again.

The technician returns that evening with a new trap and replaces the old one by isolating it with valves, shown in Figure 19-2, and allowing the fan to run for a few minutes to reduce the pressure in the coil and trap. The system is opened to the steam main and allowed to start heating. All seems well, so the technician leaves.

Summary

- Technicians should follow orderly, logical troubleshooting procedures.
- Start with the boiler and make sure it is operatable.
- Check pumps to make sure they are running with hot-water circulating systems.
- Check for air in the circulating water system.
- Check automatic air bleed system.
- Check for proper steam condensate trap operation with steam systems.

Review Questions

1. How does air affect the heat in a circulating water system and why?
2. How is air eliminated from a hot-water heating system?
3. How would a technician locate a no heat problem in a multi-story building when one floor has no heat?
4. How would a technician determine whether the pump or motor is tight with a circulating hot-water pump that will not turn when power is applied?
5. Where are the air bleed valves located in a multi-story building?
6. What simple tool can a technician use to listen to a steam condensate trap to see if it is allowing condensate to flow?
7. What is the result of a condensate trap that is not allowing condensate to flow?
8. Where is the condensate trap located on a steam coil?
9. Why would a condensate trap be located on a steam main?
10. What could cause a float type condensate trap to not function?

20 Air Humidification and Filtration

Objectives

Upon completion of this unit, you should be able to

- **explain relative humidity.**
- **list reasons for providing humidification in winter.**
- **discuss the differences between evaporative and atomizing humidifiers.**
- **describe bypass and under-duct-mount humidifiers.**
- **describe disc, plate, pad, and drum humidifier designs and the media used in each.**
- **explain the operation of the infrared humidifier.**
- **explain why a humidifier used with a heat pump or an electric furnace may have its own independent heat source.**
- **describe the spray-nozzle and centrifugal atomizing humidifiers.**
- **state the reasons for installing self-contained humidifiers.**
- **list general factors used to size humidifiers.**
- **describe general procedures for installing humidifiers.**
- **explain why cleaning the air in buildings is necessary.**
- **list five materials or devices used to filter or purify the air.**

20.1 Air Quality

The quality of the air in an environment is an important aspect of HVAC. If the air is too dry, the environment can be a health hazard. The same holds true for an environment in which the humidity is too high. When the humidity is too high, dangerous mold may be produced.

Mold is a plant and a member of the fungi family. It is one of the catalysts responsible for the breakdown of organic matter. However, unlike most plants, mold produces spores instead of seeds. These spores then become airborne particles that, when inhaled, can produce adverse health effects.

More than 100,000 species of mold can currently be found growing in a variety of different materials (food, soil, etc.). However, according to the Centers for Disease Control and Prevention (CDC), in the United States, the most common species of mold are cladosporium, penicillium, aspergillus, and alteraria. Out of these species, aspergillus and penicillium are considered toxin-producing species. One of the causes of sick building syndrome is the toxin-producing mold stachybotrys chartarum.

20.2 Relative Humidity

In fall and winter, homes are often dry because cold air from outside infiltrates the conditioned space. The infiltration air in the home is artificially dried out when it is heated because it expands and this expansion spreads the moisture. The amount of moisture in the air is expressed in terms of *relative humidity.* Relative humidity is the percentage of moisture in the air in comparison with the capacity of the air to hold the moisture. In other words, if the relative humidity is 50%, each cubic foot of air is holding one-half the moisture that it is capable of holding. The relative humidity of the air decreases as the temperature increases because the hotter the air, the more humidity the air can hold. When a cubic foot of 20°F outside air at 50% relative humidity is heated to room temperature (75°F), the relative humidity of the air drops.

This dry, warm air draws moisture from everything in the conditioned space, including carpets, furniture, woodwork, plants, and people. Furniture joints loosen, nasal and throat passages dry out, and skin becomes dry. Dry air causes more energy consumption than necessary because the air gets moisture from the human body through evaporation from the skin. The person then feels cold and sets the thermostat a few degrees higher to become more comfortable. When more humidity is in the air, a person is more comfortable at a lower temperature.

Static electricity is also more prevalent in dry air. It produces discomfort because of the small electrical shock that a person receives upon touching something after walking across a room.

Therefore, dried-out air should have its moisture replenished. The recommended relative humidity for a home is between 40% and 60%. Studies have shown that bacteria, viruses, fungi, and so forth become more active when the relative humidity varies above or below these limits. As a result, indoor air quality (IAQ) has become increasingly important.

20.3 Air Humidification

Years ago, to humidify the air, people placed pans of water on top of radiators or stoves. They even boiled water on the stove. The water evaporated into the air and raised the relative humidity. Although these methods of humidification may still be used in some homes, more efficient and effective equipment called *humidifiers* produce moisture and make it available to the air by *evaporation.* The evaporation process is accelerated by using power or heat, or by passing air over large areas of water. The area of the water can be increased by spreading it over pads or by atomization.

20.4 Types of Humidifiers

Evaporative Humidifiers

Evaporative humidifiers work on the principle of providing moisture on a surface called *media* and exposing it to the dry air. This is normally done by forcing the air through or around the media and picking up the moisture from the media as a vapor or a gas. Three types of evaporative humidifiers are discussed next.

The *bypass humidifier* relies on the difference in pressure between the supply (warm) side of the furnace and the return (cold) side. It can be mounted on either the supply plenum or duct or the cold-air return plenum or duct. Piping must be run from the plenum or duct where the humidifier is mounted to the other plenum or duct. In other words, if mounted in the supply duct, the humidifier must be piped to the cold-air return, Figures 20-1 and 20-2. The difference in pressure between the two plenums creates air flow through the humidifier and this air is distributed throughout the house.

The *plenum-mount humidifier* is mounted in the supply plenum or in the return air plenum. The furnace fan forces air through the media where the air picks up moisture. The air and moisture are then distributed throughout the

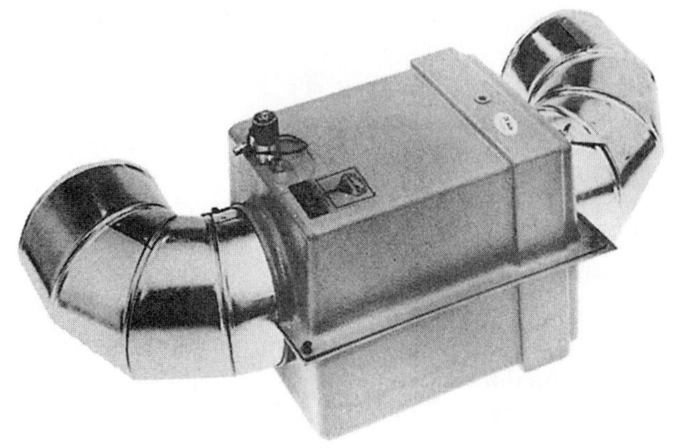

Figure 20-2. Bypass humidifier used between plenums with a pipe to each. *Courtesy Aqua-Mist, Inc.*

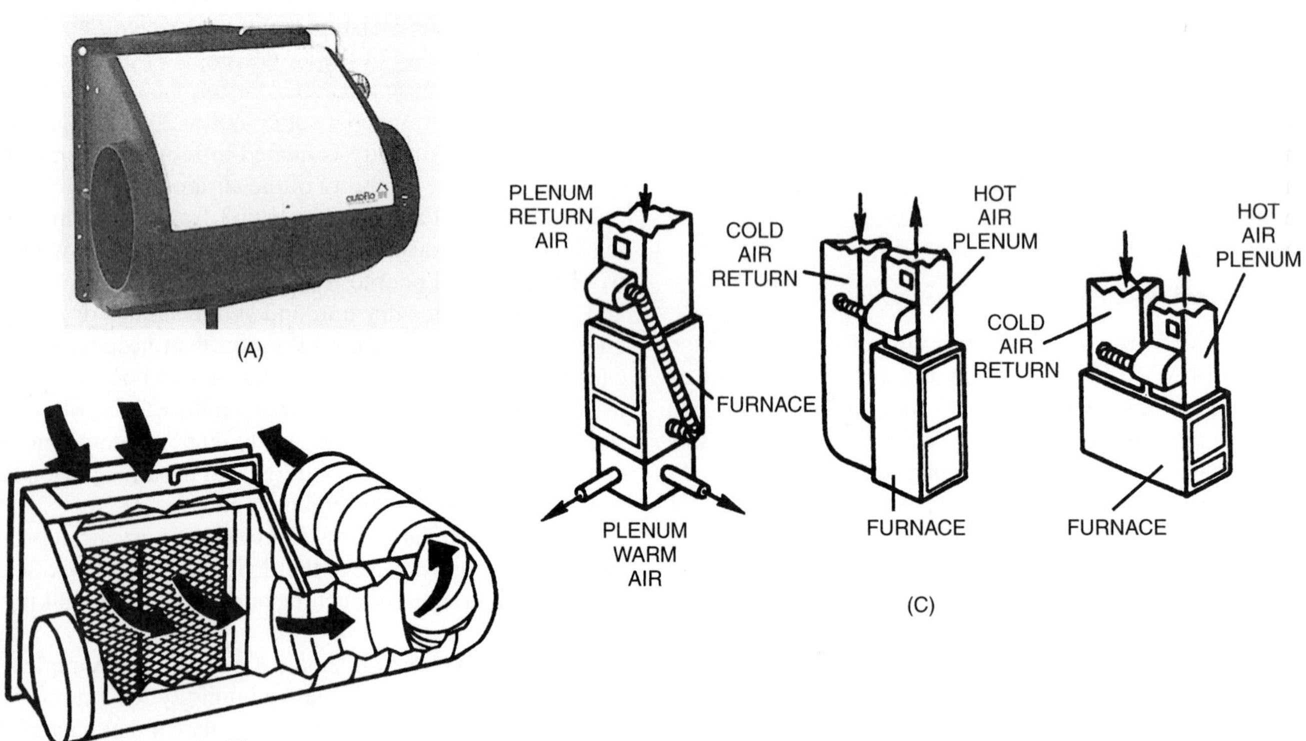

Figure 20-1. (A) Bypass humidifier. (B) Cutaway showing airflow from plenum through media to the return. (C) Typical installations. *Courtesy AutoFlo Company*

Figure 20-3. A plenum-mounted humidifier with a plate media that absorbs water from the reservoir. This water evaporates into the air in the plenum. *Courtesy AutoFlo Company*

conditioned space. Figure 20-3 shows a typical plenum-mounted humidifier.

The *under-duct-mount humidifier* is mounted on the underside of the supply duct so that the media extends into the airflow, where moisture is picked up. Figure 20-4 illustrates an under-duct-mount humidifier.

Humidifier Media

Humidifiers are available in several designs with various kinds of media. Figure 20-4 is a photograph of a humidifier using disc screens mounted at an angle. These slanted discs are mounted on a rotating shaft that allows them to pick up moisture from the reservoir. The moisture is then evaporated into the moving airstream. The discs are separated to prevent electrolysis, which causes the minerals in the water to form on the media. The wobble from the discs mounted at an angle washes the minerals off and into the reservoir. The minerals can then be drained from the bottom of the reservoir.

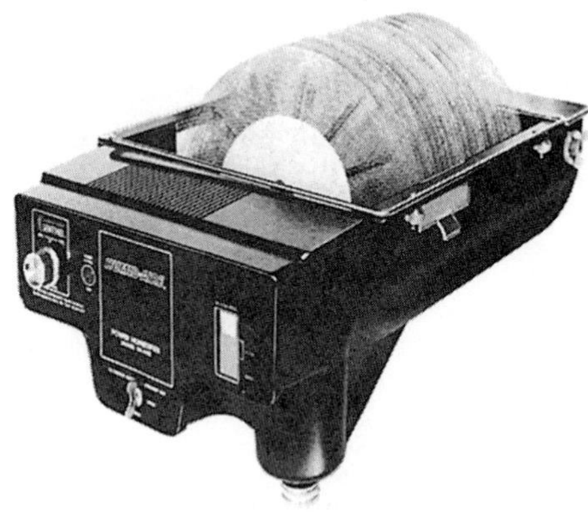

Figure 20-4. Under-duct-mount humidifier using disc screens as the media. *Courtesy Humid Aire Division, Adams Manufacturing Company*

Figure 20-5. Under-duct-mount humidifier using drum media. *Courtesy Herrmidifier Company, Inc.*

Figure 20-5 illustrates a media in a drum design. A motor turns the drum, which picks up moisture from the reservoir. The moisture is then evaporated from the drum into the moving airstream. The drums can be screen or sponge types.

A plate- or pad-type media is shown in Figure 20-3. The plates form a wick that absorbs water from the reservoir. The airstream in the duct or plenum causes the water to evaporate from the wicks or plates.

Figure 20-6 shows an electrically heated water humidifier. In an electric-furnace or a heat-pump installation, the temperature of the air in the duct is not as high as that in other types of hot-air furnaces. Media evaporation is not as easy with lower temperatures. The electrically heated

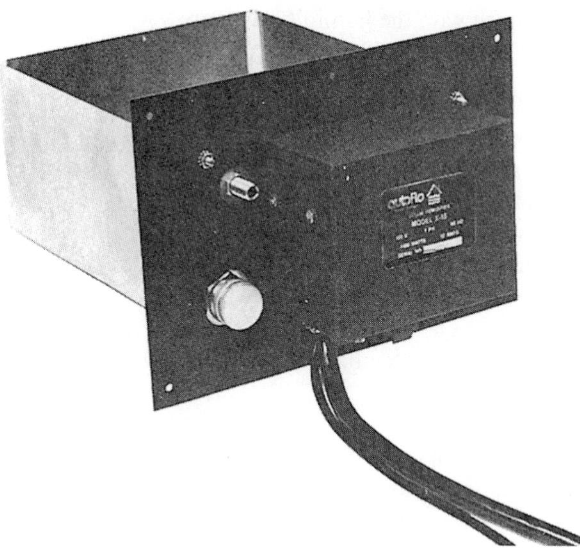

Figure 20-6. Humidifier with electric heating elements. *Courtesy AutoFlo Company*

Figure 20-7. Infrared humidifier. *Courtesy Humid Aire Division, Adams Manufacturing Company*

humidifier heats the water with an electric element, causing it to evaporate and be carried into the conditioned space by the airstream in the duct.

Another type of humidifier is the infrared humidifier, as shown in Figure 20-7. This humidifier is mounted in the duct and has infrared lamps with reflectors to reflect the infrared energy onto the water. The water thus evaporates rapidly into the duct airstream and is carried throughout the conditioned space. This action is similar to the sun's rays shining on a large lake and evaporating the water into the air.

Humidifiers are usually controlled by a *humidistat,* Figure 20-8. The humidistat controls the motor and/or the heating elements in the humidifier. The humidistat has a moisture-sensitive element, often made of hair or nylon ribbon. This material is wound around two or more bobbins and shrinks or expands, depending on the humidity. Dry air causes the element to shrink, which activates a snap-action switch and starts the humidifier.

Many other devices are used to monitor the moisture content of air, including electronic components that vary in resistance with the humidity.

Atomizing Humidifiers

Atomizing humidifiers discharge tiny water droplets (mist) into the air, which evaporate rapidly into the duct airstream or directly into the conditioned space. These humidifiers can be spray-nozzle or centrifugal types, but they should not be used with hard water because it contains minerals (lime, iron, etc.) that leave the water vapor as dust and are distributed throughout the house. Eight to 10 grains of water hardness is the maximum recommended for atomizing humidifiers.

The *spray-nozzle humidifier* sprays water through a metered bore of a nozzle into the duct airstream, which distributes the humidity to the occupied space. Another type sprays the water onto an evaporative media where it is absorbed by the airstream as a vapor. Both types can be mounted in the plenum, under the duct, or on the side of the duct. It is generally recommended that atomizing humidifiers be mounted on the hot-air or supply side of the furnace. Figures 20-9 and 20-10 illustrate two types of spray-nozzle humidifiers.

(A)

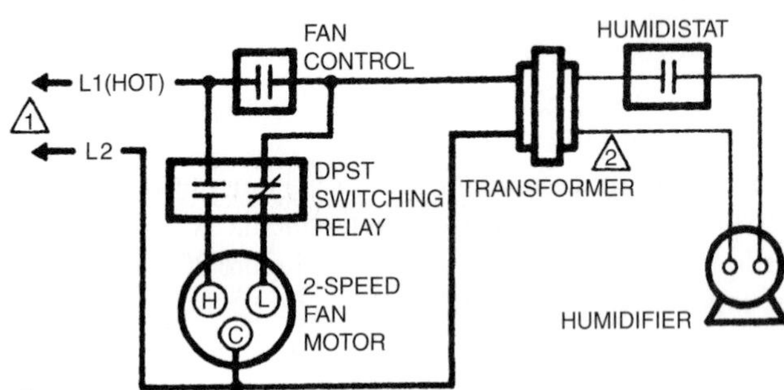

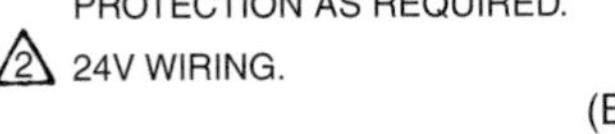

1 POWER SUPPLY. PROVIDE DISCONNECT MEANS AND OVERLOAD PROTECTION AS REQUIRED.

2 24V WIRING.

(B)

Figure 20-8. (A) Humidistat. (B) Wiring diagram. *Courtesy Honeywell, Inc., Residential Division*

Figure 20-9. Combination spray-nozzle and evaporative pad humidifier. *Courtesy Aqua-Mist, Inc.*

Figure 20-10. Atomizing humidifier. *Courtesy AutoFlo Company*

Figure 20-11. Centrifugal humidifier. *Courtesy Herrmidifier Company, Inc.*

The *centrifugal humidifier* uses an impeller or a slinger to throw the water and break it into particles that are evaporated in the airstream. Figure 20-11 shows this type of humidifier.

Caution: Atomizing humidifiers should operate only when the furnace is operating, or moisture will accumulate and cause corrosion, mildew, and a major moisture problem where they are located.

One type of centrifugal humidifier operates with a thermostat that controls a solenoid valve that turns the unit on and off. The furnace must be on and heating before this type of unit will operate. Another type of centrifugal humidifier is wired in parallel with the blower motor and operates with it. Most centrifugal humidifiers are also controlled with humidistats.

Pneumatic atomizing systems use air pressure to break up the water into a mist of tiny droplets and disperse them.

Caution: These systems as well as other atomizing systems should be applied only when the atmosphere does not have to be kept clean or when the water has a low mineral content, because the minerals in the water are also dispersed in the mist. The minerals fall out and accumulate on surfaces in the area. These systems are often used in manufacturing areas, such as textile mills.

Self-Contained Humidifiers

Many residences and light commercial buildings do not have heating equipment with duct work through which the heated air is distributed. Hydronic heating systems, electric baseboard systems, and unit heaters, for example, do not use duct work. Self-contained humidifiers can be installed to provide humidification where these systems are used and generally employ the same processes as those used with forced-air furnaces. They may use the evaporative, atomizing, or infrared humidification processes. These units may include an electric heating device to heat the water, or the water may be distributed over an evaporative media. A fan must be incorporated in the unit to distribute the moisture throughout the room or area. Figure 20-12 illustrates a drum-type self-contained humidifier. A design using steam is shown in Figure 20-13. In this system the electrodes heat the water and convert it to steam. The steam passes through a hose to a stainless steel duct. Steam humidification is also used in large industrial applications in which steam boilers are available. The steam is distributed through a duct system or directly into the air.

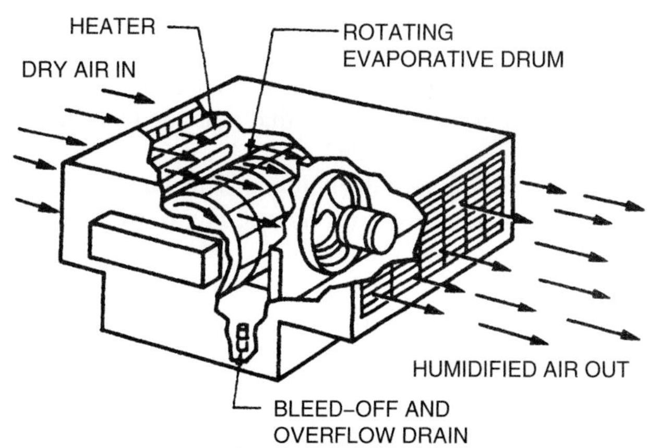

Figure 20-12. Drum-type self-contained humidifier.

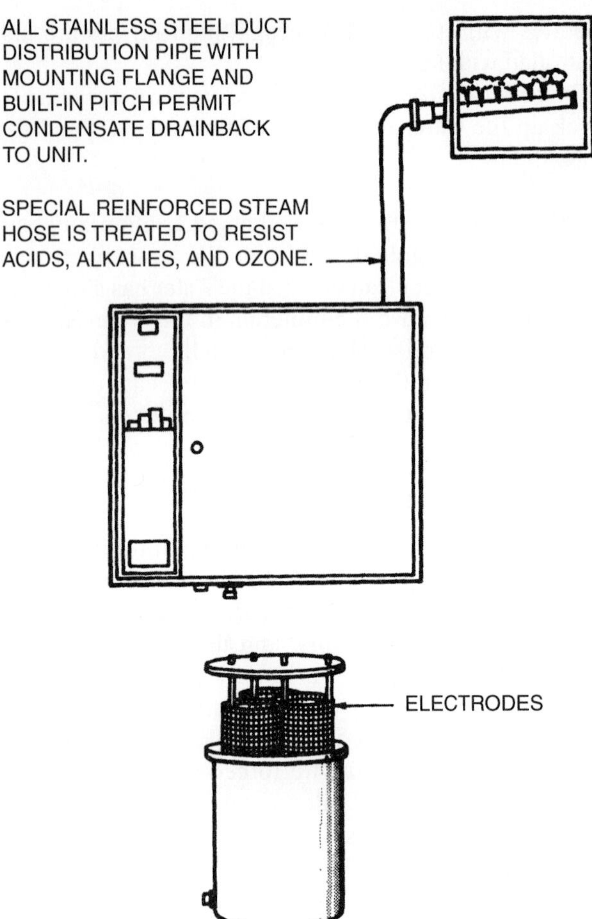

Figure 20-13. Self-contained steam humidifier.

20.5 Sizing Humidifiers

The proper-sized humidifier should be installed. This text emphasizes installation and service, so the details for determining the size, or capacity, of humidifiers is not covered. However, the technician should be aware of some general factors involved in the sizing process:

1. The number of cubic feet of space to be humidified. This is determined by multiplying the number of heated square feet of the house by the ceiling height. A 1,500-square-foot house with an 8-foot ceiling has 12,000 cubic feet of space that must be humidified (1,500 $ft^2 \times 8$ ft = 12,000 ft^3).
2. The construction of the building. This includes quality of insulation, storm windows, fireplaces, building tightness, and so forth.
3. The amount of air change per hour and the approximate lowest outdoor temperature.
4. The level of relative humidity desired.

20.6 Installing Humidifiers

The most important factor regarding installation of humidifiers is to follow the manufacturer's instructions. Evaporative humidifiers are often operated independently of the furnace. It is normally recommended that they be controlled by a humidistat, but they do no harm if they operate continuously, even when the furnace is not operating. Atomizing humidifiers, however, should not operate when the furnace blower is not operating. Moisture will accumulate in the duct if the humidifier is allowed to do so.

Particular attention should be given to clearances within the duct or plenum. The humidifier should not vent directly onto air-conditioning coils, air filters, electronic air cleaners, blowers, or turns in the duct.

If you mount a humidifier on a supply duct, choose a humidifier that serves the largest space in the house. The humid air will spread throughout the house, but the process will be more efficient when given the best distribution possible.

Plan the installation carefully, including determining the location of the humidifier and providing the wiring and the plumbing (with drain). A licensed electrician and/or plumber must provide the service where required by code or law.

20.7 Servicing, Troubleshooting, and Preventive Maintenance for Humidifiers

Proper servicing, troubleshooting, and preventive maintenance are important in keeping humidifying equipment operating efficiently. Cleaning the components that are in contact with the water is the most important factor. The frequency of cleaning depends on the hardness of the water: The harder the water, the more minerals in the water. In evaporative systems these minerals collect on the medium, on other moving parts, and in the reservoir. In addition, algae, bacteria, and virus growth can cause problems, even to the extent of blocking the output of the humidifier. Algicides can be used to help neutralize algae growth. The reservoir should be drained regularly if possible, and components, particularly the media, should be cleaned or changed periodically.

Following are some problems typically encountered with humidifiers:

HUMIDIFIER NOT RUNNING. When the humidifier does not run, the problem is usually electrical, or a component is locked because of mineral buildup. A locked condition can cause a thermal overload protector to cut out. Check the overload protection, circuit breakers, humidistat, and low-voltage controls (if there are any). Check whether the motor is burned out. Clean all components and disinfect them.

EXCESSIVE DUST. If excessive white dust is caused by the humidifier, there is mineral buildup on the media. Clean or replace the media. If excessive dust occurs in an atomizing humidifier, the wrong equipment has been installed.

WATER OVERFLOW. Water overflow indicates a defective float valve assembly. It may need cleaning, adjusting, or replacing.

MOISTURE IN OR AROUND DUCTS. Moisture in ducts is found only in atomizing humidifiers. Check the control to see if it operates at other times, such as during *cool* and *fan on* modes. A restricted airflow may also cause this problem.

LOW OR HIGH HUMIDITY LEVELS. If the humidity level is too high or too low, check the calibration of the humidistat by using a sling psychrometer. If it is out of calibration, you may be able to adjust it. Ensure that the humidifier is clean and operating properly.

20.8 Air Filtration

Concerns about energy conservation and air pollution, and other concerns, have resulted in building construction that allows as little air infiltration from the outside as possible. This means that air within a building is recirculated many times and picks up dust, dirt, smoke, and other contaminants. These impurities may include calcium carbonate, sodium chloride, and copper oxide. These contaminants must be removed or the air will become progressively more dirty and impure, and the result will be an unhealthy indoor environment.

If reservoir-type humidifiers are not maintained properly, fungi, bacteria, algae, and viruses can collect and be discharged into the air. This is especially true if the system is not used for a long time and the water stagnates.

Steam humidifiers, in which the water is heated to the point of vaporization, often alleviate many of these problems. Figure 20-14 illustrates a steam humidifier with a microprocessor control and information source. Many of these humidifiers use water as the conductor between electrodes (see Figure 20-13). As the current passes through the water, it heats the water to the boiling point, and steam is produced. In this type of humidifier, many of the waterborne minerals are left behind, and the heat kills the bacteria and viruses.

Dilution is another method that can be used to control indoor air pollution. This air dilution can be achieved by allowing more natural infiltration of outside air into the building or by using mechanical equipment to force outside air in. However, these methods reduce the energy efficiency of the heating and cooling equipment.

One design of mechanical equipment that has been developed to provide more dilution consists of two insulated ducts run to an outside wall. One duct allows fresh outside air to enter the building, and the other vents stale polluted air to the outside. A desiccant-coated heat-transfer disc rotates between the two airstreams. In the winter, this disc recovers heat and moisture from the exhaust air and transfers it to the incoming fresh air. In the summer, heat and moisture are removed from the incoming air and transferred to the exhaust air. Figure 20-15 illustrates how this equipment can be installed.

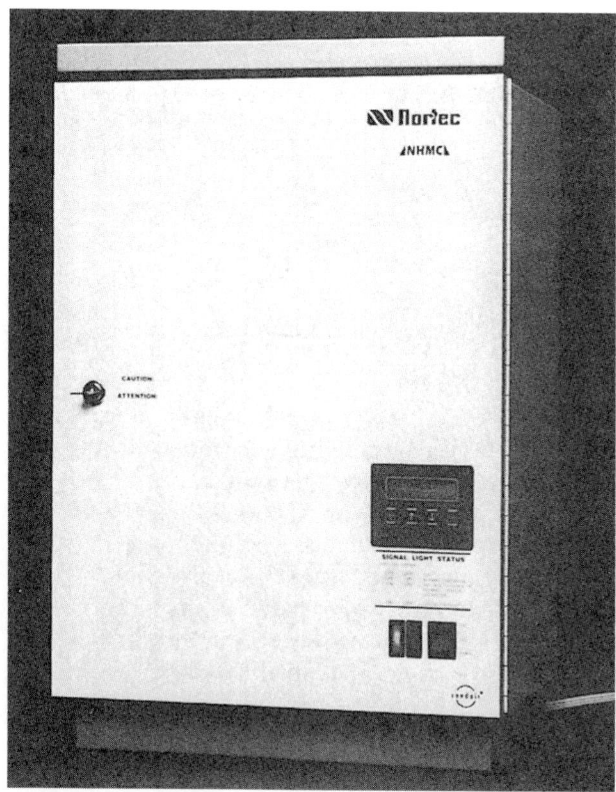

Figure 20-14. Steam humidifier with microprocessor control. *Courtesy Nortec*

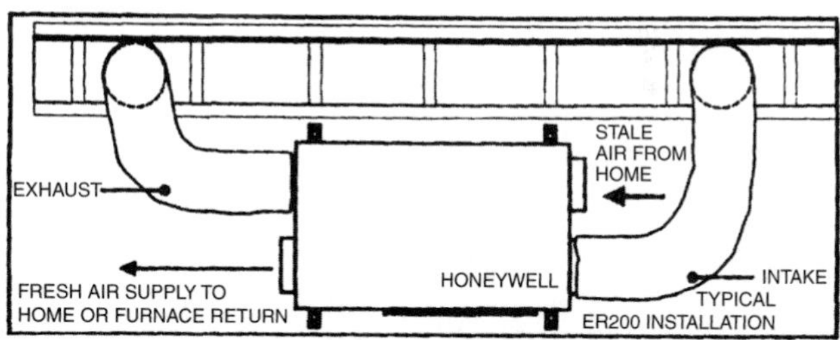

Figure 20-15. Air-dilution system with energy recovery, *Courtesy Honeywell, Inc.*

Other methods used to help control indoor air pollution are to filter the air with one or more types of filters; to use electrostatic precipitators, often called *electronic air cleaners;* or to use an electronic air-purification system.

Air Filters

The *fiberglass air filter media* can be purchased in bulk, as shown in Figure 20-16, or in frames, as shown in Figure 20-17. This material is usually 1 inch thick and coated with a special nondrying, nontoxic adhesive on each fiber. Many high-density types are designed to remove as much as 90% of dust and pollen in the air. This filter material is designed so that it becomes progressively dense as the air passes through it, as shown in Figure 20-18. This type of filter must be placed in the filter rack correctly, with

Figure 20-16. Fiberglass filter media purchased in bulk. *Courtesy W.W. Grainger, Inc.*

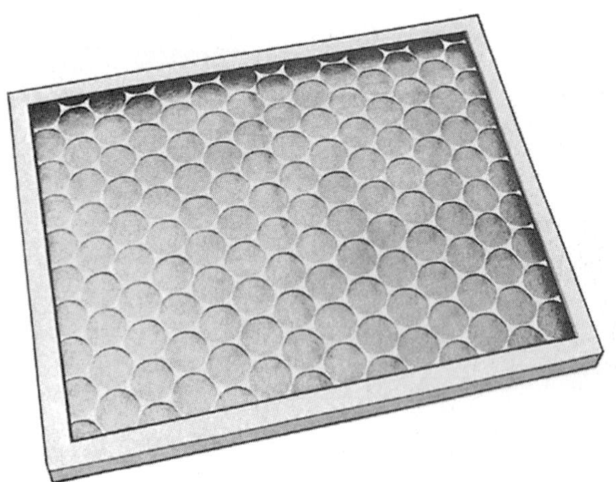

Figure 20-17. Fiberglass filter media purchased in a frame. *Courtesy W.W. Grainger, Inc.*

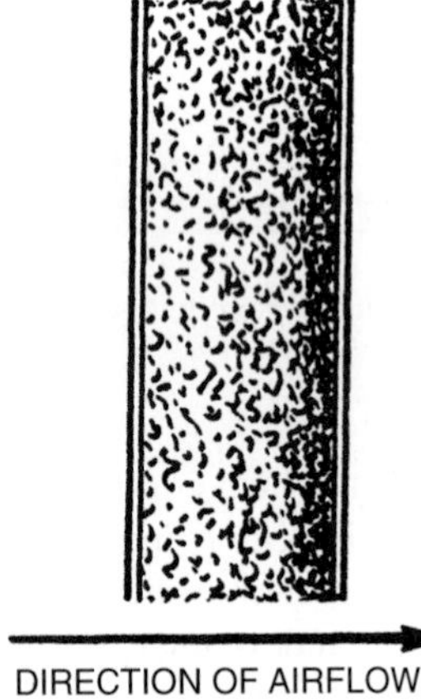

Figure 20-18. Section of fiberglass media with air flowing through it.

the arrow on the filter pointing in the direction of the airflow. Proper installation allows the larger particles to be caught first so that the filter is more efficient. These filters must be replaced regularly.

Some applications do not permit the use of fiberglass as a filter media or require a higher air velocity than fiberglass allows. *Extended-surface air filters,* Figure 20-19, are often made of nonwoven cotton and produce air-cleaning efficiencies of as much as three times more than fiberglass. This type of filter is often used in computer and electronic equipment rooms.

Figure 20-19. Extended-surface air filter.

Washable steel filters are permanent and are washed rather than replaced. Figure 20-20 shows one of these filters. They are usually used in commercial applications, such as in restaurants, hotels, and schools.

Figure 20-21 illustrates another type of filter, *the bag filter,* which is highly efficient. This filter uses a fine fiberglass media within the bags and removes microscopic particles. These filters may be used in hospital operating rooms, electronic equipment assembly rooms, and computer equipment rooms.

Electrostatic Precipitators (Electronic Air Cleaners)

There are several types of electrostatic precipitators. These cleaners can be designed to be mounted at the furnace, can be throw-away filter frames, can be installed within duct

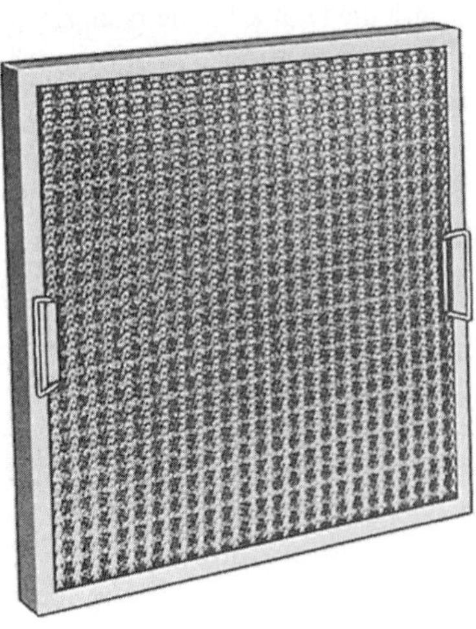

Figure 20-20. A washable steel air filter. *Courtesy W.W. Grainger, Inc.*

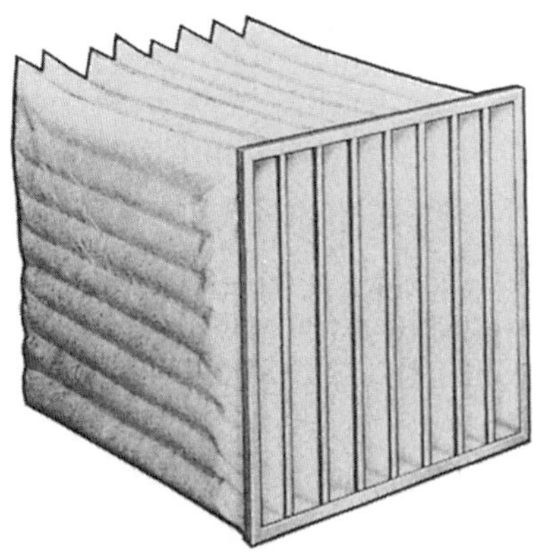

Figure 20-21. Bag air filter. *Courtesy W.W. Grainger, Inc.*

systems, or can be portable stand-alone systems. Each system usually has a prefilter section, an ionizing section, and sometimes a charcoal section.

The prefilter traps larger particles and airborne contaminants. In the ionizing section, particles are charged with a positive charge. These particles then pass through a series of negative- and positive-charged plates. The charged contaminants are repelled by the positive plates and attracted to the negative plates. The air may then pass through a charcoal filter, on those systems that have them, to remove many of the odors. Figure 20-22 shows some of these electrostatic filtering units.

Electronic Air-Purification Devices

Other devices have been developed to help clean the air and reduce pollution, Figure 20-23. One type uses an

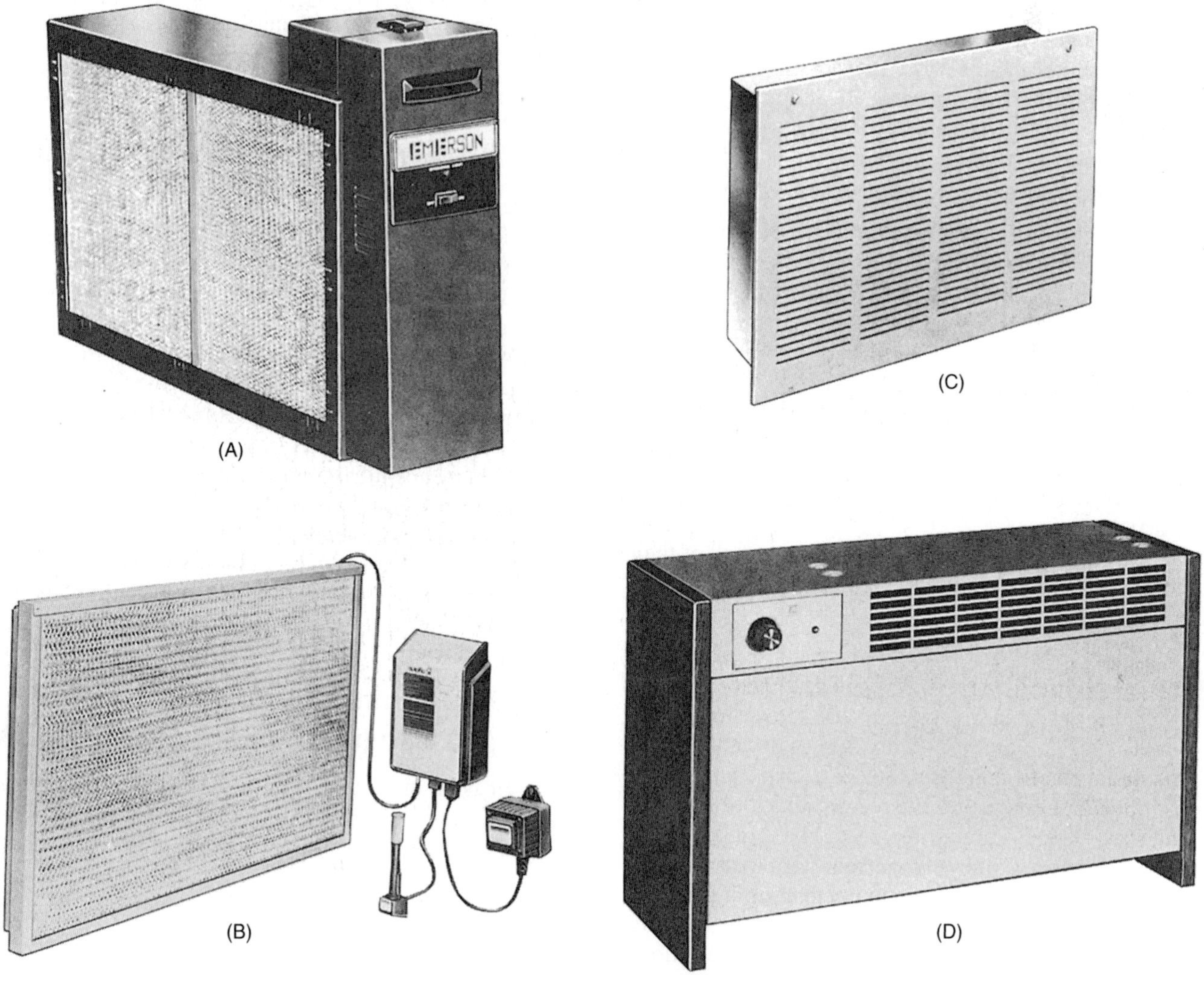

Figure 20-22. Electrostatic precipitators. (A) Furnace mount. (B) Throw-away filter frame. (C) Single-intake return system. (D) Stand-alone portable system. *Courtesy W.W. Grainger, Inc.*

Figure 20-23. An electronic air-purification system. *Courtesy Aqua-Mist, Inc.*

electronic air-purification process. This device produces trivalent oxygen, which helps to eliminate air-borne particulates and odors. Trivalent oxygen is an unstable oxidant that readily decomposes into ordinary oxygen.

HEPA Air Filters

A High Efficiency Particulate Air (HEPA) filter is a type of air filter that can remove approximately 99.97% of the airborne particles. Most air filters can remove air particles 20-30 microns; HEPA filters can remove particles down to 0.3 microns in size. A micron is equal to one-millionth of a meter. For comparison, the width of a human hair is about 50 to 100 microns.

HEPA filters are rated by the percentage of particulate matter they can capture as well as the size of the particles they are intended to capture. To be classified as a "true" HEPA filter, it must be lab tested and labeled with information including the serial number of the filter, the percentage of particles captured, and the size of the particles.

Manufacturers of true HEPA filters must test all of their filters and clearly affix the results of this test to the filter itself.

20.9 Replacing the Filters on an HVAC System

The most important thing you can do to keep an air conditioner operating efficiently is to recommend that air filters be checked, cleaned, and/or replaced on a regular schedule. Clogged, dirty filters restrict normal airflow, which can cause unfiltered air to enter the system through leaks. In addition, restricted airflow affects the equipment's efficiency. When replacing the filters on a system, the following guidelines should be followed:

- Inspect the channel that holds the filters to be sure that the channels are in good shape and that the filter is supported on at least two sides.
- Make certain that the replacement filter is the same size as the filter channel, not necessarily the size of the filter that came out of the unit.
- Install the filter in the channel, with the arrow on the filter pointing in the direction of airflow, which is toward the blower.
- Mark the edge of the filter with the date and your initials.
- Once the filter has been installed, inspect the filter and the channel to be certain that no air can bypass the filter.
- Seal any and all air leaks to prevent/eliminate air bypass.

20.10 Air Duct and Air-Handling Equipment Cleaning and Maintenance

When a structure becomes so badly polluted that an air filtration system does not work properly, oftentimes a thorough cleaning of the duct is necessary to correct the problem. Duct cleaning generally refers to the cleaning of the supply and return ducts as well as registers, grills, diffusers, heating and cooling coils on heat exchangers, condensate drain pans, fan motor and fan housing, and the air handling unit housing. When cleaning a duct system, you need to clean all the system's components. Failure to do so will negate the benefits of having the system cleaned by allowing the system to become re-contaminated. Also, the presence of mold in the air system is a good indication that moisture is present. If this is the case, the source of that moisture should be identified and corrected. Failure to correct moisture problems in a forced air system will result in the mold returning, thus requiring the system to be re-cleaned at a future time.

When a system is cleaned, special brushes are used to dislodge dirt, dust, and other debris. This is followed by a high-powered vacuum cleaning. Chemical biocides that are designed to kill microorganisms can be used. However, before using any chemical cleaning agents, always read and follow all instructions. Failure to carefully read and follow the instructions on a chemical cleaning agent can result in damage to the air system, ineffective cleaning, and/or personal injury. Also, always check all local building codes before attempting to clean an air system.

20.11 Air Duct Cleaning Associations and Regulations

Clean air is an important topic in HVAC and is often debated and argued. Over time it has become a very large industry that is establishing its own requirements, standards, and regulations. Some local and state governments have adapted these regulations for their own locales. Regardless, a good understanding of the building code governing HVAC is critical. In addition, if a technician is going to start cleaning duct systems, it is strongly recommended that the technician become acquainted with the National Air Duct Cleaning Association (NADCA). Formed in 1989, this non-profit organization was originally founded to promote the removal of contaminated HVAC equipment as the only acceptable method of cleaning. Today its mission has changed to be the number one source for HVAC cleaning and restoration. In other words, removal of the contaminated equipment is no longer the accepted method of cleaning contaminated equipment but instead is now the last line of defense.

Summary

- In fall and winter, homes are dry because colder air infiltrates the home and is heated. When heated, the air is artificially dried out by expansion.
- The relative humidity is the percentage of moisture in the air compared with the capacity of the air to hold the moisture.
- Dry air draws moisture from carpets, furniture, wood-work, plants, and people and frequently has a detrimental effect.
- Humidifiers put moisture back into the air.
- The evaporative humidifier provides moisture to a media and forces air through it. The air evaporates the moisture.
- An evaporative humidifier can be a bypass type that is mounted outside the duct work on a forced-air furnace with piping from the hot air through the humidifier to the cold-air side of the furnace. Or, an evaporative humidifier can be mounted in the plenum or under the duct.
- Evaporative humidifier media are discs, drums, and plates.
- For heat-pump or electric-heat installations, the humidifier may have an independent electric heater to increase the temperature of the water for evaporation.
- Infrared humidifiers use infrared lamps and a reflector to cause the water to be evaporated into the airstream.
- There are two types of atomizing humidifiers: spray nozzle and centrifugal. They discharge a mist (tiny water droplets) into the air. They should not be used in hard-water conditions and should be used only where there is forced-air movement.
- Self-contained humidifiers are used with hydronic, electric baseboard, or unit heaters, where forced air is not available. They have their own fans and often heaters.
- The number of cubic feet in the house, the type of construction, the amount of air change per hour, the lowest outdoor temperature, and the relative humidity level desired determine the size of the humidifier.
- The manufacturer's instructions should be followed carefully when the humidifier is installed.
- Modern buildings are being constructed to allow little air infiltration. Filtering and purification devices are used to clean the air.
- Indoor air quality (IAQ) has become a major factor in many buildings.
- Air-filtration devices include a fiberglass filter media, extended-surface air filters, washable steel air filters, bag air filters, electronic air cleaners, and electronic air-purification devices.

Review Questions

1. Why are homes drier in the winter than in the summer?
2. What is relative humidity?
3. Why does relative humidity decrease as the temperature increases?
4. How does dry warm air dry out household furnishings?
5. Why does dry air cause a cool feeling?
6. How is moisture added to the air in a house?
7. What are the differences between evaporative and atomizing humidifiers?
8. How does a bypass humidifier operate?
9. Is a plenum-mounted humidifier installed in the supply or return air plenum?
10. What is the purpose of the media in an evaporative humidifier?
11. How are disc, plate, and drum humidifiers designed? Describe their designs and the medium that can be used on each.
12. Why do some humidifiers have their own water-heating devices?
13. How does an infrared humidifier operate?
14. What are the two types of atomizing humidifiers?
15. How does each type of atomizing humidifier operate?
16. Why is it essential for the furnace to be running when an atomizing humidifier is operating?
17. Why are self-contained humidifiers used?
18. What is a self-contained humidifier?
19. What general factors are used to determine the size, or the capacity, of a humidifier?
20. In general, how are humidifiers that are used with forced-air furnaces installed?
21. Why are systems of cleaning air more necessary in modern buildings than in older ones?
22. What are five materials or devices used to clean or purify the air?
23. Why does the design of fiberglass filters require you to consider the direction of the airflow in the installation?

21 System Design

Objectives

Upon completion of this unit, you should be able to

- **explain why a well-designed system is better than a system that is merely installed without regard to design.**
- **choose the correct design temperatures, both inside and outside, for a structure.**
- **use tables to determine the various resistances for a composite wall, combine them into a total resistance, and convert this to a *U* factor.**
- **use the heat-transfer formula to find the heat-transfer rate through a wall.**
- **choose the correct size unit for a heating system.**

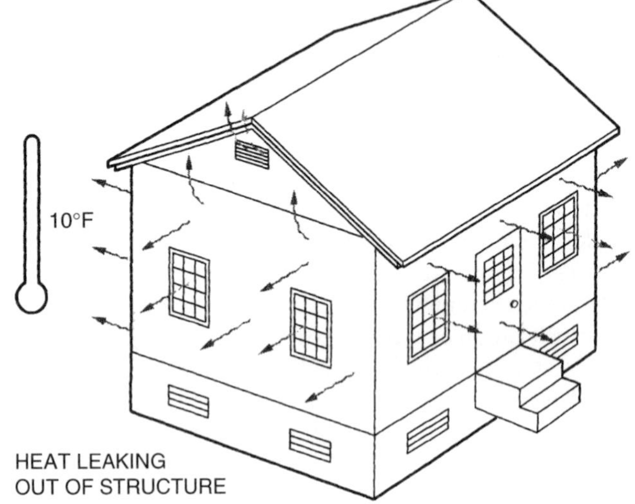

Figure 21-1. Heat must be put back into the structure as fast as it leaks out.

21.1 System Design

Designing a heating system for a structure is important if the system is to successfully heat the structure. To successfully heat the structure is to provide a heating system that keeps the space temperature comfortable during cold weather. To be comfortable, the air temperature must be within the comfort range and the air-distribution system must not create drafts, either hot or cold, in the conditioned space. Designing such a system is not complicated when the rules are followed.

The design sequence starts with a load calculation of the structure. This calculation tells the technician how much heat is leaking out of the structure. A heating system that puts the heat back into the structure as fast as it leaks out is ideal from a space temperature standpoint, Figure 21-1. Some tolerances may be allowed and are discussed in this unit. Air distribution is discussed in Unit 22.

The load calculation is important from another standpoint. Many lending institutions do not loan money for new construction unless they are ensured that the heating system will be adequate. These institutions do not want a poor heating system in a building that they will have to sell if the building owner defaults on the loan.

This text can help you understand the *methods* used for determining the load calculation for a structure. It is not meant to be used to run the actual load calculation. Designers use other methods for the actual load calculation. The most common of these methods is to use the Air Conditioning Contractors of America *ACCA Manual J.* It has a detailed account of how the load calculation is determined. Another method that many designer–estimators use is to complete manufacturers' forms. If a company installs a particular brand of equipment, it will likely have a short form for calculating the heat loss.

Whatever the method used, it should be nationally recognized. A nationally recognized method provides the designer–estimator with evidence of doing the job professionally. This evidence may be important if a legal dispute occurs. It is not uncommon for a customer to take a contractor to court over the choice of equipment. When you have evidence that your procedure was correct, you have an advantage.

There are times when the technician needs to know why the designer–estimator made certain decisions in choosing a particular system for a structure. The technician may also be caught in the middle of a dispute over why a system does not work correctly. For example, suppose that the technician is called out to work on a heating system that

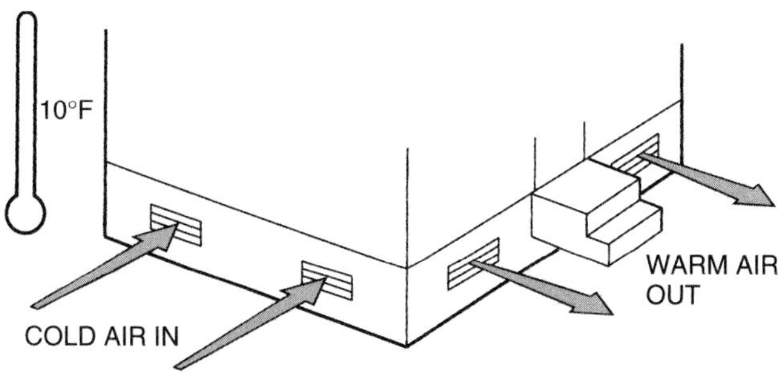

Figure 21-2. The vent covers were removed from the crawl space openings, and too much heat is being lost from the floor of the house.

will not heat a structure. The technician should first check the heating system to make sure that it is performing up to capacity. If not, the heat-loss calculation on the structure may be the problem. Perhaps the designer–estimator did not account for something, or another problem occurred. For example, it is not uncommon for a home owner to leave the vent covers off the crawl space in the winter, in which case the winter winds will blow under the house, as shown in Figure 21-2. The designer–estimator would have assumed that these vents would be covered.

The observant technician can uncover other problems and correct them without calling the designer–estimator into the problem. The designer–estimator may work for another company after the job is out of warranty because the owner may change service companies. The technician's job is to know the complete system and procedure.

21.2 Design Conditions

The technician must be familiar with two temperature design conditions before choosing a heating system for a structure: the inside and the outside design temperatures. The inside design temperature is typically 70°F for a residence and for typical offices. Not all people will be comfortable at this temperature, but it is a starting point. Typically, you will find that women are a little more cold natured than men and desire temperatures at a higher level. Older people, both men and women, typically desire warmer temperatures. If you know that the occupants of a structure will not be comfortable at this temperature, the inside design temperature can be designed around 75°F so that the capacity is available if necessary.

The size of the heating system in a residence or a small building is important to the operation of the system. In the winter, heat transfers from the inside of the building to the outside. This is why the load calculation is often called a *heat-loss calculation.* Remember that heat transfers from hot to cold. The heat in a building is much like water in a city water tank. Water is continuously being used by the city residences. Sometimes, such as during summer, people use large amounts of water while watering yards. In the winter, the usage may be much less. Water must be added at least as fast as it is being used during the high-usage summer months, or the town will run short of water (see Figure 21-3).

Likewise, the temperature in a structure must be maintained by adding heat as fast as it leaks out on the highest demand day, which is the coldest part of the winter. The greater the temperature difference between the inside of the structure and the outside, the faster the heat will transfer. For example, heat will transfer much faster on a day when the outdoor temperature is 0°F than on a day when it is 40°F, Figure 21-4. The difference in temperature is the driving force that causes heat to transfer. This temperature difference can be compared with the rate of city water usage. On cold days, more heat is required and used to maintain the same temperature in the structure.

The building materials of the structure are also a factor. Some building materials allow less heat transfer and some allow more. For example, the Styrofoam used in the coolers you use to keep cold drinks is a good insulator. You can keep drinks cold much longer in a Styrofoam cooler than in a metal washtub because Styrofoam is an insulator and metal is a conductor of heat, Figure 21-5. Choosing the best insulation for a home is worthwhile for the home owner when the home owner has input into the decision. Often homes are built by speculators, builders who put the homes up for sale after they are built. The buyer has no input into the choice of materials. When a future home owner has input into the building materials, a better choice can be made. Building materials and their application are discussed later in this unit, in Section 21.3.

Choosing a heating system that is the correct size is like choosing the pump for the city water tank. If the pump is not large enough, the water system will not have enough capacity on days of heavy usage, and the city will run out of water. In a house, the heating system would fall behind, and the house temperature would begin to drop.

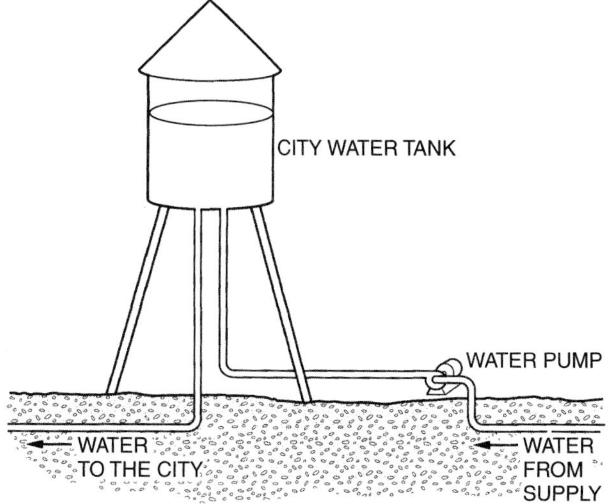

Figure 21-3. Water must be added to the city water tank as fast as it leaks out, or the tank level will drop.

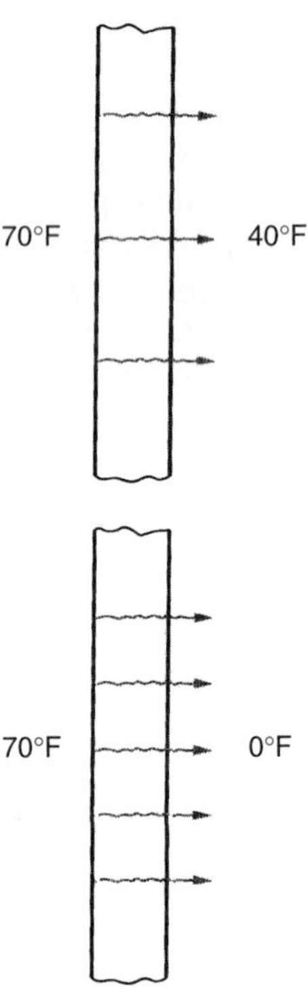

Figure 21-4. More heat is lost from a structure on a cold day than on a mild day.

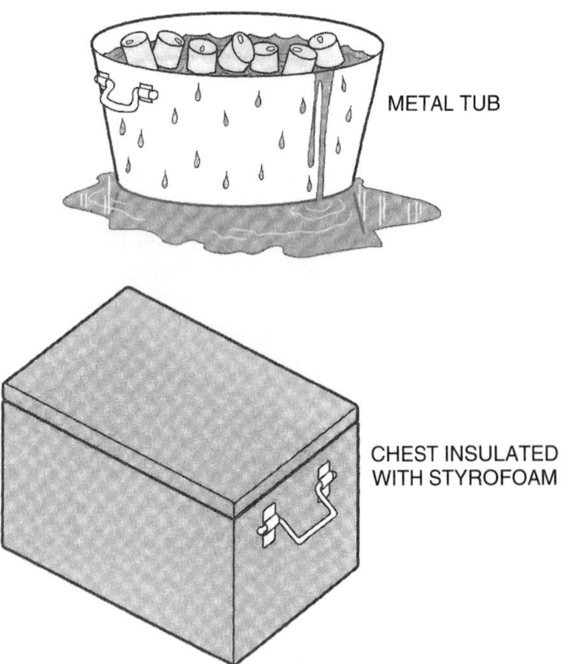

Figure 21-5. Cold drinks stored in a metal tub warm up much faster than drinks stored in a Styrofoam cooler.

If the system is much too small, the house temperature could become cold enough to cause health problems, not to mention discomfort. If the city buys a water pump that is too large, the city will spend extra money on equipment that has more power than needed and that does not operate efficiently. Likewise, if the system in a house is too large, the extra expense is unnecessary and the system may not produce the desired efficiency.

In the past, many technicians chose the furnace size by merely looking at the house and guessing. This is called the *big house, big furnace method* of choosing a furnace. If the furnace uses a fossil fuel (oil or gas), oversizing is not efficient for the furnace. Remember that a furnace requires approximately 15 minutes to get up to the proper operating temperature at the beginning of the cycle. At the end of the cycle, many natural-draft furnaces still allow heat to escape up the flue as the fan dissipates the heat at the end of the cycle. If the furnace is short cycling because it is oversized, it does not have the long running times that produce the best efficiency. The structure to be heated usually loses most of its heat in the middle of the night, so it is not known exactly how much efficiency is lost to short cycling in an oversized furnace. Many experts claim that a fossil fuel furnace, such as a gas furnace, which should have a steady-state efficiency of 80%, may have a seasonal efficiency of as low as 50% because of short cycling as a result of being oversized. Therefore, choosing a furnace that is matched to the load is especially important.

Another variable that affects the heating load on a house is the amount of daylight hours. If there is any sun during the day, there is less demand on the heating system. A heating system sized for the worst days of the worst winters would be oversized for the job at any other time of the season.

The technician who chooses the size of the heating system for a structure must balance many factors while choosing the correct fuel and the correct size for a system. This job must not be taken lightly. It is no longer wise to choose a big furnace just because it will heat the house on the coldest day with a lot of reserve capacity. Your competitor will win the bid for the job with a smaller and more efficient—and less costly—system.

The starting point for choosing the correct size for a heating system is to know the actual heating load for a structure on the coldest day. Then you can improvise by choosing the correct heating system or by suggesting some home improvements that will more closely match the needs of the structure to the heating system size.

To calculate the heat load for a structure, you must know the kinds of weather conditions that are encountered in the winter. Using the known lowest outdoor temperatures that are encountered for extended time periods will tell the technician the design temperature difference. For example, suppose that the outside design temperature is 0°F and the inside design temperature is 70°F. The design temperature difference is 70°F, Figure 21-6. Remember that the temperature difference is the driving force for heat exchange.

The American Society of Heating, Refrigerating and Air-Conditioning Engineers (ASHRAE) has a listing of the weather data for cities around the world that includes the data needed to determine the temperature difference

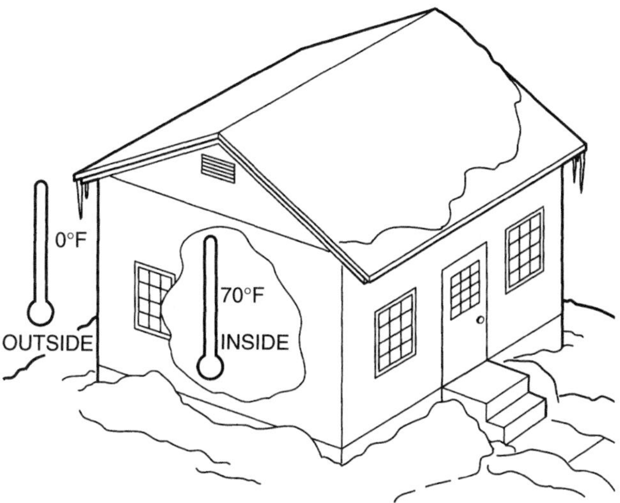

Figure 21-6. Temperature difference is the driving force for heat exchange.

for any major city. These design data have been compiled for many years, calculated to within one-tenth of a degree Fahrenheit, and rounded off to the nearest degree for practical use.

We limit our discussion to U.S. cities, but the same discussion can be expanded to any foreign country where weather data are available. Figure 21-7 is a page of the ASHRAE table showing several states—Ohio, Oklahoma, and Oregon. This particular page lists states of very different climates, with Oklahoma being western, Ohio being northern, and Oregon northwestern in climate nature. The table lists weather data gathered at first-order weather stations, U.S. Air Force bases, and airports from around the country and includes both winter and summer data. We are interested only in the winter data in column 5. Figure 21-8 shows data for column 5. Note that column 5 lists weather data for two conditions, 99% and 97½% data. Weather data are gathered in the United States during the months of December, January, and February for a total of 2,160 hours (24 h × 90 days = 2,160 h). When you use the 99% value, the temperature will not drop below the design value for more than 1% of the time during a typical winter. This amounts to 22 hours per average winter (2,160 × .01 = 21.6, or 22). The number of hours will be more during unusually cold winters. Keep in mind that all these hours are not likely to happen on any single day. Normally they occur for 1 or 2 hours at a time, in the early morning, usually between 6 and 8 A.M. sun time, on clear days. The heat loss is not instantaneous and probably does not affect the space temperature for 2 to 4 hours, or not until 10:00 A.M. to 12:00 P.M. By this time, the sun has canceled the chilling effect on sunny days by warming the space through the windows. On cloudy days, however, the effect would be felt. If our heating system did not keep up for a few hours in midmorning (10:00 A.M. to 12:00 P.M.), we probably would not notice if we were an early rising person. We would be up and about and to some extent be self-warming with activity.

The 97½% value means that the temperature will be above the design value for 97½% of the time during an average winter and below this temperature 2½% of the time, or only 54 hours (2,160 h × .025 = 54). This is an acceptable design method for many designers, particularly in the southern United States. If a designer uses this column for the outside design temperature, it will occur only in the early morning hours during normal winters for a few hours. The building may be a retail store, which may not need the extra heat in the mid-morning hours because by then the lights are on and creating internal heat. By the time customers start to arrive, the store temperature is normal. If the building is residential, spot heat may be more desirable than to oversize the system for the worst case or even the 1% conditions.

Sizing a system for 97½% weather data is quite common among many designers. The only drawbacks are uncommon winters and the fact that the system has little reserve capacity for times when it may be shut down for repairs or power problems. For example, if the system is shut off for any reason on a day or a night when the temperature is close to the design value, the structure will take a long time to get back up to design temperature. For this reason, many designers prefer to use a larger heating system.

When a system is designed, the furnace or boiler size is rarely the same value as the actual heat loss of the structure. For example, the house heating load may be 78,000 Btu per hour. A heating system with a capacity of 80,000 Btu per hour output is likely to be chosen as the closest size. This is an oversize of 2,000 Btu per hour.

However, the exact design conditions and equipment choice are regional. It is easy to write specifications and descriptions as to how to design systems, but it is much more practical to use the experience of successful local contractors. Different types of systems are popular in different parts of the country. For example, hot-water heating systems are seldom used in the southern states for heating homes because air-conditioning cannot be easily added to these systems. The northern states use much more hot-water heat. A customer in the southern states will want a system that is typical of what has been known to work in the area.

Determining the heat loss for a room or a whole house requires the same basic calculations. The only difference is that the heat losses for all the rooms are added to produce the whole-house calculation. To make this calculation, the technician must take into consideration all the places in a house from which heat can leak. These places involve any walls where temperature differences occur and any place in the house where cold air enters the structure and has to be heated, as shown in Figure 21-9. Following is a list of typical places where heat leaks occur:

1. through the outside walls
2. through partitions to rooms that are unheated, such as garages or storage rooms

TABLE 1 CLIMATIC CONDITIONS FOR THE UNITED STATES *(Continued)*

Col. 1	Col. 2		Col. 3		Col. 4	Col. 5 Winter, °F		Col. 6 Summer, °F			Col. 7	Col. 8			Col. 9 Prevailing Wind		Col. 10 Temp. °F	
State and Station[a]	Lat.		Long.		Elev.	Design Dry-Bulb		Design Dry-Bulb and Mean Coincident Wet-Bulb			Mean Daily	Design Wet-Bulb			Winter	Summer	Median of Annual Extr.	
	°	'	°	'	Feet	99%	97.5%	1%	2.5%	5%	Range	1%	2.5%	5%	Knots[d]		Max.	Min.
OHIO																		
Akron-Canton AP	40	55	81	26	1208	1	6	89/72	86/71	84/70	21	75	73	72	SW 9	SW	94.4	−4.6
Ashtabula	41	51	80	48	690	4	9	88/73	85/72	83/71	18	75	74	72				
Athens	39	20	82	06	700	0	6	95/75	92/74	90/73	22	78	76	74				
Bowling Green	41	23	83	38	675	−2	2	92/73	89/73	86/71	23	76	75	73			96.7	−7.3
Cambridge	40	04	81	35	807	1	7	93/75	90/74	87/73	23	78	76	75				
Chillicothe	39	21	83	00	640	0	6	95/75	92/74	90/73	22	78	76	74	W 8	WSW	98.2	−2.1
Cincinnati Co	39	09	84	31	758	1	6	92/73	90/72	88/72	21	77	75	74	W 9	SW	97.2	−.2
Cleveland AP (S)	41	24	81	51	777	1	5	91/73	88/72	86/71	22	76	74	73	SW 12	N	94.7	−3.1
Columbus AP (S)	40	00	82	53	812	0	5	92/73	90/73	87/72	24	77	75	74	W 8	SSW	96.0	−3.4
Dayton AP	39	54	84	13	1002	−1	4	91/73	89/72	86/71	20	76	75	73	WNW 11	SW	96.6	−4.5
Defiance	41	17	84	23	700	−1	4	94/74	91/73	88/72	24	77	76	74				
Findlay AP	41	01	83	40	804	2	3	92/74	90/73	87/72	24	77	76	74			97.4	−7.4
Fremont	41	20	83	07	600	−3	1	90/73	88/73	85/71	24	76	75	73				
Hamilton	39	24	84	35	650	0	5	92/73	90/72	87/71	22	76	75	73			98.2	−2.8
Lancaster	39	44	82	38	860	0	5	93/74	91/73	88/72	23	77	75	74				
Lima	40	42	84	02	975	−1	4	94/74	91/73	88/72	24	77	76	74	WNW 11	SW	96.0	−6.5
Mansfield AP	40	49	82	31	1295	0	5	90/73	87/72	85/72	22	76	74	73	W 8	SW	93.8	−10.7
Marion	40	36	83	10	920	0	5	93/74	91/73	88/72	23	77	76	74				
Middletown	39	31	84	25	635	0	5	92/73	90/72	87/71	22	76	75	73				
Newark	40	01	82	28	880	−1	5	94/73	92/73	89/72	23	77	75	74	W 8	SSW	95.8	−6.8
Norwalk	41	16	82	37	670	−3	1	90/73	88/73	85/71	22	76	75	73			97.3	−8.3
Portsmouth	38	45	82	55	540	5	10	95/76	92/74	89/73	22	78	77	75	W 8	SW	97.9	1.0
Sandusky Co	41	27	82	43	606	1	6	93/73	91/72	88/71	21	76	74	73			96.7	−1.9
Springfield	39	50	83	50	1052	−1	3	91/74	89/73	87/72	21	77	76	74	W 7	W		
Steubenville	40	23	80	38	992	1	5	89/72	86/71	84/70	22	74	73	72				
Toledo AP	41	36	83	48	669	−3	1	90/73	88/73	85/71	25	76	75	73	WSW 8	SW	95.4	−5.2
Warren	41	20	80	51	928	0	5	89/71	87/71	85/70	23	74	73	71				
Wooster	40	47	81	55	1020	1	6	89/72	86/71	84/70	22	75	73	72			94.0	−7.7
Youngstown AP	41	16	80	40	1178	−1	4	88/71	86/71	84/70	23	74	73	71	SW 10	SW		
Zanesville AP	39	57	81	54	900	1	7	93/75	90/74	87/73	23	78	76	75	W 6	WSW		
OKLAHOMA																		
Ada	34	47	96	41	1015	10	14	100/74	97/74	95/74	23	77	76	75				
Altus AFB	34	39	99	16	1378	11	16	102/73	100/73	98/73	25	77	76	75	N 10	S		
Ardmore	34	18	97	01	771	13	17	100/74	98/74	95/74	23	77	77	76				
Bartlesville	36	45	96	00	715	6	10	101/73	98/74	95/74	23	77	77	76				
Chickasha	35	03	97	55	1085	10	14	101/74	98/74	95/74	24	78	77	76				
Enid, Vance AFB	36	21	97	55	1307	9	13	103/74	100/74	97/74	24	79	77	76				
Lawton AP	34	34	98	25	1096	12	16	101/74	99/74	96/74	24	78	77	76				
McAlester	34	50	95	55	776	14	19	99/74	96/74	93/74	23	77	76	75	N 10	S		
Muskogee AP	35	40	95	22	610	10	15	101/74	98/75	95/75	23	79	78	77				
Norman	35	15	97	29	1181	9	13	99/74	96/74	94/74	24	77	76	75	N 10	S		
Oklahoma City AP (S)	35	24	97	36	1285	9	13	100/74	97/74	95/73	23	78	77	76	N 14	SSW		
Ponca City	36	44	97	06	997	5	9	100/74	97/74	94/74	24	77	76	76				
Seminole	35	14	96	40	865	11	15	99/74	96/74	94/73	23	77	76	75				
Stillwater (S)	36	10	97	05	984	8	13	100/74	96/74	93/74	24	77	76	75	N 12	SSW	103.7	1.6
Tulsa AP	36	12	95	54	650	8	13	101/74	98/75	95/75	22	79	78	77	N 11	SSW		
Woodward	36	36	99	31	2165	6	10	100/73	97/73	94/73	26	78	76	75			107.1	−1.3
OREGON																		
Albany	44	38	123	07	230	18	22	92/67	89/66	86/65	31	69	67	66			97.5	16.6
Astoria AP (S)	46	09	123	53	8	25	29	75/65	71/62	68/61	16	65	63	62	ESE 7	NNW		
Baker AP	44	50	117	49	3372	−1	6	92/63	89/61	86/60	30	65	63	61			97.5	−6.8
Bend	44	04	121	19	3595	−3	4	90/62	87/60	84/59	33	64	62	60			96.4	−5.8
Corvallis (S)	44	30	123	17	246	18	22	92/67	89/66	86/65	31	69	67	66	N 6	N	98.5	17.1
Eugene AP	44	07	123	13	359	17	22	92/67	89/66	86/65	31	69	67	66	N 7	N		
Grants Pass	42	26	123	19	925	20	24	99/69	96/68	93/67	33	71	69	68	N 5	N	103.6	16.4
Klamath Falls AP	42	09	121	44	4092	4	9	90/61	87/60	84/59	36	63	61	60	N 4	W	96.3	.9
Medford AP (S)	42	22	122	52	1298	19	23	98/68	94/67	91/66	35	70	68	67	S 4	WNW	103.8	15.0
Pendleton AP	45	41	118	51	1482	−2	5	97/65	93/64	90/62	29	66	65	63	NNW 6	WNW		
Portland AP	45	36	122	36	21	17	23	89/68	85/67	81/65	23	69	67	66	ESE 12	NW	96.6	18.3
Portland Co	45	32	122	40	75	18	24	90/68	86/67	82/65	21	69	67	66			97.6	20.5
Roseburg AP	43	14	123	22	525	18	23	93/67	90/66	87/65	30	69	67	66			99.6	19.5
Salem AP	44	55	123	01	196	18	23	92/68	88/66	84/65	31	69	68	66	N 6	N	98.9	15.9
The Dalles	45	36	121	12	100	13	19	93/69	89/68	85/66	28	70	68	67			105.1	7.9

Figure 21-7. Weather data for some U.S. cities. *Courtesy American Society of Heating, Refrigerating and Air-Conditioning Engineers*

Col. 1	Winter, °F Col. 5		Col. 1	Winter, °F Col. 5	
State and Station[a]	Design Dry-Bulb		State and Station[a]	Design Dry-Bulb	
	99%	97.5%		99%	97.5%
OHIO			OKLAHOMA		
Akron-Canton AP	1	6	Ada	10	14
Ashtabula	4	9	Altus AFB	11	16
Athens	0	6	Ardmore	13	17
Bowling Green	–2	2	Bartlesville	6	10
Cambridge	1	7	Chickasha	10	14
Chillicothe	0	6	Enid, Vance AFB	9	13
Cincinnati Co	1	6	Lawton AP	12	16
Cleveland AP (S)	1	5	McAlester	14	19
Columbus AP (S)	0	5	Muskogee AP	10	15
Dayton AP	–1	4	Norman	9	13
Defiance	–1	4	Oklahoma City AP (S)	9	13
Findlay AP	2	3	Ponca City	5	9
Fremont	–3	1	Seminole	11	15
Hamilton	0	5	Stillwater (S)	8	13
Lancaster	0	5	Tulsa AP	8	13
Lima	–1	4	Woodward	6	10
Mansfield AP	0	5			
Marion	0	5	OREGON		
			Albany	18	22
Middletown	0	5	Astoria AP (S)	25	29
Newark	–1	5	Baker AP	–1	6
Norwalk	–3	1	Bend	–3	4
Portsmouth	5	10	Corvallis (S)	18	22
Sandusky Co	1	6	Eugene AP	17	22
Springfield	–1	3			
			Grants Pass	20	24
Steubenville	1	5	Klamath Falls AP	4	9
Toledo AP	–3	1	Medford AP (S)	19	23
Warren	0	5	Pendleton AP	–2	5
Wooster	1	6	Portland AP	17	23
Youngstown AP	–1	4	Portland Co	18	24
Zanesville AP	1	7			
			Roseburg AP	18	23
			Salem AP	18	23
			The Dalles	13	19

Figure 21-8. Weather data for column 5, 99% and 97½%. *Courtesy American Society of Heating, Refrigerating and Air-Conditioning Engineers*

3. through floors above vented or unvented crawl spaces or unheated basements
4. through floors that are on grades and through concrete floors on the ground (these floors leak heat to the ground below and out the end of the slab)
5. through windows by conduction
6. around windows and doors by infiltration; air that leaks in because of pressure differences between the inside and outside of the structure; or ventilation, which is planned infiltration.

21.3 Heat Transfer and Building Materials

The materials used to build a structure are one of the determining factors as to how much heat leaks out of a building. As mentioned previously, some materials are conductors, some are insulators, and all are used for building structures. For example, steel is a conductor, and Styrofoam is an insulator. A solid steel door would conduct a lot of heat. The amount of heat transferred through the steel door could be reduced if the door were constructed with steel skin on the outside and polyurethane inside, as shown in Figure 21-10.

The design technician must know how much heat will transfer through the building material in order to calculate the heat that will be lost from a structure. The amount of heat transferred through a building material is known as *heat of transmission.* Actually, this is heat by conduction. This heat of transmission is calculated by using a heat-transfer coefficient called the *U* factor. The *U* factor is expressed in Btu per hour per degree Fahrenheit temperature difference per square foot of building material. The *U* factor is the reciprocal of the R-value of a material.

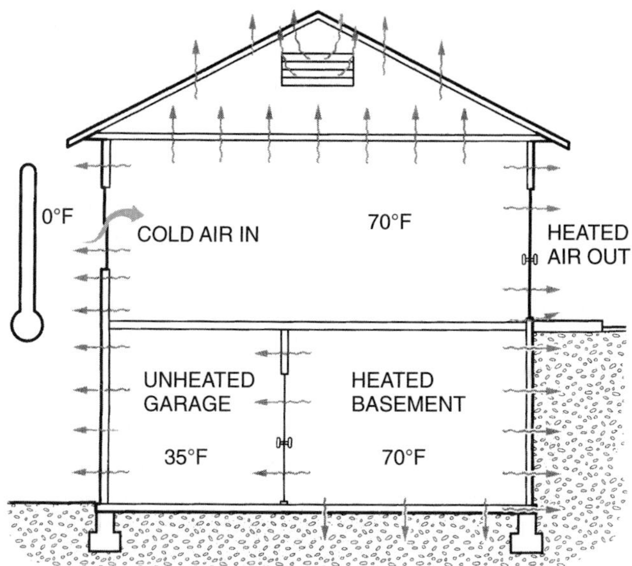

Figure 21-9. Places where heat leaks from a structure.

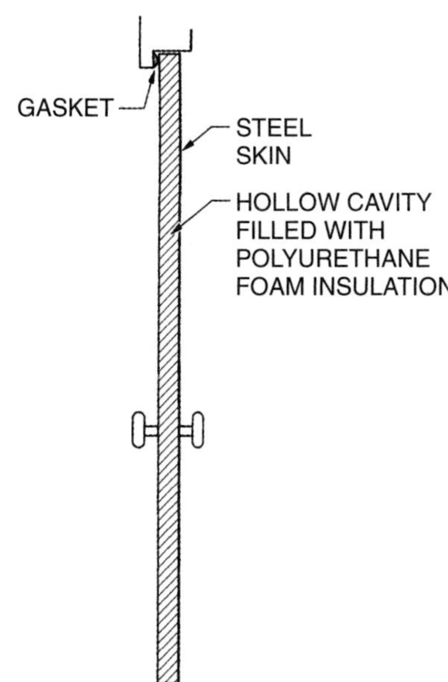

Figure 21-10. Steel door with insulation inside.

The R-value is the materials resistance to the flow of heat. Therefore for a material having an R-value of 19, its U factor can be calculated as follows:

$$U = \frac{1}{R}$$

$$U = \frac{1}{19}$$

$$U = 0.0526$$

The following equation is used when calculating the heat transfer through a particular building material.

$$q = A \times U \times \text{TD}$$

where q = Heat transfer in Btu/hr.
A = Area of the building material in square feet.
U = Overall heat-transfer coefficient.
TD = Temperature difference from one side to the other in °F.

For example, if a building material has a 1°F temperature difference from one side to the other, a U factor of 0.5, and an area of 1 square foot, it will transfer 0.5 Btu of heat per hour, Figure 21-11:

$$q = A \times U \times \text{TD}$$
$$= 1\text{sq ft} \times (0.5\ \text{Btu/hr} \times \text{°F TD} \times \text{sq ft}) \times 1\text{°F TD}$$
$$= 0.5\ \text{Btu/hr}$$

Finding the area of the building material is easy; an existing structure can be measured, or the area can be found on the blueprint for a structure that is to be built.

The temperature difference is found by subtracting the outdoor design temperature from the inside design temperature. For example, suppose that the inside design temperature is 75°F and the outside 1% design temperature is 5°F. The TD (temperature difference) is 70°F: (75 – 5 = 70°F).

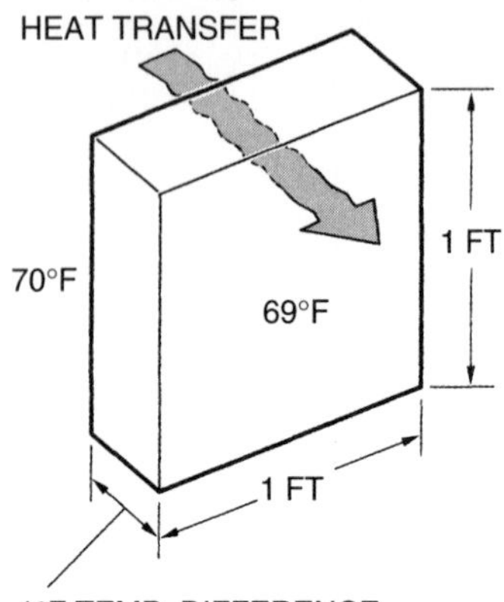

Figure 21-11. Overall heat-transfer coefficient, or U factor.

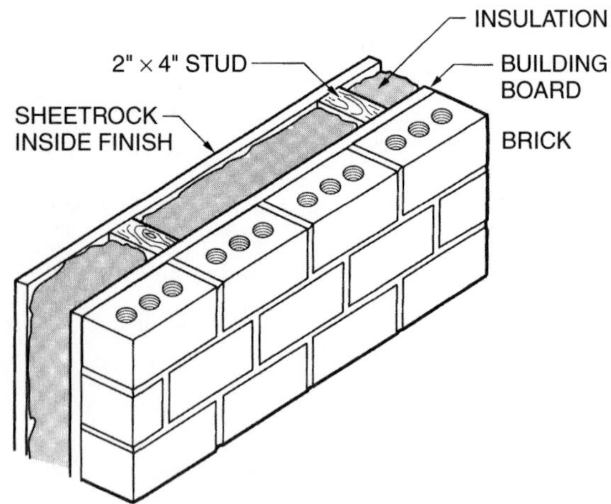

Figure 21-12. Composite stud wall.

The U factor, known as the overall coefficient of heat transfer, is the most difficult part of the formula to find. The U factor is found by testing materials and determining the rate of heat transfer and by calculation of known values of heat transfer. It is often necessary to construct, or develop a U factor when one is not published for a particular composite wall. A composite wall is one constructed of several different materials, such as a wall in a house that may have brick on the outside and a stud wall with Sheetrock fastened to it, as shown in Figure 21-12. There are several materials in this wall, each with a different rate of heat transfer. The overall U factor must be determined for use in the above formula.

21.4 Heat Loss Through Walls

The walls of a structure make up the greatest area of the structure and a large amount of heat is lost through them. As you read the following explanations, remember that the walls contain the doors and windows. When the heat loss of the wall alone is calculated, the area of the windows and doors must be subtracted from the wall area. This calculation is called the *net wall area.* Figure 21-13 illustrates this concept.

The wall of a structure may be above grade or below grade. Grade is ground level, Figure 21-14. Below-grade walls are in basements, and the heat loss for them is just as important as the heat loss for above-grade walls.

When a wall is made of several materials, the U factor can be found by determining the resistance to heat flow of each material in the wall, then adding them to find the total resistance (R_t) to heat transfer. R_t can then be converted to the U value by dividing it into 1. The U value is the reciprocal of R_t: $U = 1/R_t$.

The R_t for a wall is fairly easy to calculate by using the table in Figure 21-15. Notice that this table is quite detailed for many materials. For example, brick seems like a simple building material, but when we look at the resistances

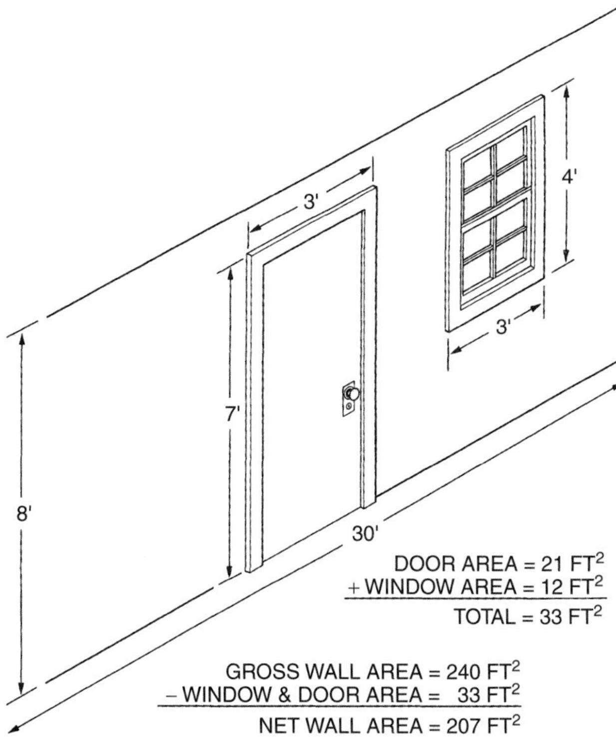

Figure 21-13. Net wall area calculation.

listed for brick, we find many combinations. The density column explains the difference. Brick density can be from 80 to 130 pounds per cubic foot of the brick material. Different bricks come from different parts of the country and are used for different applications. For this text, we use a resistance value for brick of 0.20 per inch of thickness. This was found by using the average resistance for 120 pounds per cubic foot of brick ($0.23 + 0.16 = 0.39 \div 2 = 0.195$, rounded to 0.20).

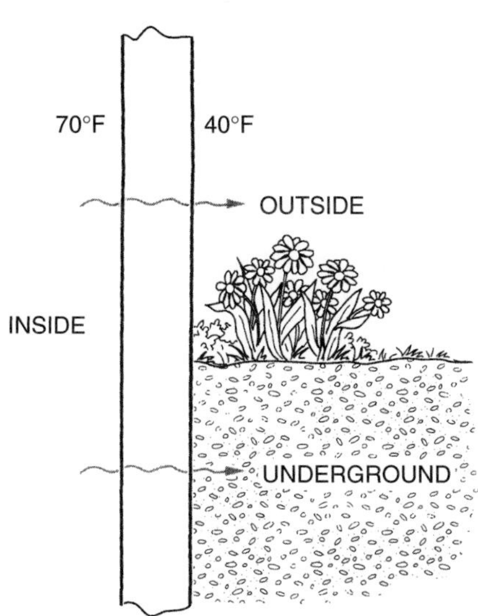

Figure 21-14. A wall may be above or below grade. Grade is ground level.

Concrete blocks are also not so easy to calculate because of the variety. In this text, we use a resistance of 1.04 ($1.11 + 0.97 = 2.08 \div 2 = 1.04$).

Before we calculate the total resistance to heat flow for a wall, we must consider two other factors. These factors have nothing to do with building materials.

An air film on each side of a wall offers resistance to the heat flow. This film is very still on the inside of the house, where the air in the room is not moving very fast. Heat does not readily transfer through still air. The resistance to heat flow from the film on the outside depends on the wind speed. This resistance is called the *film factor.* These two film factors are expressed in the form of resistances in this text and are as follows:

Inside film resistance = $0.68 + 1.47 = 2.15 \div 2 = 1.1$

Outside film resistance for 15-mph wind speed = 0.25

An example of how to calculate R_t for a wall can be seen in Figure 21-16. This is a typical masonry wall using brick, polystyrene insulation between, and concrete block with plaster on the inside. The U value for this wall is 0.107.

If a house has a wall of this material in a climate where the outside design temperature is 0°F and the inside design temperature is 75°F, the temperature difference is 75°F. If the wall has no windows, is 30 feet long, and has an 8-foot ceiling, it has an area of 240 square feet ($30 \times 8 = 240$). All this can be used to calculate the total heat loss from this wall:

$$
\begin{aligned}
q &= A \times U \times \text{TD} \\
&= 240 \times 0.107 \times 75 \\
&= 1{,}926 \text{ Btu/hr}
\end{aligned}
$$

The R-value can also be used to determine the best combination of materials that will provide the best insulation for the money. This is called the *what-if method* of building a wall. For example, suppose that a home owner is building a house and wonders, "What if 2 inches of polystyrene are used instead of 1 inch?" Notice that the polystyrene offers more resistance by far than any of the other building materials in the structure. Adding another inch would significantly improve the efficiency of the structure. R_t for the added insulation is 15.34 ($9.34 + 6 = 15.34$). When this is converted to a U factor, it becomes 0.07. Notice that the larger the R_t, the smaller the U factor. The smaller the U factor, the less heat loss from the wall. These relationships allow the public to easily understand the significance of the R-value for an insulation: Bigger is better.

The wall will now lose less heat:

$$
\begin{aligned}
q &= A \times U \times \text{TD} \\
&= 240 \times 0.07 \times 75 \\
&= 1{,}260 \text{ Btu/hr}
\end{aligned}
$$

With the addition of 1 inch of polystyrene, the wall loses only 65% of the original heat (1,260/1,926 = .654,

Table 4 Typical Thermal Properties of Common Building and Insulating Materials—Design Values[a]

Description	Density, lb/ft³	Resistance[c] (R): Per inch thickness (1/k), °F · ft² · h / Btu · in.	Resistance[c] (R): For thickness listed (1/C), °F · ft² · h / Btu
BUILDING BOARD			
Asbestos-cement board	120	0.25	—
Asbestos-cement board ... 0.125 in.	120	—	0.03
Asbestos-cement board ... 0.25 in.	120	—	0.06
Gypsum or plaster board ... 0.375 in.	50	—	0.32
Gypsum or plaster board ... 0.5 in.	50	—	0.45
Gypsum or plaster board ... 0.625 in.	50	—	0.56
Plywood (Douglas Fir)[d]	34	1.25	—
Plywood (Douglas Fir) ... 0.25 in.	34	—	0.31
Plywood (Douglas Fir) ... 0.375 in.	34	—	0.47
Plywood (Douglas Fir) ... 0.5 in.	34	—	0.62
Plywood (Douglas Fir) ... 0.625 in.	34	—	0.77
Plywood or wood panels ... 0.75 in.	34	—	0.93
Vegetable Fiber Board			
Sheathing, regular density[e] ... 0.5 in.	18	—	1.32
... 0.78125 in.	18	—	2.06
Sheathing intermediate density[e] ... 0.5 in.	22	—	1.09
Nail-base sheathing[e] ... 0.5 in.	25	—	1.06
Shingle backer ... 0.375 in.	18	—	0.94
Shingle backer ... 0.3125 in.	18	—	0.78
Sound deadening board ... 0.5 in.	15	—	1.35
Tile and lay-in panels, plain or acoustic	18	2.50	—
... 0.5 in.	18	—	1.25
... 0.75 in.	18	—	1.89
Laminated paperboard	30	2.00	—
Homogeneous board from repulped paper	30	2.00	—
Hardboard[e]			
Medium density	50	1.37	—
High density, service temp. service underlay	55	1.22	—
High density, std. tempered	63	1.00	—
Particleboard[e]			
Low density	37	1.41	—
Medium density	50	1.06	—
High density	62.5	0.85	—
Underlayment ... 0.625 in.	40	—	0.82
Waferboard	37	1.59	—
Wood subfloor ... 0.75 in.	—	—	0.94
BUILDING MEMBRANE			
Vapor—permeable felt	—	—	0.06
Vapor—seal, 2 layers of mopped 15-lb felt	—	—	0.12
Vapor—seal, plastic film	—	—	Negl.
FINISH FLOORING MATERIALS			
Carpet and fibrous pad	—	—	2.08
Carpet and rubber pad	—	—	1.23
Cork tile ... 0.125 in.	—	—	0.28
Terrazzo ... 1 in.	—	—	0.08
Tile—asphalt, linoleum, vinyl, rubber vinyl asbestos ceramic	—	—	0.05
Wood, hardwood finish ... 0.75 in.	—	—	0.68
INSULATING MATERIALS			
Blanket and Batt[f, g]			
Mineral Fiber, fibrous form processed from rock, slag, or glass			
approx. 3–4 in	0.3–2.0	—	11
approx. 3.5 in	0.3–2.0	—	13
approx. 5.5–6.5 in	0.3–2.0	—	19
approx. 6–7.5 in	0.3–2.0	—	22
approx. 9–10 in	0.3–2.0	—	30
approx. 12–13 in	0.3–2.0	—	38
Board and Slabs			
Cellular glass	8.5	2.86	—
Glass fiber, organic bonded	4.0–9.0	4.00	—
Expanded perlite, organic bonded	1.0	2.78	—
Expanded rubber (rigid)	1.5	4.55	—

Figure 21-15. This table shows the R values for many common building materials. *Courtesy American Society of Heating, Refrigerating and Air-Conditioning Engineers*

Table 4 Typical Thermal Properties of Common Building and Insulating Materials—Design Values[a] (*Continued*)

Description	Density, lb/ft³	Resistance [c](*R*) Per inch thickness (1/*k*), °F · ft² · h / Btu · in.	Resistance [c](*R*) For thickness listed (1/*C*), °F · ft² · h / Btu
Expanded polystyrene, extruded (smooth skin surface) (CFC-12 exp.)	1.8–3.5	5.00	—
Expanded polystyrene, molded beads	1.0	3.85	—
	1.25	4.00	—
	1.5	4.17	—
	1.75	4.17	—
	2.0	4.35	—
Cellular polyurethane/polyisocyanurate[h] (CFC-11 exp.) (unfaced)	1.5	6.25–5.56	—
Cellular polyisocyanurate[h] (CFC-11 exp.) (gas-permeable facers)	1.5–2.5	6.25–5.56	—
Cellular polyisocyanurate[i] (CFC-11 exp.) (gas-impermeable facers)	2.0	7.20	—
Cellular phenolic (closed cell) (CFC-11, CFC-113 exp.)	3.0	8.20	—
Cellular phenolic (open cell)	1.8–2.2	4.40	—
Mineral fiber with resin binder	15.0	3.45	—
Mineral fiberboard, wet felted			
Core or roof insulation	16–17	2.94	—
Acoustical tile	18.0	2.86	—
Acoustical tile	21.0	2.70	—
Mineral fiberboard, wet molded			
Acoustical tile[j]	23.0	2.38	—
Wood or cane fiberboard			
Acoustical tile[j] 0.5 in.	—	—	1.25
Acoustical tile[j] 0.75 in.	—	—	1.89
Interior finish (plank, tile)	15.0	2.86	—
Cement fiber slabs (shredded wood with Portland cement binder)	25–27.0	2.0–1.89	—
Cement fiber slabs (shredded wood with magnesia oxysulfide binder)	22.0	1.75	—
Loose Fill			
Cellulosic insulation (milled paper or wood pulp)	2.3–3.2	3.70–3.13	—
Perlite, expanded	2.0–4.1	3.7–3.3	—
	4.1–7.4	3.3–2.8	—
	7.4–11.0	2.8–2.4	—
Mineral fiber (rock, slag, or glass)[g]			
approx. 3.75–5 in.	0.6–2.0		11.0
approx. 6.5–8.75 in.	0.6–2.0		19.0
approx. 7.5–10 in.	0.6–2.0		22.0
approx. 10.25–13.75 in.	0.6–2.0		30.0
Mineral fiber (rock, slag, or glass)[g]			
approx. 3.5 in. (closed sidewall application)	2.0–3.5	—	12.0–14.0
Vermiculite, exfoliated	7.0–8.2	2.13	—
	4.0–6.0	2.27	—
Masonry Units			
Brick, common	80	0.45–0.31	—
	90	0.37–0.27	—
	100	0.30–0.23	—
	110	0.29–0.18	—
	120	0.23–0.16	—
	130	0.19–0.11	—
Clay tile, hollow			
1 cell deep 3 in.	—	—	0.80
1 cell deep 4 in.	—	—	1.11
2 cells deep 6 in.	—	—	1.52
2 cells deep 8 in.	—	—	1.85
2 cells deep 10 in.	—	—	2.22
3 cells deep 12 in.	—	—	2.50
Concrete blocks[k]			
Limestone aggregate			
8 in., 36 lb, 138 lb/ft³ concrete, 2 cores	—		—
Same with perlite filled cores	—		2.1
12 in., 55 lb, 138 lb/ft³ concrete, 2 cores	—		—
Same with perlite filled cores	—		3.7

Figure 21-15. (*Continued*)

Table 4 Typical Thermal Properties of Common Building and Insulating Materials—Design Values[a] (*Continued*)

Description	Density, lb/ft³	Resistance [c](*R*) Per inch thickness (1/*k*), °F · ft² · h / Btu · in.	Resistance [c](*R*) For thickness listed (1/*C*), °F · ft² · h / Btu
Normal weight aggregate (sand and gravel)			
8 in., 33–36 lb, 126–136 lb/ft³ concrete, 2 or 3 cores	—	—	1.11–0.97
Same with perlite filled cores	—	—	2.0
Same with verm. filled cores	—	—	1.92–1.37
12 in., 50 lb, 125 lb/ft³ concrete, 2 cores	—	—	1.23
Medium weight aggregate (combinations of normal weight and lightweight aggregate)			
8 in., 26–29 lb, 97–112 lb/ft³ concrete, 2 or 3 cores	—	—	1.71–1.28
Same with perlite filled cores	—	—	3.7–2.3
Same with verm. filled cores	—	—	3.3
Same with molded EPS (beads) filled cores	—	—	3.2
Same with molded EPS inserts in cores	—	—	2.7
Lightweight aggregate (expanded shale, clay, slate or slag, pumice)			
6 in., 16–17 lb, 85–87 lb/ft³ concrete, 2 or 3 cores	—	—	1.93–1.65
Same with perlite filled cores	—	—	4.2
Same with verm. filled cores	—	—	3.0
8 in., 19–22 lb, 72–86 lb/ft³ concrete..........	—	—	3.2–1.90
Same with perlite filled cores	—	—	6.8–4.4
Same with verm. filled cores	—	—	5.3–3.9
Same with molded EPS (beads) filled cores	—	—	4.8
Same with UF foam filled cores	—	—	4.5
Same with molded EPS inserts in cores	—	—	3.5
12 in., 32–36 lb, 80–90 lb/ft³ concrete, 2 or 3 cores	—	—	2.6–2.3
Same with perlite filled cores	—	—	9.2–6.3
Same with verm. filled cores	—	—	5.8
Stone. lime, or sand	—	0.08	—
Gypsum partition tile			
3 by 12 by 30 in., solid	—	—	1.26
3 by 12 by 30 in., 4 cells	—	—	1.35
4 by 12 by 30 in., 3 cells	—	—	1.67
METALS			
(See Chapter 39, Table 3)			
ROOFING			
Asbestos-cement shingles	120	—	0.21
Asphalt roll roofing	70	—	0.15
Asphalt shingles	70	—	0.44
Built-up roofing .... 0.375 in.	70	—	0.33
Slate....................................... 0.5 in.	—	—	0.05
Wood shingles, plain and plastic film faced	—	—	0.94
Spray Applied			
Polyurethane foam	1.5–2.5	6.25–5.56	—
Ureaformaldehyde foam	0.7–1.6	4.55–3.57	—
Cellulosic fiber	3.5–6.0	3.45–2.94	—
Glass fiber	3.5–4.5	3.85–3.70	—
PLASTERING MATERIALS			
Cement plaster, sand aggregate	116	0.20	—
Sand aggregate ....0.375 in.	—	—	0.08
Sand aggregate ....0.75 in.	—	—	0.15
Gypsum plaster:			
Lightweight aggregate .... 0.5 in.	45	—	0.32
Lightweight aggregate ....0.625 in.	45	—	0.39
Lightweight agg. on metal lath ....0.75 in.	—	—	0.47
Perlite aggregate	45	0.67	—
Sand aggregate	105	0.18	—
Sand aggregate ....0.5 in.	105	—	0.09
Sand aggregate .... 0.625 in.	105	—	0.11
Sand aggregate on metal lath .... 0.75 in.	—	—	0.13
Vermiculite aggregate...........................	45	0.59	—
MASONRY MATERIALS			
Concretes			
Cement mortar	105–135	0.20–0.10	—
Gypsum-fiber concrete 87.5% gypsum, 12.5% wood chips.............................	51	0.60	—
Lightweight aggregates including expanded	120	0.18–0.09	—
shale, clay or slate; expanded	100	0.27–0.17	—
slags; cinders; pumice; vermiculite;	80	0.40–0.29	—
also cellular concretes	60	0.63–0.56	—
	40	1.08–0.90	—

Figure 21-15. (*Continued*)

Table 4 Typical Thermal Properties of Common Building and Insulating Materials—Design Values[a] (*Continued*)

Description	Density, lb/ft³	Resistance [c](*R*) Per inch thickness (1/*k*), °F · ft² · h / Btu · in.	For thickness listed (1/*C*), °F · ft² · h / Btu
	30	1.33–1.10	—
	20	1.59–1.20	—
Perlite, expanded	50	0.71–0.56	—
	40	1.08	—
	30	1.41	—
	20	2.00	—
Sand and gravel or stone aggregate (oven dried)	140	0.13–0.06	—
Sand and gravel or stone aggregate (not dried)	140	0.10–0.05	—
Stucco	116	0.20	—
SIDING MATERIALS (on flat surface)			
Shingles			
Asbestos-cement	120	—	0.21
Wood, 16 in., 7.5 exposure	—	—	0.87
Wood, double, 16-in., 12-in. exposure	—	—	1.19
Wood, plus insul. backer board, 0.3125 in	—	—	1.40
Siding			
Asbestos-cement, 0.25 in., lapped	—	—	0.21
Asphalt roll siding	—	—	0.15
Asphalt insulating siding (0.5 in. bed.)	—	—	1.46
Hardboard siding, 0.4375 in	—	—	0.67
Wood, drop, 1 by 8 in	—	—	0.79
Wood, bevel, 0.5 by 8 in., lapped	—	—	0.81
Wood, bevel, 0.75 by 10 in., lapped	—	—	1.05
Wood, plywood, 0.375 in., lapped	—	—	0.59
Aluminum or Steel[l], over sheathing			
Hollow-backed	—	—	0.61
Insulating-board backed nominal 0.375 in	—	—	1.82
Insulating-board backed nominal 0.375 in., foil backed			2.96
Architectural glass	—	—	0.10
WOODS (12% Moisture Content)[e,m]			
Hardwoods			
Oak	41.2–46.8	0.89–0.80	—
Birch	42.6–45.4	0.87–0.82	—
Maple	39.8–44.0	0.92–0.84	—
Ash	38.4–41.9	0.94–0.88	—
Softwoods			
Southern Pine	35.6–41.2	1.00–0.89	—
Douglas Fir-Larch	33.5–36.3	1.06–0.99	—
Southern Cypress	31.4–32.1	1.11–1.09	—
Hem-Fir, Spruce-Pine-Fir	24.5–31.4	1.35–1.11	—
West Coast Woods, Cedars	21.7–31.4	1.48–1.11	—
California Redwood	24.5–28.0	1.35–1.22	—

Figure 21-15. (*Continued*)

rounded to .65, the decimal equivalent of 65%). This is a savings of 35% on the heating bill as applied to this wall (100% − 65% = 35%). The whole house can be evaluated in this manner by calculating the values for one wall at a time.

If the wall that we just discussed was between an unheated but enclosed garage and the heated space of the house, it would be treated as a partition, as shown in Figure 21-17. The same temperature difference would not be used because the inside of the garage would not be quite as cold as the outside air. The inside temperature of the garage can be calculated as the average temperature between inside and outside, or 37.5°F (75/2 = 37.5). If the temperature is expected to be below freezing in a garage, supplemental heat can be added to prevent any items in the garage (such as the cars) from freezing. It is very nice in the northern climates to have a car that is not covered with ice in the morning before the trip to work. The film factor for an outside wall is not used. Changing the film factor would make little difference.

When the wall is below grade, such as a wall in a heated basement, determining the temperature difference becomes more complicated. More heat transfers through the wall where it is closer to the surface than the part of the wall that is far below the surface, Figure 21-18. The table in Figure 21-19 expresses the heat loss per running foot of wall at each foot of depth. Notice that all values are averages. If you are not sure, use the largest value. The calculation is accomplished by adding the heat transfers and arriving at a total heat transfer for 1 running foot of

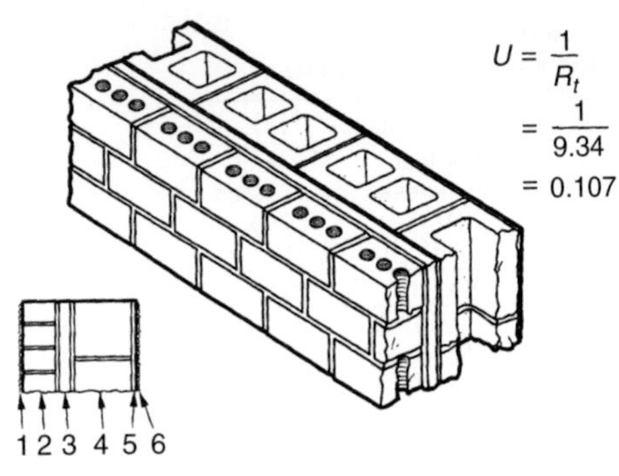

1. OUTSIDE AIR FILM 0.25
2. 4" BRICK (AVERAGE) 0.2 PER IN. × 4 0.80
3. 1" POLYSTYRENE INSULATION 6.00
4. 8" CONCRETE BLOCK 1.04
5. ¾" PLASTER 0.15
6. INDOOR AIR FILM 1.10

$R_t = 9.34$

Figure 21-16. Calculating R_t for a masonry wall.

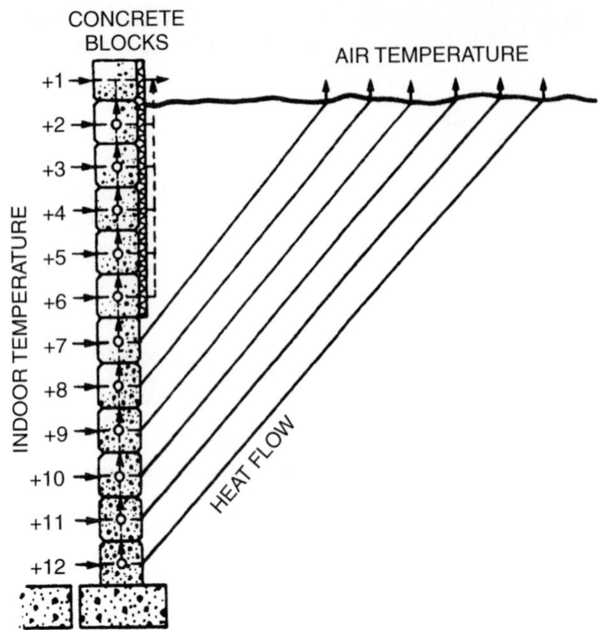

Figure 21-18. Heat transfer through a wall below grade. *Courtesy American Society of Heating, Refrigerating and Air-Conditioning Engineers*

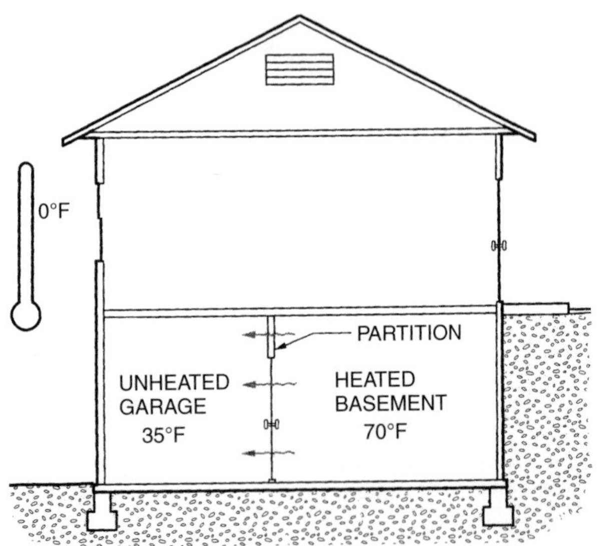

Figure 21-17. Partition wall between the home and the unheated garage.

Table 3 Heat Loss Below Grade in Basement Walls

Depth ft	Path Length Through Soil, ft	Heat Loss, Btu/hr · ft · °F Uninsulated		R = 4.17		R = 8.34		R = 12.5	
0–1	0.68	0.410		0.152		0.093		0.067	
1–2	2.27	0.222	0.632	0.116	0.268	0.079	0.172	0.059	0.126
2–3	3.88	0.155	0.787	0.094	0.362	0.068	0.240	0.053	0.179
3–4	5.52	0.119	0.906	0.079	0.441	0.060	0.300	0.048	0.227
4–5	7.05	0.096	1.002	0.069	0.510	0.053	0.353	0.044	0.271
5–6	8.65	0.079	1.081	0.060	0.570	0.048	0.401	0.040	0.311
6–7	10.28	0.069	1.150	0.054	0.624	0.044	0.445	0.037	0.348

Table 4 Heat Loss Through Basement Floors, Btu/hr · ft² · °F

Depth of Foundation Wall Below Grade	Shortest Width of House, ft 20	24	28	32
5 ft	0.032	0.029	0.026	0.023
6 ft	0.030	0.027	0.025	0.022
7 ft	0.029	0.026	0.023	0.021

Note: $\Delta F = (t_a - A)$

Figure 21-19. These tables show the heat transfer rate for a wall below grade and through basement floors. *Courtesy American Society of Heating, Refrigerating and Air-Conditioning Engineers*

wall and the total depth. For example, suppose that the wall has $R = 4.17$ insulation from the ground level to the bottom of the wall, and the wall is 7 feet underground, Figure 21-20. The total heat per running foot of wall is 2.927 Btu per hour per degree Fahrenheit of temperature difference. If the wall was 30 feet long with a temperature difference of 70°F, the total heat loss for the wall would be 6,146 Btu per hour (2.927 × 30 × 70 = 6,146).

The concrete floor in a basement also has a heat loss. This may be calculated using the bottom portion of Figure 21-19. This calculation uses the depth of the concrete floor and the width distance of the house. The table states the heat loss in Btu/hr/ft²/ of temperature difference.

For example, suppose the basement of a house is 32' × 60' and the floor is 7' below the foundation. The temperature difference is 70°F. The heat loss for the floor would be 2822. Btu/hr (32 × 60 × 0.021 × 70 = 2822.4).

21.5 Roof and Ceiling Heat Loss

Heat is also lost through the top of the structure by conduction. The ceiling of the structure may either have a vented

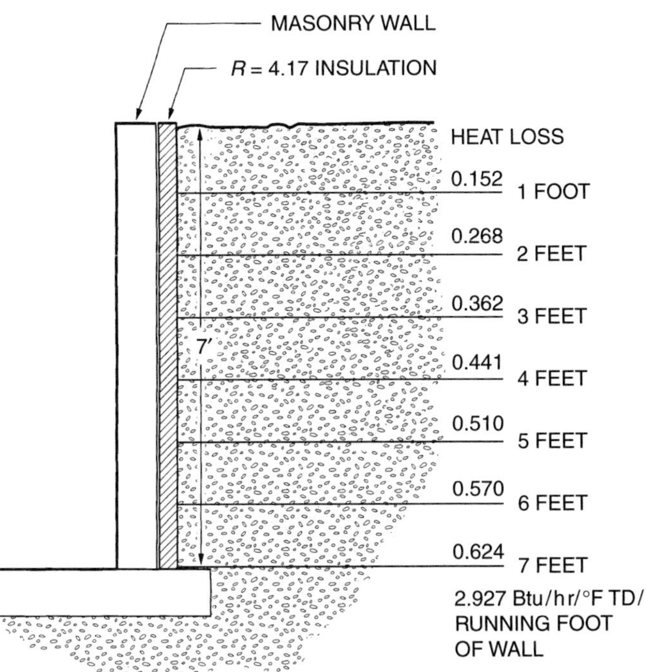

Figure 21-20. Calculating the heat loss through a wall that is below grade.

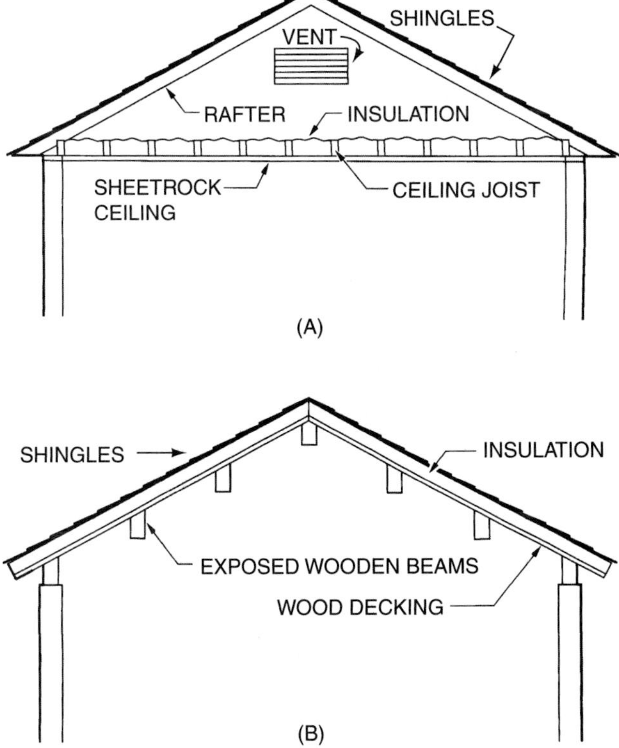

Figure 21-21. Roof and ceiling combinations. (A) Ceiling with attic above. (B) Ceiling is part of roof.

attic above or be part of the roof, as shown in Figure 21-21. Either type of ceiling will transfer heat. The building materials in the two types of structures do not have much resistance to heat transfer. The insulation used is the key to an efficient ceiling or roof for a structure. It commonly has a very high R_t (30 to 50) because the heat in the house rises. A higher ceiling temperature can transfer more heat from the room.

The ceiling loss is calculated just as the wall loss is. Suppose that the house has a floor area of 30 feet by 40 feet and a ceiling with an R_t of 30. The design temperature difference is 70°F. What is the heat loss through the ceiling? The U value is 0.033 ($U = 1/R_t$).

$$\begin{aligned} q &= A \times U \times \text{TD} \\ &= 1{,}200 \times 0.033 \times 70 \\ &= 2{,}772 \text{ Btu/hr} \end{aligned}$$

If the R_t was only 15 for the ceiling, twice as much heat would be required to overcome the loss. Insulating a ceiling is a home improvement job that the home-owner can usually do.

21.6 Heat Loss Through Floors

Many houses are located over crawl spaces or basements. If the crawl space is not vented (the vents closed in winter), it is treated like an unheated basement. It is not as cold as the outside but not as warm as inside. The temperature difference can be calculated like that for the garage; use an average between the inside and outside temperatures.

When a house is on a slab that is on top of the ground, heat is lost around the perimeter of the slab as well as through the bottom of the slab to the ground beneath (see Figure 21-22). Most of the heat leaves the perimeter of the slab. Insulating the perimeter down to about 16 inches below the slab floor is customary in cold climates. When a slab is insulated with $R = 5.4$ insulation in this manner, the heat loss is 0.49 to 0.54 Btu per hour per running foot of slab perimeter per degree Fahrenheit of temperature difference, as shown in Figure 21-23. For example, suppose that a house is on a slab with $R = 5.4$ insulation, and the slab size is 30 feet by 40 feet. Suppose that the house is in a cold climate where the temperature difference is 70°F, and the perimeter is 140 feet (30 + 40 + 30 + 40 = 140). The

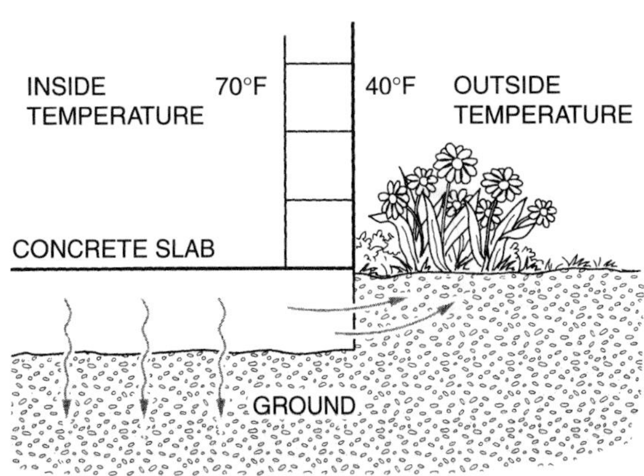

Figure 21-22. Heat loss through a concrete slab.

maximum loss of 0.54 can be used, so the slab will lose an estimated 5,292 Btu per hour (140 × 0.54 × 70 = 5,292).

Often in the colder climates the heating system is located in the perimeter of the slab. When this is the case, there is a much larger temperature difference, so insulation is essential, or excess heat will be lost. Figure 21-24 shows an example of a system with perimeter heating. The heat loss for the slab would be between 0.64 and 0.90 Btu per hour per running foot of slab per degree Fahrenheit temperature difference. This range is not a complete version as presented in the ASHRAE handbook but is a summation that should be used for reference only. To be complete, the degree days for the locality would enter into the calculation. A degree day is a method of averaging the temperatures that fell below 65°F for the day. Figure 21-23 mentions this in the table.

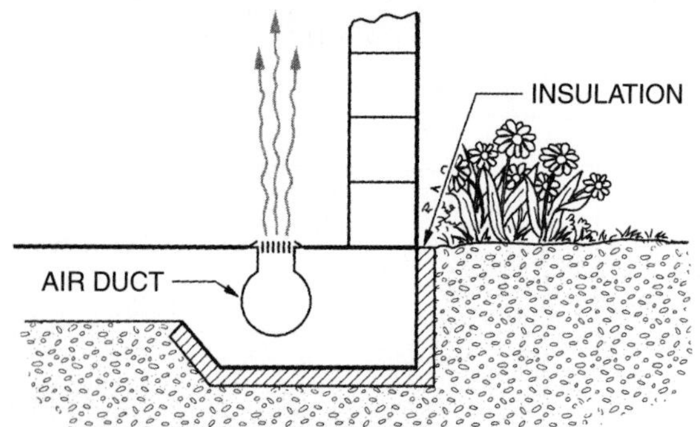

Figure 21-24. Heating system in the perimeter of a slab floor.

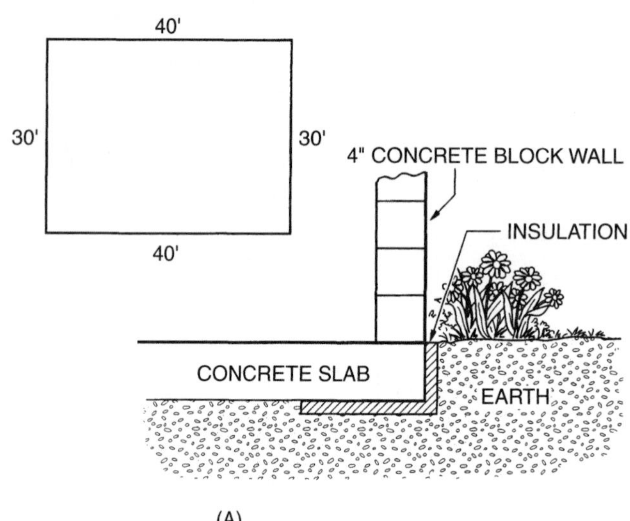

(A)

Table 5 Heat Loss Coefficient F_2 of Slab Floor Construction, Btu/hr · °F per ft of perimeter

Construction	Insulation[a]	Degree Days (65°F Base) 2950	5350	7433
8-in. block wall, brick facing	Uninsulated	0.62	0.68	0.72
	R = 5.4 from edge to footer	0.48	0.50	0.56
4-in. block wall, brick facing	Uninsulated	0.80	0.84	0.93
	R = 5.4 from edge to footer	0.47	0.49	0.54
Metal stud wall, stucco	Uninsulated	1.15	1.20	1.34
	R = 5.4 from edge to footer	0.51	0.53	0.58
Poured concrete wall with duct near perimeter[b]	Uninsulated	1.84	2.12	2.73
	R = 5.4 from edge to footer, 3 ft under floor	0.64	0.72	0.90

[a]R-value units in °F · ft² · h/Btu · in.

[b]Weighted average temperature of the heating duct was assumed at 110°F during the heating season (outdoor air temperature less than 65°F).

(B)

Figure 21-23. (A) Calculating the heat loss through a concrete slab. (B) Table of heat-loss coefficients for different concrete slabs. *Courtesy American Society of Heating, Refrigerating and Air-Conditioning Engineers*

21.7 Heat Loss Through Windows and Doors

Heat loss through windows and doors is a major factor in a structure. These openings are connected directly to the outside air and are on hinges or tracks. They can be readily opened, and it is difficult to seal the heat into the structure.

The basic window has a single glaze, commonly called the *pane.* There is not much resistance to heat transfer in a single-glaze window, so variations use more layers of glass. Double-glaze windows use two pieces of glass separated by a dead vapor space, as shown in Figure 21-25. The vapor in the space is dry nitrogen or some other inert gas, such as argon. Triple-glaze windows are also available. They are the most efficient and the most expensive. If a structure has single-glaze windows, replacing them would be expensive, so home owners often add storm windows to improve the efficiency, Figure 21-26.

When the heat loss for windows is calculated, the *U* factor is based on the rough window opening, including the frame, Figure 21-27. Because there are different materials in different frames, each type of window has a different *U* value. For example, Figure 21-28 shows that a double-glaze window with one-quarter inch Argon filled space between the glazes and a wood or vinyl frame has an U_t value of 0.35. The same window with a common aluminum frame has an U_t value of 0.72. If the aluminum-frame window has a thermal break (insulation in the frame), the R_t value is .50. You can see from this comparison that better windows perform better, especially in extremely cold climates.

An example of a window calculation follows: Suppose that a room has two wood-framed double-glaze windows that are 5 feet high and 2.5 feet wide. The design outside temperature is 0°F and the design inside temperature is 70°F. How much heat will conduct to the outside through the windows? The area is 25 square feet: (5 × 2.5 × 2 = 25). The TD is 70°F (70 − 0 = 70). Therefore, 945 Btu per hour are lost:

$$\begin{aligned} q &= A \times U \times \text{TD} \\ &= 25 \times .54 \times 70 \\ &= 945 \text{ Btu/hr} \end{aligned}$$

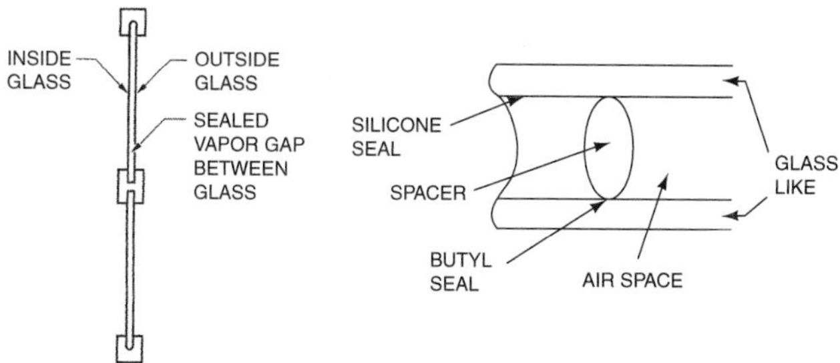

Figure 21-25. Double-glaze window.

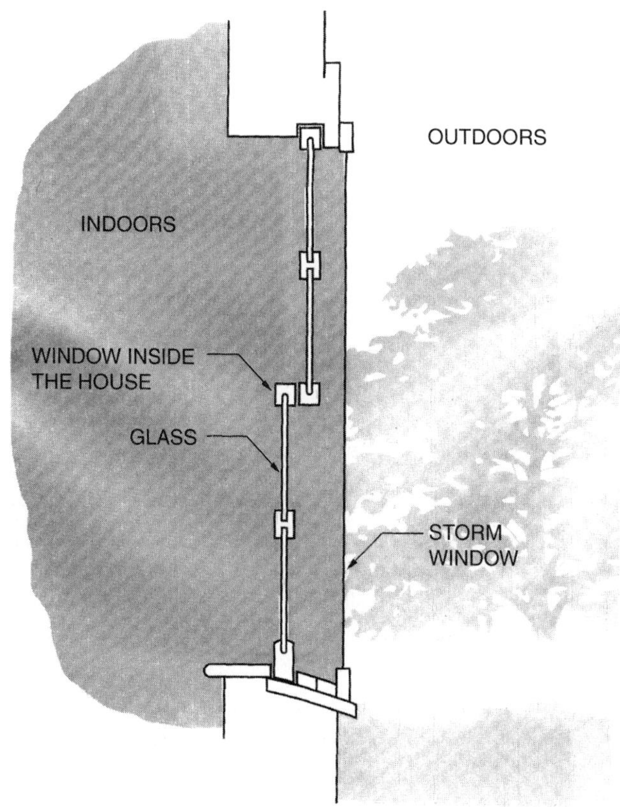

Figure 21-26. Single-glaze window with a storm window added.

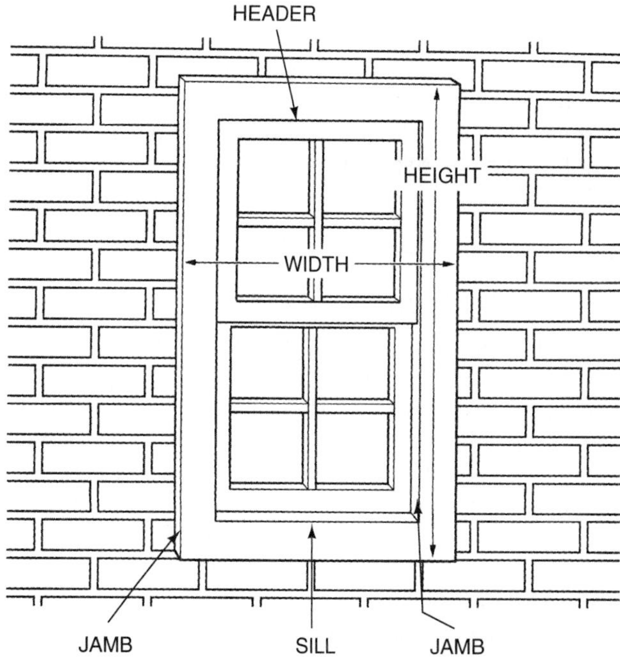

Figure 21-27. Measuring a window for the heat-loss calculation.

Table 13 Overall Coefficients of Heat Transmission of Various Fenestration Products (*Concluded*)

Part A: U-Values for Vertical Installation[a], Btu / hr · ft² · °F

	Glass Only		Aluminum Frame no thermal break ($U_f = 1.9$)		Aluminum Frame thermal break ($U_f = 1.0$)		Wood or Vinyl Frame ($U_f = 0.4$)	
	Center of Glass	Edge[c] of Glass	Product[d] type		Product[d] type		Product[d] type	
Glazing Type[b]			R	C	R	C	R	C
Triple glass or double glass with polyester film suspended in between, ε = 0.15 on surface 2, 3, 4, or 5								
1/4 in. argon spaces	(0.27)	(0.46)	0.72	0.54	0.50	0.41	0.35	0.32
3/8 in. argon spaces	(0.22)	(0.45)	0.69	0.51	0.47	0.37	0.33	0.29
1/2 in. and greater argon spaces	(0.20)	(0.44)	0.68	0.50	0.46	0.36	0.31	0.28
Triple glass or double glass with polyester film suspended in between, ε ≐ 0.15 on surfaces 2 or 3 and 4 or 5								
1/4 in. argon spaces	(0.22)	(0.45)	0.69	0.51	0.47	0.37	0.32	0.29
3/8 in. argon spaces	(0.17)	(0.43)	0.66	0.47	0.44	0.34	0.30	0.25
1/2 in. and greater argon spaces	(0.15)	(0.43)	0.65	0.46	0.43	0.32	0.29	0.24

Part B: U-Value Conversion Table for Sloped and Horizontal Glazing for Upward Heat Flow

Slope	U-Value, Btu/hr · ft² · °F												
90° (vertical)	0.10	0.20	0.30	0.40	0.50	0.60	0.70	0.80	0.90	1.00	1.10	1.20	1.30
45°	0.14	0.25	0.36	0.47	0.57	0.68	0.79	0.90	1.00	1.11	1.22	1.33	1.44
0 (horiz.)	0.19	0.29	0.40	0.51	0.61	0.72	0.82	0.93	1.04	1.14	1.25	1.35	1.46

[a]All U-values are based on standard ASHRAE winter conditions of 70°F indoor and 0°F outdoor air temperature with 15 mph outdoor air velocity and zero solar flux. The outside surface coefficient at these conditions is approximately 5.1 Btu/hr · ft² · °F, depending on the glass surface temperature. With the exception of single glazing, small changes in the interior and exterior temperatures do not significantly affect overall U-values.

[b]Glazing layer surfaces are numbered from the outside to the inside. Double and triple refer to the number of glazing lites. All data are based on 1/8 in. glass unless otherwise noted. Thermal conductivities are: 0.53 Btu/hr · ft · °F for glass, and 0.11 Btu/hr · ft · °F for acrylic and polycarbonate.

[c]Based on aluminum spacers data (see Figure 6). Edge of glass effect assumed to extend over the 2.5 in. band around perimeter of each glazing unit as seen in Figure 7.

[d]Product types described in Figure 7.

Figure 21-28. *U* values compared for wood and mental windows. *Courtesy American Society of Heating, Refrigerating and Air-Conditioning Engineers*

This does not seem like much heat, until you compare it with the same area of wall. Remember that the wall with 2 inches of polystyrene has a *U* value of 0.07. If we put 25 square feet of wall into the formula, we find a heat loss of only 122.5 Btu per hour:

$$q = A \times U \times \text{TD}$$
$$= 25 \times 0.07 \times 70$$
$$= 122.5 \text{ Btu/hr}$$

This example shows the thermal difference in a wall and a window. The window transfers 7.7 times as much heat as the wall does (945/122.5 = 7.71). The window area of many houses is 15% of the wall area. It is important to have light, but it is expensive.

Many home owners have only single-glaze windows. These windows can be improved by adding storm windows outside the existing windows. It is not very expensive to install these and turns a single-glaze window into about the same value as a double-glaze window. Most home owners can perform the installation themselves.

Doors are calculated in much the same manner as windows by using the different materials from which doors are constructed. For example, outside doors may be wood, either solid or with a core of insulation, metal with a core of insulation, or glass. Many times a combination of two doors is used, with the outer door called a *storm door.* It serves much the same purpose as the storm window.

21.8 Heat Loss from Infiltration or Ventilation

We have discussed heat loss from conduction. Heat can also be lost through infiltration or ventilation.

Infiltration is air that enters the structure as a result of prevailing winds or breezes. When the air impinges on one side of a structure, it creates a slight positive pressure on that side and when it passes over the structure, it creates a slight negative pressure on the other side of the structure, as shown in Figure 21-29. This causes air to pass through openings in the structure. Air enters one side and exits the other. This is called *infiltration* on the entering side and *exfiltration* on the exiting side.

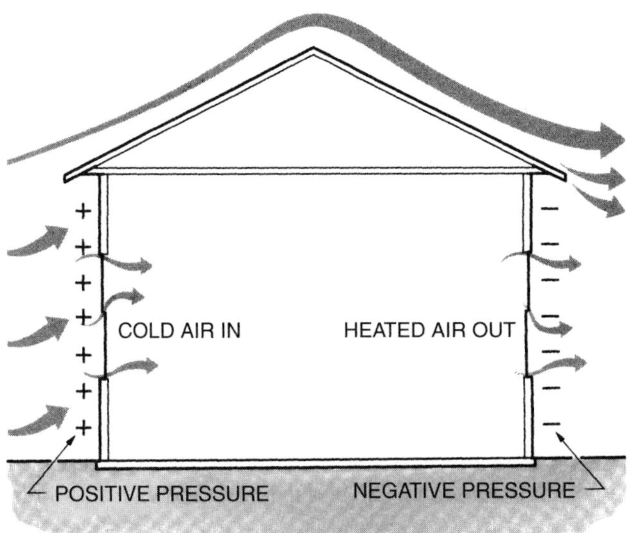

Figure 21-29. How air infiltrates a structure.

Air can enter and escape a structure through many obvious places. Some are as follows:

1. Around the windows and the doors. Some older windows and doors have very large cracks around them. You can identify an old window if it has weights to counterbalance the weight of the window.
2. Around plumbing fixtures where they enter the structure, if the fixtures are above grade. This occurs only in warmer climates. In colder climates, the plumbing is well below the frost line so that it does not freeze. If the plumbing runs in the outside walls, there is a space around the pipes or vents for air to travel by the pipes. It is not common to find plumbing in the outside walls in colder climates. However, the vent pipes penetrate the ceiling into the attic, and air can travel this route if the area where the pipes penetrate the ceiling is not sealed, Figure 21-30.
3. Where electrical wires penetrate the structure. For example, a porch light may have an inside switch. Air can enter the light fixture and exit the switch or even a plug-in receptacle on an outside wall, Figure 21-31.
4. Combustion air for fossil fuel furnaces is necessary and must enter the structure from somewhere. Combustion air inlets are discussed in the various units on the fossil fuels and must be used. It is not uncommon for an old installation to pull the combustion air through the cracks in the house. If the home owner seals the house too tightly, the furnace will back draft and spill the by-products of combustion into the house. This situation is dangerous.

Calculating the infiltration load on a structure is probably the most difficult part of a load calculation because there is no way of knowing how much air will move through the structure without an actual test. Some companies push air into the structure with a blower, then measure the entering air with a test device. The chimney and any flues must be plugged to prevent air from escaping at these points. The structure is pressurized to about 0.1 inch of water column (WC) and the entering air is measured. A well-constructed home with all precautions taken to seal the structure has about 0.3 to 0.5 air change per hour. A structure constructed with common building practices may leak from 0.5 to 1 air change per hour. Poorly constructed houses normally leak more than 1 air change per hour. An *air change* means that the total volume of the house is changed in 1 hour. For example, a house that is 30 feet by 40 feet has an area of 1,200 square feet. If the ceiling height is 8 feet, one air change is 9,600 cubic feet per hour ($30 \times 40 \times 8 = 9{,}600$). This is a leak rate of 160 cubic feet per minute ($9{,}600 \div 60$ min/h $= 160$). A very well-designed house leaks at the rate of 0.5 air change, which is 4,800 cubic feet per hour, or 80 cubic feet per minute.

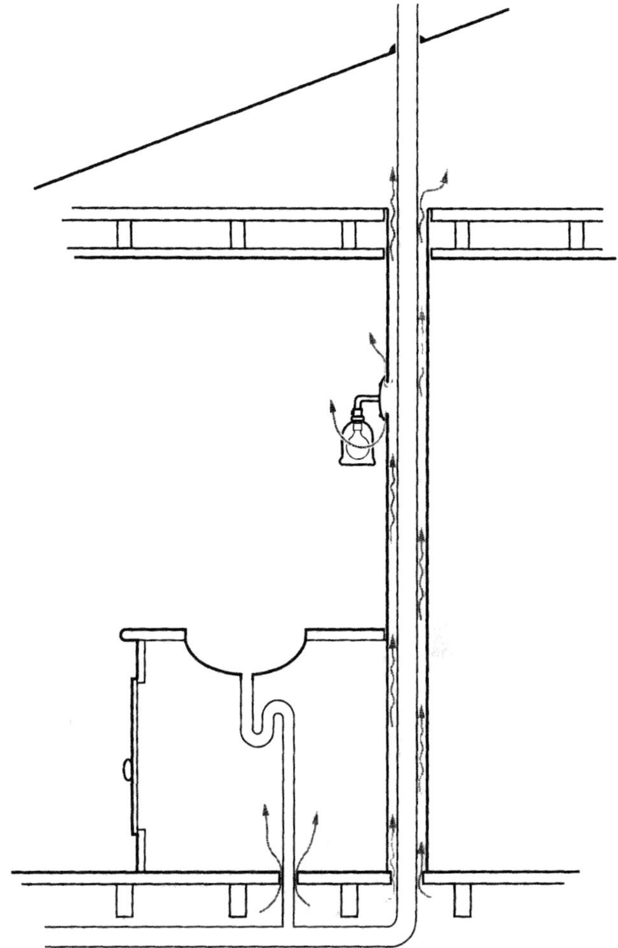

Figure 21-30. Infiltration around the plumbing fixture.

Another method of estimating infiltration is to use the known passage of air through the cracks around the windows and doors. Modern windows and doors are tested with a wind velocity across them. The typical test calls for a wind velocity of 25 miles per hour (mph) across the window or door and rates it according to how much air

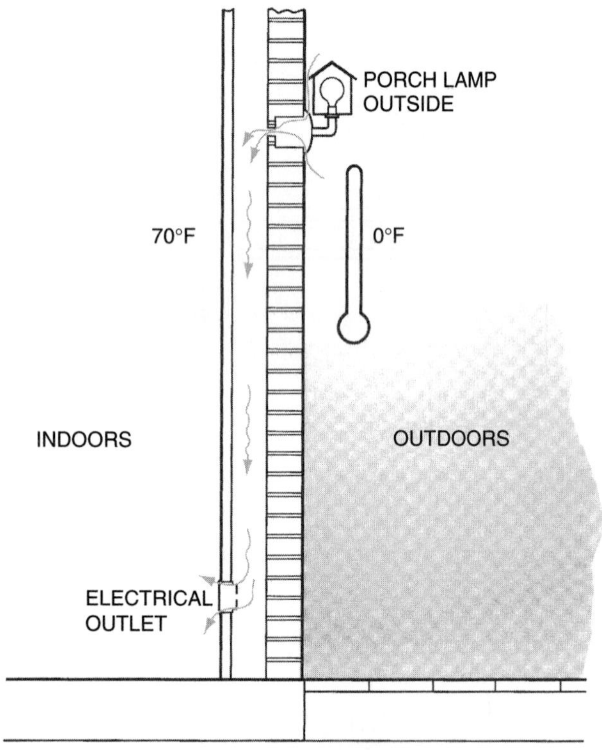

Figure 21-31. Infiltration around electrical fixtures.

leaks through per foot of crack. The best window typically leaks 0.25 cubic foot of air per minute per foot of crack. A poor window may leak 1 cubic foot per minute per foot of crack. When the leak rate per foot of crack is known, the total feet of crack for the windows in the structure can be measured and divided by 2 (remember, air leaks in one side and out the other) to determine the total cubic feet per minute (cfm) of air leaking into the structure.

Ventilation is planned induction of outside air to create a slight positive pressure on the structure. When more air is induced into the house than will leak in, the air will leak out the cracks where infiltration would occur. When the ventilation is known, there is no infiltration calculation.

After the infiltration or ventilation rate is determined in cfm, the amount of heat that must be added to the size of the heating system to offset the loss must be determined. The sensible-heat formula can be used to make this calculation. For example, suppose that the temperature difference for the structure is 70°F and 160 cfm of air are estimated to leak into the structure. The heat loss is as follows:

$$
\begin{aligned}
q &= 1.08 \times \text{cfm} \times \text{TD} \\
&= 1.08 \times 160 \times 70 \\
&= 12{,}096 \text{ Btu/hr}
\end{aligned}
$$

This is the most heat loss calculated so far. It can be seen why this house should be tightened. If the house had only a 0.5 air change, the loss would be only 6,048 Btu per hour.

Remember that a structure should not be too tight, or internal pollution will occur. The materials used to build the structure will give off chemicals for a long while after the structure is built. Carpet, drapes, upholstered furniture, and solvents used for cleaning are among some of the common indoor pollutants, Figure 21-32.

When a structure is too tight, the oxygen level in it is reduced and replaced with the CO_2 that is expelled when people exhale, Figure 21-33. Carbon dioxide buildup occurs in buildings with high occupancies for long time periods, such as office buildings.

Another indoor pollutant discovered recently is radon gas from the earth below the structure. This gas is known to cause cancer and is present in the soil under many houses, Figure 21-34. Many homes located in known radon gas localities must be designed for removing the gas. This is accomplished with a separate ventilation system under the house that does not affect the heating system unless it exchanges heat with the concrete slab or the floor of the house, Figure 21-35.

21.9 Heat Loss Through Duct Work

Another source for losing heat in a structure is the duct work when it is routed through an unconditioned space.

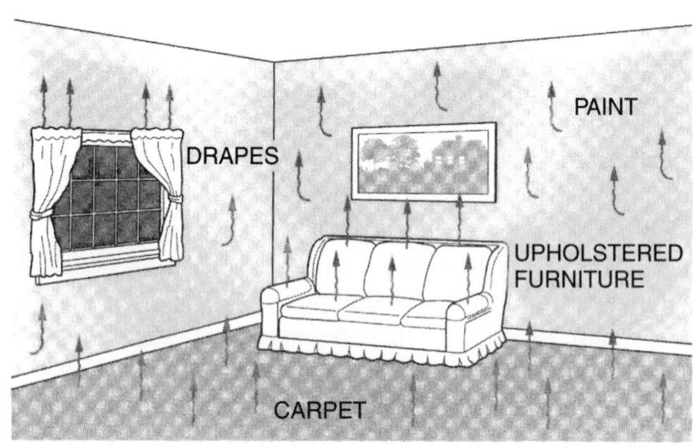

Figure 21-32. Indoor pollutants.

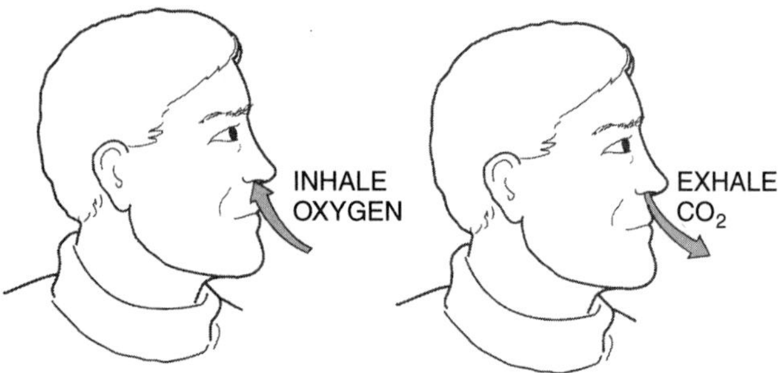

Figure 21-33. People breathe in oxygen (O) and exhale carbon dioxide (CO_2).

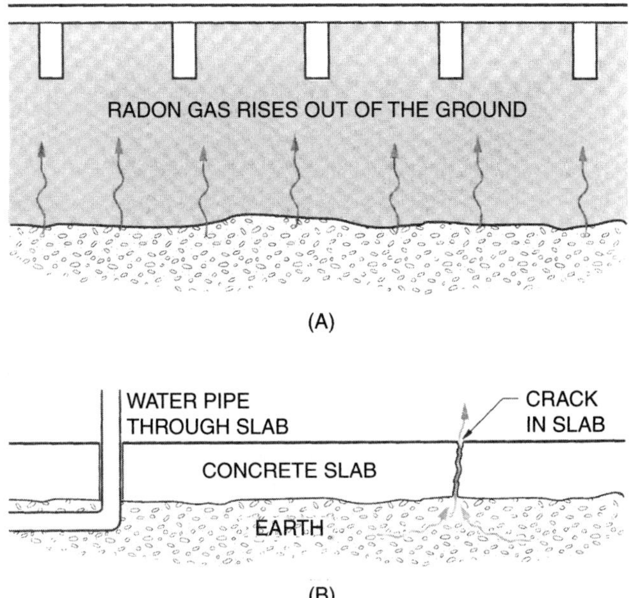

Figure 21-34. (A) Home with crawl space. Radon gas enters with infiltration. (B) Concrete slab poured on dirt. Radon gas enters through cracks in the concrete and around unsealed pipes running through the slab.

A fossil fuel furnace may operate at a supply air temperature of 130°F or more. If it is routed under a house where the temperature is 30°F, there is a 100°F temperature difference. If the duct is not insulated, it will lose a lot of heat under the house. If the vents are closed, the heat will rise and most of it can be recovered, as shown in Figure 21-36. If the duct is in a ventilated attic, the heat is lost. Typically, most designers count a 10% duct loss as a rule of thumb where the duct is routed through the unconditioned space. This percentage is usually fairly accurate and can be used to size the system. For example, suppose that the total heat loss for the structure is 73,000 Btu per hour. Adding the duct loss pushes the furnace size up to 80,300 (73,000 × 1.1 = 80,300).

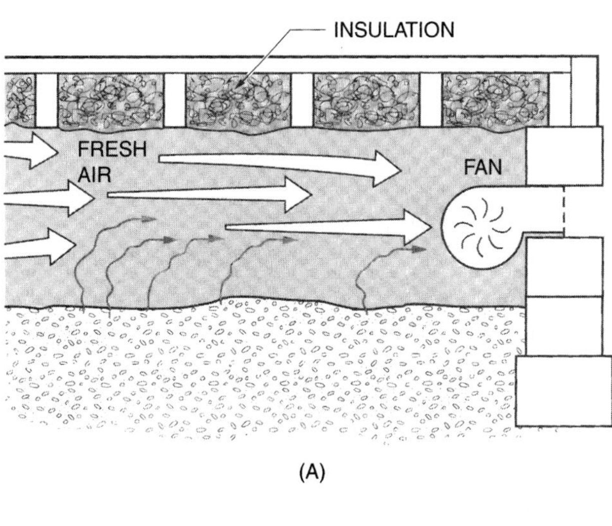

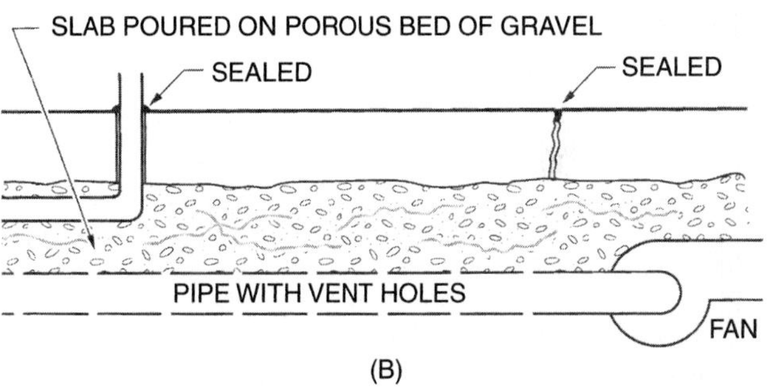

Figure 21-35. Preventing radon gas poisoning. (A) Diluting radon gas with fresh air. (B) Concrete slab poured on bed of gravel.

21.10 Total Heat Loss

Calculating the heat loss of a whole house by using the methods just discussed involves many numbers. Consequently, a bookkeeping system is important. Each manufacturer has a fill-in-the-blank form for calculating the heat loss for a structure. The areas of the various portions of the house that lose heat can be entered on the form. The technician is prompted with questions to

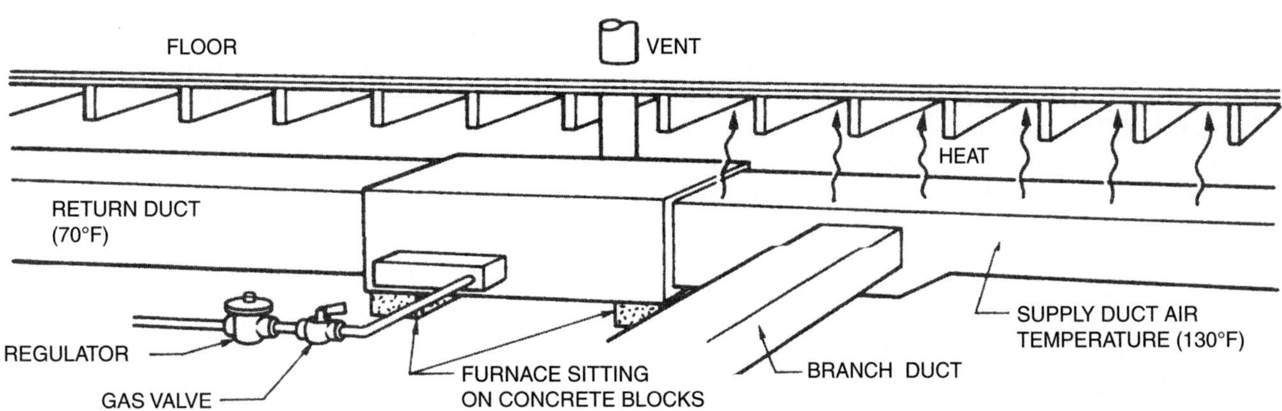

Figure 21-36. Heat loss from ducts routed under a house.

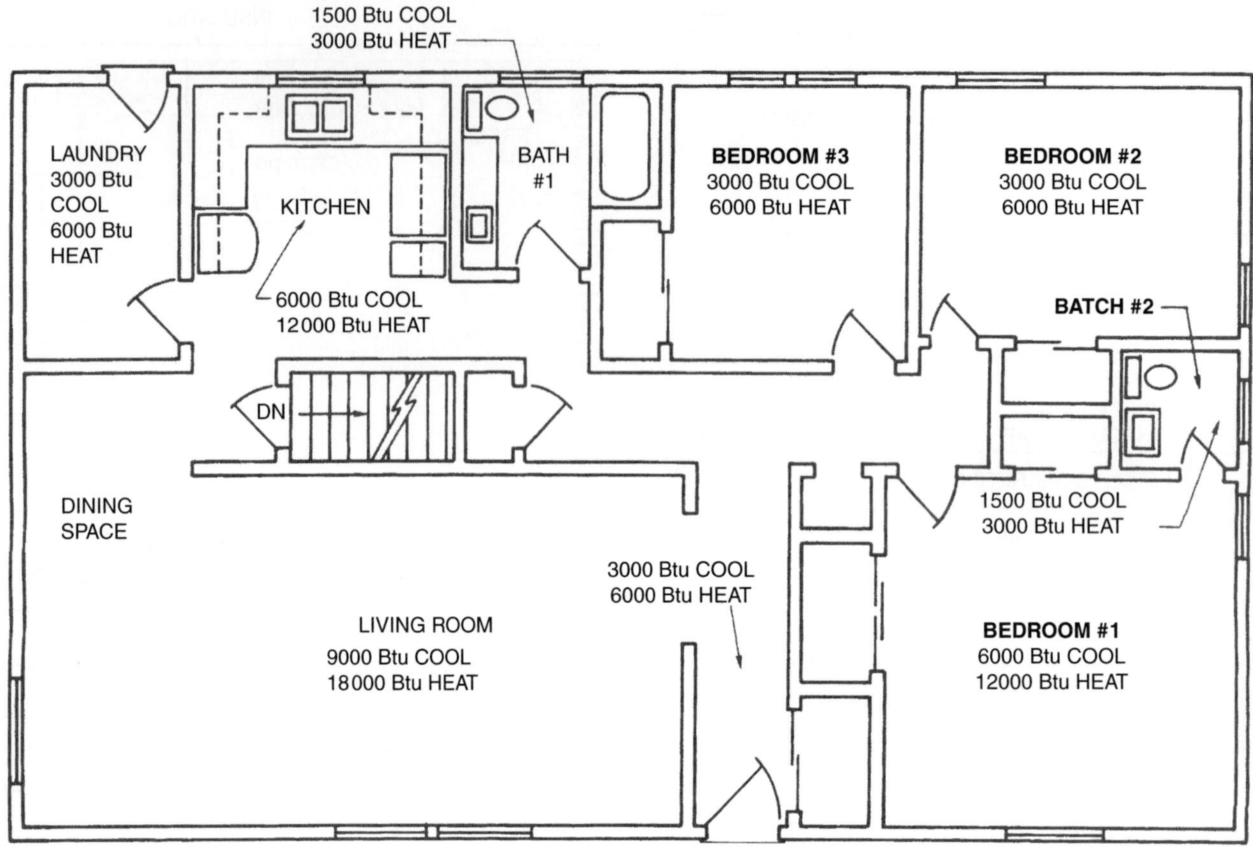

Figure 21-37. Heat distribution in a structure.

help cover all the aspects of heat loss. The totals are then summed for the entire house so that the heating equipment can be chosen.

For some heating systems, room-by-room load calculations are important in determining how much heat is to be released in each room. If the structure is large or if hydronic heat is used, a room-by-room calculation should be computed.

Room-by-room loads are calculated the same as a whole-house calculation, except each room is treated as a stand-alone heat loss. A room-by-room calculation for a house may distribute the heat as shown in Figure 21-37. Room-by-room calculations involve much detail, so many contractors take shortcuts. One shortcut that is often used is to perform a whole-house calculation and divide the total by each room area. For example, if a house has 1,200 square feet and the living room has 200 square feet, the living room is 16.7% of the total area of the house (200/1,200 = .1666, or 16.7%). If the total heat loss for the house is 80,300 Btu per hour, the heat added to this room must be 13,410 (80,300 × .167 = 13,410).

This shortcut is not as accurate as the room-by-room calculation, but it is often used. If the estimator has experience and has been using this method for a while, it works well. Experience would tell the estimator to add a little more heat to any room with extra windows.

Many designer–estimators use computer programs to estimate the heat loss for a structure. These programs are handy because the types of construction for a particular part of the country may be similar. After the designer–estimator enters the data for several structures into the computer, the similarities can be used in all other structures. This is commonly done by entering a construction number for a typical construction. The next time the heat loss for a house with the same construction must be calculated, most of the data are already entered; only the areas need to be entered. Even the temperature difference is entered in the computer. The computer also enables the designer–estimator to experiment with the building materials for a particular structure to determine the most efficient structure for the least amount of money.

21.11 Performing Room-by-Room Heating Load Calculation

Up to this point, the focus has been on the formulas and tables used to perform a heating load calculation. In this section, however, you will see that putting it all together and performing an actual heating load calculation is nothing more than calculating the heat loss

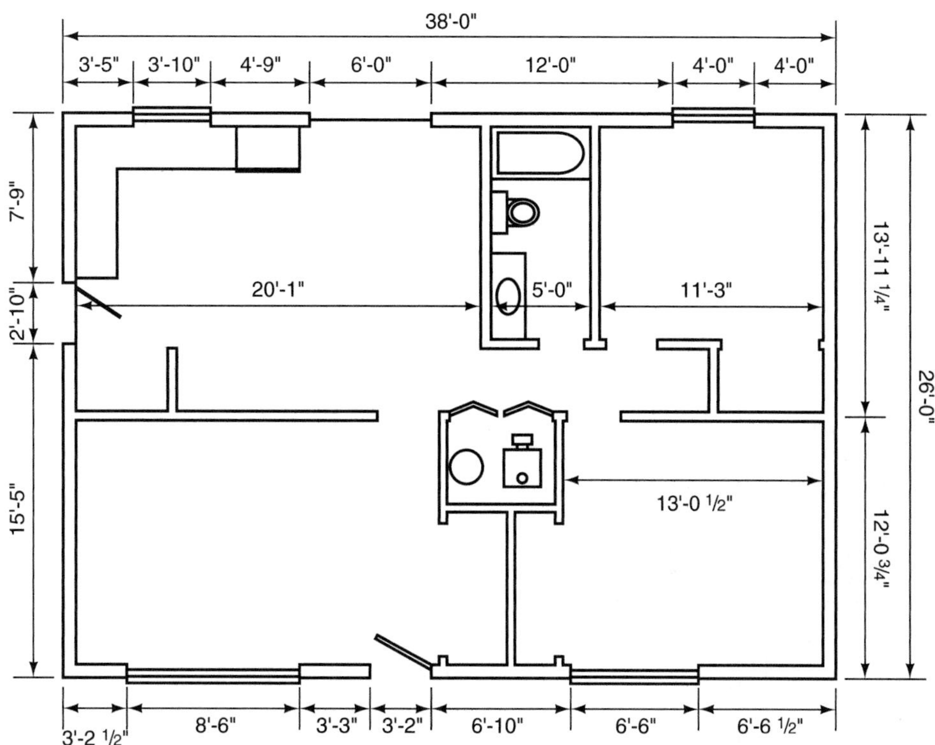

Figure 21-38. Structure floor plan.

via conduction and combining the heat loss due to infiltration. This is illustrated in the following example in which the heating load is required for the structure shown in Figure 21-38. In the following example, we will be determining the heating load for the structure shown in Figure 21-38. In this example, it will be assumed the exterior walls are constructed as shown in Figure 21-39. Also, it will be assumed that the outdoor design temperature is –5°F. Finally, for the design indoor temperature, we will be using 70°F.

The parameters for the structure are as follows:

- Unit R-value for all windows: R-30
- Unit R-value for exterior door: R-5
- R-value of foundation edge insulation: R5.4
- Rate of air infiltration in vestibule and mechanical room: 1 air change/hr
- Rate of air infiltration in all other rooms: 0.5 air changes/hr
- Outside design temperature: –5°F
- Desired indoor temperature 70°F
- Window height: 4 feet
- Exterior door height: 6 feet 8 inches
- Wall height: 8 feet

Determining the Total R-value of the Thermal Envelope Surfaces

It really doesn't matter if the technician starts with the walls, ceiling, floors, or anywhere else—the process is still the same. You must determine the R-value of the surface in question. Therefore, in this example, we will start with the walls. The walls are constructed using 2 × 6 studs framed on 24-inch centers with *R*-21 fiberglass batt insulation in the stud cavities. In addition, the walls are covered using ½-inch drywall (on the inside) and ½-inch plywood, a Tyvek infiltration barrier, and vinyl siding. The calculation is started by creating a simple three-column table that lists the material type, the R-value between the framing, and the R-value of the framing. When determining the R-value of a surface, it is necessary to determine the R-value of the cavity of the surface as well as the framing between the cavities. This is accomplished using the formula

$$R_{\text{effective}} = R_{\text{i}} \times R_{\text{f}} / P_{\text{f}}(R_{\text{i}} - R_{\text{f}}) + R_{\text{f.}}$$

where

$R_{\text{effective}}$ = effective R-value of the surface
R_{i} = R-value of the surface cavity
R_{f} = R-value of the framing
P_{f} = percentage of solid framing (for residential and light commercial structures, this is typically between 10% and 20%)

For the walls in this example, the effective R-value is calculated as follows. The R-values for the wall materials were obtained from Table 4.

Material	R-value Between Framing	R-value at Framing
Inside air film	0.68	0.68
½-inch drywall	0.45	0.45
Stud cavity	21.0	5.5
½-inch plywood sheathing	0.62	0.62
Tyvek infiltration barrier	0	0
Vinyl siding	0.61	0.61
Outside air film	0.17	0.17
	23.53	8.03

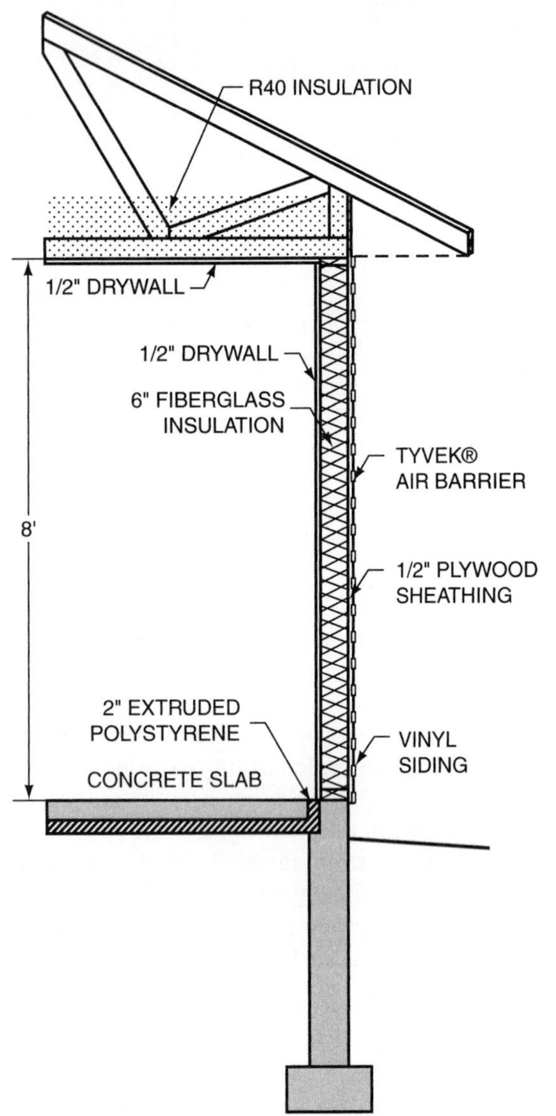

Figure 21-39. Typical wall detail.

The effective R-value for the walls is calculated as follows:

Note: assuming that the walls have a 15% solid framing the effective R-value

$(23.53 \times 8.03) / 0.15(23.53 - 8.03) + 8.03$

$188.9/10.355 = 18.3$

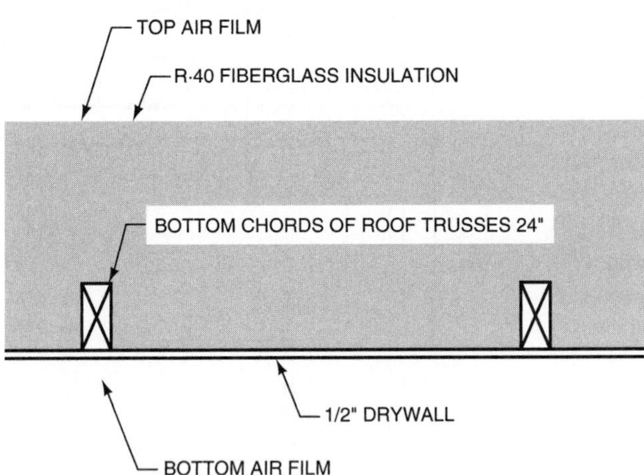

Figure 21-40. Typical ceiling detail.

The ceiling in this example consists of ½-inch drywall with approximately 12 inches of blown fiberglass insulation, Figure 21-40. Determining the effective R-value for the ceiling is accomplished as follows:

Material	R-value Between Framing	R-value at Framing
Bottom air film	0.61	0.61
½-inch drywall	0.45	0.45
Insulation	40.0	28.3
Framing	0	3.5
Top air film	0.61	0.61
	41.67	33.47

The effective R-value for the walls is calculated as follows:

Note: assuming that the ceiling has a 10% solid framing the effective R-value

$(41.67 \times 33.47) / 0.10\,(41.67 - 33.47) + 33.47$

$1394.7/34.29 = 40.7$

Now that the R-value of the walls and ceiling has been determined, the actual heating loads for the structure can be calculated. This is done by using the formula $q = A \times U \times \Delta T$ for ceiling and walls, $q = 1.08 \times \text{CFM} \times \Delta T$ for infiltration heat loss. For this house, the formulas that will be used are as follows:

$q = A \times U \times \Delta T$ for the walls and ceiling, windows and doors

$q = 1.08 \times \text{CFM} \times \Delta T$ for the heat loss due to infiltration

or

$q = 0.018 \times N \times V \times \Delta T$

where

0.018 = heat capacity of air

N = number of air changes per hour

V = interior volume of the heated space

ΔT = inside air temperature minus the outside air temperature

q = Perimeter $\times U \times \Delta T$ for the slab flooring system of the structure

To start the room-by-room calculation process, we must first select a starting point. In this example, we will start with the bedroom located in the upper right corner of the structure, as seen in Figures 21-41, 21-42, 21-43 and 21-44. The calculations are performed as follows:

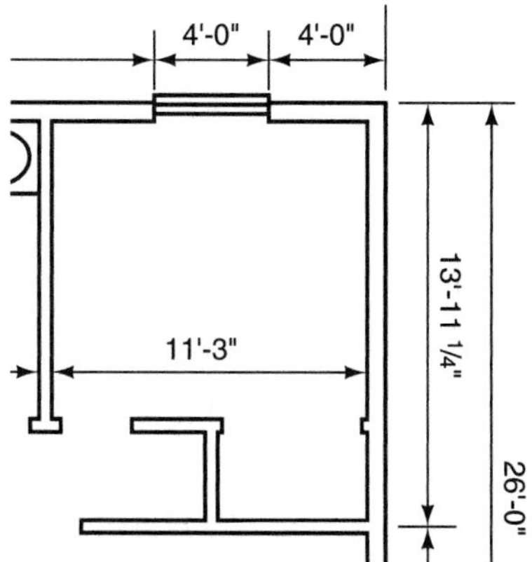

Figure 21-41. Room One.

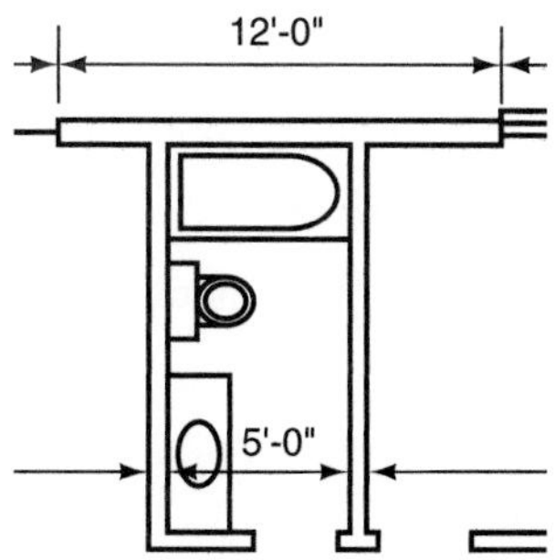

Figure 21-42. Room Two.

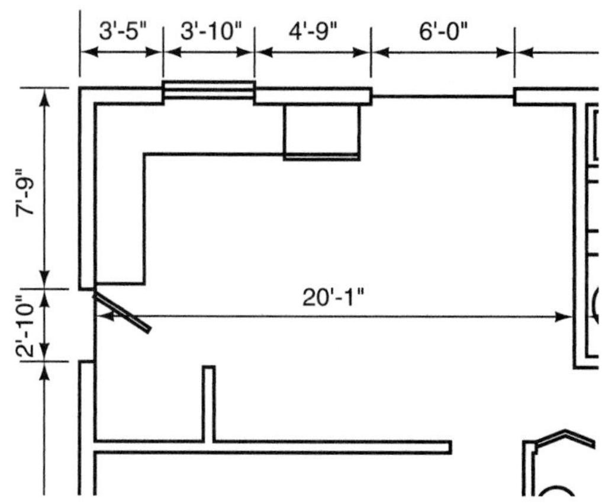

Figure 21-43. Room Three.

Room One Calculations

1. Calculate the volume of the room.

   $13.9375 \times 11.25 \times 8 = 1254.375$ sq ft

2. Calculate the area of the ceiling.

   $13.9375 \times 11.25 = 156.80$ sq ft

3. Calculate the exposed slap edge.

   $13.9375 + 11.25 = 25.1875$ ft

4. Calculate the area for the first wall.

   $13.9375 \times 8 = 111.5$ sq feet

5. Calculate the area for the second wall.

   $11.25 \times 8 = 90$ sq feet

6. Calculate the area for the first window.

   $4 \times 4 = 16$ sq ft

7. Calculate the subtotal of the wall for this room.

   $111.5 + 90 = 201.5$ sq ft

8. Calculate the total wall (subtracting the window and door area)

   $201.5 - 16 = 185.5$ sq ft

9. Perform the load calculations.

   Exposed walls

   $q = \mathrm{A} \times \mathrm{U} \times \Delta T$

   $q = 185.5 \times 1/18.3 \times 75$

   $q = 760.24$ btu/hr

   Windows

   $q = \mathrm{A} \times \mathrm{U} \times \Delta T$

   $q = 16 \times 1/3 \times 75$

   $q = 400$ btu/hr

   Exposed ceiling

   $q = A \times U \times \Delta T$

   $q = 155.80 \times 1/40.7 \times 75$

   $q = 288.94$ btu/hr

   Infiltration loss

   $q = (0.018) \times N \times V \times \Delta T$

   $q = 0.018 \times 0.5 \times 1254.3 \times 75$

   $q = 846.65$ btu/hr

   Exposed slab edge

   $q = L \times U \times \Delta T$

   $q = 25.18 \times 0.54 \times 75$

   $q = 944.25$ btu/hr

   Room one total

   $760.24 + 400 + 288.94 + 846.65 + 944.25 = q_{\text{total}}$

   $3240.08 = q_{\text{total}}$

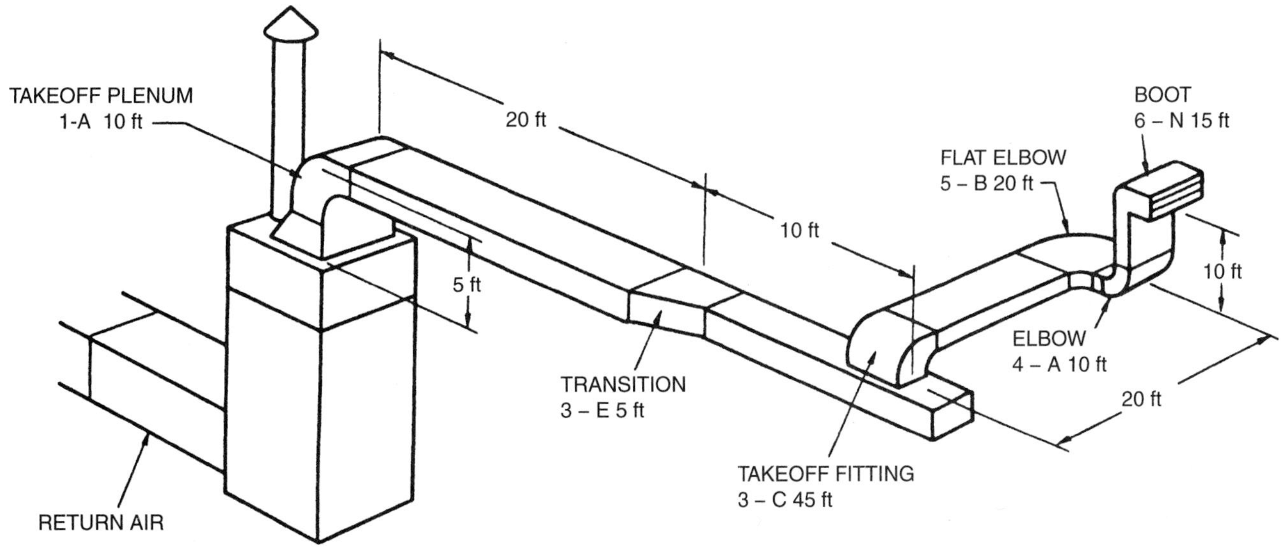

TAKEOFF PLENUM 1-A 10 ft
20 ft
10 ft
5 ft
FLAT ELBOW 5 – B 20 ft
BOOT 6 – N 15 ft
10 ft
ELBOW 4 – A 10 ft
20 ft
TRANSITION 3 – E 5 ft
TAKEOFF FITTING 3 – C 45 ft
RETURN AIR

FITTING LIST	EQUIVALENT FEET	
TAKEOFF PLENUM	10 ft	
TRANSITION	5 ft	
TAKEOFF FITTING	45 ft	
FLAT ELBOW	20 ft	
ELBOW	10 ft	
BOOT	15 ft	
	105 ft	105 ft EQUIVALENT LENGTH
ACTUAL DUCT LENGTH	5 ft	
	20 ft	
	10 ft	
	20 ft	
	10 ft	
	65 ft	65 ft TOTAL ACTUAL LENGTH
		170 ft TOTAL EFFECTIVE LENGTH FOR DESIGN PURPOSES

Figure 22-67. This duct run is more than 100 feet in length.

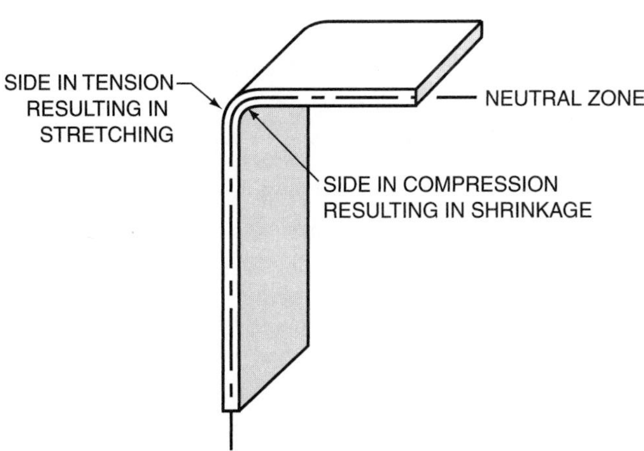

Figure 22-68. Forces experienced by a piece of steel subjected to a 90° bend.

soft steel, the formula is allowance = (0.64 × thickness) + (0.5 × 3.14 × radius of bend). For bronze, hard copper, cold-rolled steel, and spring steel, the formula is allowance = (0.71 × thickness) + (0.5 × 3.14 × radius of bend). In all three of these formulas, the calculations are based on a bend angle of 90°. When an angle other than 90° is to be used, the modifier (angle°/90°) must be applied to these formulas. For example, using soft steel and a bend angle of 56°, the formula now becomes allowance = [(0.64 × thickness) + (0.5 × 3.14 × radius of bend)] × (56°/90°). If the same sheet of soft metal has a thickness of 0.125 inches and a bend radius of 0.25 at an angle of 56°, the complete equation would be allowance = [(0.64 × 0.125) + (0.5 × 3.14 × 0.25)] × (56°/90°). After plugging the information into a calculator, a bend allowance of 0.2871 is computed.

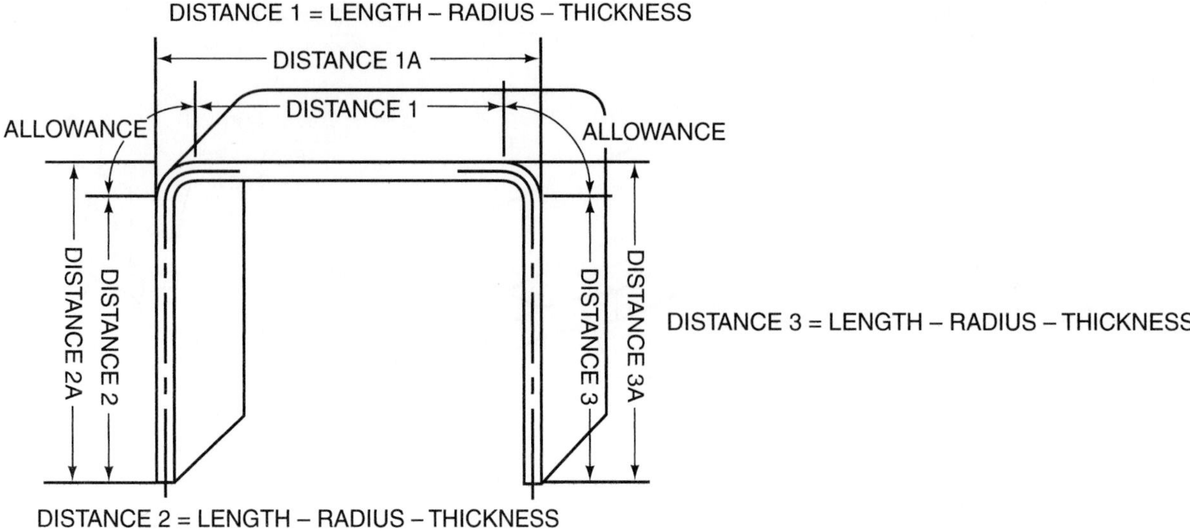

Figure 22-69. Boundaries used to measure the distances and allowances for stock length formulas.

Once the bend allowance is found, it is then used to calculate the overall stock length needed to manufacture the duct. This is accomplished by adding the inside distance from either allowance to allowance (Distance$_1$ in Figure 22-69), or from edge-of-part to allowance (Distance$_2$ in Figure 22-69). This particular duct would yield the formula stock length = distance$_1$ + distance$_2$ + distance$_3$ + allowance$_1$ + allowance$_2$. Naturally, the complexity of the duct determines the number of distances and allowances in the formula. For this equation to work, the distances supplied must be minus the bends. If the distances supplied do contain the bends, the equation would become (distance$_{1a}$ − radius − thickness) + (distance$_{2a}$ − radius − thickness) + (distance$_{3a}$ − radius − thickness) + allowance$_1$ + allowance$_2$.

$$\text{distance}_1 + \text{distance}_2 + \text{distance}_3 + \text{allowance}_1 + \text{allowance}_2$$

or

$$\text{stock length} = [(\text{distance}_{2a} - \text{radius} - \text{thickness}) + (\text{distance}_{1a} - \text{radius} - \text{thickness}) + \text{allowance}_1 + \text{allowance}_2]$$

In Figure 22-69, the formula could have been written as distance + distance + distance + allowance − (2 radius + 2 thickness). This practice is not recommended for technicians with little experience, because it can lead to confusion and possible mistakes when calculating large problems.

Another variable that must be taken into consideration when measuring for stock length is the method used to fasten sheets together. The most common methods used are **welding**, **brazing**, **soldering**, **adhesive bonds**, and **mechanical joining**. For each of these methods, a different amount of material must be allowed to make an adequate connection. The method used is determined by the application of the final product, strength of material, and the forces to which the duct will be subjected. Among the most common mechanical joining methods used today are seaming and hemming, examples of this method are illustrated in Figure 22-70. When calculating the length of stock required to construct a seam, the formulas discussed earlier must be used for each bend or fold in a particular seam. This is shown in Figure 22-71.

Material = soft steel.

Calculation for allowance$_1$:

$$\text{allowance}_1 = [(0.64 \times \text{thickness}) + (0.5 \times 3.14 \times \text{radius of bend})] \times \text{angle}°/90°$$

$$\text{allowance}_1 = [(0.64 \times 0.125) + (0.5 \times 3.14 \times 0.125)] \times 180°/90°$$

$$\text{allowance}_1 = [0.08 + 0.1963] \times 2$$

$$\text{allowance}_1 = 0.5525 \text{ inches}$$

Calculation for allowance$_2$:

$$\text{allowance}_2 = [(0.64 \times \text{thickness}) + (0.5 \times 3.14 \times \text{radius of bend})] \times \text{angle}°/90°$$

$$\text{allowance}_2 = [(0.64 \times 0.125) + (0.5 \times 3.14 \times 0.125)] \times 90°/90°$$

$$\text{allowance}_2 = [0.08 + 0.19625] \times 1$$

$$\text{allowance}_2 = 0.2763 \text{ inches}$$

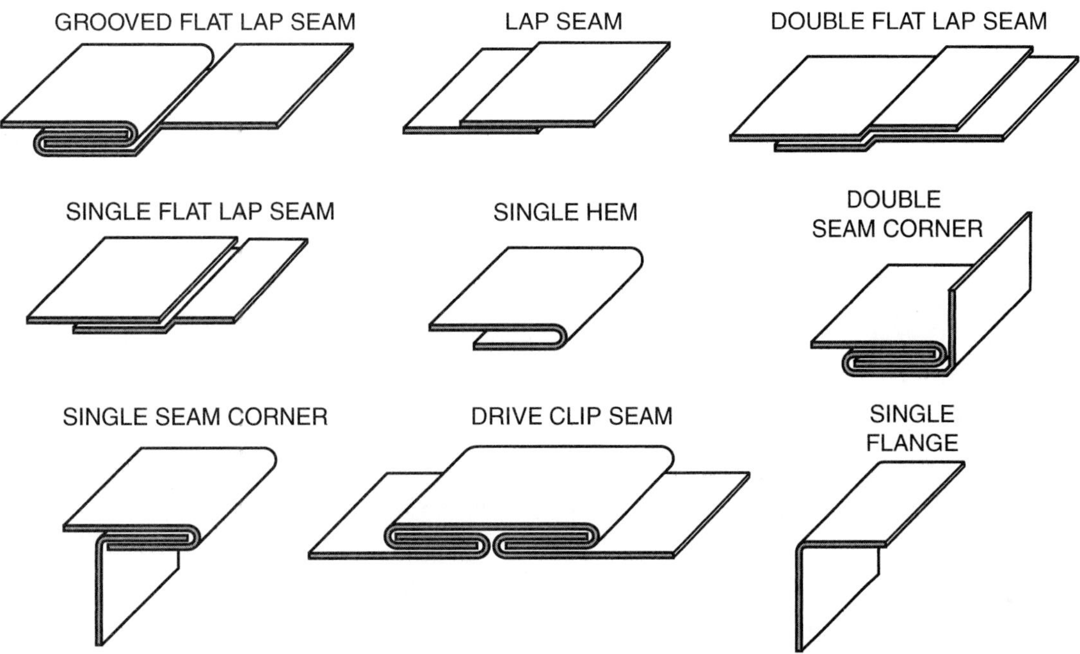

Figure 22-70. Common seams and hems used to connect sheet metal.

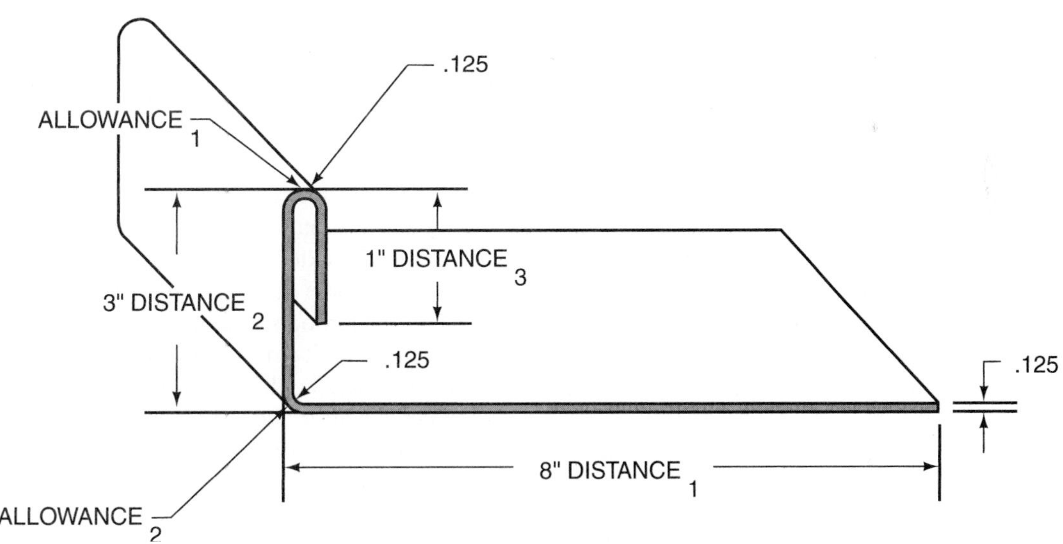

Figure 22-71. Example of a bend allowance.

Calculation for stock length:

$$\text{stock length} = (\text{distance} - (0.125 + 0.125)) + (\text{distance} - (0.125 + 0.125)) + (\text{distance} - (0.125 + 0.125)) + \text{allowance} + \text{allowance}$$

$$\text{stock length} = 7.75 + 4.25 + 0.75 + 0.5525 + 0.2763$$

$$\text{stock length} = 12.8288 \text{ inches}$$

22.22 Duct Installation Practices

The installation of the duct system can make a difference. The installers should protect the duct during the construction of the job. Stray material can make its way into the ducts and block the air. Figure 22-72 shows causes of obstructed airflow. The installers must not collapse the duct and then insulate over the collapsed part. Duct work that is run below a concrete slab can easily

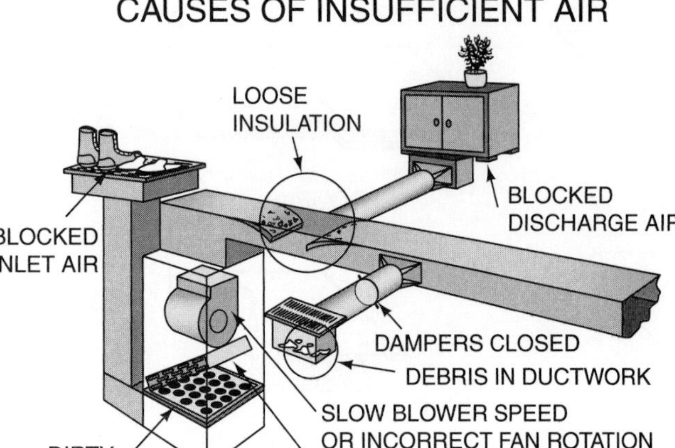

Figure 22-72. Causes of obstructed airflow. *Courtesy Carrier Corporation*

be damaged so that it does not pass the correct amount of air. When the insulation is applied to the inside of the duct, it must be fastened correctly or it may loosen and fall into the airstream.

Proper airflow techniques and design are a product of experience. The ability to design a system, install it, and get it working within a prescribed budget takes time. Working with an experienced person is an invaluable help.

Summary

- Air is passed through conditioning equipment and recirculated into the room.
- The conditioning equipment can heat, cool, humidify, dehumidify, or clean, or do a combination of these to make air comfortable.
- Infiltration is air leaking into a structure.
- Ventilation is air being induced into the conditioning equipment and conditioned before it is allowed to enter the conditioned space.
- The duct system distributes air to the conditioned space. It consists of the blower (or fan), the supply duct, and the return duct.
- The blower, or fan, uses energy to move the air.
- Propeller and centrifugal fans are commonly used in residential and small commercial systems.
- The propeller fan is used to move a lot of air against a small pressure. It can be noisy.
- The centrifugal fan is used to move large amounts of air in duct work, which offers resistance to the movement of air.
- Fans are turned either by belt-drive or direct-drive motors.
- A fan is not a positive-displacement device.
- Small centrifugal fans use energy in proportion to the amount of air that they move.
- Duct systems, supply and return, are large pipes or tunnels through which air flows.
- Duct can be made from aluminum, galvanized steel, flexible tubes, and fiberglass board.
- The pressure that the fan creates in the duct is very small and is measured in inches of water column (wc).
- One inch of wc is the amount of pressure needed to raise a column of water in a water manometer 1 inch.
- The atmosphere's pressure of 14.696 psia will support a column of water 34 feet high.
- One psig will support a column of water 27.1 inches, or 2.31 feet, high.
- Moving air in a duct system creates static pressure, velocity pressure, and total pressure.
- Static pressure is the pressure pushing outward on the duct.
- Velocity pressure is moving pressure created by the velocity of the air in the duct.
- Total pressure is the velocity pressure plus the static pressure.
- The Pitot tube is a probe device used to measure the air pressures.
- Air velocity (fpm) in a duct can be multiplied by the cross-sectional area (sq ft) of the duct to obtain the amount of air passing a particular point in cubic feet per minute (cfm).
- Typical supply duct systems are plenum, extended plenum, reducing extended plenum, and perimeter loop.
- The plenum system is economical and can easily be installed.
- The extended plenum system takes the trunk duct closer to the farthest outlets but is more expensive.
- The reducing extended plenum system reduces the trunk duct when some of the air volume has been reduced.
- The perimeter loop system supplies an equal pressure to all the outlets. It is useful in a concrete slab floor.
- Branch ducts should always have balancing dampers to balance the air to the individual areas.
- When the air is distributed in the conditioned space, it should be distributed on the outside wall to cancel the load.
- Warm air distributes better from the floor because it rises.
- Cold air distributes better from the ceiling because it falls.
- The amount of throw is how far the air from a diffuser will reach into the conditioned space.
- Return air systems normally used are individual room return and common, or central, return.
- Each foot of duct, supply or return, has a friction loss that can be plotted on a friction chart for round duct.
- Round duct sizes can be converted to square or rectangular equivalents for sizing and friction readings.

Review Questions

1. What five changes are made in air to condition it?
2. In what two ways can a structure obtain fresh air?
3. What two types of blowers move the air in a forced-air system?
4. Which type of blower is used to move large amounts of air against low pressures?
5. Which type of blower is used to move large amounts of air in duct work?
6. Why does duct work resist airflow? Name two reasons.
7. What are four types of duct distribution systems?
8. What are two types of blower drives?
9. What is the pressure in duct work expressed in?
10. What is a common instrument used to measure pressure in duct work?
11. Why is pounds per square inch (psi) not used to measure the pressure in duct work?
12. What three types of pressure are created in moving air in duct work?
13. What are two types of return air systems?
14. What component distributes the air in the conditioned space?
15. Where is the best place to distribute warm air? Why?
16. What chart is used to size duct work?
17. What four materials are used to manufacture duct?
18. What fitting leaves the trunk duct and directs the air into the branch duct?
19. The duct sizing chart is expressed in round duct. How can this be converted to square or rectangular duct?

23 Introduction to Blueprint Reading for Heating Technicians

Objectives

Upon completion of this unit, you should be able to

- **determine the length of objects presented on a blueprint by using an Architectural Scale and/or tape measure.**
- **determine the angle of a line on a blueprint by using a protractor.**
- **use sine, cosine, and tangent to determine the angle and length of lines contained on a blueprint.**
- **understand standard abbreviations and symbols used on a blueprint.**
- **identify the various views of an orthographic drawing.**
- **identify the various symbols used on plumbing plans to represent piping types, fittings, and symbols.**
- **identify the various symbols used on HVAC plans to represent HVAC linetypes, ducts, and equipment.**

23.1 Basic Blueprint Reading for HVAC-R Technicians

Because technology is evolving at an ever increasing rate and HVAC-R systems are becoming more and more complex, it is becoming ever more important that the technician have a basic understanding of blueprint reading. This is especially true when working on any system. Blueprints are drawn plans of homes and other buildings that are used to build the structure. The installing contractor must understand the portion of the prints that applies to their individual trade, electrical, piping, duct work, and location of equipment.

23.2 Linear Measurement

Linear measurement is defined as the measurement of two points along a straight line, see Figure 23-1. All objects, whether they are made by people or are the result of natural conditions and/or forces, consist of points and lines. In the HVAC industry, a point can be the center location of a gas line, as shown in Figure 23-2 and a point can be one of the edges of a rectangular duct, as seen in Figure 23-3. Whatever the application of points and/or lines, the fact remains

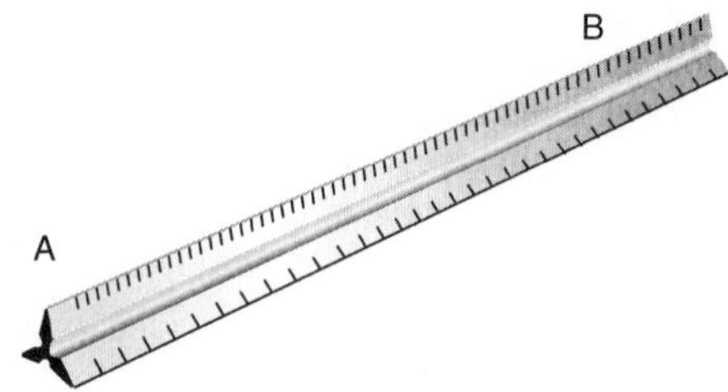

Figure 23-1. Linear measurement.

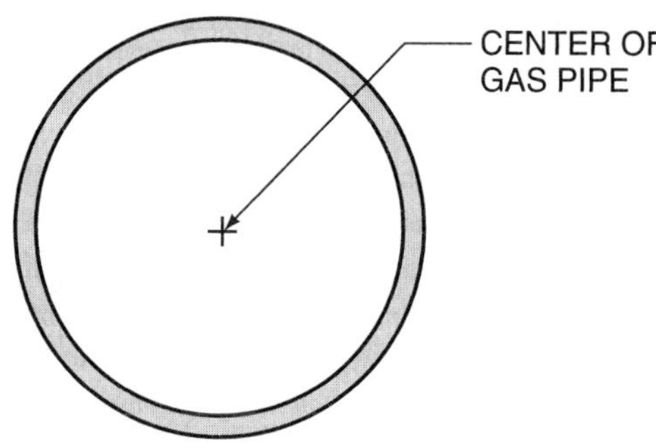

Figure 23-2. The proposed center line of a gas line.

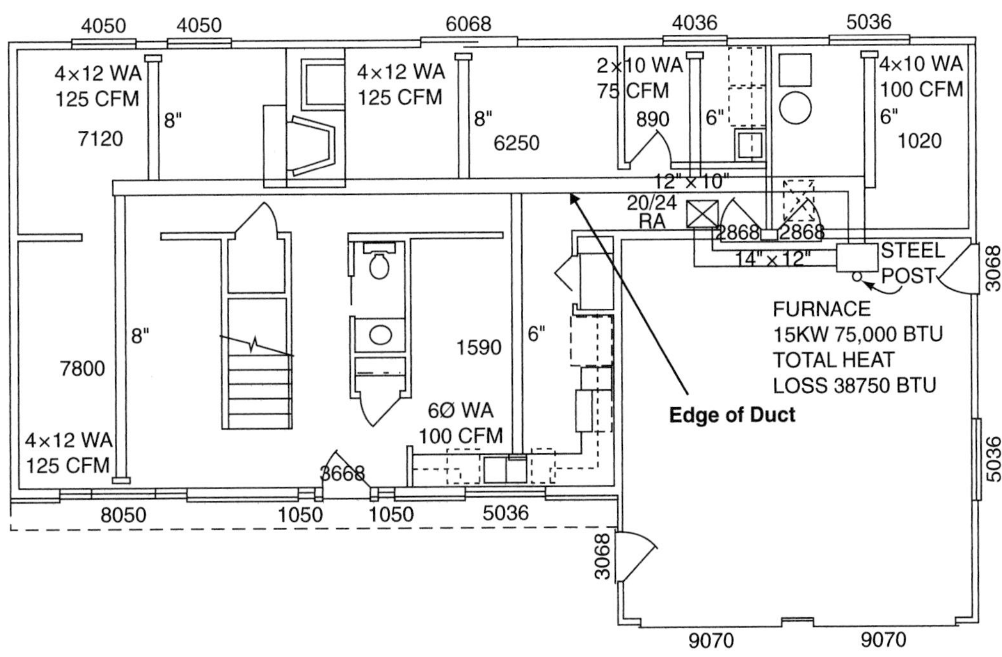

Figure 23-3. Edge of a rectangular duct supplying air to a room in a facility.

the same: any person entering a technical field must have a good understanding of how to locate them.

Currently there are two basic systems used to make measurements in the world today. They are the **English system** and the **metric system**. The English system of measure is currently used in the United States by most HVAC technicians. Its base units are inch (in), foot (ft), yard (yd), and mile (mi).

Table 1 English conversion factors

Unit	Divisions
1 inch (in)	
1 foot (ft)	12 in
1 yard (yd)	3 ft
1 mile (mi)	5,280 ft

Reading Linear Measurement on a Blueprint

When a mechanical drawing is created on the computer, typically it is drawn at a scale of 1:1 (also known as full scale). However, in most cases when the file is printed, it is printed at a reduced scale so that it can fit onto a single sheet of paper (typically 8 ½ × 11, 24 × 36, or 30 × 42). When a drawing is printed at full scale, determining the length of a line is easy to accomplish. This is done by simply placing a ruler along the length of the line and reading the dimensions, Figure 23-4. However, if the drawing is printed at a reduced scale (a scale

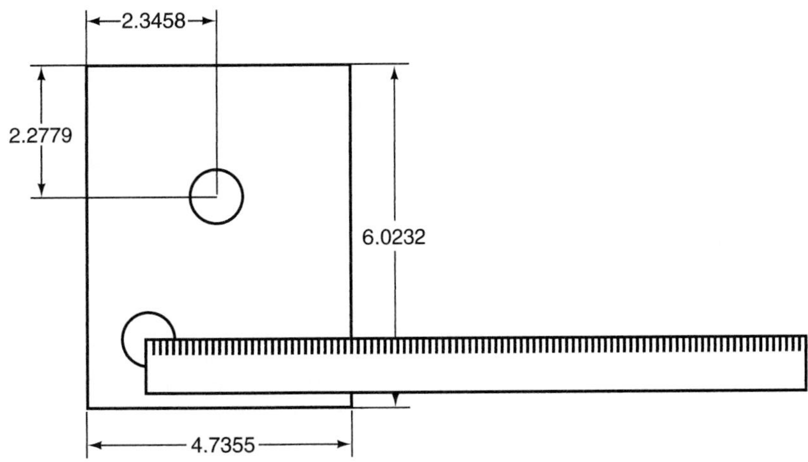

Figure 23-4. Reading the length of a line drawn at full scale.

smaller than 1:1), then a scale factor must be applied to determine the length of a line not dimensioned as shown in Figures 23-5, 23-6, and 23-7.

When a drawing is printed to a scale ¼ times smaller than its actual size, that drawing is said to be drawn at ¼ scale and is therefore read as every ¼" is equal to 1'-0". For example, if a portion of a building footing plan measures 2" using a ruler, then the actual length of the footing can be determined by multiplying the measured length by the inverse (meaning swap the top and bottom numbers) of the drawing's scale factor. For a scale factor of ¼, the inverse or multiplier would be 4, and therefore the true length of the portion of the footing would be 8 feet as illustrated in Figure 23-8.

The following table gives the conversion factors for various scales commonly used on engineering drawing.

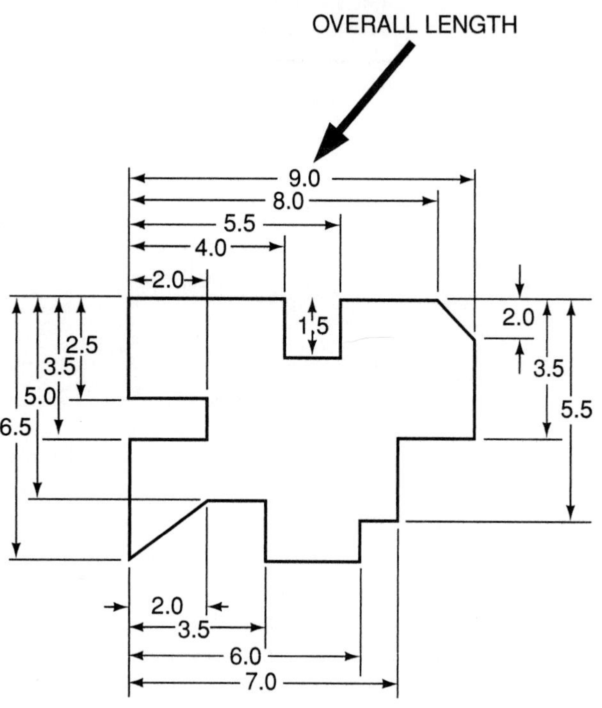

Figure 23-5. Determining the overall length of the object.

Table 2 Drawing scale conversion factors

Drawing Scale	Multiplier
1/16" = 1'-0"	16
3/32" = 1'-0"	10.66
1/8" = 1'-0"	8
3/16" = 1'-0"	5.33
¼" = 1'-0"	4
3/8" = 1'-0"	2.66
½" = 1'-0"	2
¾" = 1'-0"	1.33
1" = 1'-0"	1
1 ½" = 1'-0"	1.5
3" = 1'-0"	0.33
Half Size	2
Full Size	1

Determining Linear Measure by Using an Architectural Scale

When a drawing is created, one of the responsibilities of the drafter is to note the scale at which the drawing was created. Typically this information is listed in the drawing

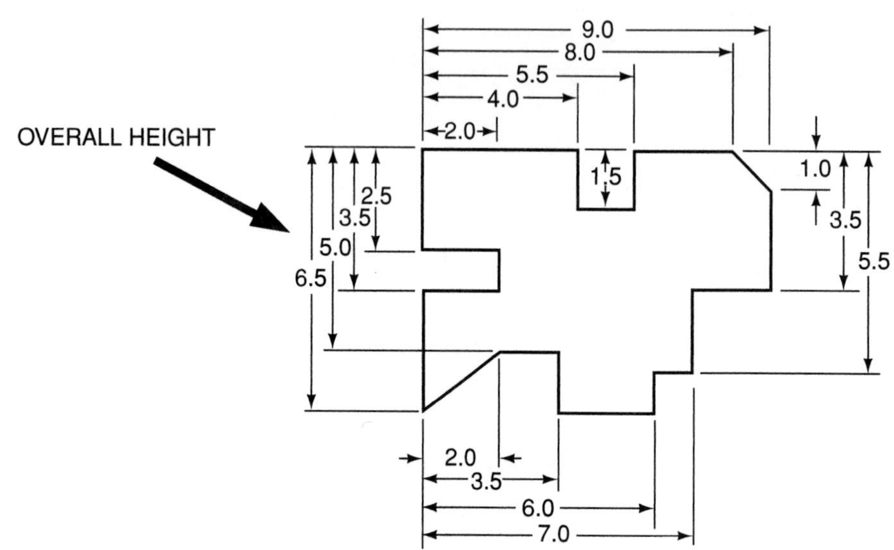

Figure 23-6. Determining the overall height of the object.

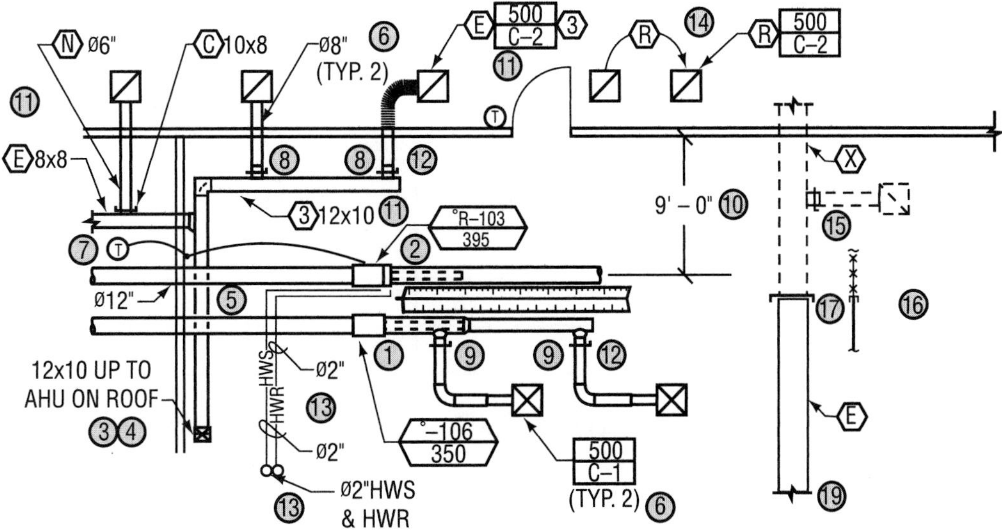

Figure 23-7. Reading the length of a line that is not dimensioned.

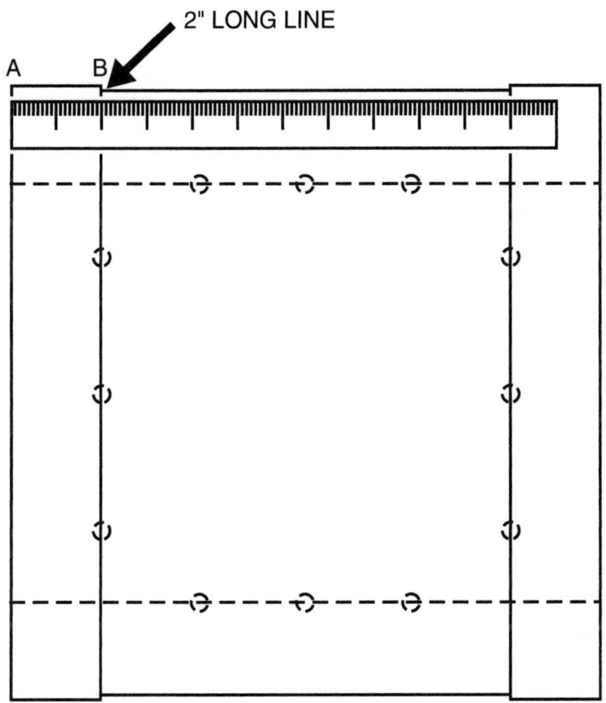

Figure 23-8. Determining the length of a portion of a footing by using a ruler.

title block region. The title block region is typically located along the bottom edge or lower right hand corner of the drawing, see Figure 23-9. The length and position of items not dimensioned on a drawing can be determined by using a ruler and the scale factor, as discussed in the previous section. However, an easier method of determining the length and position of objects not dimensioned is to measure them on the drawing by using either an Engineer's or Architectural scale, Figures 23-10 and 23-11. The scale most commonly used in residential and light commercial is the Architectural scale.

23.3 The Architectural Scale

The architect scale is a type of ruler in which a range of pre-calibrated ratios are illustrated. These scales can be made of a variety of different materials and contain as few as two scales (ratio) and as many as eleven (on 10 of them, each 1" represents a foot, which is subdivided into multiples of 12). The most commonly used architectural scale is the triangular scale. This scale received its name because its cross section is in the shape of a triangle, see Figures 23-11 and 23-12. In the United States, Architectural scales are commonly marked as *x* inches to the foot.

Reading the Architectural Scale

If you understand how to read one of the ratios (scales) on an Architect Scale, then reading the remaining scales will be simple. Because these scales are used to represent feet and inches, the scale is divided into two sections. The first section represents feet and the second section represents inches and fractions of an inch graduated in $\frac{1}{16}$", $\frac{1}{8}$", $\frac{1}{4}$", and $\frac{1}{2}$" increments. This is different from an engineer's scale, which is graduated in $\frac{1}{1000}$", $\frac{1}{100}$", and $\frac{1}{10}$" of an inch measured with caliper type devices and is used for manufacturing machine products. For example, look at the scale in Figure 23-13—the marks on the right side of

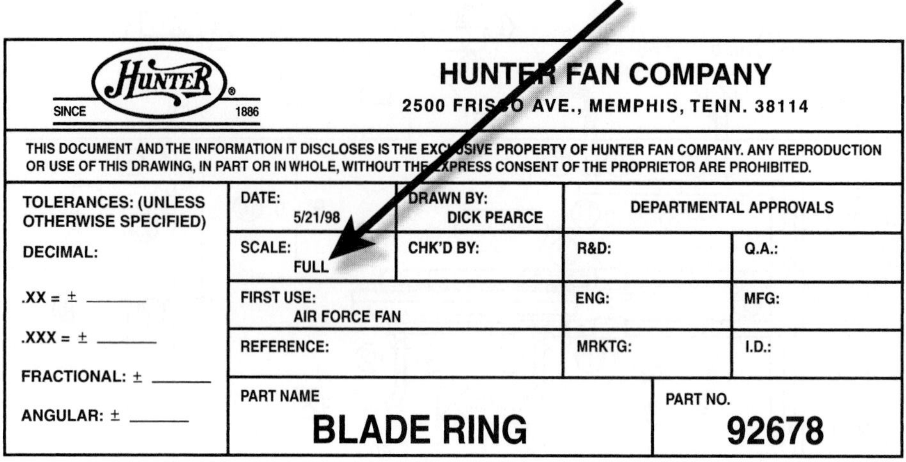

HUNTER® SINCE 1886

HUNTER FAN COMPANY
2500 FRISCO AVE., MEMPHIS, TENN. 38114

THIS DOCUMENT AND THE INFORMATION IT DISCLOSES IS THE EXCLUSIVE PROPERTY OF HUNTER FAN COMPANY. ANY REPRODUCTION OR USE OF THIS DRAWING, IN PART OR IN WHOLE, WITHOUT THE EXPRESS CONSENT OF THE PROPRIETOR ARE PROHIBITED.

TOLERANCES: (UNLESS OTHERWISE SPECIFIED)	DATE: 5/21/98	DRAWN BY: DICK PEARCE	DEPARTMENTAL APPROVALS	
DECIMAL:	SCALE: FULL	CHK'D BY:	R&D:	Q.A.:
.XX = ±	FIRST USE: AIR FORCE FAN		ENG:	MFG:
.XXX = ±	REFERENCE:		MRKTG:	I.D.:
FRACTIONAL: ±	PART NAME		PART NO.	
ANGULAR: ±	BLADE RING		92678	

Figure 23-9. Typical title block of an engineering drawing.

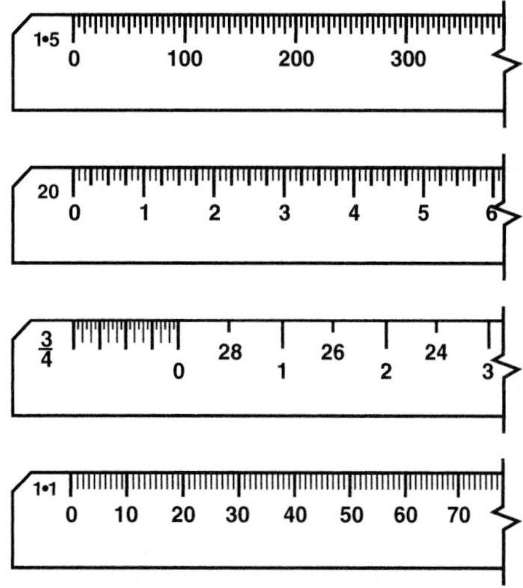

Figure 23-10. Examples of Engineers, Architectural, and Metric scales.

the zero line indicate 1-foot increments. The marks on the left side represent inches and fractions of an inch.

To find the length of a line using the 1-inch scale, the following procedure should be used:

1. Position the scale's zero mark indicator on the beginning of the line or object to be measured, Figure 23-14.
2. Moving from left to right, count the number of feet marks until you reach the end of the line or the foot mark nearest to the end, Figure 23-15.
3. Slide the scale to the right until the end of the line is on the foot mark indicated in step 2, Figure 23-16.
4. Starting at the zero, count the number of inch mark to the end of the line, Figure 23-17.

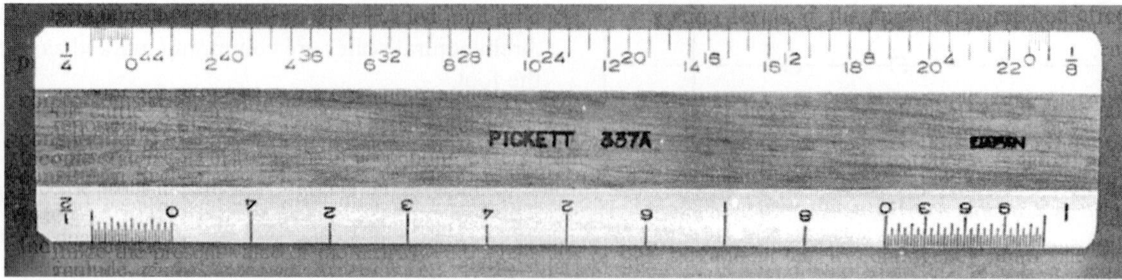

Figure 23-11. Six-inch architectural scale.

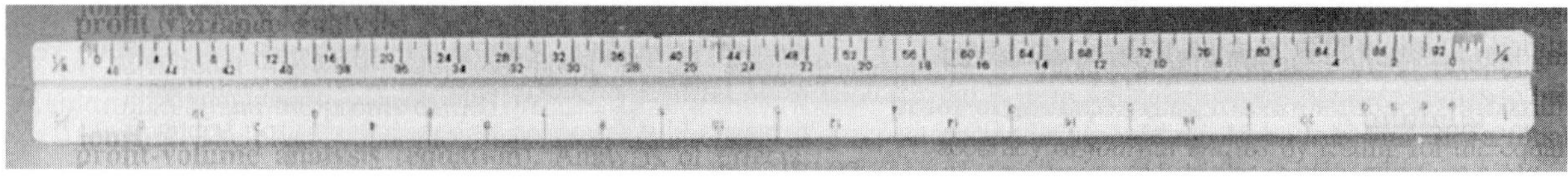

Figure 23-12. Twelve-inch triangular architectural scale.

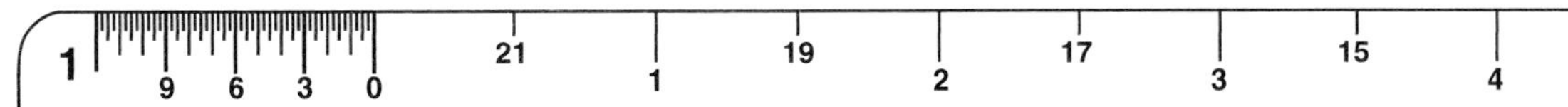

Figure 23-13. Architect Scale 1" = 1'-0".

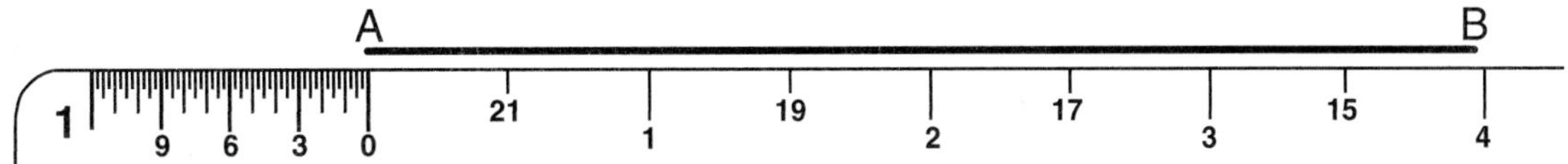

Figure 23-14. Placing the architectural scale on the beginning of the line to be measured.

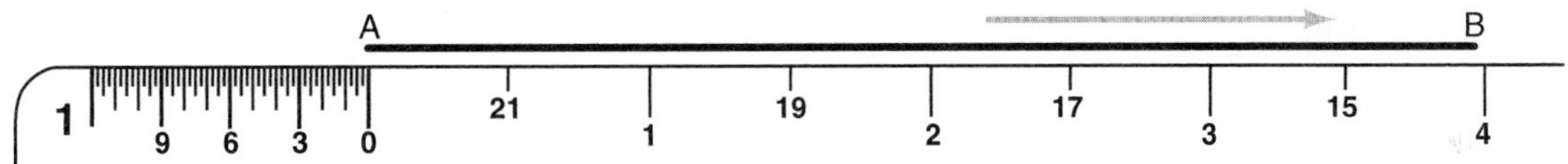

Figure 23-15. Reading the Architectural scale moving from left to right.

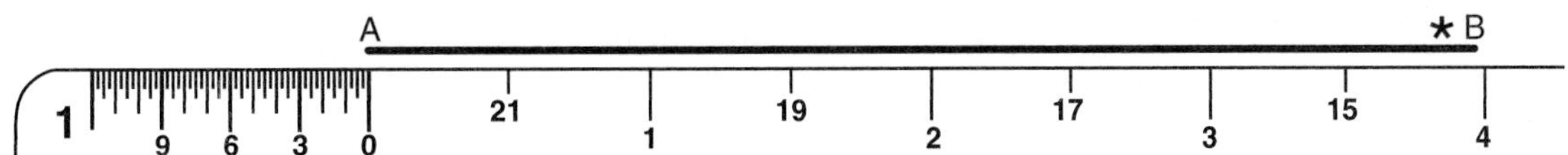

Figure 23-16. Finding the foot mark nearest to the end of the line to be measured.

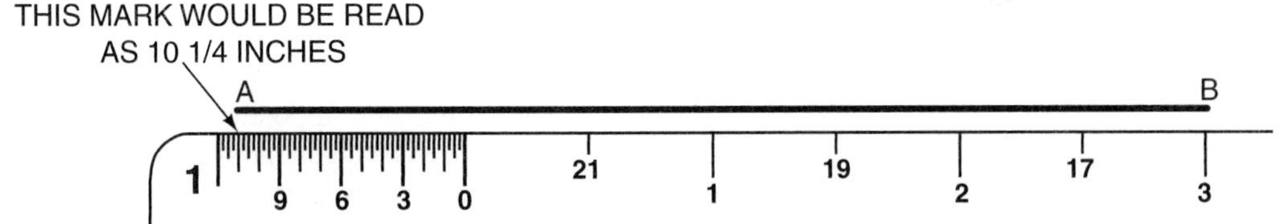

Figure 23-17. Determining the inches and fraction of an inch.

23.4 Angular Measurement

Understanding and reading linear measurements are only a part of the skill set needed to interpret a blueprint. The technician must also be able to read and understand angular measurements (i.e., they must be able to measure angles) as well.

Angles

Three types of units are used to express an angle: angular degrees, radians, and gradients. Typically, though, only angular degrees are used on blueprints. Because the circumference of any circle contains 360 degrees, an angular degree is equal to $1/360$ of the circumference of a circle. This means that by drawing a circle (of any size) and dividing its circumference into 360 equal segments (called arc lengths), the angle formed by constructing a line from the center of the circle to the endpoint of one arc length would produce a wedge equal to one degree, Figure 23-18.

Reading and Measuring Angles

Angles are typically given on a blueprint with either a dimension or a leader line. However, from time to time it may be necessary to determine the angle of something on a blueprint by using a protractor. A protractor is an instrument consisting of a half circle with a midpoint (center) marked on the horizontal position (base) of the protractor. This midpoint, also called the reference point, is marked on the protractor. The half circle is divided into degrees (180). The degrees are

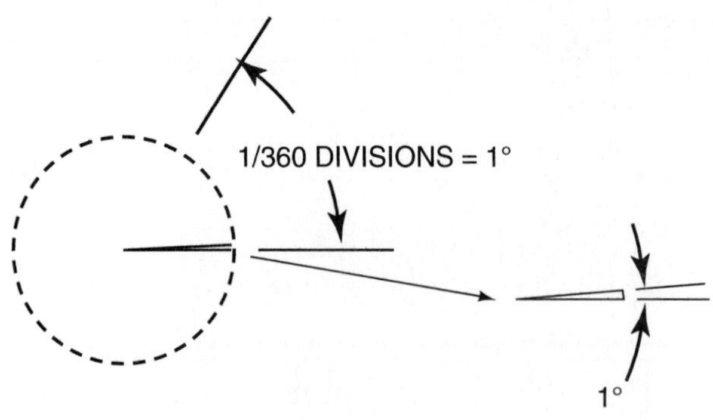

Figure 23-18. Wedge produced by dividing a circle 360 times.

labeled from right to left and from left to right, allowing for angles to be measured from either direction, Figure 23-19.

To measure an angle by using a protractor, first place the end of the protractor base angle on the vertex of the angle to be measured, Figure 23-20. Next align the baseline with one of the sides of the angle to be measured,

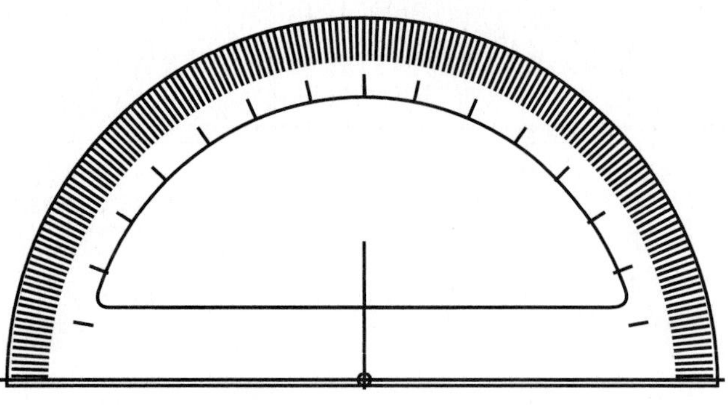

Figure 23-19. Protractor.

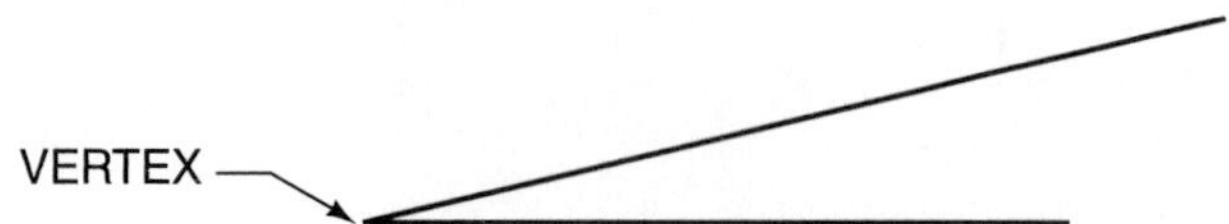

Figure 23-20. Vertex of angle to be measured.

Figure 23-21. Finally, count the number of degrees of the adjacent to the baseline, Figure 23-21.

23.5 Using Sine, Cosine, and Tangent to Determine Angles and Lengths of Lines of a Triangle

As indicated by its name, a triangle contains three angles. When these three angles are added, their sum is equal to 180 degrees, or angle A + angle B + angle C = 180°. By knowing this fact, a missing angle can be calculated if the other two angles are given. For example, the missing angle of a triangle containing a 70° and a 55° angle is found by adding the two known angles, and then subtracting their product from 180° (180° − [70° + 55°]). After performing the calculation, the missing angle is found to be 55°.

Each side of a triangle has a name: the hypotenuse, opposite, and adjacent sides. In a right triangle, the hypotenuse is always the longest side. The other two sides, opposite and adjacent, are labeled relative to the acute angle (any angle less than 90°) being focused on in a given calculation, Figures 23-22 and 23-23. The adjacent side is the side adjacent to the angle in question. The opposite is the side opposite to that angle.

In a right triangle, there is a direct relationship between the angles and the lengths of the sides. This relationship can be summed up in three fundamental trigonometric functions: sine, cosine, and tangent. The sine function is defined as the ratio of the side opposite to an acute angle divided by the hypotenuse (or sine A = a/c), as shown in Figures 23-22 and 23-23. The cosine function is defined as the ratio of the side adjacent to an acute angle divided by the hypotenuse (or cosine A = b/c), and the tangent function is defined as the ratio of the side opposite to an acute angle divided by the side adjacent (tangent A = a/b). Another important concept in trigonometry is the Pythagorean theorem. It states that the square of the hypotenuse of a right triangle is equal to the sum of the square of the other two sides (or $R^2 = X^2 + Y^2$). In this equation, *R* is equal to the hypotenuse, and X and Y are equal to the adjacent and opposite sides of the

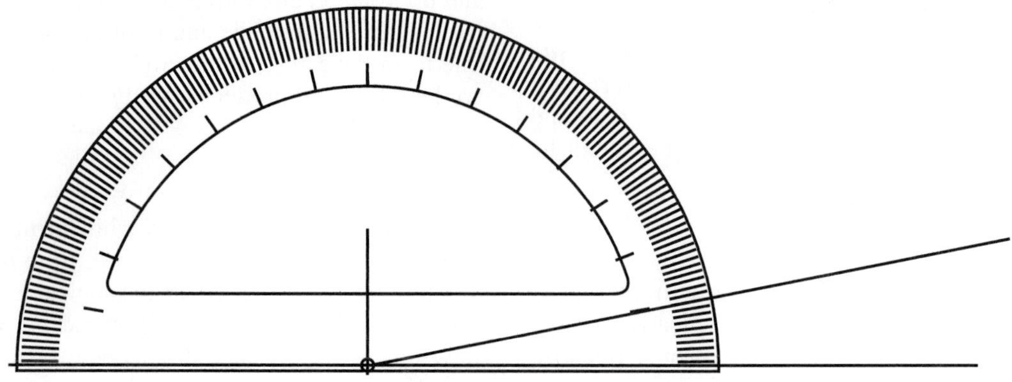

Figure 23-21. Reading the protractor.

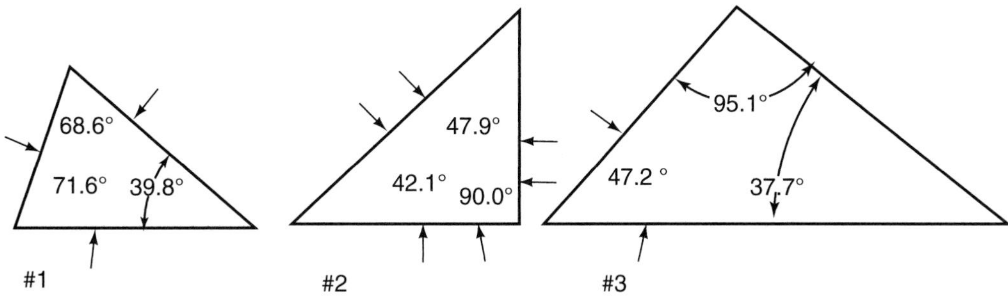

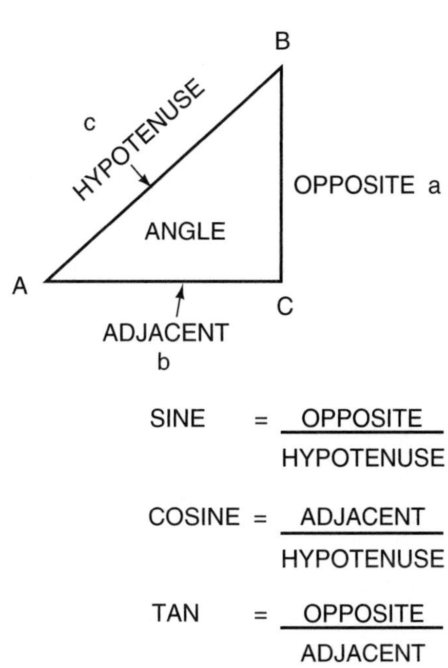

$$\text{SINE} = \frac{\text{OPPOSITE}}{\text{HYPOTENUSE}}$$

$$\text{COSINE} = \frac{\text{ADJACENT}}{\text{HYPOTENUSE}}$$

$$\text{TAN} = \frac{\text{OPPOSITE}}{\text{ADJACENT}}$$

Figure 23-22. #1 Acute triangle—all angles are less than 90°. #2 Right triangle—one angle is 90°. #3 Obtuse triangle—one angle is greater than 90°.

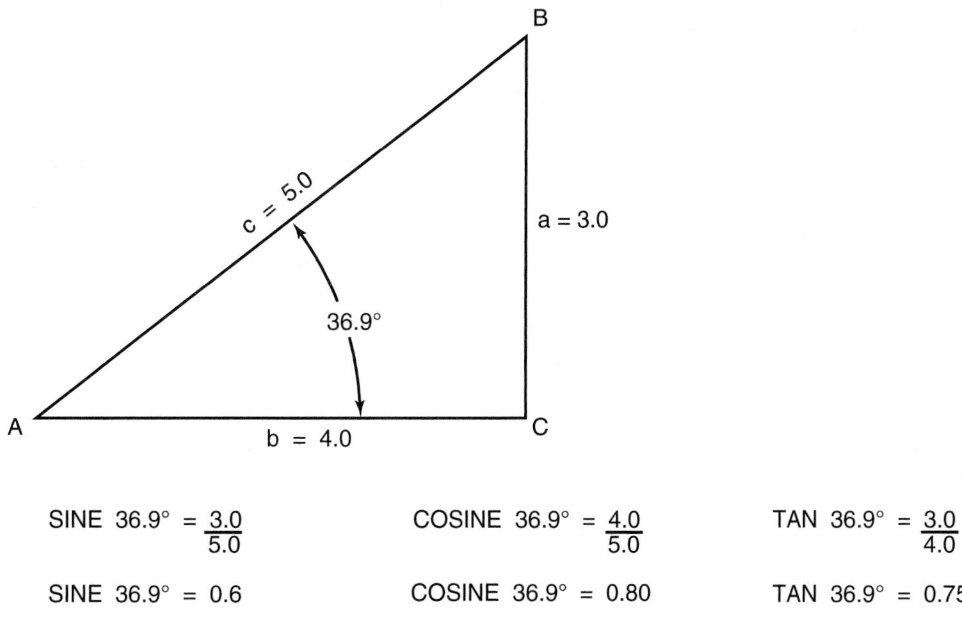

$\text{SINE } 36.9° = \frac{3.0}{5.0}$ $\text{COSINE } 36.9° = \frac{4.0}{5.0}$ $\text{TAN } 36.9° = \frac{3.0}{4.0}$

SINE 36.9° = 0.6 COSINE 36.9° = 0.80 TAN 36.9° = 0.75

Figure 23-23. Examples of sine, cosine, and tangent.

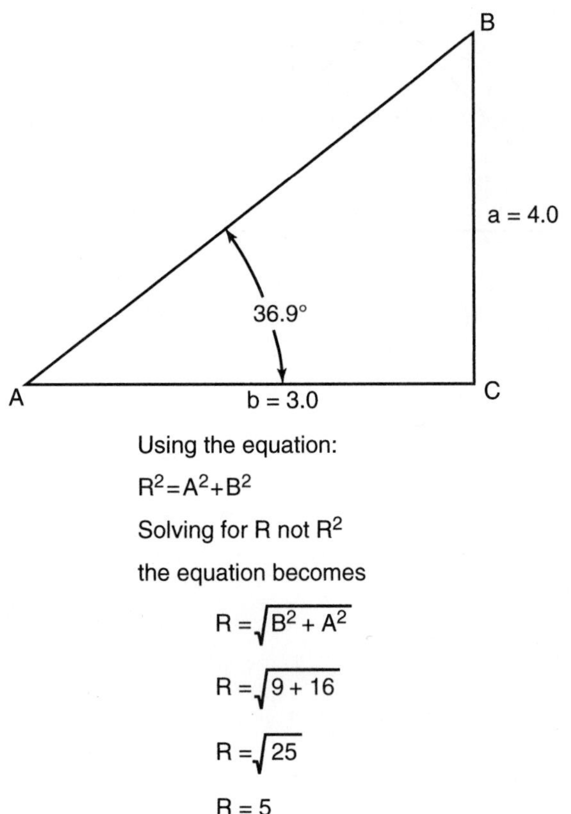

Figure 23-24. Example of the Pythagorean theorem.

triangle. With this equation, if the other two sides are given, the missing side of a right triangle can be determined by applying the Pythagorean theorem. To calculate the hypotenuse, algebra would have to be used to isolate the variable *R*. In other words, the hypotenuse is found by solving for *R*, not R^2. To isolate *R*, the square root of both sides of the equation must be taken, and doing so would yield the equation $R = \sqrt{(X^2 + Y^2)}$. For example, given a right triangle that contains an adjacent side (b) equal to three inches and an opposite side (a) equal to four inches, the hypotenuse can be found by using the formula $R = \sqrt{(23 + 42)}$, $R = \sqrt{25}$, $R = 5$, as shown in Figure 23-24.

23.6 Standard Abbreviations and Symbols

Standard abbreviations and symbols have been developed for the engineering and architectural community that not only facilitate the development of blueprints but also ensure consistency in their interpretation. Therefore, a good understanding of what these symbols are is critical in developing blueprint reading skills. An example of some of the abbreviations used in the HVAC industry is shown in the following table. A complete set of abbreviations is provided in the appendix of this book.

Table 3 HVAC abbreviations

Abbreviations	Definition
A/D	Analog to Digital
AC	Air Conditioning - Alternating current
ACH	Air Changes per Hour
ACM	Asbestos Containing Material
BACnet	A Data Communication Protocol for Building Automation and Control Networks
CFC	ChloroFluoroCarbon
CFM	Cubic Feet per Minute
CFU	Colony Forming Units
CHWP	Chilled Water Pump

Like abbreviations, symbols are used to speed up the drawing process by using a shorthand method of representing commonly used equipment. Typically, an engineering or architectural firm will use a standard set of symbols commonly used and accepted in the HVAC industry. However, if a nonstandard symbol is used, it is typically identified in the drawing legend. A sample of commonly used symbols is presented in Figure 23-25. A more complete set of symbols used in this industry can be found in the appendix of this book.

23.7 Two-Dimensional Views

The way in which an object is presented or viewed is extremely important. If a part is to be manufactured and the angle or view shown does not provide the necessary information, then the part could never be manufactured to design expectations. The following section examines the most common method of representing objects on engineering and architectural drawings.

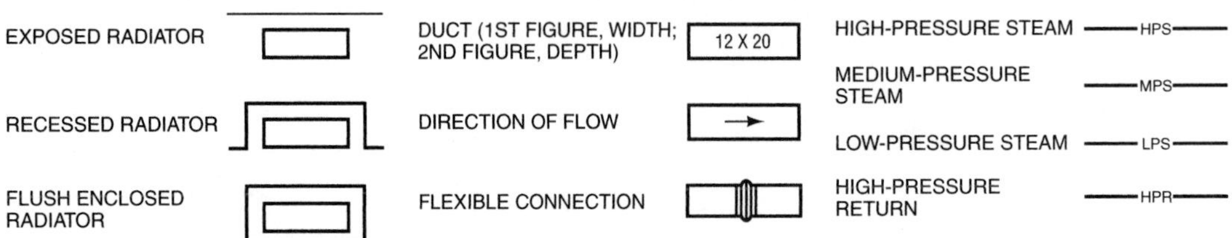

Figure 23-25. HVAC symbols.

Orthographic Projection and Engineering Drawings

When something is to be constructed or created using a blueprint, it is the job of the drafting person to supply all the necessary information regarding the object's dimensions and the location of its features. All this can be easily accomplished when the object is relatively simple, and all the essential data can be provided in one view. For example, suppose that a 2-inch square piece of steel plating that is ½" thick is to be fabricated with a ½"-diameter hole drilled at its center. All the required information about this part can be contained on one drawing consisting of a single view, Figure 23-26. When a more complex part is produced, a single view is not sufficient to clearly show all its features. In other words, a series of grooves on the opposite side of this same part would be shown as hidden features, resulting in a drawing that is difficult to interpret, Figure 23-27. To solve this problem, additional views must be created that will reveal all the hidden attributes of the part. In this instance, an additional view of the side is required, Figure 23-28A. Very

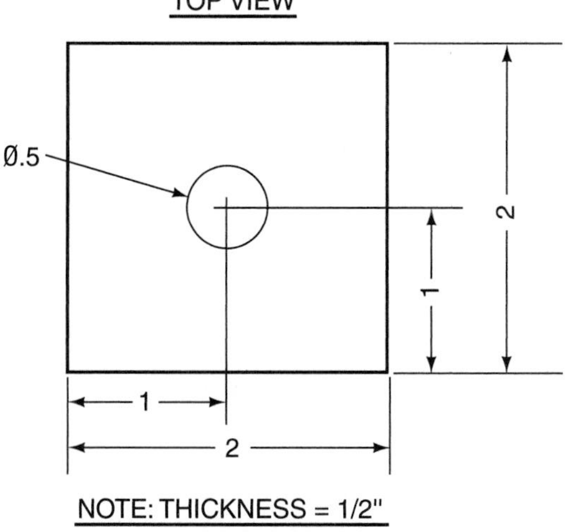

Figure 23-26. Single view of a 2" × 2" × 1/2" steel plate with a 1/2"-hole drilled in the center.

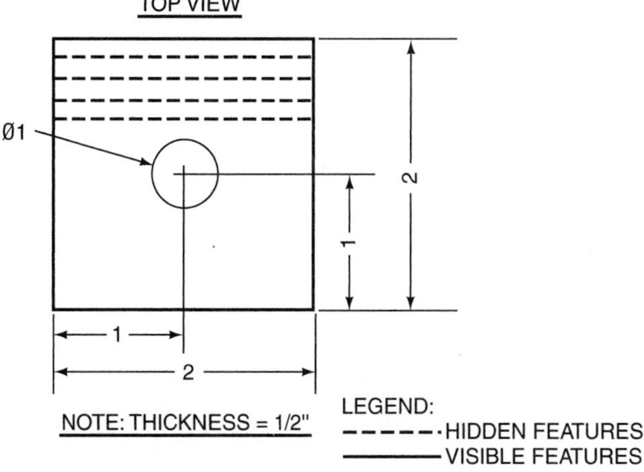

Figure 23-27. Single view of a 2" × 2" × 1/2" steel plate with a 1/2"-hole drilled in the center and containing a series of grooves as hidden features.

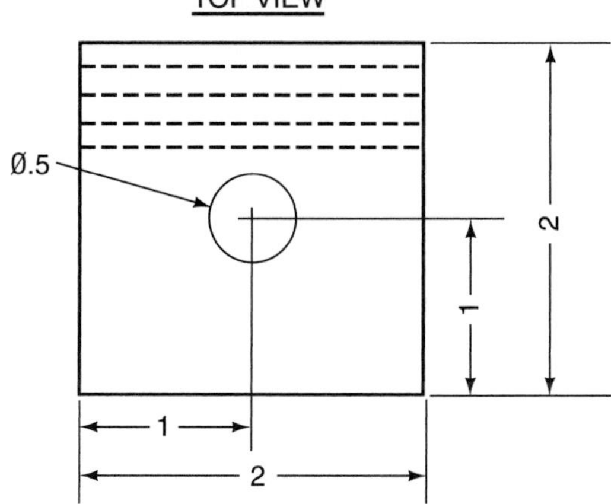

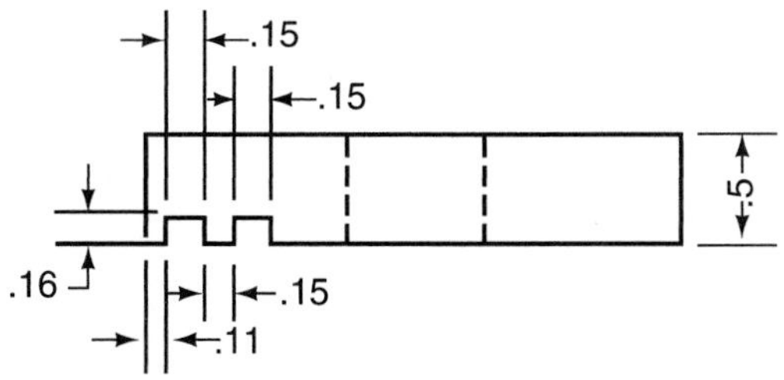

Figure 23-28A. Orthographic projection of a 2" × 2" × ½" steel plate with a ½"-hole drilled in the center and containing a series of grooves.

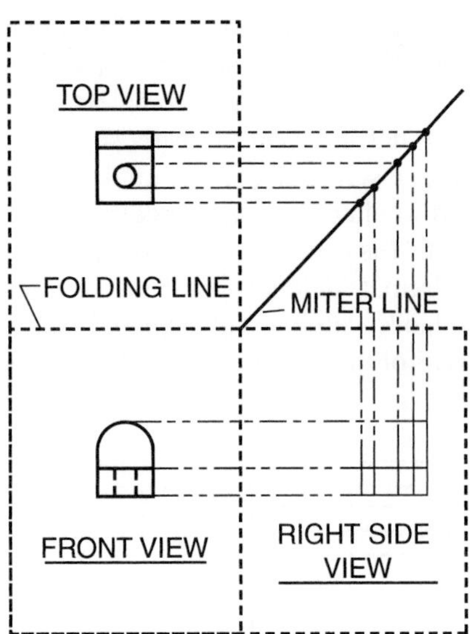

Figure 23-28B. Orthographic projection showing how the right side view was created.

complex parts might contain views showing the front, sides, back, bottom, and top of the object. A drawing containing two or more of these views is called a multiview drawing or orthographic projection. The different views in an orthographic projection are created by projecting the lines from one view to another. In other words, by projecting the edge of the object down from the top view, the width of the object can be established in the front view, Figure 23-28B. Another way of looking at this is to suppose that the object is placed in a glass box.

In a multiview drawing, the object can be visualized as being encased in a glass cube, as shown in Figure 23-29, with the object's surfaces and features being projected onto the surface of the glass, as illustrated in Figure 23-30. Once the object's features have been transferred to the glass box, the cube is then unfolded to reveal six different views of the object, Figure 23-31. The primary views used on an engineering drawing are: Front View, Right Side View, Left Side View, Top View, Bottom View, and Rear View, Figure 23-31.

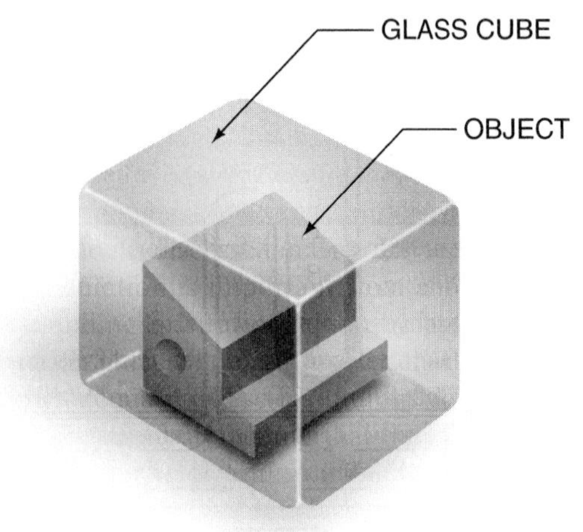

Figure 23-29. Object encased in a glass cube.

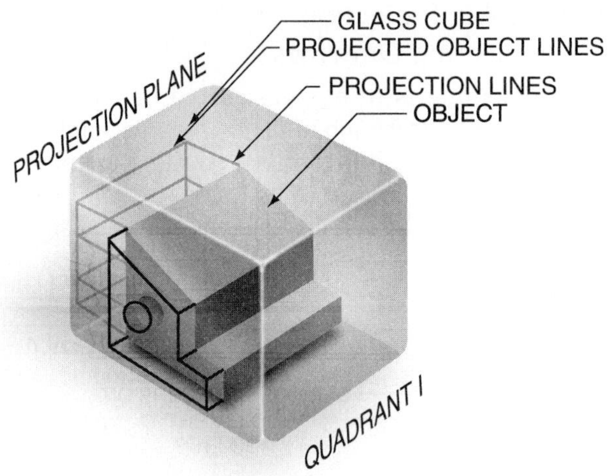

Figure 23-30. Object encased in glass cube with object lines projected onto surface of glass cube.

Orthographic Projection and Architectural Drawings

Architectural drawings are typically not labeled the same as engineering drawings. However, in reality they are still created based on the same principles. In an architectural drawing, the top view is not referred to as a top view but instead is called a plan. In addition, architectural drawings do not refer to left and right side views as left and right side views; instead, they are referred to as elevations.

23.8 Floor Plans

The floor plan provides a representation of where to locate the major items of a home. The plan shows the location of walls, doors, windows, cabinets, appliances, and plumbing fixtures. The drawing enables the owner to evaluate the project to ensure that the building will meet the current and future needs of the business once constructed. The floor plan also serves as a key tool in the communication process between the design team and the building team.

Types of Plans

The floor plan is the skeleton or framework for the development of other drawings required to complete the construction of the structure. These drawings include electrical plans, fire protection plans, framing plans, plumbing plans, and HVAC or mechanical plans.

TOP VIEW
REAR VIEW
LEFT SIDE VIEW
FRONT VIEW
RIGHT SIDE VIEW
BOTTOM VIEW

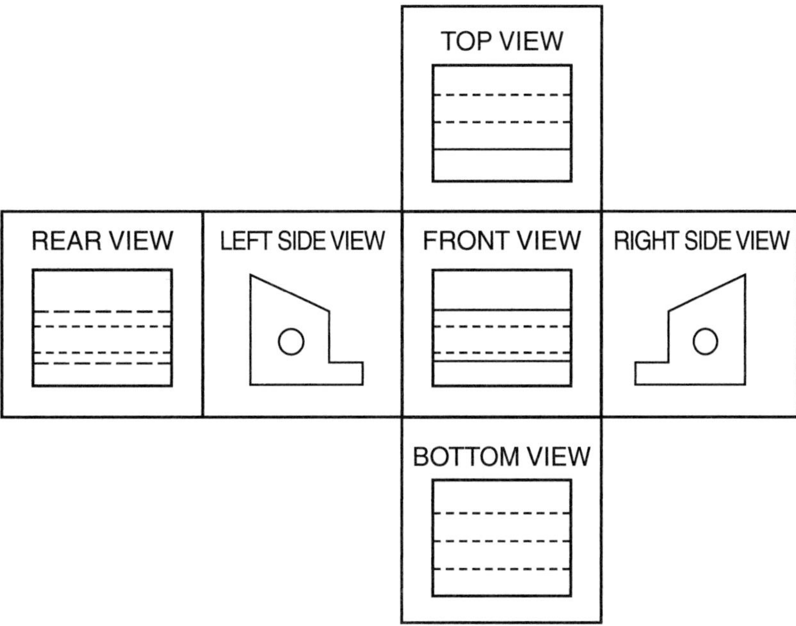

Figure 23-31. Glass cube unfolded to reveal the six different views.

Plumbing Plans

The plumbing plan contains the size and location of each piping system contained in the proposed structure. Typically piping on a plumbing plan is represented using a single line in which the different functions of that piping system are represented using different linetypes, Figure 23-32. For example, drain lines are typically shown on plumbing plans by using a heavier line, whereas the domestic cold water lines are presented as light lines having long dashes and short dashes. Domestic hot water, on the other hand, is shown as a series of long dashes followed by two short dashes. All vent piping is shown as a series of dashes.

In addition to recognizing the types of plumbing lines used in a facility, the technician must also be able to recognize various plumbing fittings and valves used in plumbing drawings. The most commonly used symbols are shown below in Figures 23-33 and 23-34.

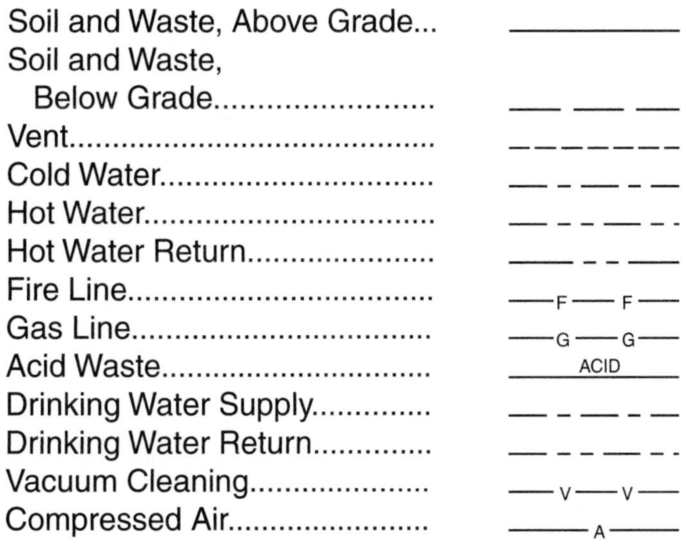

Soil and Waste, Above Grade...
Soil and Waste, Below Grade........................
Vent..
Cold Water...............................
Hot Water.................................
Hot Water Return.....................
Fire Line..................................
F F
Gas Line..................................
G G
Acid Waste...............................
ACID
Drinking Water Supply.............
Drinking Water Return.............
Vacuum Cleaning.....................
V V
Compressed Air.......................
A

Figure 23-32. Plumbing linetypes.

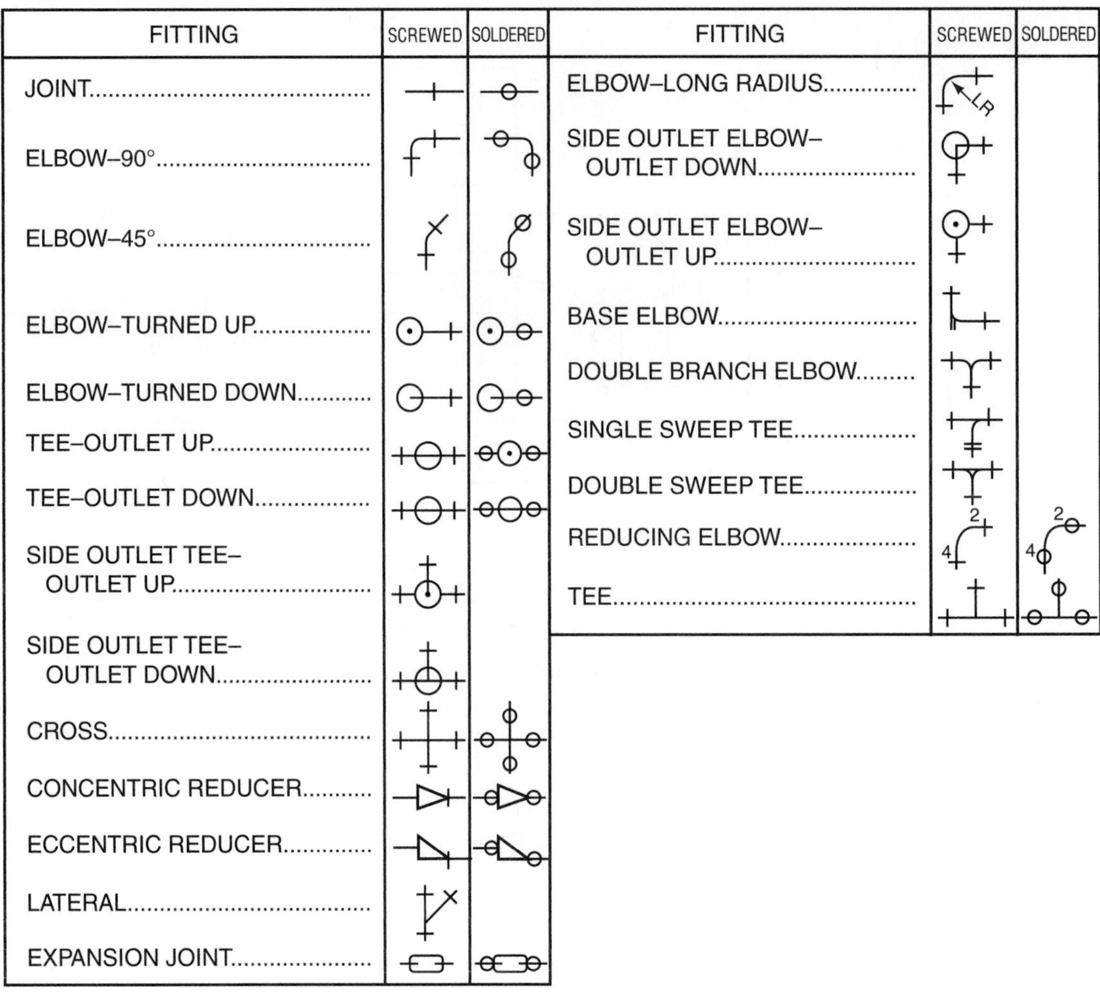

Figure 23-33. Common plumbing fittings.

HVAC Plans

The HVAC plan contains the size and location of each HVAC system contained in the proposed structure. Typically piping on a HVAC plan is represented using a single line in which the different functions of that piping system are represented using different linetypes, as shown in Figures 23-35 and 23-36. For example, drain lines are typically shown on HVAC plans by using a single line with the letter *D* inserted evenly throughout the line.

In addition to recognizing the types of HVAC lines used in a facility, the technician must also be able to recognize equipment and duct symbols used in HVAC drawings. The most commonly used symbols are shown below in Figures 23-37 and 23-38.

23.9 Elevations

An elevation is an orthographic drawing that shows one side of a building. In true orthographic projection, the elevations would be displayed as shown in Figure 23-39. The true projection is typically modified as shown in Figure 23-40 to ease viewing. No matter how they are displayed, it is important to realize that between each elevation projection and the plan view is an imaginary 90° fold line. An imaginary 90° fold line also exists between elevations in Figure 23-40. Elevations are drawn to show exterior shapes and finishes, as well as the vertical relationships of the building levels. By using the elevations, sections, and floor plans, the exterior shape of a building can be determined.

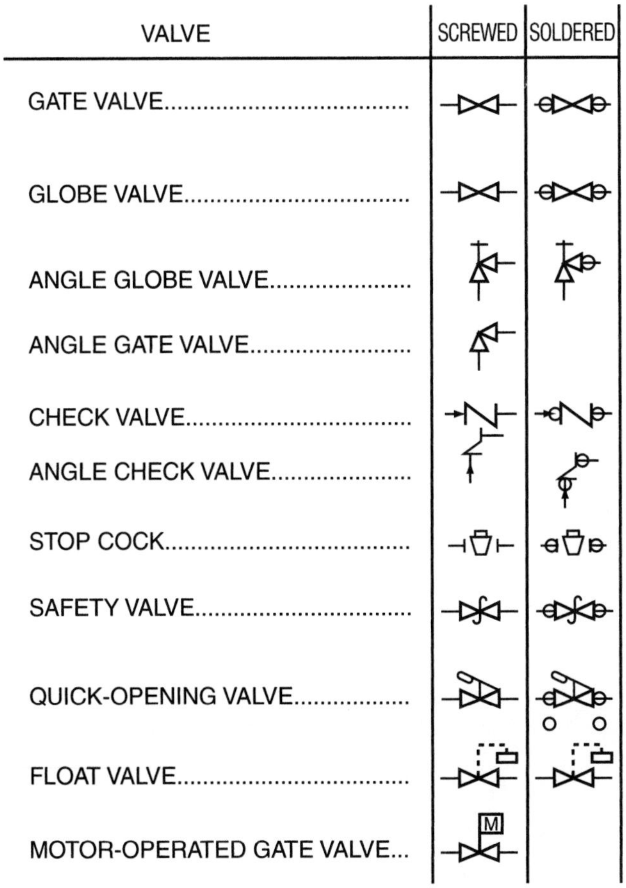

VALVE	SCREWED	SOLDERED
GATE VALVE		
GLOBE VALVE		
ANGLE GLOBE VALVE		
ANGLE GATE VALVE		
CHECK VALVE		
ANGLE CHECK VALVE		
STOP COCK		
SAFETY VALVE		
QUICK-OPENING VALVE		
FLOAT VALVE		
MOTOR-OPERATED GATE VALVE	M	

Figure 23-34. Common plumbing valves.

HIGH-PRESSURE STEAM	HPS
MEDIUM-PRESSURE STEAM	MPS
LOW-PRESSURE STEAM	LPS
HIGH-PRESSURE RETURN	HPR
MEDIUM-PRESSURE RETURN	MPR
LOW-PRESSURE RETURN	LPR
BOILER BLOW OFF	BO
CONDENSATE OR VACCUUM PUMP DISCHARGE	VPD
FEEDWATER PUMP DISCHARGE	FPD
MAKEUP WATER	MU
AIR RELIEF LINE	V
FUEL OIL SUCTION	FOS
FUEL OIL RETURN	FOR
FUEL OIL VENT	FOV
COMPRESSED AIR	A
HOT WATER HEATING SUPPLY	HW
HOT WATER HEATING RETURN	HWR

Figure 23-35. HVAC heating linetypes.

REFRIGERANT LIQUID	RL
REFRIGERANT DISCHARGE	RD
REFRIGERANT SUCTION	RS
CONDENSER WATER SUPPLY	CWS
CONDENSER WATER RETURN	CWR
CHILLED WATER SUPPLY	CHWS
CHILLED WATER RETURN	CHWR
MAKEUP WATER	MU
HUMIDIFICATION LINE	H
DRAIN	D

Figure 23-36. HVAC air-conditioning linetypes.

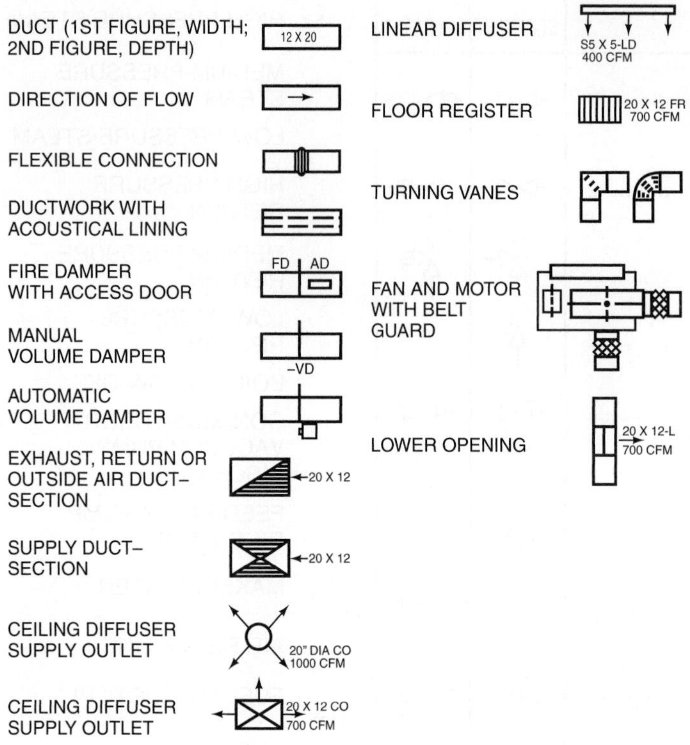

DUCT (1ST FIGURE, WIDTH; 2ND FIGURE, DEPTH)
12 X 20
DIRECTION OF FLOW
FLEXIBLE CONNECTION
DUCTWORK WITH ACOUSTICAL LINING
FIRE DAMPER WITH ACCESS DOOR
FD
AD
MANUAL VOLUME DAMPER
–VD
AUTOMATIC VOLUME DAMPER
EXHAUST, RETURN OR OUTSIDE AIR DUCT–SECTION
20 X 12
SUPPLY DUCT–SECTION
20 X 12
CEILING DIFFUSER SUPPLY OUTLET
20" DIA CO
1000 CFM
CEILING DIFFUSER SUPPLY OUTLET
20 X 12 CO
700 CFM
LINEAR DIFFUSER
S5 X 5-LD
400 CFM
FLOOR REGISTER
20 X 12 FR
700 CFM
TURNING VANES
FAN AND MOTOR WITH BELT GUARD
LOWER OPENING
20 X 12-L
700 CFM

Figure 23-37. HVAC duct symbols.

EXPOSED RADIATOR
RECESSED RADIATOR
FLUSH ENCLOSED RADIATOR
PROJECTING ENCLOSED RADIATOR
UNIT HEATER (PROPELLER) – PLAN
UNIT HEATER (CENTRIFUGAL) – PLAN
UNIT VENTILATOR – PLAN
STEAM
DUPLEX STRAINER
PRESSURE-REDUCING VALVE
AIR LINE VALVE
STRAINER
THERMOMETER
PRESSURE GAUGE AND COCK
RELIEF VALVE
AUTOMATIC 3-WAY VALVE
AUTOMATIC 2-WAY VALVE
SOLENOID VALVE
S

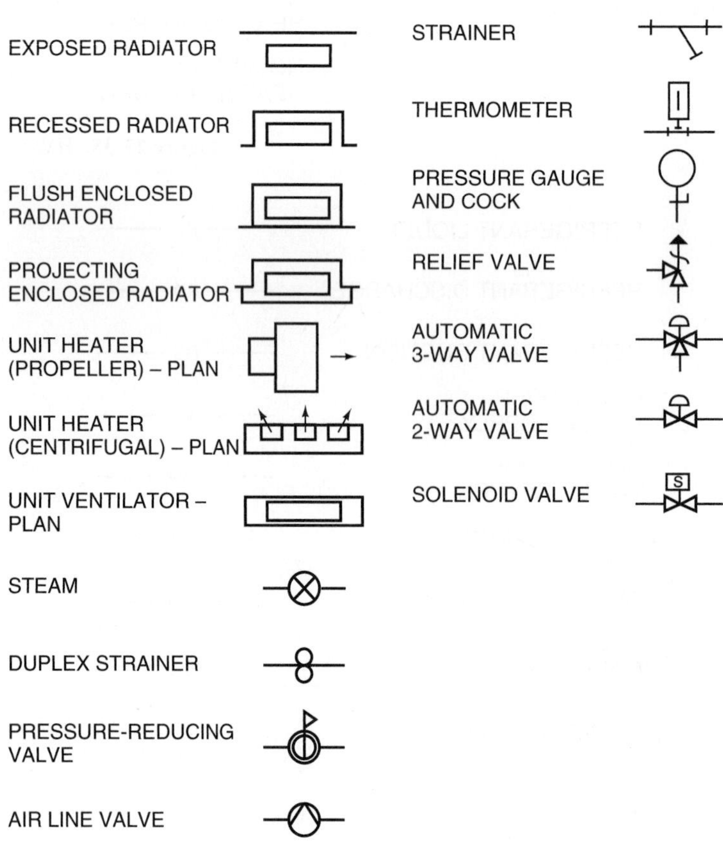

Figure 23-38. HVAC equipment symbols.

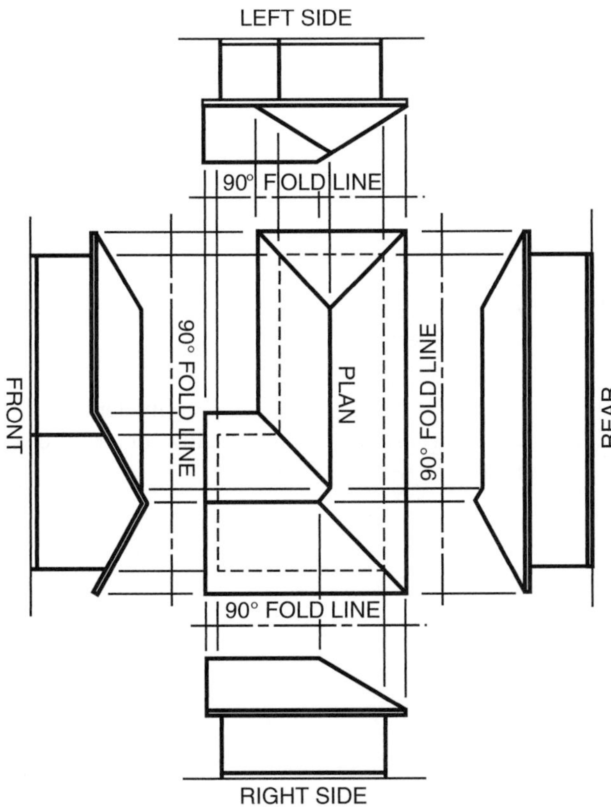

Figure 23-39. Elevations are orthographic projections showing each side of a structure.

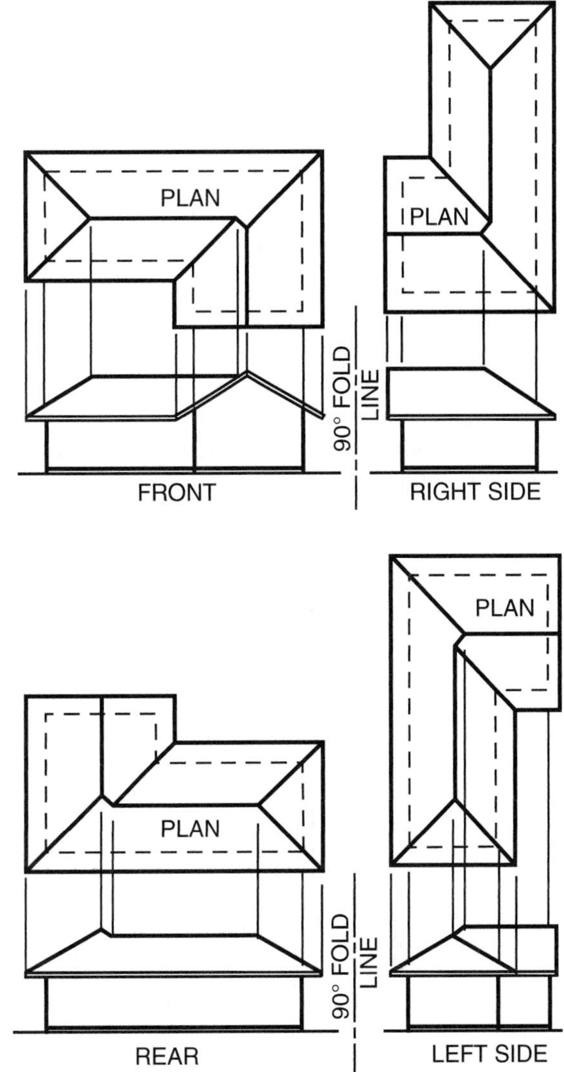

Figure 23-40. The placement of elevations is usually altered to ease viewing. Group elevations so that a 90° rotation exists between views.

Summary

- Linear measurement is defined as the measurement of two points along a straight line.
- All objects, whether they are made by people or the result of natural conditions and/or forces, consist of points and lines.
- Currently there are two basic systems used to make measurements: the English system and the metric system.
- The English system's base units are inch (in), foot (ft), yard (yd), and mile (mi).
- When a mechanical drawing is created on the computer, typically it is drawn at a scale of 1:1 (also known as full scale).
- When a drawing is created, one of the responsibilities of the drafter is to note the scale at which the drawing was created.
- The title block region is typically located along the bottom edge or lower right corner of the drawing.
- The length and position of items not dimensioned on a drawing can be determined using a ruler.
- The architect scale is a type of ruler in which a range of pre-calibrated ratios are illustrated.
- Three types of units are used to express an angle: angular degrees, radians, and gradients.
- Angles are typically given on a blueprint with either a dimension or a leader.
- When these three angles are added, their sum is equal to 180 degrees.
- Each side of a triangle has a name: the hypotenuse, opposite, and adjacent sides.
- In a right triangle, the hypotenuse is always the longest side.
- In a right triangle, there is a direct relationship between the angles and the lengths of the sides.
- The sine function is defined as the ratio of the side opposite to an acute angle divided by the hypotenuse.
- The cosine function is defined as the ratio of the side adjacent to an acute angle divided by the hypotenuse.
- The tangent function is defined as the ratio of the side opposite to an acute angle divided by the side adjacent.
- Standard abbreviations and symbols have been developed for the engineering and architectural community that not only facilitate the development of blueprints but also, ensure consistency in their interpretation.

- The primary views used on an engineering drawing are: Front View, Right Side View, Left Side View, Top View, Bottom View, and Rear View.
- Architectural drawings are typically not labeled the same as engineering drawings. However, in reality, are still created based on the same principles.
- The floor plan provides a representation of where to locate the major items of a home.
- The plan shows the location of walls, doors, windows, cabinets, appliances, and plumbing fixtures.
- The plumbing plan contains the size and location of each piping system contained in the proposed structure.
- The HVAC plan contains the size and location of each HVAC system contained in the proposed structure.
- An elevation is an orthographic drawing that shows one side of a building.

Review Questions

Convert the following:

1. 144 feet = ______________ inches
2. 234 inches = ______________ feet
3. 234,539 inches = ______________ miles
4. 5 miles = ______________ inches
5. 23 miles = ______________ yards
6. 100 feet = ______________ meters
7. 245.9 centimeters = ______________ feet
8. 123 millimeters = ______________ feet
9. 256 yards = ______________ meters
10. 452.1 inches = ______________ centimeters
11. and 12. Using the scale shown below, determine the length of lines A, B, C, G, and H.

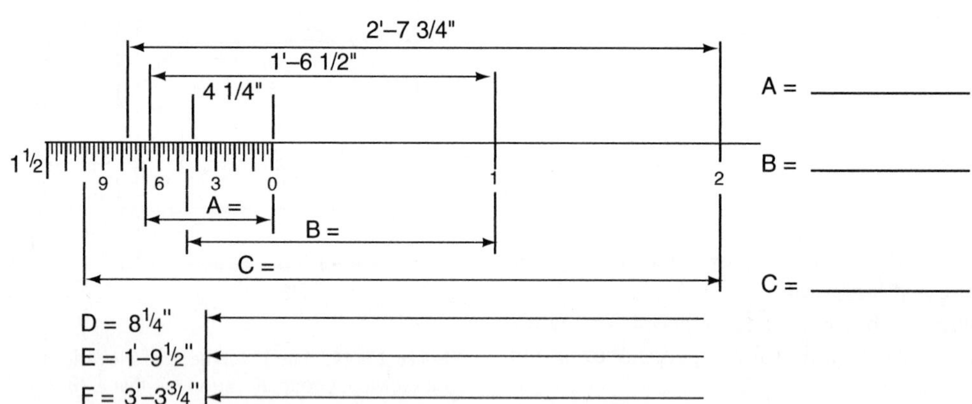

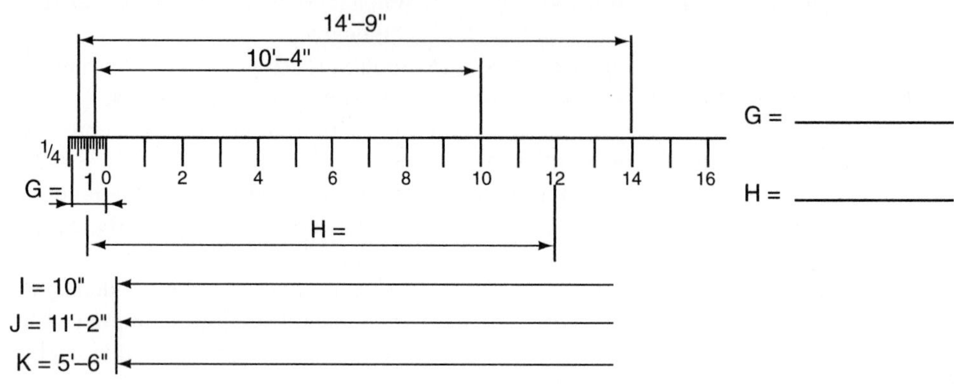

13–16. Determine the angles of the following:

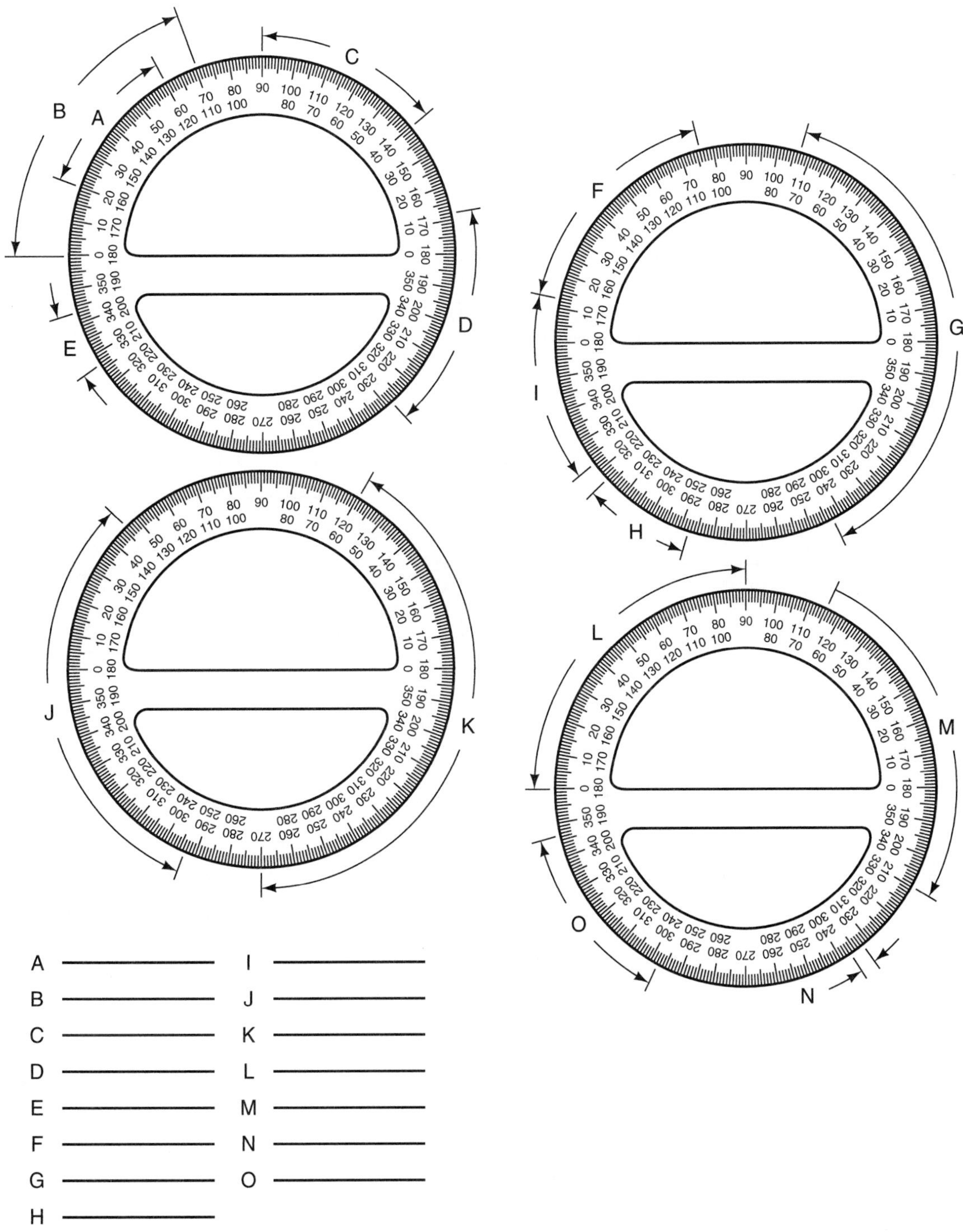

A ____________ I ____________

B ____________ J ____________

C ____________ K ____________

D ____________ L ____________

E ____________ M ____________

F ____________ N ____________

G ____________ O ____________

H ____________

17. Identify the following valve.

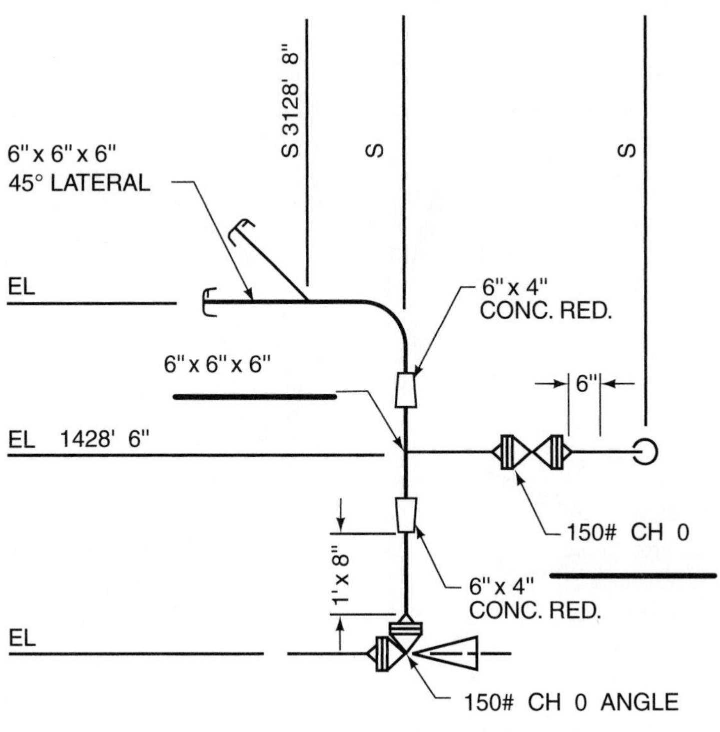

18–26. Identify the following linetypes

______________________ —— RL ——

______________________ —— RD ——

______________________ —— RS ——

______________________ —— CWS ——

______________________ —— CWR ——

______________________ —— CHWR ——

______________________ —— MU ——

______________________ —— H ——

______________________ —— D ——

27–36. Find the missing components of the right triangles shown below by using the sine, cosine, and tangent functions.

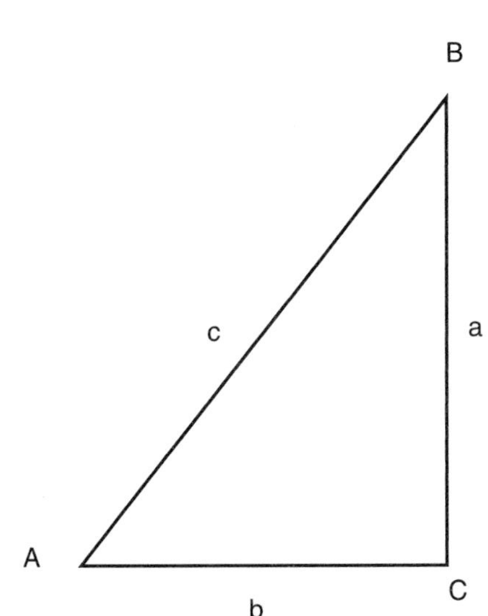

Right Triangle Chart						
Problem #	Angle A	Angle B	Angle C	Side a	Side b	Side c
27	53.13°			4	3	5
28				2	10	
29	45°			3	3	
30		35.7535°		4.1667		5.1344
31				10.783	3	
32	29°					2.2245
33		30°			1.5675	
34	76°					2.7517
35		77°		0.1574	0.6752	
36	82°			3.5142		

Index